Subcellular Biochemistry

Volume 30

Fat-Soluble Vitamins

SUBCELLULAR BIOCHEMISTRY

Recent Volumes in This Series:

Volume 21	**Endoplasmic Reticulum** Edited by N. Borgese and J. R. Harris
Volume 22	**Membrane Biogenesis** Edited by A. H. Maddy and J. R. Harris
Volume 23	**Physicochemical Methods in the Study of Biomembranes** Edited by Herwig J. Hilderson and Gregory B. Ralston
Volume 24	**Proteins: Structure, Function, and Engineering** Edited by B. B. Biswas and Siddhartha Roy
Volume 25	**Ascorbic Acid: Biochemistry and Biomedical Cell Biology** Edited by J. Robin Harris
Volume 26	***myo*-Inositol Phosphates, Phosphoinositides, and Signal Transduction** Edited by B. B. Biswas and Susweta Biswas
Volume 27	**Biology of the Lysosome** Edited by John B. Lloyd and Robert W. Mason
Volume 28	**Cholesterol: Its Functions and Metabolism in Biology and Medicine** Edited by Robert Bittman
Volume 29	**Plant–Microbe Interactions** Edited by B. B. Biswas and H. K. Das
Volume 30	**Fat-Soluble Vitamins** Edited by Peter J. Quinn and Valerian E. Kagan

---

A Continuation Order Plan is available for this series. A continuation order will bring delivery of each new volume immediately upon publication. Volumes are billed only upon actual shipment. For further information please contact the publisher.

Subcellular Biochemistry

Volume 30

Fat-Soluble Vitamins

Edited by

Peter J. Quinn
King's College London
University of London
London, United Kingdom

and

Valerian E. Kagan
University of Pittsburgh
Pittsburgh, Pennsylvania

PLENUM PRESS • NEW YORK AND LONDON

The Library of Congress cataloged the first volume of this title as follows:

Sub-cellular biochemistry.
London, New York, Plenum Press.
v. illus. 23 cm. quarterly.
Began with Sept. 1971 issue. Cf. New serial titles.
1. Cytochemistry—Periodicals. 2. Cell organelles—Periodicals.
QH611.S84 574.8'76 73-643479

ISSN 0306-0225

ISBN 0-306-45846-2

This series is a continuation of the journal *Sub-Cellular Biochemistry,* Volumes 1 to 4 of which were published quarterly from 1972 to 1975

A Division of Plenum Publishing Corporation
233 Spring Street, New York, N.Y. 10013

http://www.plenum.com

10 9 8 7 6 5 4 3 2 1

Printed in the United States of America

Contributors

Stefania Bergamini Department of Environmental and Occupational Health, University of Pittsburgh, Pittsburgh, Pennsylvania 15238; Visiting Researcher from the Department of Biomedical Science, University of Modena, Modena, Italy

Ian Byrne W. Alton Jones Cell Science Center, Lake Placid, New York 12946; and University College Dublin, Belfield, Ireland

Gianfranca Carta Department of Environmental and Occupational Health, University of Pittsburgh, Pittsburgh, Pennsylvania 15238; Visiting Researcher from the Dipartimento di Biolgia Sperimentale, Sezione di Patologia Sperimentale, Università di Cagliari, Cagliari, Italy

Andrew R. Collins Rowett Research Institute, Bucksburn, Aberdeen AB21 9SB, Scotland

Claude Conri Service de Médecine Interne et Pathologie Vasculaire, Hôpital Saint-André, 33075 Bordeaux, France

Joël Constans Service de Médecine Interne et Pathologie Vasculaire, Hôpital Saint-André, 33075 Bordeaux, France

Billy W. Day Department of Environmental and Occupational Health, and Department of Pharmaceutical Sciences, University of Pittsburgh, Pittsburgh, Pennsylvania 15238

Garry G. Duthie Rowett Research Institute, Bucksburn, Aberdeen AB21 9SB, Scotland

Susan J. Duthie Rowett Research Institute, Bucksburn, Aberdeen AB21 9SB, Scotland

Louise Flanagan W. Alton Jones Cell Science Center, Lake Placid, New York 12946; and University College Dublin, Belfield, Ireland

Lars Gille Institute of Pharmacology and Toxicology, Veterinary University of Vienna, A-1210 Vienna, Austria

Bernhard Hennig Department of Nutrition and Food Science, University of Kentucky, Lexington, Kentucky 40506-0054

R. Scott Jamison Department of Biochemistry, Vanderbilt University School of Medicine, Nashville, Tennessee 37232-0146

Valerian E. Kagan Department of Environmental and Occupational Health, and Department of Pharmacology, University of Pittsburgh, Pittsburgh, Pennsylvania 15238

Andrey V. Kozlov Institute of Pharmacology and Toxicology, Veterinary University of Vienna, A-1210 Vienna, Austria

Norman I. Krinsky Gastrointestinal Nutrition Laboratory, Jean Mayer USDA Human Nutrition Research Center on Aging at Tufts University, and Department of Biochemistry, School of Medicine, Tufts University, Boston, Massachusetts 02111

Daniel C. Liebler Department of Pharmacology and Toxicology, College of Pharmacy, University of Arizona, Tucson, Arizona 85721-0207

Maria A. Livrea Istituto di Farmacologia e Farmacognosia, Università di Palermo, 90134 Palermo, Italy

Malcolm Maden Developmental Biology Research Centre, King's College London, London WC2B 5RL, England

Colin Masters The School of Biomolecular and Biomedical Science, Griffith University, Brisbane QLD 4111, Australia

Carmen J. Narvaez W. Alton Jones Cell Science Center, Lake Placid, New York 12946

Marcia E. Newcomer Department of Biochemistry, Vanderbilt University School of Medicine, Nashville, Tennessee 37232-0146

Paul Newman Vitamin K Laboratory, The Haemophilia Centre, St. Thomas's Hospital, London SE1 7EH, England

Richard M. Niles Department of Biochemistry and Molecular Biology, Marshall University School of Medicine, Huntington, West Virginia 25755

Hans Nohl Institute of Pharmacology and Toxicology, Veterinary University of Vienna, A-1210 Vienna, Austria

Elizabeth Nolan W. Alton Jones Cell Science Center, Lake Placid, New York 12946

David E. Ong Department of Biochemistry, Vanderbilt University School of Medicine, Nashville, Tennessee 37232-0146

Evelyne Peuchant Laboratoire de Biochimie, Hôpital Saint-André, 33075 Bordeaux, France

Peter J. Quinn Division of Life Sciences, King's College London, London W8 7AH, England

William B. Rhoten Department of Anatomy, Cell and Neurobiology, Marshall University School of Medicine, Huntington, West Virginia 25704-9388

Claire Sergeant Centre d'Etudes Nucléaires de Bordeaux-Gradignan, 33075 Bordeaux, France

Igor N. Sergeev Department of Anatomy, Cell and Neurobiology, Marshall University School of Medicine, Huntington, West Virginia 25704-9388

Martin J. Shearer Vitamin K Laboratory, The Haemophilia Centre, St. Thomas's Hospital, London SE1 7EH, England

Vladimir B. Spirichev Institute of Nutrition, Russian Academy of Medical Sciences, Moscow 109240, Russia

R. Stocker Heart Research Institute, Camperdown, Sydney, NSW 2050, Australia

Luisa Tesoriere Istituto di Farmacologia e Farmacognosia, Università di Palermo, 90134 Palermo, Italy

Michal Toborek Department of Surgery, University of Kentucky Medical Center, Lexington, Kentucky 40536

Maria Tomoaia-Cotisel Molecular/Structural Biotechnologies, Rockville, Maryland 20847-2539

Vladimir A. Tyurin Department of Environmental and Occupational Health, University of Pittsburgh, Pittsburgh, Pennsylvania 15238; on leave from the Institute of Evolutionary Physiology and Biochemistry, Russian Academy of Science, St. Petersburg, 194223 Russia

Yulia Y. Tyurina Department of Environmental and Occupational Health, University of Pittsburgh, Pittsburgh, Pennsylvania 15238; on leave from the Institute of Evolutionary Physiology and Biochemistry, Russian Academy of Science, St. Petersburg, 194223 Russia

J. M. Upston Heart Research Institute, Camperdown, Sydney, NSW 2050, Australia

Shiro Urano Tokyo Metropolitan Institute of Gerontology, 35-2 Sakaecho, Itabashi-ku, Tokyo 173, Japan; *present address:* Department of Biological Chemistry, Faculty of Industrial Chemistry, Shibaura Institute of Technology, 3-9-14 Shibaura, Minato-ku, Tokyo 108, Japan

Kathryn VanWeelden W. Alton Jones Cell Science Center, Lake Placid, New York 12946; and Clarkson University, Potsdam, New York 13699

Amandio V. Vieira Department of Cell Biology, The Scripps Research Institute, La Jolla, California 92037; *present address:* Department of Biochemistry, National Public Health Institute-KTL, FIN-00300 Helsinki, Finland

Xiang-Dong Wang Gastrointestinal Nutrition Laboratory, Jean Mayer USDA Human Nutrition Research Center on Aging at Tufts University, and the Department of Biochemistry, School of Medicine, Tufts University, Boston, Massachusetts 02111

JoEllen Welsh W. Alton Jones Cell Science Center, Lake Placid, New York 12946

Eric Witt Department of Environmental and Occupational Health, University of Pittsburgh, Pittsburgh, Pennsylvania 15238

P. K. Witting Heart Research Institute, Camperdown, Sydney, NSW 2050, Australia

Preface

The subject for a volume on the fat-soluble vitamins needs no justification considering the importance of this group of nutrients and the rate of expansion of our knowledge of its role in cell biology, genetics, and disease. The level of our understanding has clearly moved from knowing what fat-soluble vitamins do to how they perform their functions. Hand in hand with a knowledge of their molecular mechanisms of action is the recognition that vitamins are used sparingly, and regeneration processes operate in certain cases to recycle vitamins from their metabolites.

We have divided the volume into alphabetical sections beginning with vitamin A and the carotenoids through vitamins D, E, F, and K, and ending with coenzyme Q. The contributors are all acknowledged experts in their particular fields and have made significant contributions to published research results. All have worked assiduously to deliver the product of their labors on a restricted time scale and to provide the most up-to-date information on their respective topics. We are truly grateful for their indulgence.

Part I: Vitamin A commences with a consideration by Niles of the different types of retinoic acid receptors in the nucleus and how they mediate the action of retinoids on gene expression by their ability to bind to specific DNA sequences in the 5′ regulatory region of target genes and activate transcription of these genes in a ligand-dependent manner. Vieira provides an account of how the various retinoid-binding proteins participate in cellular uptake, metabolism, and signaling functions of retinoids, and Newcomer *et al.* examine the detailed structure and conformation of these proteins. Maden illustrates the importance of retinoid concentration during embryo development by describing the effects on developing and regenerating limbs and the head-to-tail axis of the early embryo; when an excess of vitamin A is present, for example, extra structures are produced. The antioxidant activity of vitamin A and retinoids against oxidation of unsaturated lipids is the subject of the chapter by Livrea and Tesoriere, and Constans *et al.* consider the effects of HIV and AIDS on micronutrient deficiencies including vitamins A and E and how this may affect immune functions.

Part II: Carotenoids begins with a chapter by Wang and Krinsky on how β-carotene is converted into vitamin A via the central cleavage pathway and to apocarotenoids and retinoic acid via the excentric pathway. The controversial question of the role of carotenoids in protection against oxidative stress is addressed in two chapters; one, by Duthie *et al.*, reviews the epidemiological evidence and animal studies, and the other, by Day *et al.*, uses tissue culture methods. It is clear that more information is needed before views on the antioxidant action of carotenoids can be reconciled. The section ends with a chapter by Tomoaia-Cotisel and Quinn on the molecular conformation of carotenoids and their interaction with phospholipid model membrane systems.

The two chapters in Part III: Vitamin D are devoted to the roles of the vitamin in cell proliferation and differentiation and in calcium homeostasis and signaling. Welsh *et al.* consider the mechanism by which the $1,25(OH)_2$ D_3 form of vitamin D interacts with the vitamin D_3 receptor to modulate the growth and apoptosis of breast cancer cells. Sergeev *et al.* provide an account of how $1,25(OH)_2$ D_3 regulates calcium-mediated signal transduction by increasing intracellular calcium concentration.

Vitamin E is generally regarded as an effective antioxidant that protects unsaturated lipids from radical attack. To begin Part IV, Liebler reviews the antioxidant reactions of tocopherols, how the cellular level of α-tocopherol is maintained by redox reactions with other antioxidants, and the role of quinone oxidoreductase in maintenance of cellular antioxidant defenses. Quinn summarizes the biophysical evidence concerning the location of vitamin E in phospholipid model membranes and concludes that rather than being distributed randomly throughout the hydrophobic domain, it tends to be arranged in vitamin-enriched clusters. Witting *et al.* give an account of the role of vitamin E in lipid oxidation in lipoproteins and complex lipid emulsions. The final chapter of this section is by Urano, who deals with the role of vitamin E in maintaining antioxidant defense and arresting apoptosis and aging.

The essential fatty acids, vitamin F, are shown by Toborek and Hennig in the first chapter of Part V to play a critical role in gene expression and vascular function as it relates to the pathogenesis of atherosclerosis. One aspect of the process is said to be the influence of linoleic acid on cytokine-mediated inflammatory responses. The other chapter of this section is by Masters, who characterizes the central role of peroxisomes in lipid-mediated intracellular signaling processes.

Part VI consists of a chapter by Newman and Shearer on the metabolism and function of vitamin K. They review the absorption, transport, and metabolism of vitamin K and the molecular basis of its role in protein carboxylation.

Finally, Part VII is devoted to coenzyme Q. While this group of compounds are not strictly vitamins, they have been included because of their ancillary role in recycling vitamin E and because of their antioxidant and prooxidant functions.

We are pleased to be associated with the production of this volume, which we believe will provide a useful update on many contemporary aspects of the fat-soluble vitamins. We are grateful to all the contributors who lent their wealth of knowledge in these endeavors in order to ensure our objectives were achieved.

Peter Quinn
Valerian Kagan

London
Pittsburgh, Pennsylvania

Contents

Part I: Vitamin A

Chapter 1

Control of Retinoid Nuclear Receptor Function and Expression
Richard M. Niles

1. Introduction 3
2. Nuclear Retinoid Receptors 5
   2.1. The RAR Family 5
   2.2. The RXR Family 7
3. Control of Nuclear Retinoid Receptor Function 8
   3.1. Corepressors 8
   3.2. Coactivators 11
   3.3. Phosphorylation 12
4. Control of Nuclear Retinoid Receptor Expression 14
   4.1. Differential Splicing and Promoter Usage 14
   4.2. Control of RAR Transcription by Retinoic Acid 16
   4.3. Control of Transcription by Cyclic AMP 16
   4.4. Other Transcriptional Regulators 17
5. Conclusions 18
6. References 20

Chapter 2

Retinoid Endocrinology: From Metabolism to Cellular Signaling
Amandio V. Vieira

1. Introduction 29
2. Retinoid Transport and Metabolism 30
 2.1. Structure–Function Relations of the Retinol–RBP–TTR Complex 33
 2.2. Cellular Uptake of Retinol 33
 2.3. Biogenesis of Retinoic Acids 36
3. Retinoid Signaling Pathways 38
 3.1. Molecular Phylogeny of Retinoid Signaling 38
 3.2. Nuclear Retinoid Receptors and Regulation of Gene Expression 39
 3.3. Retinoylation and Other Functions of Retinoids 40
4. Physiological Effects of Retinoids 41
5. Future Prospects 44
6. References 44

Chapter 3

Structure and Function of Retinoid-Binding Proteins
Marcia E. Newcomer, R. Scott Jamison, and David E. Ong

1. Introduction 53
2. Extracellular Retinoid-Binding Proteins 54
 2.1. Family Members 54
 2.2. Ligand Binding 57
 2.3. Conformational Changes Induced by Ligand Binding 59
3. Intracellular Retinoid-Binding Proteins 60
 3.1. Family Members 60
 3.2. Conformational Changes upon Ligand Binding 68
 3.3. Ligand Specificity 70
4. The Nuclear Receptors for All-*trans* and 9-*cis* Retinoic Acid 71
 4.1. The Basic Structural Motif 71
 4.2. Conformational Changes 72
5. Conclusion 75
6. References 77

Chapter 4

The Role of Retinoids in Developmental Mechanisms in Embryos
Malcolm Maden

1. Introduction 81
2. Retinoids and Regeneration 82
   2.1. The Regenerating Amphibian Limb 82
   2.2. The Regenerating Amphibian Tail 85
   2.3. Endogenous Retinoids and Their Receptors 85
   2.4. Downstream Effector Genes of RA 87
3. Vitamin A and Limb Development 88
   3.1. The Chick Limb Bud 88
   3.2. Endogenous Retinoids and Their Inhibition 90
   3.3. Downstream Effector Genes of RA 91
4. Retinoids and the Anteroposterior Axis of the Embryo 92
   4.1. A Model of Patterning in the AP Axis 93
   4.2. The Node 94
   4.3. The AP Axis in the Mesoderm 96
   4.4. The AP Axis in the Central Nervous System 97
   4.5. Associated CNS Defects: The Neural Crest and Cranial Ganglia 100
   4.6. Downstream Effector Genes of RA 101
   4.7. Endogenous RA in the Early Embryo 102
   4.8. Disrupting RA Signaling 102
   4.9. Development in the Absence of RA 104
5. Conclusion 104
6. References 105

Chapter 5

Antioxidant Activity of Vitamin A within Lipid Environments
Maria A. Livrea and Luisa Tesoriere

1. Introduction 113
2. Oxidative Stress and Defense 115
   2.1. Introduction 115
   2.2. Oxidation of Unsaturated Lipids and Monitoring of Antioxidant Activity 116
3. Vitamin A and Retinoids as Antioxidants 117
   3.1. Lipoperoxyl Radical-Scavenging Activity of Vitamin A 120

3.2. *In Vitro* Studies 125
3.3. *In Vivo* Studies 130
4. Interactions with Vitamin E 133
5. Conclusions 136
6. References 137

Chapter 6

Plasma Vitamins A and E in HIV-Positive Patients
Joël Constans, Evelyne Peuchant, Claire Sergeant, and Claude Conri

1. Introduction 145
2. Plasma Vitamin A Levels in HIV-Positive Patients 146
3. Vitamin A Dietary Intakes in HIV-Positive Patients 147
4. Plasma β-Carotene in HIV-Positive Subjects 147
5. Plasma Vitamin E in HIV-Positive Subjects 148
6. Mechanisms of Vitamin A and β-Carotene Depletion in HIV-Positive Subjects 149
7. Effects of Vitamin A Deficiency in HIV-Positive Patients 150
8. Vitamin A, HIV Infection, and Pregnancy 151
9. Trials of Supplementation with β-Carotene 152
10. Conclusion 152
11. References 153

Part II: Carotenoids

Chapter 7

The Bioconversion of β-Carotene into Retinoids
Xiang-Dong Wang and Norman I. Krinsky

1. Introduction 159
2. Conversion of β-Carotene into Retinoids 161
2.1. Central Double-Bond Cleavage of β-Carotene: A Major Pathway Leading to Vitamin A Formation 161
2.2. Excentric Double-Bond Cleavage of β-Carotene: Biosynthesis of Retinoic Acid and Other Oxidative Products 164
2.3. Differentiation and Regulation between Central and Excentric Cleavage Pathway of β-Carotene 169

2.4. Biological Significance of the Central and Excentric Cleavage of β-Carotene 172
3. Conclusion 176
4. References 177

Chapter 8

The Role of Carotenoids in Modulating DNA Stability and Lipid Peroxidation: Importance for Human Health
Susan J. Duthie, Andrew R. Collins, and Garry G. Duthie

1. Carotenoid Structure and Antioxidant Activity 181
2. Human Studies 184
   2.1. Conventional Epidemiology 184
   2.2. Molecular Epidemiology 187
3. Animal Studies 194
4. Carotenoids and Cultured Cells 197
   4.1. Human Cells (Transformed and Nontransformed) 197
   4.2. Animal Cells (Transformed and Nontransformed) 200
5. Conclusions 202
6. References 203

Chapter 9

β-Carotene: An Antioxidant or a Target of Oxidative Stress in Cells?
Billy W. Day, Stefania Bergamini, Yulia Y. Tyurina, Gianfranca Carta, Vladimir A. Tyurin, and Valerian E. Kagan

1. Introduction 209
2. Methods 211
3. Results and Discussion 212
4. References 216

Chapter 10

Biophysical Properties of Carotenoids
Maria Tomoaia-Cotisel and Peter J. Quinn

1. Introduction 219
2. Monomolecular Film Techniques 221
3. Surface-Active Properties of Carotenoids: Single-Component Monolayer at the Air/Water Interface 223

3.1. Surface Pressure versus Molecular Area: Compression Isotherms 223
3.2. Structure of Carotenoid Monolayers and Molecular Packing 224
4. Mixed Monolayers of Carotenoids and Lipids at the Air/Water Interface 227
4.1. Compression (π versus A) Isotherms 228
4.2. Thermodynamics of Mixed Monolayers 228
5. Conclusions: Carotenoids in Membranes 237
6. References 238

Part III: Vitamin D

Chapter 11

The Role of Vitamin D_3 and Antiestrogens in Modulating Apoptosis of Breast Cancer Cells and Tumors
JoEllen Welsh, Kathyrn VanWeelden, Louise Flanagan, Ian Byrne, Elizabeth Nolan, and Carmen J. Narvaez

1. Overview of Apoptosis 245
1.1. Morphological and Biochemical Features 245
1.2. Role of Apoptosis in Cancer Therapy 246
2. Vitamin D_3: Negative Growth Regulator of Breast Cancer Cells 247
2.1. The Vitamin D_3 Endocrine System 247
2.2. Role of Vitamin D_3 in Cell Proliferation and Differentiation 247
2.3. Effects of Vitamin D_3 on Mammary Epithelial Cells 248
3. Interactions between Vitamin D_3 and Estrogens in Breast Cancer Cells 249
3.1. Complementary Effects of Vitamin D_3 and Antiestrogens in Estrogen-Dependent Breast Cancer Cells 249
3.2. Effects of Vitamin D_3 on Estrogen-Independent Breast Cancer Cells 249
3.3. Selection and Characterization of MCF-7 Cell Lines Resistant to Antiestrogens and Vitamin D_3 254
3.4. Mechanisms of Action in Mammary Epithelial Cells: Antiestrogens and Vitamin D_3 257
4. Synthetic Vitamin D_3 Analogs 258
4.1. Development, Description, and Characteristics 258

4.2. Preclinical *in Vivo* Trials of Vitamin D_3 Analogs in Breast Cancer 260
5. Summary and Perspectives 265
6. References 267

Chapter 12

Vitamin D and Intracellular Calcium
Igor N. Sergeev, William B. Rhoten, and Vladimir B. Spirichev

1. Introduction 271
2. Vitamin D-Mediated Signal Transduction Pathways 272
3. Cellular Ca^{2+} Homeostasis and Signaling 273
4. Effects of 1,25-Dihydroxyvitamin D_3 on Intracellular Ca^{2+} 274
5. 1,25-Dihydroxyvitamin D_3 and Ca^{2+} Fluxes 278
6. Calbindins and Intracellular Ca^{2+} 281
7. Regulation of Cell Functions via Vitamin D–Intracellular Ca^{2+} Signaling 285
8. Conclusions 288
9. References 289

Part IV: Vitamin E

Chapter 13

Antioxidant Chemistry of α-Tocopherol in Biological Systems: Roles of Redox Cycles and Metabolism
Daniel C. Liebler

1. Introduction 301
2. Antioxidant Chemistry 302
   2.1. Peroxyl Radical Scavenging 302
   2.2. Chemistry of 8a-Substituted Tocopherones 305
   2.3. Chemistry of Epoxytocopherones 307
   2.4. Other Antioxidant Reactions of Tocopherols 308
3. Redox Cycles in α-Tocopherol Antioxidant Function 309
   3.1. One-Electron Redox Cycle 309
   3.2. Two-Electron Redox Cycle 311
4. Redox Regulation of α-Tocopherolquinone and α-Tocopherolhydroquinone 312
5. Conclusion 313
6. References 314

Chapter 14

Localization of Vitamin E in Membranes
Peter J. Quinn

1. Introduction 319
2. Uptake and Transport of Vitamin E 320
   2.1. Distribution of Vitamin E in the Body 321
   2.2. Vitamin E-Binding Proteins 321
   2.3. Lipoproteins 322
3. Interaction with Phospholipid Bilayers 323
   3.1. Effect of Acyl Chain Configuration 325
   3.2. Phospholipid Class 327
   3.3. Binary Phospholipid Mixtures 327
   3.4. Orientation and Motion of Vitamin E in Bilayer Membranes 329
4. Tocopherol–Lipid Complexes 331
   4.1. Fatty Acids and Lysophospholipids 332
   4.2. Diacylphospholipids 332
5. Implications and Conclusions 338
6. References 339

Chapter 15

The Molecular Action of α-Tocopherol in Lipoprotein Lipid Peroxidation: Pro- and Antioxidant Activity of Vitamin E in Complex Heterogeneous Lipid Emulsions
P. K. Witting, J. M. Upston, and R. Stocker

1. Introduction 345
2. The Conventional Molecular Action of Vitamin E (Exclusive Antioxidant Action) 348
3. α-Tocopherol-Mediated Peroxidation (TMP) of LDL Lipid (Anti- and Prooxidant Actions of Vitamin E) 350
   3.1. Chain-Transfer Activity of LDL's Vitamin E 352
   3.2. Phase-Transfer Activity of LDL's Vitamin E 352
   3.3. Antioxidant Activity of LDL's Vitamin E under Conditions of High Radical Flux 355
4. Role of Vitamin E in Enzyme-Mediated LDL Lipid Peroxidation 357
   4.1. Horseradish Peroxidase-Induced LDL Lipid Peroxidation 357
   4.2. MPO/H_2O_2/Cl^-/Tyrosine-Induced LDL Lipid Peroxidation 359

4.3. Oxidation of LDL by 15-Lipoxygenase (15-LO) 361
5. Oxidation of LDL by Nucleophilic Oxidants 366
5.1. ^{-}OCl 367
5.2. $ONOO^{-}$ 368
6. Coantioxidants for α-TOH Are Inhibitors of TMP and Protect LDL against Lipid Peroxidation 369
6.1. $CoQ_{10}H_2$ 370
6.2. α-Tocopheryl Hydroquinone 373
6.3. Ascorbate 374
6.4. Bilirubin 374
6.5. 3-Hydroxyanthranilic Acid 375
7. Possible Roles of TMP and Anti-TMP in Atherogenesis 377
7.1. Putative Oxidants Involved in *in Vivo* LDL Oxidation 378
7.2. Frequency of Radical Hits of Intimal LDL 380
7.3. Extent of Intimal LDL Oxidation 381
8. References 382

Chapter 16

Vitamin E: Its Role in Aging
Shiro Urano

1. Appearance and Cause of Aging 391
2. Oxidative Damage to Biological Systems during Aging 393
3. Deterioration of Function in Vitamin E Deficiency 396
4. Vitamin E and Aging Phenomena 400
5. Vitamin E and Life Span Elongation 405
6. Conclusions 406
7. References 407

Part V: Vitamin F

Chapter 17

The Role of Linoleic Acid in Endothelial Cell Gene Expression: Relationship to Atherosclerosis
Michal Toborek and Bernhard Hennig

1. Introduction: Fatty Acids and Atherosclerosis 415
2. Role of NF-κB in Fatty Acid-Mediated Gene Expression 418

3. Signal Transduction Mechanisms of Fatty Acid-Induced NF-κB 420
4. Fatty Acid-Induced Activation of NF-κB and NF-κB-Related Transcription 422
5. Role of Oxidized Fatty Acids in Endothelial Cell Gene Expression 423
6. Interaction between Fatty Acids and Cytokines and Endothelial Cell Dysfunction 425
7. Role of Linoleic Acid in Proteoglycan Metabolism of the Vascular Endothelium 427
8. Summary 430
9. References 430

Chapter 18

Lipidic Signaling and Peroxisomal Function
Colin Masters

1. Introduction 437
2. Lipid Metabolism and the Peroxisome 438
3. Unique Signaling Features of the Peroxisome 440
4. Further Regulatory Signals 441
   4.1. Phospholipase A_2 Hydrolysis 442
   4.2. The Phosphatidylcholine Cycle 444
5. Essential Fatty Acids 445
6. Involvement of Oxygen Free Radicals and the Peroxisome in Cellular Signaling 447
7. Concluding Comments 448
8. References 451

Part VI: Vitamin K

Chapter 19

Vitamin K Metabolism
Paul Newman and Martin J. Shearer

1. Discovery and Early History of Vitamin K 455
   1.1. The Discovery of a Novel Essential Nutrient 455
   1.2. Structural Determination and Organic Synthesis 456
   1.3. Dietary Sources of Vitamin K and Nutritional Requirements 457

2. Absorption, Transport, and Metabolism 458
2.1. Intestinal Absorption 458
2.2. Plasma Transport 459
2.3. Storage 459
2.4. Catabolism and Turnover 459
3. The Molecular Basis of Vitamin K Action 461
3.1. Vitamin K-Dependent Protein Modification 461
3.2. Vitamin K-Dependent γ-Glutamyl Carboxylase 462
3.3. The Carboxylation Reaction Mechanism 464
4. Vitamin K Antagonists 467
4.1. The Vitamin K Reductase Inhibitors 467
4.2. Vitamin K Derivatives 470
5. Vitamin K-Dependent Proteins 470
5.1. Blood Coagulation Proteins 471
5.2. Gla-Proteins of Extracellular Matrix 475
5.3. Gas6 477
5.4. Conotoxins and Snake Venoms 478
5.5. Other Gla-Proteins 478
5.6. Vitamin K and the Sphingolipid Pathway 479
6. Noncarboxylation Functions of Vitamin K 479
6.1. Vitamin K_2 and Bone 480
6.2. Vitamin K and Leukemia 480
7. References 481

Part VII: Coenzyme Q

Chapter 20

Role of Coenzyme Q and Superoxide in Vitamin E Cycling
Valerian E. Kagan, Yulia Y. Tyurina, and Eric Witt

1. Introduction and Summary 491
2. Coenzyme Q as an Antioxidant 492
2.1. Intracellular Antioxidant Activity (Membranes) 492
2.2. Extracellular Antioxidant Activity (Low-Density Lipoproteins) 493
3. Does Coenzyme Q Recycle Vitamin E? 494
3.1. Direct Radical Scavenging Effects of Coenzyme Q 494
3.2. Reduction of Vitamin E Radicals by Ubiquinols 495
4. The Role of Superoxide in the Antioxidant Action of Ubiquinol 496

4.1. Ubisemiquinone Protects Vitamin E against Superoxide-Driven Oxidation 498
4.2. Superoxide-Dependent Reduction of Coenzyme Q Protects Cell Phospholipids against Oxidative Stress 499
5. Antioxidants, Oxidative Stress, and Human Health 501
5.1. General Health Significance of Coenzyme Q 501
5.2. Special Considerations with Low-Density Lipoprotein 501
6. References 503

Chapter 21

Prooxidant Functions of Coenzyme Q
Hans Nohl, Lars Gille, and Andrey V. Kozlov

1. Biological Functions and Side Functions of Coenzyme Q ... 509
2. Is Reduced Coenzyme Q Autoxidizable? 511
3. Conditions Allowing Autoxidation of Mitochondrial Semiquinones 514
4. Prooxidant Activity of Coenzyme Q Used as Antioxidant 518
5. Autoxidation of Ubisemiquinones Derived from Radical-Scavenging Ubiquinol 521
6. Ubisemiquinones Donate Electrons to Peroxides 522
7. The Ambiguous role of Coenzyme Q when Used as Antioxidant 523
8. References 525

Index 527

Part I

Vitamin A

Chapter 1

Control of Retinoid Nuclear Receptor Function and Expression

Richard M. Niles

1. INTRODUCTION

Vitamin A and its physiological metabolites (collectively known as retinoids) are important regulators of embryogenesis, cell growth and differentiation, vision, and reproduction (Hofmann and Eichele, 1994; Eskild and Hansson, 1994; DeLuca, 1991; Wald, 1968; Thompson *et al.*, 1964). Deficiency of vitamin A leads to well-described defects in vision, fertility, and, in animals, increased susceptibility to carcinogenesis (Chambon, 1994; Sporn *et al.*, 1976; Wilson *et al.*, 1953). Retinol (vitamin A) is metabolized by cells to form a number of biologically active compounds (Napoli *et al.*, 1993). These include all-*trans* retinoic acid, 9-*cis* retinoic acid, and didehydroretinoic acid. Retinol is first converted to retinal, which is then metabolized by a retinal dehydrogenase leading to production of all-*trans* retinoic acid (Posch *et al.*, 1992, 1991). The metabolic pathway which generates 9-*cis* retinoic acid has yet to be elucidated. Didehydroretinoic acid is produced from 3,4-didehydroretinol by a pathway that is similar to that which converts retinol to all-*trans* retinoic acid (Thaller and Eichele,

---

Richard M. Niles Department of Biochemistry and Molecular Biology, Marshall University School of Medicine, Huntington, West Virginia 25755.

Subcellular Biochemistry, Volume 30: Fat-Soluble Vitamins, edited by Quinn and Kagan. Plenum Press, New York, 1998.

1990). Dietary sources of vitamin A are β-carotene from plant sources and retinyl ester from animal sources. Vitamin A from these sources is either stored in the liver as retinyl esters, or packaged with retinol-binding protein together with transthyretin for export to the circulatory system (Blaner, 1989).

Retinol uptake by cells may be mediated by plasma membrane receptors for retinol-binding protein (Bavik *et al.*, 1992, 1991). Within cells retinol is bound by cellular retinol-binding protein (CRBP), while its major metabolite, all-*trans* retinoic acid, is bound by cellular retinoic acid-binding protein (CRABP). These proteins are members of a gene family of low-molecular-weight proteins which are carriers of hydrophobic ligands such as fatty acids (Clarke and Armstrong, 1989). There are two forms of both CRBP and CRABP, CRBP I and II, and CRABP I and II (Giguere *et al.*, 1990a; Ong, 1987; M. S. Levin *et al.*, 1987; Ong and Chytil, 1978). CRBP I is ubiquitously expressed, while CRBP II expression is retricted to the small intestine (MacDonal and Ong, 1988). It is thought that CRBP serves as a carrier to direct retinol to various sites of metabolism within the cell, resulting in its conversion to all-*trans* retinoic acid and perhaps to 9-*cis* retinoic acid (Posch *et al.*, 1992, 1991; Bernstein *et al.*, 1987). The CRABPs are expressed by all vertebrates and are highly conserved. CRABP I has a higher affinity for all-*trans* retinoic acid than CRABP II (Astrom *et al.*, 1991; Bailey and Siu, 1988). Neither of these two binding proteins has a high affinity for 9-*cis* retinoic acid (Fogh *et al.*, 1993; Fiorella *et al.*, 1993). In addition, the expression of CRABP II, but not CRABP I, can be induced by retinoic acid (Durand *et al.*, 1992). The exact function of the CRABPs is not known. It has been postulated that they (i) act as a buffer by sequestering retinoic acid in the cytoplasm and preventing excess ligand from reaching the nucleus, (ii) carry and present retinoic acid to different enzyme systems for metabolism, thus regulating its intracellular concentration, and (iii) shuttle retinoic acid to the nuclear receptors (Napoli, 1994; Ong *et al.*, 1994; Ruberte *et al.*, 1992; Takase *et al.*, 1986). However, recent experiments in which both CRABP I and II genes were inactivated demonstrated that the transgenic mice developed to full term and were essentially normal (Lampron *et al.*, 1995). Possibly under these unusal conditions (i.e., inactivation of both CRABP genes), other hydrophobic ligand-binding proteins are able to substitute, to a certain degree, for the CRABPs.

Until 1987, it was not clear how retinoic acid (RA) regulated the expression of genes in cells which differentiated and/or had decreased proliferation in response to retinoid treatment. In that year two laboratories independently identified a retinoic acid receptor (now known as RARα1) which exhibited a high degree of sequence homology to the well-known steroid hormone receptors (Giguere *et al.*, 1987; Petkovich *et al.*,

1987). It is now well accepted that these "nuclear" retinoid receptors mediate the action of retinoids on gene expression by their ability to bind to specific DNA sequences in the 5′ regulatory region of target genes (retinoic acid response elements) and activate transcription of these genes in a ligand-dependent manner.

2. NUCLEAR RETINOID RECEPTORS

The nuclear retinoid receptors consist of two distinct subfamilies, those which specifically bind all-*trans* retinoic acid and 9-*cis* retinoic acid (RAR), and those which bind only 9-*cis* retinoic acid (RXR) (Allenby *et al.*, 1993; A. A. Levin *et al.*, 1992; Mangelsdorf *et al.*, 1990). As described below, each of these subfamilies has receptor subtypes and most of these subtypes can generate different isotypes by alternative splicing and/or alternative promoter usage (Leroy *et al.*, 1991a,b; Zelent *et al.*, 1991; Giguere *et al.*, 1990). It is assumed that once these receptors are synthesized, they remain in the nucleus until they are degraded. However, little is known about the rate of synthesis or half-life of these receptors. This is due in part to the exceedingly small amount of these receptors expressed in cells. Only recently have receptor subtype antibodies been produced which have the sensitivity and specificity to detect these RAR/RXRs in some cells (Rochette-Egly *et al.*, 1991). It is not clear if these retinoid receptors reside in a specific nuclear location. Some investigators believe that these receptors are always bound to their DNA response element (Dalman *et al.*, 1991; Yang *et al.*, 1991) and that the ligands cycle on and off to activate gene transcription. It is clear that these retinoid receptors are tightly bound in the nucleus and that extraction with 6M KCl does not remove all of the receptor (R. M. Niles and A. Traish, unpublished observations). Lastly, it is still not known how the ligands (all-*trans* and 9-*cis* retinoic acid) are transported to the nucleus and delivered to the nuclear retinoid receptors.

2.1. The RAR Family

As described previously, the first nuclear receptor for retinoic acid was discovered in 1987 by two independent laboratories. The existence of additional receptor genes was suggested by demonstration of hybridization of the newly discovered receptor cDNA to other genome fragments under low-stringency conditions. At the present time three genes encoding RARs (RARα, β, and γ) have been cloned from both mouse and human. These three RAR genes map to different chromosomes in both of these species (Mattei *et al.*, 1991; Ishikawa *et al.*, 1990). The RARs are highly conserved among mammalian species. For example, there is more similarity between RARα from mouse and human than between human RARα and RARβ (Zelent *et al.*, 1989; Ragsdale *et al.*, 1989). Although many species express RARs, they have not been found in invertebrates. Each RAR gene can generate multiple isoforms by using two different promoters, alternative splicing of exons, and initiation of translation at an internal CUG codon (Nagpal *et al.*, 1992b; Leroy *et al.*, 1991a; Zelent *et al.*, 1991; Kastner *et al.*, 1990; Giguere *et al.*, 1990b). These RAR isoforms differ in the amino-terminal region of the encoded protein and could potentially add to the diversity of the retinoic acid signal.

The RARs share the modular and functional domains described for other members of the steroid hormone receptor gene family (Fig. 1). The N-terminal A/B region has been shown to function as a ligand-independent transcriptional activation domain (Folkers *et al.*, 1993; Nagpal *et al.*, 1993, 1992a). The B part of this region is well conserved among the different RAR subtypes, but the A segment is divergent. Region C contains two zinc-binding, cysteine-rich segments (so-called zinc fingers). This domain is responsible for binding to the DNA response element, usually a direct repeat of the nucleotide sequence PuGGTCA separated by 5 base pairs (bp). This region is highly conserved among the three RAR subtypes. Between this DNA-binding domain and the ligand-binding domain (E) is a so-called "hinge" region (D). This segment is also well conserved among RAR subtypes and even more conserved between a given receptor subtype in different species. These findings suggest that this region may have an RAR-subtype-specific function, which is not yet known. Segment E is the ligand-binding domain of the RARs. This region is also highly conserved between RAR subtypes and is quite complex. It not only mediates binding of all-*trans* and 9-*cis* retinoic acid, but also has ligand-dependent transcriptional

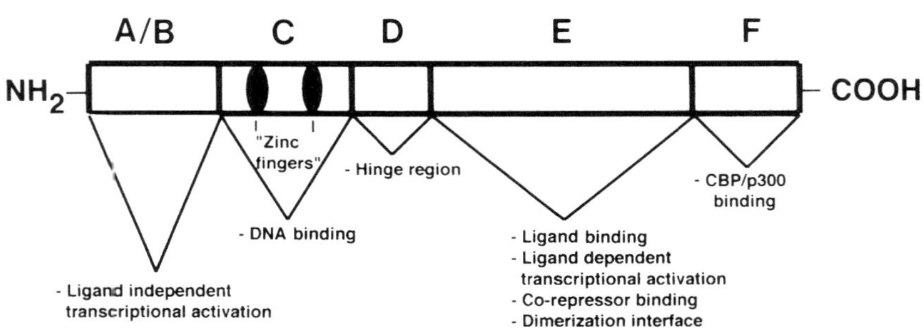

FIGURE 1. Functional domains of retinoid nuclear receptors. Currently known functions for the different regions of these receptors are listed under the appropriate segments (capital letters). RXRs do not contain region F.

activation and receptor dimerization functions (Nagpal *et al.*, 1992a; Basiahmad *et al.*, 1992; Forman *et al.*, 1989). The carboxy-terminal F region is highly divergent among RAR subtypes and its function has yet to be determined. This region has been suggested to bind the coactivator/integrator CBP/p300.

Different RAR subtypes have distinct tissue-specific expression. RARα is ubiquitously expressed, with very high levels observed in certain regions of the brain (de The *et al.*, 1989). RARβ is primarily found in kidney, prostate, pituitary, adrenal glands, spinal cord, and cerebral cortex (Benbrook *et al.*, 1988). Many tumor cells lack RARβ expression, although in some cells its expression can be induced by treatment of the cells with retinoic acid (Xu *et al.*, 1994; Gebert *et al.*, 1991; Hu *et al.*, 1991). This has led some investigators to conclude that RARβ may act as a tumor suppressor gene (Houle *et al.*, 1993). RARγ has the most restricted distribution, being predominantly expressed in skin and lungs (Kastner *et al.*, 1990; Zelent *et al.*, 1989; Noji *et al.*, 1989).

2.2. The RXR Family

In 1990 another family of nuclear retinoid receptors was discovered (Mangelsdorf *et al.*, 1990). Using low-stringency hybridization, it was found that a member of the orphan receptor family cross-reacted with the RARs. Sequencing of this receptor gene revealed similarity to the RARs especially in the DNA-binding domain. However, the ligand-binding domains of these two receptors were very divergent. This receptor, named RXR, could be activated by high concentrations of all-*trans* retinoic acid; however, this retinoid did not bind with high affinity to the receptor. Thus it was postulated that another retinoid metabolite was the true agonist. Several groups, using different biochemical strategies, identified 9-*cis* retinoic acid as the ligand for RXR (A. A. Levin *et al.*, 1992; Heyman *et al.*, 1992).

Analogous to the RAR, three distinct RXR genes (α, β, and γ) have been isolated. They map to different chromosomes in the mouse (Hoopes *et al.*, 1992). Only RXRγ has been demonstrated to have different isoforms. RXRγ1 and γ2 differ in their amino-terminal regions due to alternative usage of two different promoters (Liu and Linney, 1993). RXRα and β are widely expressed in tissues. However, the expression of RXRγ is more restricted. RNA for this receptor subtype has been observed in liver, kidney, adrenal gland, lung, and brain (Mangelsdorf *et al.*, 1992, 1990).

Although RAR, at high concentrations, can bind to its DNA response element as a homodimer, a variety of studies have provided strong evidence that the physiologically active receptor unit is a heterodimer between RAR and RXR (Leid *et al.*, 1992; Yu *et al.*, 1991; Bugge *et al.*, 1992; Kliewer *et al.*,

1992a; Marks *et al.*, 1992; Zhang *et al.*, 1992; Yang *et al.*, 1991). In most instances RXR occupies the 5′-upstream site on the DNA-response element, while RAR binds to the 3′-downstream site (Predki *et al.*, 1994; Kurokawa *et al.*, 1993). In addition to forming heterodimers with RAR, the RXRs also form heterodimers with thyroid hormone receptors (Leid *et al.*, 1992; Bugge *et al.*, 1992), vitamin D_3 receptors (Carlberg *et al.*, 1993; Kliewer *et al.*, 1992a), a number of orphan receptors (Kliewer *et al.*, 1992b; Widom *et al.*, 1992), and peroxisomal proliferator activated receptors (PPAR) (Keller *et al.*, 1993; Kliewer *et al.*, 1992c). In all of these instances, the heterodimer is thought to be the physiologically active form of the receptor. A somewhat controversial point is whether RXR functions as a silent partner in these heterodimers, or whether binding of the ligand 9-*cis* retinoic acid is required for full functional activity of the different heterodimers (Forman *et al.*, 1995; Apfel *et al.*, 1995; Kurokawa *et al.*, 1994).

3. CONTROL OF NUCLEAR RETINOID RECEPTOR FUNCTION

A variety of studies have documented that there is interference or "squelching" between the transcriptional activation functions of different steroid receptors (Tasset *et al.*, 1990; Bocquel *et al.*, 1989; Meyer *et al.*, 1989). These findings have led to the hypothesis that there exist transcriptional intermediary factors. These proteins would be involved in the mechanism by which receptor transcriptional activation domains communicate with the basal transcription machinery. Furthermore, these cofactors would be shared by different steroid hormone receptors and be present in limiting amounts such that the steroid hormone receptors would compete for these proteins (Onate *et al.*, 1995; Halachmi *et al.*, 1994). To test this hypothesis, investigators searched for proteins that would interact with receptors in the presence or absence of ligands. Using "Far Western" blotting and yeast two-hybrid screens, a variety of proteins were identified. These putative coregulators are nuclear proteins that seem to be ubiquitously expressed. As described below, these proteins act as either coactivators or corepressors. It should also be remembered that the putative activity of these factors has been tested by *in vitro* binding assays or overexpression studies and thus the physiological significance/role of many of these proteins remains to be established.

3.1. Corepressors

Unlike other steroid hormone receptors (including RXR), RAR and thyroid hormone receptor (THR) can repress the transcription of target genes in the absence of ligand. This effect is likely mediated by the carboxy-terminal region that contains the ligand-binding domain of these receptors. Studies using a mutant RAR provided some evidence for a corepressor protein that interacted with the ligand-binding domain in the absence of ligand (Perlmann and Vennstrom, 1995; Martin *et al.*, 1994). Using a biochemical approach, a 270-kDa protein was isolated that interacted with DNA-bound RAR and THR in the absence, but not in the presence of ligand (Horlein *et al.*, 1995). The protein was named N-CoR (nuclear receptor corepressor). Recent studies have found that the N-terminal region of this protein contains the repressive domain, while the C-terminus is responsible for the interaction with RAR and THR (Zamir *et al.*, 1996; Chen and Evans, 1995; Kurokawa *et al.*, 1994). Mutations in the N-terminal half of the ligand-binding domain, as well as in the hinge region (domain D, Fig. 1), abolish N-CoR binding and repression of target gene transcription in the absence of ligand.

An independent laboratory, using a different screen for proteins that interacted with unliganded receptor, cloned a different corepressor. This protein interacts with the THR and RAR ligand-binding domains, and therefore was named SMRT (silencing mediator of RAR and THR) (Chen *et al.*, 1996; Chen and Evans, 1995). These two proteins, though distinct, are closely related, especially in the C-terminal region. SMRT also contains an N-terminal repression domain, like N-CoR, but there is little sequence homology between these two regions. Nothing is known about the mechanism by which this repressive region in either protein performs its function. Some evidence has been obtained that upon binding of ligand, the receptor undergoes a conformational change which causes the corepressor to dissociate from the receptor and at the same time allows coactivators to bind to the liganded receptor.

A third corepressor protein has been described (Burris *et al.*, 1995). This protein, called TRUP (THR uncoupling protein), binds to a region of the human THR containing the hinge region and the N-terminal section of the ligand binding domain in a ligand-independent manner. TRUP was found to inhibit the transactivation by THR and RAR, but had no effect on the estrogen receptor (ER) or RXRs in mammalian cells. It was found that TRUP exerted its action by interfering with the ability of these receptors to interact with their DNA response elements. This protein apparently is unrelated to the N-CoR and SMRT corepressors described above.

Lastly, it was recently reported that the well-known transcription factor E2F can indirectly inhibit the transcriptional activity of the RAR (Costa *et al.*, 1996). E2F does not directly interact with the RAR proteins, nor does it bind to their DNA response elements. It is suggested that E2F reduces expression from retinoic acid-inducible promoters by sequestering a coactivator that is required by the RARs. Thus this factor does not fit the definition of a corepressor (directly binds to the receptor in the absence of ligand).

AP-1 and Calreticulin

The transcription factor known as AP-1 (activator protein-1) is a complex of Jun family (c-Jun, JunB, JunD) homodimers, or Jun family–Fos family (c-Fos, FosB, Fra-1, Fra-2) heterodimers. The transcriptional activity of AP-1 can be stimulated by growth factors and the tumor promoter tetradecanoyl phorbol acetate (TPA) (Lamph *et al.*, 1988; Angel *et al.*, 1988, 1987; W. Lee *et al.*, 1987). It has been shown by several laboratories that RAR (as well as other steroid hormone receptors, but not RXR) and AP-1 mutually antagonize each other's transcriptional activity. There are two mechanisms that might account for this action. In the human osteocalcin gene, the promoter contains a composite retinoic acid response element and an AP-1-binding site, which confers both retinoic acid and TPA-responsivenes (Schule *et al.*, 1990). Thus occupancy of the AP-1 site by this transcription factor may sterically hinder the binding of RAR to its response element. In the case of other genes, such as collagenase and stromelysin, this mechanism cannot be invoked, since RAR does not bind to the promoters of these genes. Instead it has been proposed that RAR and c-Jun form nonproductive complexes in solution (Schule *et al.*, 1991; Yang Yen *et al.*, 1991; Lafyatis *et al.*, 1990; Nicholson *et al.*, 1990), resulting in mutual inhibition of their respective transcriptional capabilities.

Recently another plausible mechanism has been put forth to explain the mutual antagonism between these two transcription factors. Investigators have identified a 265-kDa protein which binds to phosphorylated CREB (cyclic AMP response element binding protein) (Kwok *et al.*, 1994; Chrivia *et al.*, 1993). This protein, refered to as CBP (CREB-binding protein), is required for transcriptional activity of CREB. A protein closely related to CBP, called p300, was independently isolated through its ability to interact with the transcription factor E1A (Eckner *et al.*, 1994). These two proteins have extensive homology and they both bind to CREB, TFIIB (Kwok *et al.*, 1994; Chrivia *et al.*, 1993) E1A (Lundblad *et al.*, 1995), and c-Fos (Bannister and Kouzarides, 1995). Several experimental approaches have shown that CBP/p300 is required for transcriptional activity of AP-1 (Kwok *et al.*, 1994; Arias *et al.*, 1994). Recently, Kamei *et al.* (1996) demonstrated that the ligand-binding domains of many nuclear receptors, including RARs, have high-affinity binding with a conserved domain in the N-terminus of CBP/p300. The authors suggest that competition for limiting amounts of CBP/p300 may account for the inhibitory action of RAR on

AP-1 transcriptional activation. CBP/p300 has been found to also have histone acetyltransferase activity (Ogryzko *et al.*, 1996). This suggests that CBP/p300 is not just an adaptor between RARs or other transcription factors and the basal transcription machinery. It is likely that it directly contributes to transcriptional regulation by altering chromatin structure to allow binding of these transcription factors.

Calcium is a second messenger that regulates a variety of cellular functions (Carafoli, 1987). Because of its multiple actions there is a stringent requirement for the maintenance of cytosolic calcium by transport and storage proteins. The calcium-binding protein calreticulin fulfills some of these functions by sequestering calcium in the endoplasmic reticulum (Opas *et al.*, 1991; Tharin *et al.*, 1991). However, calreticulin has also been localized to the nucleus (Opas *et al.*, 1991), leading to speculation that the protein may also influence gene expression. Calreticulin specifically binds to the synthetic peptide KLGFFKR (Berg, 1990), which is similar to the amino acid sequence in the DNA-binding domain of the superfamily of steroid hormone receptors (Laudet *et al.*, 1992; Beato, 1989; Evans, 1988). Burns *et al.* (1994) and Dedhar *et al.* (1994) showed that calreticulin inhibited the action of glucocorticoid and androgen receptors, respectively. The N-terminal region of calreticulin was shown to interact with the DNA-binding domain of these receptors and prevent binding to their respective DNA response elements. Our laboratory has found that calreticulin can also bind to RAR and inhibits its binding to an RARE. Further, we showed that calreticulin expression is regulated through the cyclic AMP second messenger system and that both cyclic AMP and overexpression of calreticulin inhibited the retinoic acid induction of PKCα expression in B16 mouse melanoma cells (Desai *et al.*, 1996).

The material discussed above makes it clear that mammalian cells have evolved a variety of means to repress the function of nuclear retinoid receptors, especially in the absence of ligand. This suggests that it is critical to control precisely receptor function. The cell achieves this control by both negative and positive (see below) regulation.

3.2. Coactivators

A variety of studies have yielded data to suggest a requirement for coactivators in order for nuclear retinoid receptors to achieve full transcriptional activation (Goodrich *et al.*, 1993; Berkenstam *et al.*, 1992; Yoshinaga *et al.*, 1992). In addition, it is now clear that a highly conserved region in the ligand-binding domain, AF-2 (transcriptional activation function-2), undergoes a marked conformational change during ligand binding (Wurtz *et al.*, 1996; Renaud *et al.*, 1995). It was hypothesized that this conformational change allowed proteins to associate or dissociate from the receptor, altering its transcriptional activity. Using this information, several laboratories employed both biochemical techniques and the yeast two-hybrid system to identify proteins which bound to the AF-2 region in a ligand-dependent manner. Two proteins were isolated (140 and 160 kDa) which interacted with the estrogen receptor and the RAR only in the presence of the appropriate ligand (Kurokawa *et al.*, 1995; Halachmi *et al.*, 1994; Cavailles *et al.*, 1994). Cloning of genes from the yeast two-hybrid system revealed several proteins such as RIP1 (Cavailles *et al.*, 1995), TIF1 (Le Douarin *et al.*, 1995), TRIP/Sug1 (J. W. Lee *et al.*, 1995), and SRC1 (Onate *et al.*, 1995) that bind to multiple nuclear receptors in a ligand-dependent fashion. The function of all of these presumptive coactivators and their physiological significance is yet to be determined.

Another activator of RARs is CBP/p300. This activator also binds to RAR and RXR in a ligand-dependent fashion. This implies that CBP/p300 recognizes a region in the AF-2 domain exposed due to conformational change induced by the ligand. The binding site of CBP/p300 is located within its N-terminal 100 amino acid residues (Kamei *et al.*, 1996). These interactions occur on DNA as demonstrated by the ability of an N-terminal fragment of CBP to supershift an RAR–RXR/DNA complex (Kamei *et al.*, 1996). Expression of CBP together with an RARE (retinoic acid response element) reporter gene led to a threefold enhancement in retinoid-dependent transcription, but had little effect on promoters lacking an RARE (Kamei *et al.*, 1996).

The adenovirus transcription factor E1A has also been found to function as a cofactor for the RARβ receptor. A direct interaction between the ligand-binding domain (AF-2) of RARβ and the C-terminal part of CRII of E1A was observed. It is hypothesized that E1A forms a bridge between the AF-2 domain of RARβ and the TATA-box-binding protein, since E1A binds to both of these proteins (Folkers and van der Saag, 1995; Kruyt *et al.*, 1993). Since many cells have E1A-like activity, this protein is likely a significant coactivator for at least RARβ and likely for other RAR (and RXR?) subtypes.

3.3. Phosphorylation

All RAR subtypes have been shown to be phosphoproteins (Rochette-Egly *et al.*, 1992, 1991; Gaub *et al.*, 1992). Whether all RXR subtypes are also phosphoproteins remains to be determined. The phosphorylation of RARβ1 and β3 is enhanced by ligand, while RARγ and RARβ2 do not exhibit this response (Rochette-Egly *et al.*, 1992). These results suggest that either there are amino acid sites present in the RARβ1 and 3, but not in the β2 splice/differential promoter variants that undergo ligand-dependent phosphorylation, or that the conformation of the β1 and 3 variants is such that amino acids are available for phosphorylation. Interestingly, only RARβ has been shown to contain a phosphotyrosine site (Rochette-Egly *et al.*, 1992). The significance of this finding is not known. Serine-369 in RARα1 is phosphorylated in response to stimulators of protein kinase A (PKA) activity in COS cells (Rochette-Egly *et al.*, 1995). The existence of additional phosphopeptides suggest that several other amino acid sites are phosphorylated when RARα1 is overexpressed in COS cells.

The question of the role of phosphorylation on receptor function has been addressed in several studies. Lefebvre *et al.* (1995) used okadaic acid to inhibit protein phosphatases 1 and 2A and showed that phosphorylation increases the transcriptional and DNA-binding activities of nuclear retinoid receptors. They further showed that alkaline phosphatase treatment of RARα or RXRα inhibited their DNA-binding activity. Since bacterially expressed RAR and RXR were less efficient at forming heterodimers than the same receptors extracted from eukaryotic cells, it was concluded that receptor phosphorylation, which is absent in bacterially expressed receptors, enhanced heterodimer formation. This in turn would increase the efficiency of DNA binding. Transfection of the catalytic subunit of PKA and a reporter gene to measure RAR transcriptional activity showed that this kinase subunit increased ligand-independent and ligand-dependent activity of the endogenous nuclear retinoid receptors in both CV-1 and HeLa cells (Huggenvik *et al.*, 1993). The same result was found when RARα was overexpressed in these cells. Although the mechanism for ligand-independent transcriptional activation of retinoid receptors by PKA phosphorylation is not known, the investigators suggest that a phosphorylation site in the ligand-binding domain could induce a conformation change in the receptor to make it competent (bind coactivators?) for transcriptional activation.

Our laboratory has found that manipulating protein kinase C (PKC) enzyme activity in B16 mouse melanoma cells can alter the activity of the RARs (D. Desai and R. M. Niles, unpublished data). A short-term treatment (1 hr) with phorbol dibutyrate, which increases PKC enzyme activity, resulted in increased RA-induced transcription of a reporter gene. The same treatment also enhanced the transcriptional activation of coexpressed RARα, β, or γ. Long-term treatment of the cells with phorbol dibutyrate depleted the cells of PKC and inhibited RA-induced reporter gene expression by cotransfected RARα, β, or γ. This same treatment also resulted in markedly decreased DNA binding of the endogenous RAR. A previous report also found that long-term phorbol ester treatment inhibited RAR–DNA binding (Tahayato *et al.*, 1993). Yang *et al.* (1994) found that activation of T cells increased PKC enzyme activity and enhanced retinoid-induced gene transcription. They also showed that this action of PKC was independent of AP-1, since transfection and expression of a dominant-negative c-Jun did not prevent enhanced RAR transcriptional activation by PKC. Thus, signal transduction pathways that lead to increased activity of protein kinases likely alter the phosphorylation of the nuclear retinoid receptors and provide a means for fine tuning the biological actions of retinoids.

4. CONTROL OF NUCLEAR RETINOID RECEPTOR EXPRESSION

In addition to regulating receptor function by the means described in the previous sections, it is becoming increasingly evident that the expression of the nuclear retinoid receptors is highly regulated. Since these receptors are produced in limiting quantities, altering their synthesis should have a significant effect on retinoid responsiveness.

4.1. Differential Splicing and Promoter Usage

There are two RARα isoforms, at least four RARβ isoforms, and two RARγ isoforms (Fig. 2). In contrast, among the RXR subtypes, only RXRγ has been found to generate an additional isoform (Liu and Linney, 1993). As discussed earlier, these isoforms are due to alternative splicing sites and alternative promoter usage. Since some of these isoforms can generate receptor proteins with altered N-termini, these isoforms may provide another means of regulating the retinoid signal. Some differences in isoform expression have been found. RARα2 is more abundant in lung, liver, and intestine than RARα1 (Leroy *et al.*, 1991b). The distributions of RARβ 1 and of 3 are quite similar (perhaps since they are expressed from a common promoter), while the distributions of RARβ2 and of 4 are similar, but differ from those of RARβ1 and of 3. These latter isoforms use a common promoter that is different from that used by RARβ2 and 4 (Nagpal *et al.*, 1992b). RXRγ1 and 2 are expressed in both skin and lungs; however, a recent report has demonstrated that the expression of RXRγ1 in the anterior pituitary is restricted to the thyrotroph cells, where it may mediate inhibition of thyroid-stimulating hormone (TSH) expression (Haugen *et al.*,

FIGURE 2. RAR and RXR isotypes. Different retinoid receptor isoforms generated by differential splicing and/or promoter usage. The numbers represent the approximate length of the protein encoded by these mRNAs. The solid boxes represent different N-terminal regions generated by alternative splicing. Also note that RARβ4 and RXR γ2 are missing the A domain.

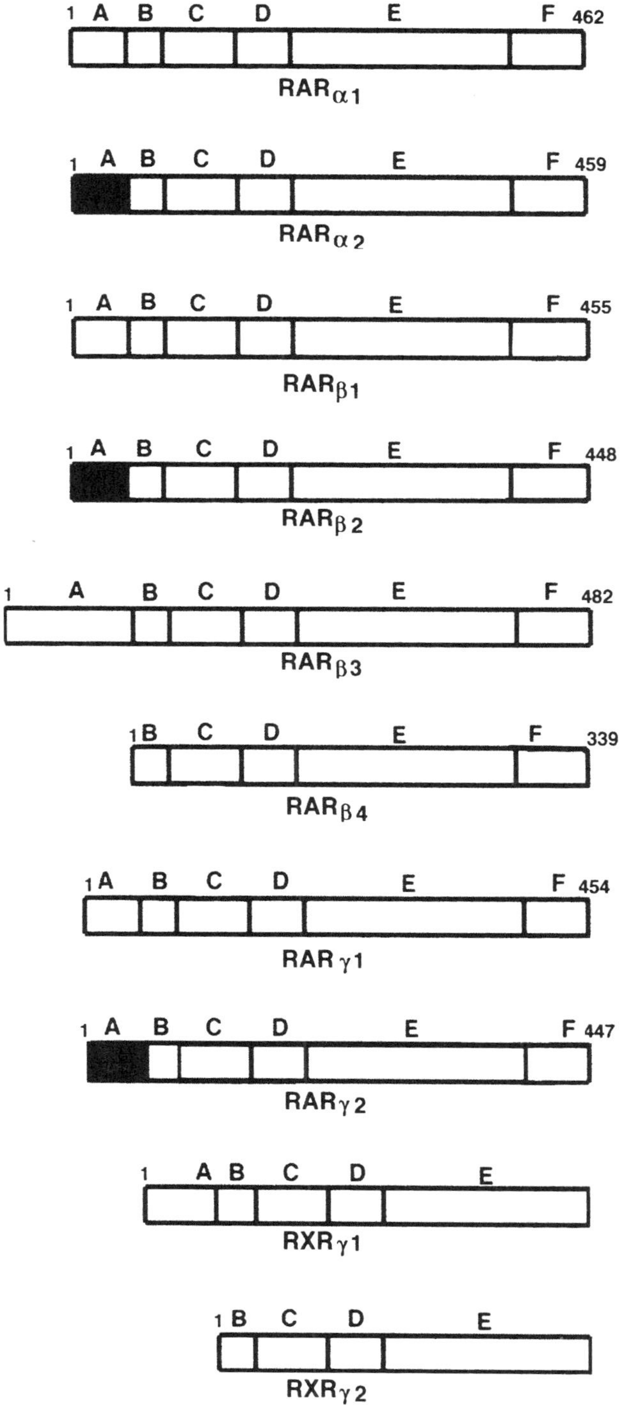

1 A B C D E F 462
RARα1
1 A B C D E F 459
RARα2
1 A B C D E F 455
RARβ1
1 A B C D E F 448
RARβ2
1 A B C D E F 482
RARβ3
1 B C D E F 339
RARβ4
1 A B C D E F 454
RAR γ1
1 A B C D E F 447
RARγ2
1 A B C D E
RXRγ1
1 B C D E
RXRγ2

1997). Taken together these findings indicate that tissue-specific transcription factors regulate expression of not only retinoid receptor subtypes, but also of isoforms within these subtypes.

4.2. Control of RAR Transcription by Retinoic Acid

The first RARE was found in the promoter region of the RARβ gene (de The *et al.*, 1990; Hoffman *et al.*, 1990; Sucov *et al.*, 1990). Further analysis of RAR 5′ regulatory regions has revealed that all three RAR genes contain an RARE in one of their two promoters (Lehmann *et al.*, 1992; Leroy *et al.*, 1991b; Sucov *et al.*, 1990; de The *et al.*, 1990). Thus, retinoids can amplify their own signal transduction pathway. Since retinoids are important regulators of growth and differentiation, defects in this amplification could have deleterious consequences to cells. Perhaps the best example of this is the absence of RARβ expression and its inducibility by retinoids in a variety of cancer cells. Many studies implicate RARβ as the RAR subtype which likely regulates growth and differentiation. Thus the inability of cells to amplify the expression of this receptor could be a contributing factor to carcinogenesis (i.e., RARβ acts like a tumor suppressor gene).

4.3. Control of Transcription by Cyclic AMP

Transfection and overexpression of the catalytic subunit of PKA into CV-1 and HeLa cells was shown to increase ligand-independent transcriptional activation of the RARs (Huggenvik *et al.*, 1993). In addition to these actions, several laboratories have shown that treatment of cells with cyclic AMP analogs, or hormones which increase intracellular cyclic AMP levels, can lead to altered expression of the RARs (Xiao *et al.*, 1996; Hu and Gudas, 1990; Martin *et al.*, 1990). F9 teratocarcinoma cells treated with dibutyryl cyclic AMP had decreased expression of RARα and γ (Hu and Gudas, 1990). The cells first had to be treated with RA in order for cyclic AMP to affect RAR expression. Although it was not possible to measure transcription directly, due to the low level of RAR expression, all other measurements indicated that the effect of cyclic AMP was indeed at the transcriptional level. Our laboratory has reported similar effects of cyclic AMP in B16 mouse melanoma cells (Xiao *et al.*, 1996). Treatment of these cells with 8-bromocyclic AMP or melanocyte-stimulating hormone (MSH) (which increases intracellular cyclic AMP levels) drastically decreased the expression of RARα and γ, and decreased the induction of RARβ by RA. These cyclic AMP-induced effects were independent of new protein synthesis, occurred within 2 hr of treatment, and did not affect the stability of the RAR mRNAs. Analogous to the case for F9 teratocarcinoma cells, we were not able to measure directly transcription rates of RAR in the B16 cells, but all of the other data suggest that cyclic AMP directly decreases the transcriptional rate of RAR genes. Whether these effects of cyclic AMP are mediated through cyclic AMP response elements (CRE) which have been found in the 5′-regulatory regions of some of the RARs is not known.

The studies reported above involved treating mouse cells with cyclic AMP. In preliminary experiments our laboratory has found that treatment of the human leukemic cell line HL-60 with 8-bromo-cyclic AMP *increases* the expression of RARα and induces the expression of RARγ, which is not normally found in these cells (D. S. Desai and R. M. Niles, unpublished data). These results may partially explain previous reports which demonstrated that cyclic AMP enhanced the RA-induced differentiation of HL-60 cells (Olsson *et al.*, 1992). They also demonstrate that the effect of cyclic AMP on RAR expression is species- and perhaps cell-type-specific. Our laboratory has also reported that 8-bromo-cyclic AMP inhibits the expression of RXRβ, but not RXRα, mRNA in a time-dependent manner (Desai and Niles, 1995). We could not determine if this was a direct effect of cyclic AMP, since incubation of the cells with cycloheximide to block new protein synthesis resulted in a large increase in RXRβ mRNA expression (see below). This combined effect of cyclic AMP on inhibition of both RAR and RXR expression (RXRβ is the major subtype expressed in B16 melanoma cells) would ensure that the retinoid signal would not be effective in altering the phenotype of the cells. Indeed we have reported (Xiao *et al.*, 1996) that cyclic AMP prevents the RA induction of PKCα and antagonizes RA-induced differentiation of B16 mouse melanoma.

4.4. Other Transcriptional Regulators

Estradiol was found to increase the amount of mRNA and protein for RARα in ER^+ human breast cancer cells (van der Leede *et al.*, 1995; Roman *et al.*, 1993). This steroid did not alter the expression of either RARβ or γ. The induction of RARα by estradiol was rapid (within 1 hr) and was maximal after 6 hr of treatment with 10^{-10} M. Antiestrogens, such a tamoxifen, block the induction of RARα by estradiol, demonstrating the involvement of ER. In contrast to estradiol, progestins markedly decreased the expression of RARα and γ mRNA in four different PR^+ human breast cancer cell lines (Roman *et al.*, 1992). The effect was maximal 3–6 hr after treatment. RARγ mRNA levels returned to normal by 24 hr, while the RARα mRNA was maintained at around 50% of the control until at least 48 hr.

While PKC is thought to alter RAR function possibly by receptor phosphorylation (see above), there is at least one report that TPA can decrease RARα, γ and RXRα mRNA within 3.5 hr after treatment of mouse skin with this tumor promoter (Kumar *et al.*, 1994). This effect is accompanied by a decrease in ^{3}H-RA binding to RAR isolated from mouse epidermal nuclear extracts.

Lastly, our laboratory has found that the expression of RXRα and β mRNAs can be regulated by inhibitors of protein synthesis (Desai and Niles, 1995). Cycloheximide treatment resulted in an increased half-life of RXRα mRNA, but the half-life of RXRβ was unchanged. Transcriptional run-on assays suggested that cycloheximide increased the transcription rate of the RXRβ gene. Thus these two RXR genes are differentially regulated in B16 mouse melanoma cells. Based on these results, we developed the hypothesis that there is a negative transcriptional regulator of the RXRβ gene that has a very short half-life. Inhibiting protein synthesis would allow the level of this negative regulatory protein to decrease, resulting in an increase in the transcription of the RXRβ gene.

5. CONCLUSIONS

Since the discovery of the nuclear retinoid receptors some 10 years ago, there have been numerous advances in our understanding of how these receptors function. Most of these studies have focused on the control of receptor activity by ligand binding, DNA response elements, and formation of heterodimers. However, it is now clear that control of receptor function involves many other players. All of the RARs are phosphoproteins, and signal transduction pathways that lead to phosphorylation of these receptors could alter their function (Fig. 3). In addition, the amount of receptor

FIGURE 3. Schematic illustrating the different mechanisms by which RAR activity is controlled. Hormones which stimulate adenylate cyclase (A.C.) increase intracellular cyclic AMP. This second messenger binds cyclic AMP-dependent protein kinase and converts it from an inactive (PKA_i) to an active form (PKA_a). We hypothesize that PKA then phosphorylates cyclic AMP response element binding protein (CREB), which makes CREB competent to inhibit RAR gene expression (downward arrow). Alternatively, phosphorylated CREB could sequester a coactivator for another transcription factor, preventing it from stimulating RAR gene expression. Also shown is the ability of estradiol (E) through the estrogen receptor (ER) to stimulate RAR (specifically RARα) gene transcription (upward arrow). Hormones which activate phosphoinositide turnover produce diacylglycerol and IP_3. These would in turn convert protein kinase C from an inactive (PKC_i) to an active (PKC_a) form. We hypothesize that PKC can either directly or indirectly increase the phosphorylation of RAR, resulting in an increase in receptor activity. CREB-binding protein (CBP)/p300 is shown as a transcriptional integrator since it stimulates the activity of different transcription factors. These transcription factors all compete for a limited pool of CBP/p300. Lastly, this diagram illustrates that unliganded RAR binds corepressor molecules. Upon binding of ligand there is a conformational change in the RAR that causes the corepressor to dissociate and allows the binding of coactivator proteins.

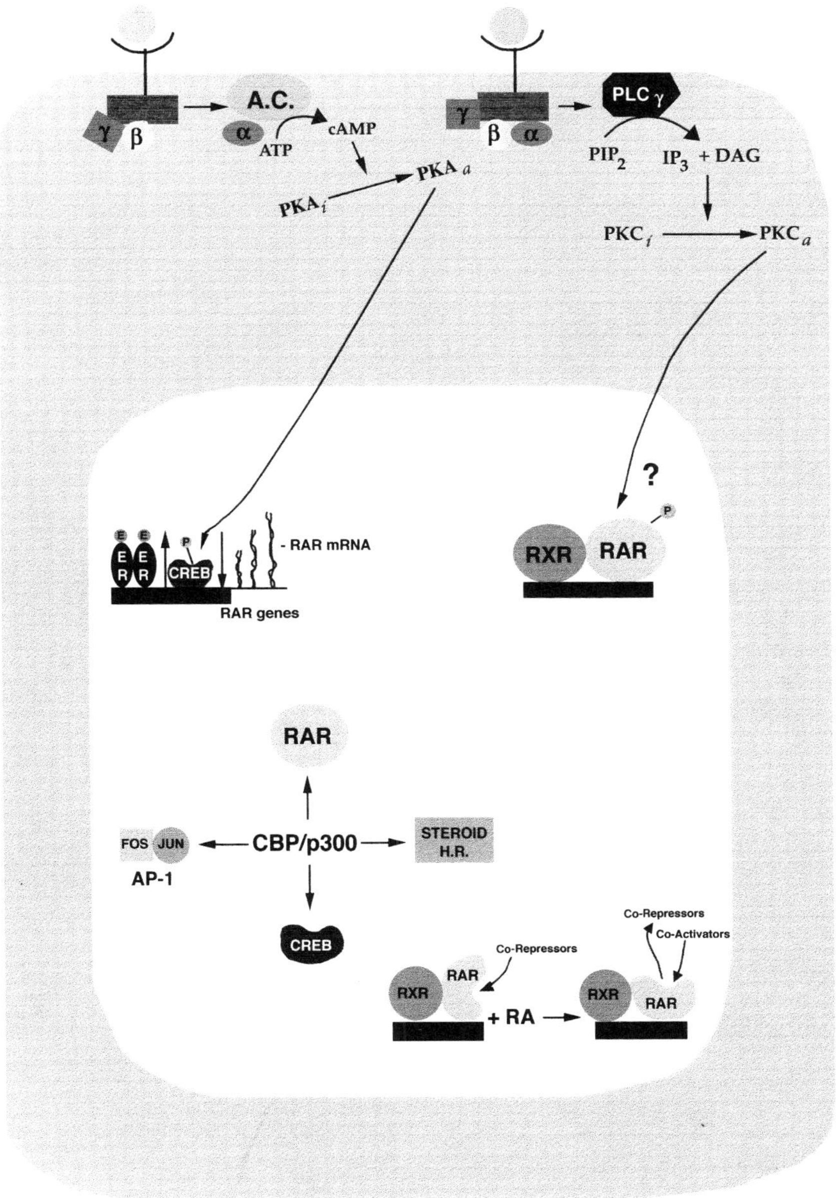
A.C.
γ
β
α
ATP
cAMP
PKA a
PKA i
PLC γ
PIP2
IP3 + DAG
PKC i
PKC a
?
P
RXR
RAR
E
E
E
R
E
R
CREB
- RAR mRNA
RAR genes
RAR
FOS
JUN
AP-1
CBP/p300
STEROID
H.R.
CREB
Co-Repressors
Co-Activators
Co-Repressors
RXR
RAR
+ RA
RXR
RAR

present at any given time can affect functionality, since we now know that many different receptors compete for common coactivators. The level of RAR expression is controled by several different regulators including cyclic AMP, estradiol, and perhaps tumor promoters. Therefore, there are many layers of control for nuclear retinoid receptor function. Since retinoids, through activation of nuclear receptors, have potential application as therapeutics for cancer and dermatological diseases, these additional control mechanisms provide targets for enhancing nuclear receptor function.

6. REFERENCES

Allenby, G., Bocquel, M.-T., Saunders, M., Kazmer, S., Speck, J., Rosenberger, M., Lovey, A., Kastner, F., Grippo, J. F., Chambon, P., and Levin, A. A., 1993, Retinoic acid receptors and retinoid X receptors: Interaction with endogenous retinoic acids, *Proc. Natl. Acad. Sci. USA* **90:**30–34.

Angel, P., Imagawa, M., Chiu, R., Stein, B., Imbra, P. J., Rahmadorf, H. J., Jonat, C., Herrlich, P., and Karin, M., 1987, Phorbol ester-inducible genes contain a common *cis* element recognized by a TPA-modulated *trans*-acting factor, *Cell* **49:**724–739.

Angel, P., Allegrato, E. A., Okino, S. T., Hattori, K., Bolye, W. J., Hunter, T., and Karin, M., 1988, Oncogene *jun* encodes a sequence-specific *trans*-activator similar to AP-1, *Nature* **332:**166–171.

Apfel, C. M., Kamber, M., Klaus, M., Mohr, P., Keidel, S., and LeMotte, P., 1995, Enhancement of HL-60 differentiation by a new class of retinoids with selective activity on retinoid X receptor, *J. Biol. Chem.* **270:**30765–30772.

Arias, J., Alberts, A. S., Brindle, P., Claret, F. X., Smeal, T., Karin, M., Feramisco, J., and Montiminy, M., 1994, Activation of cAMP and mitogen responsive genes relies on a common nuclear factor, *Nature* **370:**226–229.

Astrom, A., Tavakkol, A., Pettersson, U., Cromie, M., Elder, J. T., and Voorhees, J. J., 1991, Molecular cloning of two human cellular retinoic acid-binding proteins (CRABP). Retinoic acid-induced expression of CRABP-II in adult human skin *in vivo* and in skin fibroblasts *in vitro*, *J. Biol. Chem.* **266:**17662–17666.

Bailey, J. S., and Siu, C. H., 1988, Purification and partial characterization of a novel binding protein for retinoic acid from neonatal rat, *J. Biol. Chem.* **263:**9326–9332.

Bannister, A. J., and Kouzarides, T., 1995, CBP-induced stimulation of c-Fos activity is abrogated by E1A, *EMBO J.* **14:**4758–4762.

Basiahmad, A., Kohne, A. C., and Renkawitz, R., 1992, A transferable silencing domain is present in the thyroid hormone receptor, in the v-*erb* A oncogene product and in the retinoic acid receptor, *EMBO J.* **11:**1015–1023.

Bavik, C.-D., Ericksson, U., Allen, R. A., and Peterson, P. A., 1991, Identification and partial characterization of a retinal pigment epithelial membrane receptor for plasma retinol-binding protein, *J. Biol. Chem.* **266:**14078–14985.

Bavik, C.-D., Busch, C., and Ericksson, U., 1992, Characterization of a plasma retinol-binding membrane receptor expressed in the retinal pigment epithelium, *J. Biol. Chem.* **267:**23035–23042.

Beato, M., 1989, Gene regulation by steroid hormones, *Cell* **56:**335–344.

Benbrook, D., Lernhardt, E., and Pfahl, M., 1988, A new retinoic acid receptor identified from a hepatocellular carcinoma, *Nature* **333:**669–672.

Berg, J., 1990, Zinc fingers and other metal-binding domains. Elements for interactions between macromolecules, *J. Biol. Chem.* **265:**6513–6516.

Berkenstam, A., Ruiz, d. M. V., Barettino, D., Horikoshi, M., and Stunnenberg, H. G., 1992, Cooperativity in transactivation between retinoic acid receptor and TFIID requires an activity analogous to E1A, *Cell* **69:**401–412.

Bernstein, P. S., Law, W. C., and Rando, R. R., 1987, Isomerization of all-*trans* retinoids to 11-*cis* retinoids *in vitro*, *Proc. Natl. Acad. Sci. USA* **84:**1849–1853.

Blaner, W. S., 1989, Retinol-binding protein: The serum transporter protein for vitamin A, *Endocrine Rev.* **10:**305–316.

Bocquel, M.-T., Kumar, V., Stricker, C., Chambon, P., and Gronemeyer, H., 1989, The contribution of the N- and C-terminal regions of steroid receptors to activation of transcription is both receptor and cell-specific, *Nucleic Acids Res.* **17:**2581–2595.

Bugge, T. H., Pohl, J., Lonnoy, O., and Stunnenberg, H. G., 1992, RXRγ, a promiscuous partner of retinoic acid and thyroid hormone receptors, *EMBO J.* **11:**1409–1410.

Burns, K., Duggan, B., Atkinson, E. A., Fanulski, K. S., Nemer, M., Bleackley, R. C., and Michalak, M., 1994, Modulation of gene expression by calreticulin binding to the glucocorticoid receptor, *Nature* **367:**476–480.

Burris, T. P., Nawaz, Z., Tsai, M.-J., and O'Malley, B. W., 1995, A nuclear hormone receptor associated protein that inhibits transactivation by the thyroid hormone and retinoic acid receptors, *Proc. Natl. Acad. Sci. USA* **92:**9525–9529.

Carafoli, E., 1987, Intracellular calcium homeostasis, in *Annual Review of Biochemistry*, Vol. 55 (C. C. Richardson, P. D. Boyer, I. B. Dawid, and A. Meister, eds.), pp. 395–434, Annual Reviews, Palo Alto, California.

Carlberg, C., Bendik, I., Wyss, A., Meier, E., Sturzenbecker, L. J., Grippo, J. F., and Hunziker, W., 1993, Two nuclear signalling pathways for vitamin D, *Nature* **361:**657–660.

Cavailles, V., Dauvois, S., Danielian, P. S., and Parker, M. G., 1994, Interaction of proteins with transcriptionally active estrogen receptor, *Proc. Natl. Acad. Sci. USA* **91:**10009–10013.

Cavailles, V., Dauvois, S., L'Horset, F., Lopez, G., Hoare, S., Kushner, P. J., and Parker, M. G., 1995, Nuclear factor RIP 140 modulates transcriptional activation by the estrogen receptor, *EMBO J.* **14:**3741–3751.

Chambon, P., 1994, The retinoid signaling pathway: Molecular and genetic analysis, *Semin. Cell Biol.* **5:**115–125.

Chen, J. D., and Evans, R. M., 1995, A transcriptional co-repressor that interacts with nuclear hormone receptors, *Nature* **377:**454–457.

Chen, J. D., Umesono, K., and Evans, R. M., 1996, SMRT isoforms mediate repression and antirepression of nuclear receptor heterodimers, *Proc. Natl. Acad. Sci. USA* **93:**7567–7571.

Chrivia, J. C., Kwok, R. P. S., Lamb, N., Hagiwara, M., Montminy, M. R., and Goodman, R. H., 1993, Phosphorylated CREB binds specifically to the nuclear protein CBP, *Nature* **365:**855–859.

Clarke, S. D., and Armstrong, M. K., 1989, Cellular lipid binding proteins: Expression, function and nutritional regulation, *FASEB J.* **3:**2480–2487.

Costa, S. L., Pratt, M. A. C., and McBurney, M. W., 1996, E2F inhibits transcriptional activation by the retinoic acid receptor, *Cell Growth Diff.* **7:**1479–1485.

Dalman, F. C., Sturzenbecker, L. J., Levin, A. A., Lucas, D. A., Perdew, G. H., Petkovitch, M., Chambon, P., Grippo, J. F., and Pratt, W. B., 1991, Retinoic acid receptor belongs to a

subclass of nuclear receptors that do not form "docking" complexes with hsp 90, *Biochemistry* **30:**5605–5608.
Dedhar, S., Rennie, P. S., Shago, M., Hagesteljn, E.-Y. L., Yang, H., Filmus, J., Hawley, R. G., Bruchovsky, N., Cheng, H., Matuski, R. J., and Giguere, V., 1994, Inhibition of nuclear hormone receptor activity by calreticulin, *Nature* **367:**480–483.
DeLuca, L. M., 1991, Retinoids and their receptors in differentiation, embryogenesis and neoplasia, *FASEB J.* **5:**2924–2933.
Desai, D. S., and Niles, R. M., 1995, Expression and regulation of retinoid X receptors in B16 melanoma cells, *J. Cell. Physiol.* **165:**349–357.
Desai, D., Michalek, M., Singh, N. K., and Niles, R. M., 1996, Inhibition of retinoic acid receptor function and retinoic acid-regulated gene expression in mouse melanoma cells by calreticulin, *J. Biol. Chem.* **271:**15153–15159.
de The, H., Marchio, A., Tollais, P., and Dejean, A., 1989, Differential expression and ligand regulation of retinoic acid receptor α and β genes, *EMBO J.* **8:**429–433.
de The, H., del Mar Vivanco-Ruiz, M., Tiollais, P., Stunneberg, H., and Dejean, A., 1990, Identification of a retinoic acid responsive element in the retinoic acid receptor β gene, *Nature* **343:**177–180.
Durand, B., Saunders, M., Leroy, P., Leid, M., and Chambon, P., 1992, All *trans* and 9-*cis* retinoic acid induction of CRABP II transcription is mediated by RAR–RXR heterodimers bound to DR1 and DR2 repeated motifs, *Cell* **71:**73–85.
Eckner, R., Ewen, M. E., Newsome, D., Gerdes, M., DeCaprio, J. A., Lawrence, J. B., and Livingston, D. M., 1994, Molecular cloning and functional analysis of the adenovirus E1A-associated 300-kD protein (p300) reveals a protein with properties of a transcriptional adaptor, *Genes Dev.* **8:**869–884.
Eskild, W., and Hansson, V., 1994, Vitamin A functions in the productive organs, in *Vitamin A in Health and Disease* (R. Blomhoff, ed.), pp. 531–559, Marcel Dekker, New York.
Evans, R. M., 1988, The steroid and thyroid hormone receptor superfamily, *Science* **240:**889–895.
Fiorella, P. D., Giguere, V., and Napoli, J. L., 1993, Expression of cellular retinoic acid binding protein (type II) in *Escherichia coli*: Characterization and comparison to cellular retinoic acid binding protein (type I), *J. Biol. Chem.* **268:**21545–21552.
Fogh, K., Voorhees, J. J., and Astrom, A., 1993, Expression, purification and binding properties of human cellular retinoic acid-binding protein type I and type II, *Arch. Biochem. Biophys.* **300:**751–755.
Folkers, G. E., and van der Saag, P. T., 1995, Adenovirus E1A functions as a cofactor for retinoic acid receptor β (RARβ) through direct interaction with RARβ, *Mol. Cell. Biol.* **15:**5868–5878.
Folkers, G. E., van der Leede, B.-jM., and van der Saag, P. T., 1993, The retinoic acid receptor-β2 contains two separate cell-specific transactivation domains at the N terminus and in the ligand-binding domain, *Mol. Endocrinol.* **7:**616–627.
Forman, B. M., Umesono, K., Chen, J., and Evans, R. M., 1995, Unique response pathways are established by allosteric interactions among nuclear hormone receptors, *Cell* **81:**541–550.
Forman, B. M., Yange, C.-R., Au, M., Casanova, J., Ghysdael, J., and Samuels, H. H., 1989, A domain containing leucine-zipper-like motifs mediate novel in vivo interaction between the thyroid hormone and retinoic acid receptors, *Mol. Endocrinol.* **3:**1610–1626.
Gaub, M.-P., Rochette-Egly, C., Lutz, Y., Ali, S., Matthes, H., Scheuer, I., and Chambon, P., 1992, Immunodetection of multiple species of retinoic acid receptor α: Evidence for phosphorylation, *Exp. Cell Res.* **201:**335–346.
Gebert, J. F., Mogkal, N., Fragioni, J. V., Sugarbaker, D. J., and Neel, B. G., 1991, High frequency of retinoic acid receptor β abnormalities in human lung cancer, *Oncogene* **6:**1859–1869.

Giguere, V., Ong, S., Segui, P., and Evans, R., 1987, Identification of a receptor for the morphogen retinoic acid, *Nature* **330:**624–629.

Giguere, V., Lyn, S., Yip, P., Siu, C. H., and Amin, S., 1990a, Molecular cloning of a cDNA encoding a second cellular retinoic acid-binding protein, *Proc. Natl. Acad. Sci. USA* **87:**6233–6237.

Giguere, V., Shago, M., Zirngibl, R., Tate, P., Rossant, J., and Varmuza, S., 1990b, Identification of a novel isoform of the retinoic acid receptor γ expressed in the mouse embryo, *Mol. Cell. Biol.* **10:**2335–2340.

Goodrich, J. A., Hoey, T., Thut, C. J., Admon, A., and Tijian, R., 1993, *Drosophila* TAF_{II} 40 interacts with both a VP16 activation domain and the basal transcription factor TFIIB, *Cell* **75:**519–530.

Halachmi, S., Marden, E., Martin, G., MacKay, H., Abbondanza, C., and Brown, M., 1994, Estrogen receptor associated protein: Possible mediators of hormone-independent transcription, *Science* **264:**1455–1458.

Haugen, B. R., Brown, N. S., Wood, W. M., Gordon, D. F., and Ridgway, E. C., 1997, The thyrotrope-restricted isoform of the retinoid-X receptor-γ1 mediates 9-*cis*-retinoic acid suppression of the thyrotropin-β promoter activity, *Mol. Endocrinol.* **11:**481–489.

Heyman, R. A., Mangelsdorf, D. J., Dyck, J. A., Stein, R. B., Eichele, G., Evans, R. M., and Thaller, C., 1992, 9-*cis* Retinoic acid is a high affinity ligand for the retinoid X receptor, *Cell* **68:**397–406.

Hoffman, B., Lehmann, J. M., Zhang, X.-K., Hermann, T., Hausmann, M., Graupner, G., and Pfahl, M., 1990, A retinoic acid receptor-specific element controls the retinoic acid receptor-β promoter, *Mol. Endocrinol.* **4:**1727–1726.

Hofmann, C., and Eichele, G., 1994, Retinoids in development, in *The Retinoids, Biology, Chemistry and Medicine*, 2nd ed. (M. B. Sporn, A. B. Roberts, and D. S. Goodman, eds.), pp. 387–441, Raven Press, New York.

Hoopes, C. W., Taketo, M., Ozato, K., Liu, Q., Howard, T. A., Linney, E., and Seldin, M. F., 1992, Mapping of the Rxr loci encoding nuclear retinoid X receptors RXRα, RXRβ, and RXRγ, *Genomics* **14:**611–617.

Horlein, A. J., Naar, A. M., Heinzel, T., Torchia, J., Gloss, B., Kurokawa, R., Ryan, A., Kamei, Y., Soderstrom, M., Glass, C. K., and Rosenfeld, M. G., 1995, Ligand-independent repression by the thyroid hormone receptor mediated by a nuclear receptor co-repressor, *Nature* **377:**397–404.

Houle, B., Rochette-Egly, C., and Bradley, W. E. C., 1993, Tumor-suppressive effect of the retinoic acid receptor β in human epidermoid lung cancer cells, *Proc. Natl. Acad. Sci. USA* **90:**985–989.

Hu, L., and Gudas, L. J., 1990, Cyclic AMP analogs and retinoic acid influences the expression of retinoic acid receptor α, β and γ mRNAs in F9 teratocarcinoma cells, *Mol. Cell Biol.* **10:**391–396.

Hu, L., Crowe, D. L., Rheinwald, J. G., Chambon, P., and Gudas, L. J., 1991, Abnormal expression of retinoic acid receptors and keratin 19 by human oral and epidermal squamous cell carcinoma cell lines, *Cancer Res.* **51:**3972–3981.

Huggenvik, J. I., Collard, M. W., Kim, Y.-W., and Sharma, R. P., 1993, Modification of the retinoic acid signaling pathway by the catalytic subunit of protein kinase A, *Mol. Endocrinol.* **7:**543–550.

Ishikawa, A., Umesono, K., Mangelsdorf, D. J., Aburtatani, H., Stanger, B. Z., Shibaski, Y., Imawari, M., Evans, R. M., and Takaku, F., 1990, A functional retinoic acid receptor encoded by a gene on human chromosome 12, *Mol. Endocrinol.* **4:**837–844.

Kamei, Y., Xu, L., Heinzel, T., Torchia, J., Kurokawa, R., Gloss, B., Lin, S.-C., Heyman, R. A., Rose, D. W., Glass, C. K., and Rosenfeld, M. G., 1996, A CBP integrator complex mediates transcriptional activation and AP-1 inhibition by nuclear receptors, *Cell* **85:**403–414.

Kastner, P., Krust, A., Mendelsohn, C., Garnier, J. M., Zelent, A., Leroy, P., Staub, A., and Chambon, P., 1990, Murine isoforms of the mouse retinoic acid receptor γ with specific patterns of expression, *Proc. Natl. Acad. Sci. USA* **87:**2700–2704.

Keller, H., Dreyer, C., Medin, J., Mahfoudi, A., Ozato, K., and Wahli, W., 1993, Fatty acids and retinoids control lipid metabolism through activation of peroxisome proliferator-activated receptor–retinoid X receptor heterodimers, *Proc. Natl. Acad. Sci. USA* **90:**2160–2164.

Kliewer, S. A., Umensono, K., Mangelsdorf, D. J., and Evans, R. M., 1992a, Retinoid X receptor interacts with nuclear receptors in retinoic acid, thyroid hormone, and vitamin D_3 signalling, *Nature* **335:**446–449.

Kliewer, S. A., Umesono, K., Heyman, R. H., Mangelsdorf, D. J., Dyck, J. A., and Evans, R. M., 1992b, Retinoid X receptor-COUP-TF interactions modulate retinoic acid signaling, *Proc. Natl. Acad. Sci. USA* **89:**1448–1452.

Kliewer, S. A., Umesono, K., Noonan, D. J., Heyman, R. A., and Evans, R. M., 1992c, Convergence of 9-*cis* retinoic acid and peroxisome proliferator signalling pathways through heterodimer formation of their receptors, *Nature* **358:**771–774.

Kruyt, F. A. E., Folkers, G. E., Walhout, A. J. M., van der Leede, B.-J. M., and van der Saag, P. T., 1993, E1A functions as a coactivator of retinoic acid-dependent retinoic acid receptor-β2 promoter activation, *Mol. Endocrinol.* **7:**604–616.

Kumar, R., Shoemaker, A. R., and Verma, A. K., 1994, Retinoic acid nuclear receptors and tumor promotion: Decreased expression of retinoic acid nuclear receptors by the tumor promoter 12-O-tetradecanoylphorbol-13-acetate, *Carcinogenesis* **15:**701–705.

Kurokawa, R., Yu, V., Naar, A., Kyakumoto, S., Han, Z., Siverman, S., Rosenfeld, M. G., and Glass, C. K., 1993, Differential orientations of the DNA binding domain and C-terminal dimerization interface regulate binding site selection by nuclear receptor heterodimers, *Genes Dev.* **7:**1423–1435.

Kurokawa, R., DiRenzo, J., Boehm, M., Sugarman, J., Gloss, B., Rosenfeld, M. G., Heyman, R. A., and Glass, C. K., 1994, Regulation of retinoid signalling by receptor polarity and allosteric control of ligand binding, *Nature* **371:**528–531.

Kurokawa, R., Soderstrom, M., Horlein, A., Halachmi, S., Brown, M., Rosenfeld, M. G., and Glass, C. K., 1995, Polarity-specific activities of retinoic acid receptors determined by a co-repressor, *Nature* **377:**451–454.

Kwok, R. P., Lundblad, J. R., Chrivia, J. C., Richards, J. P., Bachinger, H. P., Brennan, R. G., Roberts, S. G., Green, M. R., and Goodman, R. H., 1994, Nuclear protein CBP is a coactivator for the transcription factor CREB, *Nature* **370:**223–226.

Lafyatis, R., Kim, S.-J., Angel, P., Roberts, A. B., Sporn, M. B., Karin, M., and Wilder, R. L., 1990, Interleukin-1 and all-*trans*-retinoic acid inhibit collagenase gene expression through its 5′ activator protein-1-binding site, *Mol. Endocrinol.* **4:**971–980.

Lamph, W. W., Wamsley, P., Sassone-Corsi, P., and Verma, I., 1988, Induction of proto-oncogene JUN/AP-1 by serum and TPA, *Nature* **334:**629–631.

Lampron, C., Rochette-Egly, C., Gorry, P., Dolle, P., Mark, M., Lufkin, T., LeMeur, M., and Chambon, P., 1995, Mice deficient in cellular retinoic acid binding protein II (CRABP II) or in both CRABP I and CRABP II are essentially normal, *Development* **121:**539–548.

Laudet, V., Hanni, C., Coll, J., Catzefles, F., and Stehelin, D., 1992, Evolution of the nuclear receptor gene superfamily, *EMBO J.* **11:**1003–1013.

Le Douarin, B., Zechel, C., Garnier, J.-M., Lutz, Y., Tora, L., Pierrat, B., Heery, D., Gronemeyer, H., Chambon, P., and Losson, R., 1995, The N-terminal part of TIF1, a putative mediator of the ligand-dependent activation function (AF-2) of nuclear receptors, is fused to B-raf in the oncogenic protein T18, *EMBO J.* **14:**2020–2033.

Lee, J. W., Ryan, F., Swaffield, J. C., Johnston, S. A., and Moore, D. D., 1995, Interaction of thyroid-hormone receptor with a conserved transcriptional mediator, *Nature* **374:**91–94.

Lee, W., Mitchell, P., and Tijian, R., 1987, Purified transcription factor AP-1 interacts with TPA-inducible enhancer elements, *Cell* **49:**741–752.

Lefebvre, P., Gaub, M.-P., Tahayato, A., Rochette-Egly, C., and Formstecher, P., 1995, Protein phosphatases 1 and 2A regulate the transcriptional and DNA binding activities of retinoic acid receptors, *J. Biol. Chem.* **270:**10806–10816.

Lehmann, J. M., Zhang, K.-K., and Pfahl, M., 1992, RARγ2 expression is regulated through a retinoic acid response element embedded in Sp1 sites, *Mol. Cell Biol.* **12:**2976–2985.

Leid, M., Kastner, P., Lyons, R., Nakshari, H., Saunders, M., Zacharewski, T., Chen, J.-Y., Staub, A., Garnier, J.-M., Mader, S., and Chambon, P., 1992, Purification, cloning, and RXR identity of the HeLa cell factor with which RAR or TR heterodimerizes to bind target sequences efficiently, *Cell* **68:**377–395.

Leroy, P., Krust, A., Zelent, A., Mendelsohn, C., Garnier, J. M., Kastner, P., Dierich, A., and Chambon, P., 1991a, Multiple isoforms of the mouse retinoic acid receptor α are generated by alternative splicing and differential induction by retinoic acid, *EMBO J.* **10:**59–69.

Leroy, P., Nakshatri, H., and Chambon, P., 1991b, Mouse retinoic acid receptor α2 isoform is transcribed from a promoter that contains a retinoic acid response element. *Proc. Natl. Acad. Sci. USA* **88:**10138–10142.

Levin, A. A., Sturgenbecker, L. J., Kazmer, S., Bosakowski, T., Huselton, C., Allenby, G., Speck, J., Kratzelsen, C., Rosenberger, M., Lovey, A., and Grippo, J. F., 1992, 9-*cis* retinoic acid stereoisomer binds and activates the nuclear receptor RXRα, *Nature* **355:**359–361.

Levin, M. S., Li, E., Ong, D. E., and Gordon, J. I., 1987, Comparison of the tissue-specific expression and developmental regulation of two closely linked rodent genes encoding cytosolic retinol-binding proteins, *J. Biol. Chem.* **262:**7118–7124.

Liu, Q., and Linney, E., 1993, The mouse X receptor-γ gene: Evidence for functional isoforms, *Mol. Endocrinol.* **7:**651–658.

Lundblad, J. R., Kwok, R. P., Laurance, M. E., Harter, M. L., and Goodman, R. H., 1995, Adenoviral E1A-associated protein p300 as a functional homologue of the transcriptional co-activator CBP, *Nature* **374:**85–88.

MacDonald, P. N., and Ong, D. E., 1988, Evidence for a lecithin-retinol acyltransferase activity in the rat small intestine, *J. Biol. Chem.* **263:**12478–12482.

Mangelsdorf, D. J., Ong, E. S., Dynk, J. A., and Evans, R. M., 1990, Nuclear receptor that identifies a novel retinoic acid response pathway, *Nature* **345:**224–229.

Mangelsdorf, D. J., Borgmeyer, U., Heyman, R. A. Zhou, J. Y., Ong, E. S., Oro, A. E., Kakizuka, A., and Evans, R. M., 1992, Characterization of three RXR genes that mediate the action of 9-*cis*-retinoic acid, *Genes Dev.* **6:**329–344.

Marks, M. S., Hallenbeck, P. I., Nagata, T., Segars, J. H., Apella, E., Nikodem, V. M., and Ozato, K., 1992, H-2 RIIBP (RXRβ) heterodimerization provides a mechanism for combinatorial diversity in the regulation of retinoic acid and thyroid hormone responsive genes, *EMBO J.* **11:**1419–1435.

Martin, C. A., Ziegler, L. M., and Napoli, J. L., 1990, Retinoic acid, dibutyryl-cyclic AMP and differentiation affect the expression of retinoic acid receptors in F9 cells, *Proc. Natl. Acad. Sci. USA* **87:**4804–4808.

Martin, R., Renakawitz, R., and Muller, M., 1994, Two silencing subdomains of v-*erbA* synergize with each other, but not with RXR, *Nucleic Acids Res.* **22:**4898–4905.

Mattei, M.-G., Riviere, M., Krust, A., Ingvarsson, S., Vennstrom, B., Islam, M. Q., Levan, G., Kastner, P., Zelent, A., Chambon, P., Szpirer, J., and Szpirer, C., 1991, Chromosomal assignment of retinoic acid receptor (RAR) genes in the human, mouse and rat genomes, *Genomics* **10:**1061–1069.

Meyer, M. F., Gronemeyer, H., Turcotte, B., Bocquel, M. T., Tasset, D., and Chambon, P., 1989, Steroid hormone receptors compete for factors that mediate their enhancer function, *Cell* **57:**433–442.

Nagpal, S., Saunders, M., Kastner, P., Durand, B., Nakshatri, H., and Chambon, P., 1992a, Promoter context- and response element-dependent specificity of the transcriptional activation and modulating functions of retinoic acid receptors, *Cell* **70:**1007–1019.

Nagpal, S., Zelent, A., and Chambon, P., 1992b, RAR-β4, a retinoic acid receptor isoform, is generated from RAR-β2 by alternative splicing and usage of a CUG initiator codon, *Proc. Natl. Acad. Sci. USA* **89:**2718–2722.

Nagpal, S., Frinat, S., Nakshatri, H., and Chambon, P., 1993, RARs and RXRs: Evidence for two autonomous transactivation functions (AF-1 and AF-2) and heterodimerization *in vivo*, *EMBO J.* **12:**2349–2360.

Napoli, J. L., 1994, Retinoic acid homeostasis. Prospective roles of β-carotene, retinol, CRBP and CRABP, in *Vitamin A in Health and Disease* (R. Blomhoff, ed.), pp. 135–188, Marcel Dekker, New York.

Napoli, J. L., Posch, K. C., Fiorella, P. D., Boerman, M. H. C. M., Salerno, G. J., and Burns, P. D., 1993, Roles of cellular retinol-binding protein and cellular retinoic acid-binding protein in the metabolic channeling of retinoids, in *Retinoids: Progress in Research and Clinical Applications* (M. A., Liverea and L. Packer, eds.), pp. 29–48, Marcel Dekker, New York.

Nicholson, R. C., Mader, S., Nagpal, S., Leid, M., Rochette-Egly, C., and Chambon, P., 1990, Negative regulation of the rat stromelysin gene promoter by retinoic acid is mediated by an AP-1 binding site, *EMBO J.* **9:**4441–4454.

Noji, S., Yamsai, T., Koyama, E., Nohno, T., Fujimoto, W., Arata, J., and Taniguchi, S., 1989, Expression of retinoic acid receptor genes in keratinizing front skin, *FEBS Lett.* **259:**86–90.

Ogryzko, V. V., Schiltz, R. L., Russanova, V., Howard, B. H., and Nakatani, Y., 1996, The transcriptional coactivators p300 and CBP are histone acetyltransferases, *Cell* **87:**953–959.

Olsson, I., Breitman, T., and Gallo, R., 1992, Priming of human myeloid leukemic cell lines HL-60 and U-937 with retinoic acid for differentiating effects of cyclic adenosine 3′–5′ monophosphate-inducing agents and a T-lymphocyte derived differentiation factor. *Cancer Res.* **42:**3928–3934.

Ong, D. E., 1987, Cellular retinoid-binding proteins, *Arch. Dermatol.* **123:**1693–1695.

Ong, D. E., and Chytil, F., 1978, Cellular retinoic acid-binding protein from rat testis, *J. Biol. Chem.* **253:**4551–4554.

Ong, D. E., Newcomer, M. E., and Chytil, F., 1994, Cellular retinoid-binding proteins, in *The Retinoids: Biology, Chemistry, and Medicine*, 2nd ed. (M. B. Sporn, A. B. Roberts, and D. S. Goodman, eds.), pp. 288–317, Raven Press, New York.

Opas, M., Dziak, E., Fliegel, L., and Michalak, M., 1991, Regulation of expression and intracellular distribution of calretinculin, a major calcium binding protein of nonmuscle cells, *J. Cell. Physiol.* **112:**160–171.

Onate, S. A., Tsai, S. Y., Tsai, M.-J., and O'Malley, B. W., 1995, Sequence and characterization of a coactivator for the steroid hormone receptor superfamily, *Science* **270:**1354–1357.

Perlmann, T., and Vennstrom, B., 1995, The sound of silence, *Nature* **377:**387–388.

Petkovich, M., Brand, N. J., Krust, A., and Chambon, P., 1987, A human retinoic acid receptor which belongs to the family of nuclear receptors, *Nature* **330:**444–450.

Posch, K. C., Boerman, M. H. E. M., Burns, R. D., and Napoli, J. L., 1991, Holocellular retinol binding protein as a substrate for microsomal retinal synthesis, *Biochemistry* **30:**6224–6230.

Posch, K. C., Burns, R. D., and Napoli, J. L., 1992, Biosynthesis of all-*trans* retinoic acid from retinol. Recognition of retinal bound to cellular retinol binding protein (type I) as substrate by a purified cytosolic dehydrogenase, *J. Biol. Chem.* **267:**19676–19682.

Predki, P. F., Zamble, D., Sarkar, B., and Giguere, V., 1994, Ordered binding of retinoic acid and retinoid X receptors to asymmetric response elements involves determinants adjacent to their DNA binding domain, *Mol. Endocrinol.* **8:**31–39.

Ragsdale, Jr., C. W., Petkoviche, M., Gates, P. B., and Chambon, P., 1989, Identification of a novel retinoic acid receptor in regenerative tissue of the newt, *Nature* **341:**654–657.

Renaud, J. P., Rochel, N., Ruff, M., Vivat, V., Chambon, P., Gronemeyer, H., and Moras, D., 1995, Crystal structure of the RARγ ligand-binding domain bound to all-*trans* retinoic acid, *Nature* **377:**454–457.

Rochette-Egly, C., Lutz, Y., Saunders, M., Scheuer, I., Gaub, M. P., and Chambon, P., 1991, Retinoic acid receptor γ: Specific immunodetection and phosphorylation, *J. Cell Biol.* **115:**535–545.

Rochette-Egly, C., Gaub, M.-P., Lutz, Y., Ali, S., Scheuer, I., and Chambon, P., 1992, Retinoic acid receptor-β: Immunodetection and phosphorylation on tyrosine residues, *Mol. Endocrinol.* **6:**2197–2209.

Rochette-Egly, C., Oulad-Abdelghani, M., Staub, A., Pfister, V., Scheuer, I., Chambon, P., and Gaub, M.-P., 1995, Phosphorylation of the retinoic acid receptor-α by protein kinase A, *Mol. Endocrinol.* **9:**860–871.

Roman, D. S., Clarke, C. L., Hall R. E., Alexander, I. E., and Sutherland, R. L., 1992, Expression and regulation of retinoic acid receptors in human breast cancer cells, *Cancer Res.* **52:**2236–2242.

Roman, S. D., Ormandy, C. J., Manning, D. L., Blaney, R. W., Nicholson, R. I., Sutherland, R. L., and Clarke, C. L., 1993, Estradiol induction of retinoic acid receptors in human breast cancer cells, *Cancer Res.* **53:**5940–5945.

Ruberti, E., Friedrich, V., Morriss-Kay, G., and Chambon, P., 1992, Differential distribution patterns of CRABPI and CRABPII transcripts during mouse embryogenesis, *Development* **115:**973–987.

Schule, R., Umesono, K., Mangelsdorf, D. J., Bolado, J., Pike, J. W., and Evans, R. M., 1990, Jun-Fos and receptors for vitamins A and D recognize a common response element in the human osteocalcin gene, *Cell* **61:**497–504.

Schule, R., Ranjarajan, P., Yang, N., Kliewer, S., Ransome, L. J., Bolado, J., Verma, I. M., and Evans, R. M., 1991, Retinoic acid is a negative regulator of AP-1 responsive genes, *Proc. Natl. Acad. Sci. USA* **88:**6092–6096.

Sporn, M. B., Dunlop, N. M., Newton, D. L., and Smith, J. M., 1976, Prevention of chemical carcinogenesis by vitamin A and its synthetic analogs (retinoids), *Fed. Proc.* **35:**1332–1338.

Sucov, H. M., Murakami, K. K., and Evans, R. M., 1990, Characterization of an autoregulated response element in the mouse retinoic acid receptor type β gene, *Nature* **343:**177–180.

Tahayato, A, Lefebvre, P., Formstecher, P., and Dautrevaux, M., 1993, A protein kinase C-dependent activity modulates retinoic acid-induced transcription, *Mol. Endocrinol.* **7:**1642–1653.

Takase, S., Ong, D. E., and Chytil, F., 1986, Transfer of retinoic acid from its complex with cellular retinoic acid-binding protein to the nucleus, *Arch. Biochem. Biophys.* **247:**328–334.

Tasset, D., Tora, L., Fromental, C., Scheer, E., and Chambon, P., 1990, Distinct classes of transcriptional activating domains function by different mechanisms, *Cell* **62:**1177–1187.

Thaller, C., and Eichele, G., 1990, Isolation of 3,4-didehydroretinoic acid, a novel morphogenetic signal in the chick wing bud, *Nature* **345:**815–819.

Tharin, S., Dziak, E., Michalak, M., and Opas, M., 1991, Widespread tissue distribution of rabbit calreticulin, a non-muscle functional analogue of calsequestrin, *Cell Tiss. Res.* **269:**29–37.

Thompson, J. N., Howell, J. M., and Pitt, G. A. J., 1964, Vitamin A and reproduction in rats, *Proc. R. Soc. Lond. B* **159:**510–535.

van der Leede, B.-J. M., Folders, G. E., van den Brink, C. E., van der Saag, P. T., and van der Burg, B., 1995, Retinoic acid receptor α1 isoform is induced by estradiol and confers retinoic acid sensitivity in human breast cancer cells, *Mol. Cell. Endocrinol.* **109:**77–86.

Wald, G., 1968, The molecular basis of visual excitation, *Nature* **219:**800–807.

Widom, R. L., Rhee, M., and Karathanasis, S. K., 1992, Repression by ARP-1 sensitizes apolipoprotein A1 gene responsiveness to RXRα and retinoic acid, *Mol. Cell. Biol.* **12:**3380–3389.

Wilson, J. G., Roth, C. B., and Warkany, J., 1953, An analysis of the syndrome of malformations induced by maternal vitamin A deficiency, *Am. J. Anat.* **92:**189–217.

Wurtz, J. M., Bourguet, W., Renaud, J. P., Vivat, V., Chambon, P., Moras, D., and Gronemeyer, H., 1996, A canonical structure for the ligand binding domain of nuclear receptors, *Nature Struct. Biol.* **3:**87–94.

Xiao, Y., Desai, D., Quick, T. C., and Niles, R. M., 1996, Control of retinoic acid receptor expression in mouse melanoma cells by cyclic AMP, *J. Cell. Physiol.* **167:**413–421.

Xu, X.-C., Ro, J. Y., Lee, J. S., Shin, D. M., Hong, W. K., and Lotan, R., 1994, Differential expression of nuclear retinoid receptors in normal, premalignant and malignant head and neck tissues, *Cancer Res.* **54:**3580–3587.

Yang, N., Schule, R., Mangelsdorf, D. J., and Evans, R. M., 1991, Characterization of DNA binding and retinoic acid binding properties of retinoic acid receptor, *Proc. Natl. Acad. Sci. USA* **88:**3559–3563.

Yang, Y., Minucci, S., Zand, D. J., Ozato, K., and Ashwell, J. D., 1994, T cell activation and increases in protein kinase C activity enhance retinoic acid-induced gene transcription, *Mol. Endocrinol.* **8:**1370–1376.

Yang Yen, H.-F., Zhang, X.-K., Graupner, G., Tzukerman, M., Karin, M., and Pfahl, M., 1991, Antagonism between retinoic acid receptors and AP-1: Implication for tumor promotion and inflammation, *Nature New Biol.* **3:**1216–1219.

Yoshinaga, S. K., Peterson, C. L., Harskowitz, I., Yamamoto, K. R., 1992, Roles of SW11, SW12, and SW13 proteins for transcriptional enhancement of steroid receptors, *Science* **258:**1598–1604.

Yu, V. C., Deisert, C., Anderson, B., Holloway, J. M., Devary, D. V., Naar, A. M., Kim, S. Y., Boutin, J. M., Glass, C. K., and Rosenfeld, M. G., 1991, RXRβ, a coregulator that enhances binding of retinoic acid, thyroid hormone, and vitamin D receptors to their cognate response elements, *Cell* **67:**1251–1266.

Zamir, I., Harding, H. P., Arkins, G. B., Horlein, A., Glass, C. K., Rosenfeld, M. G., and Lazar, M. A., 1996, A nuclear hormone receptor corepressor mediates transcriptional silencing by receptors with distinct repression domains, *Mol. Cell. Biol.* **16:**5458–5465.

Zelent, A., Krust, A., Petkovich, M., Kastner, P., and Chambon, P., 1989, Cloning of murine α and β retinoic acid receptors and a novel receptor γ predominantly expressed in skin, *Nature* **339:**714–717.

Zelent, A., Mendelsohn, C., Kastner, P., Garnier, J. M., Ruffenach, F., Leroy, P., and Chambon, P., 1991, Differentially expressed isoforms of the mouse retinoic acid receptor β are generated by usage of two promoters and alternative splicing, *EMBO J.* **10:**71–81.

Zhang, X.-K., Hoffmann, B., Tran, P. B.-V., Graupner, G., and Pfahl, M., 1992, Retinoid X receptor is an auxilliary protein for thyroid hormone and retinoic acid receptors, *Nature* **335:**441–446.

Chapter 2

Retinoid Endocrinology

From Metabolism to Cellular Signaling

Amandio V. Vieira

1. INTRODUCTION

An interplay of differentiative and proliferative factors is necessary for the normal development and maintenance of complex multicellular organisms. Aberrations in this regulatory balance are the basis of developmental abnormalities and diseases such as cancer. One important group of such regulatory factors is the retinoids, a family of structurally related isoprenoid lipids that includes vitamin A alcohol, retinol—the physiological precursor of many bioactive retinoids. Biological activity of retinoids is regulated by retinoid-binding proteins present in extracellular, cytosolic, and nuclear compartments. These retinoid–protein complexes participate in cellular uptake, metabolism, signaling, and homeostasis of retinoids. The nuclear retinoid-binding proteins, called retinoid receptors, are ligand-regulated transcription factors. The retinoid receptors can modulate the expression of genes via a complex network of interactions with corepressors, coactivators, other nuclear hormone receptors, or other components of distinct signaling

---

Amandio V. Vieira Department of Cell Biology, The Scripps Research Institute, La Jolla, California 92037; *present address:* Department of Biochemistry, National Public Health Institute-KTL, FIN-00300 Helsinki, Finland.

Subcellular Biochemistry, Volume 30: Fat-Soluble Vitamins, edited by Quinn and Kagan. Plenum Press, New York, 1998.

pathways. Such a network provides a molecular basis for the pleiotropic physiological effects of retinoids.

2. RETINOID TRANSPORT AND METABOLISM

The natural retinoids are derived ultimately from specific carotenoid products of the terpenoid biosynthetic pathway (Fig. 1) present in photosynthetic organisms. Provitamin A carotenoids such as β-carotene along with fatty acid retinyl esters are the major dietary sources of retinoids. Retinyl esters are hydrolyzed in the intestinal lumen to free retinol, which (along with carotenoids and other lipids) is absorbed by the enterocytes of the gut (reviewed in Blomhoff *et al.*, 1992). The dietary retinol in the enterocytes is bound by a cellular (cytosolic) retinol-binding protein (CRBP II), reesterified with a fatty acid by the enzyme lecithin:retinol acyltransferase (LRAT), and incorporated into lipoprotein particles (chylomicrons) for delivery to peripheral tissues, mainly to the hepatic parenchymal cells (reviewed in Ross, 1993; Herr *et al.*, 1993).

In the liver, the retinyl esters are hydrolyzed and the retinol product interacts with CRBP I (Ross, 1993; Boerman and Napoli, 1991). The hepatic retinol may then have several fates, depending on the nutritional or health status of the organism: possibilities include transfer from parenchymal to stellate cells for storage or, as shown in Fig. 2, mobilization into the circulation as a complex with plasma retinol-binding protein (RBP). The liver is the central organ for control of retinol homeostasis, but other organs such as kidney or adipose tissue may also contribute to regulation of retinol levels (Gerlach and Zile, 1991; Tsutsumi *et al.*, 1992).

Circulatory RBP is a component of a specific extracellular transport system for retinol (Kanai *et al.*, 1968). In the plasma of some animals such as mammals, the 21-kDa RBP is found associated with a 55-kDa homotetrameric protein, transthyretin (TTR). In these animals, the molar ratio of TTR to RBP in the circulation is approximately two to one; and virtually all of the RBP is complexed to TTR (Goodman, 1984). The RBP–TTR interaction (see below) is important for establishing or maintaining normal plasma RBP and retinol levels (Episkopou *et al.*, 1993; Wei *et al.*, 1995).

Secretion of RBP from hepatocytes is affected by the availability of retinol. RBP is synthesized on the rough endoplasmic reticulum membrane, and the retinol–RBP complex, but not apoRBP or retinoic acid–RBP, is efficiently transported out of the endoplasmic reticulum (Kaji and Lodish, 1993, and references therein). The mechanisms responsible for apoRBP retention in the endoplasmic reticulum are not well understood. The role of a possible intracellular complexation (e.g., in the endoplasmic reticulum)

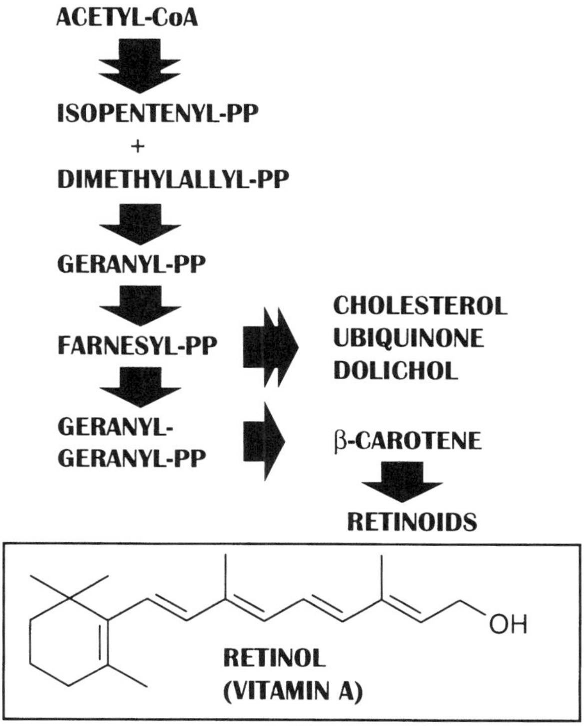

FIGURE 1. Terpenoid biosynthesis and retinoid formation. Acetyl-CoA from glycolysis and fatty acid β-oxidation acts as the carbon source for synthesis (via intermediates such as HMG-CoA and mevalonate) of the 5-carbon (C5) isoprene "building block" isopentenyl pyrophosphate (PP) and its isomer dimethylallyl-PP. The former is a nucleophile, the latter an electrophile, and their condensation gives rise to the geranyl-PP (C10). Geranyl-PP is referred to as a monoterpene and maintains electrophilic properties; subsequent additions of isopentenyl-PP generate the sesquiterpene farnesyl-PP (C15) and the diterpene geranylgeranyl-PP (C20). A step that appears to be specific for photosynthetic organisms is the formation of β-carotene (C40) from two geranylgeranyl-PPs. In animals, the β-carotene is cleaved and the cleavage products can give rise to retinol, C20 (the structure of the all-*trans* isomeric form is shown), and other retinoids. The three major retinoids in animal cells, retinol, retinoic acid, and retinal, differ in the oxidation state of the functional group (hydroxyl, carboxyl, or aldehyde) and can also differ in the *cis/trans* isomerization of the conjugated double-bond system. The terpenoid biosynthetic pathway is a source of many other important structural and regulatory molecules such as dolichol, cholesterol, and steroid hormones in animals, and giberellins and phytol in plants.

between RBP and TTR and its effect on hepatocytic secretion of RBP is also not well understood. In organisms with hepatic synthesis of RBP, but without hepatic synthesis of TTR (Achen *et al.*, 1993), any such cooperative role of TTR would have to be bypassed.

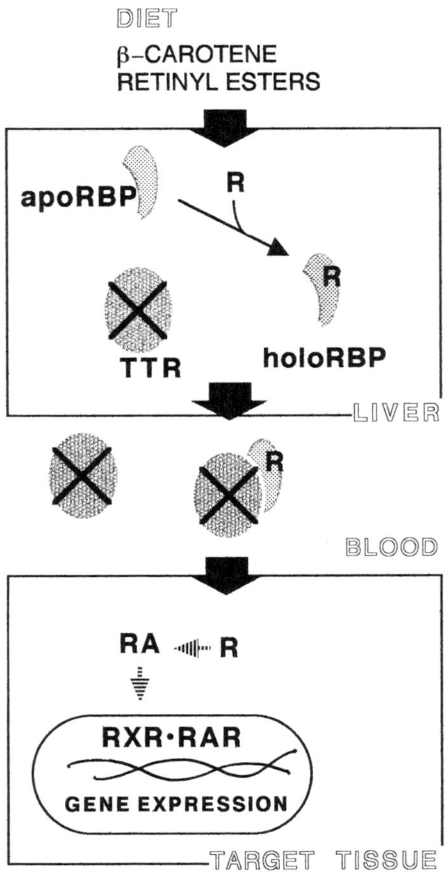

FIGURE 2. Selected steps in the transport and metabolism of retinol (R). The physiological transport protein for retinol in the circulation is retinol-binding protein (RBP). One or two RBP molecules can interact with the homotetrameric serum protein transthyretin (TTR). Circulatory retinyl esters (whose R moiety is derived from dietary retinyl esters and β-carotene breakdown) are transported by lipoproteins to the liver and converted to R. In the hepatocytes, an apoRBP is loaded with one molecule of R and the holoRBP is then secreted along with TTR into the cirulation. The R–RBP–TTR complex in the blood delivers R to target tissues, where the retinol is converted to retinoic acids (RA) such as all-*trans* and 9-*cis* retinoic acid. Interaction of RAs with nuclear hormone receptors (RAR or RXR) can modulate gene expression and thereby contribute to tissue homeostatis or to changes in the proliferation and differentiation state of the cells in that tissue.

2.1. Structure–Function Relations of the Retinol–RBP–TTR Complex

RBP and TTR belong to different protein structural families. RBP belongs to the lipocalin family, whose members are characterized by specific short-sequence motifs at the primary structure level and, at the tertiary level, by a β-barrel structure involving two orthogonal β-sheets (reviewed in Flower, 1996). TTR belongs to the Greek key β-barrel protein family, and its four identical subunits interact to form a tetrahedrally symmetric quaternary structure (Blake *et al.*, 1978; see also Terry *et al.*, 1993, and references therein). Besides participating in retinol transport, TTR can bind two thyroid hormone (e.g., T4) molecules in a central channel of the quaternary structure (Blake and Oatley, 1977). TTR also has an important pathological role in the formation of amyloid fibrils in diseases such as familial amyloidic polyneuropathy (see Terry *et al.*, 1993, and references therein).

The complexation of RBP with TTR in the circulatory system of some animals is an interaction that has been conserved among different species throughout their evolution. Chicken and human RBPs and TTRs, for example, can cross-interact (Kopelman *et al.*, 1976; Vieira and Schneider, 1993). The reasons for the conservation of this interaction may include prevention of circulatory RBP excretion during plasma filtration in the kidney (Goodman, 1984) and efficient delivery of retinoids to cells via the RBP–TTR complex (see Section 2.2.1).

Both apo- and holo-RBP can interact with TTR, but the latter has about a fivefold higher affinity (Malpeli *et al.*, 1996). This suggests that there may be interaction of the RBP-bound retinol with TTR. Retinol interacts with RBP in such a way that the β-ionone ring (see retinol structure in Fig. 1) lies deep inside the β-barrel structure of RBP; the terminal hydroxyl group at the other end of the retinol conjugated double-bond system is more exposed to solvent and is available to interact with TTR (see Monaco *et al.*, 1995). The RBP regions close to the retinol-binding pocket are also important in the interaction with TTR (Monaco *et al.*, 1995). As expected from these structures, retinoids with bulky substitutions of the terminal functional group interfere with RBP–TTR interactions (Berni and Formelli, 1992).

2.2. Cellular Uptake of Retinol

Of particular importance for retinoid physiology and pharmacology is an understanding of how cells take up retinol, the precursor of biologically active retinoids, from the circulation. It has been difficult to delineate such mechanisms because of the complexity of the transport systems and of the various possible modes of uptake (Fig. 3). The role of the retinol–RBP–

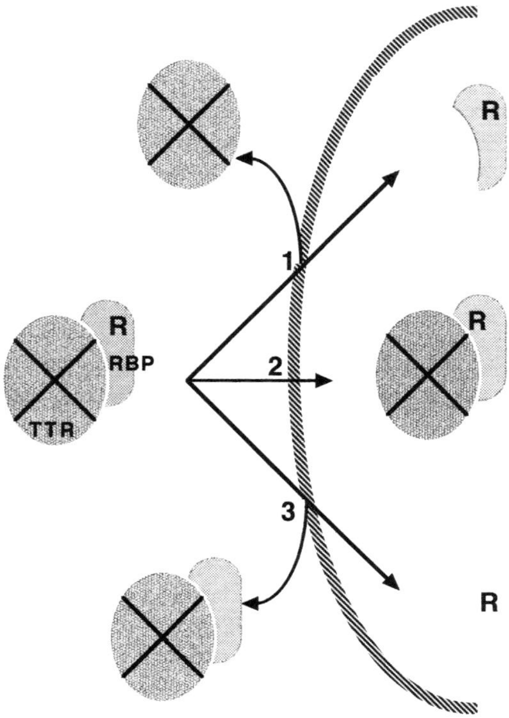

FIGURE 3. Possible cellular retinol (R) uptake mechanisms involving the circulatory RBP–TTR complex. (1) A cell surface RBP receptor disrupts the RBP–TTR interaction and endocytoses the holoRBP. About half of the TTR in the circulation is not complexed with RBP; this free TTR may also be endocytosed independently of any R delivery mechanisms. (2) The ternary R–RBP–TTR complex is endocytosed by an RBP or TTR receptor. (3) An RBP or TTR cell surface docking protein allows release of free R, which then enters the cell and is sequestered (not shown) by a cellular retinol-binding protein. Alternatively, this third form of transport may occur without a docking protein, simply by the dissociation of R from RBP. Examples representing these three putative modes of R delivery are given in the text (Section 2.2).

TTR complex in mediating cellular retinol uptake is not well defined. Based on a comparison of *in vitro* RBP–retinol dissociation rates and cellular retinol uptake rates, it has been suggested that retinol can dissociate from RBP sufficiently fast to account for its uptake into cells (Noy and Xu, 1990; Bennekum *et al.*, 1993; Levin, 1993). This is compatible with an uptake mechanism where retinol dissociates from RBP, partitions into the cell membrane, and is eventually sequestered by a CRBP in the cell. However, some determinants of the specificity of retinol delivery other than CRBPs likely exist (cf. Wahlberg *et al.*, 1989), for example, a cell surface receptor or docking protein for RBP (Heller, 1975; Rask and Peterson, 1976; Sivipradasadarao and Findlay, 1988; Bavik *et al.*, 1993). In addition, internalization of RBP or internalization of TTR has been reported in different cell types (for example, Divino and Schlussler, 1990; Malaba *et al.*, 1995; reviewed in Vieira *et al.*, 1995b). In the case of at least one cell type, the avian oocyte (Section 2.2.1), it is likely that there is endocytosis of the complete circulatory retinol transport complex.

The Avian Oocyte as a Model for Retinol Endocytosis

Chicken oocyte growth involves extensive plasma-to-oocyte transport and storage of nutritional and regulatory molecules in the oocytic yolk (reviewed in Schneider *et al.*, 1993). Retinol–RBP and TTR are known to be abundant in oocytic yolk, and active retinoid metabolites, likely derived from retinol stored in the egg yolk, are known to be important during avian embryonic development (Thompson *et al.*, 1969; Joshi *et al.*, 1973; Bermudez *et al.*, 1993; Dong and Zile, 1995). The avian oocyte is a model system which has proven to be amenable for the definition of various micro- and macronutrient systems, and represents a good system for delineating cellular retinol uptake mechanisms.

In the avian oocyte, there is evidence (Vieira, 1994) that (i) both RBP and TTR are localized to the same yolk subfraction as visualized by immuno-electron microscopy, (ii) the relative RBP and TTR yolk levels are similar in the various stages of oocyte growth, (iii) RBP: albumin and TTR: albumin ratios are at least 15-fold higher in yolk relative to serum, and (iv) RBP and TTR are present in oocytic clathrin-coated vesicles (CVs) (see also Vieira and Schneider, 1993). Chicken oocytes, but not the somatic follicular cells of the ovary, are known to have a very high density of membrane clathrin-coated pits and CVs, which reflects a high rate of clathrin-dependent, receptor-mediated endocytosis (Perry *et al.*, 1978, 1984; Woods *et al.*, 1978). In this context, the above observations suggest that there is a plasma-to-oocyte transfer of retinol–RBP–TTR, and that the endocytic mechanism is specific (relative to serum albumin) and may be receptor-mediated via the clathrin-dependent endocytic pathway. In contrast, in rat hepatocytes, internalized RBP did not appear to colocalize with an endocytic marker for clathrin-coated vesicles (Malaba *et al.*, 1995). Thus, it may be that different cell types employ different RBP uptake mechanisms, or that TTR complexation influences the RBP uptake mechanism, or both.

By ligand blotting and cross-linking techniques, a TTR receptor on oocyte membranes (but not on membrane preparations from the somatic ovarian cells) has been identified (Vieira *et al.*, 1995a); analyses with the same techniques did not reveal any holoRBP-binding activity on the membranes. The oocytic TTR receptor could bind native TTR in ligand blots when the TTR was presented as a complex with holo-RBP (Vieira *et al.*, 1995a). These results provide support for the existence of a cell membrane TTR receptor that is involved in the uptake of the retinol. This TTR-binding component is the first candidate for mediating the endocytosis of the physiological circulatory retinol-transport complex, retinol–RBP–TTR.

2.3. Biogenesis of Retinoic Acids

The conversion of retinol to biologically active retinoic acids is a critical step in retinoid endocrinology. A summary of the biogenesis of retinoic acids is shown in Fig. 4 (for a more detailed recent review see Napoli, 1996). Cellular (cytosolic) retinoid-binding proteins such as CRBP and cellular retinoic acid-binding protein (CRABP) are likely to participate in controlling specific and nonspecific changes such as oxidation and isomerization to the retinoid structure. By sequestering retinoids, these cellular binding proteins also serve to prevent excessive levels of retinoids from accumulating in the plasma membrane and in organellar membranes. Although Fig. 4 shows CRBP, CRABP, and all other components of retinoic acid (RA) biogenesis to be present in the same cell, this may not be the case; biogenesis and further metabolism of retinoic acid may be better defined within the context of a network of cells. Retinoid metabolites may function in a paracrine manner within this network; this would allow another level of control to be exerted over the differential production and accumulation of specific retinoids in a given cell type.

It is known that some tissues can obtain up to one quarter of their all-*trans* RA from the pool in the circulation (Kurlandsky *et al.*, 1995). In these tissues the control of RA uptake may be an important event in the biogenesis of active retinoids. The extracellular retinol that is complexed with RBP is likely to be the major source of RA in most tissues. Direct evidence for the importance of RBP-mediated delivery of retinoid for RA biogenesis has been obtained with mouse embryos (Bavik *et al.*, 1996). Finally, it should be mentioned that, in addition to retinoic acids, biologically active retinols in cells have been described, e.g., 13,14-dihydroxyretinol present in a variety of vertebrate and invertebrate cell lines (Derguini *et al.*, 1995).

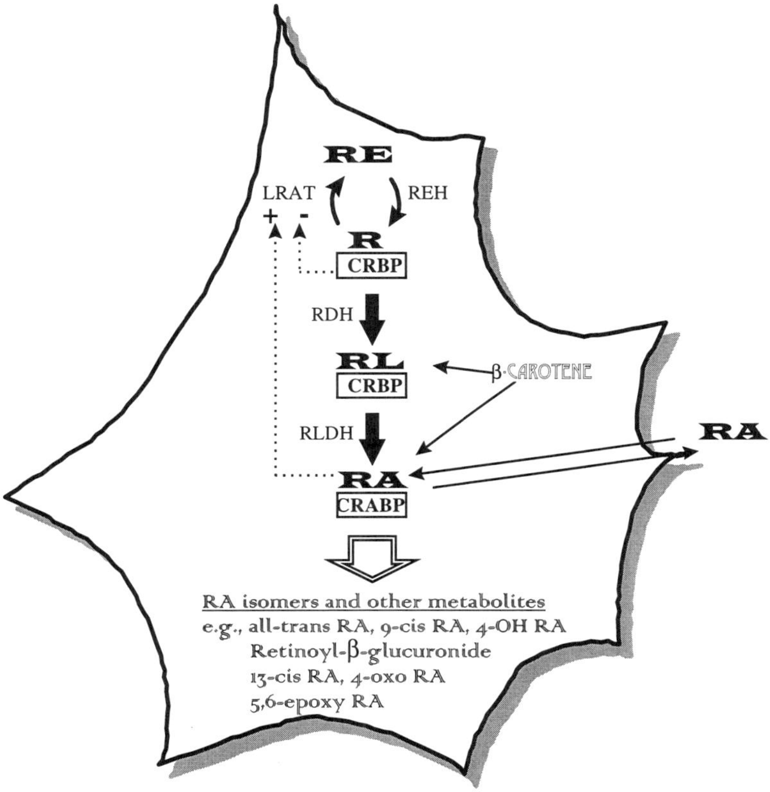

FIGURE 4. Biogenesis of retinoic acids. Cellular retinol (R) obtained primarily from the circulatory transport complex (see Fig. 3) is bound by a cellular retinol-binding protein (CRBP) and can be esterified with fatty acids to produce retinyl esters (RE) via the activity of lecithin : retinol acyltransferase (LRAT). Low retinol, high CRBP levels inhibit (−) LRAT activity and promote RE hydrolysis by retinyl ester hydrolase (REH) and the oxidation of R, bound to CRBP, to retinal (RL) by the enzyme retinol dehydrogenase (RDH). RL bound to CRBP is further oxidized by retinal dehydrogenase (RLDH) to retinoic acid (RA). β-Carotene can act as a source of RL and apocarotenals (Wang *et al.*, 1996) that are then converted to RA. Excess RA formation can promote RE formation by activation (+) of LRAT (Kurlandsky *et al.*, 1996). In some cells, RA may also be obtained from extracellular sources. In the cell, RA bound to CRABP can be isomerized or further metabolized to other retinoic acids as indicated by the large arrow. Not all of the indicated reactions occur necessarily in the same cell type.

3. RETINOID SIGNALING PATHWAYS

3.1. Molecular Phylogeny of Retinoid Signaling

The fact that retinoids and other isoprenoids appear to be part of signaling systems in a wide range of species from bacteria to mammals argues for an early evolution of these isoprenoid signaling systems. Indeed, the major neutral lipids of the primitive archaebacteria are 20-, 25-, and 30-carbon acyclic isoprenoids (Tornabene *et al.*, 1979). Isoprenoids with these numbers of carbons are also abundant in organic deposits of ancient sediments and may have been contributed by the ancient bacteria (Tornabene *et al.*, 1979). It has been suggested that cyclization of isoprenoids (i.e., formation of rings such as those in sterols and carotenes) occured later with the evolution of non-archaebacteria (Tornabene *et al.*, 1979, and references therein). Interestingly, a synthetic 20-carbon acyclic isoprenoid, geranylgeranoic acid, is active in cells and can have similar effects on gene expression as retinoic acid (Araki *et al.*, 1995).

An example of one type of retinoid signaling is represented by the rhodopsins. These proteins contain covalently bound retinal (in a Schiff-base linkage to an internal lysine) and can function as light-activated ion pumps such as the prokaryotic bacteriorhodopsin or photosensory pigments in a variety of animals. Covalent linkage (ester) of retinoic acid to proteins has also been shown to occur in a variety of human cell lines (for example, Takahashi and Breitman, 1989). Other isoprenoids can also form covalent linkages to proteins: geranylgeranylation and farnesylation of proteins are well-established regulatory modifications.

With the emergence of extracellular and intracellular binding proteins to solubilize and protect the signaling molecule or to effect ligand-mediated signaling, isoprenoids including retinoids could begin to function in a noncovalent manner as chemical signals between different organisms (e.g., pheromones) or between cells of the same organism (e.g., hormones). In addition to retinoids, examples of signaling molecules of terpenoid origin from a wide spectrum of organisms include differentiation-inducing factor in the slime mold, abscisic acid in plants, estrogen steroids in vertebrates, and juvenile hormone in insects. The signaling isoprenoids may be endogenously synthesized or, as is the case with retinoids in vertebrates, derived from dietary precursors. Various isoprenoids including retinoids and steroids (see Section 3.2), and farnesol (Forman *et al.*, 1995) and phytol (Kitareewan *et al.*, 1996) metabolites can interact with a superfamily of nuclear receptors and thereby modulate gene expression. Interestingly, nonisoprenoid lipids such as fatty acid derivatives, which can also function as intracellular signals and as pheromones, are able to form covalent linkages (esters) with proteins, and they can also interact with members of the nuclear receptor superfamily and with cellular fatty acid-binding proteins related to cellular retinoid-binding proteins (Bass, 1993, and references therein).

Conservation of specific retinoid signaling pathways during evolution is exemplified by the fact that chordates, from tunicates to vertebrates, all appear to exhibit a similar retinoic acid-dependent response in gene expression: the activation of the Hox homeobox genes (for example, see the review by Shimeld, 1996). Hox genes code for transcription factors that have a fundamental role in embryonic anterior–posterior axis formation, particularly in the central nervous system. Besides chordates, retinoid and thyroid hormone nuclear receptor superfamily members have been detected in arthropods and nematodes (Kostrouch *et al.*, 1995).

3.2. Nuclear Retinoid Receptors and Regulation of Gene Expression

With the reported identification a decade ago of nuclear retinoic acid receptors that could interact with specific DNA sequences (Petkovich *et al.*, 1987; Giguere *et al.*, 1987), a basis was provided for an important, direct function of retinoids: modulation of gene transcription. The nuclear receptor superfamily includes the retinoid receptors as well as steroid and thyroid hormone receptors. There are also structurally related receptors, the orphans, for which no ligand is known. All members of this family share a similar organization of structural domains, including a DNA-binding domain and two transcriptional activation domains.

Two families of nuclear retinoid receptors exist, termed retinoic acid receptors (RARs) and retinoid X receptors (RXRs), both with multiple isoforms (reviewed in Chambon, 1996). RARs can bind all-*trans* or 9-*cis* retinoic acid with high affinity; RXRs selectively interact with the latter ligand (Allenby *et al.*, 1993). RARs function as heterodimers with RXRs in the regulation of gene transcription (for example, Nagpal *et al.*, 1993). RXRs are also interaction partners for other nuclear hormone receptors such as vitamin D_3 receptors, thyroid hormone T_3 receptors, and several orphan receptors.

Considering the possible combinations of interactions involving different receptors and receptor isoforms, different ligands, the involvement of repressor and activator proteins that interact with the receptors, and different *cis*-acting DNA response elements to which the receptor–ligand complexes can bind, it is evident that a large set of signaling complexes involving retinoids and retinoid receptors can be generated to establish different arrays of gene expression.

Modulation of AP-1 Transcription Factor Activity by Retinoids

Direct modulation of gene transcription by the interaction of retinoid nuclear hormone receptors and DNA is a major retinoid signaling mechanism. However, effects of retinoid nuclear receptors due to interference with AP-1 activity are also likely to be important retinoid signaling events (see original references in Chen *et al.*, 1995). AP-1s are transcription factor complexes composed of members of the Jun and Fos protein families. Transrepression of AP-1 by RA and RARs may have important consequences for skin photoaging (Fisher *et al.*, 1996) as well as for various cell proliferation and transformation events (Desbois *et al.*, 1991; Fanjul *et al.*, 1994; Chen *et al.*, 1995). For example, the transformation of mouse epidermal cells by a phorbol ester tumor promoter can be inhibited by retinoids specific for AP-1-blocking activity and not by those specific for transactivation of gene expression (Li *et al.*, 1996). In contrast, a synthetic retinoid selective for RARγ activation was recently shown to upregulate AP-1 and lead to apoptosis of melanoma cells (Schadendorf *et al.*, 1996). Thus, interaction of retinoid receptors with transcription factors, such as AP-1, that can be activated by other signal transduction mechanisms is likely to have an important role in integrating different extracellular signals and cellular differentiation states at the level of gene expression.

3.3. Retinoylation and Other Functions of Retinoids

Effects of retinoids which do not lead directly to changes in transcription factor activity and gene transcription may also be important in some cell types. Covalent modification of proteins by retinoic acid may be one such mechanism (Takahashi and Breitman, 1989). Others may include retinoid-induced changes in protein and lipid glycosylation (reviewed in Roberts and Sporn, 1984), membrane structure/microviscocity (DeBoeck and Raphael, 1988; Mehta *et al.*, 1989; Urano *et al.*, 1992), activity of phospholipases (Hope *et al.*, 1990; Halter *et al.*, 1993), and intracellular vesicular trafficking (Paulik *et al.*, 1995). Retinol is a primary alcohol; and regarding the possible effects on vesicular trafficking and phospholipases mentioned above, it is of interest to note that various primary alcohols which inhibit phospholipase D-dependent production of phosphatidic acid can also affect the formation of coatamer-coated intracellular vesicles (see Ktistakis *et al.*, 1996).

4. PHYSIOLOGICAL EFFECTS OF RETINOIDS

Through regulation of gene transcription, retinoids can affect their functions during physiological events such as embryonic development, spermatogenesis, the immune system response, and control of cell proliferation (see Section 4.1).

In immunity, vitamin A deficiency has been correlated in some systems with a poor antibody-mediated immune response; this may be due at least in part to lack of normal retinoid control over interferon-γ gene transcription (Cantorna *et al.*, 1996). There is also evidence that vitamin A can stimulate directly immunoglobulin production by B cells (Wang *et al.*, 1993). Retinol is know to be necessary for growth of activated B cells (Buck *et al.*, 1990).

The involvement of retinoids in various aspects of embryogenesis has suggested a correlation between retinoid function and differential expression of particular RAR and RXR isoforms. For example, the RXRα isoform has been shown to be an important mediator of retinoid-dependent cardiac and ocular morphogenesis (Kastner *et al.*, 1994). Developmental studies such as these which involve genetic analysis of mutant mice deficient in specific RAR and/or RXR isoforms have established that RXR–RAR heterodimers are important physiological mediators of the RA signal during development (for example, see Kastner *et al.*, 1994). These investigations have also suggested that there is partial functional redundancy among several of the RAR subtypes: a null mutation in an RARγ isoform, for example, does not give rise to gross malformations of cartilage or epithelia even though its transcript is restricted to these types of tissues (or their precursors) at specific developmental stages (Lohnes *et al.*, 1993). Whether the apparent lack of RARγ function in this case is rescued by a different RAR isoform (i.e., true redundancy) or by another type of compensatory mechanism is not known. A particularly important effect of retinoid signaling during development is the regulation of Hox genes, as previously mentioned (Section 3.1), and control of the anterior–posterior embryonic axis.

Retinoid signaling has also been show to be involved in the regulation of viral genes. Retinoic acid can regulate the HIV-1 long terminal repeat gene in, for example, human brain cell lines (Sawaya *et al.*, 1996), and in human embryonal cell lines, it can enhance human cytomegalovirus replication (Angulo *et al.*, 1995).

Retinoids and Oncogenesis

Numerous studies have provided evidence that retinoid signaling events are important in controlling cell proliferation; thus, these events are likely to be significant factors in carcinogenesis and may be modulated as part of cancer chemoprevention strategies (Lotan, 1996).

The first example of a link between carcinogenesis and a member of the nuclear hormone receptor superfamily was that of leukemia with v-*erbA*, a mutated version of the thyroid hormone receptor gene *erbA* (Vennstrom and Bishop, 1982). V-*erbA* acts by suppressing erythroblast differentiation and this suppression involves interference with normal RAR and thyroid hormone receptor signaling (Schroeder *et al.*, 1992; Desbois *et al.*, 1991). Moreover, v-*erbA* studies have revealed the importance of cooperation between oncogenes derived from nuclear hormone receptors with those derived from tyrosine kinases in the establishment of cancer (for example, see Schroeder *et al.*, 1992). Besides mutant thyroid hormone receptor, the expression of mutant forms of RAR members of the nuclear hormone receptor family has also been shown to be involved in some cancers, e.g., acute promyelocytic leukemia (APL) and hepatocellular carcinoma (reviewed in de The, 1996). In APL, RA treatment (differentiation therapy) inhibits proliferation of the malignant promyelocytes and promotes their differentiation (reviewed in de The, 1996). This is remarkable because the addition of one factor, the retinoid, can change the whole oncogenic process. In HL-60 myelomonocytes, the presence of a mutant RAR which cannot bind RA inhibits the normal differentiative response of this cell line to RA (Robertson *et al.*, 1992). Thus, one of the mechanisms of how retinoid signaling can contribute to oncogenesis is through the cellular expression of mutant RARs.

Abnornal expression levels of RARs or RXRs is another mechanism that may account for oncogenesis. RAR and RXR levels are subject to a complex control mediated by, among other factors, retinoids themselves, AP-1, protein kinase C, cAMP, and the signal transduction pathways initiated by growth factors such as epidermal growth factor (EGF) and transforming growth factor β (TGFβ) (for some recent examples, see Chen *et al.*, 1996; Darwiche *et al.*, 1996; Xiao *et al.*, 1996, Roulier *et al.*, 1996). RARβ, which has been shown to have tumor suppressor activity (Berard *et al.*, 1994; Houle *et al.*, 1993; Lotan, 1996), is underexpressed, for example, in mouth and lung epithelial precancerous lesions (reviewed in Lotan, 1996). RA treatment of small lung cell carcinoma cells leads to enhanced RARβ levels and to cellular growth inhibition and neuroendocrine differentiation (Ou *et al.*, 1996). Other RARs and RXRs have also been found to be underexpressed in sqamous cell carcinomas (reviewed in Lotan, 1996) and in breast cancers (Liu *et al.*, 1996). Expression of an oncogenic v-*ras* in mouse keratinocytes was found to lead to decreased levels of RARα and γ (Darwiche *et al.*, 1996). Increasing the expression of specific RARs or RXRs (e.g., by retinoid treatment) may be a route to cancer chemoprevention. It should be mentioned, however, that in some cases retinoid treatment or the presence of a specific retinoid nuclear receptor is positively correlated with enhanced oncogenic potential (Berard *et al.*, 1994; Lotan, 1996; Tsao *et al.*, 1996). In addition, retinoids may in some cases promote tumor vascularization by inducing specific angiogenic factors such as IL-8 (see Harant *et al.*, 1996).

It is conceivable that yet another way by which defective retinoid signaling could contribute to oncogenesis is through defects in retinoid metabolism. Excessive or deficient quantities of a given retinoid may alter the activation of specific genes and of cellular proliferation.

Retinoids can interfere with the activation or signaling events of specific enzymes and growth factors, respectively, which are known to be associated with control of cell proliferation or carcinogenesis. In the following paragraphs, some of the retinoid-affected polypeptides are presented.

Ornithine decarboxylase (ODC) is the rate-limiting enzyme involved in polyamine biosynthesis and its activation is often correlated with carcinogenesis (for example, see Giardiello *et al.*, 1997). Retinoids have been shown to inhibit the ODC induction that occurs upon treatment with carcinogens or growth factors. For example, pretreatment with 13-*cis* RA inhibits the ODC activation that occurs on mouse skin upon application of several algal metabolites that can promote tumorigenesis (Fujiki *et al.*, 1981), and it has been shown that RA can suppress ODC mRNA levels (Mao *et al.*, 1993).

Protein kinase C (PKC) is an enzyme that can be activated by several different upstream signal transduction pathways. Upregulation of specific PKC isozymes can occur during carcinogenesis (for example, see Kahl-Rainer *et al.*, 1996). PKC activation, for example, can lead to decreased cellular levels of RARα and γ (Darwiche *et al.*, 1996).

Epidermal growth factor (EGF) and transforming growth factor α (TGFα) are ligands for the epidermal growth factor receptor (EGFR, also called c-*erbB*). This receptor has been found to be overexpressed in several cancer cell lines. Often, with concomitant expression of an EGFR ligand, an autocrine stimulation of growth can occur (for examples see Grandis *et al.*, 1996, and references therein). In several cell types, retinoids have been found to downregulate EGFR, EGF, or TGFα levels, primarily by decreasing the transcription rates of the genes (Grandis *et al.*, 1996, and references therein). In tracheobronchial epithelial cells (Miller *et al.*, 1996) and teratocarcinoma cells (Dmitrovsky *et al.*, 1990), for example, retinoids suppress TGFα mRNA levels and lead to suppression of proliferation and transformation. Retinoids can also interfere with EGF signaling; for example, RA can inhibit the EGF-induced increase in ODC activity in keratinocytes (Zheng *et al.*, 1995).

Other regulators of cell proliferation, transformation, or metastasis which are affected by retinoid signaling include cyclin E (Langenfeld *et al.*, 1996), transforming growth factor β (reviewed in Roberts and Sporn, 1984), protein kinase A (Rochette-Egly *et al.*, 1995; Huggenvik *et al.*, 1993), Hox transcription factors (Boylan *et al.*, 1993; Simeone *et al.*, 1990), and components of extracellular matrix assembly and disassembly (Roberts and Sporn, 1984; Fisher *et al.*, 1996; Pan *et al.*, 1995; Becker *et al.*, 1996). The influence of retinoids on these regulators is also important for embryogenesis and normal adult tissue function. Lack of, or aberrant, control of such regulators by retinoids is likely to contribute to the mechanisms of carcinogenesis in different cell types.

5. FUTURE PROSPECTS

A molecular analysis of the integration of retinoid signaling within the cellular signal transduction network is a major goal of current and future research in retinoid endocrinology, for example, direct modulation of retinoid receptor activity by associative proteins or covalent modifications imposed by other hormonal signals. This long-term objective, while facilitative, is not necessarily a prerequisite for the development of therapeutic retinoids. The development of synthetic retinoids, cellular and subcellular screening, and clinical testing of retinoids for activities such as control of carcinogenesis is already underway. Other areas of retinoid endocrinology that are not well understood include the nature of cellular retinoid uptake mechanisms, molecular reorganizations of the DNA transcription machinery induced by retinoids, and metabolic control of cellular biosynthesis and breakdown of active retinoids. While these issues need to be addressed by direct functional studies, a comparative molecular analysis of retinoid transport, metabolism, and signaling in organisms with different levels of organizational complexity (for example, among the various chordate and protochordate groups) may also provide insights into the evolution and function of these signaling molecules.

6. REFERENCES

Achen, M. G., Duan, W., Pettersson, T. M., Harms, P. J., Richardson, S. J., Lawrence, M. C., Wettenhall, R. E., Aldred, A. R., and Schreiber, G., 1993, Transthyretin gene expression in choroid plexus first evolved in reptiles, *Am. J. Physiol.* **265:**R982–989.

Allenby, G., Bocquel, M., Saunders, M., Kazmer, S., Speck, J., Rosenberger, M., Lovey, A., Kastner, P., Grippo, J. F., Chambon, P., and Levin, A. A., 1993, Retinoic acid receptors and retinoid X receptors: Interactions with endogenous retinoic acids, *Proc. Natl. Acad. Sci. USA* **90:**30–34.

Angulo, A., Suto, C., Boehm, M. F., Heyman, R. A., and Ghazal, P., 1995, Retinoid activation of retinoic acid receptors but not of retinoid X receptors promotes cellular differentiation and replication of human cytomegalovirus in embryonal cells, *J. Virol.* **69:**3831–3837.

Araki, H., Shidoji, Y., Yamada, Y., Moriwaki, H., and Muto, Y., 1995, Retinoid agonist activities of synthetic geranyl geranoic acid derivatives, *Biochem. Biophys. Res. Commun.* **209:**66–72.

Bass, N. M., 1993, Cellular binding proteins for fatty acids and retinoids: Similar or specialized functions? *Mol. Cell. Biochem.* **123:**191–202.

Bavik, C., Levy, F., Hellman, U., Wernstedt, C., and Eriksson, U., 1993, The retinal pigment epithelial membrane receptor for retinol-binding protein, *J. Biol. Chem.* **268:**20540–20546.

Bavik, C., Ward, S. J., and Chambon, P., 1996, Developmental abnormalities in cultured mouse embryos deprived of retinoic by inhibition of yolk-sac retinol binding protein synthesis, *Proc. Natl. Acad. Sci. USA* **93:**3110–3114.

Becker, P. S., Li, Z., Potselueva, T., Madri, J. A., Newburger, P. E., and Berliner, N., 1996, Laminin promotes differentiation of NB4 promyelocytic leukemia cells with all-*trans* retinoic acid, *Blood* **88:**261–267.

Bennekum, A., Blaner, W., Seifert-Bock, I., Moukides, M., Brouwer, A., and Hendriks, H. F. J., 1993, Retinol uptake from retinol binding protein by liver parenchymal cells *in vitro* does not specifically depend on its binding to retinol binding protein, *Biochemistry* **32:**1727–1733.

Berard, J., Gaboury, L., Landers, M., De Repentigny, Y., Houle, B., Kothary, R., and Bradley, W. E. C., 1994, Hyperplasia and tumors in lung, breast and other tissues in mice carrying a RARβ4-like transgene, *EMBO J.* **13:**5570–5580.

Bermudez, A. J., Swayne, D. E., Squires, M. W., and Radin, M. J., 1993, Effects of vitamin A deficiency on the reproductive system of mature White Leghorn hens, *Avian Dis.* **37:**274–283.

Berni, R., and Formelli, F., 1992, *In vitro* interaction of fenretinide with RBP and its functional consequences, *FEBS Lett.* **308:**43–45.

Blake, C. C., and Oatley, S. J., 1977, Protein–DNA and protein–hormone interactions in prealbumin, *Nature* **268:**115–120.

Blake, C. C., Geisow, M. J., Oatley, S. J., Rerat, B., and Rerat, C., 1978, Structure of prealbumin: Secondary, tertiary and quaternary interactions determined by Fourier refinement at 1.8 Å, *J. Mol. Biol.* **121:**339–356.

Blomhoff, R., Green, M. H., and Norum, K. R., 1992, Vitamin A: Physiological and biochemical processing, *Annu. Rev. Nutr.* **12:**37–57.

Boerman, M., and Napoli, J., 1991, Cholate-independent retinyl ester hydrolase, *J. Biol. Chem.* **266:**22273–22278.

Boylan, J. F., Lohnes, D., Taneja, R., Chambon, P., and Gudas, L. J., 1993, Loss of RAR γ function by gene disruption results in aberrant Hoxa-1 expression and defective cell differentiation, *Proc. Natl. Acad. Sci. USA* **90:**9601–9605.

Buck, J., Ritter, G., Dannecker, L., Katta, V., Cohen, S. L., Chait, B. T., and Hammerling, U., 1990, Retinol is essential for growth of activated human B cells, *J. Exp. Med.* **171:**1613–1624.

Cantorna, M. T., Nashold, F. E., Chun, T. Y., and Hayes, C. E., 1996, Vitamin A downregulation of IFN-gamma synthesis in cloned mouse Th1 lymphocytes depends on the CD28 costimulatory pathway, *J. Immunol.* **156:**2674–2679.

Chambon, P., 1996, A decade of molecular biology of retinoic acid receptors, *FASEB J.* **10:**940–954.

Chen, J.-Y., Penco, S., Ostrowski, J., Balaguer, P., Pons, M., Starrett, J. E., Reczek, P., Chambon, P., and Gronemeyer, H., 1995, RAR-specific agonist/antagonists which dissociate transactivation and AP1 transrepression inhibit anchorage-independent cell proliferation, *EMBO J.* **14:**1187–1197.

Chen, Y., Takeshita, A., Ozaki, K., Kitano, S., and Hanazawa, S., 1996, Transcriptional regulation by transforming growth factor beta of the expression of retinoic acid and retinoid X receptor genes in osteoblastic cells is mediated through AP-1, *J. Biol. Chem.* **271:**31602–31606.

Darwiche, N., Scita, G., Jones, C., Rutberg, S., Greenwald, E., Tennenbaum, T., Collins, S. J., De Luca, L. M., and Yuspa, S. H., 1996, Loss of retinoic acid receptors in mouse skin and skin tumors is associated with activation of the *ras*(*Ha*) oncogene and high risk for premalignant progression, *Cancer Res.* **56:**4942–4949.

DeBoeck, H., and Raphael, Z., 1988, NMR study of the interaction of retinoids with phospholipid bilayers, *Biochim. Biophys. Acta* **946:**244–252.

Derguini, F., Nakanishi, K., Hämmerling, U., Chua, R., Eppinger, T., Levi. E., and Buck, J., 1995, 13,14-Dihydroxy-retinol, a new bioactive retinol metabolite, *J. Biol. Chem.* **270:**18875–18880.

Desbois, C., Aubert, D., Legrand, C., Pain, B., and Samarut, J., 1991, A novel mechanism of action for v-*ErbA*: Abrogation of the inactivation of transcription factor AP-1 by retinoic acid and thyroid hormone receptors, *Cell* **67:**731–740.

de The, H., 1996, Altered retinoic acid receptors, *FASEB J.* **10:**955–960.

Divino, C. M., and Schlussler, G., 1990, Receptor-mediated uptake and internalization of transthyretin, *J. Biol. Chem.* **265:**1425–1429.

Dmitrovsky, E., Moy, D., Miller, W. H., Jr., Li, A., and Masui, H., 1990, Retinoic acid causes a decline in TGF-alpha expression, cloning efficiency, and tumorigenicity in a human embryonal cancer cell line, *Oncogene Res.* **5:**233–239.

Dong, D., and Zile, M. H., 1995, Endogenous retinoids in the early avian embryo, *Biochem. Biophys. Res. Commun.* **217:**1026–1031.

Episkopou, V., Maeda, S., Nishiguchi, S., Shimada, K., Gaitanaris, G. A., Gottesman, M. E., and Robertson, E. J., 1993, Disruption of the transthyretin gene results in mice with depressed levels of plasma retinol and thyroid hormone, *Proc. Natl. Acad. Sci. USA* **90:**2375–2379.

Fanjul, A., Dawson, M. I., Hobbs, P. D., Jong, L., Cameron, J. F., Harlev, E., Graupner, G., Lu, X.-P., and Pfahl, M., 1994, A new class of retinoids with selective inhibition of AP-1 inhibits proliferation, *Science* **372:**107–109.

Fisher, G. J., Datta, S. C., Talwar, H. S., Wang, Z.-Q., Varani, J., Kang, S., and Voorhees, J. J., 1996, Molecular basis of sun-induced premature skin aging and retinoid antagonism, *Nature* **379:**335–339.

Flower, D., 1996, The lipocalin protein family: Structure and function, *Biochem J.* **318:**1–14.

Forman, B. M., Goode, E., Chen, J., Oro, A. E., Bradley, D. J., Perlmann, T., Noonan, D. J., Burka, L. T., McMorris, T., Lamph, W. W., Evans, R. M., and Weinberger, C., 1995, Identification of a nuclear receptor that is activated by farnesol metabolites, *Cell* **81:**687–693.

Fujiki, H., Mori, M., Nakayasu, M., Terada, M., Sugimura, T., and Moore, R. E., 1981, Indole alkaloids: Dihydroteleocidin B, teleocidin, and lyngbyatoxin A as members of a new class of tumor promoters, *Proc. Natl. Acad. Sci. USA* **78:**3872–3876.

Gerlach, T. H., and Zile, M. H., 1991, Effect of retinoic acid and apo-retinol binding protein on serum retinol concentration in acute renal failure, *FASEB J.* **5:**86–92.

Giardiello, F. M., Hamilton, S. R., Hylind, L. M., Yang, V. W., Tamez, P., and Casero, R. A., Jr., 1997, Ornithine decarboxylase and polyamines in familial adenomatous polylosis, *Cancer Res.* **57:**199–201.

Giguere, V., Ong, E. S., Segui, P., and Evans, R. M., 1987, A human retinoic acid receptor which belongs to the family of nuclear receptors, *Nature* **330:**624–629.

Goodman, D. S., 1984, Plasma retinol-binding protein, in *The Retinoids*, Vol. 2 (D. S. Goodman, ed.), pp. 41–88, Academic Press, Orlando, Florida.

Grandis, J. R., Zeng, Q., and Tweardy, D. J., 1996, Retinoic acid normalizes the increased gene transcription rate of TGF-alpha and EGFR in head and neck cancer cell lines, *Nature Med.* **2:**237–240.

Halter, S. A., Winnier, A. R., and Arteaga, C. L., 1993, Pretreatment with vitamin A inhibits transforming growth factor alpha stimulation of human mammary carcinoma cells, *J. Cell Physiol.* **156:**80–87.

Harant, H., de Martin, R., Andrew, P. J., Foglar, E., Dittrich, C., and Lindley, I. J. D., 1996, Synergistic activation of interleukine-8 gene transciption by all-*trans*-retinoic acid and tumor necrosis factor-α involves the transcription factor NF-κB, *J. Biol. Chem.* **271:**26954–26961.

Heller, J., 1975, Interactions of retinol-binding protein with its receptor, *J. Biol. Chem.* **250:**3613–3619.

Herr, F. M., Wradlaw, S. A., Kakkad, B., Albrecht, A., Quick, T. C., and Ong, D. E., 1993, Intestinal vitamin A metabolism: Coordinate distribution of enzymes, *J. Lipid Res.* **34:**1545–1554.

Hope, W. C., Patel, B. J., Fiedler-Nagy, C., and Wittreich, B. H., 1990, Retinoids inhibit phospholipase A2 in human synovial fluid and arachidonic acid release from rat peritoneal macrophages, *Inflammation* **14:**543–559.

Houle, B., Rochette-Egly, C., and Bradley, W. E., 1993, Tumor suppressive effect of retinoic acid receptor β in human epidermoid lung cancer cells, *Proc. Natl. Acad. Sci. USA* **90:**985–989.

Huggenvik, J., Collard, M., Kim, Y., and Sharma, R., 1993, Modification of the retinoic acid signalling pathway by the catalytic subunit of protein kinase A, *Mol. Endocrinol.* **7:**543–550.

Joshi, P. S., Mathur, S. N., Murthy, S. K., and Ganguly, J., 1973, Vitamin A economy of the developing chick embryo and of the freshly hatched chick, *Biochem J.* **136:**757–761.

Kahl-Rainer, P., Sedivy, R., and Marian, B., 1996, Protein kinase C tissue localization in human colonic tumors suggests a role for adenoma growth control, *Gastroenterology* **110:**1753–1759.

Kaji, E. H., and Lodish, H. F., 1993, Unfolding of newly made retinol-binding protein by dithiothreitol. Sensitivity to retinoids, *J. Biol. Chem.* **268:**22188–22194.

Kanai, M., Raz, A., and Goodman, D. S., 1968, Retinol-binding protein: The transport protein for vitamin A in human plasma, *J. Clin. Invest.* **47:**2025–2044.

Kastner, P., Grondona, J. M., Mark, M., Gansmuller, A., LeMeur, M., Decimo, D., Vonesch, J.-L., Dolle, P., and Chambon, P., 1994, Genetic analysis of RXRα developmental functions: Convergence of RXR and RAR signaling pathways in heart and eye morphogenesis, *Cell* **78:**987–1003.

Kitareewan, S., Burka, L. T., Tomer, K. B., Parker, C. E., Deterding, L. J., Stevens, R. D., Forman, B. M., Mais, D. E., Heyman, R. A., McMorris, T., and Weinberger, C., 1996,

Phytol metabolites are circulating dietary factors that activate the nuclear receptor RXR, *Mol. Biol. Cell* **7:**1153–1166.

Kopelman, M., Cogan, U., Mokady, S., and Shinitzky, M., 1976, The interaction between retinol-binding proteins and prealbumins studied by fluorescence polarization, *Biochim. Biophys. Acta* **439:**449–460.

Kostrouch, Z., Kostrouchova, M., and Rall, J. E., 1995, Steroid/thyroid hormone receptor genes in *Caenorhabditis elegans*, *Proc. Natl. Acad. Sci. USA* **92:**156–159.

Ktistakis, N. T., Brown, H. A., Waters, M. G., Sternweis, P. C., and Roth, M. G., 1996, Evidence that phospholipase D mediates ADP ribosylation factor-dependent formation of Golgi coated vesicles, *J. Cell Biol.* **134:**295–306.

Kurlandsky, S. B., Gamble, M. V., Ramakrishnan, R., and Blaner, W. S., 1995, Plasma delivery of retinoic acid to tissues in the rat, *J. Biol. Chem.* **270:**17850–17857.

Kurlandsky, S. B., Duell, E. A., Kang, S., Voorhees, J. J., and Fisher, G. J., 1996, Auto-regulation of retinoic acid biosynthesis through regulation of retinol esterification in human keratinocytes, *J. Biol. Chem.* **271:**15346–15352.

Langenfeld, J., Lonardo, F., Kiyokawa, H., Passalaris, T., Ahn, M. J., Rusch, V., and Dmitrovsky, E., 1996, Inhibited transformation of immortalized human bronchial epithelial cells by retinoic acid is linked to cyclin E down-regulation, *Oncogene* **13:**1983–1990.

Levin, M. S., 1993, Cellular retinol-binding proteins are determinants of retinol uptake and metabolism in stably transfected Caco-2 cells, *J. Biol. Chem.* **268:**8267–8276.

Li, J. J., Dong, Z., Dawson, M. I., and Colburn, N. H., 1996, Inhibition of tumor promoter-induced transformation by retinoids that transrepress AP-1 without transactivating retinoic acid response element, *Cancer Res.* **56:**483–489.

Liu, Y., Lee, M. O., Wang, H. G., Li, Y., Hashimoto, Y., Klaus, M., Reed, J. C., and Zhang, X., 1996, Retinoic acid receptor beta mediates the growth-inhibitory effect of retinoic acid by promoting apoptosis in human breast cancer cells, *Mol. Cell. Biol.* **16:**1138–1149.

Lohnes, D., Kastner, P., Dierich, A., Mark, M., LeMeur, M., and Chambon, P., 1993, Function of retinoic acid receptor γ in the mouse, *Cell* **73:**643–658.

Lotan, R., 1996, Retinoids in cancer chemoprevention, *FASEB J.* **10:**1031–1039.

Malaba, L., Smeland, S., Senoo, H., Norum, K. R., Berg, T., Blomhoff, R., and Kindberg, G. M., 1995, Retinol-binding protein and asialo-orosomucoid are taken up by different pathways in liver cells, *J. Biol. Chem.* **270:**15686–15692.

Malpeli, G., Folli, C., and Berni, R., 1996, Retinoid binding to retinol-binding protein and the interference with the interaction with transthyretin, *Biochim. Biophys. Acta* **1294:**48–54.

Mao, Y., Gurr, J. A., and Hickok, N. J., 1993, Retinoic acid regulates ornithine decarboxylase gene expression at the transcriptional level, *Biochem. J.* **295:**641–644.

Mehta, P., Bertram, J., and Loenstein, W. R., 1989, The actions of retinoids on cellular growth correlate with their actions on gap junctions, *J. Cell Biol.* **108:**1053–1065.

Miller, L. A., Zhao, Y. H., and Wu, R., 1996, Inhibition of TGF-alpha gene expression by vitamin A in airway epithelium, *J. Clin. Invest.* **97:**1429–1435.

Monaco, H. L., Rizzi, M., and Coda, A., 1995, Structure of a complex of two plasma proteins: Transthyretin and retinol-binding protein, *Science* **268:**1039–1041.

Nagpal, S., Friant, S., Nakshatri, H., and Chambon, P., 1993, RARs and RXRs: Evidence for two autonomous transactivation functions and heterodimerization *in vivo*, *EMBO J.* **12:**2349–2360.

Napoli, J., 1996, Retinoic acid biosynthesis and metabolism, *FASEB J.* **10:**993–1001.

Noy, N., and Xu, Z., 1990, Interactions of retinol with binding proteins: Implications for cellular uptake mechanisms, *Biochemistry* **29:**3878–3883.

Ou, X., Campau, S., Slusher, R., Jasti, R. K., Mabry, M., and Kalemkerian, G. P., 1996, Mechanism of all-*trans*-retinoic acid-mediated L-*myc* gene regulation in small cell lung cancer, *Oncogene* **13:**1893–1899.

Pan, L., Eckhoff, C., and Brinckerhoff, C. E., 1995, Suppression of collagenase gene expression by all-*trans* and 9-*cis* retinoic acid is ligand dependent and requires both RARs and RXRs, *J. Cell. Biochem.* **57:**575–589.

Paulik, M., Morre, D. J., and Morre, D. M., 1995, A 55 kDa protein of transitional endoplasmic reticulum from rat liver binds retinol, *Biochim. Biophys. Acta* **1266:**273–277.

Perry, M. M., Gilbert, A. B., and Evans A. J., 1978, Electron microscope observation on the ovarian follicle of the domestic fowl during the rapid growth phase, *J. Anat.* **125:**481–497.

Perry, M. M., Griffin H. D., and Gilbert A. B., 1984, The binding of very low density and low density lipoproteins to the plasma membrane of the hen's oocyte, *Exp. Cell Res.* **151:**433–446.

Petkovich, M., Brand, N. J., Krust, A., and Chambon, P., 1987, A human retinoic acid receptor which belongs to the family of nuclear receptors, *Nature* **330:**444–450.

Rask, L., and Peterson, P. A., 1976, *In vitro* uptake of vitamin A from the retinol-binding plasma protein to mucosal epithelial cells from the monkey' s small intestine, *J. Biol. Chem.* **251:**6360–6366.

Roberts, A. B., and Sporn, M. B., 1984, Cellular biology and biochemistry of the retinoids, in *The Retinoids*, Vol. 2 (D. S. Goodman, ed.), pp. 209–286, Academic Press, Orlando, Florida.

Robertson, K., Emami, B., and Collins, S., 1992, Retinoic acid resistant HL60 cells harbour a point mutation in the RAR ligand binding domain that confers dominant negative activities, *Blood* **80:**1885–1892.

Rochette-Egly, C., Oulad-Abdelghani, M., Staub, A., Pfister, V., Scheuer I., Chambon, P., and Gaub, M. P., 1995, Phosphorylation of the retinoic acid receptor-alpha by protein kinase A, *Mol. Endocrinol.* **9:**860–871.

Ross, A., 1993, Cellular metabolism and activation of retinoids: Roles of cellular retinoid-binding proteins, *FASEB J.* **7:**317–327.

Roulier, S., Rochette-Egly, C., Alsat, E., Dufour, S., Porquet, D., and Evain-Brion, D., 1996, EGF increases retinoid X receptor-alpha expression in human trophoblastic cells in culture: Relationship with retinoic acid induced human chorionic gonadotropin secretion, *Mol. Cell. Endocrinol.* **118:**125–135.

Sawaya, B. E., Rohr, O., Aunis, D., and Schaeffer, E., 1996, Regulation of HIV-1 gene transcription by nuclear receptors in human brain cells, *J. Biol. Chem.* **271:**22895–22900.

Schadendorf, D., Kern, M. A., Artuc, M., Pahl, H. L., Rosenbach, T., Fichtner, I., Nrnberg, W., Stting, S., Stebut, E., Worm, M., Makki, A., Jurgovsky, K., Kolde, G., and Henz, B. M., 1996, Treatment of melanoma cells with the synthetic retinoid CD437 induces apoptosis via activation of AP-1 *in vitro*, and causes growth inhibition in xenografts *in vivo*, *J. Cell Biol.* **135:**1889–1898.

Schneider, W. J., Vieira, A. V., MacLachlan, I., and Nimpf, J. N., 1993, Lipoprotein receptor-mediated oocyte growth, in *Cellular Metabolism of the Arterial Wall and Central Nervous System* (G. Schettler, H. Greten, and A. Habenicht, eds.), pp. 105–115, Springer-Verlag, Berlin.

Schroeder, C., Gibson, L., and Beug, H., 1992, The v-*erbA* oncogene requires cooperation with tyrosine kinases to arrest erythroid differentiation induced by ligand-activated endogenous c-*erbA* and retinoic acid receptor, *Oncogene* **7:**203–216.

Shimeld, S. M., 1996, Retinoic acid, HOX genes and the anterior–posterior axis in chordates, *Bioessays* **18:**613–616.

Simeone, A., Acampora, D., Arcioni, L., Andrews, P. W., Boncinelli, E., and Malvillo, F., 1990, Sequencial activation of HOX2 homeobox genes by retinoic acid in human embryonal carcinoma cells, *Nature* **346:**763–767.

Sivaprasadarao, A., and Findlay, J., 1988, The mechanism of uptake of retinol by plasma membrane vesicles, *Biochem. J.* **255:**571–579.

Takahashi, N., and Breitman, T., 1989, Retinoic acid acylation (retinoylation) of a nuclear protein in the human acute myeloid leukemia cell line HL60, *J. Biol. Chem.* **264:**5159–5163.

Terry, C. J., Damas, A. M., Oliveira, P., Saraiva, M. J., Alves, I. L., Costa, P. P., Matias, P. M., Sakaki, Y., and Blake, C. C., 1993, Structure of Met30 variant of transthyretin and its amyloidogenic implications, *EMBO J.* **12:**735–741.

Thompson, J., McHowell, J., Pitt, G., and McLaughlin, C., 1969, The biological activity of retinoic acid in the domestic fowl and the effects of vitamin A deficiency on the chick embryo, *Br. J. Nutr.* **23:**471–490.

Tornabene, T. G., Langworthy, T. A., Holzer, G., and Oro, J., 1979, Squalenes, phytanes and other isoprenoids as major neutral lipids of methanogenic and thermoacidophilic "Archaebacteria," *J. Mol. Evol.* **13:**73–83.

Tsao, Y.-P., Li, S.-F., Kuo, S.-W., Liu, J.-C., and Chen, S.-L., 1996, Reversal of the temperature-shift-induced growth restriction of a temperature-sensitive simian virus 40 T-antigen-transformed human fibroblast cell line by treatment with retinoic acid, *Biochem. J.* **317:**707–711.

Tsutsumi, C., Okuno, M., Tannous, L., Piantedossi, R., Allan, M., Goodman, D. S., and Blaner, W. S., 1992, Retinoids and retinol-binding protein expression in rat liver adipocytes, *J. Biol. Chem.* **267:**1805–1810.

Urano, S., Inomori, Y., Sugawara, T., Kato, Y., Kitahara, M., Hasegawa, Y., Matsuo, M., and Mukai, K., 1992, Vitamin E: Inhibition of retinol-induced hemolysis and membrane-stabilizing behavior, *J. Biol. Chem.* **267:**18365–18370.

Vennstrom, B., and Bishop, J. M., 1982, Isolation and characterization of chicken DNA homologous to the two putative oncogenes of avian erythroblastosis virus, *Cell* **28:**135–143.

Vieira, A. V., 1994, Characterization of the avian retinol-binding protein–transthyretin complex and its role in oocytic endocytosis of retinol, *Ph.D. Thesis*, University of Alberta, Edmonton, Alberta, Canada.

Vieira, A. V., and Schneider, W. J., 1993, Transport and uptake of retinol during chicken oocyte growth, *Biochim. Biophys. Acta* **1169:**250–256.

Vieira, A. V., Sanders, E. J., and Schneider, W. J., 1995a, Transport of serum transthyretin into chicken oocytes, a receptor-mediated mechanism, *J. Biol. Chem.* **270:**2952–2956.

Vieira, A. V., Schneider, W. J., and Vieira, P. M., 1995b, Retinoids: Transport, metabolism, and mechanisms of action, *J. Endocrinol.* **146:**201–207.

Wahlberg, P., Fex, G., and Wennerberg, J., 1989, A relationship between retinol and cellular retinol-binding protein concentrations in human squamous cell carcinomas, *Biochim. Biophys. Acta* **1010:**287–293.

Wang, W., Napoli, J. L., and Ballow, M., 1993, The effects of retinol on *in vitro* immunoglobulin synthesis by cord blood and adult peripheral blood mononuclear cells, *Clin. Exp. Immunol.* **92:**164–168.

Wang, X. D., Russell, R. M., Liu, C., Stickel, F., Smith, D. E., and Krinsky, N. I., 1996, Beta-oxidation in rabbit liver *in vitro* and in the perfused ferret liver contributes to retinoic acid biosynthesis from beta-apocarotenoic acids, *J. Biol. Chem.* **271:**26490–26498.

Wei, S., Episkopou, V., Piantedosi, R., Maeda, S., Shimada, K., Gottesman, M. E., and Blaner, W. S., 1995, Studies on the metabolism of retinol and retinol-binding protein in transthyretin-deficient mice produced by homologous recombination, *J. Biol. Chem.* **270:**866–870.

Woods, J., Woodward, M., and Roth, T., 1978, Selective protein uptake, *J. Cell Sci.* **30:**87–97.

Xiao, Y., Desai, D., Quick, T. C., and Niles, R. M., 1996, Control of retinoic acid receptor expression in mouse melanoma cells by cyclic AMP, *J. Cell. Physiol.* **167:**413–421.

Zheng, Z. S., Xue, G. Z., and Prystowsky, J. H., 1995, Regulation of the induction of ornithine decarboxylase in keratinocytes by retinoids, *Biochem. J.* **309:**159–165.

Chapter 3

Structure and Function of Retinoid-Binding Proteins

Marcia E. Newcomer, R. Scott Jamison, and David E. Ong

1. INTRODUCTION

Biological processes which require compounds derived from vitamin A (retinol) include morphogenesis, spermatogenesis, the maintenance of epithelial tissue, and vision. For the last process the light-activated *cis–trans* isomerization of rhodopsin-bound retinaldehyde triggers a cascade of events which ultimately results in the hydrolysis of cyclic GMP. For morphogenesis, spermatogenesis, and the maintenance of epithelial tissue it is presumably retinoic acid that is the active metabolite. Both 9-*cis* and all-*trans* retinoic acid are transcriptional activators which exert their effects through binding to nuclear proteins that are members of the superfamily that includes the steroid hormone receptors. Various proteins mediate the transport and metabolism of the hydrophobic vitamin and consequently modulate the concentration of retinoic acid available to the nucleus. Both extracellular and intracellular retinoid-binding proteins have been identi-

Marcia E. Newcomer, R. Scott Jamison, and David E. Ong Department of Biochemistry, Vanderbilt University School of Medicine, Nashville, Tennessee 37232-0146.

Subcellular Biochemistry, Volume 30: Fat-Soluble Vitamins, edited by Quinn and Kagan. Plenum Press, New York, 1998.

fied in numerous cell types. In addition, the eye has retinoid binding proteins which are distinct for that organ: a cellular retinaldehyde-binding protein (CRalBP) and a interstitial/interphotoreceptor retinoid-binding protein (I-RBP).

A number of X-ray crystallographic structures of retinoid-binding proteins have been solved and it is clear from the X-ray structures that structural classes of retinoid-binding proteins correspond to functional classes: the extracellular, intracellular, and nuclear receptors for retinoids each have their distinct structural motif. [The cellular retinaldehyde-binding protein unique to the eye is homologous not to the ubiquitous intracellular retinol- and retinoic acid-binding proteins, but to proteins which facilitate the exchange of α-tocopherol (human liver α-tocopherol transfer protein) and phospholipid (yeast sec14) between membranes.] In this review we will describe the structural motif characteristic of the three functional classes of retinoid-binding proteins for which structural information is available: the extracellular and intracellular binding proteins, and the nuclear receptors.

2. EXTRACELLULAR RETINOID-BINDING PROTEINS

2.1. Family Members

The transport of vitamin A to the target tissues is mediated by plasma retinol-binding protein (RBP). This specific carrier protein for all-*trans* retinol is synthesized primarily in the liver, where it requires the binding of retinol to trigger its secretion (Ronne *et al.*, 1983). Other sites of synthesis are known (Soprano and Blaner, 1994) and include the kidney, peritubular, and Sertoli cells of the testis (Davis and Ong, 1992), and retinal pigment epithelium (Ong *et al.*, 1994; Jaworowski *et al.*, 1995). Circulating in the plasma, RBP is found bound to transthyretin (TTR, previously referred to as thyroxine-binding prealbumin). The formation of this protein complex prevents the low-molecular-weight retinol-binding protein (21,000) from being filtered by the kidney glomeruli (Kanai *et al.*, 1968) and also may restrict free partitioning into the intercellular space outside the vascular system. Whether ultimately utilized as retinoic acid, retinal, or retinol, vitamin A is transported to the target cells by RBP as all-*trans* retin*ol*, and only retin*ol* triggers secretion of RBP (Ronne *et al.*, 1983). However *in vitro*, RBP binds all-*trans* retinoic acid with an affinity comparable to that for all-*trans* retinol (Goodman and Blaner, 1984). The X-ray structure of RBP was determined to 2.1-Å resolution (Newcomer *et al.*, 1984; Cowan *et al.*, 1990; Zanotti *et al.*, 1993a,b). The basic structural framework is an eight-stranded up-and-down β-barrel with a C-terminal α-helix which form a coffee-mug-like structure. The ligand is bound within the barrel, or mug, in a hand-in-glove-like fit. Access to the binding site is via an opening in the open end of the barrel, at the top of the mug. The backbone trace and space-filling model of human serum retinol-binding protein are presented in Fig. 1. RBP is one of a large superfamily of extracellular proteins known as "lipocalins" which transport hydrophobic/labile ligands. These proteins share the structural framework described for the retinol-binding protein.

Extracellular transport proteins for retinoic acid (Ong and Chytil, 1988) and retinal (Lepperdinger *et al.*, 1996) have also been described, although for these two proteins only *in vitro* binding studies have been reported and the endogenous ligands have not yet been determined. The first of these proteins, an epididymal retinoic acid-binding protein (E-RABP), is an androgen-regulated protein abundant in the epididymis, the site of sperm maturation. Retinol has long been known to be required for functional sperm, as adult rats fed retinoic acid (which cannot be reduced to retinal or retinol *in vivo*) are blind and sterile but "otherwise healthy." However, retinoic acid appears to be important as well for sperm matura-

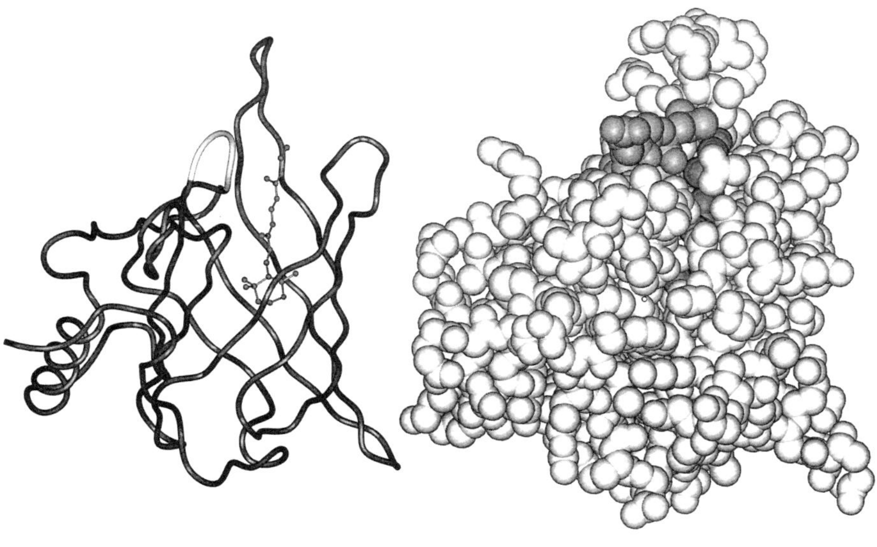

FIGURE 1. Backbone ribbon drawing and space-filling model of retinol-binding protein. The darkened spheres of the space-filling model correspond to the terminal carbons and hydroxyl of the retinol as well as the side chains which move into the binding cavity in the apo form (Cowan *et al.*, 1990; Zanotti *et al.*, 1993a).

tion, since high concentrations of cellular and extracellular retinoic acid-binding proteins are abundant in rat epididymis (Kato *et al.*, 1985; Porter *et al.*, 1985; Ong and Chytil, 1988; Newcomer and Ong, 1990), as well as transcripts for the retinoic acid receptors α and β (Wan *et al.*, 1992; Akmal *et al.*, 1996). At the very least retinoic acid is required for the maintenance of the lining epididymal epithelia and consequently normal epididymal function. In contrast to RBP, which can bind either retinoic acid or retinol *in vitro*, E-RABP is specific for retinoic acid. Although the proteins have a common structural motif and overlapping ligand specificity, important differences between the structures are thought to confer functional differences. These differences are presumably necessary to allow E-RABP to be specific for retinoic acid (versus retinol or retinal) and to accommodate both the 9-*cis* and all-*trans* isomers of the ligand in the binding site (Newcomer, 1993). The backbone trace and space-filled rendering of E-RABP are presented in Fig. 2.

Recently, a retinal-binding protein homologous to RBP and E-RABP has been identified (Lepperdinger *et al.*, 1996). The protein, *Xenopus laevis* choroid plexus lipocalin 1 (Xlcpl1), is expressed in the early embryo in the anteriormost part of the neural plate. It is also a major component of the choroid plexus of the adult frog. This protein has been shown to bind retinal with nanomolar affinity and retinoic acid with micromolar affinity, but it

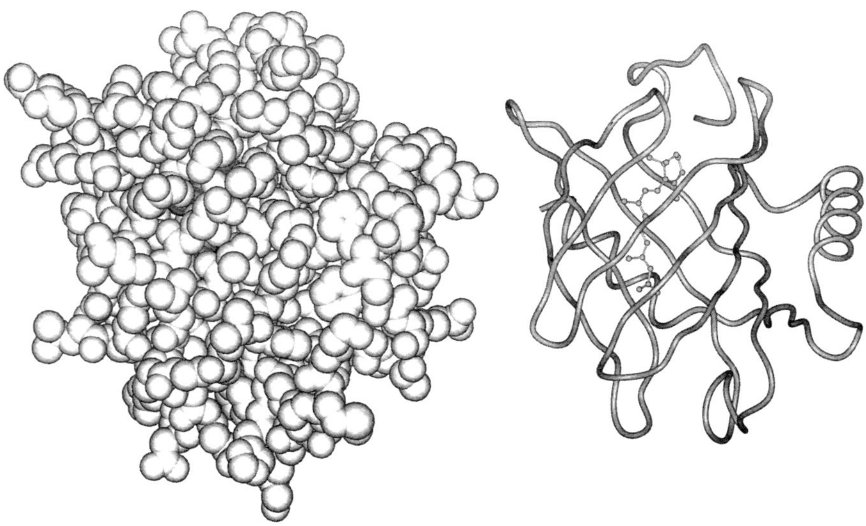

FIGURE 2. Backbone ribbon drawing and space-filling model of the epididymal retinoic acid-binding protein (Newcomer, 1993).

does not bind retinol. It has been suggested that the protein mediates the extracellular transport of retinal. Retinal is then oxidized to the morphogen retinoic acid once inside the target cell. Although no X-ray structure has been determined for the protein, the level of sequence identity with E-RABP and RBP indicates the protein has their same overall structure.

Despite a wealth of biochemical data, the function of the lipocalin β-lactoglobulin, a ruminant milk protein which binds retinol *in vitro*, is unknown. Although an earlier structure report placed the retinol-binding site on the surface of the protein rather than inside the β-barrel (Monaco *et al.*, 1987), site-directed mutagenesis experiments (Cho *et al.*, 1994) are consistent with an internal binding site. In the report of the 1.8-Å resolution structure of β-lactoglobulin (Brownlow *et al.*, 1997) the authors point out that retinol can be modeled in the binding cavity. The comparison of the retinoid-binding lipocalins which follows will be limited to those for which holo structures are available.

2.2. Ligand Binding

RBP and E-RABP bind their respective retinoids with the hand-in-glove-like fit in a deep hydrophobic cavity. Access to the cavity, which in the apo form is otherwise solvent-accessible, appears to be relatively unrestricted. The ligands bind with the β-ionone ring innermost. In RBP the all-*trans* isoprene tail is fully extended, whereas in E-RABP the isoprene tail of all-*trans* retinoic acid appears to have a *cis* configuration at C8, so that the all-*trans* retinoic acid resembles 9-*cis* retinoic acid, which also binds to E-RABP. The ligand-binding sites of both RBP and E-RABP encompass approximately 25 (RBP 25, E-RABP 23) amino acid residues. As one would predict, the binding cavities contain predominantly aromatic and hydrophobic side chains. The ligand of E-RABP is negatively charged, and consequently the protein has significantly more charged residues in the binding cavity. Furthermore, the amphipathic nature of the ligand is clearly evident in the disposition of the amino acid side chains in the binding site: aromatic amino acids circle the β-ionone ring, aliphatic amino acids flank the isoprene tail, and the negatively charged carboxylate is encompassed by a web of three charged amino acid pairs. A similar arrangement of aromatic rings around the perimeter of the β-ionone ring is observed in RBP, where aromatic rings flank a portion of the isoprene tail of the ligand as well. The strict specificities of RBP and E-RABP are presumably a result of the high degree of complementarity of the van der Waals surfaces of protein and ligand. In RBP the retinol hydroxyl is found at the surface of the protein. This positioning of the tail of the ligand allows the protein to easily accommodate the acid or aldehyde form of the vitamin, as well as retinyl esters. In contrast, the deeper placement of the ligand in E-RABP, and the positioning of two charged pairs and an unpaired positive charge within the β-barrel, appear to restrict ligand binding to the negatively charged retinoic acid.

None of the amino acid side chains that form the ligand-binding cavity in RBP is conserved in E-RABP. In RBP, with the retinol hydroxyl positioned at the surface of the protein, the β-ionone ring is found about halfway into the barrel. In E-RABP the binding site is placed about 8 Å deeper into the barrel (the ligand itself is about 15 Å long) so that the binding site penetrates more than two-thirds of the way into the barrel. This large invagination which is observed in E-RABP serves as both a binding site and a large binding site entrance. In RBP, side chains fill up the closed end of the barrel so that the binding cavity in the apo form is not so deep. In contrast to what is observed in RBP, the deeper placement of the binding site in E-RABP allows for protein–ligand interactions that clearly discriminate at the carboxylate as well the β-ionone ring end of the ligand. In Fig. 3 the structures of RBP and E-RABP are superimposed so that the alignment of their β-barrels is optimized. The structural homology between the two proteins is mostly localized to the "closed" end of the barrel, as this is the area in the tertiary structures where the amino acids that are conserved throughout the protein family cluster. There is more variation between the two structures at the entrance end of the barrels, where the proteins differ in the extent of the β-sheets and in the conformations and lengths of the loop regions. Note the difference in the placement of the binding site relative to the structural framework.

It is useful to compare the overlay of RBP and E-RABP (Fig. 3) with an overlay of a more distantly related lipocalin with RBP. In Fig. 4, the structure of odorant-binding protein (Bianchet *et al.*, 1996, Tegoni *et al.*, 1996) is superimposed on RBP. Odorant-binding protein (OBP) is a protein of mammalian nasal mucosa which is thought to function in olfactory response. The sequence identity between OBP and RBP is roughly 18% if one does a blind pairwise alignment of the two sequences. However, if one superimposes the two structures and uses this superposition to line up the sequences, the sequence identity now is remarkably less—roughly 10%. Notice also that the odorant-binding protein is a "strand-exchanged" dimer. In this type of oligoimerization the interface between monomers corresponds to an intramolecular interface in a monomer. In the OBP dimer the helix, which in RBP packs onto the β-barrel, packs onto the barrel of its mate in an intermolecular embrace.

As has been shown for other protein families, only nominal sequence identity is necessary for proteins to adopt a given structural framework. Furthermore, novel variations on the framework, such as domain swapping, can expand nature's structural repertoire.

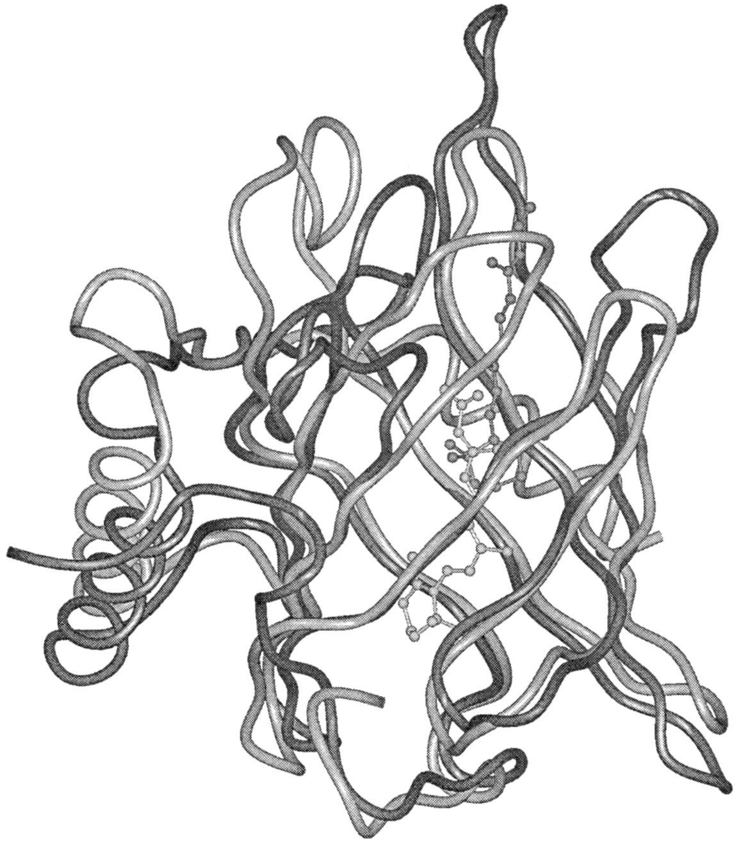

FIGURE 3. Superposition of the backbone ribbon drawings of retinol-binding protein (dark gray) and epididymal retinoic acid-binding protein (light gray).

2.3. Conformational Changes Induced by Ligand Binding

The structures for both retinol- and retinoic acid-binding proteins have been determined in both the apo and holo forms (Zanotti *et al.*, 1993a,b; Newcomer, 1993). In RBP, the changes between the apo and holo forms are limited to the movement of amino acids 33–36, which form a turn at the open end of the barrel. The side chains of Leu-35 and Phe-36 adopt strikingly different positions in the apo versus holo forms: the aromatic ring of Phe-36 is located about halfway into the binding cavity in the holo protein, but in the apo protein it is found toward the surface of the protein and occupies the space where C13 of the retinol in the holo protein is positioned. The same conformational change was described for the apo versus holo bovine protein, for which there is better than 90% identity with

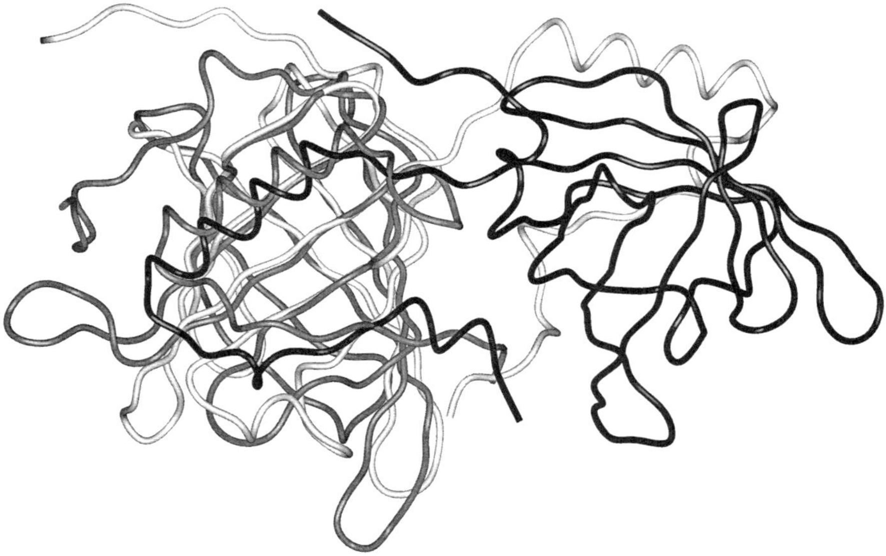

FIGURE 4. Superposition of retinol-binding protein (gray) and the dimer of odorant-binding protein (black and white).

human RBP. The region of the conformational change is highlighted in Fig. 1. In E-RABP it is a methionine (Met-39) within the binding cavity which must move to allow the binding site to conform to the contours of the ligand.

Several RBP X-ray structures have been determined and by comparison of these structures it is possible to propose possible regions of structural flexibility. The loops at 63–67 and 93–97, which are at the open (or "entrance") end of the barrel cavity, adopt alternate conformations in the different crystal structures. The corresponding "entrance loops" in E-RABP are also flexible, as evidenced by the fact that the loops adopt different conformations in crystallographically independent molecules in the same crystal unit cell (Newcomer, 1993).

3. INTRACELLULAR RETINOID-BINDING PROTEINS

3.1. Family Members

Before vitamin A can fulfill its several functions within the body, both in vision and in differentiation, it must necessarily move through a number of different cells on its journey from the lumen of the small intestine to the various sites of action. This process may begin with metabolism to convert the provitamin carotenes into retinol. This retinol, or preformed dietary retinol, must be esterified for export from the absorptive cell in chylomicrons. Retinol is also esterified for storage in other cells of the body. The retinyl esters are subsequently hydrolyzed to recover the retinol. Further metabolism is required to produce active retinoids, such as 11-*cis* retinal or retinoic acid. In the various cells that carry out the metabolic transformations one finds specific retinoid-binding proteins which mediate the intracellular transport and metabolism of the hydrophobic retinoids. Four such proteins have been well characterized: two bind retinol or retinal [cellular retinol-binding proteins type I and II, CRBP and CRBP(II)] and two bind retinoic acid [cellular retinoic acid binding proteins type I and II, CRABP and CRABP(II)].

In both the CRBP and CRABP pairs, substitutions in the amino acid sequence result in differences in ligand specificity and affinity. In the CRBP/CRBP(II) pair the level of conservation of binding-site amino acids is not significantly different from the level of overall sequence conservation: only 13 of the 23 amino acids which line the binding cavity are present in both the type I and type II proteins. Presumably this difference in binding-site amino acid side chains contributes to the fact that the proteins bind the same spectrum of ligands, but differ in the relative affinities for the two biological ligands. These differences in affinity may be due to functional differences. The putative function of CRBP(II), which is restricted to the intestinal absorptive cell in the adult animal, is to mediate the absorption of dietary retinoids and intestinal metabolism of carotenes. Both retinal and retinol occur in these processes, and CRBP(II) binds the two forms of the vitamin with similar affinities and the presence of these retinoids is transient in the absorptive cell. In contrast, CRBP, which is present in a variety of cells in the organism, has a substantially higher affinity for retinol than does CRBP(II). CRBP–retinol is a constant component of the cell and the increased affinity CRBP has for its ligand may be necessary to ensure that free retinol does not partition into the cell membrane.

The CRABPs are homologous to the CRBPs in function as well as structure. By facilitating the metabolism of retinoic acid, or simply by sequestering the retinoic acid so that it is no longer free to bind the nuclear receptors, the CRABPs are thought to modulate the amount of the vitamin/hormone available to the nuclear receptors. Holo-CRABP is a substrate for retinoic acid metabolism in rat testes microsomes (Fiorella and Napoli, 1991); the sensitivity of F9 teratocarcinoma cells to retinoic acid is inversely correlated to the amount of CRABP(I) expressed (Boylan and Gudas, 1991); and the rate of metabolism of ^{3}H-retinoic acid to 4-oxo-retinoic acid is increased with increased expression of CRABP(I) (Boylan and Gudas, 1992). In contrast to the CRBPs, both CRABPs are restricted to the cytoplasmic compartments (despite their low molecular weights), perhaps to enhance their availability to block the appearance of retinoic acid in the nucleus (Bucco *et al.*, 1995, Zheng *et al.*, 1996). The fact that two isoforms of the protein are found may be a result of different requirements for retinoic acid in different tissues, as the proteins bind the same spectrum of retinoids, but with differences in the affinities (Fiorella *et al.*, 1993). A further difference is that CRABP(II) is found in the uterine epithelial cells only at a time when cells are actively synthesizing retinoic acid, presumably for paracrine action (Bucco *et al.*, 1997).

Both CRABPs are absolutely specific for a carboxylate, as neither retinol nor retinal is a ligand. CRABP(I) and CRABP(II) bind 9-*cis* retinoic acid, but with an affinity threefold lower than that observed for the all-*trans* isomer (Fiorella *et al.*, 1993). The cellular retinol-binding proteins do not bind the 9-*cis* isomers (MacDonald and Ong, 1987). There are only three conservative substitutions in the binding-site amino acids between the type I and type II isoforms of CRABP.

The cellular retinoid-binding proteins are members of a larger protein family, as defined by sequence homology, of small (about 130 amino acids) intracellular carriers for hydrophobic ligands (Banaszak *et al.*, 1994). Other members of the family are fatty acid-binding proteins from liver, heart, muscle, and intestine, as well as myelin P2 and adipocyte P2, proteins which also bind fatty acids. CRBP and CRABP have approximately 40% sequence identity; this is not significantly more identity than there is for the CRBP/aP2 pair. The sequence identity for the functionally related pairs CRBP/CRBP(II) and CRABP/CRABP(II) is higher: 56% and 78%, respectively. Among species the individual proteins are highly conserved. For example, the sequences of CRABP from rat, mouse, and cow are identical, and human CRABP has only a single amino acid change. Such a high degree of conservation among species, in spite of the fact that it is not necessary to conserve the overall fold of the protein or the ligand specificity, is consistent with a multifunctional protein that not only recognizes ligand, but is also recognized by, or interacts with, other cellular components such as enzymes and membrane transporters.

X-ray structures of the holo cellular retinol-binding protein types I and II (Cowan *et al.*, 1993, Winter *et al.*, 1993), as well as holo cellular retinoic acid-binding protein types I and II (Kleywegt *et al.*, 1994), have been determined. Apo structures have been determined for CRBP II (Winter *et al.*, 1993) and CRABP I (Thompson *et al.*, 1995) as well. As one would expect from the level of sequence identity observed for the cellular retinoid-binding proteins, the tertiary structures of these proteins are very similar. The structures are also topologically equivalent to those described for the fatty acid-binding proteins.

The intracellular proteins are globular proteins with the approximate dimensions 40 × 40 × 30 Å (Fig. 5). Ten strands of antiparallel β-sheet (referred to as strands A–J for the sake of discussion to indicate their relative positions in the linear sequence) fold into a flattened barrel to form the structural framework of the protein. Neighboring strands B–J are connected by short reverse turns, and hydrogen bonding between strands A and J makes the barrel a continuous structure. A helix-turn-helix segment is located between β strands A and B. The ligand-binding site is sandwiched inside the flattened barrel, with the isoprene tail of the molecule innermost. As is invariably the case for this protein superfamily, the ligand-binding site is a large interior cavity considerably larger than the ligand itself. In addition to the ligand, bound water molecules fill these cavities in the holo structures. This inclusion of water molecules in the liganded binding sites is in contrast to what is observed for the extracellular retinol/retinoic acid-

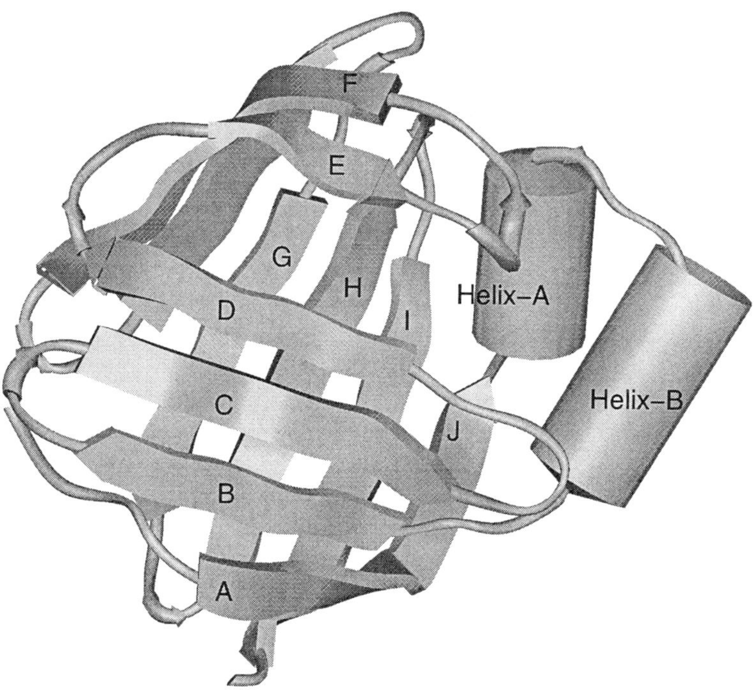

FIGURE 5. A backbone drawing of CRBP in which strands are shown as flattened ribbons and helices as cylinders. The structural elements are labeled for reference.

binding proteins, where the ligands dovetail in their binding sites with a hand-in-glove-like fit.

In the intracellular binding proteins the amino terminus wraps around the back half of the barrel, where by the "back half" is meant that end which does not hold the retinol. The front half of the barrel is capped by the two α-helices. In the backbone drawing shown in Fig. 5 another structural characteristic of the family is apparent: the separation between β-strands D and E appears to provide a gap, or access to the interior cavity. However, despite the fact that the strands are too far apart for hydrogen bonds to form between the carbonyls and amides of neighboring β-strands, amino acid side chains fill the gap and provide for side-chain : main-chain as well as side-chain : side-chain hydrogen bonds. Backbone and space-filling renderings of CRBP–retinol and CRABP–retinoic acid are given in Fig. 6. In

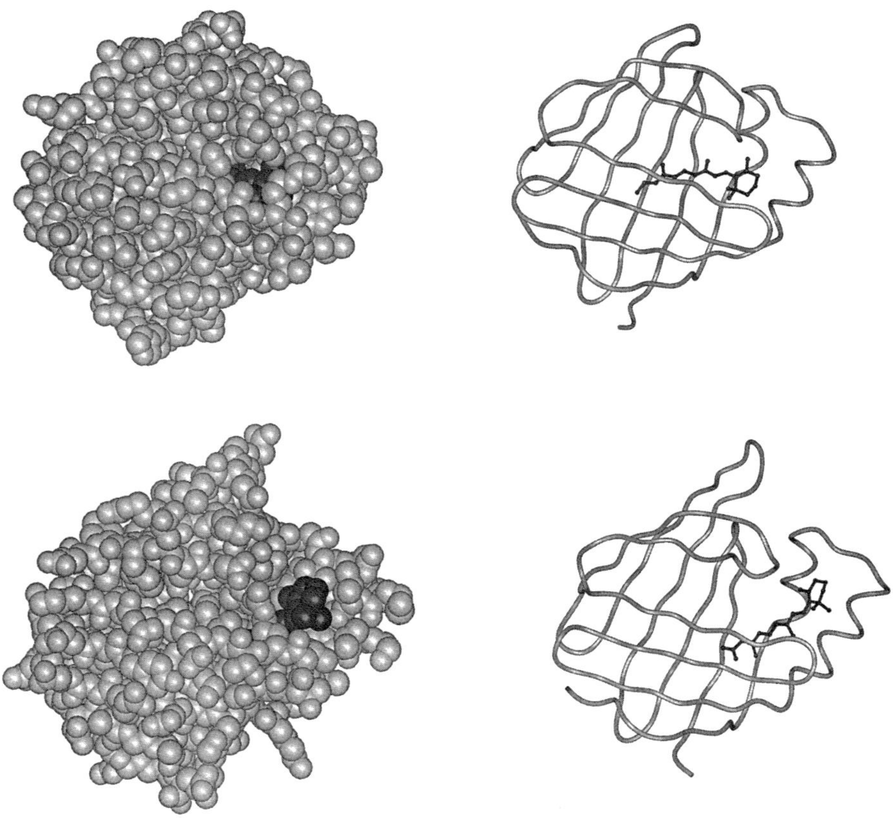

FIGURE 6. Backbone and space-filling renderings of CRBP(I) (top) and CRABP(I) (bottom). Ligand is shown in dark gray.

CRBP, the ligand is encapsulated within the core of the protein and only a small portion of the ligand, which appears as dark, gray spheres in the figure, is visible, but recessed. This portion is part of the β-ionone ring. The backbone trace for the protein is drawn in the same relative orientation. A comparable pair of representations is depicted for CRABP(I). Notice that in holo-CRABP(I) much more of the ligand is visible. In CRBP the helical cap fully encapsulates the ligand in a solvent-inaccessible cavity. However, in the CRABPs the β-ionone ring is lodged between the helical cap and the β-barrel and protrudes from the surface of the protein. As a result, the ligand-binding site of CRABP encompasses fewer amino acids (19) than that of CRBP (23).

When the backbones of CRABP and CRBP are superimposed, as in Fig. 7, one can see that just as in the RBP/E-RABP pair, the retinoid-binding site has shifted in its placement with respect to the overall structural motif. Whereas in E-RABP the retinoic acid-binding cavity is deeper than the retinol-binding cavity in RBP, in the intracellular pair of homologous proteins the retinoic acid-binding site is displaced toward the surface. This displacement of the bound retinoid results in a cavity which would have clear access to bulk solvent in the absence of ligand.

Since the ligand-binding sites of both the CRBPs and CRABPs include bound water molecules in addition to their ligands, the ligand-binding cavities include more-polar amino acids than the binding sites in the extracellular transport proteins, and lack the clear demarcation of aliphatic/polar aliphatic/aromatic amino acids described for RBP and E-RABP. In the CRBPs, Gln-108 participates in a hydrogen bond with the hydroxyl group of the ligand. In the CRABPs the primary protein–ligand electrostatic interaction is provided by Arg-132, although Arg-111 (which corresponds to Gln-108 of CRBPI/II) is also in the binding cavity and interacts with the ligand carboxylate via a water molecule. The first X-ray structures available for this superfamily of proteins were for two fatty acid-binding proteins (Jones *et al.*, 1988, Sacchettini *et al.*, 1988). In the fatty acid-binding protein Myelin protein P2 the ligand carboxylate is centered between the positively charged guanidino groups of Arg-106 and Arg-126. These arginines are conserved in the retinoic acid- and fatty acid-binding proteins and are glutamines in the cellular retinol-binding proteins, thus the prediction that the corresponding arginines (CRABPs) and glutamines (CRBPs) provide the key protein–ligand contacts in the cellular retinoid-binding proteins (Jones *et al.*, 1988). The prediction led to a number of site-directed mutagenesis experiments designed to verify the role of these amino acids in ligand binding as well as convert the CRABPs to CRBPs and vice versa. Although a mutant of type I CRBP in which glutamine-108 was substituted with arginine did have some affinity for retinoic acid (Stump *et al.*, 1991), no

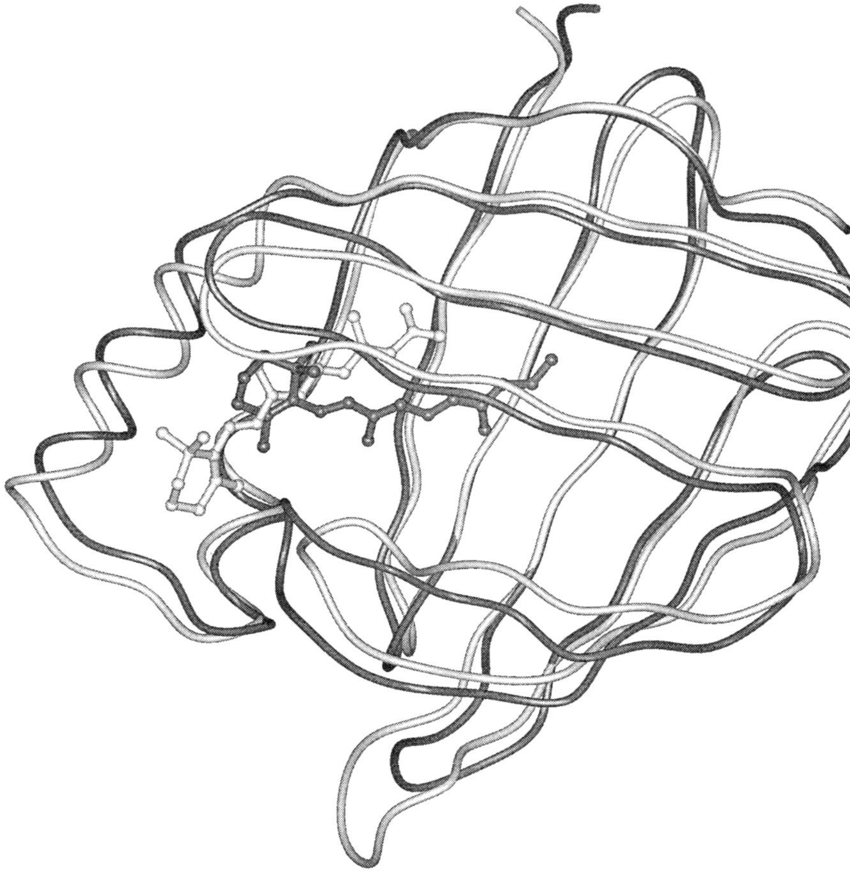

FIGURE 7. Superposition of CRBP(I) (dark gray) and CRABP(I) (white) backbones.

CRABP mutants could be engineered to bind retinol. With the structures in hand, it is easy to see why substitution of arginines 111 and 132 in the CRABPs did not produce a retinol-binding protein. Remember that the positions of retinoic acid and retinol do not superimpose when the backbones of CRBP and CRABP are overlaid: the ring of retinoic acid protrudes from the protein, whereas the retinol ring is completely encapsulated. A simple conversion of CRABP to CRBP per se is obviously not feasible. But why can one not readily engineer a CRABP mutant which binds retinol by the simple replacement of a positively charged amino acid with a hydrogen bond donor/acceptor? Clearly, any mutation at the tail end of the ligand would have to allow for the same ring protein contacts and introduce appropriate tail protein interactions. To convert CRABP to a retinol-binding protein, Arg-111 was substituted with Gln (Zhang *et al.*, 1992), a side chain which would provide a hydrogen bond donor for the retinol hydroxyl, but which is more than 2 Å shorter than an Arg, and consequently cannot reach the tail of the ligand once the retinol is positioned by protein ring interactions. Furthermore, in CRABP there is a water molecule between Arg-111 and the ligand carboxylate, so that the distance before changing to a shortened amino acid is already much too long (~4.5 Å) for the formation of a hydrogen bond. Substitution of Arg-132, which is not fully extended in the CRABP structure, with a Gln could possibly provide a hydrogen bond donor close enough to the ligand hydroxyl, but proper hydrogen bond geometry would also be required. Unlike simple electrostatic interactions which have no directional component (only a distance component), the formation of a hydrogen bond requires that the hydrogen bond donor, the hydrogen, and the hydrogen bond acceptor are colinear. This geometric constraint makes the introduction of a hydrogen bond in a protein–ligand system inherently more difficult than the introduction of an electrostatic interaction between protein and ligands (i.e., engineering a CRBP mutant which can bind retinoic acid).

Consequently, mutant CRABPs in which Arg-111 and Arg-132 are substituted with Gln lack affinity for retinoic acid, but do not gain the ability to bind retinol. Unlike the fatty acid-binding protein prototype for the family in which the ligand carboxylate is centered between two arginines, ligand interaction in CRABP appears to be primarily with Arg-132 (the C-terminal conserved Arg), as a mutant in which alanine is substituted for Arg at position 111 (the amino acid position which corresponds to the single H-bond donor of CRBP) shows only a slight decrease in retinoic acid affinity (Chen *et al.*, 1995).

It is evident from the wealth of structural data available for the fatty acid/retinoid-binding protein family (for review see Banaszak *et al.*, 1994) that very little sequence conservation is necessary for the conservation of the common structural framework. There is only 12% sequence identity between CRBP and an insect fatty acid-binding protein, yet despite this limited sequence identity, the proteins share the same overall trace (Fig. 8). The diversity of the amino acid sequences which adopt this structural motif is particularly striking, since unlike the RBP motif described above, there are no large loop regions where changes in amino acid side chains would be expected to have minimal structural consequences. (Compare Fig. 8 with Fig. 4.) However, the retinoid- and fatty acid-binding proteins have large internal cavities which include water molecules even in the holo forms, so in one sense there is both an internal and an external surface that can tolerate side-chain substitutions. Substitutions on the internal surface might be ex-

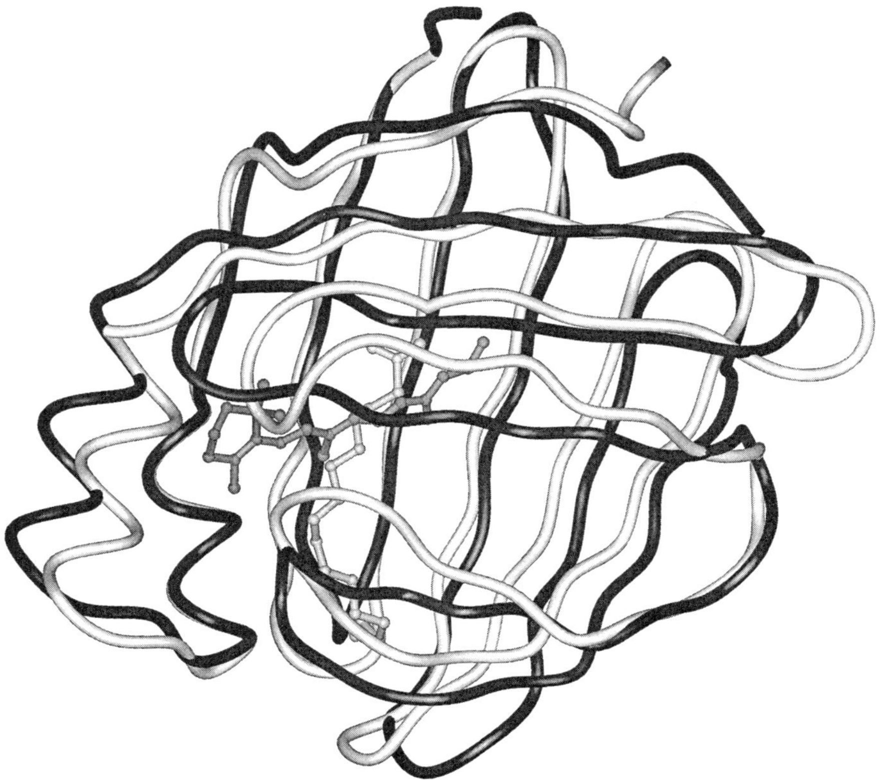

FIGURE 8. Superposition of CRBP(I) (black) and fatty acid-binding protein (white) backbone ribbons. Coordinates for the latter are from Benning *et al.* (1992).

pected to affect binding specificities, but not necessarily the overall structure. Additionally, just as with the lipocalin odorant-binding protein which provided a novel variation on the structural theme, a liver fatty acid-binding protein is in a sense an unconventional sibling in the intracellular family: liver fatty acid-binding protein binds two fatty acids in the internal binding cavity, one with the carboxylate interacting with the conserved Arg and one with the carboxylate in the vicinity of where the retinol ring is found in the CRBPs (Thompson *et al.*, 1997).

3.2. Conformational Changes upon Ligand Binding

What was not readily inferred from the structures of CRBP types I and II is how the ligand gains access to the internal binding cavity. Although the structure of CRBP(II) was determined in both the apo and holo forms, these structures are virtually identical. No portal is apparent in the apo structure. There must exist in solution a more open form of the protein which can allow entry and exit of ligand. The structure of the holo form allows speculation on how the ligand might enter the binding site. In the structure of the holo protein there are several favorable van der Waals contacts made between amino acids from the α-helices and the β-ionone ring of the retinol in holo-CRBP. It was proposed that these contacts help to secure the helix "cap" to the β-barrel and in the absence of ligand, the helix-turn-helix segment is able to move enough to provide access to the binding site (Jamison *et al.*, 1994). Movement of the helices as rigid rods would not disrupt the hydrogen bonds which stabilize them. Helical lids which move to allow entry into the catalytic site for a hydrophobic substrate have been described for human pancreatic lipase (Tilbeurgh *et al.*, 1993) and a fungal lipase (Brzozowski *et al.*, 1991).

Limited proteolysis was used to probe the conformational changes which occur upon ligand binding to CRBP and CRBP(II). It was observed that the liganded forms of both CRBP and CRBP(II) are resistant to proteolysis by a variety of proteases. However, in the absence of ligand, specific cleavage at a peptide bond in helix-B was observed (Jamison *et al.*, 1994), indicating that the helix is able to move away from the barrel and unwind (Hodson and Cistola, 1997) to allow the protease access to the scissile bond. Molecular dynamics simulations support an outward movement of helix B in apo-CRBP(I). However, this movement is reported to be accompanied by movement of the C–D and E–F loops, which lie on either side of the "gap." These movements are facilitated by hinge residues Lys-37 (in the loop between helix B and strand B) and Gly-67 (in the loop between strands D and E) (van Aalten *et al.*, 1995). Substitution of Gly-67, which is distant from the binding site, by a serine results in a protein which cannot bind retinol (van Aalten *et al.*, 1997). This glycine, found in a tight turn, is highly conserved in this protein superfamily [in only one of the 17 sequences given in the review by Banaszak *et al.* (1994) is there a substitution of alanine for Gly-67]. The high degree of conservation is consistent with an important structural role for the amino acid side chain. In holo-CRBP(I) this glycine adopts phi–psi angles accessible to nonglycine amino acids and the side chain of any other amino acid in this position would be on the surface, so it is not readily apparent why a Gly is required at this position. An X-ray or NMR structure of a CRBP-G67S mutant is necessary to understand the structural basis of the fact that this change of an invariant amino acid, removed from the ligand-binding site, results in a loss-of-function mutant.

A putative portal has also been described for the homologous fatty acid-binding protein aP2 (Xu *et al.*, 1993). It was suggested that Phe-57, which projects from the reverse turn between strands C and D and makes van der Waals contact with chains from helix B, could serve as a "hatch" to allow ligand entry/exit. This amino acid is highly conserved in the family, and in one of the fatty acid-binding proteins the "hatch" is open to allow the tail of the ligand to protrude from the cavity. Further structural support for this area of the protein providing access to the cavity in the retinoid-binding proteins comes from the subsequent structures of holo- versus apo-CRABP (Thompson *et al.*, 1995) described below. In addition, recent NMR studies on apo intestinal fatty acid-binding protein (Hodsdon and Cistola, 1997) indicate that in this protein there is discrete backbone disorder in the second helix of the α-helical "cap." A similar unwinding in the CRBP/CRBP(II) cap would be consistent with the results of the limited proteolysis experiments.

The "hatch" provided by phenylalanine-57 is not conserved in the CRABPs, in which the β-ionone ring of the ligand clearly projects from the binding cavity and is wedged between the helix and the β-barrel. In the apo-CRABP structure, β-strands C and D are pulled away as a hairpin from the body of the barrel and make an intermolecular β-sheet with the same hairpin from another CRABP molecule. The formation of this dimer dramatically increases the access to the binding cavity. Thompson *et al.* (1995) suggest that this ability to donate the C–D hairpin to form an intermolecular β-sheet may be the basis of the mechanism by which retinoic acid-metabolizing enzymes recognize holo-CRABP and gain access to the ligand. This is an intriguing hypothesis which begs to be tested experimentally.

3.3. Ligand Specificity

As described briefly above, the retinol-binding sites in CRBP and CRBP(II) are completely encapsulated by the protein, with the retinol bound such that the hydroxyl group is innermost and the isoprene tail is fully extended. The β-ionone ring abuts the α-helical cap, which blocks access to an otherwise open β-barrel. In both CRBP and CRBP(II) there are eight specifically bound water molecules which participate in a hydrogen bond network and help to form an internal binding cavity of a shape complementary to that of the ligand. The cavity, as defined without the water, is much larger than what is necessary to accommodate the ligand. The CRBPs discriminate at the tail end of the retinol, but are tolerant to changes at the β-ionone ring end of the ligand. From the structure of the ligand-occupied proteins one can understand the specificity. In both CRBP and CRBP(II) the hydroxyl group is hydrogen-bonded to Gln-108. Since Gln can serve as either a hydrogen bond donor or hydrogen bond acceptor, both retinol and retinal are ligands for the protein. However, the suggestion was made that in CRBP when Gln-108 serves as a hydrogen bond donor it does so at the expense of a favorable interaction that it would otherwise make with the phenyl ring of amino acid Phe-4. Therefore, the affinity of CRBP for retinal is markedly lower than that for retinol (Cowan *et al.*, 1993). In CRBP(II) the corresponding amino acid at position 4 is a Gln, so CRBP(II) does not discriminate strongly between retinol and retinal. However, this hypothesis was shown not to be valid, as a mutant type I protein in which the phenylalanine at position 4 was replaced by a Gln has binding affinities for retinol and retinal which are virtually indistinguishable from the wild-type protein (Jamison *et al.*, 1995). However, there is a slight, but significant difference in the positions of the oxygen atoms of retinal and retinol, and the hydrogen bond which must form between Gln-108 and the retinal oxygen would not have optimal hydrogen bonding geometry. This poor geometry alone would result in a decreased affinity of CRBP for retinal. In the retinol–CRBP(II) structure the hydrogen bonding geometry is also suboptimal, and this could account for the lower ligand binding affinity of CRBP(II) for retinol.

4. THE NUCLEAR RECEPTORS FOR ALL-*TRANS* AND 9-*CIS* RETINOIC ACID

4.1. The Basic Structural Motif

The nuclear receptors which regulate gene transcription in response to the binding of retinoic acid are members of the steroid hormone receptor family. The receptors are large multidomain proteins which can be dissected into functional modules which include a DNA-binding domain and a ligand-binding domain. Two subfamilies of retinoid receptors have been described and have been designated RAR (retinoic acid receptor, isoforms α, β, γ) and RXR (retinoid X receptor, isoforms α, β, γ). Subsequent to its identification, the ligand for RXR was determined to be 9-*cis* retinoic acid. RAR binds both all-*trans* and 9-*cis* retinoic acid. The functional nuclear receptors are dimeric and can form heterodimers with other members of the steroid hormone receptor family, hence the diversity of the biological processes affected by retinoic acid levels.

X-ray structures have been determined for the ligand-binding domains of apo-RXRα (Bourguet *et al.*, 1995) and holo-RARγ (Renaud *et al.*, 1995), as well as the homologous ligand-binding domain of the thyroxine receptor (Wagner *et al.*, 1995). The canonical ligand-binding-domain fold (Wurtz *et al.*, 1996) for the nuclear receptors (the ligand-binding domains of RXR and the thyroxine receptor have 30% and 35% sequence identity, respectively, with the RAR ligand-binding domain) can be described as an α-helical "sandwich" in which a central layer of α-helices (as well as two short β-strands) is flanked by two layers of α-helices. The helices in the flanking layers are positioned roughly orthogonally to the helices in the central layer. The resulting overall structure is a somewhat flattened cylinder composed of 12 α-helices. The ligand-binding site is at one end of the cylinder and access to the binding site appears to be permitted by the reorientation of one the helices, which in the holo form seals off the binding site. The movement of the α-helical "door" is inferred from the comparison of the apo-RXR and holo-RAR structures. In apo-RXR the C-terminal helix forms a stem at the base of a bundle of helices; in holo-RAR the corresponding helix is flipped up to join the bundle and obscure binding-site egress (see Figs. 9 and 10).

4.2. Conformational Changes

A key feature in the canonical ligand-binding domain structure is an "omega loop" between the second and third helices (the helices are numbered according to their order in the linear sequence). Omega loops are

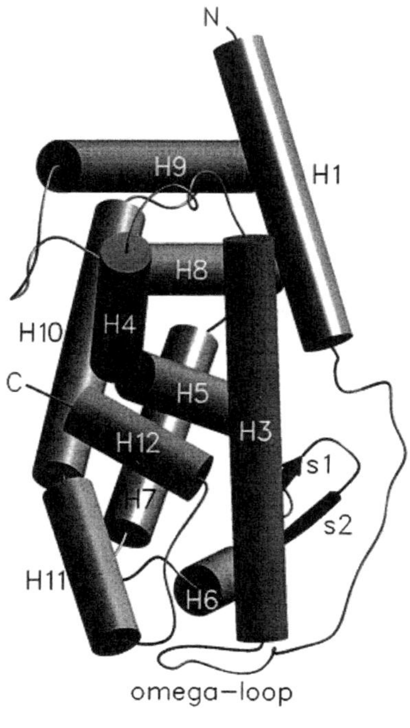

FIGURE 9. The ligand-binding domain of holo-RAR. Helices are shown as cylinders.

structural elements which are defined as loop structures which have no pattern of secondary structure, but do have intraloop hydrogen bonds which impart structural rigidity (Fetrow, 1995). The omega loop in RAR and RXR provides a hinge connecting secondary structural elements to facilitate the ligand-induced conformational change necessary for the receptor to be able to activate transcription. The omega loop in the apo structure interacts with the C-terminal helix (helix 12) which forms the "stem" of the helical bundle. In the "mouse trap" model proposed by Wurtz *et al.* (1996) ligand binding results in a shift of helices in the bundle. As a consequence, the omega loop is freed from its interaction with the C-terminal helix and is able to join the α-helical bundle.

The "mouse trap" mechanism of ligand-induced conformational change is presumed to be common to all members of the superfamily. Experimental evidence to indicate that the difference in orientation of the C-terminal helix in apo-RXR versus holo-RAR reflects an apo versus holo difference between the structures, rather than an RXR versus RAR structural difference, comes from site-directed mutagenesis experiments. In the holo-RAR structure Lys-264 and Glu-414 form a charge pair to anchor the helix (which contains Glu-414) onto the helical bundle structure. When the amino acid in RXR corresponding to Lys-264 (in mouse RXR this is Arg-307) is mutated to an alanine, the receptor binds ligand, but has a decreased transactivation capacity (Wurtz *et al.*, 1996). Additional support for the putative conformational change which occurs upon ligand binding comes from work with monoclonal antibodies. Rezcek and colleagues (Driscoll *et al.*, 1996) isolated a monoclonal antibody specific for the omega loop which recognizes apo-RARγ. Addition of ligand to the antibody:antigen complex results in the release of the monoclonal antibody. This difference in antibody recognition is consistent with the fact that in the holo-RAR structure the omega loop is not accessible to the antibody, but would be in the putative apo-RAR model based on the apo-RXR X-ray structure. Note that the antibody recognition site in the apo structure is distant from the ligand-binding site, an indication of the extent of the conformational change induced by ligand binding. Furthermore, the proposed apo versus holo structural difference is consistent with limited proteolysis experiments which indicate that the holo form of the receptors are more compact, as judged by their increased resistance to proteolysis relative to the apo receptors (e.g., Leng *et al.*, 1995).

Prior to the publication of the structures of the RAR and RXR ligand-binding domains numerous experiments were done with deletion and site-directed mutants which were designed to assign different aspects of a given function of the receptor to a particular region or "domain" of the protein. For example, nine "heptad repeats" in the sequence were proposed as

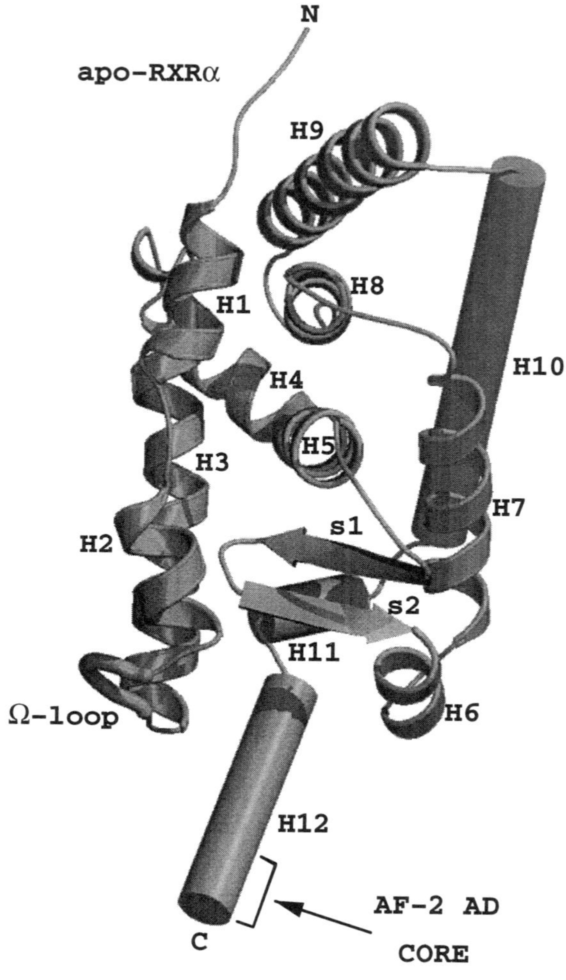

FIGURE 10. The structures of Apo-RXR and holo-RAR.

possible regions which mediate dimerization, since they are "structurally similar" to the leucine-zipper interface well characterized in other transcription factors. Deletion of these heptad repeats, an experiment designed to test their role in dimer formation, would clearly have deleterious structural consequences, as the heptad repeats span four helices in the α-helical bundle and make up a significant portion of the protein structural core. Later work narrowed the putative dimer interface to include three crucial leucine residues which were proposed to mediate dimer formation. Again, these amino acids are part of the protein hydrophobic core and they do not

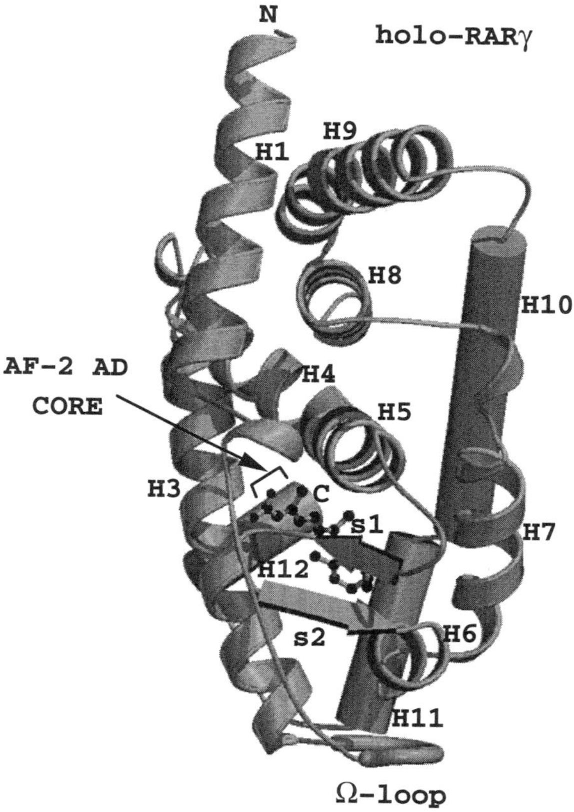

FIGURE 10. (*Continued*).

directly participate in the dimer interface. Mutations at these positions would be expected to disrupt the tertiary structure.

5. CONCLUSION

The proteins which bind retinoids fall into distinct structural classes, each with a characteristic structural motif. The extracellular proteins are eight-stranded β-barrels with a C-terminal α-helix. The ligands bind within the barrel, the ring end innermost, with a hand-in-glove-like fit into the hydrophobic binding site. Access to the binding site is via an open end in the barrel. The intracellular binding proteins are also β-barrels, but are ten-stranded and somewhat flattened. Access to the interior binding cavity, which includes polar amino acids and water molecules, appears to be blocked by two α-helices which cap the barrel. The retinol in the binding site is positioned with its ring end outermost. The third family of retinoid-binding proteins described includes the nuclear receptors which have a discrete retinoid-binding domain. These are α-helical structures in which, like the intracellular proteins, the ligand is bound with the ring end outermost in a binding site in which the exit is blocked by a helical door. The binding cavity does not include water molecules.

It is clear from the examples of these protein families (and others) that very little sequence identity is necessary to adopt equivalent structural motifs, a fact which at first might appear to circumvent the complex problem of "protein folding." One can now ask the question, "Which of the presently known protein folds might this sequence adopt?" (fold recognition), instead of, "What is the predicted tertiary structure of this sequence?" (protein folding). But, by the same token, it suggests that a given sequence might be compatible with more than one fold, and "fold recognition" techniques might give false positives [for discussions of the fold recognition problem see Fischer *et al.* (1996) and Miller *et al.* (1996)]. However, simply knowing the fold is not enough to explain the structural basis of a particular protein function. For example, both RBP and E-RABP share the lipocalin barrel. But in this functionally related pair, conserved residues are *structurally* important rather than *functionally* important, as one might infer when presented sequences of functionally related proteins. Even in functionally related proteins with well-conserved structures, such as CRBP and CRABP, small shifts in the placements of binding sites can be observed which make the extension of structural data from one member of the pair to the homologous protein not necessarily simple and direct. Furthermore, deletion or site-directed mutation experiments which result in inactive protein can be prematurely (in the absence of structural information) interpreted to indicate regions of functional importance. The mutated regions could be important for the structural integrity of the protein, but not its function per se.

It is important to emphasize that although the initial reported structure in a given superfamily will indeed establish the basic architectural framework characteristic of the superfamily, the modifications to this framework which define each member's specificity are not readily predicted from sequence comparison. In order to define accurately nature's structural repertoire, multiple structures from protein families must be determined.

ACKNOWLEDGMENTS. We thank the NIH for support, Grants GM55420 (M.E.N.), DK32642 (D.E.O.), and DK48824 (M.E.N. and D.E.O.). We thank Dino Moras for providing Figs. 9 and 10.

6. REFERENCES

Akmal, K. M., Dufour, J. M., and Kim, K. H., 1996, Region-specific localization of retinoic acid receptor-α expression in the rat epididymis, *Biol. Reprod.* **54:**1111–1119.

Banaszak, L., Winter, N., Xu, Z., Bernlohr, D. A., Cowan, S., and Jones, T. A., 1994, Lipid-binding proteins: A family of fatty acid and retinoid transport proteins, *Adv. Protein Chem.* **45:**89–151.

Benning, M. M., Smith, A. F., Wells, M. A., and Holden, H. M., 1992, Crystallization, structure determination and least-square refinements to 1.75 Å resolution of the fatty acid binding protein from *Manduca sexta*, *J. Mol. Biol.* **228:**208–219.

Bianchet, M. A., Bains, G., Pelosi, P., Pevsner, J., Snyder, S. H., Monaco, H. L., and Amzel, L. M., 1996, The three-dimensional structure of bovine odorant binding protein and its mechanism of odor recognition, *Nature Struct. Biol.* **3:**934–939.

Bourguet, W., Ruff, M., Chambon, P., Gronemeyer, H., and Moras, D., 1995, Crystal structure of the ligand-binding domain of the human nuclear receptor RXR-α, *Nature* **375:**377–382.

Boylan, J. F., and Gudas, L. J., 1991, Overexpression of the cellular retinoic acid-binding protein-I CRABP-I results in a reduction in differentiation-specific gene expression in F9 teratocarcinoma cells, *J. Cell Biol.* **112:**965–972.

Boylan, J. F., and Gudas, L. J., 1992, The level of CRABP-I expression influences the amounts and types of all-*trans*-retinoic acid metabolites in F9 teratocarcinoma stem cells, *J. Biol. Chem.* **267:**21486–21491.

Brownlow, S., Morias Cabral, J. H., Cooper, R., Flower, D. R., Yewdall, S. J., Polikarpov, I., North, A. C. T., and Sawyer, L., 1997, Bovine β-lactoglobulin at 1.8 Å resolution—Still an enigmatic lipocalin, *Structure* **5:**481–495.

Brzozowski, A. M., Derewenda, U., Derewenda, Z. S., Dodson, C. G., Lawson, D. M., Tunkerburg, J. P., Bjorkling, F., Huge-Jensen, B., Patkar, S. A., and Thim, L., 1991, A model for interfacial activation of lipases from the structure of a fungal lipase–inhibitor complex, *Nature* **351:**491–494.

Bucco, R. A., Melner, M. H., Gordon, D. S., Leers-Sucheta, S., and Ong, D. E., 1995, Inducible expression of cellular retinoic acid-binding protein II in rat ovary: Gonadotropin regulation during luteal development, *Endocrinology* **136:**2730–2740.

Bucco, R. A., Zheng, W. L., Davis, J. T., Sierra-Rivera, E., Osteen, K. G., Chaudhary, A. K., and Ong, D. E., 1997, Cellular retinoic acid-binding protein(II) presence in rat uterine epithelial cells correlates with their synthesis of retinoic acid, *Biochemistry* **36:**4009–4014.

Chen, L. X., Zhang, Z. P., Scafonas, A., Cavalli, R. G., Gabriel, J. L., Soprano, K. J., and Soprano, D. R., 1995, Arg 132 of cellular retinoic acid-binding protein (type II) is important for binding of retinoic acid, *J. Biol. Chem.* **270:**4518–4525.

Cho, Y., Batt, C. A., and Sawyer, L., 1994, Probing the retinol-binding site of bovine β-lactoglobulin, *J. Biol. Chem.* **269:**11102–11107.

Cowan, S. W., Newcomer, M. E., and Jones, T. A., 1990, Crystallographic refinement of human serum retinol binding protein at 2.0 Å resolution, *Proteins Struct. Funct. Genet.* **8:**44–61.

Cowan, S. W., Newcomer, M. E., and Jones, T. A., 1993, Crystallographic studies on a family of cellular lipophilic transport proteins. Refinement of P2 myelin protein and the structure determination and refinement of cellular retinol-binding protein in complex with all-*trans*-retinol, *J. Mol. Biol.* **230:**1225–1246.

Davis, J. T., and Ong, D. E., 1992, Synthesis and secretion of retinol-binding protein by cultured rat Sertoli cells, *Biol. Reprod.* **47:**528–533.

Davis, J. T., and Ong, D. E., 1995, Retinol processing by the peritubular cell from rat testis, *Biol. Reprod.* **52:**356–364.

Driscoll, J. E., Seachord, C. L., Lupisella, J. A., Darveau, R. P., and Reczek, P. R., 1996, Ligand-induced conformational changes in the human retinoic acid receptor detected using monoclonal antibodies, *J. Biol. Chem.* **271:**22969–22975.

Fetrow, J. S., 1995, Omega loops: Nonregular secondary structures significant in protein function and stability, *FASEB J.* **9:**708–717.

Fischer, D., Rice, D., Bowie, J. V., and Eisenberg, D., 1996, Assigning amino acid sequences to 3-dimensional protein folds, *FASEB J.* **10:**126–136.

Fiorella, P. D., and Napoli, J. L., 1991, Expression of cellular retinoic acid-binding protein CRABP in *Escherichia coli*, *J. Biol. Chem.* **266:**16572–16579.

Fiorella, P. D., Giguère, V., and Napoli, J. L., 1993, Expression of cellular retinoic acid binding protein type II in *E. coli.* Characterization and comparison to cellular retinoic acid binding protein type I, *J. Biol. Chem.* **268:**21545–21552.

Goodman, D. S., and Blaner, W. S., 1984, Plasma retinol-binding protein, in *The Retinoids* (M. B. Sporn, A. B. Roberts, and D. S. Goodman, eds.), pp. 1–39, Academic Press, New York.

Hodsdon, M. E., and Cistola, D. P., 1997, Discrete backbone disorder in the nuclear magnetic resonance structure of apo intestinal fatty acid-binding protein: Implications for the mechanism of ligand entry, *Biochemistry* **36:**1450–1460.

Jamison, R. S., Newcomer, M. E., and Ong, D. E., 1994, Cellular retinoid-binding proteins: Limited proteolysis reveals a conformational change upon ligand binding, *Biochemistry* **33:**2873–2879.

Jamison, R. S., Kakkad, B., Ebert, D. H., Newcomer, M. E., and Ong, D. E., 1995, Test of the contribution of an amino-aromatic hydrogen bond to protein function, *Biochemistry* **34:**11128–11132.

Jaworowski, A., Fang, Z., Khong, T. F., and Augusteyn, R. C., 1995, Protein synthesis and secretion by cultured retinal pigment epithelia, *Biochim. Biophys. Acta* **1245:**121–129.

Jones, T. A., Bergfors, T., Sedzik, J., and Unge, T., 1988, The three-dimensional structure of P2 myelin protein, *EMBO J.* **7:**1597–1604.

Kanai, M., Raz, A., and Goodman, D., 1968, Retinol binding protein: The transport protein for vitamin A in human plasma, *J. Clin. Invest.* **47:**2025–2044.

Kato, M., Sung, W. K., Kato, K., and Goodman, D. S., 1985, Immunohistochemical studies on the localization of cellular retinol-binding protein in rat testis and epididymis, *Biol. Reprod.* **32:**173–189.

Kleywegt, G. J., Bergfors, T., Senn, H., Le Motte, P., Gsell, B., Shudo, K., and Jones, T. A., 1994, Crystal structures of cellular retinoic acid binding proteins I and II in complex with all-*trans*-retinoic acid and a synthetic retinoid, *Structure* **2:**1241–1258.

Leng, X., Blanco, J., Tsai, S. Y., Ozato, K., O'Malley, B. W., and Tsai, M. J., 1995, Mouse retinoid X receptor contains a separable ligand-binding and transactivation domain in its E region, *Mol. Cell Biol.* **15:**255–263.

Lepperdinger, G., Strobl, B., Jilek, A., Weber, A., Thalhamer, J., Flockner, H., and Mollay, C., 1996, The lipocalin Xlcpl1 expressed in the neural plate of *Xenopus laevis* embryos is a secreted retinaldehyde binding protein, *Protein Sci.* **5:**1250–1260.

MacDonald, P. N., and Ong, D. E., 1987, Binding specificities of cellular retinol-binding protein and cellular retinol-binding protein type II, *J. Biol. Chem.* **262:**10550–10556.

Miller, R. T., Jones, D. T., and Thornton, J. M., 1996, Protein fold recognition by sequence threading: Tools and assessment techniques, *FASEB J.* **10:**171–178.

Monaco, H. L., Zanotti, G., Spadon, P., Bolognesi, M., Sawyer, L., and Eliopoulos, E. E., 1987, Crystal structure of the trigonal form of bovine β-lactoglobulin and of its complex with retinol at 2.5 Å resolution, *J. Mol. Biol.* **197:**695–706.

Newcomer, M. E., 1993, Structure of the epididymal retinoic acid binding protein at 2.1 Å resolution, *Structure* **1:**7–18.

Newcomer, M., and Ong, D. E., 1990, Purification and crystallization of a retinoic acid-binding protein from rat epididymis, *J. Biol. Chem.* **265:**12876–12879.

Newcomer, M. E., Jones, T. A., Åqvist, J., Sundelin, J., Eriksson, U., Rask, L., and Peterson, P. A., 1984, The three-dimensional structure of retinol-binding protein, *EMBO J.* **3:**1451–1454.

Ong, D. E., and Chytil, F., 1988, Presence of novel retinoic acid-binding proteins in the lumen of rat epididymis, *Arch. Biochem. Biophys.* **267:**474–478.

Ong, D. E., Davis, J. T., O'Day, W. T., and Bok, D., 1994, Synthesis and secretion of retinol-binding protein and transthyretin by cultured retinal pigment epithelium, *Biochemistry* **33:**1835–1842.

Porter, S. B., Ong, D. E., Chytil, F., and Orgebin-Crist, M.-C., 1985, Localization of cellular retinol-binding protein and cellular retinoic acid binding proteins in the rat testis and epididymis, *J. Androl.* **6:**197–212.

Renaud, J. P., Rochel, N., Ruff, M., Vivat, V., Chambon, P., Gronemeyer, H., and Moras, D., 1995, Crystal structure of the RAR-gamma ligand-binding domain bound to all-*trans* retinoic acid, *Nature* **378:**681–689.

Ronne, H., Ocklind, C., Wiman, K., Rask, L., Obrink, B., and Peterson, P. A., 1983, Ligand dependent regulation of intracellular protein transport: Effect of vitamin A on the secretion of retinol-binding protein, *J. Cell Biol.* **96:**907–910.

Sacchettini, J. C., Gordon, J. I., and Banaszak, L. J., 1988, The structure of crystalline *E. coli* derived rat intestinal fatty-acid-binding protein at 2.5 Å resolution, *J. Biol. Chem.* **263:**5815–5819.

Soprano, D. R., and Blaner, W. S., 1994, Plasma retinol binding protein, in *The Retinoids* (M. B. Sporn, A. B. Roberts, and D. S. Goodman, eds.), pp. 257–282, Raven Press, New York.

Stump, D. G., Lloyd, R. S., and Chytil, F., 1991, Site-directed mutagenesis of rat cellular retinol-binding protein. Alteration in binding specificity resulting from mutation of Glu-108 to arginine, *J. Biol. Chem.* **266:**4622–4630.

Tegoni, M., Ramoni, R., Bignetti, E., Spinelli, S., and Cambillau, C., 1996, Domain swapping creates a third putative combining site in bovine odorant binding protein dimer, *Nature Struct. Biol.* **3:**863–867.

Thompson, J. R., Bratt, J. M., and Banaszak, L. J., 1995, Crystal structure of cellular retinoic acid binding protein I shows increased access to the binding cavity due to formation of an intermolecular β-sheet, *J. Mol. Biol.* **252:**433–446.

Thompson, J., Winter, N., Terwey, D., Bratt, J., and Banaszak, L., 1997, The crystal structure of the liver fatty acid-binding protein. A complex with two bound oleates, *J. Biol. Chem.* **272:**7140–7150.

Tilbeurgh, H., Egloff, M. P., Martinez, C., Rugani, N., Verger, R., and Cambillau, C., 1993, Interfacial activation of the lipase pro-colipase complex by mixed micelles revealed by X-ray crystallography, *Nature* **362:**814–820.

van Aalten, D. M., Findlay, J. B., Amadei, A., and Berendsen, H. J., 1995, Essential dynamics of the cellular retinol-binding protein—Evidence for ligand-induced conformational changes, *Protein Eng.* **8:**1129–1135.

van Aalten, D. M., Jones, P. C., de Sousa, M., and Findlay, J. B., 1997, Engineering protein mechanics: Inhibition of concerted motions of the cellular retinol binding protein by site-directed mutagenesis, *Protein Eng.* **10:**31–37.

Wagner, R. L., Apriletti, J. W., McGrath, M. E., West, B. L., Baxter, J. D., and Fletterick, R. J., 1995, A structural role for hormone in the thyroid hormone receptor, *Nature* **378:**690–697.

Wan, Y. J., Wang, L., and Wu, T. C., 1992, Detection of retinoic acid receptor mRNA in rat tissues by reverse transcriptase–polymerase chain reaction, *J. Mol. Endocrinol.* **9:**291–294.

Winter, N., Bratt, J., and Banaszak, L. J., 1993, The crystal structures of holo and apo cellular retinol binding protein II, *J. Mol. Biol.* **230:**1247–1259.

Wurtz, J.-M., Bourget, W., Renaud, J.-R., Vivat, V., Chambon, P., Moras, D., and Gronemeyer, H., 1996, A canonical structure for the ligand binding domain of nuclear receptors, *Nature Struct. Biol.* **3:**87–94.

Xu, Z., Bernlohr, D. A., and Banaszak, L. J., 1993, The adipocyte lipid-binding protein at 1.6-Å resolution. Crystal structures of the apoprotein and with bound saturated and unsaturated fatty acids, *J. Biol. Chem.* **268:**7874–7884.

Zanotti, G., Berni, R., and Monaco, H. L., 1993a, Crystal structure of liganded and unliganded forms of bovine plasma retinol-binding protein, *J. Biol. Chem.* **268:**10728–10738.

Zanotti, G., Ottonello, S., Berni, R., and Monaco, H. L., 1993b, Crystal structure of the trigonal form of human plasma retinol-binding protein at 2.5 Å resolution, *J. Mol. Biol.* **230:**613–614.

Zhang, J., Liu, Z. P., Jones, T. A., Gierasch, L. M., and Sambrook, J. F., 1992, Mutating the charged residues in the binding pocket of cellular retinoic acid-binding protein simultaneously reduces its binding affinity to retinoic acid and increases its thermostability, *Proteins* **13:**87–99.

Zheng, W. L., Bucco, R. A., Schmitt, M. C., Wardlaw, S. A., and Ong, D. E., 1996, Localization of cellular retinoic acid-binding protein (CRABP) II and CRABP in developing rat testis, *Endocrinology* **137:**5028–5035.

Chapter 4

The Role of Retinoids in Developmental Mechanisms in Embryos

Malcolm Maden

1. INTRODUCTION

It has been known since the 1920s that the appropriate level of vitamin A (retinoids) in the maternal diet is vital for the development of the embryo (Kalter and Warkany, 1959). Both too little or too much vitamin A results in characteristic defects appearing in the embryo: defects in the central nervous system, various structures of the head and neck, the heart, the limb, and the urogenital system. These studies thus revealed that excess vitamin A was teratogenic and harmful to embryos and the usual interpretation of its effects was that it was causing cell death or an inhibition of development.

However, studies over the last 15 years on the effects of excess vitamin A on various developmental systems have revealed completely new and important stimulatory effects. That is, *extra* structures are produced in the embryo rather than deficiencies of structure and this has led to many important insights into how normal development is controlled. It has led to the proposal that retinoic acid (RA), the active metabolite of vitamin A

Malcolm Maden Developmental Biology Research Centre, King's College London, London WC2B 5RL, England.

Subcellular Biochemistry, Volume 30: Fat-Soluble Vitamins, edited by Quinn and Kagan. Plenum Press, New York, 1998.

(retinol), is responsible for organizing groups of cells in the early embryo in terms of their positional relationships so that each cell knows where it is in the embryo and knows what cell type to differentiate into. It does this by generating different concentrations of RA which cells can detect and respond to via the retinoic acid receptors (RARs) and the retinoid X receptors (RXRs) (see Chapters 1–3). As a result, differential patterns of activity of effector genes (for example, the homeobox genes) are established, which, in turn, establish differential patterns of activity of genes concerned with the final differentiated product (for example, myosin and actin in muscle cells or collagen in cartilage cells, etc).

I will exemplify these principles of different concentrations establishing patterns of effector genes resulting in different phenotypes in three embryonic systems about which we have the most information, namely the regenerating limb, the developing limb, and the head-to-tail axis (the mesoderm and the nervous system) of the early embryo. There are several other embryonic systems on which the effects of RA have been studied (for a review, see Maden and Pizzey, 1997), but not so much is known about them and so I will concentrate on the three systems about which most is known. In each case I will describe the effects of excess retinoids from the point of view of patterning mechanisms, ask whether endogenous retinoids can be detected in the system, and discuss the gene pathways which may be involved in the altered patterns of development. The clear message from these studies is that there is a remarkable similarity of retinoid-based patterning mechanisms in each of the three systems.

2. RETINOIDS AND REGENERATION

2.1. The Regenerating Amphibian Limb

The limbs of Urodele amphibians (newts) have the remarkable ability to regenerate perfectly from any level and replace precisely what has been amputated (Fig. 1a). Anurans (frogs and toads) only possess this ability at premetamorphic stages during limb development. Both these animal groups have been used for regeneration studies of the effects of retinoids. The ability to regenerate the missing part means that the cells at the amputation plane which are responsible for replacing the lost parts must have a knowledge of what is missing, they must be aware of their position in the limb as a whole. It was discovered that excess vitamin A can fool these cells into believing they are higher up the limb (toward the shoulder) with the result that duplicated limbs regenerate instead of normal ones.

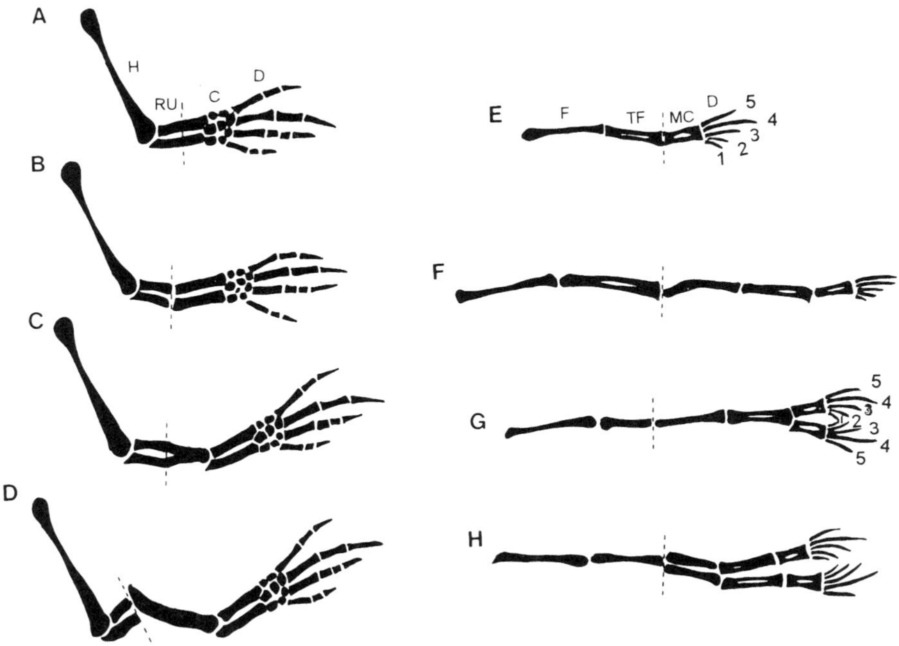

FIGURE 1. The effect of retinoids on the regenerating axolotl limb (A–D) and frog limb (E–H). Broken lines mark the amputation plane in each drawing. (A) A normal axolotl limb which has regenerated after amputation through the middle of the forearm. H, humerus; RU, radius and ulna; C, carpals; D, digits. (B) A limb amputated through the middle of the forearm and treated with a low dose of retinyl palmitate. The animal regenerated a complete radius and ulna instead of just part of it. (C) An amputated limb treated with a medium dose of retinyl palmitate, which regenerated the distal end of the humerus followed by the rest of the limb. This animal has two elbow joints. (D) An amputated limb treated with a high dose of retinyl palmitate which has regenerated a complete limb from the amputation plane. (E) A normal frog limb which has regenerated after amputation through the foot. F, femur; TF, tibia and fibula; MC, metacarpals; D, digits. (F) After a medium dose of retinyl palmitate a complete limb has regenerated from the foot level. (G) After a high dose not only is a complete limb regenerated from the foot level, but this regenerate is duplicated at the foot level. This duplication is mirror-image and of a double posterior nature, as the digital formula shows (5432112345). (H) Another high-dose regenerate which produced a complete pair of limbs from the foot level, also with double posterior structure.

Niazi and Saxena (1978) reported that when the hindlimb buds of the toad, *Bufo andersonii*, were amputated through the shank and treated with retinyl palmitate, the regenerates seemed to contain *extra* elements in the proximodistal axis and often two regenerates appeared instead of one. Investigating this phenomenon in more detail in the axolotl (Maden, 1982,

1983b), it was shown that the effect of retinyl palmitate was indeed to produce extra elements in the proximodistal axis and in a very precise and concentration-dependent fashion. Thus after amputating limbs through the mid-forearm level, at low concentrations of retinyl palmitate the regenerated radius and ulna was too long or an extra complete radius and ulna was regenerated (Fig. 1b). At a higher dose an extra elbow joint appeared (Fig. 1c) and at a higher dose still a complete limb could be regenerated from the amputation plane (Fig. 1d). The same phenomenon can be seen in hindlimbs, too. Thus what is happening is that retinoids are respecifying the cells at the amputation plane in a proximal (toward the shoulder) direction and in a gradual, concentration-dependent manner. That is, they are proximalizing the cells of the regenerating limb.

When these experiments were extended to the frog, *Rana temporaria*, a further axial effect of retinyl palmitate was observed (Maden, 1983a). Not only were the limbs proximalized, but at higher concentrations the anteroposterior axis (thumb to little finger) was also respecified. Thus after amputation of the foot, instead of just replacing the foot as controls do (Fig. 1e), treated animals produced limbs which were complete in the proximodistal axis (Fig. 1f). At higher concentrations, regenerates were not only complete in the proximodistal axis, but also duplicated in the anteroposterior (AP) axis (Figs 1g and 1h). These regenerates were always double posterior in structure, as can be seen in the figure, that is, the sequence of digits was 5, 4, 3, 2, 1, 1, 2, 3, 4, 5. This showed that the cells of the regenerate had not only been proximalized, but were also posteriorized. It is worth remembering (in relation to the results on the chick limb bud described below) that these effects appear after systemic treatment of the animals: they are dosed by placing them in solutions of retinoids.

To complete the picture, it is now apparent that retinoids can also ventralize the cells of the regenerating limb, i.e., affect the third axis of the limb, the dorsoventral (DV) axis. When axolotl limbs are cut in half in the DV axis to produce half-ventral or half-dorsal limbs, then half-ventral limbs do not regenerate at all, whereas half-dorsal limbs regenerate a complete limb from the amputation plane (Ludolph *et al.*, 1990). Thus in the regenerating amphibian limb retinoids can respecify each of the three axes of the limb: the proximodistal axis is proximalized, the anteroposterior axis is posteriorized, and the dorsoventral axis is ventralized, and the result is the regrowth of a duplicated limb.

These effects of retinoids are thus concentration dependent: the higher the concentration, the greater the degree of duplication. They are also time dependent: longer treatment times induce greater degrees of duplication; they are also stage dependent: if treatment is delayed beyond a certain stage, instead of stimulating extra growth of elements, the regenerates are phocomelic and have missing structures. Of the naturally occurring retinoids, retinoic acid (RA) is the most potent (Maden, 1983b) followed by retinol, retinyl palmitate, and then retinyl acetate, but many synthetic retinoids, particularly the benzoic acid derivatives, are far more potent than RA (Keeble and Maden, 1989; Maden *et al.*, 1991), and comparing the two RA isomers, 9-*cis*- and all-*trans*-RA, it seems that 9-*cis* is more potent (Tsonis *et al.*, 1994).

2.2. The Regenerating Amphibian Tail

An even more amazing effect of retinoids has recently been reported. When the amputated tails of tadpoles of the marbled balloon frog, *Uperodon systoma*, were treated with retinyl palmitate, instead of regenerating tails as would be expected, several hindlimbs were produced (Mohanty-Hejmadi *et al.*, 1992). This also occurs in *Rana temporaria* and shows characteristics typical of the retinoid effects described above, i.e., it is concentration dependent, time dependent, and stage dependent (Maden, 1993). Furthermore, up to nine limbs can be produced and there is a strong tendency for them to appear in even numbers (presumably in pairs) rather than in odd numbers, and they are always hindlimbs. Interestingly, the effects of retinyl palmitate are prevented by the prior administration of prolactin to *Rana* tadpoles (Maden and Corcoran, 1996). Prolactin inhibits the release of thyroid hormone, which suggests that an interaction between retinoids and thyroid hormone is involved in this transformation of tails into limbs. This is exactly the sort of interaction that can occur at the receptor level, as the RXRs can heterodimerize with the thyroid hormone receptors (see Chapter 1).

2.3. Endogenous Retinoids and Their Receptors

It is of crucial importance to demonstrate that retinoids are normally present in the cells of the regenerating limb. This would suggest that they are involved in normal pattern specification and that during the induction of limb duplication described above we are not inducing an entirely novel set of conditions, but altering the normal sequence of events. Retinoids have indeed been detected in the regenerating limb, as have the retinoic acid receptors (RARs), the nuclear transcription factors through which RA acts on the genome.

High-pressure liquid chromatography (HPLC) measurements have identified all-*trans*-retinol, 3,4-didehydroretinol, all-*trans*-retinoic acid and all-*trans*-retinaldehyde in the regenerating amphibian limb both in the mesenchyme and the epidermis (Scadding and Maden, 1994). Across the AP axis of the blastema, a gradient of all-*trans*-RA was detected with a fivefold higher concentration on the posterior side, and comparing proximal with distal blastemas gave the result that RA was 2.7 times higher in the distal blastemas than in proximal blastemas. The latter is not the result one might have expected from the proximalizing ability of RA, and it is the opposite result to another estimation of relative endogenous levels using reporter cells (see below). Since HPLC measurements estimate total retinoid content, they clearly do not take into account the relative amounts of bound and unbound ligand. Bound ligand may, for example, be unavailable for gene activation and these factors may explain this discrepancy.

The other type of detection system that has been used to determine endogenous RA involves blastemal cells transfected with a reporter construct containing a retinoic acid response element (RARE) linked to the β-galactosidase gene. When these cells were implanted under the wound epidermis of a blastema, they were activated, demonstrating the presence of endogenous RA which must have been released either from the wound epidermis or other blastemal cells (Brockes, 1992); a comparison of the level of activation in proximal and distal blastemas revealed that RA was present at 3.5-fold higher levels in proximal blastemas than in distal blastemas, exactly as one might expect since RA proximalizes distal cells. Further experiments with these reporter cells have established that the wound epidermis is a source of RA in the blastema and that this tissue synthesises and releases not all-*trans*-RA, as might be expected, but 9-*cis*-RA (Viviano *et al.*, 1995). This is the first embryonic tissue which has been shown to endogenously synthesize this particular isomer of RA.

Further evidence for a role for RA during normal regeneration comes from experiments where the synthesis of RA has been inhibited with disulphiram. Disulphiram inhibits the enzyme retinal dehydrogenase and thus prevents the synthesis of RA from retinal (McCaffery *et al.*, 1992; Stratford *et al.*, 1996). Normal tail regeneration is inhibited in the presence of disulphiram (Maden and Corcoran, 1996) and our unpublished experiments have shown that the same is true for limb regeneration.

Finally, various RARs have been identified in the cells of the regenerating newt limb. Five RAR isoforms have been detected, α_1, α_2, δ_{1a}, δ_{1b}, and δ_2 (Ragsdale *et al.*, 1989, 1992a,b). The newt δ is the counterpart of the mammalian γ receptor isoform. The major isoform expressed in the regenerating limb is δ_1 and it is expressed at sufficiently high levels to be detected by isoform-specific antibodies (Hill *et al.*, 1993). This unique study of receptor protein distribution in a developing system revealed that this receptor is expressed by 50% or more of cells in the blastema of the regenerating limb. No variation from this pattern could be detected between proximal and distal blastemas nor after RA treatment of distal blastemas. The presence of the RXRs has not been investigated in the regenerating limb.

Most excitingly, a functional analysis of each of these RAR isoforms has been performed and the results suggest that each isoform has a unique function. The strategy employed for these analyses was to generate chimeric receptors for each of the RAR isoforms present in the regenerating limb. The ligand-binding domain of each RAR isoform was swapped with that taken from a thyroid hormone receptor, thereby making a chimeric receptor which is responsive to thyroid hormone rather than RA, but which still activates the appropriate retinoic acid-responsive genes. Each chimera was then transfected one by one into regenerating newt limb blastemal cells either *in vitro*, for culture experiments, or *in vivo* for whole-animal experiments. When the RARα_1 chimera was transfected into cultured blastemal cells it was observed that these cells became growth-inhibited by thyroid hormone in the way that they would normally have responded to RA and at similar concentrations to those of RA (Schilthuis *et al.*, 1993). Thus the α_1 RAR must be responsible for retinoic acid-induced growth inhibition. By a similar procedure it was established that the δ_1 RAR chimera is responsible for changes in the secretory properties of the epidermis (Pecorino *et al.*, 1994) and, most excitingly, that the δ_2 RAR isoform is responsible for the proximalization of blastemal cells which is induced by RA (Pecorino *et al.*, 1996). This demonstration is surely a major step on the way to understanding the molecular basis of positional information in the regenerating limb and demonstrates how studies of RA and its mechanism of action have provided important insights into normal developmental processes.

2.4. Downstream Effector Genes of RA

The remarkable progress on relating particular RARs to a particular phenotype described above now needs the gap in the middle to be filled in, to answer the question of what are the effector genes in the action of RA. We know very little about this in any developmental system, but in general terms, the homeobox genes are always likely to be candidates, as they are thought to be involved in pattern specification in vertebrate embryos (Krumlauf, 1994), they are ectopically induced by RA (Conlon and Rossant, 1992), and some homeobox genes have retinoic acid response elements in their upstream sequences (Langston and Gudas, 1994). Several homeobox genes have been cloned in newts and axolotls and their expression determined in various parts of the regenerating animal such as the forelimbs, hindlimb, tail, distal limb blastema, proximal limb blastema, etc., with a view to assessing their likely role in patterning (Beauchemin and Savard, 1993; Brown and Brockes, 1991; Gardiner *et al.*, 1995; Mullen *et al.*, 1996; Savard *et al.*, 1988; Simon and Tabin, 1993; Tabin, 1989). Only three of these genes respond to RA treatment, however; *Hoxa*-13 and *Dlx*-3 are downregulated (Gardiner *et al.*, 1995; Mullen *et al.*, 1996) and *Hox*-4.5 is upregulated (Simon and Tabin, 1993). Thus, we know very little about the targets of RA action in the regenerating limb.

3. VITAMIN A AND LIMB DEVELOPMENT

3.1. The Chick Limb Bud

At the same time that these experiments were being conducted with the regenerating limb it was discovered that RA had equally profound effects on the developing chick limb bud (Tickle *et al.*, 1982). Local administration of RA to the anterior margin of the limb bud transformed the normal three-digit limb consisting of digits 2, 3, 4 (Fig. 2a) into a six-digit, mirror-image, double posterior limb consisting of digits 4, 3, 2, 2, 3, 4 (Fig. 2b). Anterior cells were thus posteriorized. Reducing the concentration of RA produced fewer additional digits; the effect is thus dose dependent and the effect is also stage dependent (Summerbell, 1983). Unlike the regenerating amphibian limb (where vitamin A is administered systemically), in the chick limb bud the effect is position dependent because applications of RA to the posterior side of the limb bud have no effect and applications to the apex of the bud have lesser effect (Tickle *et al.*, 1985). There is a minimum period of about 10–12 hr of treatment needed to induce posteriorization (Eichele *et al.*, 1985).

The remarkable feature of these results is that the effects of RA precisely mimicked the behavior of an organizing region in the chick limb bud which had been discovered more than a decade previously (Saunders and Gasseling, 1968). On the posterior side of the limb bud is a group of cells which, when removed and grafted to the anterior side of another limb bud, produces a six-digit, mirror-image, duplicated limb consisting of digits 4, 3, 2, 2, 3, 4 (e.g., Fig. 2c). This region became known as the zone of polarizing activity (ZPA) and it displays an almost identical pattern of behavior after grafting to the administration of RA: its effects are stage dependent, position dependent, concentration dependent, the grafted ZPA instructs anterior cells to change their specification, and there is a minimum time period of 12–18 hr needed to induce duplications (for review see Summerbell and Honig, 1982). On the basis of these observations a theory was developed to explain the mechanism of specification of pattern across

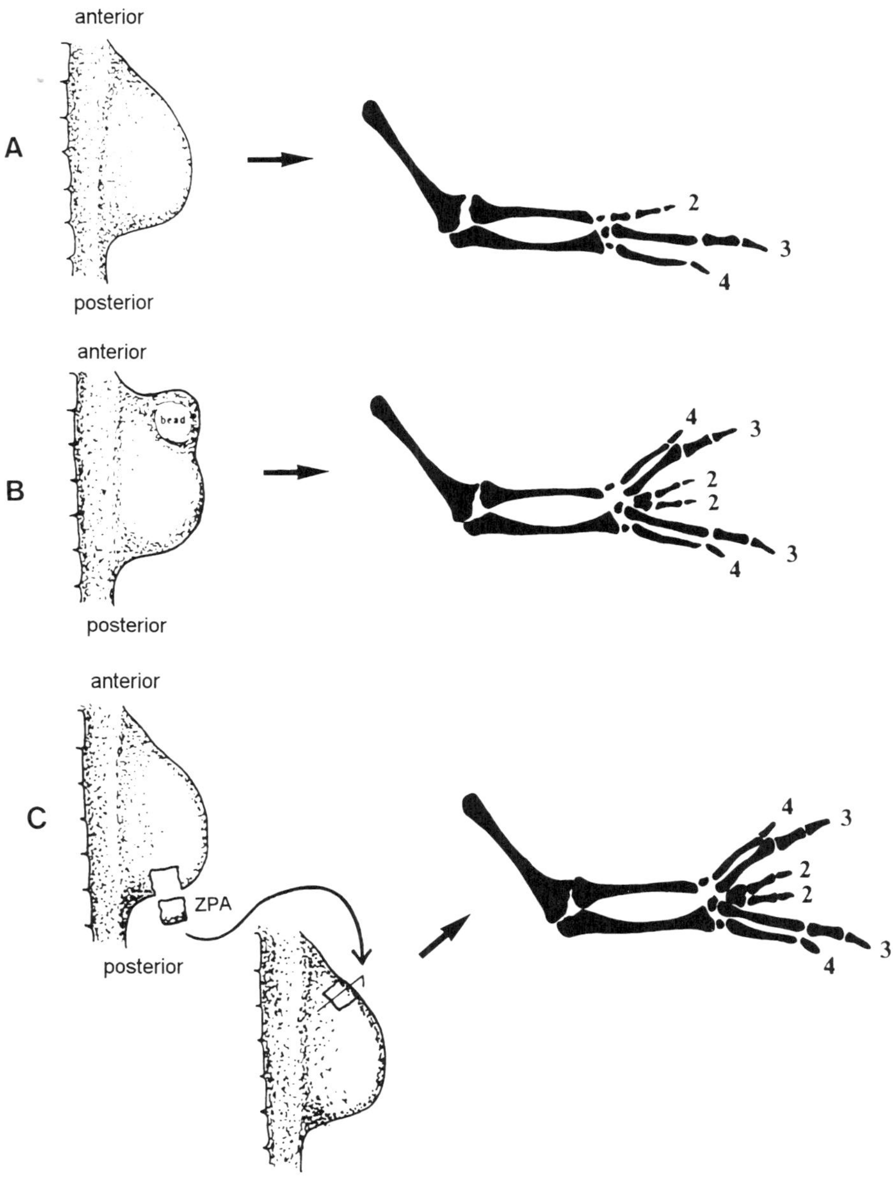

FIGURE 2. The effects of RA and a ZPA graft on the developing chick limb bud. (A) A stage 21 limb bud is drawn with its anterior and posterior margins marked, which gives rise to a three-digit limb. The digits are numbered 2, 3, and 4. (B) After implantation of a bead soaked with RA into the anterior margin of the limb bud a six-digit, mirror-image limb is produced. The sequence of digits is 4, 3, 2, 2, 3, 4, that is, double posterior, which is exactly the same as the frog limbs shown in Fig. 1. (C) The same result is produced after grafting a ZPA, which is a localized region on the posterior side of a normal limb bud. When this block of tissue is cut out of a donor bud and grafted to the anterior side of a host bud the resulting six-digit double posterior limb has exactly the same structure as the RA-treated one in panel B.

the anteroposterior axis of the chick limb bud which proposed that the ZPA released a diffusable morphogen of low molecular weight which rapidly established an exponential diffusion gradient across the limb bud (Tickle *et al.*, 1975). Cells in different positions would be exposed to different concentrations of the morphogen and would take on particular characteristics, e.g., become a digit 2, 3 or 4 according to that concentration. Grafting a ZPA to an anterior site would generate a mirror-image, U-shaped gradient leading to a mirror-imaged limb pattern.

3.2. Endogenous Retinoids and Their Inhibition

Interestingly, the generation of an exponential gradient is precisely what happens when a bead soaked with RA is placed in the limb bud (Tickle *et al.*, 1985). Furthermore, endogenous all-*trans*-RA has been detected in the limb bud with a 2.5-fold higher level on the posterior side than the anterior side (Thaller and Eichele, 1987). Subsequent measurements of this endogenous anteroposterior difference have consistently confirmed the principle, although the exact figures vary. The only other direct repeat of these endogenous measurements by HPLC has been by Scott *et al.* (1994), who found that there was a 1.38-fold excess of all-*trans*-RA on the posterior side. However, the significance of these measurements of endogenous all-*trans*-RA is dubious because it seems that the predominant retinoic acid in the chick limb bud is not all-*trans*-RA, but didehydroretinoic acid (ddRA) and there is a sixfold higher concentration of this retinoid (Scott *et al.*, 1994; Thaller and Eichele, 1990). This would thus swamp any differences in all-*trans*-RA. When measurements of ddRA are made between anterior and posterior halves of the limb bud the ratio is only 1:1.15 (Scott *et al.*, 1994). Nevertheless, measurements of the total synthesis of all retinoic acids from [^{3}H]retinol still gave an anteroposterior ratio of 1:1.3 (Stratford *et al.*, 1996), and another measurement using an F9 reporter cell system which does not distinguish between the isoforms of RA, but measures total activation of a RARE, gives a twofold difference between anterior and posterior halves of the limb bud (M. Maden, unpublished data). Further evidence for anteroposterior differences comes from the observation that the RARβ_2 isoform is enriched threefold in the posterior part of the limb bud (Smith *et al.*, 1995).

It is clear, therefore, that the effects of RA on the limb bud and these endogenous measurements fit very neatly with the theoretical concepts of a morphogen released from the ZPA as described above. If this is so, then we would expect that interference with RA synthesis or signaling would disrupt anteroposterior patterning of the limb bud. This is indeed what seems to happen in both of these instances. In the first case, the effect of inhibiting

RA synthesis with inhibitors of the metabolic pathway from retinol to RA has been examined. Disulphiram is such a compound (Vallari and Pietruszko, 1982); when the pre-limb-bud-stage chick embryo is treated with disulphiram, limb development is totally abolished in the majority of cases (Stratford *et al.*, 1996). If limb buds are treated after outgrowth has started, then the expression of certain genes is downregulated, such as *sonic hedgehog* (see below) and *fibroblast growth factor-4*, but the effect on the final form of the limb is very limited. The second experiment involves interfering with retinoid signaling using synthetic antagonists which bind to the retinoid receptors, but which then do not allow the receptors to activate gene transcription. Treatment of the pre-limb-bud embryo with both the RAR and RXR panspecific antagonists LG629 and LG754 together can prevent limb outgrowth (Helms *et al.*, 1996), and in the remaining cases where outgrowth had occurred the limbs had deletions on the posterior side. Furthermore, in both of these types of experiment if the limb buds are treated with these compounds and then a bead soaked in RA is applied to the anterior side of the bud, then the limb which grows out has a *reversed* anteroposterior polarity. Therefore, these experiments suggest that RA must be involved in both the initiation of the limb bud outgrowth and in the organization of its anteroposterior polarity, although these two may be facets of one and the same phenomenon.

3.3. Downstream Effector Genes of RA

The induction of limb duplication by identified molecules of gene products is clearly an extremely important demonstration of the role that such molecules are likely to play in normal development. This is therefore a valuable paradigm for identifying such molecules or genes, as evidenced by the discussion above concerning RA. The protein product of the gene *sonic hedgehog* (*shh*) is another such molecule, as either ectopic expression of the gene or the protein product itself can induce double posterior limbs just as RA does (Riddle *et al.*, 1993). Remarkably, the expression of this gene in the limb bud exactly colocalizes with the ZPA (Fig. 2c), as does the SHH protein (Lopez-Martinez *et al.*, 1995; Marti *et al.*, 1995). Clearly then this gene must be involved somewhere along the pathway which leads to anteroposterior patterning in the limb bud, and it is interesting to consider the relationship between RA and *shh*. Three observations suggest that *shh* is a downstream target of RA action. First, the expression of *shh* begins at stage 17/18, after the initiation of limb bud outgrowth (Riddle *et al.*, 1993) and long after the flank mesoderm has begun synthesising RA (Helms *et al.*, 1996). Second, the induction of duplicated limbs by RA involves the induction of ectopic *shh* expression on the anterior side of the limb bud 24 hr after administration of RA (Riddle *et al.*, 1993). Third, the inhibition of RA synthesis with disulphiram results in the downregulation of *shh* (Stratford *et al.*, 1996). In fact, the two molecules must be intimately involved together in the generation of the ZPA and thus the outgrowth of the limb bud because P19 cells transfected with *shh* only show low activity in terms of duplication when grafted to the anterior margin of the chick limb bud. But when the same cells are cotransfected with a constitutively active RARα, full activity is produced (Ogura *et al.*, 1996).

Another gene whose ectopic expression leads to limb duplication, at least in the mouse embryo, is a member of the Hox family of genes, namely *Hoxb*-8 (Charite *et al.*, 1994). The forelimbs of transgenic mouse embryos overexpressing this gene display perfect mirror-image duplications analogous to the chick limbs in Figs. 2b and 2c. Recent experimental data in the chick embryo also point to this gene as being an early determinant in the positioning and patterning of the forelimb bud (Lu *et al.*, 1997; Stratford *et al.*, 1997), and *Hoxb*-8 is induced within 30 min after ectopic RA application to the limb bud, demonstrating that this is likely to be a direct target of RA. Other members of the *Hox* family, particularly those of the *Hoxd* cluster (*Hoxd*-9 to *Hoxd*-13), are activated in a nested set of domains of expression across the anteroposterior axis of the chick limb bud (in a pattern analogous to Russian dolls) during normal development and a mirror-image pattern of expression is generated during the induction of limb duplication after RA administration (Izpisua-Belmonte *et al.*, 1991; Nohno *et al.*, 1991) or after ectopic *shh* expression (Riddle *et al.*, 1993).

To summarize these data, a simple scheme of events could be drawn up as follows. RA induces *Hoxb*-8, which induces *shh*; these events establish a ZPA; *shh* needing RA for continued expression establishes the nested domains of *Hoxd*-9 to *Hoxd*-13. However, it is not at all clear whether anteroposterior patterning is generated by a long-range morphogen as theory would suggest (and as RA certainly behaves, activating gene expression at a distance) or by a sequence of short-range inductions resembling a cascade of *shh*-like domains.

4. RETINOIDS AND THE ANTEROPOSTERIOR AXIS OF THE EMBRYO

The earliest patterning event in the embryo is the establishment of the head-to-tail (AP) axis. This is generated during gastrulation by an organizing center called the dorsal lip in amphibian embryos, Hensen's node in chick embryos, and the node in mouse embryos. The result of this patterning event is the production of mesoderm specified in the head-to-tail axis—

cephalic mesoderm, heart, paraxial mesoderm, intermediate mesoderm, etc. In the AP axis it is the paraxial mesoderm which is the most obviously organized, as it gives rise to the somites. The somites give rise to the muscles of the back, the vertebrae, and the ribs. Immediately after gastrulation the mesoderm induces the ectoderm above it to form the neural plate from which the nervous system will arise and the neural plate also rapidly becomes specified in the head-to-tail axis, differentiating into forebrain, midbrain, hindbrain, and spinal cord (Fig. 3a). Both the AP organization of the mesoderm and the central nervous system (CNS) as well as the node itself have been investigated with regard to RA effects. The mesoderm and the CNS react in a similar fashion to RA treatments (see below), and to understand these effects it is first pertinent to describe briefly the models which have been proposed to explain how this early patterning takes place in the embryo.

4.1. A Model of Patterning in the AP Axis

It is generally considered that AP patterning of the mesoderm and CNS are intimately connected, being different sides of the same coin, perhaps because they employ the same patterning molecules (Sasai and De Robertis, 1997). The following is considered to be a likely scheme of events (Fig. 3c). First, the mesoderm that is generated in a temporal sequence at gastrulation is induced by the interaction between various patterning molecules (chordin, noggin, bone morphogenetic proteins) and the AP sequence comes about by the gradual action of posteriorizing agents, Fig. 3c(i). Second, immediately the mesoderm is generated, it induces the overlying ectoderm to become neural plate and the same molecules (chordin, noggin, BMPs) are involved in this induction event, Fig. 3c(ii). Third, the same posteriorizing agents now act on the neural plate to gradually pattern it in the AP axis, thus generating the major divisions of the CNS (forebrain, midbrain, hindbrain, spinal cord), Fig. 3c(iii). What RA is considered to be in this scheme is one of the posteriorizing agents (fibroblast growth factor is the other one) which acts both on the initially unspecified mesoderm and CNS. Specifically, it is suggested that there is either a concentration gradient of RA from the anterior end (low point) to the posterior end (high point) or an on/off border within the mesoderm and neural plate which performs this act of posteriorization and that the anterior, low point, or border of the gradient is within the future hindbrain region, Fig. 3c(iv) (Conlon, 1995). This border is at the boundary between the expression of the *Otx* genes in the head and the expression of *Hox* genes in the rest of the embryo, Fig. 3c(iv).

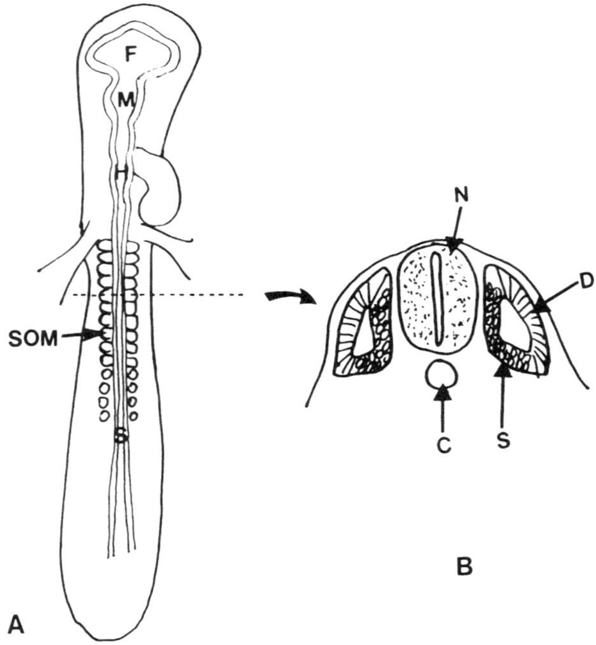

FIGURE 3. (A) Diagram of a 2-day-old chick embryo to show the basic structure of the embryo soon after the mesoderm starts to segment into the somites and the CNS has become patterned into its component parts. The CNS consists of the forebrain (F), the midbrain (M), the hindbrain (H), which consists of a series of bulges known as rhombomeres (see Fig. 4a for details), and the spinal cord (S), which begins at about the level the somites begin. The somites (SOM) are produced in pairs (one on either side of the neural tube), sequentially from the anterior end. In this diagram there are 14 pairs of somites marked and the last 4 are the most recently developed, as represented by their round shape. The dotted line marks where the transverse section drawn in panel B is taken from. (B) Transverse section though the embryo in panel A to show the structure of the somites and the relative positions of the other organs. In the center is the neural tube (N) and below that is the notochord (C). The somites to the left and right of the neural tube are spheres or cubes of cells consisting of an outermost layer, the dermamyotome (D), which subsequently splits into the dermatome (giving rise to the dermis) and the myotome (giving rise to the muscles of the back and limbs), and the sclerotome (S), which gives rise to the vertebral column.

In the discussion below it is apparent that the effects of RA on the mesoderm and the CNS are very similar and this seems to be why RA may be a posteriorizing agent for both tissues.

4.2. The Node

The first indication that RA could be involved in this early embryological event came from the demonstration that Hensen's node, when trans-

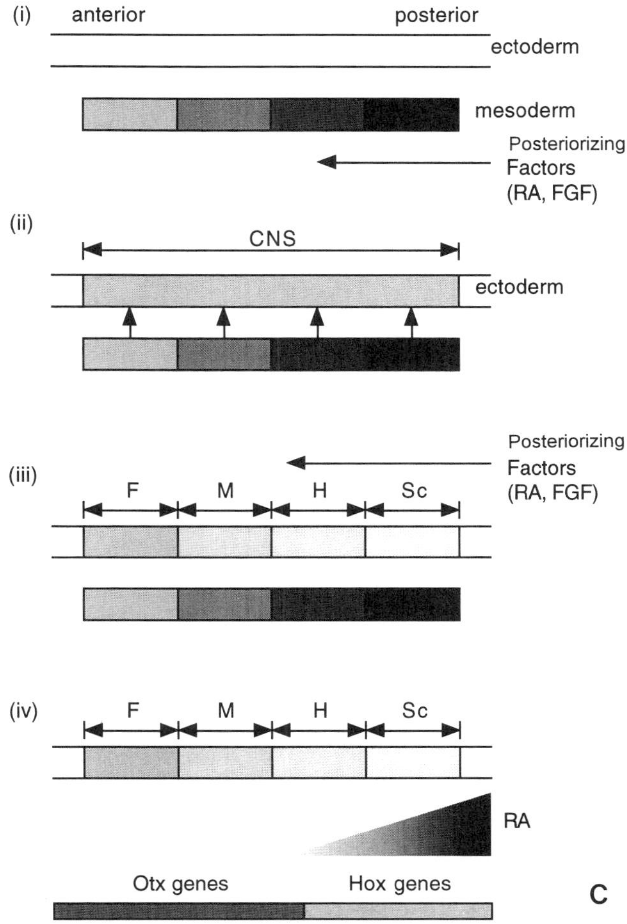

FIGURE 3. (C) A model to describe AP patterning in the mesoderm and CNS of the developing embryo. (i) First the mesoderm is generated during gastrulation, which is initially anterior in character and then becomes patterned by the action of posteriorizing factors, one of which is RA. At this stage the ectoderm still covers the embryo. (ii) Next the mesoderm induces the overlying ectoderm to become neural plate which will form the CNS. (iii) The same posteriorizing factors which earlier acted on the mesoderm now act on the neural plate to generate the subdivisions forebrain (F), midbrain (M), hindbrain (H), and spinal cord (Sc). (iv) The genes which, in part, may be responsible for this patterning event are the *Otx* genes, which are expressed from the forebrain back to the anterior hindbrain and which have a border with the *Hox* genes, which are expressed from the anterior hindbrain to the posterior end. RA in the embryo is proposed to form a gradient from the hindbrain backward and there is considerable evidence for this as described in the text. RA is known to activate *Hox* genes and repress *Otx* genes, so this fits well with the model drawn here.

planted to the anterior margin of the chick limb bud, induced mirror-image duplications in the same way that RA does (Wagner *et al.*, 1990). It was then found that RA can induce the development of a secondary axis in the early chick embryo in the same way that grafting an additional Hensen's node does (Chen and Solursh, 1992). Furthermore, using the F9 reporter cell system showed that Hensen's node itself seems to contain RA at a far higher concentration than surrounding tissues (Chen *et al.*, 1992). The equivalent structure in the mouse embryo also induced extra digits when grafted to the chick limb bud, and the mouse node can synthesize RA from its biosynthetic precursor, retinol (Hogan *et al.*, 1992). These results have led to a model whereby the AP axis is established according to the total concentration of RA seen by gastrulating mesodermal cells, with the least amount resulting in an anterior specification and the greatest amount resulting in a posterior specification (Hogan *et al.*, 1992), an idea which clearly fits with RA being a posterior signal (Fig. 3c). It is suggested that RA acts via *Hox* genes to establish this primary axis and thus shows many analogous features to the limb bud described above.

4.3. The AP Axis in the Mesoderm

The model described above (Fig. 3c) fits well with some of the effects of RA on mesodermal specification of the embryo when these are analyzed in terms of the structure of the vertebrae which develop from the somites, but does not fit with others. When mouse embryos are treated with RA during gastrulation the vertebrae exhibit anterior or posterior transformations depending on the precise time of treatment (Kessel, 1992; Kessel and Gruss, 1991). Vertebrae can easily be characterized in terms of their number within each segment (7 cervical, 13 thoracic, 6 lumbar, 4 sacral in the mouse) and certain vertebrae are readily identified individually such as the atlas, the axis, and the first and last vertebra of each segment. Early treatments with RA affected the more anterior regions of the vertebral column and the alterations were interpreted as posterior transformations. For example, the atlas often had a structure resembling more posterior cervical vertebrae, the 5th cervical vertebra resembled the 6th, the 7th cervical vertebra had ribs on it and thus resembled the 1st thoracic. These results clearly fit well with the proposed posteriorization role for RA in mesodermal specification. However, when RA was administered at a slightly later stage of embryogenesis, anterior transformation of vertebrae was generally observed and this does not obviously fit with the model. For example, the 8th thoracic vertebra, whose rib does not normally connect to the sternum, did produce such a rib and thus came to resemble the 7th thoracic vertebra, and the embryos had 14 or 15 ribs in total instead of 13, so the 1st and 2nd lumbar vertebrae came to resemble thoracic vertebrae.

These anatomical changes were preceded by a change in the expression patterns of various *Hox* genes in the mesoderm, again emphasizing the relationship between RA and *Hox* genes (Kessel, 1992; Kessel and Gruss, 1991). The expression patterns of particular *Hox* genes were anteriorized by early RA treatment and posteriorized by later RA treatment and there was a good correlation between the spread of particular genes and the level at which the anatomical abnormalities appeared. Thus the concept arose that there is a "*Hox* code" for each segment of the body, and changes in the *Hox* code, in this case induced by RA, result in anatomical changes in the AP axis of the body, here reflected in changes in vertebral structure.

Further evidence for the role of RA in mesodermal specification comes from the results of RAR knockouts in mice. In RAR double mutants, axial skeletal abnormalities are found in the cervical region which are mostly anteriorizations (Lohnes *et al.*, 1994). Thus the second cervical vertebra C2 is transformed to a C1, C6 is transformed to C5, and C7 to C6. Curiously, though, no transformations of other vertebrae (thoracic, lumbar, or sacral) were seen, although there were malformations of other vertebrae. Thus in the absence of the RA signaling machinery in these knockout embryos, these alterations fit well with the posteriorization model.

Other effects of RA on the mesoderm are also seen—for example, a loss of the complete anterior end of the embryo, loss of the heart, and effects on blood formation—and in general such affected embryos seem to have been posteriorized, thus fitting with the model. Some of these effects are described below.

4.4. The AP Axis in the Central Nervous System

Detailed studies of the effects of RA on the CNS began with the experiments of Durston *et al.* (1989) on *Xenopus*, although it was known from earlier teratogenicity studies using mammalian embryos that the CNS was most obviously affected by excess retinoids, producing defects such as exencephaly (Cohlan, 1953) and a shortened hindbrain (Morriss, 1972). But what Durston *et al.* observed most obviously fitted with the posteriorization role for RA. These researchers treated embryos with RA between the late blastula and early neurula stages and observed that they failed to develop anterior neural structures such as the forebrain, midbrain, and eyes (Fig. 4b). Identical effects are seen with axolotl embryos (Maden *et al.*, 1992). Since the total volume of CNS tissue, estimated from sections of treated embryos was the same as in controls, Durston *et al.* concluded that the missing anterior neural structures had been respecified to form posterior neural structures, exactly as the posteriorization model would predict.

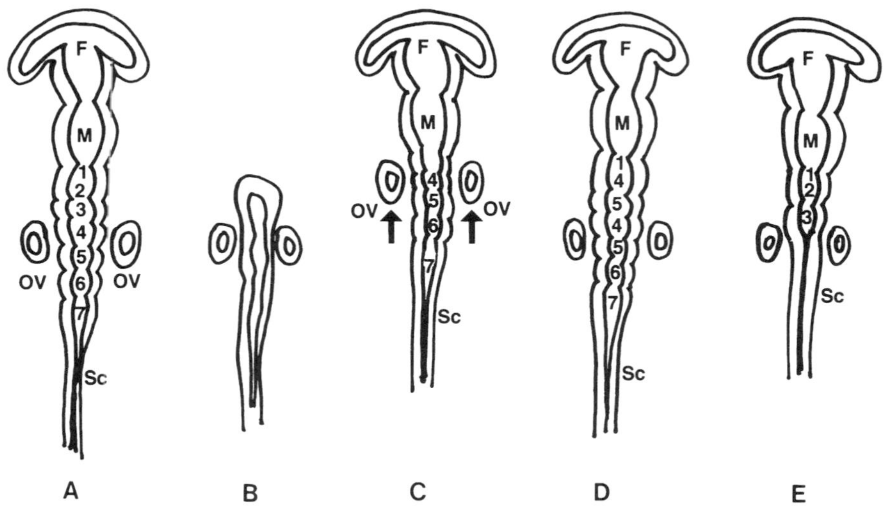

FIGURE 4. Drawings of the various CNS phenotypes found after RA treatment (B–D) or in the absence of RA (E). (A) Normal CNS of a typical embryo showing the forebrain (F) at the anterior end followed by the midbrain (M), then the hindbrain, which is segmented into seven rhombomeres (numbers 1–7) and the spinal cord (Sc). Beside rhombomeres 3 and 4 on each side are the otic vesicles, which will give rise to the inner ear and which serve as useful positional markers (see panel C). (B) Phenotype after excess RA treatment, where the forebrain and midbrain are completely lost, leaving only some hindbrain, the spinal cord, and the otic vesicles. This phenotype is seen in zebrafish after 9-*cis*-RA treatment, in amphibians at high doses of *trans*-RA, and in mice after treatment with *trans*-RA at early stages. (C) A deletion phenotype where tissue is lost at the anterior end of the hindbrain (rhombomeres 1, 2, and 3 and some posterior midbrain tissue) leaving an unidentifiable rhombomere anterior to rhombomere 4. A clear characteristic of this phenotype is the anterior shift of the otic vesicles (arrows). This is seen in all vertebrate embryos studied. (D) A respecification phenotype where rhombomeres 2 and 3 are still present, but have become respecified to resemble rhombomeres 4 and 5, which are thus repeated. This is seen in mouse emrbyos. In zebrafish a slightly different repsecification takes place and only rhombomere 2 is changed into a rhombomere 4-like structure. (E) The CNS phenotype in vitamin A-deficient quail embryos, where the embryos have developed in the absence of RA. Here rhombomeres 4–7 have completely disappeared.

When patterns of gene activity were analyzed this conclusion was reinforced because anterior-specific genes were repressed after RA treatment and posterior genes were upregulated (Cho and De Robertis, 1990; Sive *et al.*, 1990). Further experiments revealed that RA could mimic normal development by generating the correct spatial pattern of expression of two genes, *XlHbox*-6 and *XIF*-6, in isolated explants of *Xenopus* animal caps (Sharpe, 1991). This lead to the suggestion that there was an endogenous gradient of RA within the neural plate with a high point at the posterior end, precisely what had been proposed for the mesoderm. There is no doubt that embryos treated in these experiments look as if their anterior ends have been cut off and so this is one CNS phenotype of RA-treated embryos. Indeed, in the experiments of Sive *et al.* (1990) the embryos failed to develop a heart after extended RA treatment. Thus it may be asked, with reference to the model in Fig. 3c, whether RA is acting primarily on the mesoderm and thus only secondarily on the CNS, or directly on the specification of the CNS. Several experiments have been conducted using *Xenopus* to specifically answer this question. Isolated parts of the embryo have been treated with RA and then recombined or embryos have been treated before or after the induction of the nervous system, and the general conclusion from these experiments is that RA acts *both* directly on the CNS and indirectly via the underlying mesoderm (Ruiz i Altaba and Jessell, 1991a, b; Sive and Cheng, 1991; Sive *et al.*, 1990). This explains why the same general effects on anterior neural structures are seen irrespective of whether the embryos are treated with RA before or after neural induction.

Although these individual tissue and recombination experiments are most easily conducted on amphibian embryos, the "anterior amputation" phenotype (Fig. 4b) is not unique to amphibians, but is also seen in zebrafish (after 9-*cis*-RA treatments, not all-*trans*-RA) (Zhang *et al.*, 1996), and rat and mouse embryos (Cunningham *et al.*, 1994; Simeone *et al.*, 1995). In rat embryos, where the effect is seen after treatment at the late streak stage, there is a progressive loss of anterior structures with increasing length of exposure to RA (and hence an increased total dose), exactly as seen in *Xenopus* embryos.

In addition to the loss of forebrain/midbrain structures, two further CNS phenotypes are observed (Figs. 4c, 4d). These different phenotypes are due either to differences in concentration (*Xenopus*) or differences in stage of treatment (rat and mouse) or treatment with different isoforms (zebrafish), and the defects concern the hindbrain rather than the forebrain. Thus, further analyses of *Xenopus* embryos treated with lesser concentrations of RA than those used to produce the anterior amputations, revealed that rhombomeres 1–3 (rhombomeres are the individual units of the hindbrain, seven of which are normally present in vertebrate embryos; see Fig. 4a) were reduced, the associated cranial nerve tracts were abnormal, and there was no compensatory increase in posterior neural tissue (Papalopulu *et al.*, 1991). This phenotype (Fig. 4c) is also seen in the zebrafish embryo, when all-*trans*-RA is used rather than 9-*cis*, and such embryos display a compression of anterior rhombomeres and the absence of tissue at the midbrain/hindbrain border (Holder and Hill, 1991; Zhang *et al.*, 1996). The same phenotype, an abnormally rostral position of the otocyst and a shortened preotic hindbrain, had long been known to be characteristic of mammalian embryos treated with RA (Morriss, 1972) and thus is also found in rat and mouse embryos (Cunningham *et al.*, 1994; Leonard *et al.*, 1995; Morriss-Kay *et al.*, 1991; Simeone *et al.*, 1995). To complete the picture of universality the chick embryo also displays this shortened-hindbrain phenotype (Sundin and Eichele, 1992).

The third phenotype (Fig. 4d) also appears in the hindbrain and the results have prompted a return to the concept of posteriorization. Thanks to the identification of a variety of molecular and neuroanatomical markers for the individual rhombomeres of the hindbrain, the precise structure of the altered rhombomeres in RA-treated hindbrains has been deduced. In the mouse embryo, rhombomeres 2 and 3 are respecified to rhombomeres 4 and 5, giving the resultant pattern of rhombomeres of 1, 4, 5, 4, 5, 6, 7 instead of 1, 2, 3, 4, 5, 6, 7 (Marshall *et al.*, 1992) (Fig. 4d). In the zebrafish a slightly different pattern has been deduced: only rhombomere 2 is respecified to 4, giving the final pattern of 1, 4, 3, 4, 5, 6, 7 (Hill *et al.*, 1995). Whatever the precise result, it is clear that anterior hindbrain tissue has indeed been posteriorized by RA in these experiments, but only within the hindbrain region. To generate this third phenotype (Fig. 4d) rather than the second phenotype (Fig. 4c) requires a lesser concentration of RA in the zebrafish or treatment at a different stages in mouse embryos (Simeone *et al.*, 1995; Wood *et al.*, 1994).

4.5. Associated CNS Defects: The Neural Crest and Cranial Ganglia

The neural crest is a migratory group of cells which leave the dorsal surface of the newly formed neural tube and migrate in streams out into the periphery to form a wide range of tissues, including the cranial sensory nerves and ganglia. Since some of the CNS phenotypes described above involve deletion of large segments of the hindbrain, it is to be expected that abnormalities in the cranial ganglia might be observed in RA-treated embryos. This is indeed the case, and the most obvious defect in mammalian embryos is a fusion of the trigeminal and facial ganglia into one large ganglionic mass (Kessel, 1993). The original explanation of this phenomenon was that the neural crest which forms the trigeminal ganglion comes from rhombomere 2 and that forming the facial ganglion from rhombomere 4, and if rhombomere 3 was deleted (as in the phenotype in Fig. 4c), then the two streams of crest would fuse. However, it is interesting to note that under conditions when no neuroepithelium is lost, this fusion of ganglia still occurs and it is caused by the mismigration of neural crest cells (Gale *et al.*, 1996; Lee *et al.*, 1995). Thus RA can alter the gene expression patterns of neural crest cells, which results in dramatically altered routes of migration of these crest cells. This means that the cranial nerves which subsequently develop will be completely misrouted (Gale *et al.*, 1996) and presumably connect up with completely wrong innervation fields.

It seems that these CNS and neural crest defects also occur in humans (Lammer *et al.*, 1985), as some babies born to mothers who have taken Accutane during pregnancy have defects in the cerebellum (the structure which develops from the neuroepithelium at the midbrain/hindbrain border) and defects in the facial musculature (operated by the facial nerve which derives from rhombomere 4).

4.6. Downstream Effector Genes of RA

A large number of genes have been analyzed to determine the effect of RA on their expression domains in the early CNS. In the light of the experiments described in the limb bud and mesoderm, it should come as no surprise to learn that the immediate effect of RA treatment is to alter the patterns of expression of various *Hox* genes within the CNS. In general the 3′ *Hox* genes are anteriorized and the 5′ *Hox* genes are not affected in the mouse embryo (Avantaggiato *et al.*, 1996; Conlon and Rossant, 1992; Kessel, 1993; Marshall *et al.*, 1992; Morriss-Kay *et al.*, 1991; Simeone *et al.*, 1995; Wood *et al.*, 1994), the chick embryo (Sundin and Eichele, 1992; Lopez *et al.*, 1995; Gale *et al.*, 1996), and the *Xenopus* embryo (Sive and Cheng, 1991; Sive *et al.*, 1990); Papalopulu *et al.*, 1991; Leroy and De Robertis, 1992; Lopez *et al.*, 1995; Dekker *et al.*, 1992). In the zebrafish embryo, analysis of the earliest events following RA treatment has revealed that there is a change in *Hoxa*-1 and *Hoxb*-1 gene activity in the involuting cells in the axial midline during gastrulation (Hill *et al.*, 1995). Most excitingly, the phenotype of RA-treated zebrafish can be mimicked by overexpression of *Hoxa*-1 (Alexandre *et al.*, 1996), directly demonstrating the causative role that *Hox* genes play in the RA-induced phenotypes within the embryonic CNS.

Other genes that have been analysed include *goosecoid* (Cho *et al.*, 1991; Hill *et al.*, 1995), *Xlim*-1 (Taira *et al.*, 1992, 1994), *engrailed* (Hill *et al.*, 1995; Avantaggiato *et al.*, 1996), *pax*-2 (Hill *et al.*, 1995; Avantaggiato *et al.*, 1996), *Wnt*-1 (Avantaggiato *et al.*, 1996), *Krox*-20 (Conlon and Rossant, 1992; Morriss-Kay *et al.*, 1991; Hill *et al.*, 1995), and *Otx*-1 and 2, *Emx*-1 and 2, and *Dlx*-1 (Bally-Cuif *et al.*, 1995; Simeone *et al.*, 1995). It is interesting to note the relationship between the expression domains of the *Hox* genes and the *Otx* genes in the light of the posteriorization model described above, Fig. 3c(iv). There is a reciprocal relationship between these two classes of genes; the *Otx* genes are expressed in the forebrain and midbrain with a posterior border in the anterior hindbrain, whereas the

Hox genes are expressed in the rest of the embryo with an anterior border in the anterior hindbrain. RA induces *Hox* genes and represses *Otx* genes, which gives credence to the idea that RA is a posteriorizing factor operating from the hindbrain to the tail.

4.7. Endogenous RA in the Early Embryo

The postulated role for RA as a posteriorizing agent with an anterior border in the hindbrain, Fig. 3c(iv), is a hypothesis which can clearly be tested by measuring endogenous levels of RA in the embryo; the measurements indeed provide strong support. Horton and Maden (1995) determined the levels of various retinoids in the early mouse embryo by HPLC. All-*trans*-RA was present in the mesoderm of the day 10.5 embryo. In the neural tube, retinoid levels were by far the highest, up to tenfold higher than the rest of the embryo, but there was a sudden drop in level in the hindbrain and RA levels were virtually undetectable in the midbrain and forebrain. This fits precisely with the *Otx*/*Hox* expression boundary, Fig. 3c(iv). A similar result in the mouse embryo was revealed with a transgenic RARE–*lacZ* reporter line (Rossant *et al.*, 1991). At the neural plate stage strong expression of the transgene was observed in the posterior half of the embryo, and after somite formation and neural tube closure had begun there was a sharp anterior boundary of expression in the hindbrain. Again this precisely fits the theory. In the chick embryo identical results have been obtained using the F9 reporter cell system (M. Maden and E. Gale, unpublished data) and in the zebrafish embryo using the RARE transgene a sharp border of transgene expression was seen at the anterior end of the trunk (Marsh-Armstrong *et al.*, 1995). In the *Xenopus* embryo a 12-fold gradient of RA with the high point at the posterior end was detected in neurula stage using the F9 reporter cell system (Chen *et al.*, 1994). However, using HPLC, rather different and more complex results have been obtained in *Xenopus* (Creech-Kraft *et al.*, 1994). In this case at the neurula stage, a gradient of all-*trans*-RA was detected, but it was the opposite to that found by Chen *et al.*, with the high point at the anterior end. 9-*cis*-RA was detected at the anterior and posterior ends, but not in the middle. So this latter report is at odds with the general consensus of results from zebrafish, chick, and mouse embryos that the anterior end of the embryo is devoid of RA and the posterior end is rich and the switch point is in the region of the hindbrain, an excellent fit with the posteriorization model (Fig. 3c).

4.8. Disruption of RA Signaling

The expression patterns of the RARs and RXRs in the early mesoderm and neural tube have been well studied in the mouse (Dolle *et al.*,

1994; Ruberte *et al.*, 1991, 1993) and *Xenopus* embryo (Blumberg *et al.*, 1992; Ellinger-Ziegelbauer and Dreyer, 1991; Marklew *et al.*, 1994; Pfeffer and De Robertis, 1994; Sharpe, 1992; Sharpe and Goldstone, 1997), but less so in the chick embryo (Rowe *et al.*, 1991; Smith and Eichele, 1991). Interestingly, the zebrafish has the usual three RARs, but in addition to the normal three RXRs two novel ones, RXRδ and RXRε, are present, making five in all (Jones *et al.*, 1995). These two novel RXRs do not bind 9-*cis*-RA as the normal RXRs do. As might be expected, representative receptors are expressed both in the mesenchyme and neural tube of embryos. For example, in the mouse, RXRγ is particularly expressed in the somitic mesoderm (Dolle *et al.*, 1994), and RARα and RARβ are expressed in specific domains within the neural tube (Ruberte *et al.*, 1993), with RARβ particularly found in the motorneurons of the neural tube (Muto *et al.*, 1991; Ruberte *et al.*, 1993). In the zebrafish, RARα is expressed in the neural epithelium and RARγ in the anterior lateral mesenchyme (Joore *et al.*, 1994). In the chick RXRα and RARβ are expressed in the neural epithelium and RXRγ in the neural crest (Rowe *et al.*, 1991; Smith and Eichele, 1991). In *Xenopus* RXRβ is expressed in the anterior neurectoderm and the cranial neural crest (Sharpe and Goldstone, 1997).

It is very surprising, therefore, that experiments to test the role of the RARs and RXRs in mouse embryos by the creation of either single or double null mutants or compound (RAR/RXR) null mutants for these genes have failed to generate embryos in which neural development is perturbed (Kastner *et al.*, 1995, 1997). Does this suggest that the RARs play no role in neural development? In light of the experiments described above in the regenerating limb where single isoform functions have been elaborated (Section 2.3), it would be unwise to come to this negative conclusion. Indeed, a role for specific receptors in AP patterning and primary neurogenesis in *Xenopus* has been demonstrated in the following two experiments. Blumberg *et al.* (1997) microinjected either a constitutively active RARα or a dominant negative RARα into one cell of the two-cell embryo. Injection of the constitutively active receptor lead to a phenocopy of RA treatment such that anterior structures were reduced or absent. There was a corresponding alteration in AP patterning markers in the neural tube: levels of *Otx*-2 were decreased and *engrailed* and *Krox*-20 were shifted anteriorly, with no change in posterior markers. The dominant negative receptor gave the opposite effect whereby embryos showed enlarged anteriors, shorter tails, and reduced overall length. *Otx*-2 expression was expanded, *engrailed* was shifted posteriorly, and the posterior markers *Xlim*-1 and *Hoxb*-9 were either ablated or downregulated.

In the second experiment Sharpe and Goldstone (1997) followed a similar strategy, also in *Xenopus*, but injected normal or mutated RARα along with normal RXRβ to promote heterodimerization. The injection of

RARα and RXRβ produced embryos with ectopic neurons adjacent to the neural tube, whereas either injected alone had no effect. The dominant-negative RARα receptor produced embryos which lacked sensory neurons along their flanks. Clearly these experiments provide evidence for a role of RA in first AP patterning and second neural differentiation within the patterned neural plate.

4.9. Development in the Absence of RA

Finally, one can ask what happens to the patterning of the mesoderm and CNS when embryos develop in the absence of RA. This has recently been performed using quail embryos by feeding the adults on a vitamin A-deficient diet and observing the resultant embryos. Such embryos fail to produce omphalomesenteric veins and so cannot connect up the heart to the yolk circulation and the heart often loops to the wrong side (Dersch and Zile, 1993; Heine *et al.*, 1985). After about 3 days of development these A− embryos die, but not before severe defects in the AP patterning of the mesoderm and the CNS appear. Soon after gastrulation when the domains of expression of various patterning genes such as the *Hox* genes first appear it is apparent that the anterior border of these genes is shifted posteriorly (Maden *et al.*, 1996). There is clearly an AP patterning defect in the absence of RA, and as the embryos develop it seems that a segment of both mesoderm and CNS in the posterior hindbrain region has been lost and the gap closed up. The mesoderm tissue which has been lost is some cephalic mesoderm and the first few somites, and as a result the somites extend further anteriorly than normal. The CNS tissue which is lost is comprised of rhombomeres 4–7, which together make up the myelencephalon, and as a result of this the spinal cord connects directly onto the anterior hindbrain (Fig. 4e). It is interesting to note that these defects appear at the anterior boundary of the hypothesized RA gradient, Fig. 3c(iv).

5. CONCLUSION

In the above review I have picked out those systems where experiments have revealed a role for RA in patterning mechanisms in the embryo—in the regenerating limb and tail, the developing limb bud, and the AP axis of the mesoderm and CNS. Superficially, these systems seem to have little in common, but as the genes involved in the molecular mechanisms of patterning are beginning to be revealed, it is becoming apparent that the same gene pathways are used over and over again. It is also apparent that RA may be one of the molecules used early in the patterning process of each of these systems to establish patterns of gene activity. This conclusion is, in my view, a rather risky evolutionary strategy—to use a vitamin as a primary patterning molecule in the early embryo. It will be interesting in the future to determine when, and possibly how, this phenomenon evolved by examining whether these signaling systems based on RA operate in the invertebrates.

6. REFERENCES

Alexandre, D., Clarke, J. D. W., Oxtoby, E., Yan, Y.-L., Jowett, T., and Holder, N., 1996, Ectopic expression of *Hoxa*-1 in the zebrafish alters the fate of the mandibular arch neural crest and phenocopies a retinoic acid-induced phenotype, *Development* **122:**735–746.

Avantaggiato, V., Acampora, D., Tuorto, F., and Simeone, A., 1996, Retinoic acid induces stage-specific repatterning of the rostral central nervous system, *Dev. Biol.* **175:**347–357.

Bally-Cuif, L., Gulisano, M., Broccoli, V., and Boncinelli, E., 1995, c-*otx*2 is expressed in two different phases of gastrulation and is sensitive to retinoic acid treatment in chick embryos, *Mech. Dev.* **49:**49–63.

Beauchemin, M., and Savard, P., 1993, Expression of five homeobox genes in the adult newt appendages and regeneration blastemas, in *Limb Regeneration and Development* (J. F. Fallon, P. F. Goetinck, R. O. Kelley, and D. L. Stocum., eds.), Part A, pp. 41–50, Wiley-Liss, New York.

Blumberg, B., Mangelsdorf, D. J., Dyck, J. A., Bittner, D. A., Evans, R. M., and De Robertis, E. M., 1992, Multiple retinoid-responsive receptors in a single cell: Families of retinoid "X" receptors and retinoic acid receptors in the *Xenopus* egg, *Proc. Natl. Acad. Sci. USA* **89:**2321–2325.

Blumberg, B., Bolado, J., Moreno, T. A., Kintner, C., Evans, R. M., and Papalopulu, N., 1997, An essential role for retinoid signalling in anteroposterior neural patterning, *Development* **124:**373–379.

Brockes, J. P., 1992, Introduction of a retinoid reporter gene into the urodele limb blastema, *Proc. Natl. Acad. Sci. USA* **89:**11386–11390.

Brown, R., and Brockes, J. P., 1991, Identification and expression of a regeneration-specific homeobox gene in the newt limb blastema, *Development* **111:**489–496.

Charite, J., de Graff, W., Shen, S., and Deschamps, J., 1994, Ectopic expression of *Hoxb*-8 causes duplication of the ZPA in the forelimb and homeotic transformations of axial structures, *Cell* **78:**589–601.

Chen, Y.-P., and Solursh, M., 1992, Comparison of Hensen's node and retinoic acid in secondary axis induction in the early chick embryo, *Dev. Dynam.* **195:**142–151.

Chen, Y.-P., Huang, L., Russo, A. F., and Solursh, M., 1992, Retinoic acid is enriched in Hensen's node and is developmentally regulated in the early chick embryo, *Proc. Natl. Acad. Sci. USA* **89:**10056–10059.

Chen, Y.-P., Huang, L., and Solursh, M., 1994, A concentration gradient of retinoids in the early *Xenopus laevis* embryo, *Dev. Biol.* **161:**70–76.

Cho, K. W. Y., and De Robertis, E. M., 1990, Differential activation of *Xenopus* homeobox genes by mesoderm-inducing growth factors and retinoic acid, *Genes Dev.* **4:**1910–1916.

Cho, K. W. Y., Blumberg, B., Steinbeisser, H., and De Robertis, E. M., 1991, Molecular nature of Spemann's organiser: The role of the *Xenopus* homeobox gene *goosecoid*, *Cell* **67:**1111–1120.

Cohlan, S. Q., 1953, Excessive intake of vitamin A as a cause of congenital abnormalities in the rat, *Science* **117:**535–536.
Conlon, R. A., 1995, Retinoic acid and pattern formation in vertebrates, *Trends Genet.* **11:**314–319.
Conlon, R. A., and Rossant, J., 1992, Exogenous retinoic acid rapidly induces anterior ectopic expression of murine *Hox*-2 genes *in vivo*, *Development* **116:**357–368.
Creech-Kraft, J., Schuh, T., Juchau, M., and Kimelman, D., 1994, The retinoid X receptor ligand, 9-*cis*-retinoic acid, is a potential regulator of early *Xenopus* development, *Proc. Natl. Acad. Sci. USA* **91:**3067–3071.
Cunningham, M. L., Mac Auley, A., and Mirkes, P. E., 1994, From gastrulation to neurulation: Transition in retinoic acid sensitivity identifies distinct stages of neural patterning in the rat, *Dev. Dynam.* **200:**227–241.
Dekker, E.-J., Pannese, M., Houtzager, E., Boncinelli, E., and Durston, A., 1992, Colinearity in the *Xenopus laevis Hox*-2 complex, *Mech. Dev.* **40:**3–12.
Dersch, H., and Zile, M. H., 1993, Induction of normal cardiovascular development in the vitamin A-deprived quail embryo by natural retinoids, *Dev. Biol.* **160:**424–433.
Dolle, P., Fraulob, V., Kastner, P., and Chambon, P., 1994, Developmental expression of murine retinoid X receptor (RXR) genes, *Mech. Dev.* **45:**91–104.
Durston, A. J., Timmermans, J. P. M., Hage, W. J., Hendricks, H. F. J., de Vries, N. J., Heideveld, M., and Nieuwkoop, P. D., 1989, Retinoic acid causes an anteroposterior transformation in the developing nervous system, *Nature* **340:**140–144.
Eichele, G., Tickle, C., and Alberts, B. M., 1985, Studies on the mechanism of retinoid-induced pattern duplications in the early chick limb bud: Temporal and spatial aspects, *J. Cell Biol.* **101:**1913–1920.
Ellinger-Ziegelbauer, H., and Dreyer, C., 1991, A retinoic acid receptor expressed in the early development of *Xenopus laevis*, *Genes Dev.* **5:**94–105.
Gale, E., Prince, V., Lumsden, A., Clarke, J., Holder, N., and Maden, M., 1996, Late effects of retinoic acid on neural crest and aspects of rhombomere identity, *Development* **122:**783–793.
Gardiner, D. M., Blumberg, B., Komine, Y., and Bryant, S. V., 1995, Regulation of *HoxA* expression in developing and regenerating axolotl limbs, *Development* **121:**1731–1741.
Heine, U. I., Roberts, A. B., Munoz, E. F., Roche, N. S., and Sporn, M. B., 1985, Effects of retinoid deficiency on the development of the heart and vascular system of the quail embryo, *Virchow's Arch. B Cell Pathol.* **50:**135–152.
Helms, J. A., Kim, C. H., Eichele, G., and Thaller, C., 1996, Retinoic acid signalling is required during early chick limb development, *Development* **122:**1385–1394.
Hill, D. S., Ragsdale, C. W., and Brockes, J. P., 1993, Isoform-specific immunological detection of newt retinoic acid receptor d1 in normal and regenerating limbs, *Development* **117:**937–945.
Hill, J., Clarke, J. D. W., Vargesson, N., Jowett, T., and Holder, N., 1995, Exogenous retinoic acid causes specific alterations in the development of the midbrain and hindbrain of the zebrafish embryo including positional respecification of the Mauthner neuron, *Mech. Dev.* **50:**3–16.
Holder, N., and Hill, J., 1991, Retinoic acid modifies development of the midbrain–hindbrain border and affects cranial ganglion formation in zebrafish embryos, *Development* **113:**1159–1170.
Horton, C., and Maden, M., 1995, Endogenous distribution of retinoids during normal development and teratogenesis in the mouse embryo, *Dev. Dynam.* **202:**312–323.

Izpisua-Belmonte, J.-C., Tickle, C., Dolle, P., Wolpert, L., and Duboule, D., 1991, Expression of the homeobox *Hox*-4 genes and the specification of position in chick wing development, *Nature* **350:** 585–589.

Jones, B. B., Ohno, C. K., Allenby, G., Boffa, M. B., Levin, A. A., Grippo, J. F., and Petkovich, M., 1995, New retinoid X receptor subtypes in zebra fish (*Danio rerio*) differentially modulate transcription and do not bind 9-*cis* retinoic acid, *Mol. Cell Biol.* **15:**5226–5234.

Joore, J., van der Lans, G. B. L. J., Lanser, P. H., Vervaart, J. M. A., Zivkovic, D., Speksnijder, J. E., and Kruijer, W., 1994, Effects of retinoic acid on the expression of retinoic acid receptors during zebrafish embryogenesis, *Mech. Dev.* **46:**137–150.

Kalter, H., and Warkany, J., 1959, Experimental production of congenital malformations in mammals by metabolic procedure, *Physiol. Rev.* **39:**69–115.

Kastner, P., Mark, M., and Chambon, P., 1995, Nonsteroid nuclear receptors: What are genetic studies telling us about their role in real life? *Cell* **83:**859–869.

Kastner, P., Mark, M., Ghyselinck, N., Kretzel, W., Dupe, V., and Grondona, J. M., 1997, Genetic evidence that the retinoid signal is transduced by heterodimeric RXR/RAR functional units during mouse development, *Development* **124:**313–326.

Keeble, S., and Maden, M., 1989, The relationship among retinoid structure, affinity for retinoic acid-binding protein and ability to respecify pattern in the regenerating axolotl limb, *Dev. Biol.* **132:**26–34.

Kessel, M., 1992, Respecification of vertebral identities by retinoic acid, *Development* **115:**487–501.

Kessel, M., 1993, Reversal of axonal pathways from rhombomere 3 correlates with extra *Hox* expression domains, *Neuron* **10:**379–393.

Kessel, M., and Gruss, P., 1991, Homeotic transformations of murine vertebrae and comcomitant alteration of *Hox* codes induced by retinoic acid, *Cell* **67:**89–104.

Krumlauf, R., 1994, *Hox* genes in vertebrate development, *Cell* **78:**191–201.

Lammer, E. J., Chen, D. T., Hoar, R. M., Agnish, A. D., Benke, P. J., Braun, J. T., Curry, C. J., Fernhoff, P. M., Grix, A. W., Lott, I. T., Richard, J. M., and Sun, S. C., 1985, Retinoic acid embryopathy, *N. Engl. J. Med.* **313:**837–841.

Langston, A. W., and Gudas, L. J., 1994, Retinoic acid and homeobox gene regulation, *Curr. Opin. Gen. Dev.* **4:**550–555.

Lee, Y. M., Osumi-Yamashita, N., Ninomiya, Y., Moon, C. K., Eriksson, U., and Eto, K., 1995, Retinoic acid stage-dependently alters the migration pattern and identity of hindbrain neural crest cells, *Development* **121:**825–837.

Leonard, L., Horton, C., Maden, M., and Pizzey, J. A., 1995, Anteriorization of CRABP-I expression by retinoic acid in the developing mouse central nervous system and its relationship to teratogenesis, *Dev. Biol.* **168:**514–528.

Leroy, P., and De Robertis, E. M., 1992, Effects of lithium chloride and retinoic acid on the expression of genes from the *Xenopus Hox* 2 complex, *Dev. Dynam.* **194:**21–32.

Lohnes, D., Mark, M., Mendelsohn, C., Dolle, P., Dierich, A., Gorry, P., Gansmuller, A., and Chambon, P., 1994, Function of the retinoic acid receptors (RARs) during development (I) Craniofacial and skeletal abnormalities in RAR double mutants. *Development* **120:**2723–2748.

Lopez, S. L., Dono, R., Zeller, R., and Carrasco, A. E., 1995, Differential effects of retinoic acid and a retinoid antagonist on the spatial distribution of the homeoprotein *Hoxb*-7 in vertebrate embryos, *Dev. Dynam.* **204:**457–471.

Lopez-Martinez, A., Chang, D. T., Chiang, C., Porter, J. A., Ros, M. A., Simandl, B. K., Beachy, P. A., and Fallon, J. F., 1995, Limb-patterning activity and restricted posterior localization of the amino-terminal product of sonic hedgehog cleavage, *Curr. Biol.* **5:**791–796.

Lu, H.-C., Revelli, J.-P., Goering, L., Thaller, C., and Eichele, G., 1997, Retinoid signalling is required for the establishment of a ZPA and for the expression of *Hoxb*-8, a mediator of ZPA formation, *Development* **124:**1643–1651.

Ludolph, D. C., Cameron, J. A., and Stocum, D. L., 1990, The effect of retinoic acid on positional memory in the dorsoventral axis of regenerating axolotl limbs, *Dev. Biol.* **140:**41–52.

Maden, M., 1982, Vitamin A and pattern formation in the regenerating limb, *Nature* **295:**672–675.

Maden, M., 1983a, The effect of vitamin A on limb regeneration in *Rana temporaria*, *Dev. Biol.* **98:**409–416.

Maden, M., 1983b, The effect of vitamin A on the regenerating axolotl limb, *J. Embryol. Exp. Morphol.* **77:**273–295.

Maden, M., 1993, The homeotic transformation of tails into limbs in *Rana temporaria* by retinoids, *Dev. Biol.* **159:**379–391.

Maden, M., and Pizzey, J. A., 1997, The role of retinoids in patterning fish, amphibian and chick embryos, *Adv. Organ Biol.* **3:**93–139.

Maden, M., Summerbell, D., Maignan, J., Darmon, M., and Shroot, B., 1991, The respecification of limb pattern by new synthetic retinoids and their interaction with cellular retinoic acid-binding protein, *Differentiation* **47:**49–55.

Maden, M., and Corcoran, J., 1996, Role of thyroid hormone and retinoid receptors in the homeotic transformation of tails into limbs in frogs, *Dev. Genet.* **19:**85–93.

Maden, M., Gale, E., Horton, C., and Smith, J. C., 1992, Retinoid-binding proteins in the developing vertebrate nervous system, in *Retinoids in Normal Development and Teratogenesis* (G. M. Morriss-Kay, ed.), pp. 119–134, Oxford University Press, Oxford.

Maden, M., Gale, E., Kostetskii, I., and Zile, M., 1996, Vitamin A-deficient quail embryos have half a hindbrain and other neural defects, *Curr. Biol.* **6:**417–426.

Marklew, S., Smith, D. P., Mason, C. S., and Old, R. W., 1994, Isolation of a novel RXR from *Xenopus* that most closely resembles mammalian RXRβ and is expressed throughout early development, *Biochim. Biophys. Acta* **1218:**267–272.

Marsh-Armstrong, N., McCaffery, P., Hyatt, G., Alonso, L., Dowling, J. E., Gilbert, W., and Drager, U. C., 1995, Retinoic acid in the anteroposterior patterning of the zebrafish trunk, *Roux's Arch. Dev. Biol.* **205:**103–113.

Marshall, H., Nonchev, S., Sham, M. H., Muchamore, I., Lumsden, A., and Krumlauf, R., 1992, Retinoic acid alters hindbrain *Hox* code and induces transformation of rhombomeres 2/3 into a 4/5 identity, *Nature* **360:**737–741.

Marti, E., Takada, R., Bumcrot, D. A., Sasaki, H., and McMahon, A. P., 1995, Distribution of sonic hedgehog peptides in the developing chick and mouse embryo, *Development* **121:**2537–2547.

McCaffery, P., Lee, M.-O., Wagner, M. A., Sladek, N. E., and Drager, U., 1992, Asymmetrical retinoic acid synthesis in the dorsoventral axis of the retina, *Development* **115:**371–382.

Mohanty-Hejmadi, P., Dutta, S. K., and Mahapatra, P., 1992, Limbs generated at site of tail amputation in marbled balloon frog after vitamin A treatment, *Nature* **355:**352–353.

Morriss, G. M., 1972, Morphogenesis of the malformations induced in rat embryos by maternal hypervitaminosis A, *J. Anat.* **113:**241–250.

Morriss-Kay, G. M., Murphy, P., Hill, R. E., and Davidson, D. R., 1991, Effects of retinoic acid excess on expression of *Hox*-2.9 and *Krox*-20 and on morphological segementation in the hindbrain of mouse embryos, *EMBO J.* **10:**2985–2995.

Mullen, L. M., Bryant, S. V., Torok, M. A., Blumberg, B., and Gardiner, D. M., 1996, Nerve dependency of regeneration: The role of Distal-less and FGF signalling in amphibian limb regeneration, *Development* **122:**3487–3497.

Muto, K., Noji, S., Nohno, T., Koyama, E., Myokai, F., Nishijima, K., Saito, T., and Taniguchi, S., 1991, Involvement of retinoic acid and its receptor β in differentiation of motoneurons in chick spinal cord, *Neurosci. Lett.* **129:**39–42.

Niazi, I. A., and Saxena, S., 1978, Abnormal hind limb regeneration in tadpoles of the toad, *Bufo andersoni*, exposed to excess vitamin A, *Folia Biol.* (Krakow) **26:**1–8.

Nohno, T., Noji, S., Koyama, E., Ohyama, K., Myokai, F., Kuriowa, A., Saito, T., and Taniguchi, S., 1991, Involvement of the Chox-4 chicken homeobox genes in determination of anteroposterior axial polarity during limb development, *Cell* **64:**1197–1205.

Ogura, T., Alvarez, I. S., Vogel, A., Rodriguez, C., Evans, R. M., and Izpisua-Belmonte, J. C., 1996, Evidence that Shh cooperates with a retinoic acid inducible cofactor to establish ZPA-like activity, *Development* **122:**537–542.

Papalopulu, N., Clarke, J. D. W., Bradley, L., Wilkinson, D., Krumlauf, R., and Holder, N., 1991, Retinoic acid causes abnormal development and segmental patterning of the anterior hindbrain in *Xenopus embryos*, *Development* **113:**1145–1158.

Pecorino, L. T., Lo, D. C., and Brockes, J. P., 1994, Isoform-specific induction of a retinoid-responsive antigen after biolostic transfection of chimaeric retinoic acid/thyroid hormone receptors into a regenerating limb, *Development* **120:**325–333.

Pecorino, L., Entwistle, A., and Brockes, J. P., 1996, Activation of a single retinoic acid receptor isoform mediates proximodistal respecification, *Curr. Biol.* **6:**563–569.

Pfeffer, P. L., and De Robertis, E. M., 1994, Regional specificity of RARg isoforms in *Xenopus* development, *Mech. Dev.* **45:**147–153.

Ragsdale, C. W., Petkovich, M., Gates, P. B., Chambon, P., and Brockes, J. P., 1989, Identification of a novel retinoic acid receptor in regenerating tissues of the newt, *Nature* **341:**654–657.

Ragsdale, C. W., Gates, P. B., and Brockes, J. P., 1992a, Identification and expression pattern of a second isoform of the newt alpha retinoic acid receptor, *Nucleic Acids Res.* **20:**5851.

Ragsdale, C. W., Gates, P. B., Hill, D. S., and Brockes, J. P., 1992b, Delta retinoic acid receptor isoform d1 is distinguished by its exceptional N-terminal sequence and abundance in the limb regeneration blastema, *Mech. Dev.* **40:**99–112.

Riddle, R. D., Johnson, R. L., Laufer, E., and Tabin, C., 1993, Sonic hedgehog mediates the polarizing activity of the ZPA, *Cell* **75:**1401–1416.

Rossant, J., Zirngibl, R., Cado, D., Shago, M., and Giguere, V., 1991, Expression of a retinoic acid response element-*hsplacZ* transgene defines specific domains of transcriptional activity during mouse embryogenesis, *Genes Dev.* **5:**1333–1344.

Rowe, A., Eager, N. S. C., and Brickell, P. M., 1991, A member of the RXR nuclear receptor family is expressed in neural-crest-derived cells of the developing chick peripheral nervous system, *Development* **111:**771–778.

Ruberte, E., Dolle, P., Chambon, P., and Morriss-Kay, G., 1991, Retinoic acid receptors and cellular retinoid binding proteins. II Their differential pattern of transcription during early morphogenesis in mouse embryos, *Development* **111:**45–60.

Ruberte, E., Friederich, V., Chambon, P., and Morriss-Kay, G., 1993, Retinoic acid receptors and cellular retinoid binding proteins. III Their differential transcript distribution during mouse nervous system development, *Development* **118:**267–282.

Ruiz i Altaba, A., and Jessell, T., 1991a, Retinoic acid modifies mesodermal patterning in early *Xenopus* embryos, *Genes Dev.* **5:**175–187.

Ruiz i Altaba, A., and Jessell, T. M., 1991b, Retinoic acid modifies the pattern of cell differentiation in the central nervous system of neurula stage *Xenopus* embryos, *Development* **112:**945–958.

Sasai, Y., and De Robertis, E. M., 1997, Ectodermal patterning in vertebrate embryos, *Dev. Biol.* **182:**5–20.

Saunders, J. W., and Gasseling, M. T., 1968, Ectodermal–mesenchymal interactions in the origin of limb asymmetry, in *Epithelial–Mesenchymal Interactions* (R. Fleischmajer and R. E. Billingham, eds.), pp. 78–97, Williams & Wilkins, Baltimore, Maryland.

Savard, P., Gates, P. B., and Brockes, J. P., 1988, Position dependent expression of a homeobox gene transcript in relation to amphibian limb regeneration, *EMBO J.* **7:**4275–4282.

Scadding, S., and Maden, M., 1994, Retinoic acid gradients during limb regeneration, *Dev. Biol.* **162:**608–617.

Schilthuis, J. G., Gann, A. A. F., and Brockes, J. P., 1993, Chimeric retinoic acid/thyroid hormone receptors implicate RAR a1 as mediating growth inhibition by retinoic acid, *EMBO J.* **12:**3459–3466.

Scott, W. J., Walter, R., Tzimas, G., Sass, J. O., Nau, H., and Collins, M. D., 1994, Endogenous status of retinoids and their cytosolic binding proteins and limb buds of chick vs mouse embryos, *Dev. Biol.* **165:**397–409.

Sharpe, C. R., 1991, Retinoic acid can mimic endogenous signals involved in transformation of the *Xenopus* nervous system, *Neuron* **7:**239–247.

Sharpe, C. R., 1992, Two isoforms of retinoic acid receptor a expressed during *Xenopus* development respond to retinoic acid, *Mech. Dev.* **39:**81–93.

Sharpe, C. R., and Goldstone, K., 1997, Retinoid receptors promote primary neurogenesis in *Xenopus*, *Development* **124:**515–523.

Simeone, A., Avantaggiato, V., Moroni, M. C., Mavilio, F., Arra, C., Cotelli, F., Nigro, V., and Acampora, D., 1995, Retinoic acid induces stage-specific anteroposterior transformation of rostral central nervous system, *Mech. Dev.* **51:**83–98.

Simon, H.-G., and Tabin, C. J., 1993, Analysis of *Hox*-4.5 and *Hox*-3.6 expression during newt limb regeneration: Differential regulation of paralogous *Hox* genes suggest different roles for members of different Hox clusters, *Development* **117:**1397–1407.

Sive, H. L., and Cheng, P. F., 1991, Retinoic acid perturbs the expression of *Xhox.lab* genes and alters mesodermal determination in *Xenopus laevis*, *Genes Dev.* **5:**1321–1332.

Sive, H. L., Draper, B. W., Harland, R. M., and Weintrub, H., 1990, Identification of a retinoic acid-sensitive period during primary axis formation in *Xenopus laevis*, *Genes Dev.* **4:**932–942.

Smith, S. M., and Eichele, G., 1991, Temporal and regional differences in the expression pattern of distinct retinoic acid receptor-b transcripts in the chick embryo, *Development* **111:**245–252.

Smith, S. M., Kirstein, I. J., Wang, Z.-S., Fallon, J. F., Kelley, J., and Bradshaw-Rouse, J., 1995, Differential expression of retinoic acid receptor-b isoforms during chick limb ontogeny, *Dev. Dynam.* **202:**54–66.

Stratford, T., Horton, C., and Maden, M., 1996, Retinoic acid is required for the initiation of outgrowth in the chick limb bud, *Curr. Biol.* **6:**1124–1133.

Stratford, T. H., Kostakopoulou, K., and Maden, M., 1997, *Hoxb*-8 has a role in establishing early anterior–posterior polarity in chick forelimb but not hindlimb, *Development* **124:**4225–4234.

Summerbell, D., 1983, The effects of local application of retinoic acid to the anterior margin of the developing chick limb, *J. Embryol. Exp. Morphal.* **78:**269–289.

Summerbell, D., and Honig, L. S., 1982, The control of pattern across the anteroposterior axis of the chick limb bud by a unique signalling region, *Am. Zool.* **22:**105–116.

Sundin, O., and Eichele, G., 1992, An early marker of axial pattern in the chick embryo and its respecification by retinoic acid, *Development* **114:**841–852.

Tabin, C. J., 1989, Isolation of potential vertebrate limb-identity genes, *Development* **105:**813–820.

Taira, M., Jamrich, M., Good, P. J., and Dawid, I. B., 1992, The LIM domain-containing homeo box gene *Xlim*-1 is expressed specifically in the organiser region of *Xenopus* gastrula embryos, *Genes Dev.* **6:**356–366.

Taira, M., Otani, H., Jamrich, M., and Dawid, I. B., 1994, Expression of the LIM class homeobox gene *Xlim*-1 in pronephros and CNS cell lineages of *Xenopus* embryos is affected by retinoic acid and exogastrulation, *Development* **120:**1525–1536.

Thaller, C., and Eichele, G., 1987, Identification and spatial distribution of retinoids in the developing chick limb bud, *Nature* **327:**625–628.

Thaller, C., and Eichele, G., 1990, Isolation of 3,4-didehydroretinoic acid, a novel morphogenetic signal in the chick wing bud, *Nature* **345:**815–819.

Tickle, C., Summerbell, D., and Wolpert, L., 1975, Positional signalling and specification of digits in chick limb morphogenesis, *Nature* **254:**199–202.

Tickle, C., Alberts, B., Wolpert, L., and Lee, J., 1982, Local application of retinoic acid to the limb bond mimics the action of the polarizing region, *Nature* **296:**564–566.

Tickle, C., Lee, J., and Eichele, G., 1985, A quantitative analysis of the effect of all-*trans*-retinoic acid on the pattern of chick wing development, *Dev. Biol.* **109:**82–95.

Tsonis, P. A., Washabaugh, C. H., and Rio-Tsonis, K. D., 1994, Morphogenetic effects of 9-*cis*-retinoic acid on the regenerating limbs of the axolotl, *Roux's Arch. Dev. Biol.* **203:**230–234.

Vallari, R. C., and Pietruszko, R., 1982, Human aldehyde dehydrogenase: Mechanism of inhibition by disulphiram, *Science* **216:**637–639.

Viviano, C. M., Horton, C., Maden, M., and Brockes, J. P., 1995, Synthesis and release of 9-*cis* retinoic acid by the urodele wound epidermis, *Development* **121:**3753–3762.

Wagner, M., Thaller, C., Jessell, T. M., and Eichele, G., 1990, Polarizing activity and retinoid synthesis in the floor plate of the neural tube, *Nature* **345:**819–822.

Wood, H., Pall, G., and Morriss-Kay, G., 1994, Exposure to retinoic acid before or after the onset of somitogenesis reveals separate effects on rhombomeric segmentation and 3′ *HoxB* gene expression domains, *Development* **120:**2279–2285.

Zhang, Z., Balmer, J. E., Lovlie, A., Fromm, S. H., and Blomhoff, R., 1996, Specific teratogenic effects of different retinoic acid isomers and analogs in the developing anterior central nervous system of zebrafish, *Dev. Dynam.* **206:**73–86.

Chapter 5

Antioxidant Activity of Vitamin A within Lipid Environments

Maria A. Livrea and Luisa Tesoriere

1. INTRODUCTION

New information about deleterious effects of free radicals on cell compartments, and a growing amount of associations between free radicals and various pathological conditions, have produced considerable interest in the biological defense systems against oxidative injury. The antioxidant properties of vitamin A, known for decades (Monaghan and Schmitt, 1932), have been reinvestigated in recent years in chemical as well as in biological systems. Its lipid nature and the localization within the lipophilic compartment of membranes and lipoproteins make vitamin A effective in reducing lipid peroxidation by acting as a chain-breaking antioxidant.

The term "vitamin A" describes a number of molecules exhibiting the biological activity of retinol, the progenitor of naturally occurring retinoids. A long, conjugated double-bond system that binds a trimethyl-cyclohexenyl ring to each end characterizes the structure of the provitamin β-carotene (Fig. 1). Cleavage and reduction of β-carotene in the enterocytes yields all-*trans* retinol. This is metabolized intracellularly through oxidation reactions

---

Maria A. Livrea and Luisa Tesoriere Istituto di Farmacologia e Farmacognosia, Università di Palermo, 90134 Palermo, Italy.
Subcellular Biochemistry, Volume 30: Fat-Soluble Vitamins, edited by Quinn and Kagan. Plenum Press, New York, 1998.

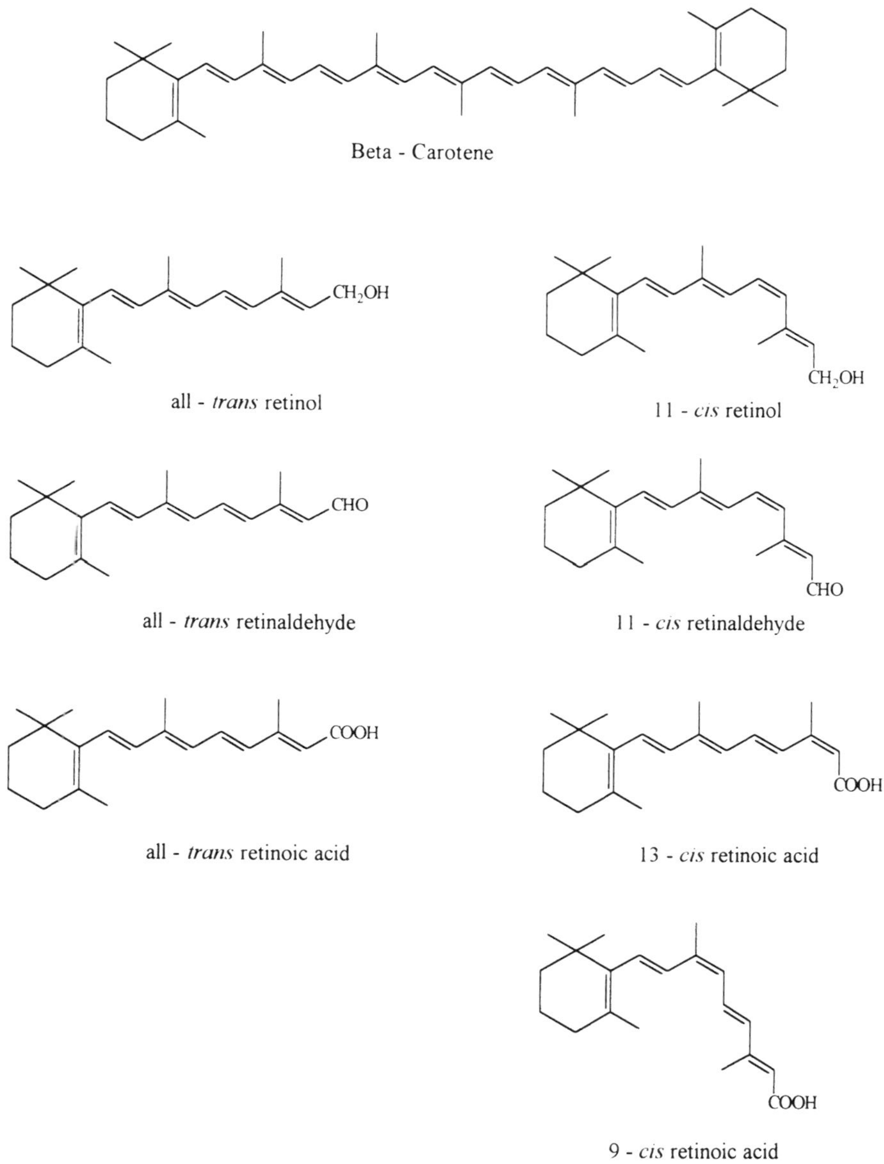

FIGURE 1. Structures of β-carotene and natural retinoids.

to form all-*trans* retinaldehyde and all-*trans* retinoic acid (Fig. 1). Isomer forms, including 11-*cis* retinol, 11-*cis* retinaldehyde, and 9-*cis* and 13-*cis* retinoic acid, have been found as products of normal metabolism of retinol in animal cells. Fatty acid esters, primarily retinyl palmitate, represent the major storage compounds. The retinol binding-protein and other specific proteins carry retinoids through plasma and the aqueous intracellular environments, deliver retinol to enzyme sites, and possibly control the level of free compounds in membranes.

Vitamin A (all-*trans* retinol) and its natural *cis* and *trans* derivatives are involved in important physiological functions, from vision and growth to immunity, embryologic development, and cell differentiation (Blomhoff, 1994). In addition, a number of apparently unrelated actions at the membrane level, from lysosomes to Golgi apparatus and plasma membrane, have been documented (Morré, 1992).

This chapter will review findings concerning the role of vitamin A and natural retinoids as antioxidants in radical-initiated processes, with particular emphasis on the scavenging activity toward lipid-derived peroxyl radicals, and attempt to provide a view into the mechanisms whereby retinoids participate in the antioxidant protection.

2. OXIDATIVE STRESS AND DEFENSE

2.1. Introduction

A number of oxygen-derived free radicals are produced through many intracellular reactions, or following exposure to various xenobiotics including drugs and pollutants, as well as in the presence of redox-active transition metal ions, mainly iron and copper, of endogenous or exogenous origin (for a review, see Halliwell and Gutteridge, 1989). These reactive species, which include the superoxide anion, O_2^{-}, hydroxyl radical, $^{\cdot}OH$, nitric oxide, $NO^{\cdot}$, hydrogen peroxide, H_2O_2, and hypochlorous acid, HOCl, can cause structural alterations to DNA, peroxidize lipids, modify the enzyme structure, depolymerize polysaccharides, and finally cause cell death (Halliwell, 1996).

A variety of protection mechanisms have evolved to protect cells against oxidative damage. These can be classified as *prevention*, *interception*, and *repair*. Enzymes such as catalase, glutathione peroxidases, and superoxide dismutase and metal chelators such as transferrin, lactoferrin, and ceruloplasmin act as preventive antioxidants. Radical scavenging antioxidants, molecules distributed into aqueous and lipid compartments, intercept and scavenge radicals to inhibit initiation and break chain propagation processes. Vitamin C, uric acid, bilirubin and albumin, vitamin E, ubiquinol, carotenoids, and vitamin A are examples of the latter. Finally, a number of enzymes are required to repair damaged DNA and to reconstitute membranes (Sies, 1993).

2.2. Oxidation of Unsaturated Lipids and Monitoring of Antioxidant Activity

The oxidation of polyunsaturated fatty acids proceeds through a free radical-mediated chain mechanism in which a hydrogen atom abstraction from lipid, LH, by peroxyl radicals, $ROO^{\cdot}$, generates a carbon-centered radical, $L^{\cdot}$, which subsequently reacts with oxygen to form a lipoperoxyl radical, $LOO^{\cdot}$ [equations (1) and (2)]. The latter acts as a chain carrier by attacking other lipids, thereby propagating lipid oxidation, and generating a nonradical lipid hydroperoxide, LOOH [equation (3)]. The chain terminates by conversion of radicals into spin-paired products [equation (4) shows one among the possible reactions]:

$$LH + R\text{—}OO^{\cdot} \rightarrow L^{\cdot} + R\text{—}OOH \tag{1}$$

chain initiation

$$L^{\cdot} + O_2 \rightarrow LOO^{\cdot} \tag{2}$$

$$LOO^{\cdot} + LH \rightarrow L^{\cdot} + LOOH \tag{3}$$

chain propagation

$$LOO^{\cdot} + LOO^{\cdot} \rightarrow \text{inactive products} \tag{4}$$

chain termination

A conventional chain-breaking antioxidant interferes with the chain propagation by donating a hydrogen atom to the chain-carrying peroxyl radical, thereby preventing the attack on other lipids. The chain propagation is inhibited as long as a sufficient amount of antioxidant is present. The faster the rate of scavenging of lipoperoxyl radicals compared to the rate of the attack of lipoperoxyl radicals on lipid, the more efficient the antioxidant.

Due to the wide variety of oxidation products, many analytical methods have been developed for detecting lipid peroxidation (Packer and Glazer, 1990; Packer, 1984). Some of the methods used to analyze the effectiveness of vitamin A and other retinoids as inhibitors of lipid peroxidation are briefly described. Measurement of conjugated diene lipid hydroperoxides is widely used to evaluate the extent of lipid peroxidation when media of well-defined composition, such as lipid solutions or liposomes, are utilized and hydroperoxides accumulate quantitatively through the peroxidation process. The thermal decomposition of an azo compound (A—N=N—A) is used in conjuction, to achieve a known and constant flux of chain-initiating peroxyl radicals ($AOO^{\cdot}$) (Niki, 1990). The production of lipid hydroperoxides, measured in the presence of antioxidant, allows calculation of kinetic parameters of the antioxidant efficiency such as the rate constant for scavenging of lipoperoxyl radicals k_{inh} and the stoichiometric factor n, i.e., the number of moles of peroxyl radicals scavenged per mole of antioxidant. The rate of production of lipid hydroperoxides in the presence of antioxidant R_{inh} and the inhibition time or lag phase t_{inh}, i.e., the time at which the propagation is inhibited before the onset of lipid peroxidation, can be calculated as well.

When more-complex systems are assayed, such as natural membranes or tissue homogenates, lipid products such as malondialdehydes and other compounds reactive with thiobarbituric acid, collectively referred to as TBARS (thiobarbituric acid-reactive substances), are usually measured (Buege and Aust, 1978), and the antioxidant efficiency may be expressed by the amount of antioxidant necessary to obtain half-maximum inhibition (IC_{50}) under a given condition. Sometimes cyclooxygenase activity has been used to assay trace amounts of peroxides in biological material (Halliwell and Chirico, 1993).

Lipid oxidation in low-density lipoproteins (LDL) has been frequently measured by continuous monitoring of the conjugated dienes (CD) production in the presence of Cu^{2+} as prooxidant (Esterbauer *et al.*, 1989b). The length of the lag phase preceding the fast increase of lipid peroxidation may be used to evaluate the effectiveness of antioxidants on the oxidation resistance of LDL. Additional indications can be obtained, e.g., from the rate of hydroperoxide production and the consumption rate of antioxidants during LDL oxidation.

3. VITAMIN A AND RETINOIDS AS ANTIOXIDANTS

Vitamin A and biological retinoids are capable of inhibiting free radical reactions, thereby acting as potential antioxidants. Retinoids do not scavenge reactive oxygen species. Although their polyenic structure displays strong similarity with that of carotenoids, and allows these molecules to act as free radical scavengers via similar mechanisms (Kennedy and Liebler, 1992; Samokyszyn and Marnett, 1990), in contrast to carotenoids they seem unable to act as preventive antioxidants. It was reported that retinol does not react with hydroxyl radicals produced during the iron/ascorbate-stimulated lipid peroxidation in heart and retinal membranes

(Livrea *et al.*, 1992). Witz *et al.* (1980), investigating the retinoid inhibition of the superoxide radical production by polymorphonuclear leukocytes, concluded that retinol does not scavenge superoxide anions, a finding in accordance with later results from other investigators (Sharma *et al.*, 1990). Retinoic acid was found unable to quench singlet oxygen (Di Mascio *et al.*, 1989), a failure that was related to the number of double bonds in the molecule, since seven or more double bonds are needed for the polyenic system to approach the energy level suitable to accept the energy transfer from the excited state of oxygen. Sahu and Das (1994) reported that high doses of vitamin A, given simultaneously with the antileprosy agent clofazimine, minimized its clastogenic effects in bone marrow cells of mice. A scavenging effect of vitamin A, removing free radicals produced by clofazimine, is assumed to be responsible for the modulation of the clastogenic effects of the drug. Although clofazimine may generate $^{\cdot}OH$, which can well be responsible for the damage to chromosomes, it is difficult to conclude that the effect of vitamin A really occurs through scavenging this radical.

The chemical reactivity of natural retinoids with various radical species, including thiyl, 1,1-diphenyl-2-picrylhydrazyl (DPPH), alkylperoxyl from the hydrophilic 2,2′-azobis(2-amidinopropane) hydrochloride (AAPH), and the lipophilic 2,2′-azobis(2,4-dimethylvaleronitrile) (AMVN), has been evaluated in either water or organic solutions and lipid bilayers by various detection methods. These studies, recently reviewed (Livrea *et al.*, 1996), provide evidence that the antioxidant potency of all-*trans* retinol, retinyl palmitate, and retinoic acid, in terms of stoichiometry and rate constants, varies considerably according to the assay system and kind of radicals generated. Table I reports on kinetic parameters measured with a number of natural retinoids under various conditions.

It may be worthwhile to mention that synthetic retinoids were shown to be not very effective as antioxidants. Acitretin, Etretinate, Temarotene, and Ro 15-1570 (arotinoid ethyl sulfone) were found unable to quench the electron spin resonance (ESR) signal of DPPH radical in ethanol (Hiramatsu and Packer, 1990) and to react with lipoperoxyl radicals from linoleic acid methyl ester in methanol (Livrea and Tesoriere, 1994). The furyl analog of retinoic acid inhibited the ESR signal of DPPH radical by 23%, while other retinoids were reported uneffective in the same system (Hiramatsu and Packer, 1990). Molecular structures of the synthetic retinoids assayed in these studies are shown in Fig. 2.

Lipid bilayers may strongly affect the reactivity of lipid-soluble antioxidants (Takahashi *et al.*, 1989; Cadenas *et al.*, 1984). For example, a stoichiometric factor lower than that measured in solution (Burton and Ingold, 1981) was found for α-tocopherol in liposomal systems, both when lipid-

Table I
Reactivity of Natural Retinoids with Radicals Generated in Chemical Systems[a]

Compound	Radical generated	Assay system	Kinetic parameters
all-*trans* Retinol	Thiyl	Spectrophotometry/ methanol[b]	$k = 1.4 \times 10^9 M^{-1} sec^{-1}$
	AAPH-peroxyl	Phycoerythrine/ aqueous[c,d,e]	$n = 0$
	AMVN-peroxyl	Parinaric acid/ hexane[e]	$n = 0$
	AMVN-peroxyl	DOPC liposomes/ parinaric acid[c]	$n = 0.3–1.0$
	AMVN-peroxyl	DOPC liposomes/ luminol[c]	$n = 0.05$
all-*trans* Retinyl palmitate	AAPH-peroxyl	Phycoerythrine/ aqueous[c,d,e]	$n = 0$
	AMVN-peroxyl	Parinaric acid/ hexane[e]	$n = 0$
	AMVN-peroxyl	DOPC liposomes/ parinaric acid[c]	$n = 0.3–1.0$
	AMVN-peroxyl	DOPC liposomes/ luminol[c]	$n = 0.05$
all-*trans* Retinoic acid	AAPH-peroxyl	Phycoerythrine/ aqueous[c,d,e]	$n = 0$
	AMVN-peroxyl	Parinaric acid/ hexane[e]	$n = 0$
	AMVN-peroxyl	DOPC liposomes/ parinaric acid[c]	$n = 0.3–1.0$
	AMVN-peroxyl	DOPC liposomes/ luminol[c]	$n = 0.05$

[a]AAPH, 2,2′-azo-bis(2-amidinopropane) hydrochloride; AMVN, 2,2′-azo-bis(2,4-dimethylvaleronitrile); DOPC, dioleoylphosphatidylcholine.
[b]D'Aquino *et al.* (1989).
[c]Freisleben and Packer (1993).
[d]Hiramatsu and Packer (1990).
[e]Tsuchiya *et al.* (1992).

soluble and water-soluble azo initiators were used (Freisleben and Packer, 1993; Barklay *et al.*, 1990; Niki *et al.*, 1986). On the other hand, studies in which the reactivity of retinol, retinyl palmitate, and retinoic acid with AMVN-derived peroxyl radicals was evaluated in solution and in lipid bilayers (Table I) are indicative that physical factors, such as interactions with lipids, as well as location and mobility in the bilayer, can substantially enhance the radical-scavenging activity of retinoids. This is interesting since, under physiological conditions, retinoids are supposed to play their

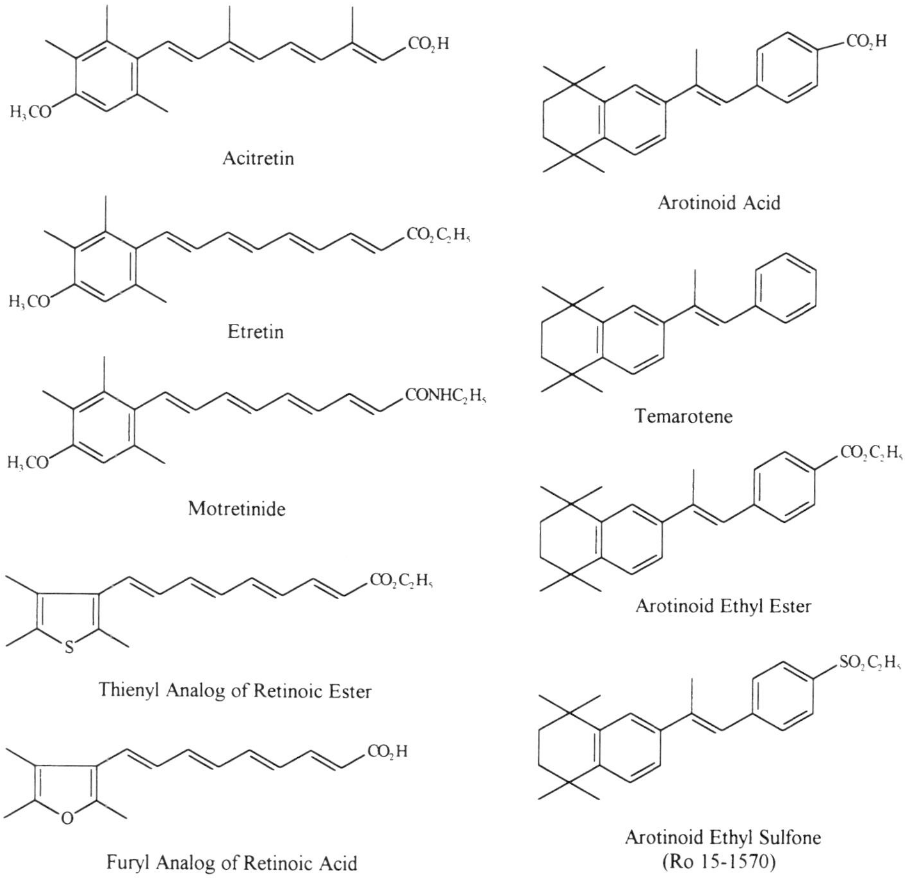

FIGURE 2. Structures of synthetic retinoids.

antioxidant role at the membrane level, by interacting with lipid-derived peroxyl radicals. Given the relevance of lipoperoxidation to biological processes, retinoids may be of great importance in the cell antioxidant defense.

3.1. Lipoperoxyl Radical-Scavenging Activity of Vitamin A

As far back as 1932, Monaghan and Schmitt demonstrated that small concentrations of vitamin A completely inhibited the oxygen uptake of linoleic acid for some hours. Later studies showed that vitamin A can reduce the extent of lipid peroxidation induced *in vitro* by oxyradical-generating systems such as NADPH oxidase (Nicotra *et al.*, 1975), ascorbate (Das, 1989; Hunter *et al.*, 1964; Green *et al.*, 1961), and transition metals (Hunter *et al.*, 1963) in tissue homogenates or isolated membranes. In addition, studies with animal models provided evidence that administration of vitamin A increased the resistance of various tissue preparations to lipid peroxidation induced *in vitro* (Kartha and Krishnamurthy, 1977). However, only in the last decade have appropriately designed experiments brought about major advances in our understanding of the chemical mechanisms and biological interactions by which vitamin A may intervene in protecting membranes and other natural lipid aggregates.

As in the case of other conjugated polyenes (Stocker *et al.*, 1987; Burton and Ingold, 1984), the inhibition of peroxidation is attributed to the ability of such compounds to act as a trap for chain-carrying peroxyl radicals, rather than acting by hydrogen donation, which characterizes conventional chain-breaking antioxidants such as tocopherol. In a lipid environment, the peroxidation of which has been induced by oxyradical-generating systems, the radical-trapping activity of vitamin A competes with LH for the chain-propagating peroxyl radical and hence inhibits the formation of hydroperoxides (Section 2.1). The highly unsaturated structure of conjugated double bonds (Fig. 1) is essential for the antioxidant activity. Since unpaired electrons can be delocalized through the polyenic chain, the extended conjugated double-bond system of the retinoid allows the formation of quite stable radicals, which may break radical chain reactions.

Peculiar features characterize the dynamics and mechanism of the reaction of polyenes with peroxyl radicals. The mechanism underlying the reaction of retinoids with lipoperoxyl radicals is outlined in Fig. 3. On the basis of the oxidation products which are formed during its action as an antioxidant, Samokyszyn and Marnett (1990, 1987) showed that 13-*cis* retinoic acid preferentially reacts with peroxyl radicals, including lipoperoxyl radicals, by addition to the double bond adjacent to the ring, thereby generating a retinoid-derived carbon-centered radical. This species, stabilized by resonance, is poorly prone to attack either another retinoic acid molecule or other oxidizable substrates. The retinoid-derived carbon-centered radical may undergo various pathways which bring about either an antioxidant effect or consume the carbon-centered radical in nonantioxidant reactions. It may react with O_2 to form a new retinoid-derived peroxyl radical, which could oxidize another retinoid molecule and/or other oxidizable substrate, leading to autocatalytic oxidation and/or eventually to prooxidant activity. The retinoid-derived carbon-centered radical can undergo unimolecular decomposition, yielding an alkoxyl radical from the added peroxyl radical and a 5,6-retinoid epoxide. This product,

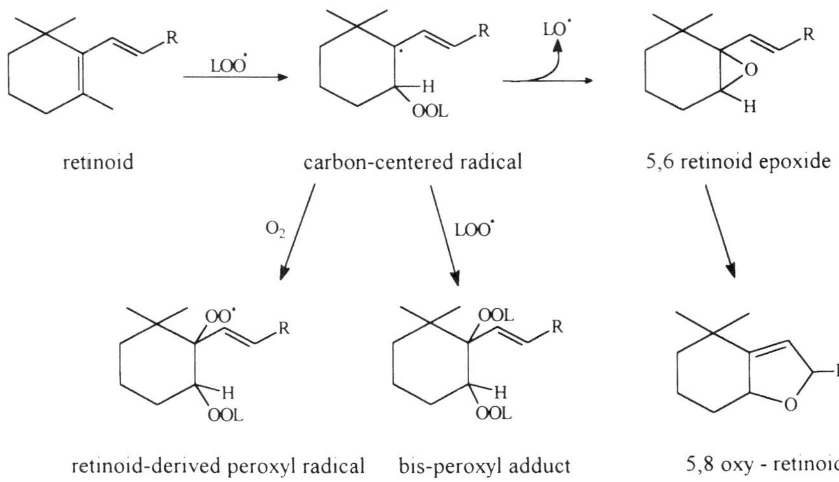

FIGURE 3. Reaction pathways of retinoids during lipid peroxidation (for explanation see text). LOO˙, lipoperoxyl radical; LO˙, alkoxyl radical; R = $C_9H_{15}O$, retinol; $C_9H_{13}O_2$, retinoic acid; $C_9H_{13}O$, retinaldehyde.

quantified after rearrangement to the more stable 5,8-oxy derivative, is one of the metabolites recovered during the ascorbate-dependent, iron-catalyzed microsomal lipid peroxidation in the presence of 13-*cis* retinoic acid (Samokyszyn and Marnett, 1990). This reaction, however, cannot entirely explain the antioxidant behavior of the retinoid, since the radical-dependent epoxidation, while consuming one peroxyl radical, results in the formation of a very reactive alkoxyl radical (Pryor, 1986). A net antioxidant behavior may arise as a consequence of cross-termination reactions where the polyene carbon-centered radical would scavenge a second peroxyl radical. Other peroxyl radicals may add to double bonds of the polyene to generate yet unidentified adducts. The complexity of the reactions consuming the retinoid-derived carbon-centered radical is supported by the observation that, besides the 5,8-oxy metabolite, unstable polar products have been detected (Samokyszyn and Marnett, 1990).

An unavoidable consequence of such a mechanism is that a number of factors may intervene to modify the antioxidant effect. This raises the question of the extent to which retinoids may be considered effective antioxidants in biological contexts under physiological conditions. According to recent findings, the antioxidant activity of all-*trans* retinol varies with varying the oxygen partial pressure, radical flux, and retinol concentration (Tesoriere *et al.*, 1997).

Effect of Oxygen Partial Pressure, Radical Flux, and All-*trans* Retinol Concentration on the Lipoperoxyl Radical-Scavenging Activity

The crucial event which drives a retinoid to an antioxidant effect involves the partition of the retinoid-derived carbon-centered radical among one radical trapping and two oxidation reactions (Fig. 3). The predominance of one or the other of these competing pathways may result in efficient radical scavenging or even prevent any antioxidant effect. Oxygen tension, by controlling the formation of retinoid-derived peroxyl radicals, modulates the autocatalytic oxidation or may even induce prooxidant reactions, thereby contributing to enhancing or minimizing the antioxidant behavior. In addition, retinoid concentrations and radical fluxes can be critical in determining the fate of the retinoid-derived carbon-centered radical.

As reported (Tesoriere *et al.*, 1994a), all-*trans* retinol clearly inhibited methyl linoleate oxidation in methanol under air until its concentration exceeded 16 μM. Then the antioxidant effect declined, as evident from the lack of correlation between duration of the inhibition time and amount of retinoid. Clear inhibition periods were no longer evident at retinol concentrations higher than 25 μM. Under reduced oxygen tension (pO_2 = 15 torr), which somewhat prevented the autocatalytic oxidation of retinol, the antioxidant effect was reduced when retinol concentration exceeded 200 μM. However, the efficacy of retinol in terms of stoichiometry with lipoperoxyl radicals was unchanged at low pO_2, suggesting that epoxidation reactions that do not result in a net radical consumption were favored over cross-termination reactions. Later results from experiments performed with soybean phosphatidylcholine liposomes (Tesoriere *et al.*, 1997) further indicated that 5,6-retinol epoxide is a major product from the retinoid-derived carbon-centered radical under air as well as at 15 torr O_2. A higher fraction of the consumed retinol is transformed into 5,6-retinol epoxide at 15 torr O_2 than under air (Tesoriere *et al.*, 1997).

The rate of consumption of retinol increases with increasing concentrations of itself and oxygen, providing evidence that a concentration- and oxygen-dependent autocatalytic oxidation plays a major role in the retinol consumption during lipid peroxidation (Tesoriere *et al.*, 1997). A decreased autocatalytic oxidation makes retinol a better lipoperoxyl radical scavenger at low, rather than at high partial pressure of oxygen.

Finally, the antioxidant effectiveness of retinol increases with the increase of the radical flux. As investigated at low oxygen tension, an increase of the peroxyl radical concentration can shift the route of the retinol activity toward radical scavenging, and brings about an increasingly more efficient radical trapping (Tesoriere *et al.*, 1997).

The above results reported in pure chemical systems may provide indications about the behavior of retinoids under physiological conditions. Relevant questions about the activity in biological microenvironments concern oxygen concentration, retinoid concentration, kind and amounts of radicals, and interactions with the lipid environment and possibly with other antioxidants. At the membrane level, where retinoids are supposed to play their antioxidant role, pO_2 is very low ($<10\,mm\,Hg$) and retinoids are usually at nanomolar concentrations (Ciaccio *et al.*, 1993). Indeed, results from experiments *in vivo* and in microenvironments such as cells and membranes clearly indicate that biological contexts favor the antioxidant behavior. A strong protection against the high levels of doxorubicin-elicited peroxyl radicals in heart and other tissues has been reported in rats pretreated with very low doses of vitamin A (Tesoriere *et al.*, 1994b; Ciaccio *et al.*, 1993). Concentrations of all-*trans* retinol that were poorly effective in solution (Tesoriere *et al.*, 1994a), were able to inhibit effectively the ascorbate/iron-stimulated lipid peroxidation in membrane preparations incubated in air (Livrea and Packer, 1994). This evidence suggests that the membrane structure may somehow protect vitamin A from autocatalytic oxidation, thus enhancing its antioxidant activity and preventing prooxidant effects. Interactions with other antioxidants in membranes may also be important (see below). Things may be different if pharmacological amounts of vitamin A are given, since prooxidant activity could account for at least part of the toxicity of retinoid excess.

Although the increased retinol concentrations did result in a fast autocatalytic oxidation, which was detrimental for the antioxidant activity, all-*trans* retinol at $10–40\,\mu M$ did not act as prooxidant in liposomal systems (Tesoriere *et al.*, 1997). This would indicate that the retinoid-derived peroxyl radical is more reactive toward retinol than toward the lipid substrate. However, prooxidant activity by excess of vitamin A has been observed. Potentiation of the hepatotoxic effects of CCl_4 by pretreatment of rats with 350,000 IU of vitamin A excess was demonstrated to occur through an enhancement of lipid peroxidation (ElSisi *et al.*, 1993). Ciaccio *et al.* (1993, and unpublished work) observed that injecting rats with 400,000 IU of vitamin A resulted in a substantial increase of free retinol and retinyl esters, with a concomitantly increased level of lipid peroxides, in membranes from extrahepatic tissues. Under these circumstances, however, if a radical stress was produced by administrating an acute dose of doxorubicin, lipid peroxides were not generated in the membranes, indicating both complete protection by vitamin A from the doxorubicin damage and absence of vitamin A-stimulated lipid peroxidation. Conceivably, excess vitamin A was consumed, and removed, in reacting with free radicals, so avoiding their deleterious effects. This prevented prooxidant activity and toxicity.

3.2. *In Vitro* Studies

The reaction of natural retinoids with lipoperoxyl radicals has been documented in a number of *in vitro* studies including lipids in homogeneous solutions, liposomes, isolated membranes, cell homogenates, and intact cells. In most studies vitamin E, which is considered the most important chain-breaking antioxidant in lipid systems, was tested as a positive control.

3.2.1. Chemical Systems

Pure chemical systems have been used to evaluate kinetic parameters for the reaction of various retinoids with lipoperoxyl radicals. Clear evidence has been provided that retinoids can interact with lipoperoxyl radicals with a high degree of affinity, which could make them very effective antioxidants. The antioxidant effectiveness, in terms of stoichiometry of the reactions, may vary widely. In addition, the tested retinoids display a different antioxidant efficiency according to the assay systems, i.e., lipid solution or bilayers.

The stoichiometric factor *n* and the inhibition constant k_{inh} were determined for all-*trans* retinol, all-*trans* retinoic acid, and 13-*cis* retinoic acid in a methanol solution, using linoleic acid methyl ester as the peroxidable substrate and AMVN as the source of initiator radicals (Livrea and Tesoriere, 1994). Among the tested retinoids, 13-*cis* retinoic acid appeared the most effective with a stoichiometric factor of 4 and a k_{inh} of $28.8 \times 10^{5}\,\mathrm{M}^{-1}\,\mathrm{sec}^{-1}$, followed by all-*trans* retinoic acid ($n = 2.7$, $k_{inh} = 6.5 \times 10^{5}\,\mathrm{M}^{-1}\,\mathrm{sec}^{-1}$) and all-*trans* retinol ($n = 0.16$–0.36, $k_{inh} = 3.6 \times 10^{5}\,\mathrm{M}^{-1}\,\mathrm{sec}^{-1}$). Under similar assay conditions, the inhibition constant measured for α-tocopherol was $3.24 \times 10^{5}\,\mathrm{M}^{-1}\,\mathrm{sec}^{-1}$ (Tesoriere *et al.*, 1993b). Hence, natural retinoids exhibit a chemical reactivity as high as vitamin E in homogeneous solutions. Table II summarizes the reactivity of natural retinoids with lipoperoxyl radicals.

Studies carried out in water-dispersed soybean phosphatidylcholine liposomes provided evidence that the incorporation into a bilayer increases the all-*trans* retinol reactivity with lipid radicals with respect to lipid solutions. The value of the stoichiometric factor for the reaction with lipoperoxyl radicals is over ten times higher than that measured in methanol solution (Tesoriere *et al.*, 1993b). However, the antioxidant effectiveness of retinol was shown to vary considerably according to the source of initiating radicals, i.e., whether the lipid-soluble AMVN or the water-

Table II
Reactivity of Natural Retinoids with Lipoperoxyl Radicals Generated in Chemical System as Compared with Vitamin E

		Kinetic parameters	
Compound	Assay system	Vitamin A	Vitamin E
all-*trans* Retinol	Methyl linoleate/ methanol[a]	$n = 0.16–0.36$ $k_{inh} = 3.6 \times 10^5 M^{-1} sec^{-1}$	$n = 2$[b] $k_{inh} = 3.24 \times 10^5 M^{-1} sec^{-1}$,[c] $5.1 \times 10^5 M^{-1} sec^{-1}$[d]
	Soybean PC/ liposomes[c]	$n = 3.1$ $k_{inh}(A)/k_{inh}(E) = 1.5$[f]	—
	Dilinoleoyl PC/ liposomes[e]	—	— $k_{inh} = 5.8 \times 10^{-3} M^{-1} sec^{-1}$
all-*trans* Retinoic acid	Methyl linoleate/ methanol[a]	$n = 2.7$ $k_{inh} = 6.5 \times 10^5 M^{-1} sec^{-1}$	—
13-*cis* Retinoic acid	Methyl linoleate/ methanol[a]	$n = 4$ $k_{inh} = 28.8 \times 10^5 M^{-1} sec^{-1}$	—

[a]Livrea and Tesoriere (1994).
[b]Burton and Ingold (1981).
[c]Tesoriere *et al.* (1993b).
[d]Niki *et al.* (1984).
[e]Barklay *et al.* (1990).
[f]The relative antioxidant activity of all-*trans* retinol (A) is evaluated with respect to α-tocopherol (E) by the ratio $R_{inh}(A)/R_{inh}(E) = nk_{inh}(E)/nk_{inh}(A)$.

soluble AAPH was used. The highest lipoperoxyl radical-scavenging activity was measured when radicals were generated by AMVN inside the bilayer of multilamellar liposomes. Under these conditions all-*trans* retinol appeared to be more efficient than α-tocopherol. The stoichiometric factor was 3.1, and the ratio between the inhibition constants of retinol and α-tocopherol was calculated to be 1.5 (Tesoriere *et al.*, 1993b). When radicals were generated by AAPH in the aqueous phase of a unilamellar liposomal system, all-*trans* retinol was three times less effective than α-tocopherol (Tesoriere *et al.*, 1993b).

A number of studies questioned the effects of the site of radical generation on the antioxidant effectiveness of lipid-soluble or water-soluble antioxidants (Pryor *et al.*, 1993; Tsuchiya *et al.*, 1992; Yoshida and Niki, 1992). It has been shown that in the case of lipid-soluble inhibitors such as tocopherol, similar effects can be expected regardless of whether the radicals are produced in the lipid or water phase. This is because the radicals must diffuse to the lipid phase before lipid peroxidation can be initiated, and the inhibitors are not sensitive to whether radicals were formed in the lipid phase or diffused into the lipid phase from the aqueous phase. This behavior does not appear to apply to retinol.

The antioxidant effectiveness of all-*trans* retinol when incorporated in the liposomal bilayer may be explained by considering the location of the polyene chain deep inside the lipid moiety (Ortiz *et al.*, 1992; De Boek and Zidovetzki, 1988; Wassal *et al.*, 1988) and the autocatalytic nature of the lipid peroxidation. Because of its polyene chain, all-*trans* retinol mobility may be higher than that of α-tocopherol, which penetrates the lipid bilayer with its saturated phytyl tail (Shinitzky, 1984). This circumstance could favor interactions of all-*trans* retinol with lipid radicals and explain why it is more efficient than vitamin E when AMVN is the initiator radical. On the other hand, although all-*trans* retinol has been shown to react with AAPH-derived peroxyl radicals (Tesoriere *et al.*, 1997), the fast attack by hydrophilic radicals on lipids on the liposome surface produces large amounts of chain-branching initiation sites and of lipid hydroperoxides before lipoperoxyl radicals can be effectively scavenged by all-*trans* retinol. In other terms, under these circumstances the antioxidant capacity of all-*trans* retinol is exceeded and chain propagation can proceed almost unhindered.

3.2.2. Natural Membranes

All-*trans* retinol, retinaldehyde, retinoic acid, retinyl palmitate, and retinyl acetate are able to contrast production of TBARS induced by NADPH, ascorbate, or iron in either membrane preparations or homogenates from various rat tissues, including liver, heart, and nervous tissues (Livrea and Packer, 1994; Livrea *et al.*, 1992; Scheschonka *et al.*, 1990; Das, 1989; Kartha and Krishnamurthy, 1978; Nicotra *et al.*, 1975; Pokrovsky *et al.*, 1974; Hunter *et al.*, 1964; Green *et al.*, 1961). Structure and membrane lipid composition, while varying physicochemical interactions of retinoids with the bilayer (Noy, 1993), appear to affect the antioxidant behavior. Interestingly, vitamin A and its analogs were found to be very active in nervous tissues (Livrea and Packer, 1964; Livrea *et al.*, 1992; Das, 1989), which, due to their high content in polyunsaturated lipids, are easily susceptible to peroxidation and may produce high amounts of lipid hydroperoxides. The IC_{50} measured for all-*trans* retinol was in the range of 0.2–5.0 μM, according to the preparation assayed, whereas IC_{50} of α-tocopherol was 100–120 μM (Livrea and Packer, 1994; Livrea *et al.*, 1992). Table III summarizes the data concerning the quantitative evaluations of the antioxidant effectiveness of various natural retinoids as assayed under several conditions.

Table III
Lipoperoxyl Radical-Scavenging Activity of Retinoids in Natural Membranes as Compared with Vitamin E[a]

Compound	Membrane	IC_{50} (μM)	
		Vitamin A	Vitamin E
all-*trans* Retinol	Brain total membrane[b]	0.8	100
	Retina total membrane[b]	0.2	100
	Retina microsomes[c]	5	120
	Liver microsomes[d]	4.8	—
	Heart total membrane[b]	2	>200
	RPE microsomes[c]	100	—
11-*cis* Retinol	Retina microsomes[c]	2	—
	RPE microsomes[c]	65	—
13-*cis* Retinol	Liver microsomes[d]	4.9	—
all-*trans* Retinaldehyde	Retina microsomes[c]	30	—
11-*cis* Retinaldehyde	Retina microsomes[c]	4	—
all-*trans* Retinyl palmitate	Retina microsomes[c]	35	—
all-*trans* Retinoic acid	Liver microsomes[d]	6.1	—
13-*cis* Retinoic acid	Liver microsomes[e]	10	—

[a]The assays were carried out with tissue amounts as high as 100 μg of protein for retina, RPE (retinal pigment epithelium), and brain membrane preparations, 200 μg of protein for heart preparations, and 500 μg of protein for liver microsomes. Lipid peroxidation was induced by 4 mM ADP/15 μM Fe^{+3}/0.8 mM ascorbate (Livrea and Packer, 1994; Tesoriere *et al.*, 1993a; Samokyszyn and Marnett, 1990) or 1 mM ADP/10 μM Fe^{3+} and a NADPH regenerating system (Scheschonka *et al.*, 1990). Lipid peroxides were measured as thiobarbituric acid-reactive substances (TBARS). IC_{50}, half-maximum inhibition concentration.
[b]Livrea and Packer (1994).
[c]Tesoriere *et al.* (1993a).
[d]Scheschonka *et al.* (1990).
[e]Samokyszyn and Marnett (1990).

Interesting results were obtained when the antioxidant activity of 11-*cis* isomers was evaluated in comparison with that of all-*trans* retinoids in retina and retinal pigment epithelium subjected to an iron/ascorbate lipid peroxidation system (Tesoriere *et al.*, 1993a). Higher reactivity was observed for *cis* compounds. Iron-dependent lipid peroxidation in liver microsomes was also effectively inhibited by 13-*cis* retinoic acid (Samokyszyn and Marnett, 1990) and by 13-*cis* retinol (Scheschonka *et al.*, 1990). Differences in the thermodynamic properties between *cis* and *trans* double bonds could explain the higher reactivity of *cis* isomers. Compared to *trans* bonds, *cis* bonds possess a higher potential energy and therefore are less stable and more susceptible to various reactions (Graham Solomons, 1984). A major displacement of the two parts of the molecule toward one another is brought about by the introduction of a *cis* double bond into a conjugated double-bond system (Olson, 1993). This conformational change leads to a higher steric interference between the two parts of the molecule and to higher reactivity of the antioxidant.

Inhibition of the enzyme-catalyzed oxidation of arachidonic acid by retinol has been described by Halevy and Sklan (1987). Using microsomal preparations from seminal vesicles and kidney, they reported that 1 mM retinol was able to inhibit the lipoxygenase- and cyclooxygenase-catalyzed oxidation of arachidonic acid when added into the system in 10% Tween 80 in ethanol and buffer. Under the same conditions, α-tocopherol was six times less effective. It is not possible to differentiate from this study between scavenging activity of retinol toward the reactive peroxides required for cyclooxygenase and lipoxygenase reactions, and other modes of inhibition in the metabolic pathways of eicosanoids and prostaglandins.

Those studies in which α-tocopherol was concurrently tested as a positive control showed that the activity of vitamin A was comparable to (Hunter *et al.*, 1964; 1963; Green *et al.*, 1961) or even more potent than that of α-tocopherol (Livrea and Packer, 1994; Tesoriere *et al.*, 1993a; Livrea *et al.*, 1992; Das, 1989; Kharta and Krishnamurthy, 1978). This may not be surprising since, though being an excellent lipid-soluble antioxidant in homogeneous solution (Burton and Ingold, 1989), α-tocopherol is less effective in scavenging lipoperoxyl radicals in membranes (Niki and Komuro, 1989; Cadenas *et al.*, 1984). The location of the chromanol —OH group of α-tocopherol in lipid bilayers and its dynamics in inhibiting lipid peroxidation may account for the observed activities in biological membranes. As suggested, retinoids may establish more favorable interactions with lipid radicals in bilayers (Tesoriere *et al.*, 1993b).

Retinoids are usually bound to proteins in the body. Then it may be of interest that retinol bound to albumin or to retinol-binding protein (Tesoriere *et al.*, 1993a) also has antioxidant capacity. When assayed in retinal or brain membranes where peroxidation was stimulated by iron/ascorbate, the retinol concentration (as a 1:1 protein complex) useful to obtain the half-maximum inhibition was five to ten times higher with respect to the condition where free retinol was dissolved in the assay. It should be mentioned that concentrations of such an order of magnitude (1×10^{-6}M) for the complex retinol/retinol-binding protein are found in plasma and in liver.

Small amounts of free retinol and retinyl esters occur in membranes (Ciaccio *et al.*, 1993). In light of the ability of vitamin A to scavenge lipid peroxyl radicals, an effective defense of cell membranes may be expected. Interestingly, the membrane effects observed in vitamin A deficiency are similar to those of excess (Sudakaran and Kurup, 1974), although less severe (Dingle *et al.*, 1966). According to Sudhakaran and Kurup (1974), an optimum vitamin A status ensures stability of cell membranes, whereas either vitamin A deficiency or excess leads to their destabilization. Pertinent to radical scavenging-activity of vitamin A, a unifying concept may be that the antioxidant activity of optimum amounts of vitamin A account for its role in ensuring membrane stability, whereas prooxidant activity of the excess (see above) could account for its destabilizing effects.

3.2.3. Isolated Cells

A few investigations described the antioxidant effects of retinol and retinoic acid in isolated cells, PC 12 rat pheochromocytoma (Jackson *et al.*, 1991), fibroblasts (Geesin *et al.*, 1990), and B16 melanoma and C6 glioma (Gundimeda *et al.*, 1993). From the reported results, however, it is not possible to decide if or to what extent prevention of lipid peroxidation by retinoids is involved in the prevention of the oxidative modification observed in those cells, or whether other effects should be considered. Recently, retinol at a very low concentation was found to be effective in reducing lipoperoxide formation induced by adriamycin in K562 human erythroleukemia cells (Ciaccio *et al.*, 1994). Remarkably, the cytotoxic activity of the drug was entirely preserved or slightly enhanced, suggesting that new therapeutic approches with safe doses of vitamin A could be attempted in cancer chemotherapy with redox-cycling anthracyclines.

3.3. *In Vivo* Studies

Peroxyl radicals are highly toxic, due to their ability to diffuse in biological systems (Simic, 1988). By inhibiting production of organic radicals such as those involved in lipid peroxidation, retinoids can in turn prevent the attack on other cell components, including proteins, by these radicals (Dean *et al.*, 1991).

Studies with animal models, in which vitamin A was either injected (Piergiacomi *et al.*, 1996; Tesoriere *et al.*, 1994b; Ciaccio *et al.*, 1993), supplemented with diet (Sanjeev *et al.*, 1981; Kartha and Krishnamurthy, 1977), or administered in stomach by tube (Pokrovsky *et al.*, 1974), or in humans receiving single oral doses (Livrea *et al.*, 1995), provided evidence that retinoids can prevent peroxidative damage *in vivo*. In animals, the efficiency of vitamin A as an antioxidant has been evaluated in terms of protection against either membrane lipid peroxidation (Ciaccio *et al.*, 1993; Kartha and Krishnamurthy, 1977; Pokrovsky *et al.*, 1974) or oxidative modifications of proteins (Tesoriere *et al.*, 1994b), following radical stress induced *in vivo* (Ciaccio *et al.*, 1993; Sanjeev *et al.*, 1981) or produced *in vitro* (Piergiacomi *et al.*, 1996; Kartha and Krishnamurthy, 1977; Pokrovsky *et al.*,

1974). In humans, the resistance of low-density lipoproteins (Livrea *et al.*, 1995) to peroxidative stress produced *in vitro* has been evaluated.

In view of the toxicity by vitamin A excess (Hathcock *et al.*, 1990), when considering the antioxidant activity *in vivo*, the administration of high vitamin A doses should be considered cautiously. Feeding rats with 250,000 IU (Pokrovsky *et al.*, 1974) or with 100,000 IU of vitamin A for 5 days (Kartha and Krishnamurthy, 1977) resulted in a marked lowering of lipid peroxidation induced *in vitro* in tissue homogenates incubated in air. Prior administration of a single oral dose of 400,000 IU of retinyl palmitate markedly reduced the extent of diene conjugation found in liver of rats treated with carbon tetrachloride, a potent hepatotoxic agent which causes peroxidation of endoplasmic reticulum lipids (Sanjeev *et al.*, 1981). However, clear symptoms of hypervitaminosis were reported (Karta and Krishnamurthy, 1977). More recently rats have been injected with very low doses of vitamin A that did not cause any macroscopic (Ciaccio *et al.*, 1993) or microscopic (Tesoriere *et al.*, 1994b) sign of toxicity. This resulted in a greatly improved resistance of membrane lipids and proteins from brain and heart to peroxidation induced both *in vivo* by challenging rats with the redox-cycling drug doxorubicin and stimulated *in vitro*. This protection, which did not appear to be mediated by antioxidant enzyme activities such as superoxide dismutase and catalase, was related to the increase in membranes of extrahepatic tissues of very low amounts of all-*trans* retinol and all-*trans* retinyl esters after vitamin A administration (Ciaccio *et al.*, 1993). Noy and Xu (1990a,b) suggested that retinol bound to plasma retinol-binding protein will spontaneously equilibrate with unbound retinol in the circulation, and that unbound retinol may freely traverse plasma membranes, the flux of retinol from blood into cells being determined by the concentration of unbound retinol. Administration of vitamin A to vitamin A-sufficient animals may temporarily change retinol dynamics, with a modification of liver storage, plasma retinol, and uptake into and subcellular distribution in extrahepatic tissues, after saturation of the specific binding proteins (Blomhoff *et al.*, 1991). It seems of interest to report that vitamin E, given in excess before CCl_4 administration, failed to prevent liver microsome peroxidation. This has been explained in terms of the absence of vitamin E at the locus of activation of CCl_4, i.e., the endoplasmic reticulum (Ghoshal, 1976).

Low-Density Lipoproteins

One of the most important events in which lipoperoxidation can have crucial effects is the oxidation of low-density lipoproteins (LDL), in that this event is acknowledged as a major factor in the development of atherogenesis (Esterbauer *et al.*, 1993; Steinberg *et al.*, 1989; Steinbrecher *et al.*, 1984). Thus, a great number of studies addressed the question how important the antioxidants contained in the LDL particle are in preventing or limiting the oxidative damage. Surprisingly, in spite of the essential role of vitamin E in protecting against lipid peroxidation, assays aimed at correlating vitamin E content of LDL to their resistance to oxidative stress in normal subjects failed (Babiy *et al.*, 1990; Esterbauer *et al.*, 1989a). In addition, some reports show that under certain circumstances vitamin E is a rather ineffective inhibitor in that system and may behave as a prooxidant rather than a radical trap (Ziouzenkova *et al.*, 1996; Bowry and Stocker, 1993; Bowry *et al.*, 1992). On the other hand, epidemiologic studies that examined the relation between dietary antioxidant supplements and coronary heart disease showed a statistically significant protective effect of antioxidants. Some of these studies showed that, among other antioxidants, including vitamin C and E, vitamin A plasma levels were also inversely related to mortality from ischemic heart disease (Gey *et al.*, 1993, 1994; Gey and Puska, 1989). Other studies claimed decreased atherosclerosis as a result of supplement with vitamin E or vitamin A (Brattsand, 1975), and a decrease in lipid peroxide content was found in blood of patients with atherosclerosis treated with vitamin A (Kalmikova and Dmytrovsky, 1972).

Recently, Livrea *et al.* (1995) provided evidence of enhanced resistance of LDL against *ex vivo* copper-induced oxidation, due to an increase of vitamin A content in LDL. In this study subjects received a single oral dose of 20,000 IU. This produced a two- to threefold increase of retinol and retinyl esters in the LDL isolated 8 hr after the supplementation. In comparison to autologous control LDL, vitamin A-enriched LDL showed a decrease in the susceptibility to peroxidation, as expressed by the elongation of the lag phase that precedes the rapid oxidation of LDL and by the reduction of the rate of hydroperoxide production during the propagation phase. The calculated incremental increase in the lag phase determined by 1 mole retinol per mole LDL was about 1000 min. Therefore, on an equimolar basis, retinol appears to be a more potent antioxidant in LDL than α-tocopherol assayed under similar conditions (Esterbauer *et al.*, 1992).

It is difficult to reconcile the observed activity of vitamin A with the chemistry of the antioxidant mechanism of retinoids (Samokyszyn and Marnett, 1990). At the level at which retinol is found in LDL, its role solely as a direct chain-breaking antioxidant is puzzling. Simple molecular mechanisms do not seem to apply to LDL. The antioxidant response elicited by the increase of very low amounts of retinol entails that amplification mechanisms, and possibly interaction with other antioxidants, are concerned.

Synergistic interactions with α-tocopherol might occur (Tesoriere *et al.*, 1996).

4. INTERACTIONS WITH VITAMIN E

Although remarkable, the lipoperoxyl radical-scavenging activity measured for all-*trans* retinol in chemical systems cannot entirely account for the antioxidant activity evaluated in animal models and in humans when minute amounts of all-*trans* retinol are accumulated in membranes (Ciaccio *et al.*, 1993) and in lipoproteins (Livrea *et al.*, 1995). This suggests that other interactions might contribute to enhance the antiradical activity of vitamin A in biological contexts, a point that led to the study of the possibility that vitamin A could act in cooperation with other antioxidants. Vitamin E, the major lipid antioxidant in natural membranes, appeared as a candidate. Interactions between vitamin A and vitamin E in suppressing lipid peroxidation were investigated in retinal membranes (Tesoriere *et al.*, 1995) and quantitative studies were carried out by including variable proportions of all-*trans* retinol and α-tocopherol in liposomes.

Retinal membranes, which are endowed with high amounts of vitamin E (Stephens *et al.*, 1988; Farnsworth and Dratz, 1976), are a good model in that they offer the advantage of a sizable pool of α-tocopherol in its natural localization. Vitamin E was crucial in determining the antioxidant effectiveness of all-*trans* retinol. Incorporation into native retinal membranes of an amount of all-*trans* retinol comparable with that of endogenous α-tocopherol caused a significant increase of the resistance to peroxidative stress induced by the water-soluble AAPH. On the other hand, membranes that had been deprived of the endogenous vitamin E by ultraviolet light exposure did not exert a significant resistance to peroxidation even after incorporation of all-*trans* retinol. When the antioxidants were allowed to act in combination, their effects appeared to be cooperative rather than additive. Finally, all-*trans* retinol appeared to be consumed more rapidly when incorporated in membranes deprived of α-tocopherol than in native ones, and addition of all-*trans* retinol to the membranes slowed down the consumption of endogenous α-tocopherol. These data suggest that the antioxidant activities of all-*trans* retinol and α-tocopherol may be interrelated.

Kinetic studies have been carried out with phosphatidylcholine liposomes in which peroxidation was initiated by AAPH (Tesoriere *et al.*, 1996). In such a system, all-*trans* retinol alone is poorly effective, but the combination of all-*trans* retinol with α-tocopherol brings about an inhibition far beyond the sum of the inhibitions observed with individual antioxidants, providing evidence for synergistic interactions (Fig. 4). The extent of synergism is a function of the ratio all-*trans* retinol/α-tocopherol. A linear relation was observed within a 0.1–1.0 retinol/tocopherol ratio, whereas a drop was evident at a ratio of 2.0 (Tesoriere *et al.*, 1996).

Cooperative interactions are very common mechanisms in the action of biological antioxidants, in particular when α-tocopherol is involved. α-Tocopherol is known to interact synergistically with ascorbic acid (Bowry and Stocker, 1993; Costantinescu *et al.*, 1993; Ho and Chan, 1992; Chan *et al.*, 1991; Doba *et al.*, 1985) and coenzyme Q-10 (Bowry and Stocker, 1993; Ingold *et al.*, 1993) to protect against lipid peroxidation. This activity is due to "regeneration" of α-tocopherol from its tocopheroxyl radical. Synergistic interactions between vitamin E and peroxyl radical scavengers containing an extended system of conjugated double bonds such as bilirubin and biliverdin have been observed in liposomes (Stocker and Peterhans, 1989), and have been accounted for by the regeneration of vitamin E by hydrogen donation to the tocopheroxyl radical from bilirubin and biliverdin. On the other hand, data demonstrating some synergism between α-tocopherol and β-carotene in microsomal membranes (Palozza and Krinsky, 1992) showed that α-tocopherol may be "consumed" to avoid autoxidation of β-carotene and reduce its prooxidant effects.

The mechanisms underlying the synergism between vitamin A and vitamin E are not entirely clear. The kinetics of consumption of α-tocopherol and all-*trans* retinol in the course of lipid peroxidation did not show regeneration mechanisms for any of the antioxidants (Fig. 4). On the other hand, when α-tocopherol and all-*trans* retinol were combined, the consumption of *both* antioxidants was significantly delayed with respect to assays where they acted separately, indicating a reciprocal protection. Vitamin E may prevent vitamin A oxidation (Mejia, 1986; Robison *et al.*, 1980, 1979). So, keeping in mind the antioxidant mechanism of retinoids, a possible explanation is that, by limiting the autocatalytic oxidation of all-*trans* retinol, α-tocopherol strongly promotes its antioxidant effectiveness. The concerted radical scavenging action in turn results in a synergistic protection of the lipid system and ultimately slows down α-tocopherol consumption. In accordance with this explanation, combination of all-*trans* retinol with butylated hydroxytoluene, a synthetic antioxidant known to prevent vitamin A autoxidation (Budowski and Bondi, 1960), reduced all-*trans* retinol consumption and produced synergistic effects in a similar liposomal system (Tesoriere *et al.*, 1996).

The interactions between all-*trans* retinol and α-tocopherol observed in retina (Tesoriere *et al.*, 1995) may be of importance for the visual function. Early *in vivo* studies had showed that interrelationships between

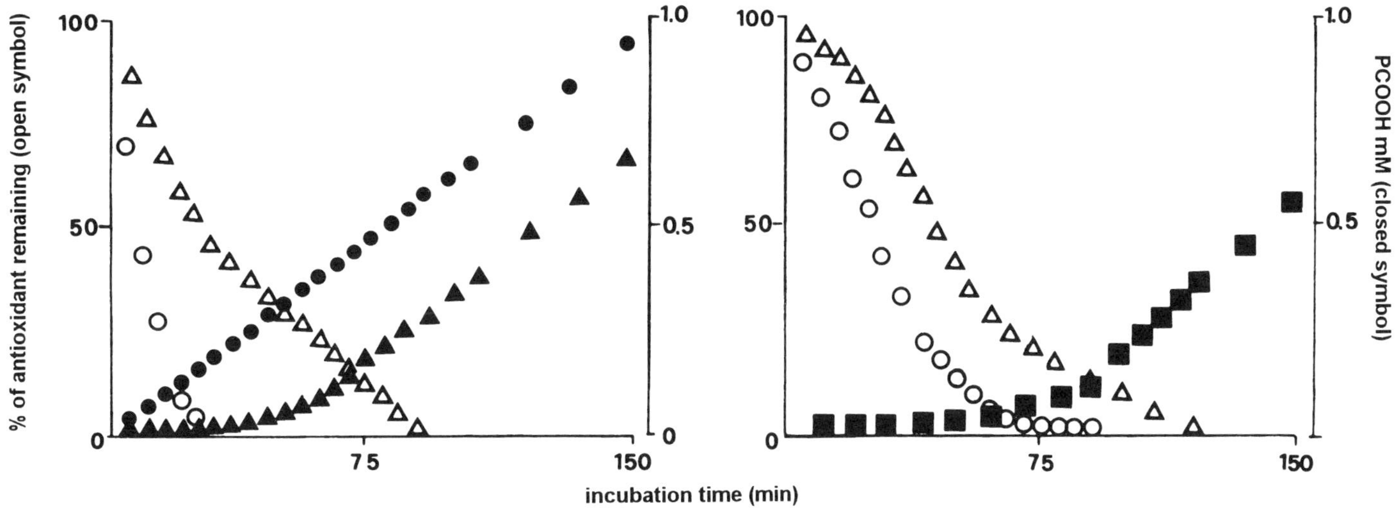

FIGURE 4. Synergism between vitamin A and vitamin E in inhibiting lipid peroxidation in phosphatidylcholine liposomes. Peroxidation of liposomes containing α-tocopherol (▲) or all-*trans* retinol (●) or both (■) was stimulated by AAPH, and evaluated as conjugated diene lipid hydroperoxides (PCOOH). In the same assays consumption of α-tocopherol (△) and all-*trans* retinol (○) was monitored. Uninhibited reaction curve was equivalent to the curve in the presence of all-*trans* retinol. Adapted from Tesoriere *et al.* (1996).

vitamins E and A were essential in preventing oxidative destruction of rod outer segments (Robison *et al.*, 1979). The amount of all-*trans* retinol measured in native retinal membranes ($0.1\,\mathrm{nmol\,mg\,prot^{-1}}$) is in a 1:7 ratio with the endogenous α-tocopherol (Tesoriere *et al.*, 1995), i.e., a ratio useful for the synergistic effects to occur (Tesoriere *et al.*, 1996). Moreover, according to the visual cycle, larger amounts of all-*trans* retinol can be released in the retina (Bongiorno *et al.*, 1991; Zimmerman, 1974), providing an effective antioxidant defense when free radical reactions consequential to light absorption may easily occur (Handelman and Dratz, 1986).

Vitamin E is not uniformly distributed in membranes, but forms clusters in the lipid bilayer (Kagan *et al.*, 1990). Lipid domains in the membrane where retinol and tocopherol can be concentrated may create conditions for a much more effective antioxidant action. Moreover, *in vivo*, complementary antioxidant systems in the aqueous compartment (Packer, 1992), by maintaining α-tocopherol status, may concur to improve their synergistic effect.

5. CONCLUSIONS

Data indicate that, by scavenging lipoperoxyl radicals, the main chain-carrying species during the peroxidation of unsaturated lipids, vitamin A and its congeners may act as chain-breaking antioxidants in biological membranes and other lipid-rich environments. This activity is unusual for several reasons. Retinoids do not donate hydrogen atoms, but rather act as a trap for lipoperoxyl radicals. A delicate balance between antioxidant and nonantioxidant, or even prooxidant, pathways establishes the effectiveness of retinol. The latter appears to depend critically on the oxygen partial pressure, extent of radical fluxes, and retinol concentration (Tesoriere *et al.*, 1997). Low oxygen tension and retinol concentrations favor the antioxidant activity in membranes *in vivo*. Increase of radical fluxes may enhance the antioxidant effectiveness. In addition, interactions with other agents seem to be critical in determining an effective antioxidant behavior of vitamin A.

In view of the growing number of associations between various pathological conditions and changes in oxidative balance, redox status, and oxidative injury, one of the goals of current research concerns whether antioxidant nutrients may be pharmacologically exploited. The importance of vitamin A in reducing the risk of certain types of cancer (Hong and Itri, 1994) and of cardiovascular diseases (Gey *et al.*, 1994; Gey and Puska, 1989), as well as the therapeutic value in skin diseases (Vahlquist, 1994; Orfanos *et al.*, 1987), raise the question of whether its beneficial role is associated with its antioxidant properties. Since carcinogenesis has been implicated with reactive oxygen species (Guyton and Kensler, 1993; Cerutti, 1985), it was suggested that the antioxidant properties of retinoids may be the basis of their anticancer action (Gey *et al.*, 1994; Gey, 1993; Gundimeda *et al.*, 1993). This concept also seems to be supported by the evidence that vitamin A may inhibit several aspects of the carcinogenesis process (Moon *et al.*, 1994). On the other hand, the uncertainties about the role of lipid peroxidation products on stimulation of tumor growth, or conversely inhibition of cell proliferation (Welsch, 1995; Burdon, 1995), make the effect of lipid antioxidants on tumorigenesis difficult to assess at the moment. Prevention of photoaging of the skin by retinoic acid (Kligman *et al.*, 1984) may be an expression of antioxidant activity, as emerging evidence suggests that free radicals are responsible for at least part of the degenerative changes leading to cutaneous aging (Black, 1987). In any event one may ask what are the safety limits wherein production of toxic radicals can be counterbalanced by vitamin A without being open to prooxidant activity. The redox character of retinoids must be considered in terms of the extracellular and intracellular oxygen microenvironments and of the actual oxidative insult.

6. REFERENCES

Babiy, A. V., Gebicki, J. M., and Sullivan, D. R., 1990, Vitamin E content and low density lipoprotein oxidizability induced by free radicals, *Atherosclerosis* **81:**175–182.

Barklay, R. L. C., Baskin, K. A., Dakin, K. A., Locke, S. J., and Vinqvist, M. R., 1990, The antioxidant activities of phenolic antioxidants in free radical peroxidation of phospholipid membranes, *Can. J. Chem.* **68:**2258–2269.

Black, H. S., 1987, Potential involvement of free radical reactions in ultraviolet light-mediated cutaneous damage, *Photochem. Photobiol.* **46:**213–221.

Blomhoff, R., 1994, Introduction: Overview of vitamin A metabolism and function, in *Vitamin A in Health and Disease* (R. Blomhoff, ed.), pp. 1–35, Marcel Dekker, New York.

Blomhoff, R., Green M. H., Green, J., Berg, T., and Norum, K. R., 1991, Vitamin A metabolism: New perspectives on absorption, transport, and storage, *Physiol. Rev.* **71:**951–990.

Bongiorno, A., Tesoriere, L., Livrea, M. A., and Pandolfo, L., 1991, Distribution of vitamin A compounds in bovine eyes after bleaching adaptation, *Vision Res.* **31:**1099–1106.

Bowry, V. W., and Stocker, R., 1993, Tocopherol-mediated peroxidation.The prooxidant effect of vitamin E on the radical-initiated oxidation of human low-density lipoprotein, *J. Am. Chem. Soc.* **115:**6029–6044.

Bowry, V. W., Ingold, K. U., and Stocker, R., 1992, Vitamin E in human low-density lipoprotein. When and how this antioxidant becomes a pro-oxidant, *Biochem. J.* **288:**341–344.

Brattsand, R., 1975, Actions of vitamins A and E and some nicotinic acid derivatives on plasma lipids and lipid infiltration of aorta in cholesterol-fed rabbits, *Atherosclerosis* **22:**47–61.

Budowski, P., and Bondi, A., 1960, Autoxidation of carotene and vitamin A. Influence of fat and antioxidants, *Arch. Biochem. Biophys.* **89:**66–73.

Buege, J. A., and Aust, S. D., 1978, Microsomal lipid peroxidation, in *Methods in Enzymology*, Vol. 51 (S. Fleischer and L. Packer, eds.), pp. 302–308, Academic Press, New York.

Burdon, R. H., 1995, Superoxide and hydrogen peroxide in relation to mammalian cell proliferation, *Free Radic. Biol. Med.* **18:**775–794.

Burton, G. W., and Ingold, K. U., 1981, Autoxidation of biological molecules. 1. The antioxidant activity of vitamin E and related chain-breaking antioxidant *in vitro*, *J. Am. Chem. Soc.* **103:**6472–6477.

Burton, G. W., and Ingold, K. U., 1984, Beta-carotene, an unusual type of lipid anti-oxidant, *Science* **224:**569–573.

Burton, G. W., and Ingold, K. U., 1989, Vitamin E as an *in vitro* and *in vivo* antioxidant, *Ann. N. Y. Acad. Sci.* **570:**7–22.

Cadenas, E., Ginsberg, M., Rabe, U., and Sies, H., 1984, Evaluation of alpha-tocopherol antioxidant activity in microsomal lipid peroxidation as detected by low-level chemiluminescence, *Biochem. J.* **223:**755–759.

Cerutti, P. A., 1985, Prooxidant states and tumor promotion, *Science* **227:**375–381.

Ciaccio, M., Valenza, M., Tesoriere, L., Bongiorno, A., Albiero, R., and Livrea, M. A., 1993, Vitamin A inhibits doxorubicin-induced membrane lipid peroxidation in rat tissues *in vivo*, *Arch. Biochem. Biophys* **302:**103–108.

Ciaccio, M., Tesoriere, L., Pintaudi, A. M., Re, R., Vallesi-Cardillo, S., Bongiorno, A., and Livrea, M. A., 1994, Vitamin A preserves the cytotoxic activity of adriamycin while counteracting its peroxidative effects in human leukemic cells *in vitro*, *Biochem. Mol. Biol. Int.* **34:**329–335.

Chan, A. C., Tran, K., Raynor, T., Ganz, P. R., and Chow, C. K., 1991, Regeneration of vitamin E in human platelets, *J. Biol. Chem.* **266:**17290–17295.

Costantinescu, A., Han, D., and Packer, L., 1993, Vitamin E recycling in human erythrocyte membranes, *J. Biol. Chem.* **268:**10906–10913.

D'Aquino, M., Dunster, C., and Willson, R. L., 1989, Vitamin A and glutathione-mediated free radical damage: Competing reactions with polyunsaturated fatty acids and vitamin C, *Biochem. Biophys. Res. Commun.* **161:**1199–1203.

Das, N. P., 1989, Effects of vitamin A and its analogs on nonenzymatic lipid peroxidation in rat brain mitochondria, *J. Neurochem.* **52:**585–588.

Dean, R. T., Hunt, J. V., Grant, A. J., Yamamoto, Y., and Niki, E., 1991, Free radical damage to proteins: The influence of the relative localization of radical generation, antioxidants, and target proteins, *Free Radic. Biol. Med.* **11:**161–168.

De Boek, H., and Zidovedzki, R., 1988, NMR study of the interaction of retinoids with phospholipid bilayers, *Biochem. Biophys. Acta* **946:**244–252.

Di Mascio, P., Kaiser, S., and Sies, H., 1989, Lycopene as the most efficient biological carotenoid singlet oxygen quencher, *Arch. Biochim. Biophys.* **274:**532–538.

Dingle, J. T., Sharman, I. M., and Moore, T., 1966, Nutrition and lysosomal activity, *Biochem. J.* **98:**476–484.

Doba, T., Burton, G. W., and Ingold, K. U., 1985, Antioxidant and co-antioxidant activity of vitamin C. The effect of vitamin C, either alone or in the presence of vitamin E or a water-soluble vitamin E analogue, upon the peroxidation of aqueous multilamellar phospholipid liposomes, *Biochem. Biophys. Acta* **835:**298–303.

ElSisi, A. E. D., Earnest, D. L., and Sipes, A. I. G., 1993, Vitamin A potentiation of carbon tetrachloride epatotoxicity: Enhanced lipid peroxidation without enhanced biotransformation, *Toxicol. Appl. Pharmacol.* **119:**289–294.

Esterbauer, H., Striegl, G., Puhl, H., Oberreither, S., Rotheneder, M., El-Saadani, M., and Jurgens, G., 1989a, The role of vitamin E and carotenoids in preventing oxidation of low density lipoproteins, *Ann. N. Y. Acad. Sci.* **570:**254–267.

Esterbauer, H., Striegl, G., Puhl, H., and Rotheneder, M., 1989b, Continuous monitoring of *in vitro* oxidation of human low density lipoproteins, *Free Radic. Res. Commun.* **6:**67–75.

Esterbauer, H., Gebicki, J., Puhl, H., and Jurgens, G., 1992, The role of lipid peroxidation and antioxidants in oxidative modification of LDL, *Free Radic. Biol. Med.* **13:**341–390.

Esterbauer, H., Wag, G., and Puhl, H., 1993, Lipid peroxidation and its role in atherosclerosis, *Br. Med. Bull.* **49:**566–576.

Farnsworth, C. C., and Dratz, E. A., 1976, Oxidative damage of retinal rod outer segment membranes and the role of vitamin E, *Biochim. Biophys. Acta* **443:**556–570.

Freisleben, J.-H., and Packer, L., 1993, Free radical scavenging activities, interactions and recycling of antioxidants, *Biochem. Soc. Trans.* **21:**325–330.

Geesin, J. C., Gordon, J. S., and Berg, R. A., 1990, Retinoids affect collagen synthesis through inhibition of ascorbate-induced lipid peroxidation in cultured human dermal fibroblasts, *Arch. Biochem. Biophys.* **278:**350–355.

Gey, K. F., 1993, Prospects for the prevention of free radical disease, regarding cancer and cardiovascular disease, *Br. Med. Bull.* **49:**679–699.

Gey, K. F., and Puska, P., 1989, Plasma vitamin E and A inversely correlated to mortality from ischemic heart disease in cross-cultural epidemiology, *Ann. N. Y. Acad. Sci.* **570:**268–282.

Gey, K. F., Moser, U. K., Jordan, P., Stahelin, H. B., Eicholzer, M., and Ludin, E., 1993, Increased risk of cardiovascular disease at suboptimal plasma concentration of essential antioxidants: An epidemiological update with special attention to carotene and vitamin C, *Am. J. Clin. Nutr.* **57**(Suppl.)**:**787S–797S.

Gey, K. F., Stahelin, H. B., Eichholzer, M, and Ludin, E., 1994, Prediction of increased cancer risk in humans by interacting suboptimal plasma levels of retinol and carotene, in *Retinoids. From Basic Science to Clinical Applications* (M. A. Livrea and G. Vidali, eds.), pp. 137–164, Birkhäuser Verlag, Basel.

Ghoshal, A. K., 1976, Antioxidant role of vitamin E and carbontetrachloride toxicity, *Exp. Mol. Pathol.* **25:**202–207.

Graham Solomons, T. W., 1984, *Organic Chemistry*, pp. 225–228, Wiley, New York.

Green, J., Diplock, A. T., Bunyan, J., Edwin, E. E., and McHale, D., 1961, Ubiquinone (coenzyme Q) and the function of vitamin E, *Nature* **190:**318–326.

Gundimeda, U., Hara, S. K., Anderson, W. B., and Gopalakrishna, R., 1993, Retinoids inhibit the oxidative modification of protein kinase C induced by oxidant tumor promoters, *Arch. Biochem. Biophys.* **300:**526–530.

Guyton, K. Z., and Kensler, T. W., 1993, Oxidative mechanisms in carcinogenesis, *Br. Med. Bull.* **49:**523–544.

Halevy, O., and Sklan, D., 1987, Inhibition of arachidonic acid oxidation by β-carotene, retinol and α-tocopherol, *Biochim. Biophys. Acta* **918:**304–307.

Halliwell, B., 1996, Cellular stress and protection mechanisms, *Biochem. Soc. Trans.* **24:**1023–1027.

Halliwell, B., and Chirico, S., 1993, Lipid peroxidation: Its mechanism, measurement and significance, *Am. J. Clin. Nutr.* **57**(Suppl.)**:**715S–725S.

Hallywell, B., and Gutteridge, J. M. C., 1989, *Free Radicals in Biology and Medicine*, 2nd ed., Clarendon Press, Oxford.

Handelman, G. J., and Dratz, E. A., 1986, The role of antioxidants in the retina and retinal pigment epithelium and the nature of prooxidant-induced damage, *Adv. Free Radic. Biol. Med.* **2:**1–89.

Hathcock, J. N., Hattan, D. G., Jenkins, H. Y., McDonald, J. T., Sundaresan, P. R., and Wilkening, V. L., 1990, Evaluation of vitamin A toxicity, *Am. J. Clin. Nutr.* **52:**183–202.

Hiramatsu, M., and Packer, L., 1990, Antioxidant activity of retinoids, in *Methods in Enzymology*, Vol. 190 (L. Packer, ed.), pp. 273–280, Academic Press, New York.

Ho, C. T., and Chan, A. C., 1992, Regeneration of vitamin E in rat polymorphonuclear leucocytes, *FEBS Lett.* **306:**269–272.
Hong, W. K., and Itri, L. M., 1994, Retinoids and human cancer, in *The Retinoids*, 2nd ed. (M. B. Sporn, A. B. Roberts, and D. W. S. Goodman, eds.), pp. 597–630, Raven Press, New York.
Hunter, F. E., Jr., Gebicki, J. M., Hoffsten, P. E., Weinstein, J., and Scott, A., 1963, Swelling and lysis of rat liver mitochondria induced by ferrous ions, *J. Biol. Chem.* **238:**828–835.
Hunter, F. E., Jr., Scott, A., Hoffsten, P. E., Guerra, F., Weinstein, J., Schneider, A., Schutz, B., Fink, J., Ford, L., and Smith, E., 1964, Studies on the mechanism of ascorbate-induced swelling and lysis of isolated liver mitochondria, *J. Biol. Chem.* **239:**604–613.
Ingold, K. U., Bowry, V. W., Stocker, R., and Walling, C., 1993, Autoxidation of lipids and antioxidation by alpha-tocopherol and ubiquinol in homogeneous solution and in aqueous dispersions of lipids: Unrecognized consequences of lipid particle size as exemplified by oxidation of human low density lipoproteins, *Proc. Natl. Acad. Sci. USA* **90:**45–49.
Jackson, G. R., Morgan, B. C., Werrbach-Perez, K., and Perez-Polo, J. R., 1991, Antioxidant effect of retinoic acid on PC 12 rat pheochromocytoma, *Int. J. Dev. Neurosci.* **9:**161–170.
Kagan, V. E., Bakalova, R. A., Serbinova, E., and Stoytchev, T. S., 1990, Fluorescence measurements of incorporation and hydrolysis of tocopherol and tocopheryl esters in biomembranes, in *Methods in Enzymology*, Vol. 186 (L. Packer and A. N. Glazer, eds.), pp. 355–367, Academic Press, San Diego, California.
Kalmikova, V. I., and Dmytrovsky, A. A., 1972, in *Proceedings of the Symposium on Structure, Biosynthesis, and Metabolism of Lipids in Man and Animals*, Leningrad, p. 59.
Kartha, V. N. R., and Krishnamurthy, S., 1977, Antioxidant function of vitamin A, *Int. J. Vitamin Nutr. Res.* **47:**394–401.
Kartha, V. N. R., and Krishnamurthy, S., 1978, Effect of vitamins antioxidants and sulfhydryl compounds on *in vitro* rat brain lipid peroxidation, *Int. J. Vitamin Nutr. Res.* **48:**38–43.
Kennedy, T. A., and Liebler, D. C., 1992, Peroxyl radical scavenging by β-carotene in lipid bilayers, Effect of oxygen partial pressure, *J. Biol. Chem.* **267:**4658–4663.
Kligman, L. H., Chen, H. D., and Kligman, A. M., 1984, Topical retinoic acid enhances the repair of ultraviolet damaged dermal connective tissue, *Connect. Tiss. Res.* **12:**139–150.
Livrea, M. A., and Packer, L., 1994, Vitamin A as an antioxidant *in vitro* and *in vivo*, in *Retinoids. From Basic Science to Clinical Applications* (M. A. Livrea and G. Vidali, eds.), pp. 293–304, Birkhäuser, Basel.
Livrea, M. A., and Tesoriere, L., 1994, Lipoperoxyl radical-scavenging activity of vitamin A and analogs in homogeneous solution, in *Methods in Enzymology*, Vol. 234 (L. Packer, ed.), pp. 401–410, Academic Press, New York.
Livrea, M. A., Albiero, A., Valenza, A., Ciaccio, M., and Tesoriere, L., 1992, *In vitro* peroxyl radical scavenging activity of vitamin A reduces iron/ascorbate-induced lipid peroxidation in subcellular membrane preparations from rat tissues, *It. Biochem. Soc. Trans.* **3:**181.
Livrea, M. A., Tesoriere, L., Bongiorno, A., Pintaudi, A. M., Ciaccio, M., and Riccio, A., 1995, Contribution of vitamin A to the oxidation resistance of human low density lipoproteins, *Free Radic. Biol. Med.* **18:**401–409.
Livrea, M. A., Tesoriere, L., and Freisleben, H.-J., 1996, Vitamin A as antioxidant, in *Handbook of Antioxidants* (E. Cadenas and L. Packer, eds.), pp. 371–405, Marcel Dekker, New York.
Mejia, L. A., 1986, Vitamin A–nutrient interrelationships, in *Vitamin A Deficiency and Its Control* (J. C. Bauernfeind, ed.), pp. 69–100, Academic Press, Orlando, Florida.
Monaghan, B. R., and Schmitt, F. O., 1932, The effects of carotene and of vitamin A on the oxidation of linoleic acid, *J. Biol. Chem.* **96:**387–385.

Moon, R. C., Metha R. G., and Rao, K. V. N., 1994, Retinoid and cancer in experimental animals, in *The Retinoids*, 2nd ed. (M. B. Sporn, A. B. Roberts, and D. W. S. Goodman, eds.), pp. 573–595, Raven Press, New York.

Morré, D. M., 1992, Intracellular actions of vitamin A, *Int. Rev. Cytol.* **135:**1–38.

Nicotra, C., Livrea, M. A., and Bongiorno, A., 1975, Vitamin A and lipid peroxidation in bovine retina, *IRCS Med. Sci.* **3:**141–142.

Niki, E., 1990, Free radical initiators as source of water- or lipid-soluble peroxyl radicals, in *Methods in Enzymology*, Vol. 186 (L. Packer, ed.), pp. 100–108, Academic Press, New York.

Niki, E., and Komuro, E., 1989, Inhibition of peroxidation of membranes, in *Oxygen Radicals in Biology and Medicine* (M. G. Simic, K. A. Taylor, J. F. Ward, and C. Sonntag, eds.), pp. 561–566, Plenum Press, New York.

Niki, E., Saito, T., Kawakami, A., and Kamiya, Y., 1984, Inhibition of oxidation of methyl linoleate in solution by vitamin E and vitamin C, *J. Biol. Chem.* **259:**4177–4182.

Niki, E., Takahashi, M., and Komuro, E., 1986, Antioxidant activity of vitamin E in liposomal membranes, *Chem. Lett.* 1573–1576.

Noy, N., 1993, Interactions of retinoids with lipid bilayers and with membranes, in *Retinoids. Progress in research and clinical applications* (M. A. Livrea and L. Packer, eds.), pp. 17–28, Marcel Dekker, New York.

Noy, N., and Xu, Z.-J., 1990a, Kinetics parameters of the interactions of retinol with lipid bilayers, *Biochemistry* **29:**3883–3888.

Noy, N., and Xu, Z.-J., 1990b, The thermodynamic parameters of the binding of retinol to binding proteins and to membranes, *Biochemistry* **29:**3888–3892.

Olson, J. A., 1993, Atwater lecture. The irresistible fascination of carotenoids and vitamin A, *Am. J. Clin. Nutr.* **57:**833–839.

Orfanos, C., Ehlert, R., and Gollnick, H., 1987, The retinoids: A review of their chemical, pharmacologic and therapeutic use, *Drugs* **34:**459–503.

Ortiz, A., Aranda, F. J., and Gomez-Fernandez, J., 1992, Interaction of retinol and retinoic acid with phospholipid membranes. A differential scanning calorimetry study, *Biochim. Biophys. Acta* **1106:**282–290.

Packer, L. (ed.), 1984, *Methods in Enzymology*, Vol. 105, Academic Press, Orlando, Florida.

Packer, L., 1992, Interactions among antioxidants in health and disease: Vitamin E and its redox cycle, *Proc. Soc. Exp. Biol. Med.* **200:**271–276.

Packer, L., and Glazer, A. N. (eds.), 1990, *Methods in Enzymology*, Vol. 186, Academic Press, San Diego, California.

Palozza, P., and Krinsky, N. I., 1992, β-Carotene and α-tocopherol are synergistic antioxidants, *Arch. Biochem. Biophys.* **297:**184–187.

Piergiacomi, V. A., Palacios, A., and Catala, A., 1996, Vitamin A inhibits lipoperoxidation ascorbate-Fe^{++} dependent of rat kidney microsomes and mitochondria, *Mol. Cell. Biochem.* **165:**121–125.

Pokrovsky, A. A., Lashneva, N. V., and Kon, I. Y., 1974, Study on vitamin A effects upon lipid peroxidation in rat liver, *Int. J. Vitamin Nutr. Res.* **44:**477–486.

Pryor, W. A., 1986, Oxy-radicals and related species: Their formation, lifetimes and reactions, *Annu. Rev. Physiol.* **48:**657–667.

Pryor, W. A., Cornicelli, J. A., Devall, L. J., Tait, B., Trivedi, B. K., Witiak, D. T., and Wu, M., 1993, A rapid screening test to determine the antioxidant potencies of natural and synthetic antioxidants, *J. Org. Chem.* **58:**3521–3532.

Robison, W. G., Jr., Kuwabara, T., and Bieri, J. G., 1979, Vitamin E deficiency and the retina: Photoreceptor and pigment epithelial changes, *Invest. Ophthalmol. Vis. Sci.* **18:**683–690.

Robison, W. G., Jr., Kuwabara, T., and Bieri, J. G., 1980, Deficiences of vitamin E and A in the rat. Retinal damage and lipofuscin accumulation, *Invest. Ophthalmol. Vis. Sci.* **19:**1030–1037.
Sahu, K., and Das, R. K., 1994, Reduction of clastogenic effect of clofazimine, an antileprosy drug, by vitamin A and vitamin C in bone marrow cells of mice, *Food Chem. Toxicol.* **32:**911–915.
Samokyszyn, V. M., and Marnett, L. J., 1987, Hydroperoxide-dependent cooxidation of 13-*cis*-retinoic acid by prostaglandin H synthase, *J. Biol. Chem.* **262:**14119–14133.
Samokyszyn, V. M., and Marnett, L. J., 1990, Inhibition of liver microsomal lipid peroxidation by 13-*cis* retinoic acid, *Free Radic. Biol. Med.* **8:**491–496.
Sanjeev Kumar, P., Vasudevan, D. M., and Jayanthi Bai, N., 1981, Antioxidant role of retinyl palmitate on carbontetrachloride toxicity, *Acta Vitaminol. Enzymol.* **3:**214–218.
Scheschonca, A., Murphy, M. E., and Sies, H., 1990, Temporal relationships between the loss of vitamin E, protein sulfhydryls and lipid peroxidation in microsomes challenged with different prooxidants, *Chem. Biol. Interact.* **74:**233–252.
Sharma, A, S., Lewandoski, J. R., and Zimmermann, J. J., 1990, Retinol inhibition of *in vitro* human neutrophil superoxide anion radical, *Pediatric Res.* **27:**574–579.
Shinitzky, M., 1984, Membrane fluidity and cellular function, in *Physiology of Membrane Fluidity*, Vol. 1 (M. Shinitzky, ed.), pp. 1–51, CRC Press, Boca Raton, Florida.
Sies, H., 1993, Strategies of antioxidant defense, *Eur. J. Biochem.* **215:**213–219.
Simic, M. G., 1988, Mechanisms of inhibition of free-radical processes in mutagenesis and carcinogenesis, *Mutat. Res.* **202:**377–386.
Steinberg, D., Parthasarathy, S., Carew, T. E., Khoo, J. C., and Witztum, J. L., 1989, Beyond cholesterol. Modification of low density lipoprotein that increase its atherogenicity, *N. Engl. J. Med.* **320:**915–924.
Steinbrecher, U. P., Parthasarathy, S., Leake, D. S., Witztum, J. L., and Steinber, D., 1984, Modification of low density lipoprotein by endothelial cells involves lipid peroxidation and degradation of low density lipoprotein phospholipids, *Proc. Natl. Acad. Sci. USA* **83:**3883–3887.
Stephens, R. J., Negi, D. S., Short, S .M., van Kuijk, F. J. G. M., Dratz, E. A., and Thomas, D. W., 1988, Vitamin E distribution in ocular tissues following long-term dietary depletion and supplementation as determined by microdissection and gas chromatography-mass spectrometry, *Exp. Eye Res.* **47:**237–245.
Stocker, R., and Peterhans, E., 1989, Synergistic interaction between vitamin E and bile pigments bilirubin and biliverdin, *Biochim. Biohys. Acta* **1002:**238–244.
Stocker, R., Yamamoto, Y., McDonagh, A. F., Glazer, A. N., and Ames, B. N., 1987, Bilirubin is an antioxidant of possible physiological importance, *Science* **235:**1043–1046.
Sudhakaran, P. R., and Kurup, P. A., 1974, Vitamin A and lysosomal stability in rat liver, *J. Nutr.* **104:**1466–1475.
Takahashi, M., Tsuchiya, J., and Niki, E., 1989, Scavenging of radicals by vitamin E in the membranes as studied by spin labeling, *J. Am. Chem. Soc.* **111:**6350–6353.
Tesoriere, L., Bongiorno, A., Ciaccio, M., and Livrea, M. A., 1993a, Kinetic evaluation of the antioxidant activity of all-*trans*- and *cis*-isomers of natural retinoids against non-enzymatic microsomal lipid peroxidation in ocular tissues, *It. J. Biochem.* **42:**99–101.
Tesoriere, L., Ciaccio, M., Bongiorno, A., Riccio, A., Pintaudi, A. M., and Livrea, M. A., 1993b, Antioxidant activity of all-*trans* retinol in homogeneous solution and in phosphatidylcholine liposomes, *Arch. Biochem. Biophys.* **307:**217–223.
Tesoriere, L., Ciaccio, M., Bongiorno, A., Freisleben, H.-J., and Livrea, M. A., 1994a, Kinetics of the antioxidant activity of all-*trans* retinol in homogeneous solution, in *Retinoids. From*

Basic Science to Clinical Applications (M. A. Livrea and G. Vidali, eds.), pp. 305–314, Birkhäuser Verlag, Basel.

Tesoriere, L., Ciaccio, M., Valenza, M., Bongiorno, A., Maresi, E., Albiero, R., and Livrea, M. A., 1994b, Effect of vitamin A administration on resistance of rat heart against doxorubicin-induced cardiotoxicity and lethality, *J. Pharmacol. Exp. Ther.* **269:**430–436.

Tesoriere, L., Bongiorno, A., Re, R., and Livrea M. A., 1995, Reciprocal protective effects of all-*trans* retinol and α-tocopherol during lipid peroxidation in retinal membrane, *Biochem. Mol. Biol. Int.* **37:**1–7.

Tesoriere, L., Bongiorno, A., Pintaudi A. M., D'Anna, R., D'Arpa, D., and Livrea, M. A., 1996, Synergistic interactions between vitamin A and vitamin E against lipid peroxidation in phosphatidylcholine liposomes, *Arch. Biochem. Biophys.* **326:**57–63.

Tesoriere, L., D'Arpa, D., Re, R., and Livrea, M.A., 1997, Antioxidant reactions of all-*trans* retinol in phospholipid bilayers. Effect of oxygen partial pressure, radical fluxes and retinol concentration, *Arch. Biochem. Biophys.* **343**:13–18.

Tsuchiya, M., Scita, G., Freisleben, H.-J., Kagan, V. E., and Packer, L., 1992, Antioxidant radical-scavenging activity of carotenoids and retinoids compared to α-tocopherol, in *Methods in Enzymology*, Vol. 213 (L. Packer, ed.), pp. 460–472, Academic Press, New York.

Vahlquist, A., 1994, Role of retinoids in normal and diseased skin, in *Vitamin A in Health and Disease* (R. Blomhoff, ed.), pp. 365–424, Marcel Dekker, New York.

Wassal, S. R., Phelps, T. M., Albrecht, M. R., Langsford, C. A., and Stillwell, W., 1988, Electron spin resonance study of the interactions of retinoids with a phospholipid model membrane, *Biochim. Biophys. Acta* **939:**393–402.

Witz, G., Goldstein, B. D., Amoruso, M., Stone, D. S., and Troll, W., 1980, Retinoid inhibition of superoxide anion radical production by human polymorphonuclear leukocytes stimulated with tumor promoters, *Bioch. Biophys. Res. Commun.* **97:**883–888.

Welsch, C. W., 1995, Review of the effects of dietary fat on experimental mammary gland tumorigenesis: Role of lipid peroxidation, *Free Radic. Biol. Med.* **18:**757–773.

Yoshida, Y., and Niki, E., 1992, Oxidation of methyl linoleate in aqueous dispersions induced by copper and iron, *Arch. Biochem. Biophys.* **295:**107–114.

Zimmerman, W. F., 1974, The distribution and proportions of vitamin A components during the visual cycle in the rat, *Vision Res.* **14:**795–802.

Ziouzenkova, O., Gieseg, S. P., Ramos, P., and Esterbauer, H., 1996, Factors affecting resistance of low density lipoproteins to oxidation, *Lipids* **31:**S71–S76.

Chapter 6

Plasma Vitamins A and E in HIV-Positive Patients

Joël Constans, Evelyne Peuchant, Claire Sergeant, and Claude Conri

1. INTRODUCTION

Malnutrition is frequent in HIV infection. Wasting may be the only AIDS-defining condition in some patients (weight loss greater than 10% body weight), but in many cases wasting occurs in patients with opportunistic infection or chronic diarrhea (Serwadda *et al.*, 1985). Even at an early stage of disease, nutritional abnormalities can be demonstrated by studies of body composition (Kotler *et al.*, 1985), and malabsorption may be present in as many as one third of patients without diarrhea (Laine *et al.*, 1993). Furthermore, a correlation has been demonstrated between lean body mass and the timing of death from AIDS (Kotler *et al.*, 1989). Wasting, together with opportunistic infections, may contribute to depletion in some micronutrients, deficits in several of which have been demonstrated in HIV-positive

Joël Constans and Claude Conri Service de Médecine Interne et Pathologie Vasculaire, Hôpital Saint-André, 33075 Bordeaux, France. **Evelyne Peuchant** Laboratoire de Biochimie, Hôpital Saint-André, 33075 Bordeaux, France. **Claire Sergeant** Centre d'Etudes Nucléaires de Bordeaux-Gradignan, 33075 Bordeaux, France.

Subcellular Biochemistry, Volume 30: Fat-Soluble Vitamins, edited by Quinn and Kagan. Plenum Press, New York, 1998.

patients: vitamins, B6, B12, D, and A, folates, zinc, copper, iron, selenium (Forest and Confalonieri, 1987; Baum *et al.*, 1991; Rule *et al.*, 1994; Haug *et al.*, 1994; Smith *et al.*, 1987; Graham *et al.*, 1991). Few studies have been conducted about vitamin E status in HIV infection, but several works have demonstrated a decrease in plasma vitamin A in HIV-positive subjects (Constans *et al.*, 1995b; Beach *et al.*, 1992; Semba *et al.*, 1993a; Karter *et al.*, 1995). Vitamin A plays a positive role in immune functions, and a deficit in vitamin A in children results in a decrease in CD4 cell count (Krishnan *et al.*, 1974). Therefore a negative role of vitamin A depletion on the course of HIV infection has been suggested.

2. PLASMA VITAMIN A LEVELS IN HIV-POSITIVE PATIENTS

All the available studies demonstrate that plasma vitamin A levels are lower in HIV-positive subjects than in controls (Table I) (Constans *et al.*, 1995b; Beach *et al.*, 1992; Semba *et al.*, 1993a; Karter *et al.*, 1995). The differences observed among the studies may reflect geographical differences, but also differences in disease stages among patients. Furthermore, a deficit in vitamin A defined as a plasma vitamin A level below 20 μg/dl is prevalent in 18–32% of HIV-positive patients (Constans *et al.*, 1995b; Beach *et al.*, 1992; Semba *et al.*, 1993a; Karter *et al.*, 1995). Karter *et al.* (1995) found that the prevalence of hyporetinemia is 73-fold higher in a HIV-positive population than in historical controls. However, clinical manifestations of retinol depletion such as corneal xerostomia, nocturnal amblyopia, or Bitot's spots are not reported in HIV-positive subjects. In fact such manifestations are unlikely unless extreme deficiencies are present. An association has been suggested between plasma vitamin A deficit and diarrheal illness or non-*Pneumocystis carinii* pneumonia (Karter

Table I
Plasma Vitamin A in HIV-Positive Patients

	Number of patients	Mean plasma vitamin A in patients (μg/dl)	Mean plasma vitamin A in controls (μg/dl)	Percentage of patients with vitamin A deficit
Beach *et al.* (1992)	100	44	52	18
Semba *et al.* (1993a)	53	40.4	55.6	15.1
Karter *et al.* (1995)	59	36.5	—	22
Constans *et al.* (1995b)	95	29.7	51	32

et al., 1995). A weak correlation has been reported between plasma vitamin A and CD4 cells, a marker of immunological impairment, in one study (Constans *et al.*, 1995b), but not in another one (Karter *et al.*, 1995). A weak correlation has also been found between the plasma levels of vitamin A and albumin, a marker of nutritional status and acute phase reaction (Constans *et al.*, 1995b).

3. VITAMIN A DIETARY INTAKES IN HIV-POSITIVE PATIENTS

HIV-positive patients' vitamin A intakes exceeded the recommended dietary allowances (RDA) in all the studies (Beach *et al.*, 1992; Tang *et al.*, 1993; Abrams *et al.*, 1993). Beach *et al.* (1992) studied 100 homosexual HIV-positive subjects and reported a mean dietary intake of 22,072 IU/day (RDA 5000) and a 3% prevalence of dietary intakes below the RDA. They found a mild correlation between vitamin A plasma levels and intakes ($r = 0.31$). Abrams *et al.* (1993) found that 296 HIV-positive men had a median vitamin A intake 200% above RDA. Tang *et al.* (1993) reported median vitamin A intakes between 12,930 and 14,291 IU/day (RDA 5000) in three groups of HIV-infected homosexuals divided according to CD4 cell count. Karter *et al.* (1995) found that 27 normoretinemic and 8 hyporetinemic patients had vitamin A intakes above the RDA, respectively 1830 and 1295 retinol equivalents (RE) (RDA 800–1000 RE). Thus HIV-positive patients' decrease in plasma vitamin A does not seem to be related to decreased intake.

4. PLASMA β-CAROTENE IN HIV-POSITIVE SUBJECTS

A deficit in plasma β-carotene has been found in all studies conducted in HIV-positive patients. Bogden *et al.* (1990) found that 30% of 30 HIV-positive patients were carotene-deficient. Ullrich *et al.* (1994) reported that 77% of subjects were carotene-deficient. In their study, mean serum carotene concentration was lower in HIV-seropositive than in HIV-seronegative subjects (0.47 μmol/L, range 0.06–1.69 versus 1.37 μmol/L, range 0.88–2.92) Moreover, they found a correlation between plasma β-carotene level and CD4 cell count. Baranowitz *et al.* (1996) found that 31% of 84 HIV-positive patients were carotene deficient (β-carotene below 50 μg/dl). The prevalence of β-carotene deficit was higher in those patients whose CD4 cell count was below 50/mm^3 (43.5%) than in those whose CD4 cells were above 500/mm^3 (20%) (Baranowitz *et al.*, 1996). The same authors defined a low-normal β-carotene level as the lower quartile of the normal range (51–100 μg/dl for this laboratory), and an additional 40% of the population was carotene deficient, so that 71% of their 84 HIV-positive patients were found to have low or low-normal plasma levels of β-carotene. Low plasma levels of β-carotene were also reported elsewhere (median levels between 30 and 50 μg/dl in three HIV-positive patient groups; (Constans *et al.*, 1996).

5. PLASMA VITAMIN E IN HIV-POSITIVE SUBJECTS

One group found mean plasma vitamin E concentration to be lower in 50 HIV-seropositive patients than in 50 HIV-seronegative individuals (Bjarnason *et al.*, 1996). In other studies, HIV-positive patients' mean plasma vitamin E levels were found to be comparable to those of controls (Abrams *et al.*, 1993; Beach *et al.*, 1992; Constans *et al.*, 1995b). Beach *et al.* (1992) reported that plasma vitamin E level was 8.9 ± 4.8 μg/dl in 100 HIV-positive subjects and 9.8 ± 4.7 μg/dl in 42 controls. However the percentage of vitamin E-deficient subjects was 27% in the HIV-seropositive group (19% overt deficiencies and 8% marginal deficiencies) and no subject was vitamin E deficient in the HIV-seronegative group. Mean dietary vitamin E intakes were far higher than RDA in the HIV-positive subjects and in the controls, but 10% of the HIV-positive and 26% of the HIV-negative subjects had a dietary vitamin E intake below the RDA (Beach *et al.*, 1992). The correlation between intakes and plasma levels was higher for vitamin E than for vitamin A ($r = 0.54$ vs. 0.31, respectively). Abrams *et al.* reported normal vitamin E intakes when the taking of supplements was included in the evaluation, but when it was excluded, mean vitamin E intakes were below the RDA in this population of 296 HIV-seropositive homosexual men. In another study the prevalence of vitamin E deficiency was 4.1% in asymptomatic HIV-positive subjects and 27.2% in AIDS patients (Bjarnason *et al.*, 1996). In four groups of 95 HIV-positive patients divided according to CD4 cell count, plasma vitamin E levels were comparable to those of 20 controls (Constans *et al.*, 1995b). A mild correlation was reported between plasma vitamin E levels and body mass index, a marker of nutritional status ($r = 0.21$), or p24 antigenemia, a marker of HIV replication ($r = -0.33$) (Constans *et al.*, 1995b). Finally little evidence exists for a reduction of plasma vitamin E concentration in selected groups of HIV-positive patients compared to control groups, but the proportion of vitamin E-deficient subjects seems to be increased in populations of HIV-seropositive patients.

6. MECHANISMS OF VITAMIN A AND β-CAROTENE DEPLETION IN HIV-POSITIVE SUBJECTS

Vitamin A depletion is not simply related to decreased intake in HIV-positive patients since most have retinol intakes far above the RDA (Abrams *et al.*, 1993; Beach *et al.*, 1992; Constans *et al.*, 1995b). In one study only 3% of patients had vitamin A intakes below the RDA. Karter *et al.* (1995) reported that 27% of those with normal retinol intakes had hyporetinemia. Moreover, they found that 87% of HIV-positive subjects with hyporetinemia had vitamin A intakes above the RDA. Other factors are implied in HIV-infection-related hyporetinemia. Fat malabsorption is a plausible mechanism. AIDS patients have been reported to have an abnormal α1-antitrypsin clearance, indicating exsudative enteropathy in 70% of cases when they are hypoalbuminemic and in 30% of cases when they are normoalbuminemic (Laine *et al.*, 1993). Recently significant intestinal inflammation (increased excretion of ^{111}indium) and malabsorption of ^{58}cobalt-labeled cyanocobalamine were demonstrated in HIV-infected subjects with or without diarrhea (Bjarnason *et al.*, 1996). However, discrepancies exist between fat-soluble vitamins, since mean plasma vitamin A, but not vitamin E levels are decreased in the series of HIV-positive patients.

Jolly *et al.* (1996) found that three of six AIDS patients had an increased excretion of retinol in urine. Stephensen *et al.* (1994) demonstrated that vitamin A was excreted in the urine during acute infections and that the retinol levels in urine were higher in those patients with fever, who were taking aminoglycosides, or who had low glomerular filtration rates. Other studies suggest a role for acute infections on vitamin A depletion: vitamin A was found to be depleted in children after an episode of chickenpox (Campos *et al.*, 1987), and the risk of xerophtalmia was increased after diarrhea or respiratory infections (Sommer *et al.*, 1987). In AIDS patients, recurrent episodes of opportunistic infections might increase retinol excretion and contribute to the depletion of vitamin A body stores as reported by Jolly *et al.* (1996). However, plasma vitamin A level is decreased in HIV-positive patients even at an early stage of HIV infection when the patients do not experience episodes of opportunistic infections (Constans *et al.*, 1995b).

Another mechanism must be involved in AIDS-related retinol depletion. One can hypothesize an increased utilization of vitamin A in metabolic reactions. Vitamin A has an antioxidant role through its action on singlet oxygen, and its precursor β-carotene also has an antioxidant role (Bates, 1995). Oxidative stress has been demonstrated in AIDS: peroxidation of fatty acids, increased production of end products of peroxidation such as malonic-dialdehyde, and decreased levels of antioxidants (selenium, vitamin A, β-carotene, glutathione) (Constans *et al.*, 1995a,b; Ullrich *et al.*, 1994; Roederer *et al.*, 1992; Sönnerborg *et al.*, 1988). However, vitamin E is a potent natural antioxidant, and mean vitamin E levels are not decreased in HIV-positive subjects (Meydani, 1995). It would be important to consider the cellular membrane levels of vitamins A and E since both components have their antioxidant function at this site, but there is a dearth of data on these membrane levels. Data about the mechanisms of β-carotene depletion are lacking. HIV-positive patients seem to have adequate intakes (Abrams *et al.*, 1993), and malabsorption may be implied in some part. An increased utilization is possible for β-carotene as for vitamin A, but data are lacking to assess this hypothesis.

7. EFFECTS OF VITAMIN A DEFICIENCY IN HIV-POSITIVE PATIENTS

During the last 10 years it has been demonstrated that vitamin A deficiency not only results in xerophtalmia and blindness, but also in mortality and susceptibility to infectious diseases (Lammer *et al.*, 1985; Reddy *et al.*, 1986). Treatment with massive intermittent dosages of vitamin A has resulted in a decrease in mortality in developing countries (Sommer *et al.*, 1986). It has been suggested that vitamin A might have a positive effect on the immune system and that marginal deficiencies in vitamin A that are unable to give rise to xerophtalmia and blindness might impair immune defenses (Bates, 1995; Sommer *et al.*, 1984). Deficiency in vitamin A clearly has effects on the immune system, and the number, and the function of natural killer cells were depressed in vitamin A-deficient rats (Bates, 1995). Supplementation with vitamin A (60 mg retinol equivalents) resulted in higher CD4/CD8 ratio, higher CD4 naive T cells, and lower proportions of CD8 and CD45 RO T cells in vitamin A-deficient children compared to placebo-treated children (Semba *et al.*, 1993b). Vitamin A deficiency might also result in depression of humoral response to proteins and alterations of mucosal surfaces (Bates, 1995). Watson *et al.* (1988) reported that high dietary vitamin A given to mice with retroviral infection increased survival and numbers of macrophages and total T lymphocytes.

It has been hypothesized that vitamin A deficiency may contribute to the loss of CD4 cells in HIV-positive subjects and accelerate the progression to AIDS. In a group of 179 HIV-positive patients, vitamin A deficiency was associated with increased mortality (relative risk 6.3, 95% confidence interval 2.1–18.6) (Semba *et al.*, 1993a). Tang *et al.* (1993) studied the relationship between vitamin A intakes and outcome, and found that the relation between total vitamin A intake and progression to AIDS was U-shaped: the lowest and highest quartiles had faster progression to AIDS than the middle two quartiles.

β-Carotene is a precursor of vitamin A, but also acts itself on immune response by increasing the proportion of natural killer cells and the expression of activation markers on mononuclear cells (Prabhala *et al.*, 1989, 1991; Bendich, 1991). Oral β-carotene specifically increased the number of CD4 cells in one study of normal volunteers (Alexander *et al.*, 1985). Serum concentration of carotene has been demonstrated to be decreased, but the relationship between serum carotene concentration and progression to AIDS has not been studied, to our knowledge (Bogden *et al.*, 1990; Ullrich *et al.*, 1994; Baranowitz *et al.*, 1996).

8. VITAMIN A, HIV INFECTION, AND PREGNANCY

Since a decrease in plasma vitamin A has been demonstrated in HIV-positive patients, and pregnancy may also contribute to vitamin A deficiency (Wallingford and Underwood, 1986), HIV-positive pregnant women's vitamin A status was studied together with the relationship between hyporetinemia and vertical transmission of HIV (Semba *et al.*, 1994; Phuapradit *et al.*, 1996). Semba *et al.* (1994) investigated plasma vitamin A levels in 567 HIV-infected mothers in Malawi. The plasma vitamin A concentration was 24.6 μg/dl in those women who transmitted the virus and 30.7 μg/dl in those who did not. There was a monotonic association between HIV vertical transmission and the mother's plasma vitamin A level. Transmission occurred in 32% of the women whose plasma vitamin A was below 20 μg/dl versus 17% in those whose vitamin A was above 40 μg/dl. Maternal CD4 cell percentage and vitamin A were both independently predictive of vertical transmission of HIV. The relative risk of vertical transmission was 4.38 in the mothers with plasma vitamin A levels below 20 μg/dl when compared to those with vitamin A above 40 μg/dl. Phuapradit *et al.* (1996) compared 74 HIV-seropositive pregnant women to 148 HIV-seronegative pregnant women, and found that the seropositive subjects had lower mean serum vitamin A and β-carotene levels (0.820 vs. 1.308 μmol/L and 1.486 vs. 2.363 μmol/L, respectively). The serum vitamin A and β-carotene levels correlated with the CD4 cell count (r = 0.772 and 0.614, respectively). These authors wondered whether supplementation in vitamin A or its precursors would reduce the risk of HIV vertical transmission, but they believed that risk versus benefit must be carefully considered because of the enhancement of birth defects reported with retinoids (Lammer *et al.*, 1985).

9. TRIALS OF SUPPLEMENTATION WITH β-CAROTENE

A potential toxicity of vitamin A and the association of high dietary intakes of vitamin A with a poorer outcome have led investigators to use the provitamin A β-carotene, which also has positive specific effects on the immune system (Semba *et al.*, 1993b; Moriguchi *et al.*, 1985). Few trials of β-carotene supplementation have been conducted in HIV-seropositive patients. In a preliminary study, CD4 cell count increased by 89% in three patients treated with 180 mg of β-carotene/day for 2 weeks (Coodley, 1991). Garewal *et al.* (1992) gave 60 mg of β-carotene/day to 11 HIV-infected patients during a 4 month period. He did not find any change in CD4 cell count, but the number of cells with natural killer markers increased. In another trial, Coodley *et al.* (1993) gave 180 mg/day or placebo to 21 HIV-seropositive patients during 4 weeks with a crossover design. They found that β-carotene supplementation resulted in a significant increase in white blood cells, CD4 cell count, and CD4/CD8 ratio. Our group supplemented 15 patients with 60 mg of β-carotene/day during 1 year (Constans *et al.*, 1996; Delmas-Beauvieux *et al.*, 1996). There was no positive effect of this supplementation on CD4 cell count or β2-microglobulin, but plasma concentrations of β-carotene and glutathione increased and malonic dialdehyde levels decreased in those patients who were supplemented. Erythrocyte glutathione peroxidase increased in the β-carotene-supplemented group, although nonsignificantly so. Finally, this study concluded on some improvement of the antioxidant system by β-carotene supplementation, although no clear effect was seen on the immune system.

10. CONCLUSION

Deficiencies in vitamin A and β-carotene are frequent in HIV-positive patients, although clinical manifestations of hyporetinemia are uncommon. Plasma vitamin A mean concentration decreases in groups of HIV-positive subjects compared to HIV-seronegative controls. Beta-carotene plasma levels are also low in HIV-positive subjects and a marginal deficiency in β-carotene is present in as many as 70% of the patients. Although a deficiency in vitamin E seems more frequent in the HIV-positive population than in than HIV-seronegative subjects, the mean plasma vitamin E concentration does not seem to decrease significantly in groups of HIV-positive patients compared to controls. HIV-positive patients' hyporetinemia is not related to a decrease in intake and other mechanisms such as increased urinary retinol or metabolic use might be implied. Hyporetinemia has been related to increased mortality in AIDS and increased vertical transmission of HIV.

Supplementation with β-carotene has not proved a beneficial effect on the immune system or on the course of HIV infection. A severe vitamin A deficiency probably has to be corrected in HIV-positive patients, but supplementation of large numbers of patients cannot be recommended. In pregnant women, the efficacy of supplementation on vertical transmission has to be evaluated.

11. REFERENCES

Abrams, B., Duncan, D., and Hertz-Picciotto, I., 1993, A prospective study of dietary intake and acquired immune deficiency syndrome in HIV-seropositive homosexual men, *J. Acquir. Immune Defic. Syndr.* **6:**949–958.

Alexander, M., Newmark, M. and Miller, R. G., 1985, Oral beta carotene can increase the number of OKT4+ cells in human blood, *Immunol. Lett.* **9:**221–224.

Baranowitz, S. A., Starrett, B., and Brookner, A. R., 1996, Carotene deficiency in HIV patients, *AIDS* **10:**115.

Bates, C. J., 1995, Vitamin A, *Lancet* **345:**31–35.

Baum, M. K., Mantero-Atienza, E., Shor-Posner, G., Fletcher, M. A., Morgan, R., Eisdorfer, C., Sauberlich, H. E., Cornwell, P. E., and Beach, R. S., 1991, Association of vitamin B6 status with parameters of immune function in early HIV-1 infection. *J. Acquir. Immune Defic. Syndr.* **4:**1122–1132.

Beach, R. S., Mantero-Atienza, E., Shor-Posner, G., Javier, J. J., Szapocznik, J., Morgan, R., Sauberlich, H. E., Cornwell, P. E., Eisdorfer, C., and Baum, M. K., 1992, Specific nutrient abnormalities in asymptomatic HIV-1 infection, *AIDS* **6:**701–708.

Bendich, A., 1991, β-Carotene and the immune response, *Proc. Nutr. Soc.* **50:**263–274.

Bjarnason, I., Sharpstone, D. R., Francis, N., Marker, A., Taylor, C., Barrett, M., Macpherson, A., Baldwin, C., Menzies, I. S., Crane, R. C., Smith, T., Pozniak, A., and Gazzard, B. G., 1996, Intestinal inflammation, ileal structure and function in HIV, *AIDS* **10:**1385–1391.

Bogden, J. D., Baker, H., Frank, O., Perez, G., Kemp, F., Bruening, K., and Lousia, D., 1990, Micronutrient status and human immunodeficiency virus (HIV) infection, *Ann. N. Y. Acad. Sci.* **587:**189–195.

Campos, F. A. C. S., Flores, H., and Underwood, B. A., 1987, Effect of an infection on vitamin A status of children as measured by the relative dose response (RDR), *Am. J. Clin. Nutr.* **46:**91–94.

Constans, J., Pellegrin, J. L., Sergeant, C., Simonoff, M., Pellegrin, I., Fleury, H., Leng, B., and Conri, C., 1995a, Serum selenium predicts outcome in HIV infection, *J. Acquir. Immune Defic. Syndr.* **10:**392.

Constans, J., Peuchant, E., Pellegrin, J. L., Sergeant, C., Hamon, C., Dubourg, L., Thomas, M. J., Simonoff, M., Pellegrin, I., Brossard, G., Barbeau, P., Fleury, H., Clerc, M., Leng, B., and Conri, C., 1995b, Fatty acids and plasma antioxidants in HIV-positive patients: Correlation with nutritional and immunological status, *Clin. Biochem.* **28:**421–426.

Constans, J., Delmas-Beauvieux, M. C., Sergeant, C., Peuchant, E., Pellegrin, J. L., Pellegrin, I., Clerc, M., Fleury, H., Simonoff, M., Leng, B., and Conri, C., 1996, One-year antioxidant supplementation with β-carotene or selenium for patients infected with human immunodeficiency virus: A pilot study, *Clin. Infect. Dis.* **23:**654–656.

Coodley, G., 1991, Beta carotene therapy in human immunodeficiency virus infection (abstract), *Clin. Res.* **39:**634A.

Coodley, G. O., Nelson, H. D., Loveless, M. O., and Folk, C., 1993, β-Carotene in HIV infection, *J. Acquir. Immune Defic. Syndr.* **6:**272–276.

Delmas-Beauvieux, M. C., Peuchant, E., Couchouron, A., Constans, J., Sergeant, C., Simonoff, M., Pellegrin, J. L., Leng, B., Conri, C., and Clerc, M., 1996, The enzymatic antioxidant system in blood and glutathione status in human immunodeficiency virus (HIV)-infected patients: Effects of supplementation with selenium or β-carotene, *Am. J. Clin. Nutr.* **64:**101–107.

Forest, V., and Confalonieri, F., 1987, Wernicke's encephalopathy in AIDS, *Lancet* **I:**1499.

Garewal, H. S., Ampel, N. M., Watson, R. R., Prabhala, R. H., and Dols, C. L., 1992, A preliminary trial of beta carotene in subjects infected with the human immunodeficiency virus, *J. Nutr.* **122:**728–732.

Graham, N. M. H., Sorensen, D., Odaka, N., *et al.*, 1991, Relationship of serum copper and zinc levels to HIV-1 seropositivity and progression to AIDS, *J. Acquir. Immune Defic. Syndr.* **4:**976–980.

Haug, C., Müller, F., Aukrust, P., and Froland, S. S., 1994, Subnormal serum concentrations of 1,25-vitamin D in human immunodeficiency virus infection: Correlation with degree of immune deficiency and survival, *J. Infect. Dis.* **169:**889–893.

Jolly, P. E., Yang, Y. L., Alvarez, J. O., and Smoot, T. M., 1996, Vitamin A depletion in HIV infection and AIDS, *AIDS* **10:**114.

Karter, D. L., Karter, A. J., Yarrish, R., Patterson, C., Kass, P. H., Nord, J., and Kislak, J. W., 1995, Vitamin A deficiency in non vitamin-supplemented patients with AIDS: A cross-sectional study, *J. Acquir. Immune Defic. Syndr.* **8:**199–203.

Kotler, D. P., Wang, J., and Pierson, R. N., 1985, Body composition studies in patients with the acquired immunodeficiency syndrome, *Am. J. Clin. Nutr.* **42:**1255–1265.

Kotler, D. P., Tierney, A. R., Wang, J., and Pierson, R. N., 1989, Magnitude of body cell mass depletion and the timing of death from wasting in AIDS, *Am. J. Clin. Nutr.* **50:**444–447.

Krishnan, S., Bjuyam, U. M., Talwar, G. P., and Ramalingaswami, V., 1974, Effects of vitamin A and protein-calorie malnutrition on immune responses, *Immunology* **27:**383–392.

Laine L., Garcia F., McGilligan K., Malinko A., Sinatra, F. R., and Thomas, D. W., 1993, Protein-losing enteropathy and hypoalbuminemia in AIDS, *AIDS* **7:**837–840.

Lammer, E. J., Chen, D. T., Hoar, R. M., Agnish, R. D., Benke, P. J., Braun, J. T., Curry, C. J., Fernhoff, P. M., Grix, A. W., Lott, I. T., Richard, J. M., and Sun, S. C., 1985, Retinoic acid embryopathy, *N. Engl. J. Med.* **313:**837–841.

Meydani, M., 1995, Vitamin E, *Lancet* **345:**170–175.

Moriguchi, S., Werner, L., and Watson, R. R., 1985, High dietary vitamin A (retinyl palmitate) and cellular immune functions in mice, *Immunology* **56:**169–177.

Phuapradit, W., Chaturachinda, K., Taneepanichskul, S., Sirivarasry, J., Khupulsup, K., and Lerdvuthisopon, N., 1996, Serum vitamin A and β-carotene levels in pregnant women infected with human immunodeficiency virus-1, *Obstet. Gynecol.* **87:**564–567.

Prabhala, R. H., Maxey, V., Hicks, M. J., and Watson, R. R., 1989, Enhancement of the expression of activation markers on human peripheral blood mononuclear cells by *in vitro* culture with retinoids and carotenoids, *J. Leukocyte Biol.* **45:**249–254.

Prabhala, R. H., Garewal, H. S., Hicks, M. J., Sampliner, R. E., and Watson, R. R., 1991, The effects of 13-*cis*-retinoic acid and beta-carotene on cellular immunity in humans, *Cancer* **67:**1556–1560.

Reddy, V., Bhaskaram, P., Raghuramulu, N., Milton, R. C., Rao, V., Madhusudan, J., and Krishna, R., 1986, Relationship between measles, malnutrition and blindness: A prospective study in Indian children, *Am. J. Clin. Nutr.* **44:**924–930.

Roederer, M., Staal, F. J. T., Anderson, M. T., and Herzenberg, L. A., 1992, Glutathione deficiency and human immunodeficiency virus infection, *Lancet* **339:**909–912.

Rule, S. A. J., Hooker, M., Costello, C., Luck, W., and Hoffbrand, A. V., 1994, Serum vitamin B12 and transcobalamin levels in early HIV disease, *Am. J. Hematol.* **47:**167–171.

Semba, R. D., Graham, N. M. H., Caiaffa, W. T., Margolick, J. B., Clement, L. and Vlahov, D., 1993a, Increased mortality associated with vitamin A deficiency during human immunodeficiency virus type 1 infection, *Arch. Intern. Med.* **153:**2149–2154.

Semba, R. D., Ward, M. B. J., Griffin, D. E., Scott, A. L., Natadisastra, G., West, K. P., Jr., and Sommer, A., 1993b, Abnormal T-cell subset proportions in vitamin-A deficient children, *Lancet* **341:**5–8.

Semba, R. D., Miotti, P. G., Chiphangwi, J. D., Saah, A. J., Canner, J. K., Dallabetta, G. A., and Hoover, D. R., 1994, Maternal vitamin A deficiency and mother-to-child transmission of HIV-1, *Lancet* **343:**1593–1597.

Serwadda, D., Sewankamboo, N. K., Carswell, J. W., Bayley, A. C., Redder, R. S., and Weiss, R. A., 1985, Slim disease: A new disease in Uganda and its association with HTLV-III infection, *Lancet* **II:**849–852.

Smith, J., Howells, D. W., Kendall, B., Levinsky, R., and Hyland K., 1987, Folate deficiency and demyelination in AIDS, *Lancet* **II:**215.

Sommer, A., Katz, J., and Tarwotjo, I., 1984, Increased risk of respiratory disease and diarrhea in children with preceding mild vitamin A deficiency, *Am. J. Clin. Nutr.* **40:**1090-1095.

Sommer, A., Tarwotjo, I., Djunaedi, E., West, K. P., Loeden, A. A., Tilden, R., Mele, L., and the ACEH study group, 1986, Impact of vitamin A supplementation on childhood mortality, *Lancet* **1:**1169–1173.

Sommer, A., Tarwotjo, I., and Katz, J., 1987, Increased risk of xerophtalmia following diarrhea and respiratory disease, *Am. J. Clin. Nutr.* **45:**977–980.

Sönnerborg, A., Carlin, G., Akerlund, B., and Jarstrand, C., 1988, Increased production of malondialdehyde in patients with HIV infection, *Scand. J. Infect. Dis.* **20:**287–290.

Stephensen, C. B., Alvarez, J. O., Kohatsu, J., Harmeier, R., Kennedy, J. I., and Gammon, Jr., R. B., 1994, Vitamin A is excreted in the urine during acute infection, *Am. J. Clin. Nutr.* **60:**388–392.

Tang, A. M., Graham, N. H., Kirby, A. J., Mc Call, D., Willett, W. C. and Saah, A. J., 1993, Dietary micronutrient intake and risk of progression to acquired immunodeficiency syndrome (AIDS) in human immunodeficiency virus type 1 (HIV-1) infected homosexual men, *Am. J. Epidemiol.* **138:**937–951.

Ullrich, R., Schneider, T., Heise, W., Schmidt, W., Averdunk, R., and Zeitz, M., 1994, Serum carotene deficiency in HIV-infected patients, *AIDS* **8:**661–665.

Wallingford, J. C., and Underwood, B. A., 1986, Vitamin A deficiency in pregnancy, in *Vitamin A Deficiency and Its Control* (J. C. Bauenfeind ed.), pp. 101–152, Academic Press, Orlando.

Watson, R. R., Yaha, M. D., Darban, H. R., and Prabhala, R. H., 1988, Enhanced survival by vitamin A supplementation during a retrovirus infection causing murine AIDS, *Life Sci.* **43:**12–18.

Part II

Carotenoids

Chapter 7

The Bioconversion of β-Carotene into Retinoids

Xiang-Dong Wang and Norman I. Krinsky

1. INTRODUCTION

Vitamin A (or retinol) is an essential nutrient for all mammals since it cannot be synthesized in the body. The most important physiologically active derivative of vitamin A is retinoic acid, since this molecule fulfills the vitamin A requirements for growth and maintenance of normal epithelial function (Goodman and Blaner, 1984). However, retinoic acid cannot fulfill all vitamin A requirements, for the visual and the reproductive systems require retinal or retinol, respectively. Retinoic acid is a well-known inducer of cell differentiation, particularly in certain carcinoma cell lines, and this property has been used in the treatment of acute promyelocytic leukemia (Warrell *et al.*, 1993; Chen *et al.*, 1991). Two families of nuclear retinoic acids receptors (RARs and RXRs) have been cloned that are active in receptor-mediated regulation of gene transcription (for reviews see Chambon, 1996, and Chapters 2 and 3 of this volume). It is now well established that cellular actions of vitamin A are mediated by the binding of

Xiang-Dong Wang and Norman I. Krinsky Gastrointestinal Nutrition Laboratory, Jean Mayer USDA Human Nutrition Research Center on Aging at Tufts University, and the Department of Biochemistry, School of Medicine, Tufts University, Boston, Massachusetts 02111.

Subcellular Biochemistry, Volume 30: Fat-Soluble Vitamins, edited by Quinn and Kagan. Plenum Press, New York, 1998.

retinoic acid to specific nuclear receptor proteins (for reviews see Pfahl and Chytil, 1996, and Chapter 1 of this volume).

Among the more than 560 carotenoids found in nature, about 50 have some provitamin A activity. β-Carotene is the most active precursor of vitamin A (Olson, 1996). It is a symmetrical molecule with 11 conjugated double bonds in its polyene chain and rings (Fig. 1). Recently, a number of studies have shown that β-carotene can generate retinoic acid, bypassing retinal formation (Hebuterne *et al.*, 1996; Wang *et al.*, 1992b; Napoli and Race, 1988). More recently, 9-*cis*-β-carotene has been demonstrated both *in vitro* (Wang *et al.*, 1994; Nagao and Olson, 1994) and *in vivo* (Hebuterne *et al.*, 1995) to be a precursor of either 9-*cis*-retinoic acid or 9-*cis*-retinal. Whereas all-*trans*-retinoic acid activates the RAR family of nuclear receptors, 9-*cis*-retinoic acid activates both the RAR and RXR receptors (Levin *et al.*, 1992). As such, these metabolites of provitamin A carotenoids play an important role in regulating cell function. Given that carotenoids provide between one third and one half of the daily vitamin A intake in the diet of developed societies and 70–90% of the total vitamin A intake in the less developed societies of Africa and Asia (FAO/WHO, 1988), the understanding of the mechanism(s) of carotenoid conversion into retinol and retinoic acid is an important and fundamental issue.

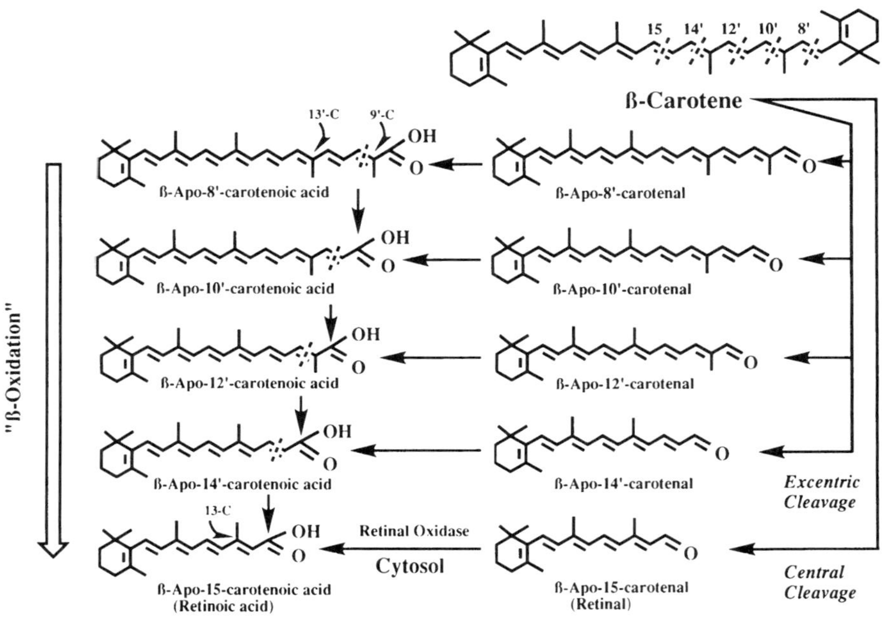

FIGURE 1. Structures of metabolites in the metabolic pathway of β-carotene. Reproduced with permission from Wang *et al.* (1996).

There are at least two proposed mechanisms for the biosynthesis of vitamin A (retinol) and retinoic acid from β-carotene (Fig. 1). The central cleavage pathway utilizes a 15,15′-dioxygenase which theoretically should yield two molecules of retinal (retinaldehyde) (Goodman and Olson, 1969; Olson and Hayaishi, 1965; Goodman and Huang, 1965), which can either be reduced to retinol or oxidized to retinoic acid (Crain *et al.*, 1967) (Fig. 1). This central cleavage mechanism has been accepted as the pathway for the formation of vitamin A from other provitamin A carotenoids. The excentric cleavage mechanism yields a series of β-apo-carotenals of different chain lengths. These β-apo-carotenals are subsequently oxidized to their corresponding β-apo-carotenoic acids. By a process analogous to fatty acid β-oxidation, the β-apo-carotenoic acids are converted stepwise to retinoic acid, where the β-oxidation-type process is stopped by the presence of the methyl substituent on carbon 13 (Fig. 1). This hypothesis was first put forth by Glover (1960) and supported by the work of Ganguly and co-workers (Ganguly and Sastry, 1985). Since we lack information about the specific enzyme(s) that carry out these important biological processes, the existence of two cleavage pathways in mammals had been a controversial issue among scientists for several decades (Olson, 1989). One of the major problems is that enzyme(s) for both pathways have not yet been purified and characterized.

Although the exact cleavage mechanism of β-carotene needs to be studied further, recent evidence clearly demonstrated that both a central and an excentric cleavage pathway are involved in biosynthesis of retinoids from β-carotene. More specifically, central cleavage is the major pathway leading to vitamin A formation, and excentric cleavage is involved in the metabolism of β-carotene into other oxidative metabolites, e.g., apo-carotenoids and retinoic acid. Furthermore, it is possible that the two pathways may be independently regulated and play a different role in either physiological or pathological states. Since there have been several very carefully reviewed papers on the early work on β-carotene conversion into retinoids (Krinsky *et al.*, 1993; Ganguly and Sastry, 1985; Goodman and Olson, 1969; Glover, 1960), we will focus on most recent research on this subject in this review.

2. CONVERSION OF β-CAROTENE INTO RETINOIDS

2.1. Central Double-Bond Cleavage of β-Carotene: A Major Pathway Leading to Vitamin A Formation

Since the mid-1960s, based on the work of Olson and Goodman (Olson and Hayaishi, 1965; Goodman and Huang, 1965), it has been widely accepted that β-carotene is converted to vitamin A via a putative dioxetane intermediate formed across the central, 15,15′ double bond of β-carotene by a soluble 15,15′-dioxygenase enzyme present in the intestinal mucosa and liver (Fig. 2). The original experiments were carried out in rats, but many other species have similar enzymatic activities (Goodman and Blaner, 1984). The product of the reaction, retinal, can then be either reduced to retinol or oxidized further to form retinoic acid. Since these workers did not find any other reaction products, they called the enzyme "β-carotene 15,15′-dioxygenase" (EC 1.13.11.21), which will be referred to as "dioxygenase" in this chapter. This observation was extended by a number of workers using different species as the source of the enzyme (Lakshmanan *et al.*, 1968; Villard and Bates, 1986). Recently, several investigators have extended these observations using enzyme preparations from either intestinal mucosa of rats (Duszka *et al.*, 1996; Nagao and Olson, 1994; van Vliet *et al.*, 1992), hamster (van Vliet *et al.*, 1992), or guinea pig (Devery and Milborrow, 1994) as the source of the β-carotene cleavage enzyme activity. These investigators confirmed that retinal is the major product of β-carotene central cleavage in an *in vitro* system. In some cases, the yield of product formed from β-carotene was determined, and these results are summarized in Table I. Among these studies, Devery and Milborrow (1994) achieved a stoichiometry of 1.5–2 moles of retinal produced per mole of β-carotene cleaved by hamster intestinal mucosal cytosol, which provides conclusive evidence for the central cleavage mechanism of β-carotene metabolism. In their study, the β-carotene dioxygenase was obtained as a soluble enzyme which was partially purified by gel filtration and ion-exchange chromatography. Their improved assay procedure included adding the antioxidants butylated hydroxytoluene to the organic solvents and

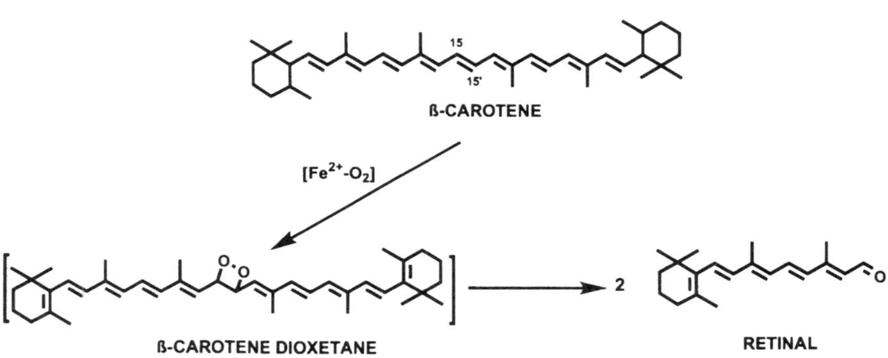

FIGURE 2. Proposed mechanism for the central double-bond cleavage of β-carotene.

glutathione to the aqueous solvents. The rate of conversion of 1 μM β-carotene to retinal reached 300 pmol/hr/mg protein (Table I). The most active dioxygenase activity is associated with the supernatant fraction after centrifugation at 9000–10,000 × *g* for rat liver, and 100,000 × *g* for guinea pig liver (Table I). It has been reported that β-apo-carotenals, the potential intermediates of excentric cleavage, are converted into retinal by the mucosal cytoplasmic dioxygenase at a markedly faster rate than that of β-carotene (Lakshmanan *et al.*, 1968). Thus, many of these studies with stoichiometry of less than 1 mole retinal/mole β-carotene cannot exclude the possibility that β-apo-carotenals were formed and very rapidly converted to retinal. The study by Wang *et al.* (1991), using a low-speed centrifugation (800 × *g*) to remove nuclei, showed that the production of β-apo-carotenals from β-carotene is linear for 30 min, whereas there was a lag in the appearance of retinal, clearly suggesting that the β-apo-carotenals are intermediate compounds in retinal formation in this system. The detection of retinal has been interpreted as only occurring via the central cleavage of β-carotene, despite the facile ability of the dioxygenase to cleave β-apo-carotenals to yield retinal.

Table I
The Formation of Retinal and Retinoic Acid from β-Carotene after Incubation with Various Intestinal Mucosal Fractions[a]

Species	Centrifugation at which enzyme fraction was prepared (×*g*)	β-Carotene concentration (μM)	Product formed (pmol/mg protein/hr)		Yield[b]	Reference
			Retinal	Retinoic acid		
Guinea pig	100,000	1.0	300	0	1.5–2.0	Devery and Milborrow (1994)
Pig	10,000	15.0	480	0	1.88	Nagao and Olson (1994)
Hamster	9,000	2.8	30	0		van Vliet *et al.* (1996)
Human	800	2.0	19	8		Wang *et al.* (1991)
Ferret	800	2.0	6	2		Wang *et al.* (1991)
Rat	105,000	2.8	9	7		Duszka *et al.* (1996)
Rat	104,000	0.8	15–30	—		Villard and Bates (1986)
Rat	100,000	1.4	0	0		Hansen and Maret (1988)
Rat	100,000	10.0	0	224		Napoli and Race (1988)
Rat	10,000	15.0	694	0		During *et al.* (1996)
Rat	9,000	2.8	30	0		van Vliet *et al.* (1996)
Rat	9,000	2.8	199	66		Duszka *et al.* (1996)
Rat	800	2.0	6	3		Wang *et al.* (1991)

[a]The enzyme preparation, purification, and incubation conditions differed in each citation.
[b]Defined as the number of moles of product/mole of substrate β-carotene.

In recent *in vivo* intestinal perfusion experiments, we have demonstrated that retinol and retinyl esters are the major metabolites formed in ferrets (Hebuterne *et al.*, 1996; Wang *et al.*, 1992a). There is a significant increase in the concentration of both retinol and retinyl esters in the ferret intestinal mucosa after perfusion of either β-carotene, β-apo-14′-carotenal, or retinal (Table II). When ^{14}C-labeled β-carotene was perfused through the ferret intestine, both retinol and retinyl esters account for about 72% of the radioactivity recovered in the lymph after 4 hr of perfusion (Table III) (Wang *et al.*, 1992a). Of particular interest in this study is the observation that the portal vein plays an important role in the transport of β-carotene metabolites, and that the differential absorption of β-carotene and its metabolites into lymph or into portal blood is dependent on the polarities of the metabolites involved. Of the total radioactivity removed from the perfusate, only 3.2% was found in the lymph after 4 hr, whereas 28% was transported via the portal vein (Table III). This observation may be due to the fact that the differential flow rate between lymph and portal vein is very large. It was noted in earlier studies that of the absorbed radioactivity from ^{14}C-β-carotene, only 3–6% could be found in the lymph in the rat (Huang and Goodman, 1965) and 8.7–16.8% in the lymph of humans (Blomstrand and Werner, 1967; Goodman *et al.*, 1966), even after prolonged lymph collection. As seen in Table III, β-carotene, retinyl esters, retinol, and the less-polar metabolites are absorbed via the lymphatic system, whereas the more-polar metabolites, which include retinoyl-β-glucuronide, retinyl-β-glucuronide, β-apo-carotenals, and retinoic acid, are absorbed directly into the portal blood (Wang *et al.*, 1993). No intact ^{14}C-β-carotene was detected in the portal blood.

2.2. Excentric Double-Bond Cleavage of β-Carotene: Biosynthesis of Retinoic Acid and Other Oxidative Products

In 1960, Glover hypothesized that β-carotene could undergo both central and excentric cleavage, but it was not until 1976 that Sharma *et al.* (1977, 1976) showed strong evidence of the excentric cleavage of β-carotene. They isolated significant amounts of β-apo-carotenals as well as retinal from the intestine of chickens given β-carotene. They showed further that when β-apo-8′-carotenal was fed to rats, it was rapidly oxidized to the β-apo-8′-, 10′-, and 12′-carotenoic acids and also led to the deposition of considerable amounts of retinyl esters in the liver (Sharma *et al.*, 1977). In a 1985 review, Ganguly and Sastry claimed that the "weight of the evidence so far is overwhelmingly in favor of random cleavage and is against any specific attack at the central double bond of the carotene molecule." An interesting paper by Hansen and Maret (1988) failed to demonstrate any enzymatic

Table II
The Concentration of Retinoic Acid, β-apo-14′-Carotenoic Acid, Retinol, and Retinyl Esters Formed in the Ferret after a 2-hr Intestinal Mucosa Perfusion of 10 μM β-Carotene, 10 μM β-apo-14′-Carotenal, or 1 μM Retinal with or without 2 mM Citral[a]

	Substrate perfused (means ± SEM)							
	β-Carotene ($n = 4$)		β-apo-14′-Carotenal ($n = 5$)		Retinal ($n = 3$)		None ($n = 1$)	
Product analyzed	−Citral	+Citral	−Citral	+Citral	−Citral	+Citral	−Citral	+Citral
Retinoic acid	36 ± 3	19 ± 5*	51 ± 4	30 ± 3	30 ± 2	ND	ND	ND
β-apo-14′-Carotenoic acid	ND	ND	218 ± 28	245 ± 37	ND	ND	ND	ND
Retinol	251 ± 36	437 ± 68*	329 ± 11	884 ± 10*	994 ± 337	2084 ± 185*	83	276
Retinyl esters	7310 ± 2882	6847 ± 2978	1523 ± 117	1655 ± 143	10,440 ± 2176	5446 ± 897	2483	3187

[a]Data from Hebuterne *et al.*, (1996), Liu *et al.* (1997), with permission. ND, Not detected. Asterisk indicates significantly different from the no-citral control ($p < .05$).

Table III
Percentage Distribution of Radioactivity in the Intestine, Lymph, and Portal Vein Blood of Ferrets Collected over 4 hr of Intestinal Perfusion of 15,15′-^{14}C-β-Carotene *in Vivo*[a]

	Percentage distribution of ^{14}C in metabolites[b]		
Metabolite in HPLC fraction	Intestine [68.6%]	Lymph [3.2%]	Portal blood [28.2%]
More polar than retinoic acid	25	0	29
Retinoic acid	10	3	4
Retinol	6	9	8
β-apo-Carotenals	11	4	16
Retinyl esters	15	63	36
β-Carotene	23	10	0
Other fraction	10	11	7

[a]Modified with permission from Wang *et al.* (1992a). The samples of intestine, lymph, and portal vein blood of ferrets were analyzed by using the HPLC described in Wang *et al.* (1992a).
[b]Brackets give percent of total distribution of radioactivity.

cleavage of β-carotene in rat intestinal mucosa cytoplasmic fractions, but their results suggested that small amount of β-apo-carotenals were formed in vitamin A-deficient rats. Dmitrovskii (1991) claimed that apo-carotenoids can be formed from β-carotene in chicken intestinal mucosal membrane in the presence of NADPH. Since the major metabolite is apo-12′-carotenal, they referred to the enzyme activity, localized in the nuclear-membrane fraction of the enterocyte, as NADPH-dependent carotene-11,12-dioxygenase. Dmitrovskii (1991) also reported that xanthophylls such as canthaxanthin, isozeaxanthin, and astaxanthin could also be metabolized; however, there was no detailed experimental procedure described. In order to assay for excentric cleavage products of β-carotene metabolism, *in vitro* experiments were carried out utilizing the postnuclear fraction (low centrifugation, 800 × *g*) of intestinal mucosa (Wang *et al.*, 1991). We have shown that when purified β-carotene is incubated with homogenates of small intestine and other tissues, a series of homologous carbonyl cleavage products are produced including β-apo-14′-, 12′-, 10′-, and 8′-carotenal, as well as β-apo-13-carotenone, retinoic acid, and retinal (β-apo-15-carotenal) (Wang *et al.*, 1991; Tang *et al.*, 1991). However, further isolation of enzymatic activity has been unsuccessful. Only minor amounts of any metabolites were detected when all-*trans*-β-carotene was incubated with isolated mitochondrial, microsomal, or cytosolic fractions of human intestinal mucosa cell (Wang *et al.*, 1994). The lower enzymatic activity of the isolated fractions may have been caused by the fractionation procedure, effects of detergent, or lack of cofactor(s) for the oxidation reaction. The postnuclear fraction consistently formed both central and excentric cleavage products when incubated with either all-*trans*- or 9-*cis*-β-carotene (Wang *et al.*, 1994, 1992b, 1991; Tang *et al.*, 1991). Human intestinal homogenates have been shown to be particularly useful, inasmuch as the conversion of β-carotene to vitamin A in intestinal mucosa fractions in different animal species is usually lower than in comparison to humans *in vitro* (Wang *et al.*, 1991).

As mentioned above, lower specific activities of the isolated cellular fractions increase the difficulty of isolating an active enzyme, and thus facilitating an understanding of the cleavage mechanism. Several possible mechanisms, including lipid peroxidation, enzymatic cooxidation by lipoxygenase, cytochrome P-450, and peroxidases, have been proposed to account for excentric cleavage (Hansen and Maret, 1988). We have demonstrated that one possible cleavage mechanism is via a coupled oxidation with lipoxygenase and unsaturated fatty acids in gastric tissue (Yeum *et al.*, 1995). However, we do not know if this coupled oxidation cleavage mechanism is a major pathway or is tissue specific, because lipoxygenase has not been described in the intestine, but was reported to occur in the stomach (Hansen and Maret, 1988; Hove, 1943).

In 1988, Napoli and Race demonstrated the ability of β-carotene to serve as a precursor of retinoic acid in rat intestine *in vitro*; retinal was not detected as an intermediate in this study. They stated that determining whether a pathway exists for retinoic acid synthesis that does not require retinol or retinal as a substrate is a prerequisite to understanding the metabolic relationship between carotenoids and retinoids. Since retinal was not detected in their studies, they suggested that β-carotene was converted to retinoic acid through a pathway that did not go through retinal. It is interesting that the retinoic acid produced from β-carotene represents one third (Wang *et al.*, 1991) of the total retinoids formed (Wang *et al.*, 1994), even when citral (an inhibitor of retinal oxidation) is added to the incubation mixture (Wang *et al.*, 1992b). Of particular interest is the observation that the retinoic acid produced by the excentric cleavage pathway of β-carotene represents at least 50% of the retinoic acid formed *in vivo* (Hebuterne *et al.*, 1996) even when citral is added to the perfusate (Table II). Both retinoic acid and β-apo-14′-carotenoic acid are formed after the intestinal perfusion of β-apo-14′-carotenal in the absence and the presence of citral (Table II). In view of these observations, we concluded that in addition to the central cleavage pathway, an excentric cleavage mechanism is involved in the metabolism of β-carotene into retinoic acid.

Since earlier studies showed that β-apo-carotenals formed from the cleavage of β-carotene are rapidly oxidized to the corresponding β-apo-

Table IV
The Formation of Metabolites from Various β-apo-Carotenoic Acids in the Mitochondrial Fraction of Rabbit Liver[a]

	β-apo-Carotenoic acid product (pmol/mg protein/hr)			
Substrate	10′-RA	12′-RA	14′-RA	15-RA
β-apo-8′-Carotenoic acid	35 ± 5	25 ± 2	16 ± 3	11 ± 2
β-apo-12′-Carotenoic acid			32 ± 4	18 ± 3
β-apo-14′-Carotenoic acid				30 ± 7

[a]The incubations were carried out at 10 μM substrate concentration. Data are means ± SEM, from at least three independent determinations. Reproduced with permission from Wang *et al.* (1996).

carotenoic acids both in animals (Sharma *et al.*, 1977, 1976) and humans (Zeng *et al.*, 1992), we investigated whether the mitochondrial fraction of rabbit liver could carry out the further metabolism of β-apo-carotenoic acids, with particular emphasis on the formation of products indicative of oxidation at the β-position of each carotenoic acid (Fig. 1). Using β-apo-8′, 12′-, and 14′-carotenoic acids as substrates, comparable rates of β-apo-10′-carotenoic acid, β-apo-14′-carotenoic acid, and β-apo-15-carotenoic acid (retinoic acid) were achieved, respectively, using the rabbit liver mitochondrial fraction as the enzyme source (Table IV). These results support the hypothesis that β-apo-carotenoic acids are oxidized in the β-position, removing successively two carbon units from the linear polyene chain. Since the rate of formation of each lower homologue is approximately the same when the individual β-apo-carotenoic acids were used as substrates, this would suggest that the methyl groups on carbons 9′ and 13′ (which are in α-position, see Fig. 1) do not interfere with the β-oxidation-like process of β-apo-carotenoic acid metabolism. However, the β-oxidation-type process would be blocked by a methyl group in the β-position, which occurs with the methyl group on carbon 13, thus preventing any further oxidation beyond carbon 15, i.e., forming retinoic acid as the terminal oxidation product of this β-oxidation-like process of β-apo-carotenoic acids. This hypothesis is supported by our results that retinoic acid formation was increased in the order β-apo-14′- > 12′- > 8′-carotenoic acid when these carotenoic acids were incubated with the mitochondrial fraction (Table IV). Additional evidence for the stepwise oxidation is that the products of β-apo-8′-carotenoic acid in mitochondria occurred in the order β-apo-10′- > 12′- > 14′- > 15-carotenoic acid (retinoic acid) (Table IV).

To study the metabolism of β-apo-carotenoic acid *in vivo*, we carried out a kinetic study after infusing β-apo-8′-carotenoic acid through the por-

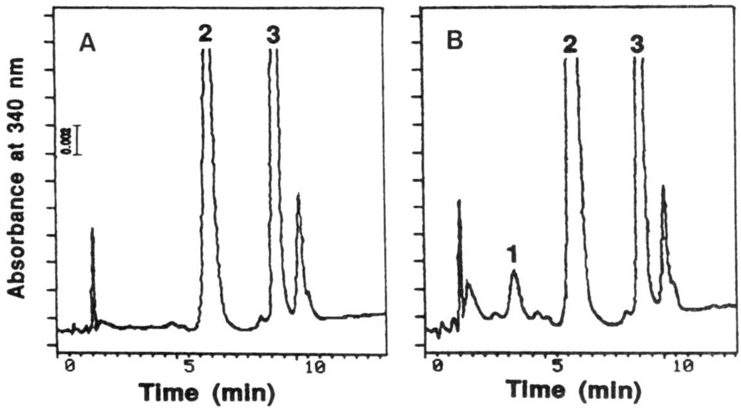

FIGURE 3. HPLC profile of the ferret liver (100 mg) before (A) and after (B) perfusion of 10 μM β-apo-8′-carotenoic acid into the portal vein. Peak 1, retinoic acid; peak 2, retinol; peak 3, retinyl acetate (internal standard). Reproduced with permission from Wang *et al.* (1996).

tal vein of ferrets, and monitoring retinoic acid secretion in bile and accumulation of retinoic acid in the liver. Since the increase of retinoic acid concentration in the ferret intestine and portal blood after the intestinal perfusion of β-carotene has been demonstrated using chemical derivatization and gas chromatography/mass spectrometry techniques (Wang *et al.*, 1993), and one of the major pathways of retinoic acid metabolism is through glucuronidation and secretion in the bile, we reasoned that retinoic acid secreted in the bile would reflect the metabolic fate of β-apo-carotenoic acids in the liver. After the perfusion of 10 μm of β-apo-8′-carotenoic acid for 180 min, we detected 150 pmol of retinoic acid/g liver sample (Fig. 3), equivalent to 4500 pmol/whole liver (based on the liver weight of 30 g/1.5 kg ferret; Fox, 1992). During this same 3-hr perfusion period, we collected 90 pmol of retinoic acid in the bile, indicating that only a very small fraction of the newly formed retinoic acid is secreted in the bile, and that the β-oxidation-like pathway of the β-apo-carotenoic acids may contribute to the steady-state concentration of retinoic acid in the body (Wang *et al.*, 1996; Hebuterne *et al.*, 1996).

2.3. Differentiation and Regulation between Central and Excentric Cleavage Pathway of β-Carotene

Based on the information discussed above, it seems that there are two different enzyme systems in cells that convert β-carotene into retinoids

(Krinsky *et al.*, 1992), namely specific cleavage at the 15,15′-double bond by cytosolic β-carotene 15,15′-dioxygenase to form retinal (central cleavage), or nonspecific cleavage at any of the linear double bonds (7′,8′-; 9′,10′-; 11′,12′-; 13′,14′-; and 15,15′-) by some unidentified, membrane-associated enzyme(s) to yield both retinal and β-apo-carotenals as products (excentric cleavage). The β-apo-carotenals formed from β-carotene may undergo additional cleavage reactions to form retinal or be oxidized to their corresponding apocarotenoic acid, which, in a process analogous to β-oxidation, ultimately yields retinoic acid.

In order to differentiate between the central and excentric cleavage pathways of β-carotene metabolism, several experiments have been reported in both *in vivo* and *in vitro* studies during which one of the pathways has been blocked by various inhibitors. An example of this type of study is shown in Fig. 4. In our previous *in vitro* study (Wang *et al.*, 1992b), we could not detect any significant decrease in retinoic acid formation when citral was added to the incubation mixture of human intestinal homogenates, suggesting that the excentric cleavage pathway contributes to the production of retinoic acid from β-carotene under our experimental conditions. What is probably more important is our observation that citral caused a 42–47% decrease in retinoic acid synthesis in the intestinal mucosa of ferrets after the *in vivo* perfusion of either β-apo-14′-carotenal or β-carotene, suggesting that *both* central and excentric cleavage pathways were operating to produce retinoic acid in approximately equal amounts (Table II). Also, there is no reduction of retinoic acid formation when citral was added to the liver perfusate of β-apo-8′-carotenoic acid *in vivo* (Fig. 4).

3-Mercaptopropionic acid is a potent inhibitor of long-chain acyl coenzyme A (CoA) dehydrogenase in coupled rat heart mitochondria (Sabbagh *et al.*, 1985). This interesting property prompted us to examine the effects of 3-mercaptopropionic acid on β-apo-carotenoic acid metabolism in perfused liver from ferrets. We have demonstrated the immediate inhibition of retinoic acid formation after the addition of 2 mM 3-mercaptopropionic acid to the perfusate (Fig. 4), strongly suggesting that β-apo-carotenoic acid metabolism goes through a step analogous to the dehydrogenase step in the β-oxidation of fatty acids. Long-chain fatty acids must first enter mitochondria by a carnitine-dependent process to be available for β-oxidation, and the carnitine palmitoyl transferase system may be rate-limiting for long-chain fatty acid oxidation. 2-Tetradecylglycidic acid inhibits oxidation of long-chain fatty acids in the liver, by irreversibly inhibiting carnitine palmitoyl transferase I (Kiorpes *et al.*, 1984). The *in vivo* addition of 2-tetradecylglycidic acid in the perfusate caused a 75% decrease in retinoic acid secretion in bile (Fig. 4), which suggests that a similar system might

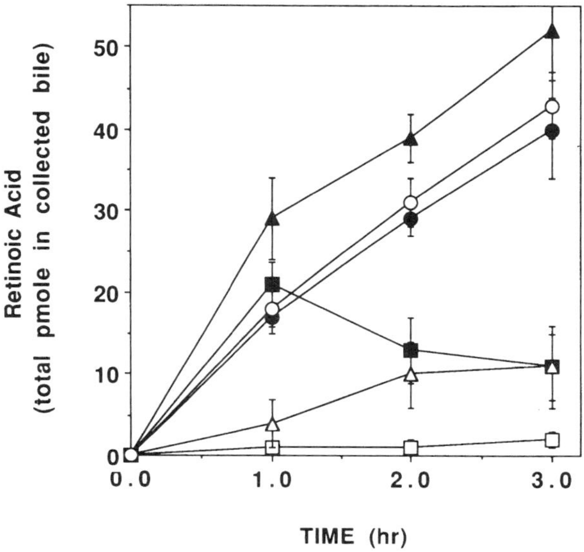

FIGURE 4. Effect of citral, 2-tetradecylglycidic acid, and 3-mercaptopropionic acid on the production of retinoic acid from β-apo-8′-carotenoic acid in the perfused ferret liver, as measured by collecting retinoic acid in bile. In each group of three ferrets, livers were perfused with a continuous infusion of β-apo-8′-carotenoic acid with varying conditions, as follows: 2 μM β-apo-8′-carotenoic acid plus 2 mM citral (closed triangles); 10 μM β-apo-8′-carotenoic acid (open circles); 2 μM β-apo-8′-carotenoic acid only (closed circles); 2 μM β-apo-8′-carotenoic acid plus 2 mM 3-mercaptopropionic acid (open squares); addition of 2 mM 3-mercaptopropionic acid after 1 hr to 2 μM β-apo-8′-carotenoic acid (closed squares); 2 μM β-apo-8′-carotenoic acid plus 2 mM 2-tetradecylglycidic acid (open triangles). Each point represents the means ± SEM. Reproduced with permission from Wang *et al.* (1996).

exist with respect to carotenoic acid transport into mitochondria. This hypothesis was also supported by our observation that the biosynthesis of retinoic acid declined markedly when oleoyl CoA, a substrate for carnitine palmitoyl transferase I, was added to the incubation mixture (Wang *et al.*, 1996). The addition of a microsomal fraction did not stimulate mitochondrial oxidation of β-apo-carotenoic acids, suggesting that whatever enzymes or cofactors are necessary to activate and facilitate carotenoic acid entry into mitochondria are contained in the mitochondrial fraction alone. Although we do not have direct evidence that β-apo-carotenoic acid converts to β-apo-carotenoyl-CoA before its oxidation, the deletion of CoA resulted in a 70% inhibition of retinoic acid formation, indicating that CoA plays an important role in the oxidation of β-apo-carotenoic acids in mitochondria (Wang *et al.*, 1996). Based on these studies, it seems that the excentric

Table V
The Concentration of α-Tocopherol, β-Carotene, Retinol, and β-apo-12′-Carotenal in Ferret Intestinal Mucosa after a 3-hr Perfusion of 5 μM β-Carotene with Varying Concentrations of α-Tocopherol[a]

	αT = 0.0 μM	αT = 0.4 μM	αT = 5.0 μM	αT = 10.0 μM
α-Tocopherol (nmol/g)	3 ± 3[a]	32 ± 11[b]	158 ± 75[c]	268 ± 75[d]
β-Carotene (nmol/g)	1.4 ± 0.4[a]	1.8 ± 0.4[a]	1.8 ± 0.8[a]	1.1 ± 0.2[a]
Retinol (pmol/g)	10 ± 3[a]	44 ± 19[ab]	73 ± 20[b]	89 ± 24[b]
β-apo-12′-carotenal (pmol/g)	106 ± 20[a]	117 ± 7[a]	87 ± 10[a]	70 ± 18[a]

[a]αT, α-Tocopherol Values are means ± SEM ($n = 3$). Data in each line not sharing a common superscript letter are significantly different at $p < .05$. Reproduced with permission from Wang *et al.* (1994).

cleavage pathway contributes at least half of retinoic acid synthesis in the liver. The major metabolic pathway of β-carotene in both intestine and liver is summarized in Fig. 5.

The conversion of β-carotene into retinoids may be regulated by several factors. Vitamin A status, such as sufficient (van Vliet *et al.*, 1996) or deficiency (Villard and Bates, 1986), may down- or upregulate β-carotene cleavage activity. The activity of the enzyme carotene dioxygenase on β-carotene conversion is dependent on intake of dietary protein (Gronowska-Senger and Wolf, 1970). Sklan (1983) suggested that dioxygenase enzyme may be zinc dependent, but there was no information about the regulation of β-carotene cleavage activity. A recent observation has show that α-tocopherol enhances the conversion of β-carotene to vitamin A (Wang *et al.*, 1995). As shown in Table V, the retinol level in the intestinal mucosa increased in a dose-dependent manner with increasing concentrations of α-tocopherol in the perfusate. There was an insignificant decrease in the β-apo-12′-carotenal level as a function of the α-tocopherol concentration. One mechanism for this α-tocopherol effect could be through an enhancement of the conversion of β-carotene into retinol. Another possible explanation is that α-tocopherol may inhibit the excentric cleavage pathways of β-carotene, either chemical or enzymatic, thereby enhancing the central cleavage pathway and generating more retinol. An understanding of the mechanism of these effects will require further study.

2.4. Biological Significance of the Central and Excentric Cleavage of β-Carotene

Retinoids play an important role in the control of cellular proliferation and differentiation. Retinoid signaling occurs through nuclear receptors,

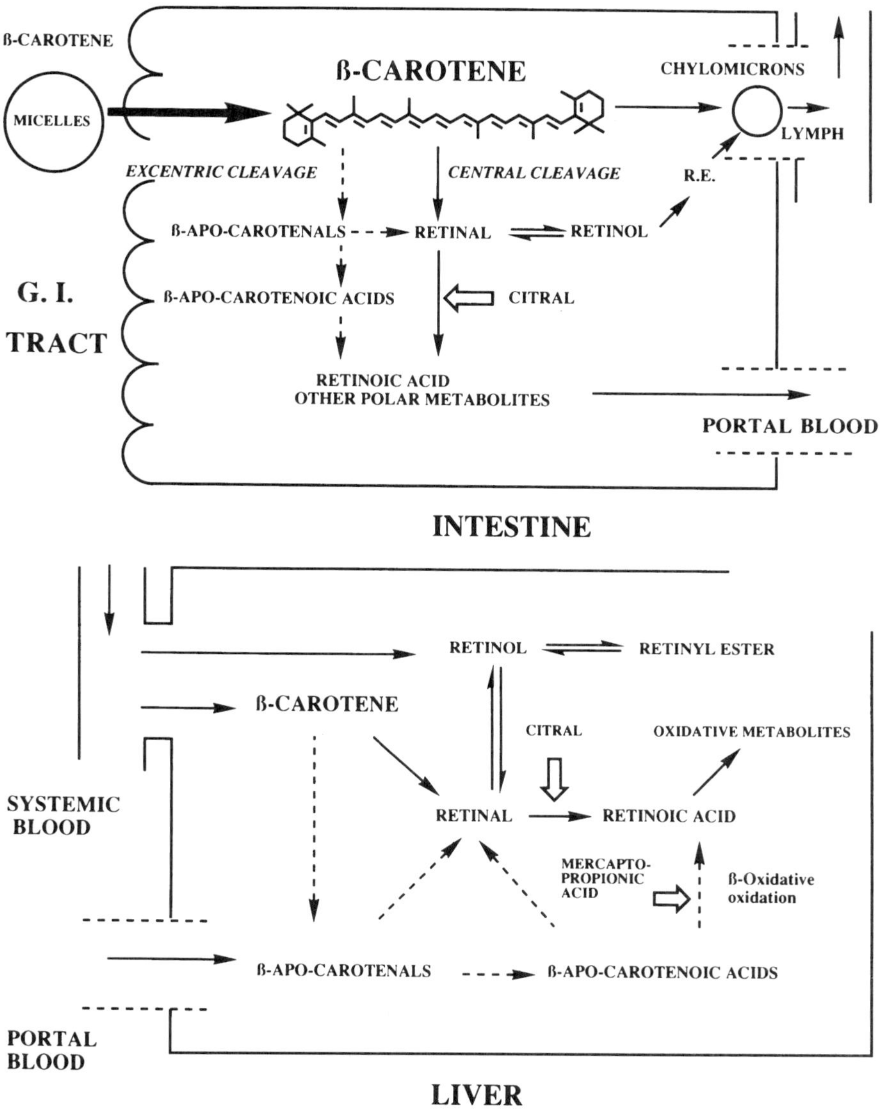

FIGURE 5. Major metabolic pathway of β-carotene in both intestine and liver.

which appear to act as transcription factors (for review see Pfahl and Chytil, 1996). There are at least two biologically active retinoic acid stereoisomers, all-*trans* and 9-*cis*-retinoic acid, and two classes of retinoid receptors, retinoic acid receptors (RARα, β, and γ) and retinoid X receptors (RXRα, β, and γ). RARs bind all-*trans*- and 9-*cis*-retinoic acid, whereas RXRs bind only 9-*cis*-retinoic acid. The identification of 9-*cis*-retinoic acid as the specific ligand for the RXR class of nuclear receptors (Heyman *et al.*, 1992; Levin *et al.*, 1992) raises the question as to the source of 9-*cis*-retinoic acid in the body. There is little information available about the formation of 9-*cis*-retinoic acid, although it is known that it is not formed from all-*trans*-retinoic acid during a normal extraction procedure (Heyman *et al.*, 1992). 9-*cis*-Retinoic acid may arise via isomerization of all-*trans*-retinoic acid in the body, by the oxidation of 9-*cis*-retinal, or through the conversion of 9-*cis*-β-carotene to this biologically active retinoid. The very rapid clearance of retinoic acid from body fluids by tissues increases the significance of the conversion of tissue 9-*cis*-β-carotene. An important question is whether the intestinal cleavage of dietary 9-*cis*-β-carotene or tissue cleavage of stored 9-*cis*-β-carotene can provide a source of 9-*cis*-retinoic acid for the human body. The enzymatic conversion of either all-*trans*-β-carotene or 9-*cis*-β-carotene into all-*trans*-retinoic acid and 9-*cis*-retinoic acid by incubation of homogenates of human intestinal mucosa *in vitro* and intestinal perfusion of ferret *in vivo* has been reported (Hebuterne *et al.*, 1995; Wang *et al.*, 1994). Since equal amounts of 9-*cis*-retinoic acid and all-*trans*-retinoic acid were formed from the 9-*cis*-β-carotene (Fig. 6), we cannot draw any conclusions as to whether the cleavage was primarily central, excentric, or a combination of both. However, the results are consistent with β-carotene functioning as a precursor of retinoic acid under physiological conditions, and therefore, through its metabolism, regulating cell function.

The biological significance of the excentric cleavage pathway is unknown, besides serving as an alternative pathway for retinoic acid formation. Recently, it has been reported that excentric cleavage metabolites other than retinoic acid have biological activity in HL-60 cells (Suzuki *et al.*, 1995). It was found that all-*trans*-β-apo-12′-carotenoic acid and 15′-*cis*-apo-12′-carotenoic acid per se have effects on the inhibition of growth and the induction of differentiation. However, it is not known if this activity is due to metabolism of retinoic acid or is a function of direct binding to nuclear receptors and transcriptionally activating genes. Certainly, these β-carotene metabolites are converted into retinoic acid as reported by Wang *et al.* (1996).

In a recent study, β-apo-14′-carotenoic acid has been evaluated as an inducer of RARβ_2 activity using a luciferase reporter gene coupled to the RARβ_2 promoter and the retinoic acid response element, in a transient transfection assay (Wang *et al.*, 1997). The results show that β-apo-14′-carotenoic acid transactivates the nuclear receptors, although its concentration had to be 100 times higher than that of retinoic acid in order to reach equivalent activity (Table VI). Since high-pressure liquid chromatography

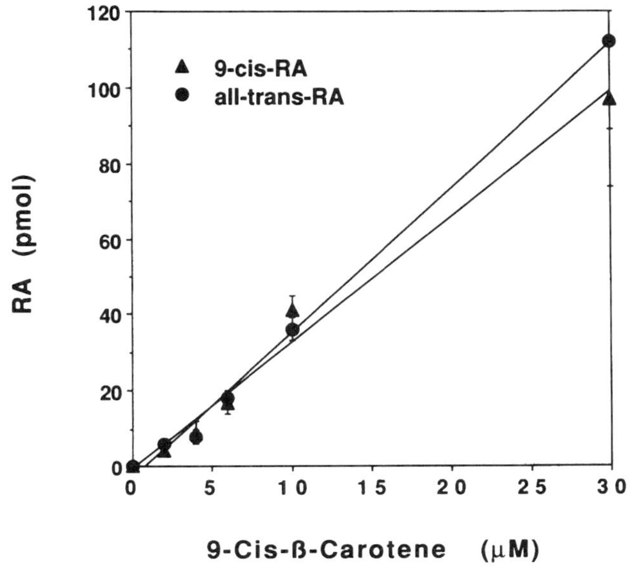

FIGURE 6. Effects of 9-*cis*-β-carotene concentration on the synthesis of retinoic acid by the human intestinal mucosa postnuclear fraction. Retinoic acid (9-*cis*-retinoic acid, filled triangles; all-*trans*-retinoic acid, filled circles) synthesis was measured from various 9-*cis*-β-carotene concentrations with 1 mg of protein for 30 min. Data are expressed as the means ± SEM. Reproduced with permission from Wang *et al.* (1994).

Table VI
Comparison of RARβ$_2$ Transactivation by 10 nM Retinoic Acid, 1000 nM β-apo-14′-Carotenoic Acid, or Both in Rat Embryo Fibroblasts Cells in the Absence and Presence of 3-Mercaptopropionic Acid (MPA)[a]

	Luciferase activity (fold increase)		Retinoic acid (pmol)	
Treatment	−MPA	+MPA	−MPA	+MPA
Retinoic acid (10 nM)	7	7	47	44
β-apo-14′-Carotenoic acid (1000 nM)	11	5	35	15
Retinoic acid (10 nM) + β-apo-14′-Carotenoic acid (1000 nM)	23	14	106	42

[a]Luciferase activities are expressed as a fold induction in treated versus control cells (DMSO only) after an 18 hr incubation at 37°C. Retinoic acid concentrations are expressed as pmol/plate. Data represent the means of two separate experiments (Wang *et al.*, 1997).

(HPLC) analysis indicated that some β-apo-14′-carotenoic acid was converted to retinoic acid, the transactivation activity of β-apo-14′-carotenoic acid could be attributed to its conversion into retinoic acid. In the presence of 3-mercaptopropionic acid, which inhibits the oxidation of β-apo-14′-carotenoic acid into retinoic acid, both the transactivation activity of β-apo-14′-carotenoic acid and the formation of retinoic acid decreased (Wang *et al.*, 1997). However, the residual transactivation activity was greater than could be accounted for by the small amount of retinoic acid formed (Table VI). These results suggest that β-apo-14′-carotenoic acid can either bind $RAR\beta_2$ directly with a lower affinity than retinoic acid, or, once bound, may have a lower transactivation potential, thus acting as a weak agonist. However, the biological significance of the excentric cleavage products of β-carotene acting as weak agonists to the retinoic acid receptor is uncertain. It is possible that they may enhance activity of retinoic acid by their conversion at physiological concentration, or interfere with retinoid signal transduction by their binding to receptor at high concentration. These observations have not been taken into consideration in the human intervention studies in which high-dose β-carotene has been administered to large populations for periods of several years.

3. CONCLUSION

The biosynthesis of retinoids from β-carotene involves both a central and an excentric cleavage pathway. The central cleavage pathway utilizes a cytosolic 15,15′-dioxygenase to yield, under optimal *in vitro* conditions, two molecules of retinal (retinaldehyde), which in turn can be readily reduced to retinol (vitamin A), or oxidized to retinoic acid. A number of studies indicate that the major pathway for the intestinal formation of vitamin A from β-carotene *in vivo* is via central cleavage. The excentric cleavage pathway yields a series of carbonyl metabolites, of shortened chain length, the β-apo-carotenals. β-Apo-carotenals subsequently are oxidized to their corresponding β-apo-carotenoic acids. By a process analogous to fatty acid β-oxidation, the β-apo-carotenoic acids are converted stepwise to retinoic acid, where the β-oxidation-type process is stopped by the presence of a methyl substituent on carbon-13. The biosynthesis of retinoic acid from β-carotene via the excentric cleavage pathway has been demonstrated both *in vitro* and *in vivo*. It has been reported that some of the excentric cleavage products of β-carotene have biological function in terms of inducing cell differentiation, at least in HL-60 cells. It seems that central cleavage is a major pathway leading to vitamin A formation, and excentric cleavage is involved in the metabolism of β-carotene into other oxidative metabolites.

The two pathways may be independently regulated and play a different role in either physiological or pathological states.

ACKNOWLEDGMENTS. This study was funded in part by grants R01CA49195-07 and R01CA66914-02 from the National Institute of Health, Bethesda, Maryland, and under Contract 53-3K06-01 from the U. S. Department of Agriculture. We are greatly indebted to the following colleagues for their valuable contributions: Drs. R. M. Russell, C. Liu, X. Hebuterne, F. Stickel, G. Tang, D. E. Smith, and J. G. Fox. The contents of this publication do not necessarily reflect the views or policies of the U. S. Department of Agriculture, nor does mention of trade names, commercial products, or organizations imply endorsement by the U. S. Government.

4. REFERENCES

Blomstrand, R., and Werner, B., 1967, Studies on the intestinal absorption of radioactive β-carotene and vitamin A in man, *Scand J. Clin. Lab. Invest.* **19:**339–345.

Bollag, W., and Holdener, E. E., 1992, Retinoids in cancer prevention and therapy, *Ann. Oncol.* **3:**513–526.

Chambon, P., 1996, A decade of molecular biology of retinoic acid receptors, *FASEB J.* **10:**940–953.

Chen, Z. X., Xue, Y. Q., Tao, R. F., Xia, X. M., Li, C., Zu, W. Y., Yao, X. Z., and Ling, B. J., 1991, A clinical and experimental study on all-*trans* retinoic acid-treated acute promyelocytic leukemia patients, *Blood* **78:**1413–1419.

Crain, F. D., Lotspeich, F. J., and Krause, R. F., 1967, Biosynthesis of retinoic acid by intestinal enzymes of the rat, *J. Lipid Res.* **8:**249–254.

Devery, J., and Milborrow, B. V., 1994, β-Carotene-15,15′-dioxygenase (EC 1.13.11.21): Isolation, reaction mechanism and an improved assay procedure, *Br. J. Nutr.* **72:**397–414.

Dmitrovskii, A. A., 1991, Metabolism of natural retinoids and their functions, in *New Trends in Biological Chemistry* (T. Ozawa, ed.), pp. 297–308, Japan Scientific Society Press, Tokyo/Springer-Verlag, Berlin.

Duszka, C., Grolier, P., Azim, E.-M., Alexandre-Gouabau, M.-C., Borel, P., and Azais-Braesco, V., 1996, Rat intestinal β-carotene dioxygenase activity is located primarily in the cytosol of mature jejunal enterocytes, *J. Nutr.* **126:**2550–2556.

FAO/WHO, 1988, Requirements of vitamin A, iron folate and vitamin B12, Report of a joint FAO/WHO expert consultation, FAO Food Nutrition Series, No. 23, FAO, Rome.

Fox, J. G., 1992, *The Biology and Disease of the Ferret*, Lea & Febiger, Philadelphia.

Ganguly, J., and Sastry, P. S., 1985, Mechanism of conversion of β-carotene into vitamin A—Central cleavage versus random cleavage, *World Rev. Nutr. Dietetics*, **45:**198–220.

Glover, J., 1960, Conversion of β-carotene into vitamin A in the intestine of the rat, *Vitamins Horm.* **18:**371–386.

Goodman, D. S., and Blaner, W. S., 1984, Biosynthesis, absorption, and hepatic metabolism, in *The Retinoids*, Vol. 2 (M. B. Sporn, A. B. Roberts, and D. S. Goodman, eds.), pp. 2–41, Academic Press, Orlando, Florida.

Goodman, D. S., and Huang, H. S., 1965, Biosynthesis of vitamin A with rat intestinal enzymes, *Science* **149:**879–880.

Goodman, D. S., and Olson, J. A., 1969, The conversion of all-*trans*-β-carotene into retinol, *Meth. Enzymol.* **15:**510–535.

Goodman, D. S., Blomstrand, R., Werner, B., Huang, H. S., and Shiratori, T., 1966, The intestinal absorption and metabolism of vitamin A and β-carotene in man, *J. Clin. Invest.* **45:**1615–1623.

Gronowska-Senger, A., and Wolf, G., 1970, Effect of dietary protein on the enzyme from rat and human intestine which converts β-carotene to retinaldehyde, *J. Nutr.* **100:**300–308.

Heyman, R. A., Mangelsdorf, D. J., Dyck, J. A., Stein, R. B., Eichele, G., Evans, R. M., and Thalle, C., 1992, 9-*cis* retinoic acid is a high affinity ligand for the retinoid X receptor, *Cell* **68:**397–406.

Hansen, S., and Maret, W., 1988, Retinal is not formed *in vitro* by enzymatic cleavage of β-carotene, *Biochemistry* **27:**200–206.

Hebuterne, X., Wang, X.-D., Johnson, E. J., Krinsky, N. I., and Russell, R. M., 1995, Intestinal absorption and metabolism of 9-*cis*-β-carotene *in vivo*: Biosythesis of 9-*cis*-retinoic acid, *J. Lipid Res.* **36:**1264–1273.

Hebuterne, X., Wang, X.-D., Smith, D., Tang, G., and Russell, R. M., 1996, *In vivo* biosynthesis of retinoic acid from β-carotene involves an excentric cleavage pathway in ferret intestine, *J. Lipid Res.* **37:**482–492.

Hove, E. L., 1943, The *in vitro* destruction of carotene by water extracts of minced rat stomachs in the presence of methyl linolate, *Science* **98:**433–434.

Huang, H. S., and Goodman, D. S., 1965, Vitamin A and carotenoids. I. Intestinal absorption and metabolism of 14C-labeled vitamin A alcohol and β-carotene in the cat, *J. Biol. Chem.* **240:**2839–2844.

Kiorpes, T. C., Hoerr, D., Ho, W., Weaner, L. E., Inman, M. G., and Tutwiler, G. F. 1984, Identification of 2-tetradecylglycidyl coenzyme A as the active form of methyl 2-tetradecylglycidate (methyl palmoxirate) and its characterization as an irreversible, active site-directed inhibitor of carnitine palmitoyltransferase A in isolated rat liver mitochondria, *J. Biol. Chem.* **259:**9750–9755.

Krinsky, N. I., Wang, X.-D., Tang, G., and Russell, R. M., 1992, Mechanism of carotenoid cleavage to retinoids, in *Carotenoids in Human Health* (L. M. Canfield, N. I. Krinsky, and J. A. Olson, eds.), *Ann. N.Y. Acad. Sci.* **691:**167–176.

Krinsky, N. I., Wang, X.-D., Tang, G., and Russell, R. M., 1993, Conversion of carotenoids to retinoids, in *Retinoids: Progress in Research and Clinical Applications* (M. A. Livrea and L. Packer, eds.), pp 1–16, Marcel Dekker, New York.

Lakshmanan, M. R., Pope, J. L., and Olson, J.A., 1968, The specificity of a partially purified carotenoid cleavage enzyme of rabbit intestine, *Biochem. Biophys. Res. Commun.* **33:**347–352.

Levin, A. A., Sturzenbecker, L. J., Kazmer, S., Bosakowski, T., Huselton, C., Allenby, G., Speck, J., Kratseisen, C., Rosenberger, M., Lovey, A., and Grippo, J. F., 1992, 9-*Cis* retinoic acid stereoisomer binds and activates the nuclear receptor RXRα, *Nature* **355:**359–361.

Liu, C., Wang, X.-D., and Russell, R. M., 1997, Biosynthesis of retinoic acid from beta-apo-14′-carotenal in ferret *in vivo*, *J. Nutr. Biochem.* **8:**652–657.

Nagao, A., and Olson, J. A., 1994, Enzymatic formation of 9-*cis*, 13-*cis*, and all-*trans*-retinals from isomers of β-carotene, *FASEB J.* **8:**968–973.

Napoli, J. L., and Race, K. R., 1988, Biogenesis of retinoic acid from β-carotene. Differences between the metabolism of β-carotene and retinal, *J. Biol. Chem.* **263:**17372–17377.

Olson, J. A., 1989, Provitamin A function of carotenoids: The conversion of β-carotene into vitamin A, *J. Nutr.* **119:**105–108.

Olson, J. A., 1996, Vitamin A, in *Present Knowledge in Nutrition* (E. E. Ziegler and L. J. Filer, Jr., eds.), pp. 109–119, ILSI Press, Washington, DC.

Olson, J. A., and Hayaishi, O., 1965, The enzymatic cleavage of β-carotene into vitamin A by soluble enzymes of rat liver and intestine, *Proc. Natl. Acad. Sci. USA* **54:**1364–1370.

Pfahl, M., and Chytil, F., 1996, Regulation of metabolism by retinoic acid and its nuclear receptor, *Annu. Rev. Nutr.* **16:**257–283.

Sabbagh, E., Cuebas, D., and Schulz, H., 1985, 3-Mercaptopropionic acid, a potent inhibitor of fatty acid oxidation on rat heart mitochondria, *J. Biol. Chem.* **260:**7337–7342.

Sharma, R. V., Mathur, S. N., and Ganguly, J., 1976, Studies on the relative biopotencies and intestinal absorption of different apo-β-carotenoids in rats and chickens, *Biochem. J.* **158:**377–383.

Sharma, R. V., Mathur, S. N., Dimitrovskii, A. A., Das, R. C., and Ganguly, J., 1977, Studies on the metabolism of β-carotene and apo-β-carotenoids in rats and chickens, *Biochim. Biophys. Acta* **486:**183–194.

Sklan, D., 1983, Carotene-cleavage activity in chick intestinal mucosa cytosol: Association with a high-molecular-weight lipid-protein aggregate fraction and partial characterization of the activity, *Br. J. Nutr.* **50:**417–425.

Suzuki, T., Matsui, M., and Murayama, A., 1995, Biological activity of (all-E)-β-apo-12′-carotenoic acid and the geometrical isomers on human acute promyelocytic leukemia cell line HL-60, *J. Nutr. Sci. Vitaminol.* **41:**575–585.

Tang, G., Wang, X.-D., Russell, R. M., and Krinsky, N. I., 1991, Characterization of β-apo-13-carotenone and β-apo-14′-carotenal as enzymatic products of the excentric cleavage of β-carotene, *Biochemistry* **30:**9829–9834.

van Vliet, T., Schaik, F., and Van Den Rerg, H., 1992, Beta-carotene metabolism: The enzymatic cleavage to retinal, *Neth. J. Nutr.* **53:**186–190.

van Vliet, T., Vlissinger, V., Schaik, F., and Van Den Rerg, H., 1996, Beta-carotene absorption and cleavage in rats is affected by the vitamin A concentration of the diet, *J. Nutr.* **126:**499–508.

Villard, L., and Bates, C. J., 1986, Carotene dioxygenase (EC 1.13.11.21) activity in rat intestine: Effects of vitamin A deficiency and of pregnancy, *Br. J. Nutr.* **56:**115–122.

Wang, X.-D., Tang, G., Fox, J. G., Krinsky, N. I., and Russell, R.M., 1991, Enzymatic conversion of β-carotene into β-apo-carotenals and retinoids by human, monkey, ferret, and rat tissues, *Arch. Biochem. Biophys.* **285:**8–16.

Wang, X.-D., Krinsky, N. I., Marini, R., Tang, G., Yu, J., Fox, J., and Russell, R. M., 1992a, Intestinal uptake and lymphatic absorption of β-carotene in ferrets: A model for human β-carotene metabolism, *Am. J. Physiol.* **263:**G480–G486.

Wang, X.-D., Krinsky, N. I., Tang, G., and Russell, R. M., 1992b, Retinoic acid can be produced from excentric cleavage of β-carotene in human intestinal mucosa, *Arch. Biochem. Biophys.* **293:**293–304.

Wang, X.-D., Russell, R. M., Marini, R., Tang, G., Fox, J. G., and Krinsky, N. I., 1993, Intestinal perfusion of β-carotene in the ferret raises retinoic acid level in portal blood, *Biochim. Biophys. Acta* **1167:**159–164.

Wang, X.-D., Krinsky, N. I., Benotti, P. N., and Russell, R. M., 1994, Biosynthesis of 9-*cis*-retinoic acid from 9-*cis*-β-carotene in human intestinal mucosa, *Arch. Biochem. Biophys.* **313:**150–155.

Wang, X.-D., Marini, R. P., Hebuterne, X., Fox, J., Krinsky, N. I., and Russell, R. M., 1995, Vitamin E can enhance lymphatic absorption of β-carotene and its conversion into vitamin A in the ferret, *Gastroenterology* **108:**719–726.

Wang, X.-D., Russell, R. M., Liu, C., Stickel, F., Smith, D. E. H., and Krinsky, N. I., 1996, β-Oxidation in rabbit liver *in vitro* and in the perfused ferret liver contributes to retinoic acid biosynthesis from β-apo-carotenoic acid, *J. Biol. Chem.* **271:**26490–26498.

Wang, X.-D., Russell, R. M., Liu, C., Peacocke, M., and Krinsky, N. I., 1997, β-Apo-14′-carotenoic acid *trans*-activates the response element of the retinoic acid receptor β2 gene via its conversion into retinoic acid, submitted.

Warrell, R. P., Jr., The, H. D., Wang, Z.-Y., and Degos, L., 1993, Acute promyelocytic leukemia, *N. Engl. J. Med.* **329:**177–189.

Yeum, K. J., Lee-Kim, Y. C., Yoon, S., Lee, K. Y., Park, I. S., Lee, K. S., Kim, B. S., Tang, G., Russell, R. M., and Krinsky, N. I., 1995, Similar metabolites formed from β-carotene by human gastric mucosa homogenates, lipoxygenase, or linoleic acid hydroperoxide, *Arch. Biochem. Biophys.* **321:**167–174.

Zeng, S., Furr, H. C., and Olson, J. A., 1992, Metabolism of carotenoid analogs in humans, *Am. J. Clin. Nutr.* **56:**433–439.

Chapter 8

The Role of Carotenoids in Modulating DNA Stability and Lipid Peroxidation

Importance for Human Health

Susan J. Duthie, Andrew R. Collins, and Garry G. Duthie

1. CAROTENOID STRUCTURE AND ANTIOXIDANT ACTIVITY

Carotenoids are red, yellow, and orange pigments that absorb light in the 400- to 500-nm region of the visible spectrum (Isler, 1971). They are synthesized by photosynthetic microorganisms and plants, but not by animals. Particularly rich sources are yellow-orange fruits and dark green, leafy vegetables. About 700 naturally occurring carotenoids have been identified, of which about 50 can serve as precursors for vitamin A (Olsen, 1992). All are primarily symmetrical, C-40, polyisoprenoid structures with an extensive conjugated double-bond system (Fig. 1). The conjugated backbone of isoprene units is usually inverted at the center of the molecule, imparting the symmetry. Changes in the geometrical configuration about the double bonds result in many *cis* and *trans* isomers. Moreover, there are many

---

Susan J. Duthie, Andrew R. Collins, and Garry G. Duthie Rowett Research Institute, Bucksburn, Aberdeen AB21 9SB, Scotland.

Subcellular Biochemistry, Volume 30: Fat-Soluble Vitamins, edited by Quinn and Kagan. Plenum Press, New York, 1998.

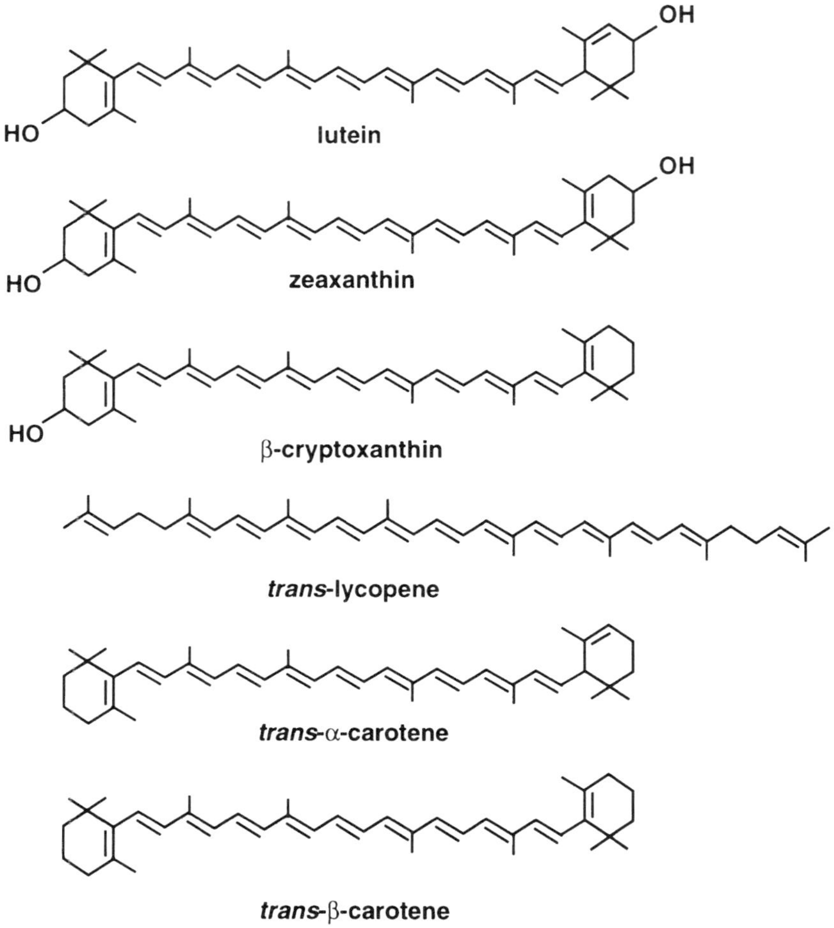

FIGURE 1. Selected carotenoid structures.

hydroxylated, oxidized, hydrogenated, and ring-containing derivatives. In general, major dietary classes of carotenoids are hydrocarbon carotenoids (carotenes), oxygen-containing carotenoids (xanthophylls), and carotenol fatty acid esters (Breecher and Khachik, 1992).

The nutritional and biochemical significance of dietary carotenoids is unclear. Most attention has been directed toward the potential antioxidant role of carotenoids and in particular β-carotene in preventing procarcinogenic and atherogenic oxidative damage to tissues. However, whether the primary roles of carotenoids are as biological antioxidants is a matter of debate. Other important biological functions have been suggested, including the regulation of gap junction communication (Bertram, 1993), immunocompetence enhancement (Schmidt, 1991), and the inhibition of non-free radical-mediated mutagenesis and cell transformation (Krinsky, 1993). These will be discussed later. However, even if an antioxidant role for carotenoids *in vivo* is merely adventitious, there is little doubt that because of their polyene structural feature many react with reactive oxygen species in chemical and biochemical systems. For example, β-carotene has long been known to be a particularly effective scavenger of singlet oxygen (Bendich and Olsen, 1989), i.e.,

$$^{1}O_2 + \beta\text{-carotene} \rightarrow {}^{3}O_2 + \beta\text{-carotene}$$

This ability is also shared by α-carotene, lycopene, zeaxanthin, lutein, and β-cryptoxanthin and appears to depend largely on physical quenching as opposed to chemical reaction. Rate constants for the reactions *in vitro* approach diffusion control, with lycopene being the most effective $^{1}O_2$ quencher (Sies *et al.*, 1992).

The antioxidant efficacy of certain carotenoids is also apparent in nonphotochemical free radical-mediated reactions. In organic solvents carotenoids produce a concentration-dependent inhibition of azo-compound-induced lipid peroxidation (Liebler *et al.*, 1997). This effectiveness declines with increasing oxygen tension and is most markedly decreased at 100% oxygen. Similarly, β-carotene can inhibit the oxidation of methyl linoleate by peroxyl radicals (Seis *et al.*, 1992). This ability is greatest at the lower concentrations and partial pressures of oxygen found in tissues, suggesting that the carotenoid could effectively protect cell membranes from lipid peroxidation *in vivo*. β-Carotene has also been shown to inhibit peroxidation of liposomal phospholipids exposed to a range of free radical generators including azo-dye-induced peroxyl radicals, superoxide, and transition metals (Liebler *et al.*, 1997). In these systems, antioxidant effects are seen over a range of oxygen tensions within the normal physiological range.

Assessment of the antioxidant activity of carotenoids in biological membranes isolated from tissues has been more difficult to evaluate because of the confounding antioxidant effect of vitamin E which is invariably present in the preparations. However, preincubation of hepatic microsomal preparations from vitamin E-deficient rats with a range of carotenoids suggests that some are effective at suppressing lipid peroxidation (G. G. Duthie and P. C. Morrice, in preparation). In this system, lycopene was the most effective antioxidant, indicating that the presence of hydroxyl groups on the terminal rings is not a criterion for antioxidant activity of carotenoids. Whether sufficient carotenoids are absorbed from the diet and then accumulate in adequate concentrations in membranes or low-density lipoprotein (LDL) to exert antioxidant effects *in vivo* is unclear. Moreover, little is known about possible prooxidant actions at high dietary intakes.

Nevertheless, there is now a substantial body of data which suggests that some carotenoids may be important nutritional antioxidants and that low intakes may predispose to the development of coronary heart disease (CHD) and certain cancers. This chapter will present evidence from human, whole-animal, and cellular studies which indicate that carotenoids are chemopreventive against both DNA damage and oxidant-induced membrane damage.

2. HUMAN STUDIES

2.1. Conventional Epidemiology

Oxidative damage to biomolecules contributes to the etiology of degenerative diseases in general, and CHD and cancer in particular. Since carotenoids act as antioxidants *in vitro* (specifically protecting lipids and DNA), it is reasonable to suggest that they may play a defensive role *in vivo*. Similarly, there is considerable epidemiological evidence linking consumption of foods rich in carotenoids with decreased risk of disease.

Epidemiological studies are of various kinds. Comparisons of cancer or cardiovascular disease incidence in countries or regions which differ in diet (such as northern Europe compared with the Mediterranean fringe) provide circumstantial evidence. Thus, the relatively low incidence of colon cancer, lung cancer, and CHD in Greece, southern Italy, Spain, etc., is commonly attributed to the "Mediterranean diet." However, while estimates of intake of specific dietary carotenoids (on a population level) may allow a negative correlation with decrease risk to be proposed (Bellizzi *et al.*, 1994), it is always possible that protection is due to the presence of some other associated food component.

Case–control studies are carried out on a single population. Subjects with the disease of interest are matched with disease-free controls, and differences in diet are sought retrospectively. In a series of studies in northern Italy, consumption of vegetables and fruit was associated with a lowered risk of myocardial infarction (Gramenzi *et al.*, 1990) and various cancers (Negri *et al.*, 1991), while serum levels of β-carotene and other carotenoids have been reported to be lower in lung cancer cases than in matched controls (Connet *et al.*, 1989).

Prospective studies, in which a large cohort of normal volunteers is followed over a period of years, allows for identification of factors distinguishing individuals who contract a disease from those who do not. Such studies are more rigorous than case–control studies, though more expensive and slower to yield results. The Prospective Basel Study (Stahelin *et al.*, 1991) showed an association between low plasma carotene and increased risk of cancer, especially of the bronchus. The California study (Paganini-Hill *et al.*, 1987) did not find evidence for protection by β-carotene or retinol. A prospective study in Finland (Knekt *et al.*, 1991), based on dietary interviews to estimate micronutrient intakes, found that carotenoids, as well as vitamins C and E, were correlated with a low incidence of lung cancer in nonsmokers. However, the evidence from these and other studies is not overwhelming (Ziegler, 1989).

A different approach is to intervene with micronutrients that have a putative protective role. In Linxian, China, almost 30,000 people were given combinations of four vitamin/mineral supplements daily for 5 years. A reduction in total morbidity and in cancer mortality and incidence (especially for cancer of the stomach) was seen in the groups receiving β-carotene, vitamin E, and selenium (Blot *et al.*, 1993). In direct contrast, there was an 8% increase in mortality for participants in the Finnish Trial who received β-carotene, resulting from higher rates of lung cancer (increased by 18%) and ischemic heart disease (Alpha-Tocopherol, Beta-Carotene Cancer Prevention Study Group, 1994). The Beta-Carotene and Retinol Efficiency Trial (CARET) was prematurely halted when it was found that supplementation with these antioxidants was associated with more cases of lung cancer (up by 28%) and a 26% increase in deaths from CHD (Omenn *et al.*, 1996). However, supplementation for 12 years with β-carotene in the Physicians' Health Study revealed no significant effects on incidence of cancer or cardiovascular disease (Hennekens *et al.*, 1996).

There are over 240 epidemiologically derived risk factors for CHD, including frontal baldness, creases on the ear lobe, short stature, and rainfall (McCormick and Skrabanek, 1988). Presumably many of these are not causal, but merely adventitious, and illustrate the need for caution in the interpretation of statistically based population studies. Moreover, even if it is assumed that the "classical" risk factors (smoking, blood cholesterol, and hypertension) are causal, they explain only 50% of the variance in the incidence of the disease (Gey, 1986). Elevated blood cholesterol concentrations are regarded by many as major indicators of CHD risk. Such a belief arose primarily from the "Seven-Countries" cross-cultural study (Keys, 1980), which found a strong relationship between consumption of saturated fatty acids and mortality from CHD. This finding subsequently evolved into the "lipid hypothesis," which suggests that the consumption of high levels of saturated fat increases blood cholesterol concentrations, which in turn are related to the number of deaths from CHD. Unfortunately most intervention trials aimed at lowering blood cholesterol by dietary and pharmacological means have not reduced the incidence of CHD or total mortality in the intervention groups (Rosenman, 1992). Although trials with the new generation of cholesterol-lowering statins appear to have found a reduction in progression of CHD and possible regression of atheromatous plaques, effectiveness may also relate to the antioxidant properties of the drug (Hussein *et al.*, 1997).

There remain many anomalies in the "lipid hypothesis." In many areas of France, a high intake of saturated fat is accompanied by a markedly low incidence of CHD. In the United Kingdom, the upper social classes have the highest fat intake and highest cholesterol levels, but, surprisingly, the lowest incidence of CHD. Similarly, British Asian immigrants have a higher rate of cardiovascular disease than Caucasians, despite the fact that, on average, Asians smoke less and have lower blood pressure and plasma cholesterol levels (James *et al.*, 1989). Moreover, in 194 consecutive autopsies at the Providence Veterans Hospital in the United States, only 10% of cases of fatal atherosclerosis had elevated serum cholesterol (Rosenman, 1992). Most cases had developed without evidence of elevated serum cholesterol, diabetes, or hypertension. CHD incidence was also unrelated to dietary fats, saturated fatty acids or percent calories from saturated fatty acids in a 5-year study of 10,000 male Israelis (James *et al.*, 1989). Similarly, the long-term incidence of CHD in a major U.K. study was not associated with either saturated or polyunsaturated fatty acid intakes (James *et al.*, 1989).

Although it is debatable whether cholesterol is a major epidemiological indicator of CHD risk in normolipidemic populations, there is considerable evidence that the susceptibility of low-density lipoprotein (LDL) to oxidative modification is a key event in the development of atherosclerosis (for recent review, see Westhuyzen, 1997). In brief, LDL is a cholesterol fraction containing a high proportion of polyunsaturated fatty acids (PUFA) in the phospholipid layer, which is susceptible to free radical-mediated peroxidation. When LDL is oxidized *in vitro* there is a loss of PUFA, an increase in toxic products such as PUFA hydroperoxides, aldehydes, and lysolecithin, and a modification of the chemical and physical properties of LDL, all of which may increase atherogenicity. The oxidized LDL is taken up by monocytes, which are attracted to a site of injury on the arterial wall. These monocytes promote chemotactic attraction by further monocytes. On transformation of monocytes to macrophages, the oxidized LDL appears to limit macrophage mobility and decreases their ability to migrate away from the artery wall. The enhanced rate of uptake of oxidized

LDL may then convert macrophages into foam cells, which are the precursors of the plaque that ultimately occludes the artery. Since the macrophage can oxidatively modify native LDL via the respiratory burst, autocatalytic progression may lead to the continuous growth of the atheroma. Many studies have shown that oxidation of LDL can be inhibited *in vitro* by the addition of antioxidants to the LDL and that susceptibility of oxidation may depend on antioxidant content (Rice-Evans *et al.*, 1996). For example, when LDL is exposed to a copper-mediated oxidation system, vitamin E in the outer phospholipid layer of the LDL is depleted first, followed by carotenoids such as lycopene and β-carotene. LDL oxidation is delayed if vitamin C is present in the external medium, presumably because vitamin C regenerates vitamin E. Increasing the vitamin E or ubiquinone content of the LDL by dietary supplementation of volunteers also inhibits oxidation of LDL to the atherogenic form, suggesting that nutritional antioxidant intakes may be important in inhibiting the pathogenesis of coronary heart disease (Esterbauer *et al.*, 1992). Such *in vitro* studies are complemented by epidemiological studies which indicate that low intakes of antioxidant nutrients increase risk of developing CHD. As with certain cancers, epidemiological studies have consistently suggested that the intake of antioxidant-rich foods such as fruit and vegetables is associated with reduced CHD (e.g., Kushi *et al.*, 1996; Singh *et al.*, 1993; Crombie *et al.*, 1990; Armstrong *et al.*, 1975). These are further corroborated by significant cross-cultural correlations relating mortality from coronary heart disease with low dietary intakes of antioxidant nutrients (Bellizzi *et al.*, 1994).

2.2. Molecular Epidemiology

While conventional epidemiology suggests an association between low consumption of fruit and vegetables or specific antioxidants with a high incidence of CHD and cancer, determining the mechanisms underlying this relationship requires a more focused experimental approach. Molecular epidemiology relies on the measurement of "biomarkers"—quantitative indicators of disease risk, or of exposure to causative agents or the presence of protective factors. In nutritional studies, molecular epidemiology avoids reliance on subjective reporting of dietary intake and instead measures indices such as plasma antioxidants. By investigating the earliest stages in the development of the disease, molecular epidemiology dispenses with the need to wait for the disease to develop or for death to occur. Studies can be completed in a relatively short time with a small number of people. Molecular epidemiology lends itself to intervention trials, where the aim is to register a significant change in a defined set of biomarkers. However, the selection and validation of biomarkers is all important. Ideally, the relevance of a biomarker representing an early stage of disease should be confirmed in a trial which correlates the marker with the clinical endpoint. In practice, we tend to rely on well-established causal connections. For example, genetic change is known to be the primary cause of cancer. Determining mutations in an oncogene would be an acceptable molecular epidemiological approach, but such changes are extremely rare events. However, genetic change results from DNA damage, and although most DNA damage does not result in mutations in the genes that are involved in cancer, and there are several steps between DNA damage and tumorigenic transformation, the former is relatively easy to measure and is widely accepted as a viable marker for cancer risk. Similarly, indices of lipid peroxidation and LDL oxidation, which are involved in the formation of plaques in arterial walls and which can be readily measured, are regarded as biomarkers for heart disease.

There are significant associations between the incidence of CHD and concentrations of antioxidant nutrients in plasma. Gey (1992) found that coronary heart disease mortality was inversely related to a cumulative antioxidant index, defined as

$$\frac{[\text{vitamin E}]\times[\text{vitamin C}]\times[\beta\text{-carotene}]\times[\text{selenium}]}{[\text{cholesterol}]}$$

where all concentrations are those in plasma. In addition, both vitamin C and vitamin E concentrations in plasma are significantly correlated with mortality from coronary heart disease, particularly in populations with comparable blood cholesterol concentrations. Similarly, case–control studies suggest that low plasma concentrations of vitamins E and C and β-carotene are related to risk of angina in men (Riemersma *et al.*, 1989).

Serum concentrations of α-carotene and β-carotene have also been correlated with reduced CHD risk, but it is possible that blood levels of these carotenoids may simply reflect fruit and vegetable consumption and act as a marker for other protective dietary factors (Stahl, 1996). However, studies on smokers (at high risk from CHD) suggest a mechanistic connection between carotenoids and disease. For example, plasma β-carotene concentrations are approximately 15–20% less in smokers than in nonsmokers (Ross *et al.*, 1995). Whether this is due to decreased intake of β-carotene by smokers or to changes in metabolic turnover ascribed to oxidative stress is unclear, although we have recently shown that low plasma carotene concentration is maintained even after adjustment for dietary intake (Duthie *et al.*, 1993). Oxidative stress may differentially influence carotenoids. Plasma concentrations of α-carotene, β-carotene, and β-cryptoxanthin are lower in smokers, whereas concentrations of other carotenoids such as lycopene and phytofluene are similar to those of nonsmokers (Ross *et al.*, 1995). Few other studies have compared a range of carotenoid concentrations between smokers and never-smokers, although Ito (1991) found that in Japanese subjects serum β-cryptoxanthin was inversely associated with cigarette consumption.

2.2.1. Dietary Intake, Plasma Carotenoid Levels, DNA Damage, and Lipid Peroxidation

The fruits and vegetables normally consumed in the human diet contain more than 40 different types of carotenoids, some of which are described in Table I. However, only 14 have been identified in human plasma, the most common being β-carotene, α-carotene, lutein, zeaxanthin, lycopene, and β-cryptoxanthin (Fig. 2). These are widely distributed in fruits and vegetables (Table II). This restricted range suggests that although carotenoids are absorbed from the gut along with dietary fat, chemical modification or the regulation of absorption of specific carotenoids may occur. Moreover, tissue uptake of carotenoids may also be highly specific, as some tissues accumulate particular carotenoids. For example, lycopene is the major carotenoid in the testes, while the macula of the retina contains relatively high concentrations of zeaxanthin and lutein (Sies *et al.*, 1992). Plasma and possibly tissue concentrations may also depend on dietary patterns relating to demographic preferences and the availability and preparation of food items. For example, differences in plasma carotenoid

Table I
Major Carotenoid Classes and Food Sources[a]

Carotenoid class	Food source	Carotenoid type
Hydrocarbon	Green, orange, red, and yellow fruits and vegetables	α-Carotene, β-carotene, γ-carotene, lycopene, phytoene, phytofluene, neurosporene
Hydroxylated	Green vegetables and some yellow-orange fruits and vegetables	α-Cryptoxanthin, β-cryptoxanthin, lutein, zeaxanthin, 3′-epilutein, 5,6-dihydroxylycopene
Epoxides	Green vegetables and some orange, red, and yellow fruits and vegetables	Auroxanthin, β-cryptoxanthin-5′,6-epoxide, flavoxanthin, luteoxanthin, lycopene-5,6-epoxide, neoxanthin

[a]There is also a range of carotenoid esters of laurate, myristate, and palmitate in foods. Data abstracted from Breecher and Khachik (1992).

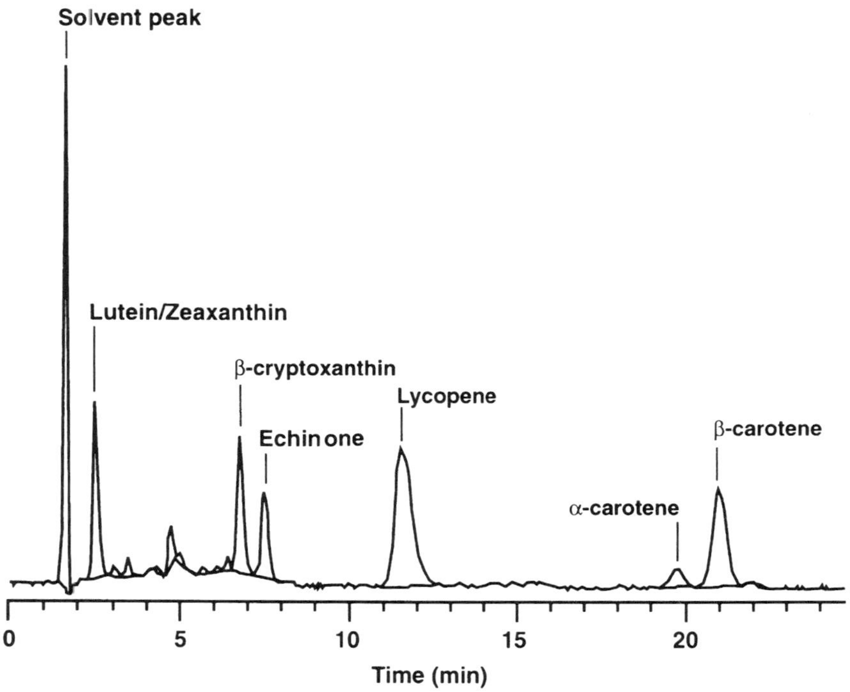

FIGURE 2. HPLC chromatogram showing the major carotenoids identified in human plasma. Echinone was used as internal standard.

Table II
Common Dietary Sources of Major Carotenoids Detected in Human Plasma[a]

Carotenoid	Dietary origin
β-Carotene	Mango, cantaloupe, carrot, pumpkin, papaya, peach, prune, apricot, cabbage, green bean, Brussel sprout, kale, kiwi fruit, lettuce, tomato, orange
α-Carotene	Apricot, pumpkin, canataloupe, green bean, cabbage, kale, lettuce, prune, peach, carrot, broccoli
Lycopene	Tomato, pink grapefruit, red grape, watermelon, apricot
Lutein	Mango, orange, papaya, peach, kiwi fruit, pea, green bean, Brussel sprout, cabbage, sweet potato
Zeaxanthin	Peach, apricot, prune, orange, mango, kale, kiwi fruit, yellow corn
β-Cryptoxanthin	Orange, mango, cantaloupe, peach, prune, papaya

[a]Data taken from several sources.

profiles in Scottish men compared with healthy subjects from southern Europe (Ross *et al.*, 1995) are presumably a reflection of varying regional dietary patterns. β-Carotene was found to be the predominant carotenoid in the plasma of Scottish men (31% of total carotenoids), but constituted only 16% of total carotenoids from Spanish men (Olmedilla *et al.*, 1992). Conversely, β-cryptoxanthin constituted only 6% of total carotenoids in the Scottish sample against 20% in Spaniards. Other factors which affect uptake and plasma and tissue concentrations of carotenoids include effects of cooking, use of oral contraceptives, smoking, alcohol consumption, and use of supplements which may competitively exclude the uptake of other carotenoids in the diet (Micozzi, 1992).

In general, plasma carotenoid concentrations reflect dietary intake, with habitually low dietary intake related to low plasma concentrations. However, there is considerable interindividual variation in serum carotenoid concentrations following consumption of single supplements or carotenoid-rich foods and drinks. Individuals whose plasma concentrations do not increase markedly following such intakes are classed as nonresponders. Whether this reflects genetically determined differences in carotenoid absorption or metabolism is unclear and it is not known whether tissue stores of nonresponders are lower than those of responders.

Oxidative damage to DNA can be measured in several ways. However, these methods give widely differing values for endogenous DNA damage in normal human cells. Gas chromatography-mass spectrometry (GC-MS) gives the highest values (several hundred oxidized based per 10^6 base pairs [bp]), which seems unlikely in terms of the genetic burden this would impose on the cell and ultimately the organism. Much of the damage is apparently artifactual. High-pressure liquid chromatography (HPLC) estimates are more reasonable (less than 1 per 10^6 bp), although spurious oxidation may still occur during isolation of the DNA. Methods based on enzymic recognition of oxidized bases give frequencies of approximately 0.1 per 10^6 bp. Here, there may be underestimation of DNA damage if oxidized bases are not all accessible to the enzyme (Collins *et al.*, 1996).

We have used both HPLC and an enzymic method (the comet assay) to investigate the hypothesis that dietary antioxidants protect against DNA damage. In a study of a semirural Scottish population, we reported weak negative correlations between endogenous oxidative DNA damage and plasma antioxidant concentrations (Duthie *et al.*, 1996) and concluded that the range in antioxidant levels in this population (where consumption of fruit and vegetables tends to be low) might be too narrow to see a significant relationship. Recently, however, looking at a population group in Spain, we observed significant correlations between oxidized pyrimidines in lympho-

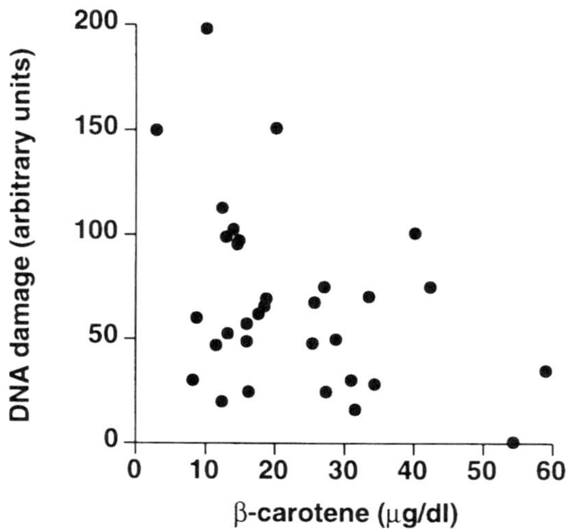

FIGURE 3. Plasma β-carotene levels and oxidative DNA base damage in human lymphocytes. Each data point represents DNA base damage (the comet assay) and plasma β-carotene levels (HPLC) for one individual ($n = 32$, $r = -0.4$, $p < .05$).

cytes DNA and plasma concentrations of β-carotene (Fig. 3), lutein, and total carotenoids (A. R. Collins *et al.*, in preparation).

2.2.2. Carotenoid Supplementation

Intervention trials may employ methods of molecular epidemiology to investigate the mechanisms of action of dietary antioxidants. In one study, groups of 50 smokers and 50 nonsmokers (healthy men in their 50s) received either a cocktail of vitamin C (100 mg), vitamin E (280 mg), and β-carotene (25 mg) or a placebo daily for 20 weeks. Blood was analyzed for antioxidant concentrations, lipid peroxidation, and DNA damage in isolated lymphocytes. At the end of the trial there was a highly significant decrease in oxidized DNA bases in the lymphocytes of subjects receiving supplement, confirming the protective role of at least one of the antioxidants *in vivo*. Supplemented lymphocytes were also more resistant to oxidative damage inflicted *in vitro* by hydrogen peroxide (Fig. 4). Similarly, antioxidant supplementation led to a decrease in lipid peroxidation in smokers. Indices of lipid peroxidation (LDL oxidation and susceptibility of erythrocytes to oxidation) decreased in response to supplementation, indicating that the oxidative stress sustained by smokers could be ameliorated, at least in part, by increased antioxidant intake. In a smaller study, lympho-

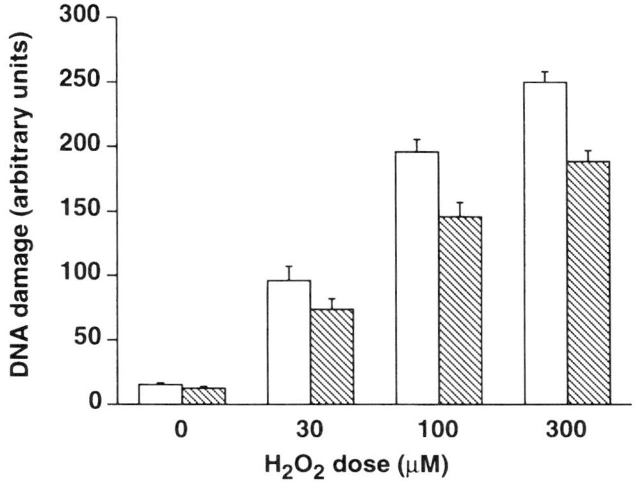

FIGURE 4. *Ex vivo* test for antioxidant protection. Lymphocytes from subjects (never-smokers) receiving placebo (open bars) or supplement (shaded bars) were challenged *in vitro* with hydrogen peroxide, and DNA strand breaks were measured by the comet assay. Bars indicate SE of the mean; $n = 24$ (placebo), $n = 17$ (supplement). Data taken from Duthie *et al.* (1996).

cytes from volunteers supplemented with β-carotene alone (30 mg/day for 6 days) and X-irradiated *ex vivo* had significantly fewer micronuclei (chromosomes or chromosomal fragments which have been excluded from the nucleus during mitosis) than cells from untreated subjects. X-rays induce DNA damage via the formation of oxygen radicals, and the antioxidant activity of β-carotene may have prevented genotoxicity in this system (Umegaki *et al.*, 1994a). We have also studied the protective effects of single antioxidants supplemented over shorter time periods. Lymphocytes were isolated from blood taken before or 24 hr after a single dose of β-carotene (45 mg) and challenged *in vitro* with hydrogen peroxide. They were subsequently incubated for up to 24 hr to allow DNA repair to proceed and DNA breaks were measured using the comet assay. The protective effect of β-carotene against DNA damage was complex and varied among subjects. Certain individuals showed very slow repair of strand breaks in the lymphocytes isolated before supplementation followed by accelerated repair after β-carotene. Carotenoid plasma levels were relatively low in these individuals, but were increased by the supplement (Fig. 5). Other individuals showed no effect of β-carotene on the rejoining of breaks, but the rate of repair was already elevated compared with the first group. Plasma β-carotene concentrations in these individuals were high and not increased by

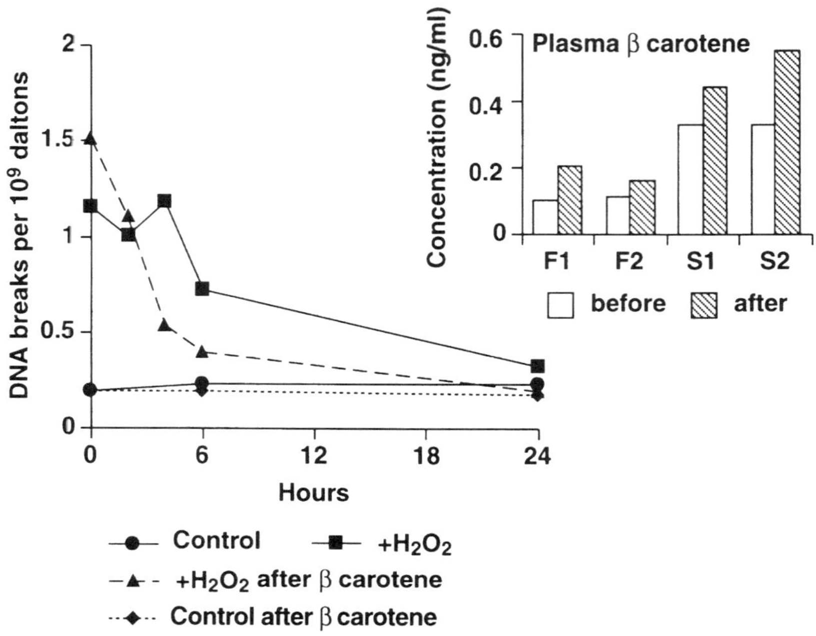

FIGURE 5. DNA repair in human lymphocytes after β-carotene supplementation. Lymphocytes, isolated from subjects (two individuals, F and S, each sampled twice) before and after a single dose of β-carotene (45 mg), were incubated with hydrogen peroxide (100 μM), then allowed to repair for up to 24 hr. DNA breaks were measured using the comet assay. The inset shows corresponding plasma β-carotene levels in these individuals.

supplementation. It appears, therefore, that β-carotene can modulate DNA repair.

It is clear from these studies that β-carotene does influence oxidative damage at the cellular level. Why this effect is not reflected in conventional intervention trials is not clear, but one possibility is that high doses of β-carotene interfere with uptake or action of other carotenoids which might have a particular antioxidant role that cannot be substituted by β-carotene.

3. ANIMAL STUDIES

Animal studies indicate that dietary carotenoids protect against DNA damage, carcinogenesis, and membrane lipid peroxidations both *in vivo* and *in vitro* (Table III). β-Carotene has been shown to decrease tumorigenesis and to inhibit conversion of papillomas to carcinomas in the skin of dimethylbenz(a)anthracene (DMBA)-treated mice (Kornhauser *et al.*, 1994; Chen *et al.*, 1993). Similarly, spontaneous mammary tumor formation in C3H (Jax) mice (Bhide *et al.*, 1994) and the incidence of preneoplastic $GST\pi^+$ liver foci in rats treated with the food mutagen, 2-amino-3-methylimidazo [4,5-f] quinoline are inhibited by β-carotene (Tsuda *et al.*, 1994). Astaxanthin and canthaxanthin inhibited cell proliferation and the number of preneoplastic and neoplastic lesions in mouse bladder following initiation by *N*-butyl-*N* (4-hydroxybutyl) nitrosamine (Tanaka *et al.*, 1994), while lycopene, lutein, and α-carotene decreased aberrant crypt foci formation in rat colon (Narisawa *et al.*, 1996). β-Carotene has also been found to inhibit micronuclei formation in the bone marrow of X- or γ-irradiated and cyclophosphamide-treated mice (Sarma *et al.*, 1994; Umegaki *et al.*, 1994b; Mukhergee *et al.*, 1991).

The carotenoids may be anticarcinogenic at different sites. β-Carotene is readily converted to retinol and it may be that vitamin A activity inhibits malignant transformation. However, this does not explain protection by non-provitamin A-active carotenoids. Many carcinogens require metabolic activation before they can exert their effect. Dietary canthaxanthin and β-carotene upregulate cytochrome P450 enzyme activities in rat liver microsomes and cause a corresponding decline in 2-acetylaminofluorene-induced hyperplastic nodules (Astorg *et al.*, 1994; Sarkar *et al.*, 1994). Conversely, β-carotene decreased glutathione-S transferase and glucuronyl transferase activities—markers of free radical exposure (Sarker *et al.*, 1994). Carotenoids may therefore inhibit carcinogenesis by increasing precarcinogen metabolism, thereby decreasing DNA binding, initiation, and the formation of secondary reactive oxygen species and radicals.

The carotenoids are also effective protective agents against oxidant-induced membrane damage. Carbon tetrachloride-induced lipid peroxidation (thiobarbituric acid-reactive products, TBARS) was completely abolished in mouse tissue by beta-carotene supplementation of the diet (Iyama *et al.*, 1996), while beta-carotene and zeaxanthin protect against cumene hydroperoxide-mediated peroxidation in chick tissues (Woodall *et al.*, 1996). Iron-mediated increases in lipid peroxidation, total cholesterol, and high-density lipoprotein cholesterol have all been shown to decrease in rats fed β-carotene (Whittaker *et al.*, 1996).

There are worries, however, that animal studies do not accurately reflect what is happening in humans, since carotenoid absorption and metabolism differ among species (Krinsky, 1989).

Table III
The Effect of Carotenoids in Animal Models (*in Vivo*)[a]

Cell type	Carotenoid	Concentration	Damaging agent	Effect of carotenoid	Reference
Swiss albino mice	β-Carotene	2.7, 27 mg/kg	Cyclophosphamide (25 mg/kg)	↓ Chromosome aberrations	Mukherjee *et al.* (1991)
SENCAR mice	β-Carotene	0.6–600 μg/g diet	DMBA (20 μg)	↓ Carcinoma	Chen *et al.* (1993)
Swiss albino mice	β-Carotene	0.5 mg/kg/day	γ-Irradiation	↓ Micronuclei	Sarma *et al.* (1994)
ICR mice	β-Carotene	50–400 mg/kg	X rays (0.3 Gy)	↓ Micronuclei	Umegaki *et al.* (1994)
C3H mice	β-Carotene	2.5 mg/day	—	↓ Tumor incidence	Bhide *et al.* (1994)
SD rats	β-Carotene	100 mg/kg diet	2-AAF	↓ Hyperplastic nodules	Sarkar *et al.* (1994)
Skh mice	β-Carotene	0.5% diet	DMBA (150 μg)	↓ Tumorigenesis	Lambert *et al.* (1994)
Fischer mice	β-Carotene	0.02% diet	IQ (100 mg/kg)	↓ GST liver foci	Tsuda *et al.* (1994)
Mice	β-Carotene	0.05% diet	Carbon tetrachloride	↓ Lipid peroxidation	Iyama *et al.* (1996)
SD rats	β-Carotene	0.05% diet	Iron (10,000 μg/g)	↓ Lipid peroxidation ↓ HDL/cholesterol	Whittaker *et al.* (1996)
SD rats	β-Carotene, α-carotene, lutein, lycopene	0.06–6 mg/day	MNU	↓ Aberrant crypt formation	Narisawa *et al.* (1996)

[a]DMPA, Dimethylbenz(a)anthracene; 2-AAF, 2-acetylaminofluorene; IQ, 2-amino-3-methylimidazo [4,5-f] quinoline; MNU, *N*-methyl nitrosourea; HDL, high-density lipoprotein; GST, glutathione-S transferase.

4. CAROTENOIDS AND CULTURED CELLS

Cultured cells provide an appropriate and simple system with which to investigate the putative protective effects of carotenoids. Proposed mechanisms include provitamin A activity, antioxidant radical quenching ability, selective toxicity toward transformed cells, and carotenoid-mediated alteration in gene expression. Studies have been carried out in normal and transformed animal and human cells using biomarkers for DNA stability, membrane integrity, and cellular viability (Tables IV and V).

4.1. Human Cells (Transformed and Nontransformed)

Murakoshi *et al.* (1989) found that α-carotene inhibited human neuroblastoma cell (GOTO) proliferation in a time- and dose-dependent manner, and that arrest of the cells in the G0–G1 phase of the cell cycle corresponded with almost total suppression of N-*myc* mRNA. Modulation of gene expression was unrelated to provitamin A activity, as β-carotene was ten times less effective than α-carotene, despite being converted to retinol more readily. β-Carotene also reduced the colony-forming ability of human keratinocytes derived from normal and preneoplastic biopsies and in human cell lines from oral cavity tumors (KB and SCC-25 cells). The frequency of micronuclei was decreased following treatment with carotenoid (Toma *et al.*, 1991). Beta-carotene was selectively toxic (decreased cell survival and ^{3}H-thymidine incorporation) toward SCC-25 tumor cells compared with foreskin-derived normal human keratinocytes. Enzyme activity (superoxide dismutase and glutathione-S transferase) was also reduced in transformed cells following exposure to carotenoid. The antiproliferative ability of carotenoids has also been demonstrated in normal mammary epithelial cells. Both β-carotene and canthaxanthin decreased the proliferation of primary human mammary epithelial cells and induced a more differentiated morphological structure in the cells. The observation that canthaxanthin was as effective as β-carotene, combined with evidence that the cells were unable to convert β-carotene to retinol, suggests that the carotenoids induce differentiation themselves (Rock *et al.*, 1995). Similarly, β-carotene increased time- and concentration-dependent cell death in transformed human colon cells (LS 180, SW 620, and HCT-15) at concentrations achievable following supplementation in humans (Iftikhar *et al.*, 1996). Differential toxicity of β-carotene to cancer cells compared with normal cells has been observed. Nontumorigenic acinar cells (2HPC8) from a pleomorphic parotid adenoma were less sensitive to β-carotene treatment (inhibition of growth) than their spontaneously transformed counterparts (2HPIG). Retinoic acid was less effective than

Table IV
The Effect of Carotenoids on Human Cells

Cell type	Carotenoid	Concentration	Damaging agent	Effect	Reference
GOTO neuroblastoma	α-Carotene, β-Carotene	0–20 μM	—	↓ N-*myc* mRNA, ↓ Cell proliferation	Murakoshi *et al.* (1989)
KB, SCC-25, keratinocytes (transformed and normal)	β-Carotene	10 μM	—	↓ Colony-forming ability, ↓ Micronuclei	Toma *et al.* (1991)
SCC-25	β-Carotene, Canthaxanthin	0–100 μM	—	↓ Thymidine uptake, ↓ Cell survival, ↑ Melphalan toxicity	Schwartz *et al.* (1992)
Human lymphocytes (*ex vivo*)	β-Carotene	30 mg/day for 6 days	X rays (0.6 Gy)	↓ Micronuclei	Umegaki *et al.* (1994a)
MCF-10	β-Carotene	7 μM	—	↑ Morphological differentiation	Rock *et al.* (1995)
Mammary epithelium	β-Carotene, Canthaxanthin	7 μM	—	↓ Cell proliferation, ↑ Morphological differentiation	Rock *et al.* (1995)
HepG2 hepatoma	β-Carotene Lutein	1.1 μM 10.9 μM	—	↓ Lipid peroxidation ↓ Lactate dehydrogenase leakage, ↑ Amino acid uptake	Martin *et al.* (1996)
Colon cancer cells (LS 180, SW 620, HCT-15)	β-Carotene	0–5.8 μM	—	↑ Cell killing	Iftikhar *et al.* (1996)
Parotid acinar cells (2HPIG, 2HPC8)	β-Carotene	20 μg/ml	—	↓ Cell proliferation	Prasad and Kumar (1996)

Table V
The Effect of Carotenoids on Animal Cells[a]

Cell type	Carotenoid	Concentration	Damaging agent	Effect of carotenoid	Reference
C3H/10T1/2	β-Carotene, Canthaxanthin	10^{-8} to 10^{-5} M	3MC (1 μg/ml, 24 hr), X rays (600 rad)	↓ Transformation	Pung *et al.* (1988)
C3H/10T1/2	β-Carotene	10^{-5} M	—	Uptake studies	Rundhaug *et al.* (1988)
C3H/10T1/2	β-Carotene, Canthaxanthin, Lycopene	3×10^{7} to 10^{-5} M	3MC (1 or 3 μg/ml, 24 hr)	↓ Transformation	Bertram *et al.* (1991)
C3H/10T1/2	β-Carotene, Canthaxanthin, α-Carotene, Lycopene	0 to 10^{-5} M	3MC (1 or 3 μg/ml, 24 hr)	↑ Cellular communication, ↓ lipid peroxidation	Zhang *et al.* (1991)
C3H/10T1/2	β-Carotene, Lycopene, Canthaxanthin	3×10^{7} to 10^{-5} M	3MC (1 or 3 μg/ml, 24 hr)	↑ Cellular communication, ↑ connexin43 expression	Zhang *et al.* (1992)
C3H/10T1/2	β-Carotene	0 to 10^{-6} M	X rays (6 Gy)	↓ Transformation	Kennedy and Krinsky (1994)
Woodchuck hepatocytes	β-Carotene	0–14.9 μM	AFB_1	↑ AFB_1 DNA binding	Yu *et al.* (1994)
BALBc 3T3 fibroblasts	β-Carotene	3 μM	5 μM 8-MOP/ UVA (3 kJ/m^2)	↑ cell proliferation	Kornhauser *et al.* (1994)
BALBc 3T3 thymocytes	β-Carotene, Canthaxanthin	0–50 μM	———	↓ Malondialdehyde	Palozza *et al.* (1996)
Chicken fibroblasts	β-Carotene	0–10 μM	Paraquat	↓ Antioxidant enzymes	Lawlor and O'Brien (1997)

[a]3MC, 3-Methylcolanthrene; AFB_1, aflatoxin B_1; 8-MOP, 8-methoxypsoralen.

β-carotene, indicating that inhibition was independent of provitamin A activity. This selective killing of tumor cells by β-carotene has been proposed to be due to prooxidant effects (Prasad and Kumar 1996). Conversely, β-carotene and lutein protected transformed human liver cells (HepG2) from tert-butyl hydroperoxide-induced membrane damage. Oxidant-induced enzyme leakage, lipid peroxidation, and disrupted amino acid uptake all decreased following incubation with carotenoids (Martin *et al.*, 1996). However, HepG2 cells retain many specialized hepatic functions and are considered a good model for normal human liver (Duthie *et al.*, 1995; Duthie and Grant 1989).

4.2. Animal Cells (Transformed and Nontransformed)

4.2.1. C3H10T1/2 Cells

Perhaps the most widely used animal cell culture system with which to investigate the ability of dietary carotenoids to protect against DNA damage (in the form of oncogenesis) is the C3H10T1/2 transformation assay, which mimics crucial aspects of carcinogenesis. First described by Reznikoff *et al.* (1973), the assay uses immortal C3H mouse embryo fibroblasts (C3H10T1/2 clone 8) to detect and study the mutagenic potential of both chemical and physical carcinogens. Normal C3H10T1/2 cells display a strong density-dependent inhibition of proliferation and cease to divide upon reaching confluency. Carcinogen-transformed cells continue to grow uncontrollably even after confluency, forming distinct multilayered foci which can be readily detected. C3H10T1/2 cells have a low rate of spontaneous transformation, while transformed cells induce tumors in irradiated C3H mice (Boreiko *et al.*, 1985; Reznikoff *et al.*, 1973). Retinoids were the first dietary compounds to be employed in this assay system as putative protective agents against 3-methylcholanthrene (3MC)-induced neoplastic transformation. Retinyl acetate, retinol, and retinal all effectively inhibited transformation of mouse fibroblasts when applied after the chemical carcinogen. Suppression of transformation was reversible following exclusion of the retinoids, indicating that protection occurred at the progressional stage of carcinogenesis (Merriman and Bertram 1979).

Subsequently, β-carotene and canthaxanthin were found to inhibit neoplastic transformation of C3H10T1/2 cells by both chemical (3MC) and physical (X rays) carcinogens. Surprisingly, canthaxanthin was more effective at suppressing oncogenesis than β-carotene, suggesting that the carotenoids protected by a mechanism other than conversion to retinol. Neither β-carotene nor canthaxanthin altered the colony-forming ability of the mouse cells, and inhibition of transformation was reversible, indicating that the carotenoids were acting to inhibit malignant transformation rather than by selectively killing initiated cells. Since mammalian cells cannot convert canthaxanthin to retinoids and C3H10T1/2 cells are incapable of converting β-carotene to retinol (Pung *et al.*, 1988; Rundhaug *et al.*, 1988), it was proposed that inhibition of transformation was due to the antioxidant ability of the carotenoids via inhibition of lipid peroxidation. Other carotenoid antioxidants prevalent in the human diet (α-carotene and lycopene) are also effective in this system (Bertram *et al.*, 1991; Kennedy and Krinsky 1994). However, the ability of dietary carotenoids to inhibit malignant transformation is linked not with their antioxidant potential (TBARS), but rather with increased intracellular communication between the mouse cells (Zhang *et al.*, 1991). Gap junctions, composed of transmembrane connexin proteins, allow the transfer of growth-controlling signals between neighboring cells. Upregulation of cell-to-cell communication, demonstrated as increased transfer of the fluorescent dye Lucifer Yellow following microinjection, had been demonstrated for retinoids, but had not been shown for carotenoids (Mehta *et al.*, 1989; Hossain *et al.*, 1989). Both carotenoids and retinoids increased cellular communication and inhibited dedifferentiation so that malignant transformation was retarded. Further evidence for carotenoid-mediated increases in cellular communication came from Zhang *et al.* (1992), who observed that dietary carotenoids induced connexin 43 mRNA and protein levels. Non-provitamin A carotenoids were as effective as those able to undergo conversion to retinol. Conversely, the potent antioxidant α-tocopherol was inactive, indicating that carotenoid-induced upregulation of connexin gene expression, rather than provitamin A or antioxidant activity, was responsible for suppression of malignant transformation. β-Carotene does, however, inhibit endogenous lipid peroxidation (TBARS) in C3H10T1/2 cells (Franke *et al.*, 1994). β-Carotene and canthaxanthin modulate the expression of cytokeratin and filaggrin (differentiation-specific genes) in human keratinocytes grown in culture (Bertram and Bortkiewicz, 1995).

It appears, therefore, that dietary carotenoids inhibit neoplastic transformation and alter the expression of gene products in both mouse and normal human cells.

4.2.2. Other Animal Cells

β-Carotene can reduce the phototoxicity (measured as inhibition of growth) of 8-methoxypsoralen in combination with UVA irradiation in BALBc 3T3 embryonic mouse fibroblasts (Kornhauser *et al.*, 1994). Similarly, in a study by Palozza *et al.* (1996) β-carotene and canthaxanthin decreased tert-butyl hydroperoxide-induced malondialdehyde formation in BALBc 3T3 mouse thymocytes. Canthaxanthin was a more potent inhibitor of oxidant-induced membrane damage than β-carotene. Modulation of antioxidant enzyme activity has also been employed as a marker for oxidative stress. Lawlor and O'Brien (1997) found that paraquat increased superoxide dismutase and catalase activity in normal chicken fibroblasts, while glutathione peroxidase activity was decreased. β-Carotene reversed these effects.

While cell-based *in vitro* assays are undoubtedly invaluable in elucidating mechanisms of carotenoid protection, it is important to note that the concentrations of carotenoids required to protect against DNA damage or lipid peroxidation in the majority of these systems are only achievable in humans after supplementation.

5. CONCLUSIONS

Epidemiological studies have indicated that dietary carotenoids may protect against cancer and heart disease. Moreover, conventional intervention trials and studies employing methods of molecular epidemiology have shown that β-carotene supplementation decreases both DNA damage and indices of lipid peroxidation, putative biomarkers for disease. *In vivo* and *in vitro* animal and cell studies confirm that the carotenoids are chemoprotective by a number of different mechanisms, only some of which have been reviewed in this chapter. They act as antioxidants to inhibit lipid peroxidation and oxidant-induced DNA damage, and are highly effective modulators of gene expression. Tumor cells are more sensitive to prooxidant-induced cell death and inhibition of proliferation associated with carotenoids than their untransformed counterparts, although cell type and method of delivery appear to be crucial. Chemoprevention appears to be independent of provitamin A activity. In addition, carotenoids may influence carcinogen activation and DNA binding.

It can be concluded, therefore, that the carotenoids are dietary chemopreventive agents of great potential importance, which protect against both coronary heart disease and specific types of cancer via a number of different pathways. However, many questions remain to be answered concerning the effects of intervention with nonphysiological concentrations of single carotenoids on the uptake, metabolism, and distribution of other essential dietary components and the activity of cellular enzyme systems before carotenoids can be endorsed as safe and effective chemoprotective agents.

6. REFERENCES

Alpha-Tocopherol, Beta-Carotene Cancer Prevention Study Group, 1994, The effect of vitamin E and beta-carotene on the incidence of lung cancer and other cancers in male smokers, *N. Engl. J. Med.* **330:**1029–1035.

Armstrong, B. K., Mann, J. I., and Eskin, F., 1975, Commodity consumption and ischaemic heart disease mortality, with special reference to dietary practices, *J. Chron. Dis.* **28:**455–469.

Astorg, P., Gradelet, S., LeClerc, J., Canivenc, M.-C., and Siess, M.-H., 1994, Effects of beta-carotene and canthaxanthin on liver xenobiotic-metabolizing enzymes in the rat, *Food Chem. Toxicol.* **32:**735–742.

Bellizzi, M. C., Franklin, M. F., Duthie, G. G., and James, W. P. T., 1994, Vitamin E and coronary heart disease: The European paradox, *Eur. J. Clin. Nutr.* **48:**822–831.

Bendich, A., and Olsen, J., 1989, Biological action of carotenoids, *FASEB J.* **3:**927–1932.

Bertram, J. S., 1993, Cancer prevention by carotenoids: Mechanistic studies in cultured cells, *Ann. N. Y. Acad. Sci.* **691:**177–191.

Bertram, J. S., and Bortkiewicz, H., 1995, Dietary carotenoids inhibit neoplastic transformation and modulate gene expression in mouse and human cells, *Am. J. Clin. Nutr.* **62**(Suppl.)**:**1327S–1336S.

Bertram, J. S., Pung, A., Churley, M., Kappock, T. J., IV, Wilkins, L. R., and Cooney, R. V., 1991, Diverse carotenoids protect against chemically-induced neoplastic transformation, *Carcinogenesis* **12:**671–678.

Bhide, S. V., Azuine, M. A., Lahiri, M., and Telang, N. T., 1994, Chemoprevention of mammary tumor virus-induced and chemical carcinogen-induced rodent mammary tumors by natural plant products, *Breast Cancer Res. Treat.* **30:**233–242.

Blot, W. J., Li, J.-Y., Taylor, P. R., Guo, W., Dawsey, S., Wang, G.-Q., Yang, C. S., Zheng, S.-F., Gail, M., Li, G.-Y., Yu, Y., Liu, B.-Q., Tangrea, J., Sun, Y.-H., Liu, F., Fraumeni, J. F., Zhang, Y.-H., and Li, B., 1993, Nutrition intervention trails in Linxian, China: Supplementation with specific vitamin/mineral combinations, cancer incidence, and disease-specific mortality in the general population, *J. Natl. Cancer Inst.* **85:**1483–1492.

Boreiko, C. J., 1985, Mechanistic aspects of initiation and promotion in C3H/10Ti/2 cells, *Carcinogenesis* **9:**153–326.

Breecher, G. R., and Khackik, F., 1992, Qualitative relationship of dietary and plasma carotenoids in human beings, *Ann. N. Y. Acad. Sci.* **669:**320–321.

Chen, L.-C., Sly, L., Jones, C. S., Tarone, R., and DeLuca, L. M., 1993, Differential effects of dietary beta-carotene on papilloma and carcinoma formation induced by initiation-promotion protocol in SENCAR mouse skin, *Carcinogenesis* **14:**713–717.

Collins, A. R., Dusinska, M., Gedik, C. M., and Stetina, R., 1996, Oxidative DNA damage to DNA: Do we have a reliable biomarker? *Environ. Health Perspect.* **104**(Suppl.)**:**465–469.

Connett, J. E., Kuller, L. H., Kjelsberg, M. O., Polk, B. F., Collins, G., Rider, A., and Hulley, S. B., 1989, Relationship between carotenoids and cancer, *Cancer* **64:**126–134.

Crombie, I. K., Smith, W. C. S., Tavendale, R., and Tunstall Pedoe, H., 1990, Geographic clustering of risk factors and lifestyle for coronary heart disease in the Scottish Heart Health study, *Br. Heart. J.* **64:**199–203.

Duthie, S. J., and Grant, M. H., 1989, The toxicity of menadione and mitozantrone in human liver-derived HepG2 hepatoma cells, *Biochem. Pharmacol.* **38:**1247–1255.

Duthie, G. G., Arthur, J. R., Beattie, J. A. G., Brown, K. B., Morrice, P. C., Robertson, J. D., Shortt, C. T., Walker, K. A., and James, W. P. T., 1993, Cigarette smoking, antioxidants, lipid peroxidation and coronary heart disease, *Ann. N. Y. Acad. Sci.* **686:**120–129.

Duthie, S. J., Melvin, W. T., and Burke, M. D., 1995, Drug toxicity mechanisms in human hepatoma HepG2 cells: Cyclosporin A and tamoxifen, *Xenobiotica* **25:**1151–1164.
Duthie, S. J., Ma, A.-G., Ross, M. A., and Collins, A. R., 1996, Antioxidant supplementation decreases oxidative DNA damage in human lymphocytes, *Cancer Res.* **56:**1291–1295.
Esterbauer, H., Gebicki, J., Puhl, H., and Jurgens, G., 1992, The role of lipid peroxidation and antioxidants in the oxidative modification of LDL, *Free Radic. Biol. Med.* **13:**341–390.
Franke, A. A., Harwood, P. J., Shimamoto, T., Lumeng, S., Zhang, L.-X., Bertram, J. S., Wilkens, L. R., Le Marchand, L., and Cooney, R. V., 1994, Effects of micronutrients and antioxidants on lipid peroxidation in human plasma and in cell culture, *Cancer Lett.* **70:**17–26.
Gey, K. F., 1986, On the antioxidant hypothesis with regard to arteriosclerosis, *Bibliotheca Nutritio Dieta* **37:**53–91.
Gey, K. F., 1992, Epidemiological correlations between poor plasma levels of essential antioxidants and the risk of coronary heart disease and cancer, in *Lipid Soluble Antioxidants* (A. S. H. Ong and L. Packer, eds.), pp. 442–456, Birkhauser, Basel.
Gramenzi, A., Gentile, A., Fasoli, M., Negri, E., Parazzini, F., and La Vecchia, C., 1990, Association between certain foods and risk of acute myocardial infarction in women, *Br. Med. J.* **300:**771–773.
Hennekens, C. H., Buring, J. E., Manson, J. E., Stampfer, M., Rosner, B., Cook, N. R., Belanger, C., LaMotte, F., Gaziano, J. M., Ridker, P. M., Willett, W., and Peto, R., 1996, Lack of effect of long-term supplementation with beta-carotene on the incidence of malignant neoplasms and cardiovascular disease, *N. Engl. J. Med.* **334:**1145–1149.
Hossain, M. Z., Wilkens, L. R., Mehta, P. P., Loewenstein, W., and Bertram, J. S., 1989, Enhancement of gap junctional communication by retinoids correlates with their ability to inhibit neoplastic transformation, *Carcinogenesis* **10:**1743–1748.
Hussein, O., Schlezinger, S., Rosenblat, M., Keidar, S., and Aviram, M., 1997, Reduced susceptibility of low density lipoprotein (LDL) to lipid peroxidation after fluvastatin therapy is associated with the hypercholesterolemic effect of the drug and its binding to the LDL, *Atherosclerosis* **128:**11–18.
Iftikhar, S., Lietz, H., Mobarhan, S., and Frommel, O., 1996, *In vitro* beta-carotene toxicity for human colon cancer cells, *Nutr. Cancer* **25:**221–230.
Isler, O., 1971, *Carotenoids*, Birkhauser, Basel.
Ito, Y., Sasaki, R., Suzuki, S., and Aoki, K., 1991, Relationship between serum xanthophyll levels and the consumption of cigarettes, alcohol or foods in healthy inhabitants of Japan, *Int. J. Epidemiol.* **20:**615–620.
Iyama, T., Takasuga, A., and Azuma, M., 1996, Beta-carotene accumulation in mouse tissues and a protective role against lipid peroxidation, *Int. J. Vitamin Res.* **66:**301–305.
James, W. P. T., Duthie, G. G., and Wahle, K. W., 1989, The Mediterranean diet: Protective or merely non-toxic? *Eur. J. Clin. Nutr.* **43**(Suppl. 2)**:**31–41.
Kennedy, A. R., and Krinsky, N. I., 1994, Effects of retinoids, beta-carotene and canthaxanthin on UV- and X-ray-induced transformation of C3H10T1/2 cells *in vitro*, *Nutr. Cancer* **22:**219–232.
Keys, A., 1980, *Seven Countries: A Multivariate Analysis of Death and Coronary Heart Disease*, Harvard University Press, Cambridge, Massachusetts.
Knekt, P., Jarvinen, R., Seppanen, R., Rissanen, A., Aromaa, A., Heinonen, O. P., Albanes, D., Heinonen, M., Pukkala, E., and Teppo, L., 1991, Dietary antioxidants and the risk of lung cancer, *Am. J. Epidemiol.* **134:**471–479.
Kornhauser, A., Wamer, W. G., Lambert, L. A., and Wei, R. R., 1994, Beta-carotene inhibition of chemically-induced toxicity *in vivo* and *in vitro*, *Food Chem. Toxicol.* **32:**149–154.

Krinsky, N. I., 1989, Carotenoids and cancer in animal models, *J. Nutr.* **119:**123–126.
Krinsky, N. I., 1993, Micronutrients and their influence on mutagenicity and malignant transformation, *Ann. N. Y. Acad. Sci.* **686:**229–242.
Kushi, L. H., Folsom, A. R., Prineas, R. J., Mink, P. J., Wu, Y., and Bostick, R. M., 1996, Dietary antioxidant vitamins and death from coronary heart disease in postmenopausal women, *N. Engl. J. Med.* **334:**1156–1162.
Lambert, L. A., Wamer, W. G., Wei, R. R., Lavu, S., Chirtel, S. J., and Kornhauser, A., 1994, The protective but nonsynergistic effect of dietary beta-carotene and vitamin E on skin tumorigenesis in Skh mice, *Nutr. Cancer* **21:**1–12.
Lawlor, S. M., and O'Brien, N. M., 1997, Modulation of paraquat toxicity by beta-carotene at low oxygen partial pressure in chicken embryo fibroblasts, *Br. J. Nutr.* **77:**133–140.
Liebler, D. C., Stratton, S. P., and Kaysen K. L., 1997, Antioxidant actions of beta-carotene in liposomal and microsomal membranes: Role of carotenoid-membrane incorporation and alpha-tocopherol, *Arch. Biochem. Biophys.* **338:**244–250.
Martin, K. R., Failla, M. L., and Smith, J. C., 1996, Beta-carotene and lutein protect HepG2 human liver cells against oxidant-induced damage, *J. Nutr.* **126:**2098–2106.
McCormick, J., and Skrabanek, P., 1988, Coronary heart disease is not preventable by population interventions, *Lancet* **ii:**839–841.
Mehta, P. P., Bertram, J. S., and Loewenstein, W. R., 1989, The actions of retinoids on cellular growth correlates with their actions on gap junctional communication, *J. Cell Biol.* **108:**1053–1065.
Merriman, R. L., and Bertram, J. S., 1979, Reversible inhibition by retinoids of 3-methylcholanthrene-induced neoplastic transformation in C3H/10T1/2 clone 8 cells, *Cancer Res.* **39:**1661–1666.
Micozzi, M. S., 1992, Plasma carotenoid responses to chronic intake of selected foods and beta-carotene supplements in man, *Am. J. Clin. Nutr.* **55:**1125–1129.
Mukherjee, A., Agarwal, K., Aguilar, M. A., and Sharma, A., 1991, Anticlastogeneic activity of beta-carotene against cyclophosphamide in mice *in vivo*, *Mutat. Res.* **263:**41–46.
Murakoshi, M., Takayasu, J., Kimura, O., Kohmura, E., Nishino, H., Iwashima, A., Okuzumi, J., Sakai, T., Sugimoto, T., Imanishi, J., and Iwasaki, R., 1989, Inhibitory effects of alpha-carotene on proliferation of the human neuroblastoma cell line GOTO, *J. Natl. Cancer Inst.* **81:**1649–1652.
Narisawa, T., Fukaura, Y., Hasebe, M., Ito, M., Aizawa, R., Murakoshi, M., Uemura, S., Khachik, F., and Nishino, H., 1996, Inhibitory effects of natural carotenoids, alpha-carotene, beta-carotene, lycopene and lutein, on aberrant crypt foci formation in rats, *Cancer Lett.* **107:**137–142.
Negri, E., La Vecchia, C., Franceschi, S., D'Avanzo, B., and Parazzini, F., 1991, Vegetable and fruit consumption and cancer risk, *Int. J. Cancer* **48:**350–354.
Olmedilla, B., Granado, F., Blanco, I., and Rojas-Hidalgo, E., 1992, Determination of nine carotenoids, retinol, retinyl palmitate and alpha-tocopherol in control human serum using two internal standards, *Food Chem.* **45:**205–213.
Olsen, J. A., 1992, Carotenoids and vitamin A: An overview, in *Lipid Soluble Antioxidants* (A. S. H. Ong and L. Packer, eds.), pp. 178–192, Birkhauser, Basel.
Omenn, G. S., Goodman, G. E., Thornquist, M. D., Balmes, J., Cullen, M. R., Glass, A., Keogh, J. P., Meyskens, F. L., Valaris, B., Williams, J. H., Barnhart, S., and Hammer, S., 1996, Effects of a combination of beta-carotene and vitamin A on lung cancer and cardiovascular diseases, *N. Engl. J. Med.* **334:**1150–1155.
Paganini-Hill, A., Chao, A., Ross, R. K., and Henderson, B. E., 1987, Vitamin A, beta-carotene and the risk of cancer: A prospective study, *J. Natl. Cancer Inst.* **79:**443–448.

Palozza, P., Luberto, C., Ricca, P., Sgarlata, E., Calviello, G., and Bartoli, G. M., 1996, Effect of beta-carotene on tert-butyl hydroperoxide-induced lipid peroxidation in murine normal and tumor thymocytes, *Arch. Biochem. Biophys.* **325:**145–151.

Prasad, K. N, and Kumar, R., 1996, Effect of individual and multiple antioxidant vitamins on growth and morphology of human nontumorigenic and tumorigenic parotid acinar cells in culture, *Nutr. Cancer* **26:**11–19.

Pung, A., Rundhaug, J. E., Yoshizawa, C. N., and Bertram, J. S., 1988, Beta-carotene and canthaxanthin inhibit chemically- and physically-induced neoplastic transformation in 10T1/2 cells, *Carcinogenesis* **9:**1533–1539.

Reznikoff, C. A., Bertram, J. S., Brankow, D. W., and Heidelberger, C., 1973, Quantitative and qualitative studies of chemical transformation of cloned C3H mouse embryo cells sensitive to postconfluence inhibition of cell division, *Cancer Res.* **33:**3239–3249.

Rice-Evans, C., Leake, D., Bruckdorfer, R., and Diplock, A. T., 1996, Practical approaches to low density lipoprotein oxidation: Why's, wherefores and pitfalls, *Free Radic. Res. Commun.* **25:**285–312.

Riemersma, R. A., Wood, D. A., MacIntyre, C. C. A., Elton, R., Gey, K. F., and Oliver, M. F., 1989, Low plasma vitamins E and C: Increased risk of angina in Scottish men, *Ann. N. Y. Acad. Sci.* **570:**291–295.

Rock, C. L., Kusluki, R. A., Galvez, M. M., and Ethier, S. P., 1995, Carotenoids induce morphological changes in human mammary epithelial cell cultures, *Nutr. Cancer* **23:**319–333.

Rosenman, R. H., 1992, Diet in haste; repent at leisure? *Biochemist* **1992**(Aug–Sept)**:**6–10.

Ross, M., Crosley, K. M., Brown, K. M., Duthie, S. J., Arthur, J. R., and Duthie, G. G., 1995, Plasma concentrations of carotenoids and antioxidant vitamins in Scottish males: Influences of smoking, *Eur. J. Clin. Nutr.* **49:**861–865.

Rundhaug, J. E., Pung, A., Read, C. M., and Bertram, J. S., 1988, Uptake and metabolism of betacarotene and retinal by C3H/10T1/2 cells, *Carcinogenesis* **9:**1541–1545.

Sarkar, A., Mukherjee, B., and Chatterjee, M., 1994, Inhibitory effect of beta-carotene on chronic 2-acetylaminofluorene induced hepatocarcinogenesis in rat: Reflection in hepatic drug metabolism, *Carcinogenesis* **15:**1055–1060.

Sarma, L., Abraham, S. K., and Kesavan, P. C., 1994, Chromosomal damage by low doses of radiation: Protection by combinations of dietary antioxidants, *Curr. Sci.* **66:**861–862.

Schmidt, K., 1991, Antioxidant vitamins and beta-carotene: Effects on immunocompetence, *Am. J. Clin. Nutr.* **53:**383–385S.

Schwartz, J. L., Tanaka, J., Khandekar, V., Herman, T. S., and Teicher, B. A., 1992, Beta-carotene and/or vitamin E as modulators of alkylating agents in SCC-25 human squamous carcinoma cells, *Cancer Chemother. Pharmacol.* **29:**207–213.

Sies, H., Stahl, W., and Sundquist, A. R., 1992, Antioxidant functions of vitamins: Vitamin E and C, beta-carotene and other carotenoids, *Ann. N. Y. Acad. Sci.* **669:**7–20.

Singh, R. B., Niaz, M. A., Ghosh, S., and Rastogi, S. S., 1993, Effect on mortality and reinfarction of adding fruits and vegetables to a prudent diet in The Indian Experiment of Infarct Survival (IEIS), *J. Am. Coll. Nutr.* **12:**255–261.

Stahelin, H. B., Gey, K. F., Eichholzer, M., Ludin, E., Bernasconoi, F., Thurneysen, J., and Brubacher, G., 1991, Plasma antioxidant vitamins and subsequent cancer mortality in the 12-year follow-up of the prospective Basel study, *Am. J. Epidemiol.* **133:**766–775.

Stahl, W., 1996, Desirable versus potentially harmful intake levels of various forms of carotenoids, in *Natural Antioxidants and Food Quality in Atherosclerosis and Cancer Prevention* (J. T. Kumpulainen and J. T. Salonen, eds.), pp. 102–109, Royal Society of Chemistry, London.

Tanaka, T., Morishita, Y., Suziu, M., Kojima, T, Okumura, A., and Mori, H., 1994, Chemoprevention of mouse urinary bladder carcinogenesis by the naturally occurring carotenoid astaxanthin, *Carcinogenesis* **15:**15–19.

Toma, S., Albanese, E., Paluma, R., Nicolo, G., Morerio, C., Mangiante, P. E., Bonatti, S., Cancedda, R., and Santi, L., 1991, *In vitro* effects of beta-carotene on human oral keratinocytes from precancerous lesions and squamous carcinoma, *Anticancer Drugs* **2:**581–589.

Tsuda, H., Uehara, N., Iwahori, Y., Asamoto, M., Iigo, M., Nagao, M., Matsumoto, K., Ito, M., and Hirono, I., 1994, Chemopreventive effects of beta-carotene, alpha-tocopherol and five naturally occurring antioxidants on initiation of hepatocarcinogenesis by 2-amino-3-methylimidazo[4,5-f]quinoline in the rat, *Jpn. J. Cancer Res.* **85:**1214–1219.

Umegaki, K., Ikegami, S., Inoue, K., Ichikawa, T., Kobayashi, S., and Soeno, N., 1994a, Beta-carotene prevents X-ray induction of micronuclei in human lymphocytes, *Am. J. Clin. Nutr.* **59:**409–412.

Umegaki, K., Takeuchi, N., Ikegami, S., and Ichikawa, T., 1994b, Effect of beta-carotene on spontaneous and X-ray-induced chromosomal damage in bone marrow cells of mice, *Nutr. Cancer* **22:**277–284.

Westhuyzen, J., 1997, The oxidation hypothesis of atherosclerosis: An update, *Ann. Clin. Lab. Med.* **27:**1–10.

Whittaker, P., Wamer, W. G., Chanderbhan, R. F., and Dunkel, V. C., 1996, Effects of alpha-tocopherol and beta-carotene on hepatic lipid peroxidation and blood lipids in rats with dietary iron overload, *Nutr. Cancer* **25:**119–128.

Woodall, A. A., Britton, G., and Jackson, M. J., 1996, Dietary supplementation with carotenoids: Effects on alpha-tocopherol levels and susceptibility of tissues to oxidative stress, *Br. J. Nutr.* **76:**307–317.

Yu., M.-W., Zhang, Y.-J., Blaner, W. S., and Santella, R. M., 1994, Influence of vitamin A, C, and E and beta-carotene on aflatoxin B_1 binding to DNA in woodchuck hepatocytes, *Cancer* **73:**596–604.

Zhang, L.-X., Cooney, R. V., and Bertram, J. S., 1991, Carotenoids enhance gap junctional communication and inhibit lipid peroxidation in C3H/10T1/2 cells: Relationship to their cancer chemopreventive action, *Carcinogenesis* **12:**2109–2114.

Zhang, L.-X., Cooney, R. V., and Bertram, J. S., 1992, Carotenoids up-regulate *connexin43* gene expression independent of their provitamin A or antioxidant properties, *Cancer Res.* **52:**5707–5712.

Ziegler, R. G., 1989, A review of epidemiologic evidence that carotenoids reduce the risk of cancer, *J. Nutr.* **119:**116–122.

Chapter 9

β-Carotene

An Antioxidant or a Target of Oxidative Stress in Cells?

Billy W. Day, Stefania Bergamini, Yulia Y. Tyurina, Gianfranca Carta, Vladimir A. Tyurin, and Valerian E. Kagan

1. INTRODUCTION

The results from numerous observational studies conducted in the 1980s have nurtured a strong opinion that the carotenoids in fruits and vegetables are responsible for the associated decreased incidence of degenerative dis-

Billy W. Day Department of Environmental and Occupational Health, and Department of Pharmaceutical Sciences, University of Pittsburgh, Pittsburgh, Pennsylvania 15238. **Stephania Bergamini and Gianfranca Carta** Department of Environmental and Occupational Health, University of Pittsburgh, Pittsburgh, Pennsylvania 15238; Visiting Researchers from the Department of Biomedical Science, University of Modena, Modena, Italy; and the Dipartimento di Biologia Sperimentale, Sezione di Patologia Sperimentale, Università di Cagliari, Cagliari, Italy. **Yulia Y. Tyurina and Vladimir A. Tyurin** Department of Environmental and Occupational Health, University of Pittsburgh, Pittsburgh, Pennsylvania 15238; on leave from the Institute of Evolutionary Physiology and Biochemistry, Russian Academy of Science, St. Petersburg, 194223 Russia. **Valerian E. Kagan** Department of Environmental and Occupational Health, and Department of Pharmacology, University of Pittsburgh, Pittsburgh, Pennsylvania 15238.

Subcellular Biochemistry, Volume 30: Fat-Soluble Vitamins, edited by Quinn and Kagan. Plenum Press, New York, 1998.

eases like cardiovascular disease and cancer (Sies *et al.*, 1992). One of the most popular among the proposed mechanisms by which carotenoids might be disease preventive centers on their putative antioxidant activity (Palozza and Krinsky, 1992a,b; Tsuchihashi *et al.*, 1995). Some recent, large-scale, double-blind epidemiological studies as well as clinical intervention trials have yielded discouraging results, however, in that β-carotene supplements do not appear to be effective in disease prevention. Moreover, the CARET Study (β-Carotene and Retinol Efficacy Trial) was discontinued 21 months prior to its planned end because of clear evidence that no benefit, and rather substantial evidence of possible harm (Omenn *et al.*, 1996), resulted from this supplementation. There were 28% more lung cancers and 17% more deaths in the active CARET intervention group (smokers and asbestos-exposed workers at high risk for lung cancer) than in controls (Omenn *et al.*, 1996).

These disappointing results raised anew the question of whether or not carotenoids act as biologically significant antioxidants in physiologically relevant environments (Liebler *et al.*, 1997). While high reactivity of carotenoids toward a variety of free radical species (e.g., peroxyl, alkoxyl, phenoxyl, nitrogen dioxide, thiyl, and sulfonyl radicals) has been well established in simple chemical systems (Mortensen and Skibsted, 1996; Everett *et al.*, 1996), reports concerning carotenoid antioxidant potency in biological model systems (e.g., subcellular fractions, cells) and *in vivo* are conjectural (Palozza and Krinsky, 1992b; Palozza *et al.*, 1996).

A number of earlier studies demonstrate that enrichment of subcellular fractions or liposomes with β-carotene *in vitro* results in resistance to oxidative stress (Palozza and Krinsky, 1992b). It was noted, however, that the antioxidant effect of carotenoids was dramatically dependent on their mode of integration and the uniformity of their partitioning in membrane lipid bilayers. Not surprisingly, the data reported are thus highly controversial, ranging from attributing high antioxidant potency (Palozza and Krinsky, 1992a) to very low or no antioxidant activity to the carotenoids (Liebler *et al.*, 1997). These incorporation-dependent effects are absent in experiments in which membrane fractions are isolated from tissues of animals fed β-carotene-rich diets, where proper integration and orientation of β-carotene in membranes occurs. Recent studies show that β-carotene exerts a very weak, if any, antioxidant effect in such membranes containing other endogenous lipid-soluble antioxidants (e.g., vitamin E) (Liebler *et al.*, 1997; Bast *et al.*, 1996).

Experiments performed with cultured cells may be a critical source of information on the antioxidant properties of β-carotene. Indeed, detailed studies with such systems have demonstrated that β-carotene can protect cells against oxidative stress and damage, especially at low (150 mm Hg pO_2)

oxygen tensions (Palozza *et al.*, 1995; Martin *et al.*, 1996). At high (760 mm HgpO_2) oxygen pressure, however, β-carotene exhibits prooxidant effects (Palozza *et al.*, 1995). One serious concern in interpreting these experiments is that high concentrations of oxidants—standardly used in such experiments—cause significant cell killing. Hence, it is not clear whether the apparent protective effects of β-carotene against oxidative stress are realized in living cells, or if these are postmortal events occurring in damaged cells. Additionally, it is not clear what contribution the non-antioxidant-driven protection afforded by β-carotene may have in its apparent "antioxidant" effect due to the fact that the constituents of living cells experience less cumulative oxidation than dead cells; i.e., dead cells have no means to avoid and do not repair oxidative lesions.

With this in mind, we conducted experiments in which cells were loaded with β-carotene, then exposed to sublethal fluxes of peroxyl radicals. We used our newly developed technique, metabolic integration of a fluorescent fatty acid (Ritov *et al.*, 1996), to measure site-specific oxidative stress in membrane phospholipids in living cells. Our results demonstrate that β-carotene does not provide any antioxidant protection to live cells under normobaric oxygen tension.

2. METHODS

HL-60 cells were grown in RPMI 1640 medium supplemented with 10% fetal bovine serum at 37°C in a 5% CO_2 atmosphere. Cells from passages 25–40 were used in the experiments. The density of cells at collection time was 1×10^6 cells/ml. Cell viability was determined microscopically by Trypan blue exclusion. Cells diluted to a density of 0.5×10^6 cells/ml in RPMI 1640 medium containing 10% fetal bovine serum were treated with β-carotene or α-tocopherol predissolved in tetrahydrofuran (THF) to give a final concentration of 10–50 mM. The cells were incubated at 37°C in a 5% CO_2 atmosphere for 24 hr. In order to remove β-carotene or α-tocopherol not incorporated into the cells, after the 24 hr incubation, cells were washed with L1210 buffer and recovered by centrifugation at $800 \times g$ for 15 min. Cells so treated were determined to be 96% viable. Aliquots were taken for determination of β-carotene and α-tocopherol content by high-pcrformance liquid chromatography (HPLC) procedures described previously (Tyurin *et al.*, 1997; Lang *et al.*, 1986).

cis-Parinaric acid (PnA; Molecular Probes, Eugene, OR) was metabolically incorporated into HL-60 cells enriched with β-carotene or α-tocopherol by addition of its human serum albumin (hSA) complex (PnA–hSA). Cell suspensions (1×10^6 cells/ml) in L1210 buffer were treated with PnA–hSA at 37°C for 2 hr, and washed twice with L1210 buffer with or without hSA (0.5 mg/ml) to give a final incorporated PnA concentration of 5 mg/10^6 cells determined as previously reported (Ritov *et al.*, 1996). HL-60 cells thus enriched with β-carotene or α-tocopherol and metabolically loaded with PnA were incubated for 2 hr in the presence of 500 μM 2,2′-azobis(2,4-dimethylvaleronitrile) (AMVN). Aliquots of cells were taken to determine cell viability and antioxidant and phospholipid content. Total lipids were extracted from cells using a slightly modified Folch procedure (Folch *et al.*, 1957), and lipid phosphorus content was determined using a modification of a previously reported molybdate-based method (Chalvardjian and Rudnicki, 1970). PnA content in individual phospholipid classes was determined by HPLC with fluorescence detection (emission at 420 nm after excitation at 324 nm) as previously reported (Ritov *et al.*, 1996). Data are expressed as mean ± SE values.

The statistical significance of differences between determinations was determined using Student's *t* test. Values of $p < .05$ were considered significant.

3. RESULTS AND DISCUSSION

We developed procedures to load HL-60 cells with levels of β-carotene or α-tocopherol that had no effect on cell viability or growth characteristics. In a previous work (Tyurin *et al.*, 1997), we used two representative phenolic compounds, phenol and a water-soluble derivative of α-tocopherol, 2,2,5,7,8-pentamethyl-6-hydroxychromane (PMC), to study interactions of their respective phenoxyl radicals toward the β-carotene-enriched and control cells. In this work, we exposed the β-carotene-enriched, α-tocopherol-enriched, and control cells to a low flux of peroxyl radicals generated by the lipid-soluble azo initiator AMVN. In both cases, experiments were performed during time periods when there was no induction of cell killing, i.e., we determined changes that occurred only in living cells.

We previously noted that β-carotene integrated into HL-60 cells is oxidized by phenol/H_2O_2, but not by either agent alone. HL-60 cells contain very high myeloperoxidase activity [see Kettle *et al.* (1995), who also show that myeloperoxidase is the most abundant hemoprotein in these cells]. Inhibition of myeloperoxidase with azide (and thus of phenol oxidation) simultaneously inhibits oxidation of β-carotene, suggesting that the oxidation of β-carotene is catalyzed by endogenous myeloperoxidase. In contrast to phenol, PMC/H_2O_2 is ineffective in oxidizing β-carotene (Fig. 1). In the absence of H_2O_2, HL-60 cells loaded with β-carotene oxidize phenol or PMC only very slowly. In the presence of H_2O_2, the oxidation of phenol

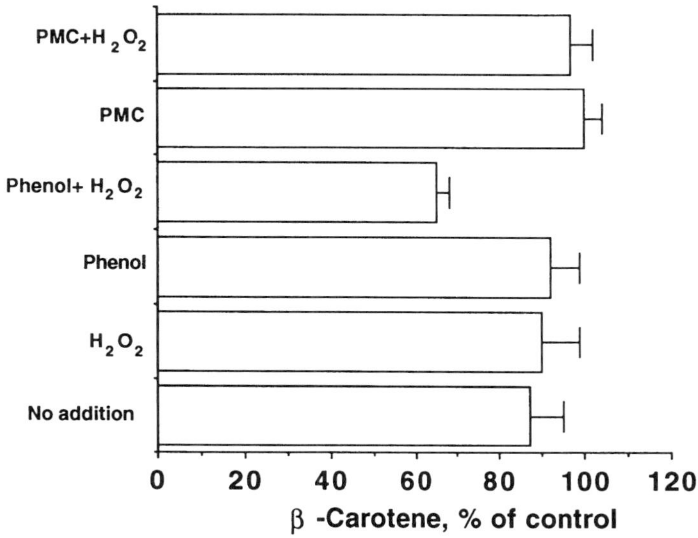

FIGURE 1. Effect of phenol and PMC on the oxidation of β-carotene in HL-60 cells in the presence or absence of H_2O_2. Reaction conditions: β-carotene-loaded HL-60 cells (1.0×10^6) were incubated in L1210 medium in the presence of phenol (300 μM) or PMC (300 μM) and H_2O_2 (400 μM) at 37°C. The initial concentration of β-carotene was 0.33 nmol/10^6 cells. Results presented are means ± standard deviations of the mean (n = 3).

oxidation is increased about 15-fold while that of PMC is increased by less than 3-fold, i.e., the rate of phenol oxidation is enchanced to twice that of PMC by addition of H_2O_2. Since, in the presence of H_2O_2, phenol and PMC are oxidized at comparable rates, the dramatic difference in the oxidation of β-carotene caused by the two phenolic compounds (lack of oxidation by PMC/H_2O_2 and significant oxidation by phenol/H_2O_2) cannot simply be attributed to differences in their rates of oxidation. Hence, hindered phenols such as vitamin E and its congeners are somewhat protective against carotenoid oxidation, while those whose one-electron oxidized free radical (phenoxyl) forms are more reactive have the ability to extend the active site of peroxidases to cause oxidation of β-carotene. This is of some likely importance in the case of smokers (and thus the CARET study), as tobacco smoke is rich in unhindered phenols and is a potent source of one-electron oxidative stress.

In the present study, the level of oxidative stress induced by the peroxyl radicals generated from the azo initiator AMVN did not cause any loss in cell viability in the β-carotene-enriched, α-tocopherol-enriched, or control HL-60 cells (Tables I and II). Even in the absence of cell killing,

Table I
Viability of β-Carotene-Enriched HL-60 Cells after Exposure to AMVN

Incorporation of β-carotene (nmol/10^6 cells)	Viability of cells (%)	
	Control	AMVN
0.0	84.3 ± 4.3	84.2 ± 4.1
0.42 ± 0.04	82.9 ± 6.6	88.0 ± 6.0
0.72 ± 0.38	92.6 ± 9.8	83.5 ± 3.5
1.50 ± 0.50	79.2 ± 4.6	77.1 ± 5.0

Table II
Viability of α-Tocopherol-Enriched HL-60 Cells after Exposure to AMVN

Incorporation of α-tocopherol (nmol/10^6 cells)	Viability of cells (%)	
	Control	AMVN
0.0	84.3 ± 4.3	84.2 ± 4.1
0.25 ± 0.06	85.0 ± 2.4	87.5 ± 3.4
0.50 ± 0.01	86.6 ± 2.6	88.8 ± 3.7
0.71 ± 0.02	86.7 ± 3.4	90.5 ± 1.4

there was a remarkable oxidation-induced loss of PnA-labeled residues in all classes of membrane phospholipids in control cells. In line with all previous literature, even low levels of integrated α-tocopherol afforded almost complete and concentration-dependent protection against radical-induced damage. In contrast, β-carotene-enriched cells, regardless of the level integrated into membranes, suffered phospholipid peroxidation no different than control cells. In other words, β-carotene provided no antioxidant protection under the conditions studied.

Not only was β-carotene ineffective at concentrations at which α-tocopherol exerted near-complete protection, but even sixfold higher levels of integrated β-carotene still had no protective, antioxidant effect (Tables III and IV).

Both α-tocopherol and β-carotene were consumed by the peroxyl radicals generated from AMVN in a concentration-dependent manner. Hence, we conclude that AMVN-derived peroxyl radicals interacted with both α-tocopherol and β-carotene, but that only the former provided any form of antioxidant protection to the cells. Since no protection was afforded by β-

Table III
Effect of β-Carotene on the Oxidation of PnA in the Major Phospholipid Classes in HL-60 Cells Induced by AMVN[a]

Incorporation of β-carotene (nmol/10^6 cells)	PnA content (% of control)		
	PEA	PS	PC
0.0	21.7 ± 6.4	28.5 ± 14.1	28.3 ± 5.9
0.42 ± 0.04	27.2 ± 5.1	35.9 ± 5.3	25.3 ± 3.9
0.72 ± 0.38	18.1 ± 1.8	60.9 ± 3.5	22.7 ± 3.0
1.50 ± 0.50	30.8 ± 3.4	44.2 ± 5.0	28.8 ± 4.0

[a]PEA, phosphatidylethanolamine; PS, phosphatidylserine; PC, phosphatidylcholine.

Table IV
Effect of α-Tocopherol on the Oxidation of PnA in the Major Phospholipid Classes in HL-60 Cells Induced by AMVN[a]

Incorporation of α-tocopherol (nmol/10^6 cells)	PnA content (% of control)		
	PEA	PS	PC
0.0	21.7 ± 6.4	28.5 ± 14.1	28.3 ± 5.9
0.25 ± 0.06	29.7 ± 9.9	79.0 ± 5.2	29.8 ± 9.2
0.50 ± 0.01	31.9 ± 4.8	92.0 ± 5.4	43.6 ± 2.3
0.71 ± 0.02	45.5 ± 2.7	67.9 ± 11.0	62.8 ± 4.9

[a]PEA, phosphatidylethanolamine; PS, phosphatidylserine; PC, phosphatidylcholine.

carotene, but rather it was consumed, it seems that β-carotenyl radicals participated in lipid peroxidation and were as effective as AMVN-derived peroxyl radicals at inducing phospholipid oxidation.

One possible explanation for the lack of protection by β-carotene is that it was not properly integrated into domains containing PnA-labeled phospholipids. This is likely not the case, however, as the same concentrations of AMVN that caused consumption of β-carotene were also responsible for significant oxidation of PnA-labeled phospholipids.

Our studies strongly suggest that β-carotene has no antioxidant activity in cells under normobaric conditions, and that it likely has in fact prooxidant properties with respect to cellular phospholipids. The seminal paper on potential antioxidant properties of β-carotene (Burton and Ingold, 1984) clearly states a warning that should be heeded by those planning to use β-carotene supplements in humans: "β-carotene and related compounds will tend to be concentrated . . . in those particular membranes and organelles that are exposed to the lowest partial pressures of oxygen. As a corollary . . . vitamin E will tend to be concentrated in those lipid regions that are exposed to the highest partial pressures of oxygen."

Clinical trials should be based on sound approaches whose mechanisms are thoroughly understood. While the results of CARET and related studies have been disappointing, they were thus not totally unexpected.

4. REFERENCES

Bast, A., van der Plas, R. M., van den Berg, H., and Haenen, G. R., 1996, β-Carotene as antioxidant, *Eur. J. Clin. Nutr.* **50**(Suppl 3):S54–S56.

Burton, G. W., and Ingold, K. U., 1984, β-Carotene, an unusual type of lipid antioxidant, *Science* **224:**569–573.

Chalvardjian, A., and Rudnicki, E., 1970, Determination of lipid phosphorus in the nanomolar range, *Anal. Biochem.* **36:**256–226.

Everett, S. A., Dennis, M. F., Patel, K. B., Maddix, S., Kundu, S. C., and Willson, R. L., 1996, Scavenging of nitrogen dioxide, thiyl, and sulfonyl free radicals by the nutritional antioxidant β-carotene, *J. Biol. Chem.* **271:**3988–3994.

Folch, J., Lees, M., and Sloan-Stanley, G. H., 1957, A simple method for isolation and purification of total lipids from animal tissues, *J. Biol. Chem.* **226:**497–509.

Kettle, A. J., Gedye, C. A., Hampton, M. B., and Winterbourn, C. C., 1995, Inhibition of myeloperoxidase by benzoic acid hydrazides, *Biochem. J.* **308:**559–563.

Lang, J. K., Gohil, K., and Packer, L., 1986, Simultaneous determination of tocopherols, ubiquinols, and ubiquinones on blood plasma, tissue homogenates, and subcellular fractions, *Anal. Biochem.* **157:**106–116.

Liebler, D. C., Stratton, S. P., and Kaysen, K. L., 1997, Antioxidant actions of β-carotene in liposomal and microsomal membranes: Role of carotenoid-membrane incorporation and α-tocopherol, *Arch. Biochem. Biophys.* **338:**244–250.

Martin, K. R., Failla, M. L., and Smith, J. C., Jr., 1996, β-Carotene and lutein protect HepG2 human liver cells against oxidant-induced damage, *J. Nutr.* **126:**2098–2106.

Mortensen, A., and Skibsted, L. H., 1996, Kinetics of parallel electron transfer from β-carotene to phenoxyl radical and adduct formation between phenoxyl radical and β-carotene, *Free Radic. Res.* **25:**515–523.

Omenn, G. S., Goodman, G. E., Thornquist, M. D., Balmes, J., Cullen, M. R., Glass, A., Keogh, J. P., Meyskens, F. L., Jr., Valanis, B., Williams, J. H., Jr., Barnhart, S., Cherniack, M. G., Brodkin, C. A., and Hammar, S., 1996, Risk factors for lung cancer and for intervention effects in CARET, the Beta-Carotene and Retinol Efficacy Trial, *J. Natl. Cancer Inst.* **88:**1550–1559.

Palozza, P., and Krinsky, N. I., 1992a, β-Carotene and α-tocopherol are synergistic antioxidants, *Arch. Biochem. Biophys.* **297:**184–187.

Palozza, P., and Krinsky, N. I., 1992b, Antioxidant effects of carotenoids *in vivo* and *in vitro*: An overview, *Meth. Enzymol.* **213:**163–170.

Palozza, P., Calviello, G., and Bartoli, G. M., 1995, Prooxidant activity of β-carotene under 100% oxygen pressure in rat liver microsomes, *Free Radic. Biol. Med.* **19:**887–892.

Palozza, P., Luberto, C., Ricci, P., Sgarlata, E., Caviello, G., and Bartoli, G. M., 1996, Effect of β-carotene and canthaxanthin on tert-butyl hydroperoxide-induced lipid peroxidation in murine normal and tumor thymocytes, *Arch. Biochem. Biophys.* **325:**145–151.

Ritov, V. B., Banni, S., Yalowich, J. C., Day, B. W., Claycamp, H. G., Corongiu, F. P., and Kagan, V. E., 1996, Non-random peroxidation of different classes of membrane in live cells detected by metabolically integrated *cis*-parinaric acid, *Biochim. Biophys. Acta* **1283:**127–140.

Sies, H., Stahl, W., and Sundquist, A. R., 1992, Vitamins E and C, β-carotene, and other carotenoids, *Ann. N. Y. Acad. Sci.* **669:**7–20.

Tsuchihashi, H., Kigoshi, M., Iwatsuki, M., and Niki, E., 1995, Action of β-carotene as an antioxidant against lipid peroxidation, *Arch. Biochem. Biophys.* **323:**137–147.

Tyurin, V. A., Carta, G., Tyurina, Y. Y., Banni, S., Day, B. W., Corongiu, F. P., and Kagan, V. E., 1997, Peroxidase-catalyzed oxidation of β-carotene in HL-60 cells and in model systems—Involvement of phenoxyl radicals, *Lipids* **32:**131–142.

Chapter 10

Biophysical Properties of Carotenoids

Maria Tomoaia-Cotisel and Peter J. Quinn

1. INTRODUCTION

Carotenoids are important biological pigments that play vital roles in nutrition, vision, cellular differentiation, growth control, photooxidative protection, cell membrane stability, and photosynthesis. A number of recent review articles aim to evaluate the literature in this field and identify areas for further research, such as human clinical trials which serve to prevent or protect against serious health disorders. It is likely that more studies attempting to relate carotenoids to therapy of cancer, coronary heart disease, HIV infections, or light-sensitive diseases will be undertaken. These are exciting areas for the study of functions and actions of carotenoids in biological systems. However, it is essential that these investigations are based on reliable biophysical and chemical principles so that the biological effects of carotenoids can be explained at the molecular level.

It is widely assumed that the probable protective effect of carotenoids against cardiovascular (Kohlmeier and Hastings, 1995; Gaziano and Hennekens, 1993) and age-related diseases in humans is mainly due to their antioxidant properties (Bohm *et al.*, 1997; Sies and Stahl, 1995).

Maria Tomoaia-Cotisel Molecular/Structural Biotechnologies, Rockville, Maryland 20847-2539. **Peter J. Quinn** Division of Life Sciences, King's College London, London W8 7AH, England.

Subcellular Biochemistry, Volume 30: Fat-Soluble Vitamins, edited by Quinn and Kagan. Plenum Press, New York, 1998.

Among carotenoids already identified in nature, only about 10% are nutritionally active as precursors of vitamin A (Olson, 1993), and therefore it is widely assumed that they have biological effects unrelated to their provitamin A action (Nikawa *et al.*, 1995; Bertram, 1993; Krinsky, 1993; Mathews-Roth, 1993) and act synergistically to protect the membrane systems (Bohm *et al.*, 1997). A growing body of literature exists regarding the anticarcinogenic (Toma *et al.*, 1995; Willett, 1994; Galligan *et al.*, 1993; Ziegler, 1993), antimutagenic, and immunomodulating (Coodley *et al.*, 1993) effects of carotenoids in humans.

Although numerous studies have suggested some possible beneficial effects of carotenoids on human health and longevity as well as in human clinical trials, only little data are available concerning their absorption, metabolism, and transport *in vivo* (Parker, 1996; Erdman *et al.*, 1993; Krinsky *et al.*, 1958). However, it is stated that carotenoids are transported in plasma by lipoproteins, with a preferential distribution among lipoprotein classes determined in large part by the structure of carotenoid molecules (Borel *et al.*, 1996). The consequences of the differential distribution in terms of tissue uptake and retention are not completely understood.

Nevertheless, it is clear that the varied biological activities of carotenoids in living organisms are determined by the biophysical and chemical properties of these molecules. These properties and particularly the surface activity of carotenoids are defined by their molecular structure and geometry. In addition, specific interactions of carotenoids with other molecules in their vicinity may ensure their correct orientation at various biological interfaces and their proper location within cellular and subcellular structures. Further, carotenoids and their metabolites may influence or modulate the properties of cellular and subcellular structures. Finally, these structures may allow carotenoids to exert their function and act efficiently.

Thus, studies of biophysical properties of carotenoids in single-component membranes or in mixture with other biologically active molecules, such as lipids, are essential to determine the orientation of carotenoid molecules at biological interfaces and to evaluate the molecular interactions in self-assembled systems. It has been shown that various carotenoids act differently among liposomal systems (Liebler *et al.*, 1997; Gabrielska and Gruszecki, 1996; Yin and Subczynski, 1996; Woodall *et al.*, 1995; Jezowska *et al.*, 1994; Strzalka and Gruszecki, 1994; Gruszecki and Sielewiesiuk, 1990; Subczynski *et al.*, 1991; Lazrak *et al.*, 1987), biological membranes (Britton, 1995; Bone *et al.*, 1992; Khomutov *et al.*, 1990; Picorel *et al.*, 1988; Bone and Landrum, 1984; Rohmer *et al.*, 1979), and biological emulsions and intracellular storage (e.g., mixtures of carotenoids and lipids) droplets (Borel *et al.*, 1996; Miller and Small, 1983). These findings suggest that carotenoids are found in a precise location in lipid membrane systems.

This chapter considers the biophysical properties of carotenoids in monomolecular membranes (e.g., monolayers) in the context of the effects of carotenoids on biological membranes. We describe the surface-active properties of some carotenoids in monolayers at fluid interfaces in the absence and the presence of phospholipids or galactolipids. The results provide a basis for understanding how carotenoids may stabilize the lipid membranes and act in biological systems. In addition, the biophysical properties of carotenoids are strongly influenced by other molecules incorporated within or adsorbed on carotenoid monolayers.

We report the biophysical studies performed on carotenoid monolayers, also referred to as monomolecular films or monomolecular membranes (Tomoaia-Cotisel *et al.*, 1990, 1987a, b, 1983a, 1980, 1975; Chifu *et al.*, 1989, 1986, 1983, 1979b, 1978, 1975; Zsako *et al.*, 1989, 1986, 1979; Chifu and Tomoaia-Cotisel, 1981; Tomoaia-Cotisel and Chifu, 1981; Tomoaia-Cotisel, 1979) and describe carotenoid molecules in self-assembled arrangements or interacting with lipid monomolecular membranes, (Tomoaia-Cotisel *et al.*, 1995, 1987c, d, 1985a; Zsako *et al.*, 1991, 1984; Tomoaia-Cotisel and Chifu, 1983, 1979; Chifu and Tomoaia-Cotisel, 1979; Chifu *et al.*, 1979a; Tomoaia-Cotisel, 1979).

Our findings are in general agreement with those obtained in liposomal systems and emulsions and show that carotenoids form molecular associations with phospholipids and galactolipids and can act synergistically to stabilize membranes. This ability of carotenoids might be significant in their potential beneficiary effects on human health. Carotenoid and lipid associations may be involved in the transfer of carotenoids from lipid droplets to the lipid structures dispersed in aqueous phase, such as mixed micelles. Also, these associations might be important in the transfer of the carotenoids between lipoprotein particles, and between lipoproteins and cellular membranes. These investigations open new perspectives for future developments in pharmaceutical and clinical applications of carotenoids.

In this chapter we provide a brief introduction to the methodology used in monomolecular film studies. Then we describe the behavior of some selected carotenoids in single-component monolayers. Lastly, we present a biophysical characterization of molecular interactions between carotenoids and diacyl polar lipids in monolayers.

2. MONOMOLECULAR FILM TECHNIQUES

The properties of carotenoid monolayers in the absence and the presence of lipids can be characterized by compression isotherms and film stabilities by using surface pressure versus mean molecular area measurements.

To study surface-active properties by Langmuir techniques, biomolecules are dissolved in an appropriate solvent and spread on a clean air/water interface in a Langmuir trough (Kuhn and Mobius, 1993; Tomoaia-Cotisel *et al.*, 1987b; Gaines, 1966). After spreading, the solvent evaporates from the surface leaving all of the nonvolatile solutes in the interfacial region; the number of molecules can be determined by the volume of solution introduced onto the surface.

The area of the interface is also known and therefore the mean area per molecule of biosurfactant can be determined fairly precisely. The presence of surface-active molecules at the air/water interface reduces the surface tension of water (about 72 mN/m at 20°C). The magnitude of the decrease in surface tension is referred to as the surface pressure of the monolayer. The surface pressure exerted by the monolayer molecules can be measured by determining the lateral force on a barrier separating the monolayer from a clean air/water surface or by the downward force exerted on a rod or plate dipping through the film. As the area of the interfacial monolayer is decreased by a movable barrier at one end of the trough, a compression isotherm of the monolayer spread on the water subphase can be obtained by recording associated changes in pressure. The rate of compression should be maintained constant. Compression isotherms are plotted in terms of the mean molecular area A in angstroms squared ($Å^2$/molecule) versus the surface pressure π in mN/m. In the experiments described the surface pressure was recorded by using the Wilhelmy dipping plate method (Gaines, 1966). Generally, the uniformity of the monolayer and the reproducibility of the results depend on the compression speed, due to the different surface mobility of the film constituents. An optimum compression speed was determined with each biosurfactant studied.

The measurements were made at constant temperature of 22 ± 2°C. The accuracy of the surface pressure (referred to also as lateral pressure) measurements was ±0.5 mN/m and the compression (π versus A) isotherms were reproducible within ±2 $Å^2$/molecule. The optimum compression speed, ranging between 2 and 4 $Å^2$/(molecule·min), was sufficiently slow to allow equilibrium to be established, as evidenced by maintenance of constant surface pressure for each value of the surface area (Tomoaia-Cotisel *et al.*, 1990, 1987a,b, 1985b, 1981; Zsako *et al.*, 1989).

To verify that there was no degradation of the carotenoids either in the spreading solution or in the monolayer during the measurements, absorption spectra were recorded and analysis by thin-layer chromatography was performed on recovered monolayers (Tomoaia-Cotisel, 1979; Chifu *et al.*, 1983; Tomoaia-Cotisel *et al.*, 1987d). Additional structural and spectroscopic studies can be performed on Langmuir–Blodgett films. In this method the biosurfactant monolayer, compressed at a particular surface pressure, is transferred onto an appropriate solid substrate by dipping through the surface of the film. Successive layers may be deposited on the support (Tomoaia-Cotisel *et al.*, 1983b).

The carotenoids examined in our studies are of both natural origin and synthetic compounds. They were all-*trans* isomers of purity established by chromatography and spectral analysis. The diacylglycerolipids were also of natural origin, e.g., egg phosphatidylcholine (EPC), synthetic, e.g., 1,2-distearoylphosphatidylcholine (DSPC), and semisynthetic, e.g., 1,2-distearoyldigalactosylglycerol (DGDG).

The spreading solvent for β-cryptoxanthin and all the esters of xanthophylls was benzene. Since zeaxanthin and astaxanthin are sparingly soluble in benzene, the spreading solvent consisted of benzene containing 2–4% absolute ethanol.

3. SURFACE-ACTIVE PROPERTIES OF CAROTENOIDS: SINGLE-COMPONENT MONOLAYER AT THE AIR/WATER INTERFACE

The biophysical properties of carotenoid monolayers have been examined with a major emphasis on the relationship between molecular structure and surface-active properties with reference to the molecular packing at fluid interfaces. Single-component monolayers offer a unique opportunity for structural studies of highly oriented systems, essentially involved in biological systems, as well as in the development of new drug delivery and nutrient systems. Within these monolayers, carotenoids have been assembled at defined concentrations and therefore their molecular area, orientation, and packing within the monolayer assemblies can be precisely determined. The monolayer features are dependent on the molecular structure of the individual carotenoid and on the long-range action of surface forces in interfacial structures.

3.1. Surface Pressure versus Molecular Area: Compression Isotherms

The selected carotenoids are all-*trans* isomers: β-cryptoxanthin (3–hydroxy β-carotene; CX), β-cryptoxanthin palmitate (CP), zeaxanthin (3,3′-dihydroxy β-carotene; ZX), zeaxanthin monopalmitate (ZP), and astaxanthin (3,3′-dihydroxy-, 4,4′-dioxo-β-carotene; AX).

As an example, Fig. 1 shows the compression isotherm for β-cryptoxanthin and β-cryptoxanthin palmitate monolayers. From the compression isotherm the surface characteristics are determined, namely the molecular area extrapolated to zero surface pressure (A_0: limiting molecu-

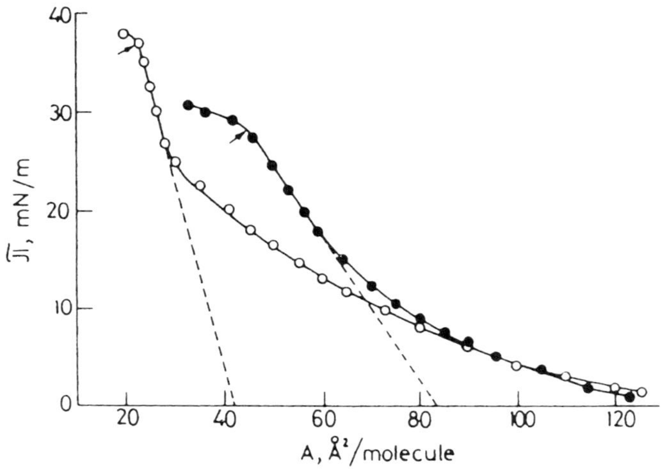

FIGURE 1. Compression (surface pressure versus area) isotherms of single-component monolayers of β-cryptoxanthin (CX) (open circles) and β-cryptoxanthin palmitate (solid circles).

lar area), the collapse pressure (π_c), and corresponding molecular area (A_c: collapse area) and surface compressional modulus (C_{s0}^{-1}) (for details see Chifu *et al.*, 1983, 1979b; Tomoaia-Cotisel and Chifu, 1981; Tomoaia-Cotisel *et al.*, 1980, 1975; Zsako *et al.*, 1979).

The term collapse pressure is used for the greatest surface pressure to which monolayers can be compressed without detectable expulsion of the molecules to form a new, three-dimensional collapsed phase (Tomoaia-Cotisel *et al.*, 1990; Chifu *et al.*, 1989; Zsako *et al.*, 1986). The collapse pressure corresponds to the slope change of the isotherm at high surface pressure (marked by an arrow in Fig. 1).

The surface compressional modulus indicates the physical state of these monolayers (Davies and Rideal, 1963; Tomoaia-Cotisel, 1979; Zsako *et al.*, 1979). These monolayer characteristics A_0, A_c, π_c, and C_{s0}^{-1} are given in Table I. It can be seen from Table I that C_{s0}^{-1} values are similar to values typical of condensed-liquid monolayers.

3.2. Structure of Carotenoid Monolayers and Molecular Packing

In interpreting the area requirements A_0 and A_c, geometric models are used according to the molecular structures of carotenoids. Their properties in the monolayer can be considered in terms of a rotating rigid plate model (Chifu *et al.*, 1983; Tomoaia-Cotisel, 1979). As for the significance of the

Table I
Calculated Molecular Areas (A_4 and A_p) and the Experimental Features of Compression Isotherms (A_0, A_c, π_c, C^{-1}_{s0})[a]

Compound	A_4	A_0	A_p	A_c	π_c	C^{-1}_{s0}
β-Cryptoxanthin (CX)	40	42	24	23	37	82
β-Cryptoxanthin palmitate (CP)	83	83	45.5	44	28	58
Zeaxanthin (ZX)	42	49	29	30	40	103
Zeaxanthin monopalmitate (ZP)	67	70	52	52	35.0	136
Astaxanthin (AX)	42	46	29	25	45	99

[a]Units for A_4, A_0, A_p, and A_c are Å^2; for π_c and C^{-1}_{s0} are mN/m.

rotation, it can either be a real rotation or assumed to be a random, dynamic orientation of the terminal polar group axes.

Briefly, in self-assembled monolayers, the carotenoid molecules are oriented at the air/water interface, with one polar head group in water and the hydrophobic domain (i.e., polyene hydrocarbon chain, with its delocalized π-electron system, carrying at its extremity another β-ionone ring with or without substituents) is extended in the air phase.

The linear dimensions of the molecule (coordinates *a*, *b*, and *c* in Fig. 2) oriented in the monolayer can be calculated from the internuclear distances, bond angles, and van der Waals radii of the atoms. With the palmitate derivatives of β-cryptoxanthin and zeaxanthin, the dimensions of the molecules are calculated for their most favorable conformation in monolayers, having the longitudinal axis of the hydrocarbon chain in the plane of the rigid carotenoid plate (Chifu *et al.*, 1983).

Further, the area requirements of vertically oriented molecules were calculated for different approximations, such as for a tetragonal packing (denoted A_4; $A_4 = a^2$), and a hexagonal compact packing [A_6; $A_6 = (3/4)^{1/2}A_4 = (3/4)^{1/2}a^2$] of molecular cylinders, obtained by rotation of oriented molecules in monolayers (Tomoaia-Cotisel, 1979). Futhermore, the area of nonrotating, $A_p = ab$, molecules was calculated for the surface lattice with closest packed parallel-oriented molecules. These calculated areas are also given in Table I.

The calculated areas are compared with experimentally determined molecular areas (Table I). As can be seen, in the case of β-cryptoxanthin, β-cryptoxanthin palmitate, and zeaxanthin monopalmitate, the experimental data are in close agreement with the calculated values. For example, A_0 is similar to A_4, and A_c is almost identical to A_p. There are three exceptions,

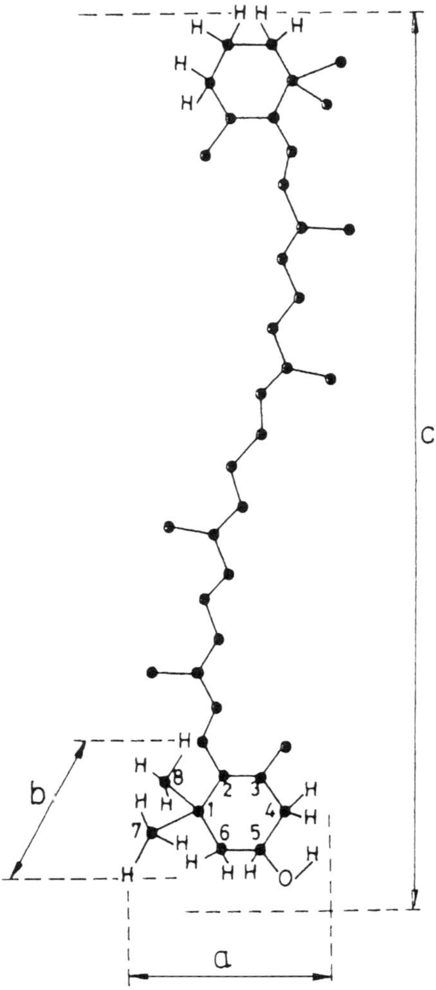

FIGURE 2. Geometric, rigid-plate molecular model for β-cryptoxanthin (CX) oriented in a monolayer at the air/water interface. The delocalized π-electron system of the polyene chain determines the S-shape of the CX molecule.

for astaxanthin, zeaxanthin monopalmitate, and especially zeaxanthin, where $A_0 > A_4$. This might be explained by the fact that these carotenoids have polar groups located in the air phase, and particularly the OH group which participates in a dimerization process. Consequently, the area requirement increases greatly if the dimer rotates and seemingly $A_0 > A_4$. This dimerization does not alter the area requirement if the rotation of molecules is restricted as is the case at film collapse. Therefore, in the case of zeaxanthin and its palmitate derivative, $A_c = A_p$. For astaxanthin,

$A_c < A_p$, due to the enhanced molecular interactions within the astaxanthin monolayers.

Since the difference between astaxanthin and zeaxanthin is the presence of a carbonyl group at each extremity on the β-ring of the astaxanthin molecule, it is reasonable to assume that the small A_c value of astaxanthin is due to its increased monolayer stability. This stability is also evidenced by the high surface pressure required for the collapse of the astaxanthin monolayer.

Therefore, at the collapse of carotenoid monolayers a surface lattice of nonrotating molecules develops, and generally A_c does not differ much from A_p. At low surface pressure a tetragonal packed molecular lattice is preferred by rotating carotenoid molecules oriented in expanded-liquid monolayers.

Exceptions are observed when the possibility exists of hydrogen bond formation (intermolecular hydrogen bridges) in the air phase. Thus, the monolayer properties and structure are strongly determined by the molecular structure of carotenoids and by molecular interactions in these self-assembled and oriented monolayers.

4. MIXED MONOLAYERS OF CAROTENOIDS AND LIPIDS AT THE AIR/WATER INTERFACE

Biophysical properties of some mixtures of carotenoids and diacylglycerolipids spread at the air/water interface have been examined. From these studies relationships have been established between carotenoid structure, surface activity, molecular interactions, and miscibility with lipids in oriented membranes.

These data suggest a fundamental structural role of carotenoids in strengthening lipid membranes and modifying their properties. Carotenoids also appear to enhance the stability of lipid membranes composed of either saturated or unsaturated lipids, suggesting an explanation of how carotenoids may protect membranes from damage. These data contribute to our knowledge of carotenoid transport, bioavailability, and biological actions.

The carotenoids used in these studies include β-cryptoxanthin (CX), zeaxanthin (ZX), and astaxanthin (AX). All possess the same polyenic chain, but have different polar groups. The polarity of carotenoid molecules is reflected in their monolayer properties. Thus, astaxanthin is more polar than zeaxanthin or β-cryptoxanthin, and consequently the cohesion of astaxanthin monolayers at the air/water interface, as judged by surface collapse pressure, is greater than is the case of zeaxanthin or β-cryptoxanthin monolayers (see Table I).

Six binary mixed monolayers were examined, namely (1) CX and EPC, (2) CP and EPC, (3) ZP and EPC, (4) ZX and DSPC, (5) AX and DSPC, and (6) AX and DGDG.

The surface properties of these mixed monolayers were derived from compression isotherms and the results were interpreted in terms of both the surface phase rule and solution theory, such as regular molecular associations in monolayers. Accordingly, the curves of surface collapse pressure versus composition are discussed and the stability constants of molecular associations of carotenoids with lipids in mixed monolayers are calculated.

For the sake of simplicity, we denote the lipid as component A and the carotenoid as component B.

4.1. Compression (π versus A) Isotherms

Compression isotherms for β-cryptoxanthin and egg phosphatidylcholine (CX and EPC) monolayers and for binary mixtures of various mole fractions of EPC are shown in Fig. 3. The surface-active characteristics, such as collapse surface pressure π_c (arrows in Fig. 3) and corresponding collapse molecular area A_c and the mean molecular area at given values of surface pressure are derived from these isotherms.

4.2. Thermodynamics of Mixed Monolayers

The miscibility of two components A and B (denoted 1 and 2) in a monolayer has been interpreted assuming an additivity rule of molecular areas (Tomoaia-Cotisel *et al.*, 1985a; Tomoaia-Cotisel and Chifu, 1983, 1979; Chifu and Tomoaia-Cotisel, 1979, Chifu *et al.*, 1979a) and by the Gibbs free energies of mixing (Tomoaia-Cotisel and Chifu, 1983), both methods being applicable at constant values of the surface pressure, but smaller than the collapse pressure of the component B (denoted π_B), which is the component that, in a monocomponent film, collapses at the lowest pressure. In all the mixtures examined π_B is less than π_A, where π_A represents the collapse pressure of the single-component lipid monolayer.

4.2.1. Mean Molecular Areas and the Additivity Rule

The mean area values were found to follow the additivity rule in condensed monolayers at high surface pressures (e.g., 20 and 35 mN/m) for mixtures of CX and EPC.

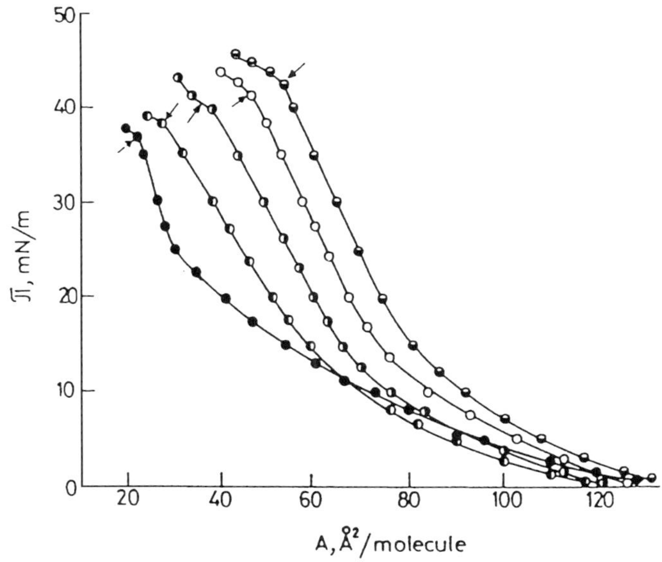

FIGURE 3. Surface pressure versus area isotherms of β-cryptoxanthin (CX) and egg phosphatidylcholine (EPC) mixed monolayers at the air/water interface at various mole fractions of EPC: (●) 0, (◐) 0.25, (◑) 0.5, (○) 0.75, (◒) 1.

The additivity rule is given by $A = x_1A_1 + x_2A_2$, where A is the mean molecular area in the two-component monolayer, A_1 and A_2 are the respective molecular areas in the two single-component monolayers at the same given surface pressure, and x_1 and x_2 are the respective mole fractions of the components in the mixed monolayer.

For the CX:EPC mixture, system 1 in Table II, negative deviations were found at low surface pressures (e.g., at 10 mN/m) in the expanded state of the monolayers. This condensing effect observed in the mixed monolayers has been interpreted in terms of special mechanisms for packing of the compounds (Tomoaia-Cotisel *et al.*, 1987d; Cadenhead and Muller-Landau, 1980) when the geometric accommodation or space-filling possibilities (Tomoaia-Cotisel and Chifu, 1983) in relation to the molecular associations (Tomoaia-Cotisel *et al.*, 1987d) are taken into account or in terms of specific interactions in the monolayers (Tomoaia-Cotisel *et al.*, 1995, 1985a; Zsako *et al.*, 1991).

The condensing effect of xanthophylls on the EPC monolayer at low surface pressure generally noted in these binary systems could be explained on the basis of the intermolecular voids or cavities caused by the presence of unsaturated fatty acyl chains in the EPC molecules.

The reduction in the mean molecular area for the ZX:EPC and lutein:EPC systems (Tomoaia-Cotisel and Chifu, 1983) even at high surface pressures can be explained, at least in part, by the geometric fit between the components in the mixed monolayers. It seems likely that this process will always be accompanied by altered molecular interactions resulting in enhanced intermolecular hydrophobic interactions or even in specific interactions between the polar groups.

These results are in substantial agreement with the excess free energies of mixing (Tomoaia-Cotisel and Chifu, 1983) and indicate a high stability of the mixed (carotenoid and lipid) monolayers. Similarly, carotenoids could be adsorbed onto the surface or inserted into cell membranes thereby modifying the biophysical properties and stability of the structure and modulating molecular processes associated with those membranes. These effects of carotenoids might be an important part of their role in human health.

4.2.2. Miscibility and the Phase Rule

Mixed monolayers have been investigated by the collapse surface pressure method. Upon collapse, the monolayer (M) is in equilibrium with a new collapsed bulk phase (C).

In order to evaluate reciprocal miscibilities of the two components in both M and C phases, the surface phase rule has been used (Tomoaia-Cotisel *et al.*, 1985a; Tomoaia-Cotisel and Chifu, 1983; Defay *et al.*, 1966).

In the case of planar interfaces at constant temperature and external pressure the phase rule takes the form

$$w = (c - \phi) - (\psi - s)$$

where w is the variance of the system, c is the number of independent chemical components, ϕ is the number of bulk phases, ψ is the number of surface phases, and s is the number of the types of surfaces. In this situation, $c = 4$ (water, the inert gas or air, and the two components that form the monolayer), $\phi = 3$ (water phase, air phase, and collapsed phase) and $s = 3$ (collapsed phase/water, collapsed phase/air, and air/water).

If all the types of surfaces carry only one surface phase, then $\psi = 3$ and $w = 1$. In this case, the system is monovariant and the two components are completely miscible in both the M and C phases.

Consequently, the collapse pressure is a unique function of the monolayer composition and is given by $\pi_c = \pi_c(x_A^M)$, which describes the monolayer curve. Some examples of monolayer curves are given in Figs. 4–6.

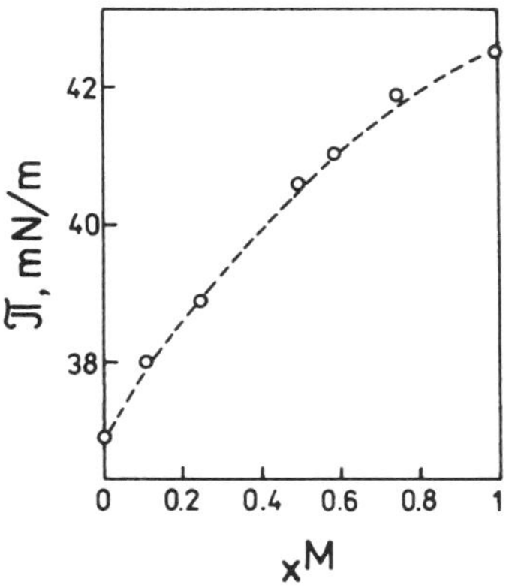

FIGURE 4. Phase diagram (experimental collapse pressure π versus composition = mole fraction of EPC, denoted x^{M}) of mixed monolayers of β-cryptoxanthin (CX) and egg phosphatidylcholine (EPC). Dashed line, theoretical curve for an ideal system computed for the interaction parameters $\xi^{\mathrm{C}} = \xi^{\mathrm{M}} = 0$, using equation (2).

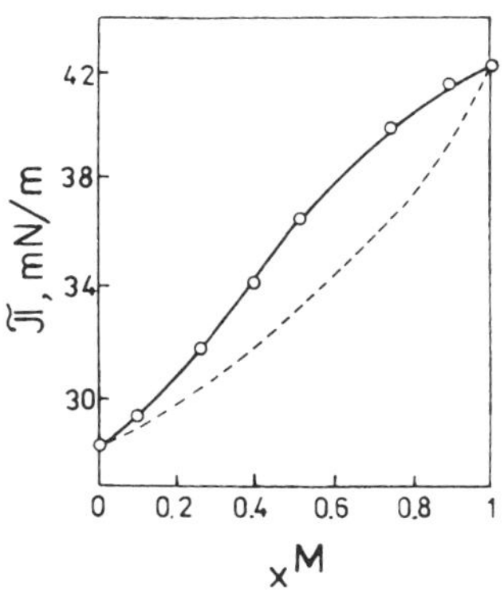

FIGURE 5. Experimental collapse pressure π of mixed monolayers of β-cryptoxanthin palmitate (CP) and egg phosphatidylcholine (EPC) as a function of mole fraction of EPC, denoted x^{M}. Solid line, theoretical monolayer curve for regular-solution approximation calculated by $\xi^{\mathrm{C}} = -1.73$ and $\xi^{\mathrm{M}} = -2.65$. Dashed line, theoretical curve for an ideal system computed for $\xi^{\mathrm{C}} = \xi^{\mathrm{M}} = 0$.

4.2.3. Collapse Pressure versus Composition of Mixtures

A thermodynamic analysis of monolayer curves of binary mixtures in terms of the surface phase rule showed that the carotenoid and lipid molecules are completely miscible in both the monolayer (M) and collapsed (C) phases.

The shape of the monolayer curve $\pi_c = \pi_c(x^{\mathrm{M}})$ is described by the following general equation obtained from surface solution theory (Tomoaia-Cotisel *et al.*, 1987d; Zsako *et al.*, 1984):

$$\sum_{i=1}^{m} \frac{f_i^{\mathrm{M}}}{f_i^{\mathrm{C}}} x_i^{\mathrm{M}} \exp\left[\left(\pi - \pi_i\right)\frac{A_i}{kT}\right] = 1 \quad (1)$$

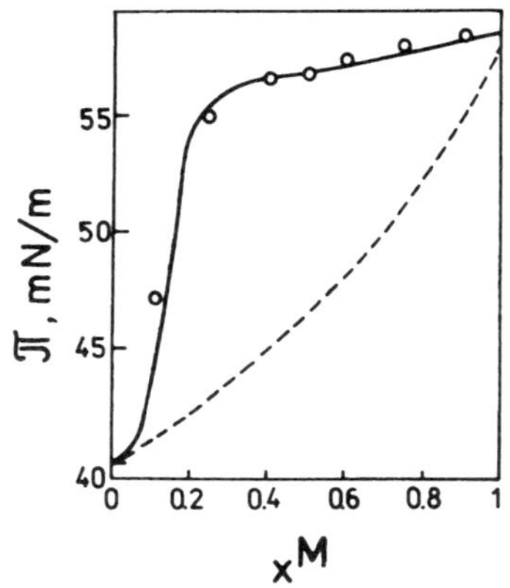

FIGURE 6. Experimental collapse pressure π of mixed monolayers of zeaxanthin (ZX) and distearoyl phosphatidylcholine (DSPC) as a function of mole fraction of DSPC, denoted x^M. Solid line, theoretical monolayer curve for regular molecular associations calculated for stability constant $K = 6.5$, increment in surface pressure $\Delta\pi = 0.35$ mN/m, and minimum standard deviation $\Delta_{m,m} = 0.291$ mN/m. Dashed line, theoretical curve for an ideal-solution approximation computed for $\xi^C = \xi^M = 0$.

where f_i^ϕ stands for the activity coefficient of component i in the phase ϕ, x_i^ϕ is the molar fraction of the component i in phase ϕ, and π is the collapse pressure of the mixed monolayer; π_i and A_i represent the collapse pressure and the collapse area, respectively, in the pure monolayer formed of the component i, and m is the total number of independent components in the system.

In the case where there is no specific interaction between components, i.e., $f_i^M = f_i^C = 1$, then both the M and C phases can be considered as ideal solutions, and equation (1) for a two-component (A and B) monolayer becomes

$$x_A{}^M \exp\left[\left(\pi - \pi_A\right)\frac{A_A}{kT}\right] + x_B{}^M \exp\left[\left(\pi - \pi_B\right)\frac{A_B}{kT}\right] = 1 \qquad (2)$$

The deviation of the experimental data from the curve given by equation (2) indicates the existence of some specific interactions between A and B components.

In the case of the mixtures examined in these studies, where $\pi_A > \pi_B$, the notation $x_A{}^\phi = x^\phi$ and $x_B{}^\phi = 1 - x^\phi$ was used, assuming the molar fraction of the subphase liquid to be negligible at the collapse of the mixed monolayer.

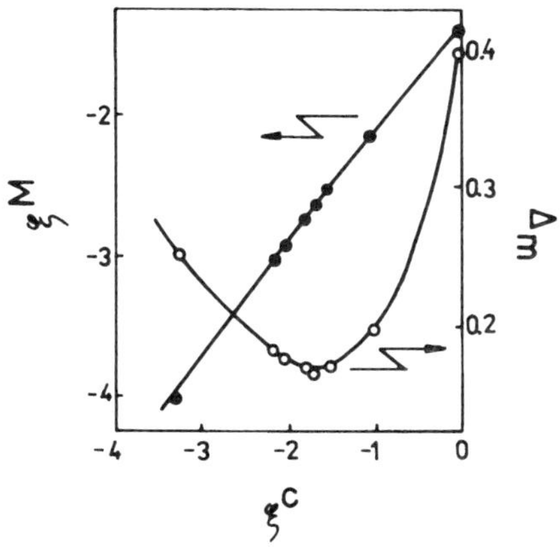

FIGURE 7. Derivation of interaction parameters for β-cryptoxanthin palmitate (CP) and egg phosphatidylcholine (EPC) mixed monolayers. For symbols see the text.

Equation (2) accurately describes the experimental curve for monolayers of CX and EPC (Fig. 4), but positive deviations from the ideal monolayer curve are observed for all other mixtures examined (e.g., Figs. 5 and 6).

An attempt to account for these deviations has been made by applying regular solution theory, which expresses the activity factors in both M and C phases as a function of interaction parameters (ξ^M and ξ^C) (Tomoaia-Cotisel *et al.*, 1995, 1985a; Zsako *et al.*, 1984):

$$\begin{aligned} f_A{}^{\phi} &= \exp\left[\xi^{\phi}\left(1 - x^{\phi}\right)^2\right] \\ f_B{}^{\phi} &= \exp\left[\xi^{\phi}\left(x^{\phi}\right)^2\right] \end{aligned} \tag{3}$$

We found that the regular solution approximation describes an experimental monolayer curve with maximum positive deviations from ideal behavior at $x^M \geq 0.5$ in mixed monolayers of CP:EPC and ZP:EPC. The interaction parameters have been obtained by double minimization of the standard deviation Δ of the experimental π data from the computed values (see Fig. 7), using the corresponding relations of regular solution theory obtained from equations (1) and (3) as shown previously (Tomoaia-Cotisel *et al.*, 1985a; Zsako *et al.*, 1984). Similarly, the behavior of mixed monolayers of ZP:EPC was adequately described by regular solution theory.

The interaction parameters are given in Table II for the CP:EPC and ZP:EPC mixed monolayers. The greater values of interaction parameters obtained with ZP:EPC as compared to CP:EPC indicate an important role of the OH groups of carotenoid molecules in the interactions between egg phosphatidylcholine and carotenoid palmitates.

Table II
Parameters Derived from Collapse Pressure versus Composition Curves of Mixed Monolayers of Carotenoids and Lipids[a]

System	Component A	Component B	n	$\xi^{M}_{m,m}$	ξ^{C}_{m}	K	$\Delta\pi$ (mN/m)	$\Delta_{m,m}$ (mN/m)
1	EPC	CX	—	0	0	—	—	—
2	EPC	CP	—	−2.65	−1.73	—	—	0.169
3	EPC	ZP	—	−9.48	−8.67	—	—	0.105
4	DSPC	ZX	4	—	—	6.5	0.35	0.291
5	DSPC	AX	6	—	—	1.45	1.00	0.213
6	DGDG	AX	6	—	—	1.17	0.05	0.172

[a]Interaction parameter ξ^{Φ}, stability constant K and surface collapse pressure increment $\Delta\pi$ of the regular molecular associations, coordination number n of lipids, and minimum standard deviation $\Delta_{m,m}$.

For mixed monolayers of ZX:DSPC, AX:DSPC, and AX:DGDG, the maximum deviation from ideal behavior is found at $x^{M} \leq 0.5$. In this situation (an example is given in Fig. 6), it was shown that regular solution theory is no longer applicable (Tomoaia-Cotisel *et al.*, 1995, 1987c,d; Zsako *et al.*, 1991).

The shape of the experimental monolayer curve suggests the formation of molecular associations of general AB_n type or interfacial complexes. Consequently, in the monolayer phase, there are $(n + 2)$ molecular species: A, B, AB, $AB_2, \ldots, AB_n$, which are characterized by n stability constants. Presumably, these new species have a higher surface activity than A and B and their formation entails the increase of the collapse pressure.

In this new situation, equation (1) for the π versus x^{M} curves becomes

$$\frac{f_A{}^{M}}{f_A{}^{C}} x_A{}^{M} \exp\left[\left(\pi - \pi_A\right)\frac{A_A}{kT}\right] + \frac{f_B{}^{M}}{f_B{}^{C}} x_B{}^{M} \exp\left[\left(\pi - \pi_B\right)\frac{A_B}{kT}\right] + \sum_{i=1}^{n} \frac{f_i{}^{M}}{f_i{}^{C}} x_i{}^{M} \exp\left[\left(\pi - \pi_i\right)\frac{A_i}{kT}\right] = 1 \tag{4}$$

in which the calculation of the unknown parameters from a single experimental curve is unreliable.

Therefore, the following simplifications for approximating ideal solutions of regular molecular associations are proposed. In this approximation, the molecular species i (A, B, AB_i; $1 \leq i \leq n$) form ideal solutions in both the M and C phases. Consequently the molecular areas are additive, and are given by

$$A_{\mathrm{AB}_i} = A_{\mathrm{A}} + iA_{\mathrm{B}} \tag{5}$$

Further, the attachment of each molecule B to the molecule A, or to the complex AB_{i-1}, is given by $\mathrm{AB}_{i-1} + \mathrm{B} \Leftrightarrow \mathrm{AB}_i$ and results in the same collapse pressure increment, $\Delta\pi$, to the pure monolayer consisting of A. That is $\pi_{\mathrm{AB}_i} = \pi_{\mathrm{A}} + i(\Delta\pi)$. The equilibrium constants K_i have the same value K for each complexing step in the formation of molecular associations:

$$K_i = x_{\mathrm{AB}_i}{}^{\mathrm{M}} \Big/ \left(x_{\mathrm{AB}_{i-1}}{}^{\mathrm{M}} x_{\mathrm{B}}{}^{\mathrm{M}} \right) = K \tag{6}$$

We have suggested a method for deriving the most probable values of n, K, and $\Delta\pi$ from the experimental collapse pressure versus composition curves (e.g., Fig. 6 for ZX and DSPC in mixed monolayers) by performing a triple minimization of the standard deviation of the experimental collapse pressure values from the theoretical values (Tomoaia-Cotisel *et al.*, 1995, 1987d; Zsako *et al.*, 1991).

In the ideal-solution approximation of regular molecular associations, the general equation (1) becomes

$$\begin{aligned} &x_{\mathrm{A}}{}^{\mathrm{M}} \exp\left[\left(\pi - \pi_{\mathrm{A}} \right) \frac{A_{\mathrm{A}}}{kT} \right] + x_{\mathrm{B}}{}^{\mathrm{M}} \exp\left[\left(\pi - \pi_{\mathrm{B}} \right) \frac{A_{\mathrm{B}}}{kT} \right] \\ &+ \sum_{i=1}^{n} x_{\mathrm{AB}_i}{}^{\mathrm{M}} \exp\left[\left(\pi - \pi_{\mathrm{A}} - i\Delta\pi \right) \frac{A_{\mathrm{A}} + iA_{\mathrm{B}}}{kT} \right] = 1 \end{aligned} \tag{7}$$

The mole fractions of all the molecular species AB_i are calculated from the relation

$$x_{\mathrm{AB}_i}{}^{\mathrm{M}} = \left(1 - x_{\mathrm{B}}{}^{\mathrm{M}} \right) \left(x_{\mathrm{B}}{}^{\mathrm{M}} K \right)^i \Big/ \left[1 + \sum_{i=1}^{n} \left(x_{\mathrm{B}}{}^{\mathrm{M}} K \right)^i \right] \tag{8}$$

The $x_{\mathrm{B}}{}^{\mathrm{M}}$ is obtained by solution of the following equation:

$$x_{\mathrm{A}}{}^{0\mathrm{M}} \left(1 - x_{\mathrm{B}}{}^{\mathrm{M}} \right) \sum_{i=1}^{n} i \left(x_{\mathrm{B}}{}^{\mathrm{M}} K \right)^i = \left(1 - x_{\mathrm{A}}{}^{0\mathrm{M}} - x_{\mathrm{B}}{}^{\mathrm{M}} \right) \left[1 + \sum_{i=1}^{n} \left(x_{\mathrm{B}}{}^{\mathrm{M}} K \right)^i \right] \tag{9}$$

where $x_{\mathrm{A}}{}^{0\mathrm{M}}$ represents the apparent mole fraction of A initially spread in the film.

Thus, in equations (7)–(9) there are three unknown parameters, namely K, n, and $\Delta\pi$, which can be calculated from the experimental data (π versus $x_{\mathrm{A}}{}^{0\mathrm{M}}$) using a triple minimization of the standard deviation of the experimental collapse pressure values from the theoretical values. The values of K and $\Delta\pi$, the coordination number of the lipid n, as well as the corresponding minimum standard deviations $\Delta_{m,m}$ are presented in Table II.

Once the value of K is known, the mole fractions of all molecular species versus the spread mixture composition (Tomoaia-Cotisel *et al.*, 1995, 1987d; Zsako *et al.*, 1991) can be computed using equations (8) and (9).

The most probable coordination number n of the lipid (DSPC or DGDG), the collapse pressure increment ($\Delta\pi$, mN/m) and the minimum standard deviation ($\Delta_{m,m}$, mN/m), as well as the equilibrium constant K for the mixed monolayers of ZX:DSPC, AX:DSPC, and AX:DGDG are given in Table II.

As seen in Table II, the probable coordination number of DSPC in the presence of ZX is 4, which is the value also determined for the surface lattice of pure ZX monolayers from geometric models of ZX molecules. It appears, therefore, that zeaxanthin molecules are able to change the preferred hexagonal packing in pure DSPC monolayers near their collapse pressure to a tetragonal mixed lattice of $(DSPC)(ZX)_4$ molecular associations in mixed monolayers. In the case of the mixed monolayers of AX and DSPC or of AX and DGDG, it seems that the hexagonal lattice preferred by these lipids at collapse pressure is preserved even in the presence of astaxanthin in monolayers. The different effect of ZX and AX molecules may be due to differences in the molecular structure of these carotenoid molecules.

Further, from Table II, it may be noted that the stability of the molecular associations is greater in ZX:DSPC monolayers and diminishes in the order of DSPC:ZX > DSPC:AX > DGDG:AX. This order suggests an important role for the carotenoid OH group in the formation of molecular associations of carotenoids and lipids, while the carbonyl groups of AX seemingly prevent these associations.

Also, it may be seen from Table II that $\Delta_{m,m}$ values for ZX:DSPC, AX:DSPC, and AX:DGDG monolayers range below the mean experimental error of π, and therefore it may be concluded that the regular-associations approximation provides a satisfactory description of the experimental data. The values of K and $\Delta\pi$ indicate the existence of specific interactions between the lipid and carotenoid molecules in monolayer assemblies at the air/water interface.

These studies of mixed films are important for understanding the structural arrangements in biomembranes and the specific complexes required for certain membrane phenomena. Also, these results could have applications in therapeutic preparations and nutritional formulations of emulsions or liposomes. The intake of specific carotenoids, for example, may be improved by their administration in liposomal preparations and emulsions.

The observed difference in association between zeaxanthin and astaxanthin with polar lipids may represent the basis of a functional differentiation in photosynthetic membranes where carotenoids act to protect against chlorophyll (membrane) photooxidation.

5. CONCLUSIONS: CAROTENOIDS IN MEMBRANES

The differential effects of carotenoids on lipid monolayer membranes result from differences in the molecular structure and polarity and consequent biophysical properties of particular carotenoids. The presence of the unsaturated double bonds in acyl chains of EPC molecules greatly attenuates the effect of some carotenoids, resulting in a surface lattice with an ideal behavior in mixed monolayers as is the case of egg phosphatidylcholine and β-cryptoxanthin (EPC:CX) monolayers.

The presence of bulky galactosyl groups attached to the polar head group of lipids, as in DGDG molecules, appears to diminish the stability constant of molecular associations with astaxanthin compared with diacyl phosphatidylcholine, with the same acyl chains, but stabilizes the compact packing in mixed monolayers.

The different behavior of carotenoids in mixtures with polar lipids might result from different locations and orientations of the carotenoids in subcellular structures and may explain the various functions that carotenoids exercise *in vivo*.

The biophysical properties of carotenoids are strongly influenced by other biomolecules in their vicinity, especially, as we have shown here, membrane polar lipids. Carotenoids could therefore be expected to modify or modulate the biophysical properties of subcellular structures. It appears that the molecular structure, shape, size, and polarity, and in turn the surface activity, are essential determinants of the ability of carotenoids to locate appropriately into their biological environment and fulfil their functions.

From the thermodynamic data, considerations of symmetry and additivity of mean molecular areas suggest that the carotenoid molecules are likely to be oriented in lipid monolayers near the collapse with their long axis perpendicular to the monolayer/water interface and the β-rings penetrating into the polar domain of lipid monolayers where they are exposed to the aqueous phase. This orientation is similar to that found in the case of zeaxanthin *in vivo* (Bone *et al.*, 1992; Bone and Landrum, 1984).

It is noteworthy that the use of monolayers offers considerable advantages over other model membrane systems for undertaking these biophysical studies. Thus the concentration of molecules in monolayer assemblies can be monitored precisely and the molecular interactions can be treated quantitatively.

Molecular interactions between carotenoid molecules themselves and their aggregates, such as dimers evidenced in monomolecular membranes, may also have significance in biological membranes. Molecular aggregations, for example, may change the physical properties of the biological membranes.

The present results are in general agreement with studies of liposomal systems and indicate that carotenoids are able to form molecular associations with lipids and stabilize membranes (Gabrielska and Gruszecki, 1996; Lazrak *et al.*, 1987; Tomoaia-Cotisel *et al.*, 1987d). These molecular associations determined in mixed monolayers (Zsako *et al.*, 1991; Tomoaia-Cotisel *et al.*, 1987c, d) could explain the behavior of carotenoids in phospholipid-stabilized triglyceride emulsions and the preferential distribution of carotenoids in the phospholipid monolayer that surrounds the triglyceride droplets in emulsions or in intracellular lipid storage droplets (Borel *et al.*, 1996). Therefore, the behavior of mixed monolayer assemblies may mimic the enhanced stability observed in biological systems and dietary emulsions containing carotenoids.

The molecular associations of carotenoids and phospholipids that are observed in mixed monolayers could be a reliable model for the transfer of carotenoids from lipid droplets to lipid structures dispersed in aqueous media, such as mixed micelles. Also, these associations might be important in the transfer of the carotenoids between lipoprotein particles and between lipoproteins and cellular membranes.

Thus, these data relating to the solubility of carotenoids in lipid structures, such as monolayers, coupled with data on micelles, bilayers, and the bulk lipid core of both collapsed phases and of the storage droplet may explain the role of carotenoids in modifying cellular membrane structure, properties, stability, and function.

ACKNOWLEDGMENT. This chapter is dedicated to the late Emil Chifu, whose support and counsel over many years inspired the work described here.

6. REFERENCES

Bertram, J. S., 1993, Cancer prevention by carotenoids: Mechanistic studies in cultured cells, in Carotenoids in Human Health, *Ann. N. Y. Acad. Sci.* **691:**177–191.

Bohm, F., Edge, R., Land, E. J., McGarvey, D. J., and Truscott, T. G., 1997, Carotenoids enhance vitamin E antioxidant efficiency, *J. Am. Chem. Soc.* **119:**621–622.

Bone, R. A., and Landrum, J. T., 1984, Macular pigment in Henle fiber membranes: A model for Haidinger's brushes, *Vision Res.* **24:**103–108.

Bone, R. A., Landrum, J. T., and Cains, A., 1992, Optical density spectra of the macular pigment *in vivo* and *in vitro*, *Vision Res.* **32:**105–110.

Borel, P., Grolier, P., Armand, M., Partier, A., Lafont, H., Lairon, D., and Azais-Braesco, V., 1996, Carotenoids in biological emulsions: Solubility, surface-to-core distribution, and release from lipid droplets, *J. Lipid Res.* **37:**250–261.

Britton, G., 1995, Structure and properties of carotenoids in relation to function, *FASEB J.* **9:**1551–1558.

Cadenhead, D. A., and Muller-Landau, F., 1980, Molecular accommodation and molecular interactions in mixed insoluble monomolecular films, *J. Colloid Interface Sci.* **78:**269–270.

Chifu, E., and Tomoaia-Cotisel, M., 1979, Insoluble monolayers of lecithin and carotenoid pigments, *Rev. Roum. Chim.* **24**(7)**:** 979–986.

Chifu, E., and Tomoaia-Cotisel, M., 1981, Carotene and protein films at the oil/water interface, *Stud. Univ. Babes-Bolyai Chem.* **26**(2)**:**3–8.

Chifu, E., Tomoaia, M., and Ioanette, A., 1975, Behavior of canthaxanthin at the benzene/water and air/water interfaces, *Gazz. Chim. Ital.* **105**(11–12)**:**1225–1232.

Chifu, E., Tomoaia, M., Nicoara, E., and Olteanu, A., 1978, Isozeaxanthin films at the oil/water and air/water interfaces, *Rev. Roum. Chim.* **23**(8)**:**1163–1169.

Chifu, E., Tomoaia-Cotisel, M., and Andrei, Z., 1979a, Mixed monolayers of canthaxanthin with lipids, *Stud. Univ. Babes-Bolyai Chem.* **24**(2)**:**63–67.

Chifu, E., Tomoaia-Cotisel, M., Andrei, Z., and Bonciu, E., 1979b, β-Apo-8′-carotenoic acid ethyl ester films at fluid interfaces, *Gazz. Chim. Ital.* **109**(6–7)**:**365–369.

Chifu, E., Zsako, J., and Tomoaia-Cotisel, M., 1983, Xanthophyll films. I. Single-component monolayers at the air/water interface, *J. Colloid Interface Sci.* **95:**346–354.

Chifu, E., Zsako, J., Tomoaia-Cotisel, M., Salajan, M., and Albu, I., 1986, Xanthophyll films. IV. Interaction of zeaxanthin and astaxanthin with electrolytes at the air/water interface, *J. Colloid Interface Sci.* **112:**241–251.

Chifu, E., Tomoaia-Cotisel, M., Salajan, M., Chifu, A., and Zsako, J., 1989, Xanthophyll films. V. Dynamics of xanthophyll monolayers at the liquid/air interface, *Stud. Univ. Babes-Bolyai Chem.* **34**(2)**:**28–34.

Coodley, G. O., Nelson, H. D., Loveless, M. O., and Folk, C., 1993, β-Carotene in HIV infection, in Carotenoids in Human Health, *Ann. N. Y. Acad. Sci.* **691:**277–278.

Davies, J. T., and Rideal, E. K., 1963, *Interfacial Phenomena*, 2nd ed., Academic Press, New York.

Defay, R., Prigogine, I., Bellemans, A., and Everett, D. H., 1966, *Surface Tension and Adsorption*, p. 77, Longmans, Green, London.

Erdman, J. W., Jr., Bierer, T. L., and Gugger, E.T., 1993, Absorption and transport of carotenoids, in Carotenoids in Human Health, *Ann. N. Y. Acad. Sci.* **691:**76–85.

Gabrielska, J., and Gruszecki, W. I., 1996, Zeaxanthin (dihydroxy-β-carotene) but not β-carotene rigidifies lipid membranes: A ^{1}H-NMR study of carotenoid-egg phosphatidylcholine liposomes, *Biochim. Biophys. Acta* **1285:**167–174.

Gaines, G. L., Jr., 1966, *Insoluble Monolayers at Liquid–Gas Interfaces*, Interscience, New York.

Galligan, L. J., Jackson, C. L., and Gerber, L. E., 1993, Carotenoids slow the growth of small cell lung cancer cells, in Carotenoids in Human Health, *Ann. N. Y. Acad. Sci.* **691:**267–269.

Gaziano, J. M., and Hennekens, C. H., 1993, The role of β-carotene in the prevention of cardiovascular disease, in Carotenoids in Human Health, *Ann. N. Y. Acad. Sci.* **691:**148–155.

Gruszecki, W. I., and Sielewiesiuk, J., 1990, Orientation of xanthophylls in phosphatidylcholine multibilayers, *Biochim. Biophys. Acta* **1023:**405–412.

Jezowska, I., Wolak, A., Gruszecki, W. I., and Strzalka, K., 1994, Effect of β-carotene on structural and dynamic properties of model phosphatidylcholine membranes. II. A ^{31}P-NMR and ^{13}C-NMR study, *Biochim. Biophys. Acta* **1194:**143–148.

Khomutov, G., Fry, I. V., Huflejt, M. E., and Packer, L., 1990, Membrane lipid composition, fluidity, and surface charge changes in response to growth of the fresh water cyanobacterium *Synechococcus* 6311 under high salinity, *Arch. Biochem. Biophys.* **277:**263–267.

Kohlmeier, L., and Hastings, S. B., 1995, Epidemiologic evidence of a role of carotenoids in cardiovascular disease prevention, *Am. J. Clin. Nutr.* **62**(Suppl)**:**1370S–1376S.

Krinsky, N. I., 1993, Actions of carotenoids in biological systems, *Annu. Rev. Nutr.* **13:**561–587.

Krinsky, N. I., Cornwell, D. G., and Oncley, J. L., 1958, The transport of vitamin A and carotenoids in human plasma, *Arch. Biochem. Biophys.* **73:**233–246.

Kuhn, H., and Mobius, D., 1993, Monolayer assemblies, in *Physical Methods of Chemistry*, Vol. 9B: *Investigations of Surfaces and Interfaces—Part B*, 2nd ed. (B. W. Rossiter and R. C. Baetzold, eds.), pp. 375–542, Wiley, New York.

Lazrak, T., Milon, A., Wolff, G., Albrecht, A.-M., Miehe, M., Ourisson, G., and Nakatani, Y., 1987, Comparison of the effects of inserted C_{40}- and C_{50}-terminally dihydroxylated carotenoids on the mechanical properties of various phospholipid vesicles, *Biochim. Biophys. Acta* **903:**132–141.

Liebler, D. C., Stratton, S. P., and Kaysen, K. L., 1997, Antioxidant actions of β-carotene in liposomal and microsomal membranes: Role of carotenoid-membrane incorporation and α-tocopherol, *Arch. Biochem. Biophys.* **338:**244–250.

Mathews-Roth, M. M., 1993, Carotenoids in erythropoietic protoporphyria and other photosensitivity diseases, in Carotenoids in Human Health, *Ann. N. Y. Acad. Sci.* **691:**127– 138.

Miller, K. W., and Small, D. M., 1983, Surface-to-core and interparticle equilibrium distributions of triglyceride-rich lipoprotein lipids, *J. Biol. Chem.* **258:**13772–13784.

Nikawa, T., Schulz, W. A., van den Brink, C. E., Hanusch, M., van der Saag, P., Stahl, W., and Sies, H., 1995, Efficacy of all-*trans*-β-carotene, canthaxanthin, and all-*trans*-, 9-*cis*-, and 4-oxoretinoic acids in inducing differentiation of an F9 embryonal carcinoma RARβ-*lacZ* reporter cell line, *Arch. Biochem. Biophys.* **316:**665–672.

Olson, J.A., 1993, Molecular actions of carotenoids, in Carotenoids in Human Health, *Ann. N. Y. Acad. Sci.* **691:**156–166.

Parker, R. S., 1996, Absorption, metabolism, and transport of carotenoids, *FASEB J.* **10:**542–551.

Picorel, R., Holt, R. E., Cotton, T. M., and Seibert, M., 1988, Surface-enhanced resonance Raman scattering spectroscopy of bacterial photosynthetic membranes. The carotenoid of *Rhodospirillum rubrum*, *J. Biol. Chem.* **263**(9)**:**4374–4380.

Rohmer, M., Bouvier, P., and Ourisson, G., 1979, Molecular evolution of biomembranes: Structural equivalents and phylogenetic precursors of sterols, *Proc. Natl. Acad. Sci. USA* **76:**847–851.

Sies, H., and Stahl, W., 1995, Vitamins E and C, β-carotene, and other carotenoids as antioxidants, *Am. J. Clin. Nutr.* **62**(Suppl)**:**1315S–1321S.

Strzalka, K., and Gruszecki, W. I., 1994, Effect of β-carotene on structural and dynamic properties of model phosphatidylcholine membranes. I. An EPR spin label study, *Biochim. Biophys. Acta* **1194:**138–142.

Subczynski, W. K., Markowska, E., and Sielewiesiuk, J., 1991, Effect of polar carotenoids on the oxygen diffusion-concentration product in lipid bilayers. An EPR spin label study, *Biochim. Biophys. Acta* **1068:**68–72.

Toma, S., Losardo, P. L., Vincent, M., and Palumbo, R., 1995, Effectiveness of β-carotene in cancer chemoprevention, *Eur. J. Cancer Prev.* **4:**213–224.

Tomoaia-Cotisel, M., 1979, Study on films of natural pigments and lecithin, Ph.D. Thesis, University of Cluj-Napoca, Cluj-Napoca, Romania.

Tomoaia-Cotisel, M., and Chifu, E., 1979, Mixed insoluble monolayers with β-apo-8′-carotenoic acid ethyl ester, *Gazz. Chim. Ital.* **109**(6–7)**:**371–375.

Tomoaia-Cotisel, M., and Chifu, E., 1981, Carotenoid pigment films at fluid interfaces, *Rev. Chim.* (Bucharest) **32**(11)**:**1063–1069.

Tomoaia-Cotisel, M., and Chifu, E., 1983, Xanthophyll films. II. Two-component monolayers of some xanthophylls and egg lecithin at the air/water interface, *J. Colloid Interface Sci.* **95:**355–361.

Tomoaia, M., Ioanette, A., and Chifu, E., 1975, Behavior of β-apo-8′-carotenal at the oil/water and air/water interfaces, in *Proceedings of the International Conference of Colloid and Surface Science* (E. Wolfram, ed.), Vol. 1, pp. 559–566, Akademiao Kiado, Budapest.

Tomoaia-Cotisel, M., Chifu, E., Tamas, V., and Marculetiu, V., 1980, Behaviour of some apocarotenoid derivatives at the air/water interface, *Rev. Roum. Chim.* **25**(2)**:**175–180.

Tomoaia-Cotisel, M., Zsako, J., and Chifu, E., 1981, Dipalmitoyl lecithin and egg lecithin monolayers at an air/water interface, *Ann. Chim.* (Rome) **71**(3–4)**:**189–200.

Tomoaia-Cotisel, M., Chifu, E., and Albu, I., 1983a, Critical micelle concentration of egg lecithin in the presence of some carotenoids, *Stud. Univ. Babes-Bolyai Chem.* **28:**42–45.

Tomoaia-Cotisel, M., Sen, A., and Quinn, P. J., 1983b, Surface-active properties of 1,2-distearoylgalactosylglycerols, *J. Colloid Interface Sci.* **94:**390–398.

Tomoaia-Cotisel, M., Chifu, E., and Zsako, J., 1985a, Mixed monolayers of egg lecithin and carotenoids, *Colloids Surf.* **14:**239–246.

Tomoaia-Cotisel, Chifu, E., and Zsako, J., 1985b, The structure of some lecithin monolayers at the air/water interface, in *Water and Ions in Biological Systems, Proceedings of the 2nd International Conference* (A. Pullman, V. Vasilescu, and L. Packer, eds.), pp. 243–250, Plenum Press, New York.

Tomoaia-Cotisel, M., Zsako, J., Chifu, E., Cadenhead D. A., and Ries, H. E., Jr., 1987a, Collapse mechanism of some carotenoid monomolecular films—membrane models, in *Photosynthesis Research*, Vol. 2 (J. Biggins, ed.), pp. 333–337, Martinus Nijhoff, Dordrecht.

Tomoaia-Cotisel, M., Zsako, J., Chifu, E., and Quinn, P. J., 1987b, Monomolecular films and collapse structures as biomembrane models, *Stud. Univ. Babes-Bolyai Chem.* **32**(1)**:**35–49.

Tomoaia-Cotisel, M., Zsako, J., Chifu, E., and Quinn, P. J., 1987c, Molecular associations in lipid–carotenoid monolayers, in *The Metabolism, Structure and Function of Plant Lipids* (P. K. Stumpf, J. B. Mudd, and W. D. Ness, eds.), pp. 131–133, Plenum Press, New York.

Tomoaia-Cotisel, M., Zsako, J., Chifu, E., and Quinn, P. J., 1987d, Intermolecular interactions in lipid/carotenoid monolayers, *Biochem. J.* **248**(3)**:**877–882.

Tomoaia-Cotisel, M., Zsako, J., Chifu, E., and Cadenhead, D. A., 1990, Relaxation phenomena in apocarotenoid monolayers, *Langmuir* **6**(1)**:**191–197.

Tomoaia-Cotisel, M., Chifu, E., Zsako, J., and Quinn, P. J., 1995, Biophysical properties of lipid and carotenoid films, in *International Symposium on Micelles, Microemulsions and Monolayers: Quarter Century Progress and New Horizons*, p. 118, University of Florida Press, Gainesville.

Willett, W. C., 1994, Micronutrients and cancer risk, *Am. J. Clin. Nutr.* **59**(Suppl)**:**1162S–1165S.

Woodall, A. A., Britton, G., and Jackson, M. J., 1995, Antioxidant activity of carotenoids in phosphatidylcholine vesicles: Chemical and structural considerations, *Biochem. Soc. Trans.* **23:**133S.

Yin, J. J., and Subczynski, W. K., 1996, Effects of lutein and cholesterol on alkyl chain bending in lipid bilayers: A pulse electron spin resonance spin labeling study, *Biophys. J.* **71:**832–839.

Ziegler, R. G., 1993, Carotenoids, cancer, and clinical trials, in Carotenoids in Human Health, *Ann. N. Y. Acad. Sci.* **691:**110–119.

Zsako, J., Chifu, E., and Tomoaia-Cotisel, M., 1979, Rotating rigid-plate model of carotenoid molecules and the behaviour of their monolayers at the air/water interface, *Gazz. Chim. Ital.* **109**(12)**:**663–668.

Zsako, J., Tomoaia-Cotisel, M., and Chifu, E., 1984, Insoluble mixed monolayers. I. Phase equilibria at the collapse of binary monolayers at gas/liquid interfaces, *J. Colloid Interface Sci.* **102:**186–205.

Zsako, J., Tomoaia-Cotisel, M., Mocanu, A., and Chifu, E., 1986, Insoluble mixed monolayers. II. Protolytic equilibria and the influence of the pH on the collapse pressure, *J. Colloid Interface Sci.* **110:**317–334.

Zsako, J., Tomoaia-Cotisel, M., and Chifu, E., 1989, Discussion of compression isotherms of some carotenoid monolayers on the basis of HMO calculations, in *Surfactants in Solution*, Volume 9 (K. L. Mittal, ed.), pp. 311–324, Plenum Press, New York.

Zsako, J., Tomoaia-Cotisel, M., and Chifu, E., 1991, Insoluble mixed monolayers. V. Molecular associations in binary films: The regular association approach, *J. Colloid Interface Sci.* **146**(2)**:**353–362.

Part III

Vitamin D

Chapter 11

The Role of Vitamin D_3 and Antiestrogens in Modulating Apoptosis of Breast Cancer Cells and Tumors

JoEllen Welsh, Kathryn VanWeelden,
Louise Flanagan, Ian Byrne, Elizabeth Nolan, and
Carmen J. Narvaez

1. OVERVIEW OF APOPTOSIS

1.1. Morphological and Biochemical Features

The death of cells can be classified as either *necrosis*, usually the result of tissue damage or insult, or *apoptosis*, an active process of cellular self-destruction. Apoptosis can be distinguished by its characteristic morphological features such as cell shrinkage, nuclear condensation, and irreversible DNA fragmentation (Tenniswood *et al.*, 1992). In most cell types, apoptosis requires gene transcription and activation of specific

JoEllen Welsh, Elizabeth Nolan, and Carmen J. Narvaez W. Alton Jones Cell Science Center, Lake Placid, New York 12946. **Kathryn VanWeelden** W. Alton Jones Cell Science Center, Lake Placid, New York 12946; and Clarkson University, Potsdam, New York 13499. **Louise Flanagan and Ian Byrne** W. Alton Jones Cell Science Center, Lake Placid, New York 12946; and University College Dublin, Belfield, Ireland.

Subcellular Biochemistry, Volume 30: Fat-Soluble Vitamins, edited by Quinn and Kagan. Plenum Press, New York, 1998.

proteases which mediate destruction of the cell without release of intracellular contents or initiation of an immune response (Vaux and Strasser, 1996; Zhivotovsky *et al.*, 1997). In addition to genes linked to activation of apoptosis, genes associated with repression of apoptosis, such as bcl-2, have been identified (Craig, 1995; Yang and Korsmeyer, 1996). Thus, the regulation of apoptosis within a cell reflects a balance between antiapoptotic and proapoptotic gene products. Extracellular influences on this balance include physiological triggers of apoptosis, cell-type-specific "survival factors," and the components of the extracellular matrix (Tenniswood *et al.*, 1992; Goberdhan *et al.*, 1996; Merio *et al.*, 1997). The recognition that mutations in critical components or regulators of the apoptotic pathway, such as p53 and bcl-2, may be linked to cancer development or therapeutic responsiveness has emphasized the importance of apoptosis in the context of tumor biology (Barry *et al.*, 1990; Lowe *et al.*, 1993; Yang and Korsmeyer, 1996).

1.2. Role of Apoptosis in Cancer Therapy

Effective therapeutic agents induce tumor regression by altering the balance between the net rate of cell proliferation and the net rate of cell death. Cytostatic agents primarily block cell cycle progression, thereby reducing the rate of tumor cell proliferation, whereas cytotoxic agents enhance the rate at which tumor cells die. It is now clear that many therapeutic strategies, including DNA-damaging agents, radiation, and antihormonal drugs, induce tumor regression through their ability to target the apoptotic pathway (Barry *et al.*, 1990; Fesus *et al.*, 1995). In estrogen-dependent mammary epithelial cells, estrogen ablation or administration of antiestrogens results in activation of apoptosis secondary to disruption of estrogen-mediated survival signals (Kyprianou *et al.*, 1991; Warri *et al.*, 1993; Guenette *et al.*, 1994; Quarrie *et al.*, 1996). Unfortunately, in the case of breast cancers, prolonged treatment with antihormonal therapies often leads to the emergence of hormone-independent cells in which the apoptotic pathway is no longer triggered by disruption of the hormone-dependent survival signals. These hormone-independent breast cancer cells either lack estrogen receptor (ER) or express nonfunctional variants which do not respond to antiestrogen therapies. Since hormone resistance is associated with acceleration of tumor growth, and is common in aggressive tumors which have acquired metastatic capacity, development of resistance frequently correlates with reduced patient survival time. Induction of apoptosis in breast tumors consisting of estrogen-independent cells would require use of agents which exploit signaling pathways that are independent from those triggered by disruption of ER function. Identification of agents which induce apoptosis in breast cancer cells which have escaped estrogen dependence is critical to the development of novel therapeutics with efficacy against later stage tumors.

2. VITAMIN D_3: NEGATIVE GROWTH REGULATOR OF BREAST CANCER CELLS

2.1. The Vitamin D_3 Endocrine System

$1,25(OH)_2D_3$ is the biologically active form of vitamin D_3, a fat-soluble vitamin identified as an antirachitic factor in the early 1920s. Since that discovery, the mechanism of action of $1,25(OH)_2D_3$ in maintenance of calcium homeostasis and bone remodeling has been intensely investigated. $1,25(OH)_2D_3$ interacts with the vitamin D_3 receptor (VDR), a member of the steroid/thyroid/retinoic acid family of nuclear receptors, and modulates the expression of target genes whose products mediate the biological responses attributed to vitamin D_3 (Studzinski *et al.*, 1993; Carlberg *et al.*, 1993). Identification of a variety of vitamin D_3 responsive elements in the promoters of numerous $1,25(OH)_2D_3$-regulated genes underscores the transcriptional mechanism of action of the occupied VDR and its similarity to other nuclear receptors (Carlberg *et al.*, 1993; Haussler *et al.*, 1995; Darwish and DeLuca, 1996).

2.2. Role of Vitamin D_3 in Cell Proliferation and Differentiation

It has recently become apparent that the effects of vitamin D_3 are not limited to the regulation of calcium and bone homeostasis (Reichel *et al.*, 1989; Pols *et al.*, 1990; Studzinski *et al.*, 1993; Bouillon *et al.*, 1995). The expression of the VDR in tissues not normally involved in calcium homeostasis, and the recognition that $1,25(OH)_2D_3$ is a potent regulator of cell proliferation and differentiation, has enormously expanded the role of the vitamin D_3 endocrine system in human physiology. Examples of normal cell types sensitive to vitamin D_3 are keratinocytes, prostatic epithelial cells, monocytes/macrophages, lymphocytes, and pancreatic β cells. Transformed cell lines which express the VDR include those derived from bone, colon, prostate, leukemias, melanoma, and breast. The first report of a negative growth regulatory effect of $1,25(OH)_2D_3$ in cancer-derived cells was in melanoma cells (Colston *et al.*, 1981). The effects of $1,25(OH)_2D_3$ on transformed cell lines which express functional VDR include induction of differentiation markers, cell cycle arrest, and/or activation of apoptosis (Tsuchiya *et al.*, 1993; Shabahang *et al.*, 1994; Elstner *et al.*, 1995, 1997; Welsh *et al.*,

1995). Preclinical studies have demonstrated significant inhibitory effects of 1,25$(OH)_2D_3$ on growth or progression of leukemias, melanoma, colon cancer, osteosarcoma, and breast cancer *in vivo* (reviewed by Bouillon *et al.*, 1995).

2.3. Effects of Vitamin D_3 on Mammary Epithelial Cells

The presence of the VDR in normal and lactating mammary gland, in breast tumors, and in cell lines derived from human breast cancers (Eisman *et al.*, 1981; Colston *et al.*, 1989; Berger *et al.*, 1991; Buras *et al.*, 1994) has generated interest in the possible functional role(s) of vitamin D_3 in the mammary gland. Early studies indicated that vitamin D_3 may function in regulation of calcium transport during lactation, or may act in concert with other hormones to regulate mammary cell differentiation and milk protein production (Battcherjee *et al.*, 1987; Mezettie *et al.*, 1988). However, in breast cancer cells, 1,25$(OH)_2D_3$ has consistently been identified as a potent negative growth regulator (Chouvet *et al.*, 1986; Colston *et al.*, 1989; Eisman *et al.*, 1989). Treatment of estrogen-dependent MCF-7 human breast cancer cells with 1,25$(OH)_2D_3$ induces cell cycle arrest in G_0/G_1, upregulation of the cell cycle inhibitors p53 and p21, and dephosphorylation of the retinoblastoma protein (Welsh *et al.*, 1994; Simboli-Campbell *et al.*, 1996, 1997; James *et al.*, 1996). Since our original demonstration that 1,25$(OH)_2D_3$ induces typical features of apoptosis in MCF-7 cells (Welsh *et al.*, 1994), several additional reports have confirmed that vitamin D_3 compounds induce cytoplasmic condensation, nuclear matrix reorganization, pyknotic nuclei, and DNA fragmentation in breast cancer cells (Elstner *et al.*, 1995; Simboli-Campbell and Welsh, 1995; Vandewalle *et al.*, 1995; James *et al.*, 1996; Simboli-Campbell *et al.*, 1996). Additional indicators of apoptosis which have been demonstrated in MCF-7 cells after treatment with vitamin D_3 compounds include upregulation of apoptosis-associated proteins, such as cathepsin B and clusterin, and downregulation of the antiapoptosis protein bcl-2 (Elstner *et al.*, 1995; James *et al.*, 1996; Simboli-Campbell *et al.*, 1996, 1997). An intriguing link between induction of apoptosis by vitamin D_3 compounds and depletion of intracellular calcium stores has been reported in MCF-7 cells (Vandewalle *et al.*, 1995).

The effects of 1,25$(OH)_2D_3$ on cell cycle and apoptosis in MCF-7 cells are illustrated in Fig. 1. Within 48 hr of treatment, 1,25$(OH)_2D_3$ (at doses greater than 10^{-9}M) significantly reduces viable cell number (Fig. 1, A). The loss of cell viability after 1,25$(OH)_2D_3$ treatment is associated with the induction of DNA fragmentation, as detected by terminal transferase-mediated fluorescein dUTP nick end labeling of DNA strand breaks (Fig. 1, B), a characteristic feature of apoptotic cell death. Flow-cytometric analysis of propidium iodide-stained MCF-7 cells (Fig. 1, C) indicates that $1,25(OH)_2D_3$ increases the percentage of cells in G_0/G_1 and reduces the percentage of cells in S phase (see also Simboli-Campbell *et al.*, 1997). Collectively, these data indicate that the reduction of viable cell number after $1,25(OH)_2D_3$ treatment of MCF-7 cells reflects both cell cycle arrest and activation of apoptosis.

3. INTERACTIONS BETWEEN VITAMIN D_3 AND ESTROGENS IN BREAST CANCER CELLS

3.1. Complementary Effects of Vitamin D_3 and Antiestrogens in Estrogen-Dependent Breast Cancer Cells

Analyses of human breast tumor biopsies have demonstrated that the ER and VDR do not necessarily colocalize within the same cells (Berger *et al.*, 1991), indicating that these two nuclear receptors are present in distinct target cells within a mixed tumor. These findings suggest that combining agents which activate vitamin D_3-dependent pathways with those that disrupt estrogen-dependent signaling might offer a therapeutic advantage over either approach alone. This concept is supported by *in vitro* studies demonstrating complementary effects of antiestrogens and vitamin D_3 compounds on growth of ER positive (ER+) breast cancer cells (Abe-Hashimoto *et al.*, 1993; James *et al.*, 1994; Vink-van Wijngaarden *et al.*, 1994; Love-Schimenti *et al.*, 1996). In our lab, we have studied the interactions between $1,25(OH)_2D_3$ and the pure steroidal antiestrogen ICI 182,780 (Zeneca Pharmaceuticals, Macclesfield, UK) in modulating growth of MCF-7 cells, which express both ER and VDR. The complementary effects of $1,25(OH)_2D_3$ and ICI 182,780 on MCF-7 cell number are presented in Fig. 2. A dose-dependent decrease in cell number is observed in MCF-7 cells treated for 72 hr with ICI 182,780 (dashed bars). Coincubation with 10 nM $1,25(OH)_2D_3$ further reduces cell number at all doses of ICI 182,780 (solid bars). Flow-cytometric analysis (data not shown) has indicated that MCF-7 cells treated with both $1,25(OH)_2D_3$ and ICI 182,780 exhibit enhanced rates of cell cycle arrest and apoptosis compared to either treatment alone.

3.2. Effects of Vitamin D_3 on Estrogen-Independent Breast Cancer Cells

In ER+ breast cancer cells, the effects of $1,25(OH)_2D_3$ are very similar to those induced by antiestrogens (Welsh, 1994) and $1,25(OH)_2D_3$-mediated apoptosis in MCF-7 cells is preceded by downregulation of the

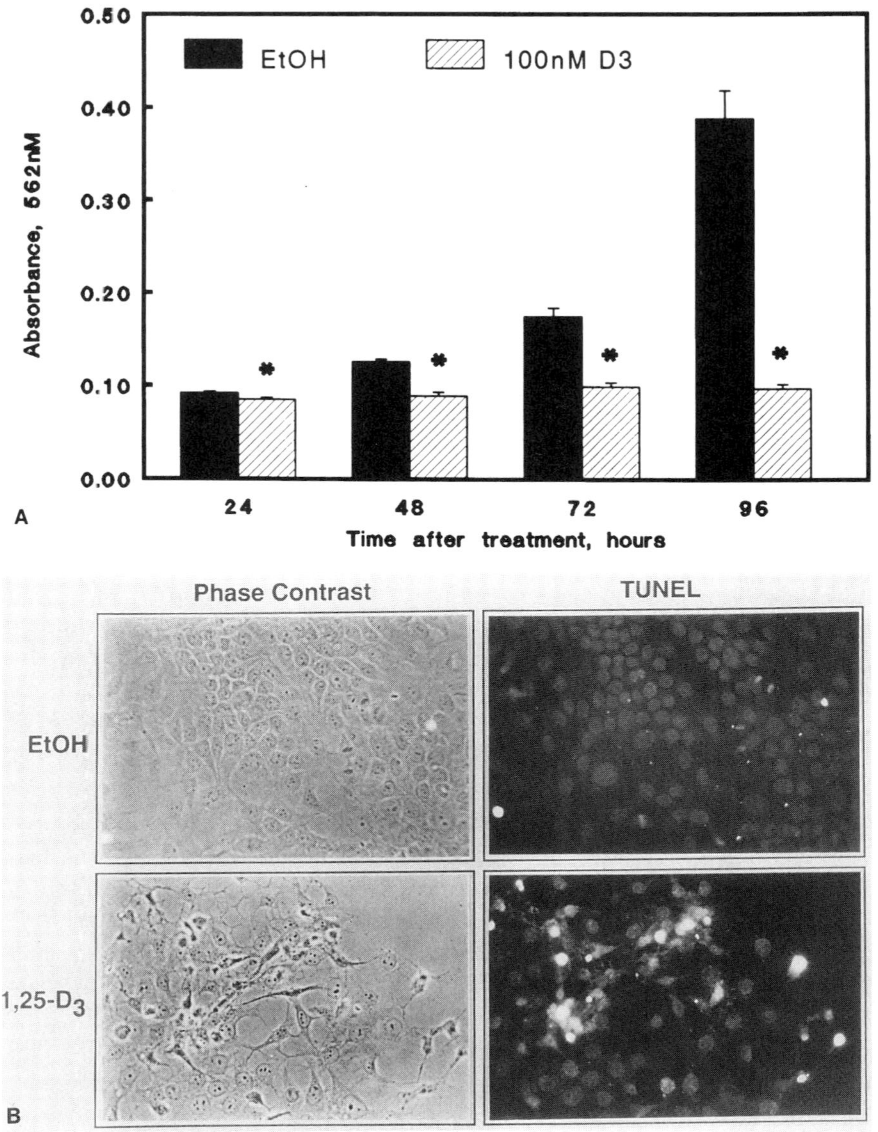

FIGURE 1. Effects of 100 nM $1,25(OH)_2D_3$ on MCF-7 cells. (A) Temporal changes in MCF-7 cell number after one dose of 100 nM $1,25(OH)_2D_3$. Cell number was quantitated by crystal violet staining after treatment with $1,25(OH)_2D_3$ (D3) or ethanol vehicle (EtOH). Data are expressed as mean ± SEM of four determinations. Asterisk indicates statistically significant difference, $1,25(OH)_2D_3$ vs. ethanol. (B) Phase contrast (left panels) and epifluorescent (right panels) images of MCF-7 cells treated for 96 hr with ethanol vehicle (EtOH) or 100 nM $1,25(OH)_2D_3$ (1,25-D_3). Cells were processed for terminal transferase-mediated fluorescein dUTP nick end labeling (TUNEL) of DNA strand breaks, a characteristic of apoptosis.

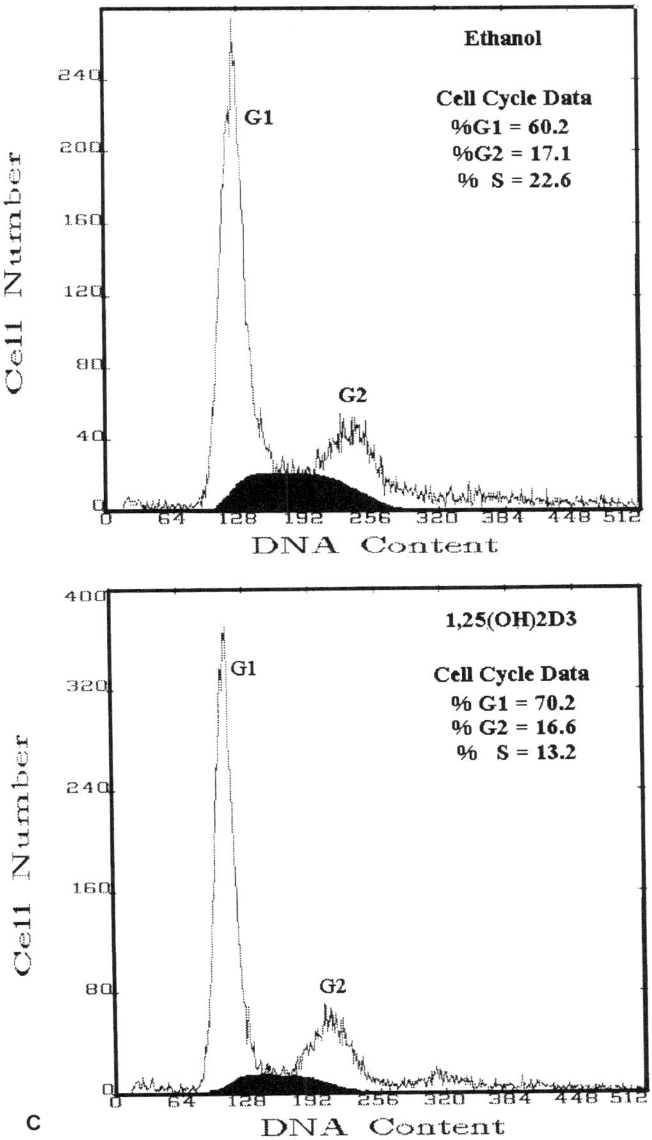

C

FIGURE 1. (C) Flow-cytometric analysis of propidium iodide-stained MCF-7 cells treated for 144 hr with ethanol vehicle or 100 nM 1,25$(OH)_2D_3$. The percentage of cells in G1, G2, and S phases of the cell cycle are indicated for each treatment.

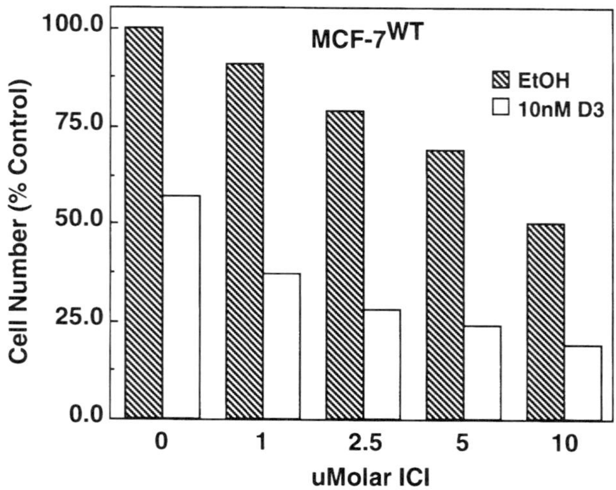

FIGURE 2. Complementary effects of 1,25(OH)$_2$D$_3$ and antiestrogen on growth of MCF-7 cells. Wild-type MCF-7 (MCF-7^{WT}) cells were treated with the antiestrogen ICI 182,780 (Zeneca Pharmaceuticals, Macclesfield, UK), in the presence or absence of 10 nM 1,25(OH)$_2$D$_3$, for 72 hr. Cell number was quantitated by crystal violet staining and expressed as percentage of control (ethanol-treated) values (mean of four determinations).

ER and decreased expression of estrogen-dependent genes such as those for the progesterone receptor and pS$_2$ (James *et al.*, 1994; Colston *et al.*, 1995; Simboli-Campbell *et al.*, 1997). These findings have led to the hypothesis that, at least in MCF-7 cells, the effects of vitamin D$_3$ may in part be related to disruption of estrogen-regulated survival signals (Colston *et al.*, 1995; Welsh *et al.*, 1994). If so, then sensitivity to 1,25(OH)$_2$D$_3$-mediated growth arrest/apoptosis could be reduced in estrogen-independent breast cancer cells. Since late-stage, aggressive breast cancers with poor outcome are primarily estrogen independent, it is clearly of interest to determine whether 1,25(OH)$_2$D$_3$ can effectively induce apoptosis in ER-negative (ER−) mammary cells. We and others have therefore examined the relationship between sensitivity to 1,25(OH)$_2$D$_3$ and estrogen independence in breast cancer cells. The presence of the VDR has been demonstrated in many established estrogen-independent human breast cancer cell lines (Buras *et al.*, 1994; Elstner *et al.*, 1995). Furthermore, 1,25(OH)$_2$D$_3$ inhibits the growth of ER− breast carcinoma cells such as MDA-MB-231 and BT-20 (Chouvet *et al.*, 1986; Elstner *et al.*, 1995; Love-Schimenti *et al.*, 1996), suggesting that the growth-inhibitory effects of 1,25(OH)$_2$D$_3$ are not completely dependent on the presence of a functional ER signaling pathway.

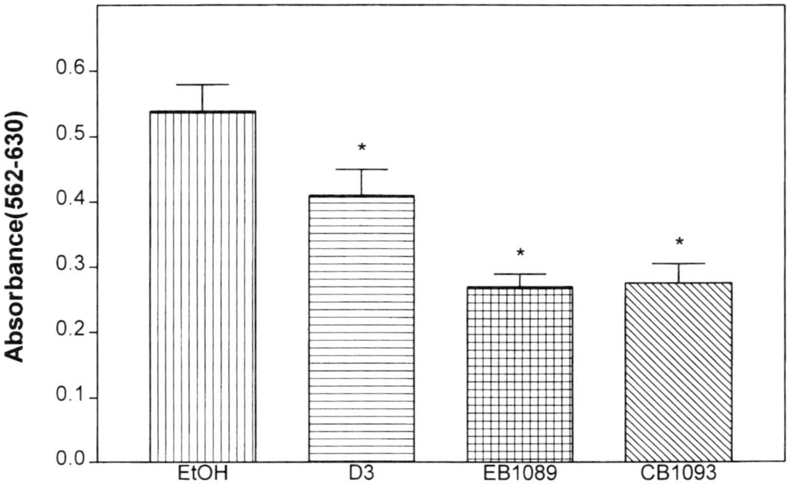

FIGURE 3. Effects of vitamin D_3 compounds on growth of estrogen-independent SUM159 human mammary cancer cells. SUM159 cells were treated for 48 hr with ethanol vehicle (EtOH), 100 nM 1,25$(OH)_2D_3$ (D3), or the synthetic analogs EB1089 or CB1093 (LEO Pharmaceuticals, Ballerup, Denmark). Cell number was quantitated by crystal violet staining and data represent mean ± SEM of four determinations. Asterisk indicates statistically significant difference, ethanol vs. 1,25$(OH)_2D_3$, EB1089, or CB1093.

We have recently characterized the vitamin D_3 signaling pathway in a newly established estrogen-independent human breast cancer cell line, SUM159 (obtained from Dr. Steve Ethier, University of Michigan, Ann Arbor, Michigan). In contrast to estrogen-dependent MCF-7 cells, which were isolated from the pleural effusions of a patient with adenocarcinoma, the estrogen-independent SUM159 cell line was established from the primary tumor of a patient with metaplastic carcinoma of the breast. The estrogen independence of SUM159 cells has been established by the lack of estradiol-binding sites, insensitivity to antiestrogens *in vitro*, and the ability to form tumors in ovariectomized nude mice (data not shown). We have established that SUM159 cells express functional VDR (based on ligand-binding assays, Western blotting, and gel shift analysis) and that 1,25$(OH)_2D_3$ potently inhibits growth of these cells *in vitro*. As demonstrated in Fig. 3, treatment with 100 nM 1,25$(OH)_2D_3$ significantly reduces SUM159 cell number within 48 hr. Characteristics of apoptosis, such as cytoplasmic and nuclear condensation and DNA fragmentation, have been observed in 1,25$(OH)_2D_3$-treated SUM159 cells, using the techniques of terminal transferase-mediated fluorescein dUTP nick end labeling, Hoechst dye fluorescence, and flow cytometry (L. Flanagan, S. Ethier, and J. E.

Welsh, in preparation). These data indicate that despite the lack of functional ER, SUM159 cells retain sensitivity to vitamin D_3-mediated apoptosis. Preliminary studies indicate that SUM159 cells exhibit invasive behavior *in vitro*, and are tumorigenic in nude mice (see Section 4.2.2). Thus, SUM159 cells/tumors represent a promising model for late-stage breast cancer, and are very responsive to the growth-inhibitory effects of $1,25(OH)_2D_3$.

3.3. Selection and Characterization of MCF-7 Cell Lines Resistant to Antiestrogens and Vitamin D_3

Although studies with ER− breast cancer cell lines, such as those reviewed above, have suggested that $1,25(OH)_2D_3$ can effectively inhibit growth independently of estrogen signaling, these data were generated by comparison of distinct cell lines with variable levels of VDR function and diverse genetic alterations. To provide more conclusive evidence for independent signaling pathways toward apoptosis by antiestrogens and vitamin D_3, we have selected variants of MCF-7 cells which display resistance to these agents. The selection and characterization of a vitamin D_3-resistant MCF-7 cell line has already been described (Narvaez *et al.*, 1996). Briefly, MCF-7 cells were selected for growth in 100 nM $1,25(OH)_2D_3$, and a stable cell line which grows equally well in the presence or absence of 100 nM $1,25(OH)_2D_3$ was generated (Fig. 4, top). Despite selective resistance to $1,25(OH)_2D_3$ and structurally related compounds (Narvaez *et al.*, 1996), these vitamin D_3-resistant MCF-7 cells express a functional VDR. We have established that the nucleic acid sequence of the VDR from the vitamin D_3-resistant MCF-7 cell line is identical to that of the VDR from parental MCF-7 cells (I. Byrne and J. E. Welsh, unpublished). Our studies have demonstrated that MCF-7 cells selected for resistance to $1,25(OH)_2D_3$ express the ER and retain sensitivity to antiestrogens such as tamoxifen (Narvaez *et al.*, 1996) and ICI 182,780 (Fig. 4, bottom). Similar results have been obtained with several distinct vitamin D_3-resistant clones generated from two different MCF-7 stock cultures, suggesting that the ability to dissociate sensitivity to vitamin D_3 compounds from that of antiestrogens is a general phenomenon.

In similar experiments, we have generated several distinct MCF-7 clonal cell lines which are resistant to the antiestrogen ICI 182,780. In contrast to the parental MCF-7 cells, which can form tumors in nude mice only in the presence of estradiol, the ICI 182,780-resistant MCF-7 cells are tumorigenic in ovariectomized nude mice (Nolan *et al.*, in press). Despite their estrogen independence, we have observed that ICI 182,780-resistant MCF-7 cell clones retain VDR expression and are growth inhibited by

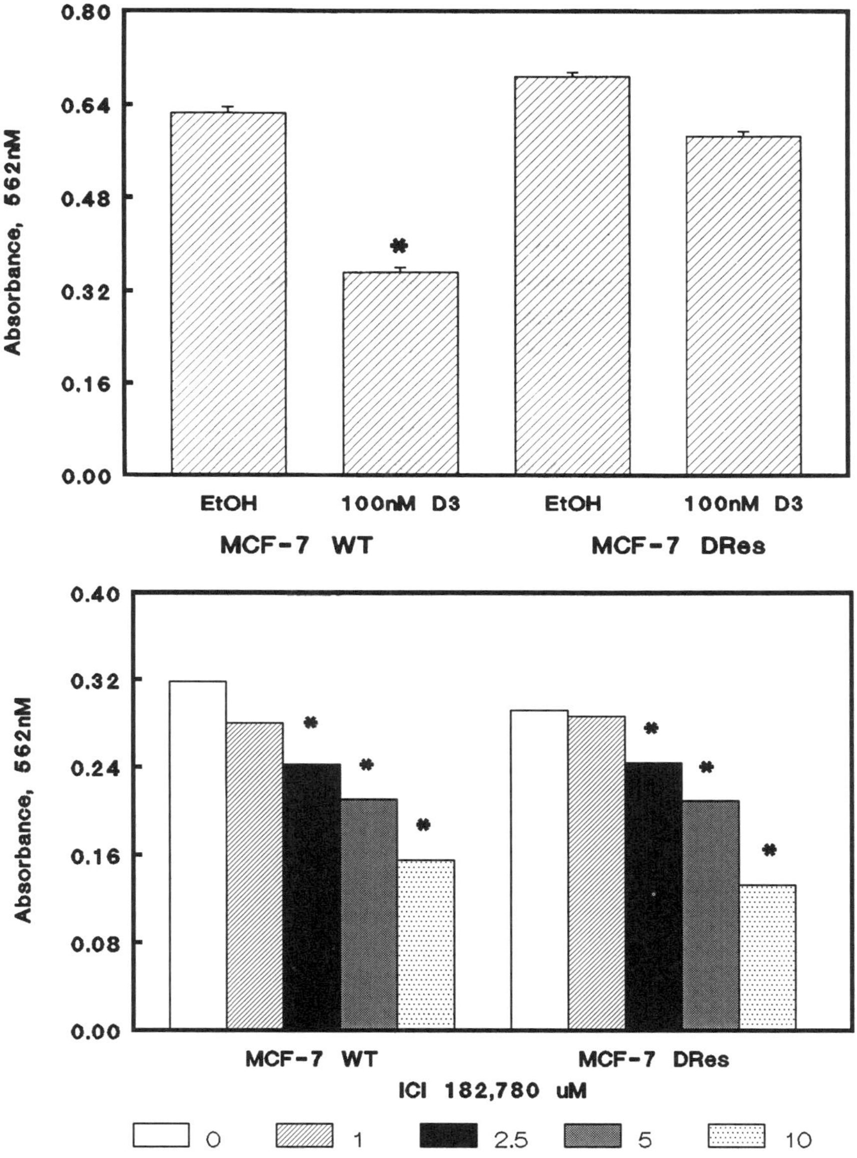

FIGURE 4. Effects of $1,25(OH)_2D_3$ and ICI 182,780 on growth of parental and vitamin D_3 resistant MCF-7 cells. *Top*: Parental ($MCF\text{-}7^{WT}$) and vitamin D_3-resistant ($MCF\text{-}7^{DRes}$) cells were grown for 48 hr in the absence of $1,25(OH)_2D_3$, and then treated with ethanol vehicle (EtOH) or 100 nM $1,25(OH)_2D_3$ (D_3) for 96 hr. Data represent cell number, quantitated by crystal violet staining, and are expressed as mean ± SEM of four determinations. Asterisk indicates statistically significant difference, ethanol vs. $1,25(OH)_2D_3$. *Bottom*: Parental ($MCF\text{-}7^{WT}$) and vitamin D_3-resistant ($MCF\text{-}7^{DRes}$) cells were grown for 24 hr in the absence of $1,25(OH)_2D_3$, and then treated with the indicated doses of the antiestrogen ICI 182,780 (Zeneca Pharmaceuticals, Macclesfield, UK) for 72 hr. Data represent cell number, quantitated by crystal violet staining, and are expressed as mean ± SEM of four determinations. Asterisk indicates statistically significant difference, ethanol vs. ICI 182,780.

vitamin D_3. As demonstrated in Fig. 5, parental MCF-7^{WT} cells are growth inhibited by $1,25(OH)_2D_3$ or ICI 182,780, and an additive effect on cell number is observed in response to combination treatment. In clone 1.10, selected for resistance to ICI 182,780, no decrease in cell number is observed in the presence of 10 μM ICI 182,780, but growth is significantly reduced by $1,25(OH)_2D_3$. Surprisingly, $1,25(OH)_2D_3$ may sensitize antiestrogen-resistant cells to ICI 182,780, as suggested by the further reduction in cell number observed when clone 1.10 cells are coincubated with $1,25(OH)_2D_3$ and ICI 182,780. In several of the ICI 182,780-resistant MCF-7 clonal cell lines, sensitivity to $1,25(OH)_2D_3$ is higher than that of the parental MCF-7 cells (not shown). These studies demonstrate that, even in MCF-7 cells, triggering of apoptosis by $1,25(OH)_2D_3$ does not require functional ER. In addition, these findings suggest that the sensitivity to vitamin D_3-mediated growth arrest/apoptosis is maintained during the progression of human breast cancer cells from an estrogen-dependent to an estrogen-independent state.

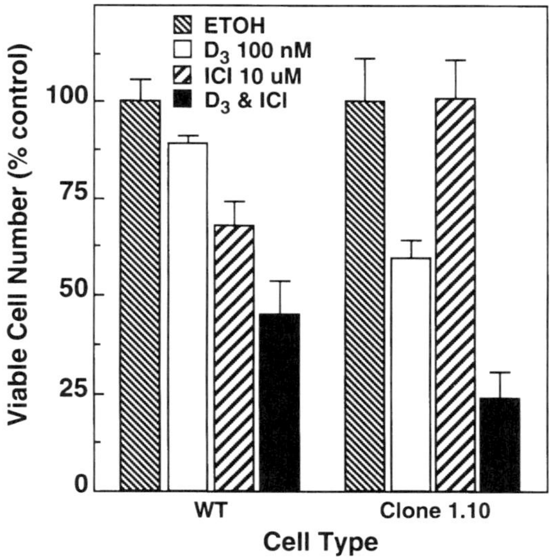

FIGURE 5. Effects of $1,25(OH)_2D_3$ and ICI 182,780 on growth of parental and antiestrogen-resistant MCF-7 cells. Parental (MCF-7^{WT}) and ICI 182,780-resistant (clone 1.10) cells were grown for 24 hr in the absence of ICI 182,780 or $1,25(OH)_2D_3$ and then treated with ethanol vehicle (ETOH) or with 100 nM $1,25(OH)_2D_3$ (D_3), 10 μM ICI 182,780 (ICI), or both for 96 hr. Data represent cell number, quantitated by crystal violet staining, and are expressed as percentage of control (ethanol-treated) values for each cell line. Bars represent mean ± SEM of four determinations.

3.4. Mechanisms of Action in Mammary Epithelial Cells: Antiestrogens and Vitamin D_3

The VDR and the ER are members of the superfamily of nuclear receptors which act as ligand-activated transcription factors. It is important to remember, however, that apoptosis is induced by *activation* of VDR signaling, or by *disruption* of ER signaling. Thus, in breast cancer cells, these two nuclear receptors act antagonistically with respect to cell survival. In breast cancer cells containing both ER and VDR, $1,25(OH)_2D_3$ and antiestrogens may independently modulate expression of one or more critical modulators of apoptosis. Identification of the specific genes involved in modulating apoptosis will be necessary to determine whether $1,25(OH)_2D_3$ and estrogens interact through their cognate receptors at the level of the target gene promoters. As previously discussed, downregulation of ER signaling may contribute to the effects of $1,25(OH)_2D_3$ in estrogen-dependent cells such as MCF-7.

In estrogen-independent cells such as SUM159 and BT-20, $1,25(OH)_2D_3$ treatment induces apoptosis despite the lack of functional ER. In addition, experiments conducted with the ICI 182,780-resistant variants of MCF-7 cells support the concept that the induction of apoptosis by vitamin D_3 can be dissociated from antiestrogen sensitivity. These findings imply that, in addition to potential interactions between VDR and ER signaling discussed above, $1,25(OH)_2D_3$ can independently activate the apoptotic program. This concept is further supported by studies demonstrating that $1,25(OH)_2D_3$ can inhibit proliferation and/or induce apoptosis in transformed cells derived from a variety of tumor types, such as osteosarcoma, colon, and leukemia, which are not estrogen dependent (Tsuchiya *et al.*, 1993; Shabahang *et al.*, 1994; Elstner *et al.*, 1997).

Although the specific target genes regulated by the $1,25(OH)_2D_3$–VDR complex which induce growth arrest and apoptosis in breast cancer cells have yet to be conclusively identified, one candidate is the cell cycle inhibitor, p21. The promoter of the human p21 gene contains a functional vitamin D_3 response element which has been linked to the antiproliferative effects of $1,25(OH)_2D_3$ in U937 leukemic cells (Liu *et al.*, 1996). We have observed that $1,25(OH)_2D_3$ induces p21 protein expression in parental MCF-7 cells, but not in the vitamin D_3-resistant variant (Narvaez and Welsh, 1997). Although upregulation of p21 has been associated with cell cycle arrest, further studies are necessary to determine whether induction of p21 is sufficient and/or necessary for MCF-7 cell apoptosis. Our own studies indicate that the phorbol ester tetradecanoylphorbol acetate (TPA) can upregulate p21 and induce growth arrest without triggering apoptosis in

MCF-7 cells (Narvaez and Welsh, 1997), supporting the concept that $1,25(OH)_2D_3$ induces additional signals which participate in activation of apoptosis. Other proteins linked to apoptosis which are modulated in breast cancer cells by $1,25(OH)_2D_3$ include bcl-2, p53, retinoblastoma, clusterin, and cathepsin B (Welsh *et al.*, 1994; Elstner *et al.*, 1995; James *et al.*, 1996; Simboli-Campbell *et al.*, 1996, 1997). However, further studies are required to determine whether the genes for any of these are transcriptionally regulated by $1,25(OH)_2D_3$ in breast cancer cells, and if so, whether they contribute to the regulation of proliferation or apoptosis by vitamin D_3 in these cells.

The similarity in the characteristics of apoptosis in response to $1,25(OH)_2D_3$ and antiestrogens (Welsh *et al.*, 1995) suggests that these two nuclear signaling pathways may converge downstream of receptor activation, ultimately activating the same or similar apoptotic mediators. Comparison of the specific mediators involved in triggering and/or completing the apoptotic process in response to $1,25(OH)_2D_3$ and antiestrogens, such as the caspases, members of the bcl-2 family, or changes in intracellular calcium will be necessary in order to clarify how and where these two pathways may overlap in the regulation of growth arrest and apoptosis of breast cancer cells.

4. SYNTHETIC VITAMIN D_3 ANALOGS

4.1. Development, Description, and Characteristics

From a clinical standpoint, the data reviewed above emphasize the potential of vitamin D_3-based therapeutics for induction of apoptosis in late-stage, estrogen-independent breast cancers. Although studies have demonstrated that $1,25(OH)_2D_3$ can significantly retard the growth of chemically induced rat mammary tumors *in vivo* (Colston *et al.*, 1992a; Saez *et al.*, 1993), its therapeutic use is limited by its inherent calcemic activity. A wide variety of synthetic vitamin D_3 analogs have recently been developed by several pharmaceutical firms, with the intent of identifying vitamin D_3-based therapeutics possessing enhanced growth-modulatory properties and minimal toxicity (Binderup *et al.*, 1991; Bouillon *et al.*, 1995). EB1089 (LEO Pharmaceuticals, Ballerup, Denmark), 22-oxacalcitriol (Chugai Pharmaceutical Co., Tokyo, Japan), and Ro24-5531 (1,25-dihydroxy-16-ene-23-yne-26,27-hexafluorocholecalciferol; Hoffmann LaRoche, Inc, Nutley, NJ) are three representatives of this novel class of vitamin D_3 compounds which have been extensively studied. All three of these compounds are more potent inhibitors of cell proliferation *in vitro* than

$1,25(OH)_2D_3$, but display less pronounced effects on bone resorption and intestinal calcium absorption *in vivo* (Abe *et al.*, 1991; Colston *et al.*, 1992b; Anzano *et al.*, 1994). Like $1,25(OH)_2D_3$, EB1089 induces cell cycle arrest and apoptosis in ER+ MCF-7 cells (James *et al.*, 1996; Simboli-Campbell *et al.*, 1997), but at doses up to 100-fold lower. The dose response of two LEO vitamin D_3 analogs, EB1089 and CB1093, compared to that of $1,25(OH)_2D_3$ on viable cell number in MCF-7 cells is shown in Fig. 6. As indicated, 72 hr treatment with EB1089 significantly reduces viable cell number at doses as low as 1 nM, and CB1093 is effective at doses as low as 0.1 nM. In contrast, significant effects of $1,25(OH)_2D_3$ are detected at doses of 10 nM or higher. In the ER− SUM159 cell line, equivalent doses of EB1089 and CB1093 have a greater effect on viable cell number than does $1,25(OH)_2D_3$ (Fig. 3). These studies indicate that vitamin D_3 analogs which lack potent calcemic action retain the ability to induce growth arrest and apoptosis in both ER+ and ER− breast cancer cells.

The mechanism by which synthetic vitamin D_3 analogs preferentially modulate cellular proliferation as opposed to calcium homeostasis is poorly understood. Factors which may impinge on the downstream efficacy of

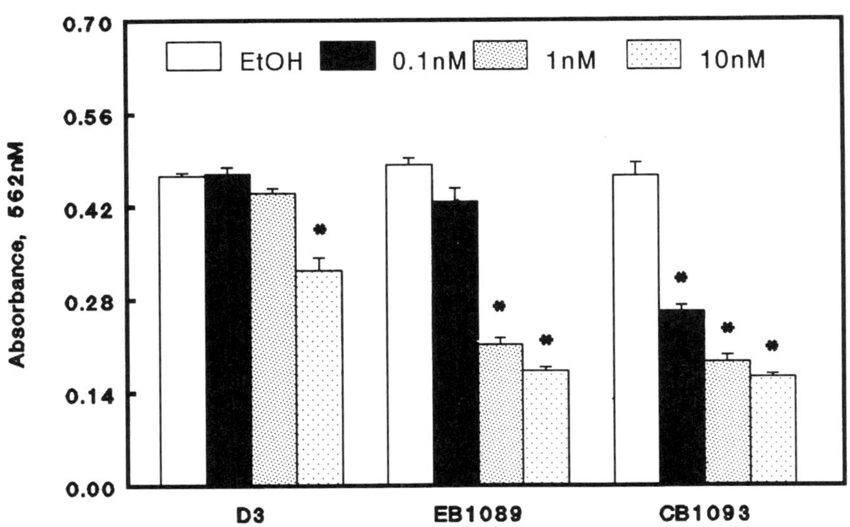

FIGURE 6. Comparative efficacy of vitamin D_3 compounds on growth inhibition of MCF-7 cells. MCF-7 cells were treated for 72 hr with ethanol vehicle (EtOH) or increasing doses of $1,25(OH)_2D_3$ (D3) or the synthetic analogs EB1089 or CB1093 (LEO Pharmaceuticals, Ballerup, Denmark). Cell number was quantited by crystal violet staining and data represent mean ± SEM of four determinations. Asterisk indicates statistically signficant difference, ethanol vs. $1,25(OH)_2D_3$, EB1089, or CB1093.

vitamin D_3 analogs include binding to transport proteins, affinity for the VDR, metabolism of analogs to products with more or less biological activity, half-life of analogs, and differential structure/function of the VDR when bound to distinct ligands (Dusso *et al.*, 1991; Bouillon *et al.*, 1995). Recent *in vitro* studies have demonstrated that, compared to $1,25(OH)_2D_3$, certain vitamin D_3 analogs can differentially alter the protease sensitivity and DNA-binding capability of the VDR (Peleg *et al.*, 1995) and exhibit promoter selectivity in transactivation assays when tested against structurally distinct types of vitamin D_3 response elements (Nayeri *et al.*, 1995; Mork-Hansen *et al.*, 1996). These data support the concept that synthetic analogs of $1,25(OH)_2D_3$ can selectively target distinct biological activities of vitamin D_3 such as growth control, immunoregulation, or calcium homeostasis, which would be of enormous therapeutic value with respect to cancer, autoimmune disease, or osteoporosis.

4.2. Preclinical *in Vivo* Trials of Vitamin D_3 Analogs in Breast Cancer

4.2.1. Effects of Vitamin D_3 Analogs on Estrogen-Dependent Mammary Tumors

Several studies have examined the efficacy of synthetic vitamin D_3 analogs to modulate the development and/or growth kinetics of estrogen-dependent mammary tumors. The first demonstration of a beneficial effect of a vitamin D_3 analog on breast tumor growth was reported by Colston *et al.* (1989). These authors demonstrated that $1\alpha(OH)D_3$, a synthetic analog which is rapidly converted to $1,25(OH)_2D_3$ *in vivo*, suppressed the growth of established *N*-nitrose-*N*-methylurea (NMU)-induced rat mammary tumors when administered intraperitoneally three times weekly at a dose of 0.1 μg per rat. The effective dose range for $1\alpha(OH)D_3$, however, was narrow, with doses of 0.25 μg per day inducing unacceptable hypercalcemia, even when rats were maintained on a low-calcium diet. In a similar study, Iino *et al.* (1992) examined the effects of the same analog, $1\alpha(OH)D_3$, given orally at doses of 0.5–1.0 μg/kg body weight daily, on 7,12-dimethylbenz(a)anthracence (DMBA)-induced rat mammary tumors. Although a significant dose-dependent reduction in tumor growth was observed in $1\alpha(OH)D_3$-treated rats, the doses used induced significant hypercalcemia and weight loss. Reducing the dosing schedule to alternate days alleviated these side effects somewhat, but not entirely.

In collaboration with LEO Pharmaceuticals (Ballerup, Denmark), Colston *et al.* (1992b, 1994, 1995) demonstrated that the vitamin D_3 analog EB1089, administered orally at doses up to 1 μg/kg body weight daily, inhibited the growth of established NMU—induced rat mammary tumors without induction of hypercalcemia. Significantly, these studies were the first to identify a vitamin D_3 analog that not only inhibited growth of tumors, but induced tumor regression. With respect to tumor regression, EB1089 was as effective as daily administration of the antiestrogen tamoxifen (given orally at 1 mg/kg body weight). In follow-up studies, Colston *et al.* (1994, 1995) made the critical observation that some vitamin D_3 analogs which display very potent antiproliferative effects against MCF-7 cells *in vitro* do not exert antitumor effects *in vivo*. These results emphasize the importance of *in vivo* tumor models for testing the therapeutic potential of vitamin D_3 analogs.

Abe-Hashimoto *et al.* (1993) demonstrated that oral administration of 22-oxacalcitriol inhibited the growth of MCF-7 tumor fragments transplanted into nude mice when given beginning 3 days after transplantation. At the effective tumor suppressive doses (0.001–1.0 μg/kg body weight, given five times weekly), 22-oxacalcitriol did not induce hypercalcemia. In these experiments, a synergistic tumor-suppressive effect of 22-oxacalcitriol and the antiestrogen tamoxifen (administered orally at 0.01 mg/kg body weight) was reported. Time course data for tumor growth were not reported, so it is not clear if this vitamin D_3 analog induced tumor regression. Due to the nature of the experimental design, the relative effects of 22-oxacalcitriol on tumor take compared to tumor growth kinetics are hard to determine.

Recently, the ability of the Roche vitamin D_3 analog Ro24-5531 to prevent breast cancer was examined by Anzano *et al.* (1994). One week after injection of NMU, rats were fed control diets or diets containing Ro24-5531 at levels of 1.25 or 2.5 nmol/kg diet, and tumor burden was examined over a 6-month period. Ro24-5531 significantly reduced the mean tumor burden and percentage of tumor-free mice at both dose levels, and no hypercalcemic side effects were detected. In addition, coadministration of Ro24-5531 with the antiestrogen tamoxifen offered a statistically significant benefit in reduction of mammary carcinogenesis compared to tamoxifen treatment alone.

In our own *in vivo* studies, we have focused on determining whether the mechanism of tumor growth inhibition by vitamin D_3 analogs involves activation of apoptosis (VanWeelden *et al.*, 1998). We have utilized the MCF-7 xenograft model system outlined in Fig. 7. During the trial, mice were maintained on a low-calcium diet to minimize calcium toxicity. In this system, the amount of estrogen is carefully controlled in order to minimize influences of physiological estradiol cycling on growth of the MCF-7 tumors, which are highly estrogen dependent (Kyprianou *et al.*, 1991). Ovariectomized mice bearing continuous-release estradiol pellets (Innovative Research, Sarasota, FL) were inoculated with MCF-7 cells and

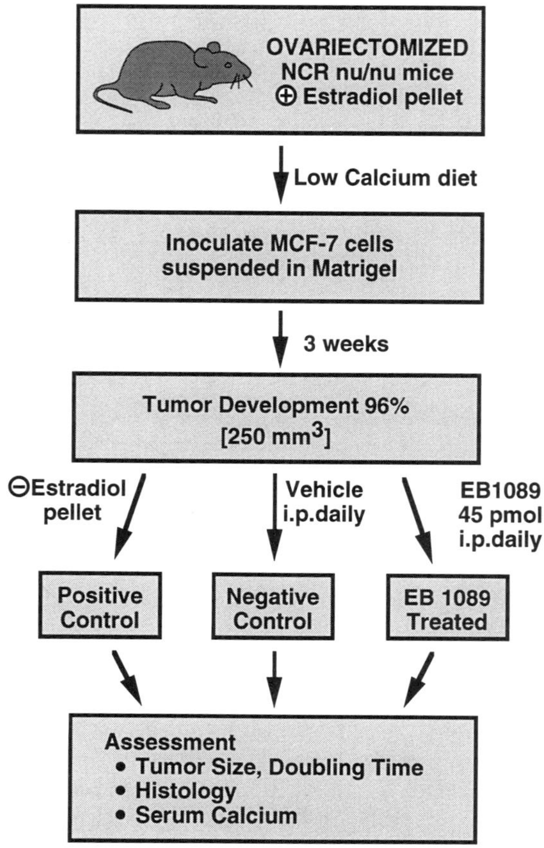

FIGURE 7. Schematic depicting MCF-7 human mammary tumor xenograft system utilized in preclinical studies of vitamin D_3 analogs. See text for details.

tumor growth was monitored by weekly caliper measurements. Cells were allowed to grow for 3 weeks in order to establish rapidly growing tumors with volumes of approximately 250 mm^3. Mice were then treated with either EB1089 (given by daily intraperitoneal injection at doses up to 45 pmol/mouse/day (approximately 0.8 µg/kg body weight) or vehicle injections. At this dose of EB1089, serum calcium was not elevated, and no weight loss was observed. In contrast, doses of 60 pmol/mouse/day or higher were associated with hypercalcemia and weight loss. Compared to control mice, tumors from mice treated daily with 45 pmol of EB1089 grew at a slower rate from week 1 on. Tumor doubling times, calculated between weeks 4 and 5, were sevenfold longer in EB1089-treated mice than in control mice. The mean final weights of tumors from EB1089-treated mice after 5 weeks

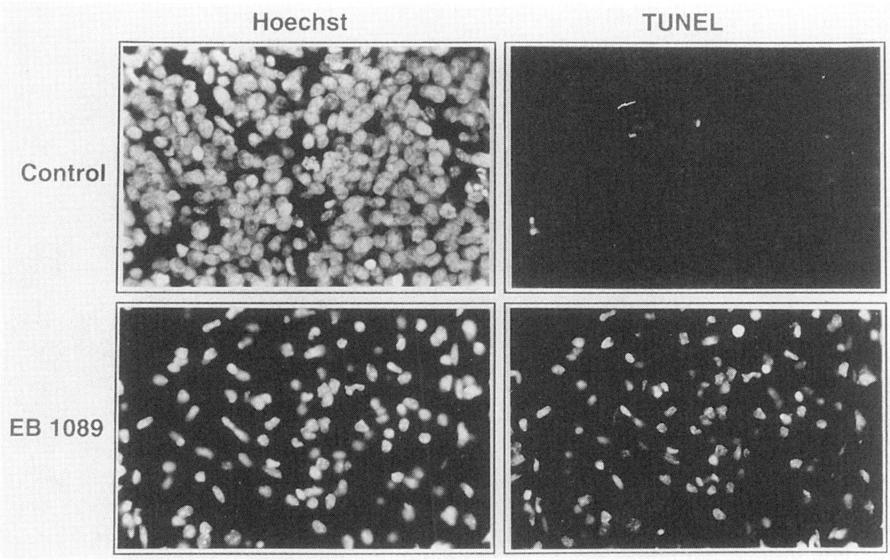

FIGURE 8. Hoechst staining and TUNEL positivity of MCF-7 tumors. Tumors derived from mice treated for 2 weeks with 60 pmol/day EB1089 or vehicle (control) were fixed, embedded in paraffin, sectioned, and processed for Hoechst staining and terminal transferase-mediated fluorescein dUTP nick end labeling of DNA strand breaks (TUNEL) to assess nuclear morphology and DNA fragmentation.

of treatment was less than 30% that of tumors from vehicle-treated mice. Histological analysis indicated that tumors from EB1089-treated mice exhibited large areas of apoptotic regression (identified by terminal transferase-mediated fluorescein dUTP nick end labeling of DNA strand breaks), which were infrequent in tumors from control mice (Fig. 8). In the positive control group, tumors regressed rapidly upon removal of the estradiol pellet, confirming the estrogen dependence of the MCF-7 xenografts. Tumors from the positive control group exhibited the same histological evidence of apoptosis as tumors from EB1089-treated mice. Thus, our *in vivo* results with EB1089 correlate well with the *in vitro* reports that vitamin D_3 compounds induce growth arrest and apoptosis in ER+ breast cancer cells.

4.2.2. Effects of Vitamin D_3 Analogs on Estrogen-Independent Mammary Tumors

Only one group has reported the effects of a vitamin D_3 analog in an *in vivo* model of estrogen-independent breast cancer. Abe *et al.* (1991) demonstrated that 22-oxacalcitriol significantly inhibited the growth of the estrogen-independent MX-1 human breast tumor. In this study, 22-oxacalcitriol was injected directly into developing tumors beginning 4 days after transplantation of the MX-1 tumor fragments. At the most effective dose (1 μg/kg body weight), 22-oxacalcitriol reduced tumor weight by 70% after 26 days without inducing hypercalcemia. In these experiments, 22-oxacalcitriol reduced tumor growth rate, but no obvious tumor regression was observed. In a follow-up study (Abe-Hashimoto *et al.*, 1993), oral administration of 22-oxacalcitriol three times weekly (at doses up to 1 μg/kg body weight) significantly reduced the growth rate of MX-1 tumors. These authors also reported that the effect of 22-oxacalcitriol on tumor growth was equivalent to that of the chemotherapeutic drug adriamycin, and that additive effects on tumor growth were observed when both agents were administered simultaneously. Because of the short time between tumor transplantation and initiation of 22-oxacalcitriol treatment, it is difficult to distinguish between effects of the vitamin D_3 analog on initial tumor take and on tumor growth kinetics.

In recent studies, we have examined the effects of EB1089 on established estrogen-independent SUM159 tumors growing in ovariectomized nude mice. The system used for SUM159 tumors was similar to those described in Section 4.2.1 for MCF-7 xenografts (Fig. 7), except that mice were not supplemented with estradiol pellets. In these experiments, EB1089 was administered via subcutaneously implanted, continuous-release pellets (custom made by Innovative Research, Sarasota, FL) at doses of 60–120 pmol/day. An 80–85% reduction in SUM159 tumor volume was observed in mice treated for 4 weeks with EB1089 compared to control mice bearing placebo pellets. In contrast to daily injections of 60 pmol of EB1089, minimal elevation of serum calcium was observed when EB1089 was delivered continuously via pellets, even at these higher doses. These data indicate that continuous-release administration may minimize the side effects of vitamin D_3 analogs and maximize the induction of tumor regression. Histological analyses of the SUM159 tumors indicated that the decrease in tumor volume after EB1089 treatment was associated with apoptotic regression, as indicated by Hoechst staining and terminal transferase-mediated fluorescein dUTP nick end labeling of DNA strand breaks.

Collectively, the preclinical studies support the concept that some synthetic vitamin D_3 analogs can effectively retard the growth of estrogen-dependent and estrogen-independent breast tumors. However, as detailed by Colston *et al.* (1994, 1995), not all vitamin D_3 analogs which are effective antiproliferative agents for breast cancer cells *in vitro* exert comparable negative growth-regulatory effects on mammary tumors *in vivo*. Comparative analysis of structurally distinct analogs may offer insight into the discrepancies between *in vivo* and *in vitro* studies as well as the apparent dissociation of growth-regulatory and calcemic actions. A more complete understanding of the molecular mechanism of action of vitamin D_3 analogs is critical to the development of second-generation analogs with lower toxicity and enhanced effectiveness in inducing regression of human breast tumors.

5. SUMMARY AND PERSPECTIVES

In summary, the VDR is present in both estrogen-dependent and estrogen-independent breast cancer cell lines, and both ER+ and ER− cells are growth inhibited by $1,25(OH)_2D_3$. Studies have demonstrated that the effects of vitamin D_3 compounds on breast cancer cells, *in vitro* and *in vivo*, reflect both cell cycle arrest and activation of apoptosis. Structure–function analysis of synthetic vitamin D_3 analogs has lead to the development of several novel analogs which display enhanced potency in mediating tumor regression with minimal calcemic side effects.

Preclinical studies have emphasized complementary effects of antiestrogens and vitamin D_3 compounds on growth of estrogen-dependent mammary tumors. Our working model for the interactions between $1,25(OH)_2D_3$ and antiestrogens in the modulation of apoptosis in breast cancer cells is presented in Fig. 9. This model emphasizes that vitamin D_3 compounds and antiestrogens trigger apoptosis in breast cancer cells by independent mechanisms. In ER+ cells, we propose that $1,25(OH)_2D_3$, through the VDR, induces a specific pattern of gene expression which induces growth arrest and facilitates apoptosis. In such estrogen-dependent cells, estrogen normally suppresses apoptosis, and disruption of estrogen signaling by treatment with antiestrogens negates the protective effect of estrogen. $1,25(OH)_2D_3$-mediated downregulation of the ER would similarly disrupt estrogen signaling, contributing to its negative growth-regulatory effect. In ER− cells, the estrogen signaling pathway is disabled, allowing for escape from estrogen dependence and resistance to antiestrogen-mediated apoptosis. Although in ER− cells, factors other than estrogens presumably repress apoptosis, the presence of VDRs renders them sensitive to activation of apoptosis in response to $1,25(OH)_2D_3$. This working model suggests that, for patients with mixed tumors containing estrogen-dependent and estrogen-independent cells, a distinct therapeutic advantage may be achieved by combining therapies that activate vitamin D_3-dependent pathways with those that disrupt estrogen-dependent signaling. Further, we have shown that breast cancer cell sensitivity to $1,25(OH)_2D_3$-mediated apoptosis is not diminished during the progression to estrogen

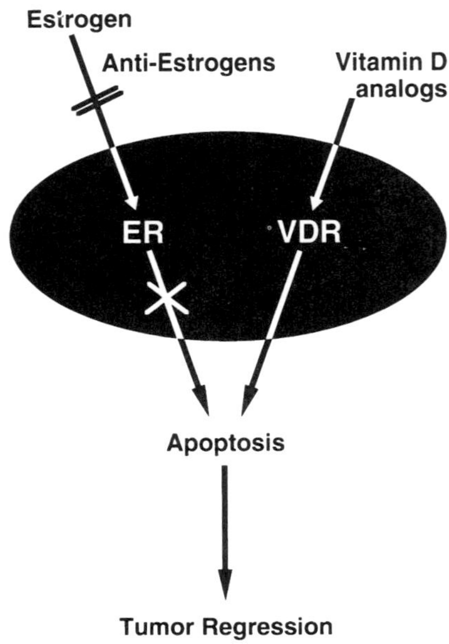

FIGURE 9. Working model for interactions between antiestrogens and vitamin D_3 in breast cancer cells. In cells containing both ER and VDR, estrogen stimulates proliferation and suppresses apoptosis, while vitamin D_3 compounds induce growth arrest and apoptosis. In estrogen-dependent mammary cells, antiestrogens disrupt estrogen signaling, negating the suppressive effect of estrogen on apoptosis. In estrogen-independent cells, loss of functional ER signaling renders cells resistant to antiestrogens. Estrogen-independent cells which express functional VDR retain sensitivity to vitamin D_3-mediated growth arrest and apoptosis. See text for further details.

independence, indicating that vitamin D_3 analogs may offer an effective therapeutic approach for aggressive, late-stage tumors which exhibit resistance to the standard antiestrogen regimens.

ACKNOWLEDGMENTS. Work in the author's laboratory is funded by operating grants from the NIH (#CA69700, CA69233) and the American Institute for Cancer Research (#95B068). I would like to thank Dr. Martin Tenniswood (W. Alton Jones Cell Science Center, Lake Placid, New York) for his integral contributions to the ideas presented in the paper, and Dr. Steve Ethier (University of Michigan, Ann Arbor, Michigan) for the SUM159 cell line.

6. REFERENCES

Abe, J., Nakano, T., Nishii, Y., Matsumoto, T., Ogata, E., and Ikeda, K., 1991, A novel vitamin D_3 analog, 22-oxa-1,25 dihydroxyvitamin D_3, inhibits the growth of human breast cancer *in vitro* and *in vivo* without causing hypercalcemia, *Endocrinology* **129:**832–837.

Abe-Hashimoto, J., Kikuchi, T., Matsumoto, T., Nishii, Y., Ogata, E., and Ikeda, K., 1993, Antitumor effect of 22-oxa-calcitriol, a noncalcemic analogue of calcitriol, in athymic mice implanted with human breast carcinoma and its synergism with tamoxifen, *Cancer Res.* **53:**2534–2537.

Anzano, M. A., Smith, J. M., Uskokovic, M. R., Peer, C. W., Mullen, L. T., Letterio, J. J., Welsh, M. C., Shrader, M. W., Logsdon, D. L., Driver, C. L., Brown, C. C., Roberts, A., and Sporn, M. B., 1994, 1α,25-Dihydroxy-16-ene-23-yne-26,27-hexafluorocholecalciferol (Ro24-5531), a new deltanoid (vitamin D analogue) for prevention of breast cancer in the rat, *Cancer Res.* **54:**1653–1656.

Barry, M. A., Behnke, C. A., and Eastman, A., 1990, Activation of programmed cell death (apoptosis) by cisplatin, other anticancer drugs, toxins and hyperthermia, *Biochem. Pharmacol.* **40:**2353–2362.

Battcharjee, M., Weintroub, S., and Vonderhaar, B. K., 1987, Milk protein synthesis of vitamin D deficient mice, *Endocrinology* **121:**865–874.

Berger, U., McClelland, R., Wilson, P., Greene, G., Haussler, M., Pike, J., Colston, K., Easton, D., and Coombes, R. C., 1991, Immunoctyochemical determination of estrogen receptor, progesterone receptor and $1,25(OH)_2$-vitamin D_3 receptor in breast cancer and relationship to prognosis, *Cancer Res.* **51:**239–244.

Binderup, L., Latini, S., Binderup, E., Bretting, C., Caverley, M., and Hansen, K., 1991, 20-Epi-vitamin D_3 analogues: A novel class of potent regulators of cell growth and immune responses, *Biochem. Pharmacol.* **42:**1569–1575.

Bouillon, R., Okamura, W. H., and Norman, A. W., 1995, Structure–function relationships in the vitamin D endocrine system, *Endocrine Rev.* **16:**200–257.

Buras, R., Schumaker, L. M., Davoodi, F., Brenner, R. V., Shabahang, M., Nauta, R., and Evans, S., 1994, Vitamin D receptors in breast cancer cells, *Breast Cancer Res. Treat.* **31:**191–204.

Carlberg, C., Bendik, I., Wyss, A., Meier, E., Sturzenbecker, L. J., Grippo, J., and Hunziker, W., 1993, Two nuclear signaling pathways for vitamin D, *Nature* **361:**657–660.

Chouvet, C., Berger, U., and Coombes, R. C., 1986, 1,25 Dihydroxyvitamin D_3 inhibitory effect on the growth of two human breast cancer cell lines (MCF-7, BT-20), *J. Steroid Biochem.* **24:**373–376.

Colston, K., Colston, M. J., and Feldman, D., 1981, 1α25-Dihydroxyvitamin D_3 inhibits the clonogenic growth of transformed cells via its receptor, *Endocrinology* **108:**1083–1086.

Colston, K., Berger, U., and Coombes, R. C., 1989, Possible role for vitamin D in controlling breast cancer cell proliferation, *Lancet* **1:**188–191.

Colston, K., Chander, S. K., Mackay, A. G., and Coombes, R. C., 1992a, Effects of synthetic vitamin D analogs on breast cancer cell proliferation *in vivo* and *in vitro*, *Biochem. Pharmacol.* **44:**693–702.

Colston, K., Mackay, A. G., James, S. Y., and Binderup, L., 1992b, EB1089: A new vitamin D_3 analogue that inhibits the growth of breast cancer cells *in vivo* and *in vitro*, *Biochem. Pharmacol.* **44:**2273–2280.

Colston, K., Mackay, A. G., and James, S. Y., 1994, The role of $1\alpha25(OH)_2D_3$ and its analogs in breast cancer, in *Proceedings of the Ninth Workshop on Vitamin D* (A. W. Norman, R. Bouillon, and M. Thomasett, eds.), pp. 477–484, DeGruyter, Berlin.

Colston, K., Mackay, A. G., and James, S. Y., 1995, Vitamin D_3 derivatives and breast cancer, in *Apoptosis in Hormone Dependent Cancers* (M. Tenniswood and H. Michna, eds.), pp. 201–224, Springer-Verlag, Berlin.

Craig, R. W., 1995, The BCL-2 gene family, *Semin. Cancer Biol.* **6:**35–43.

Darwish, H. M., and DeLuca, H. F., 1996, Recent advances in the molecular biology of vitamin D action, *Prog. Nucleic Acids Res. Mol. Biol.* **53:**321–344.

Dusso, A. S., Negrea, L., Gunawardhana, S., Lopez-Hilker, S., Finch, J., Mori, T., Nishii, Y., Slatopolsky, E., and Brown, A. J., 1991, On the mechanisms for the selective action of vitamin D analogs, *Endocrinology* **128:**1687–1692.

Eisman, J., Suva, L., Sher, E., Pearce, P., Funder, J., and Martin, T. J., 1981, Frequency of 1,25-dihydroxyvitamin D_3 receptor in human breast cancer, *Cancer Res.* **41:**5121–5124.

Eisman, J., Sutherland, R. L., McMenemy, M. L., Fragonas, J. C., Musgrove, E. A., and Pang, G., 1989, Effects of 1,25-dihydroxyvitamin D_3 on cell cycle kinetics of T47D human breast cancer cells, *J. Cell. Physiol.* **138:**611–616.

Elstner, E., Linker-Israeli, M., Said, J., Umiel, T., deVos, S., Shintaku, I. P., Heber, D., Binderup, L., Uskokovic, M., and Koeffler, H. P., 1995, 20-epi Vitamin D_3 analogues: A novel class of potent inhibitors of proliferation and inducers of differentiation of human breast cancer cells, *Cancer Res.* **55:**2822–2830.

Elstner, E., Linker-Israeli, M., Le, J., Umiel, T., Michl, P., Said, J., Binderup, L., Reed, J. C., and Koeffler, H. P., 1997, Synergistic decrease of clonal proliferation, induction of differentiation and apoptosis of acute promyelocytic leukemia cells after combined treatment with novel 20-epi vitamin D_3 analogues and 9-*cis* retinoic acid, *J. Clin. Invest.* **99:**349–260.

Fesus, L., Szondy, Z., and Uray, I., 1995, Probing the molecular program of apoptosis by cancer chemopreventive agents, *J. Cell. Biochem.* **S22:**151–161.

Goberdhan, N., Dive, C., and Streuli, C. H., 1996, The role of the extracellular matrix and lactogenic hormones in mammary gland apoptosis, *Biochem. Soc. Trans.* **24:**348S.

Guenette, R. S., Corbeil, H., Leger, J., Wong, K., Mezl, V., Mooibroek, M., and Tenniswood, M., 1994, Induction of gene expression during involution of the lactating mammary gland of the rat, *J. Mol. Endocrinol.* **12:**47–60.

Haussler, M. R., Jurukta, P. W., Hsieh, J. C., Thompson, P. D., Selznick, S. H., Haussler, C. A., and Whitfield, G. K., 1995, New understanding of the molecular mechanism of receptor mediated genomic actions of the vitamin D hormone, *Bone* **17**(2Suppl.)**:**33S–38S.

Iino, Y., Yoshida, M., Sugamata, N., Maemura, M., Ohwada, S., Yokoe, T., Ishikita, T., Ooriuchi, R., and Morishita, Y., 1992, 1α-Hydroxyvitamin D_3, hypercalcemia, and growth suppression of 7,12-dimethylbenz[a]anthracene-induced rat mammary tumors, *Breast Cancer Res. Treat.* **22:**133–140.

James, S. Y., Mackay, A., Binderup, L., and Colston, K., 1994, Effects of a new synthetic vitamin D analogue, EB1089, on the oestrogen responsive growth of human breast cancer cells, *J. Endocrinol.* **141:**555–563.

James, S. Y., Mackay, A. G., and Colston, K., 1996, Effects of 1,25 dihydroxyvitamin D_3 and its analogues on induction of apoptosis in breast cancer cells, *J. Steroid Biochem. Mol. Biol.* **58:**395–401.

Kyprianou, N., English, H., Davidson, N., and Isaacs, J., 1991, Programmed cell death during regression of the MCF-7 human breast cancer following estrogen ablation, *Cancer Res.* **51:**162–166.

Liu, M., Lee, M.-H., Cohen, M., Bommakanti, M., and Freedman, L., 1996, Transcriptional activation of the Cdk inhibitor p21 by vitamin D_3 leads to the induced differentiation of the myelomonocytic cell line U937, *Genes Dev.* **10:**142–153.

Love-Schimenti, C., Gibson, D., Ratnam, A., and Bikle, D., 1996, Antiestrogen potentiation of antiproliferative effects of vitamin D_3 analogues in breast cancer cells, *Cancer Res.* **56:**2789–2794.

Lowe, S., Ruley, H. E., Jacks, T., and Housman, D., 1993, p53-Dependent apoptosis modulates the cytotoxicity of anticancer agents, *Cell* **74:**957–967.

Merio, G. R., Cella, N., and Hynes, N. E., 1997, Apoptosis is accompanied by changes in bcl-2 and bax expression, induced by loss of attachment, and inhibited by specific extracellular matrix proteins in mammary epithelial cells, *Cell Growth Diff.* **8:**251–260.

Mezettie, G., Monti, M. G., Casolo, P., Piccinini, G., and Moruzzi, M. S., 1988, 1,25-Dihydroxycholecalciferol dependent calcium uptake by mouse mammary gland in culture, *Endocrinology* **122:**389–394.

Mork-Hansen, C., Danielsson, C., and Carlberg, C., 1996, The potent anti-proliferative effect of 20-epi analogues of 1,25 dihydroxyvitamin D_3 in human breast cancer MCF-7 cells is related to promoter specificity, *Int. J. Cancer* **67:**739–742.

Narvaez, C. J. and Welsh, J. E., 1997, Differential effects of 1,25 Dihydroxyvitamin D_3 and TPA on cell cycle and apoptosis of MCF-7 cells and a vitamin D_3 resistant variant, *Endocrinology* **138:**4690–4698.

Narvaez, C. J., VanWeelden, K., Byrne, I., and Welsh, J. E., 1996, Characterization of a Vitamin D_3 resistant MCF-7 cell line, *Endocrinology* **137:**400–409.

Nayeri, S., Danielsson, C., Kahlen, J. P., Schräder, M., Mathiasen, I., Binderup, L., and Carlberg, C., 1995, The anti-proliferative effect of vitamin D_3 analogues is not related to the AP-1 pathway, but related to promoter selectivity, *Oncogene* **11:**1853–1858.

Nolan, E., Donepudi, M., Van Weelden, K., Flanagan, L., and Welsh, J. E., 1998, Dissociation of vitamin D_3 and anti-estrogen mediated growth regulation in MCF-7 breast cancer cells, *Mol. Cell. Biochem.* in press.

Peleg, S., Sastry, M., Collins, E. D., Bishop, J. E., and Norman, A. W., 1995, Distinct conformational changes induced by 20-epi analogues of 1α,25-dihydroxyvitamin D_3 are associated with enhanced activation of the vitamin D receptor, *J. Biolo. Chem.* **270:**10551–10558.

Pols, H., Birkenhager, J. C., Foekens, J., and van Leeuwen, J., 1990, Vitamin D: A modulator of cell proliferation and differentiation, *J. Steroid Biochem. Mol. Biol.* **37:**873–876.

Quarrie, L. H., Addey, C. V., and Wilde, C. J., 1996, Programmed cell death during mammary tissue involution induced by weaning, litter removal and milk stasis, *J. Cell. Physio.* **168:**559–569.

Reichel, H., Koeffler, H. P., and Norman, A. W., 1989, The role of the vitamin D endocrine system in health and disease, *N. Eng. J. Med.* **320:**980–991.

Saez, S., Falette, N., Guillot, C., Meggouh, F., Lefebvre, M., and Crepin, M., 1993, $1,25(OH)_2D_3$ modulation of mammary tumor growth *in vitro* and *in vivo*, *Breast Cancer Res. Treat.* **27:**69–81.

Shabahang, M., Buras, R., Davoodi, F., Schumaker, L., Nauta, R., Uskokovic, M., Brenner, R., and Evans, S., 1994, Growth inhibition of HT-29 human colon cancer cells by analogues of 1,25 dihydroxyvitamin D_3, *Cancer Res.* **54:**4057–4064.

Simboli-Campbell, M., and Welsh, J., 1995, 1,25 Dihydroxyvitamin D_3: Coordinate regulator of active cell death and proliferation in MCF-7 breast cancer cells, in *Apoptosis in Hormone Dependent Cancers* (M. Tenniswood and H. Michna, eds.), pp. 181-200, Springer-Verlag, Berlin.

Simboli-Campbell, M., Narvaez, C. J., Tenniswood, M., and Welsh, J. E., 1996, $1,25(OH)_2D_3$ induces morphological and biochemical indices of apoptosis in MCF-7 breast cancer cells. *J. Steroid Biochem. Mol. Biol.* **58:**367–376.

Simboli-Campbell, M., Narvaez, C. J., VanWeelden, K., Tenniswood, M., and Welsh, J. E., 1997, Comparative effects of $1{,}25(OH)_2D_3$ and EB1089 on cell cycle kinetics and apoptosis in MCF-7 cells, *Breast Cancer Res. Treat.* **42:**31–41.

Studzinski, G., McLane, J. A., and Uskokovic, M., 1993, Signaling pathways for vitamin D induced differentiation: Implications for therapy of proliferative and neoplastic diseases, *Crit. Rev. Eukaryotic Gene Expression* **3:**279–312.

Tenniswood, M., Guenette, R. S., Lakins, J., Mooibroek, M., Wong, P., and Welsh, J. E., 1992, Active cell death in hormone dependent tissues, *Cancer Metastasis Rev.* **11:**197–220.

Tsuchiya, H., Morishita, H., Tomita, K., Ueda, Y., and Tanaka, M., 1993, Differentiating and antitumor activities of 1α,25 dihydroxyvitamin D_3 *in vitro* and 1α-hydroxyvitamin D_3 *in vivo* on human osteosarcoma, *J. Orthopaedic Res.* **11:**122–130.

Vandewalle, B., Hornez, L., Wattez, N., Revillion, F., and Lefebvre, J., 1995, Vitamin D_3 derivatives and breast tumor cell growth: effect on intracellular calcium and apoptosis, *Int. J. Cancer* **61:**806–811.

VanWeelden, K., Flanagan, L., Binderup, L., Tenniswood, M. P., and Welsh, J. E., 1998, Apoptotic regression of MCF-7 xenografts in nude mice treated with the vitamin D_3 analog, EB 1089, *Endocrinology* **139:**2102–2110.

Vaux, D., and Strasser, A., 1996, The molecular biology of apoptosis, *Proc. Natl. Acad. Sci. USA* **93:**2239–2244.

Vink-van Wijngaarden, T., Pols, H., Buurman, C. J., van den Bemd, G., Dorssers, L., Birkenhager, J., and van Leeuwen, J., 1994, Inhibition of breast cancer cell growth by combined treatment with vitamin D_3 analogues and tamoxifen, *Cancer Res.* **54:**5711–5717.

Warri, A. M., Huovinen, R. L., Laine, A. M., Martikainen, P. M., and Harkonen, P. L., 1993, Apoptosis in toremifene-induced growth inhibition of human breast cancer cells *in vivo* and *in vitro*, *J. Natl. Cancer Ins.* **85:**1412–1418.

Welsh, J. E., 1994, Induction of apoptosis in breast cancer cells in response to vitamin D and antiestrogens, *Biochem. Cell Biol.* **72:**537–545.

Welsh, J., Simboli-Campbell, M., and Tenniswood, M., 1994, Induction of apoptotic cell death by $1{,}25(OH)_2D_3$ in MCF-7 breast cancer cells. in *Proceedings of the Ninth Workshop on Vitamin D* (A. W. Norman, R. Bouillon, and M. Thomasett, eds.), pp. 526–527, DeGruyter, Berlin.

Welsh J. E., Simboli-Campbell, M., Narvaez C. J., and Tenniswood M., 1995, Role of apoptosis in the growth inhibitory effects of vitamin D in MCF-7 cells, *Adv. Exp. Biol. Med.* **375:**45–52.

Yang, E., and Korsmeyer, S., 1996, Molecular thanatopsis: A discourse on the BCL2 family and cell death, *Blood* **88:**386–401.

Zhivotovsky, B., Burgess, D., Vanags, D., and Orrenius, S., 1997, Involvement of cellular proteolytic machinery in apoptosis, *Biochem. Biophys. Res. Commun.* **230:**481–488.

Chapter 12

Vitamin D and Intracellular Calcium

Igor N. Sergeev, William B. Rhoten, and
Vladimir B. Spirichev

1. INTRODUCTION

The role of the hormonally active form of vitamin D, 1α,25-dihydroxyvitamin D_3 [$1,25(OH)_2D_3$], in the regulation of Ca^{2+} metabolism is well established (Norman, 1979; Spirichev and Sergeev, 1988). Current concepts of the vitamin D- and intracellular Ca^{2+}-mediated signaling are based on steroid and second messenger dogmas. Attempts have been made to integrate both of these concepts. In particular, the hypothesis of membrane steroid receptors continues to be a promising area of research (Norman, 1995; Szego, 1994). Understanding the relationship between the vitamin D- and intracellular Ca^{2+}-mediated signal transduction pathways is especially important due to the crucial role of $1,25(OH)_2D_3$ in maintaining Ca^{2+} homeostasis.

The goal of this chapter is to review the effects of $1,25(OH)_2D_3$ on Ca^{2+} metabolism at the cellular and subcellular levels and attempt to present the concept of interactive vitamin D–intracellular Ca^{2+} signaling. Such a review

Igor N. Sergeev and William B. Rhoten Department of Anatomy, Cell and Neurobiology, Marshall University School of Medicine, Huntington, West Virginia 25704-9388.
Vladimir B. Spirichev Institute of Nutrition, Russian Academy of Medical Sciences, Moscow 109240, Russia.

Subcellular Biochemistry, Volume 30: Fat-Soluble Vitamins, edited by Quinn and Kagan. Plenum Press, New York, 1998.

of the literature seems to be a timely undertaking because of the current significant interest in this area of research and its potential importance for discovery of new drugs for treatment of a variety of diseases related to Ca^{2+} metabolism and its regulation.

2. VITAMIN D-MEDIATED SIGNAL TRANSDUCTION PATHWAYS

Vitamin D_3 is a precursor of the seco-steroid hormone $1{,}25(OH)_2D_3$ (calcitriol; other terms sometimes used in the literature are the vitamin D hormone or hormone D). $1{,}25(OH)_2D_3$ is traditionally considered as the principal Ca^{2+} regulatory hormone, which exerts its physiological effects by increasing intestinal Ca^{2+} absorption, bone Ca^{2+} mobilization, and renal Ca^{2+} reabsorption (Spirichev and Sergeev, 1988). Generally, $1{,}25(OH)_2D_3$, like some other hormones of the steroid/thyroid/retinoid superfamilly, produces a wide spectrum of biological effects via both receptor-mediated regulation of nuclear events and rapid actions independent of the genomic pathway (Norman, 1995; Sergeev and Rhoten, 1995a; Norman *et al.*, 1992). A substantial body of evidence accumulated over the past 25 years supports the concept of the genomic mechanism of action of $1{,}25(OH)_2D_3$. Conventional nuclear/cytosolic vitamin D receptors (VDRs) have been shown to regulate a number of genes associated with Ca^{2+} homeostasis (such as vitamin D-dependent Ca^{2+}-binding proteins, or calbindins), as well as genes related to proliferation, differentiation, development, and programmed cell death (Norman, 1997, 1995; Hannah and Norman, 1994; Lowe *et al.*, 1992; Sergeev and Norman, 1992; Walters, 1992; Sergeev, 1989). Reports of nongenomic actions of $1{,}25(OH)_2D_3$ are less widespread. Rapid effects of $1{,}25(OH)_2D_3$ on Ca^{2+} transport and activity or translocation of enzymes involved in signal transduction (phospholipases, Ca^{2+}- and mitogen-activated protein kinases) have been described (Norman, 1997; Bouillon *et al.*, 1995; Norman, 1995; Walters, 1995, 1992; Nemere *et al.*, 1993; Nemere and Norman, 1991). The concept of membrane VDRs has emerged in an attempt to explain mechanisms of rapid, nongenomic effects of $1{,}25(OH)_2D_3$ (Nemere *et al.*, 1994; Norman *et al.*, 1993, 1992, 1991; Farach-Carson *et al.*, 1991a).

Significant evidence indicates that actions of $1{,}25(OH)_2D_3$ are mediated or influenced by intracellular Ca^{2+} signaling events. $1{,}25(OH)_2D_3$ rapidly (within seconds to minutes) induces Ca^{2+} uptake, activates Ca^{2+} channels, and stimulates phosphoinositide turnover in many different cell types (Walters, 1995, 1992; Nemere *et al.*, 1993; Norman *et al.*, 1992). Thus, it is appealing to suggest that the activation of intracellular Ca^{2+} signaling pathways is essential for the action of $1,25(OH)_2D_3$. In turn, Ca^{2+} can interact with VDRs via Ca^{2+} binding γ-carboxyglutamate residues of the protein, which results in downregulation of vitamin D mediated biological responses (Sergeev and Norman, 1992; Sergeev and Spirichev, 1989).

This review will present evidence for the effects of $1,25(OH)_2D_3$ on intracellular free Ca^{2+} concentration ($[Ca^{2+}]_i$), Ca^{2+} entry from extracellular space, Ca^{2+} mobilization from intracellular stores, and intracellular Ca^{2+} buffering. Interaction of $1,25(OH)_2D_3$- and intracellular Ca^{2+}-mediated signaling pathways in regulation of cellular functions will be also discussed.

3. CELLULAR Ca^{2+} HOMEOSTASIS AND SIGNALING

To better understand the significance of the effects of $1,25(OH)_2D_3$ on intracellular Ca^{2+}, we present a short summary of cellular Ca^{2+} homeostasis and signaling. The concentration of free Ca^{2+} in the cytosol, $[Ca^{2+}]_i$, is carefully regulated in most cells at the level of 10^{-7} M, whereas the concentration of free Ca^{2+} in the extracellular fluid and compartments of the endoplasmic reticulum (ER) is 10^{-3} M and 10^{-4} M, respectively (Golovina and Blaustein, 1997; Clapham and Sneyd, 1995). The transient influx of Ca^{2+} from the extracellular space or the ER compartments into the cytosol, down a steep concentration gradient, is a means of transmitting signals within the cell. In an activated cell, $[Ca^{2+}]_i$ increases up to 10^{-6} M. The plasma and ER membrane Ca^{2+}-ATPases keep the resting $[Ca^{2+}]_i$ low. Specialized Ca^{2+}-binding proteins (such as cytosolic calbindins) serve as Ca^{2+} buffers (Rhoten and Sergeev, 1994a; Christakos *et al.*, 1989), e.g., during propagation of the Ca^{2+} signal within the cell.

Pathways for Ca^{2+} entry consist of voltage-dependent Ca^{2+} channels (VDCC), typical of electrically active cells, and voltage-insensitive Ca^{2+} channels (VICC), present in both electrically excitable and nonexcitable cells. VDCC of different types have been extensively characterized; VICC have been the focus of research only for the past several years. The three main types of VICC are (1) receptor-operated Ca^{2+} channels, which are tightly coupled to a membrane receptor, (2) depletion-operated Ca^{2+} channels, which are activated by the Ca^{2+} influx factor released from ER when the ER Ca^{2+} stores are depleted (capacitative Ca^{2+} entry), and (3) second messenger-operated Ca^{2+} channels, which are regulated by Ca^{2+} itself (Ca^{2+} release activated Ca^{2+} current), inositol 1,4,5-trisphosphate (IP_3), or other second messengers released following receptor activation (Felder *et al.*, 1994). Pathways of Ca^{2+} mobilization from intracellular stores consist of two types of Ca^{2+} release channels, ligand gated and mechanically activated.

IP_3 and ryanodine receptors are Ca^{2+} release channels located in the ER and sarcoplasmic reticulum membranes; they are activated, correspondently, by IP_3 or Ca^{2+} and cyclic ADP-ribose (Meissner, 1994; Galione, 1994). Released Ca^{2+} causes more of the ryanodine receptor/Ca^{2+} release channels to open (Ca^{2+}-induced Ca^{2+} release).

Calcium ions can mediate two distinct signaling pathways: (1) in electrically active cells, depolarization of plasma membrane causes an influx of Ca^{2+} through VDCC followed by Ca^{2+} release from intracellular stores (through mechanically gated Ca^{2+} release channels/ryanodine receptors), thus initiating the biological response, e.g., hormone secretion; (2) in electrically nonexcitable cells, the binding of extracellular signaling molecules to plasmalemmal receptors causes entry of Ca^{2+} through VICC or release of Ca^{2+} from the ER. In the latter scenario, the events at the cell surface are coupled to the opening of Ca^{2+} channels in the ER membrane through IP_3 and perhaps other signaling molecules.

Specific routes of Ca^{2+} entry and/or Ca^{2+} release from intracellular stores determine both the absolute level and the duration of increased cytosolic Ca^{2+}, thus providing spatial and temporal specificity of Ca^{2+} signaling. Coordinative functioning of Ca^{2+} entry and Ca^{2+} release pathways is responsible for producing and maintaining Ca^{2+} oscillations (Clapham and Sneyd, 1995; Sergeev and Rhoten, 1995a). Amplitude and frequency of Ca^{2+} oscillations often depend on the concentration of extracellular signaling molecules and may be transferred into amplitude- and frequency-dependent cellular responses. The initial Ca^{2+} signal may also propagate as a Ca^{2+} wave through the cytosol from a localized region of the cell. Ca^{2+}-binding proteins associated with the cytoskeletal elements (such as calbindin-D_{28k}, colocalized with tubulin; Sergeev and Rhoten, unpublished observation) may influence propagation of Ca^{2+} waves and the pattern of Ca^{2+} oscillations.

The sections which follow describe the effects of $1{,}25(OH)_2D_3$ on $[Ca^{2+}]_i$ and possible mechanisms of such effects. We will emphasize the role of $1{,}25(OH)_2D_3$ in the intracellular Ca^{2+}-mediated signaling and consider effects of $1{,}25(OH)_2D_3$ on Ca^{2+} transport in this context.

4. EFFECTS OF 1,25-DIHYDROXYVITAMIN D_3 ON INTRACELLULAR Ca^{2+}

The basic observation made on many cell types is that $1{,}25(OH)_2D_3$ increases, usually very rapidly, $[Ca^{2+}]_i$. In most published reports describing this phenomenon, spectrofluorometric measurements of $[Ca^{2+}]_i$ were made in cell suspension with fluorescent Ca^{2+}-sensitive dyes (fura-2, fluo-3, or, in earlier studies, quin-2) (Lieberherr, 1987; Baran, 1994); more recently, digital video Ca^{2+} imaging has been applied to monitor simultaneously changes in $[Ca^{2+}]_i$ in a number of individual cells (Sergeev and Rhoten, 1995a, b).

Lieberherr (1987) and Baran and Milne (1986) first reported effects of $1{,}25(OH)_2D_3$ and vitamin D_3 metabolites on $[Ca^{2+}]_i$. Lieberherr *et al.* (1994) showed that $1{,}25(OH)_2D_3$ acts very rapidly (within seconds) on $[Ca^{2+}]_i$ in osteoblasts and osteoblastlike cells. The increase in $[Ca^{2+}]_i$ was transient and occurred at physiological concentrations of $1{,}25(OH)_2D_3$ (1–10 pM). Mechanisms of such an increase in $[Ca^{2+}]_i$ included Ca^{2+} influx through VDCC and Ca^{2+} mobilization from the ER. $1{,}25(OH)_2D_3$ also increased the formation of IP_3 and diacylglycerol (DAG); importantly, IP_3 levels oscillated with specific (as compared with other steroid hormones) amplitude and frequency (Grosse *et al.*, 1993).

Reports from Baran's laboratory (Baran, 1994; Baran *et al.*, 1991) described a similar rapid increase in $[Ca^{2+}]_i$ in rat osteosarcoma cells (ROS 17/2.8), which was consistent with Ca^{2+} influx through VDCC and Ca^{2+} release from intracellular stores. Importantly, $1{,}25(OH)_2D_3$ increased $[Ca^{2+}]_i$ in rat osteosarcoma cell line ROS 24/1, lacking the conventional VDR. The analogue $1\beta{,}25(OH)_2D_3$, which does not interact with the conventional VDR, blocked the rise in $[Ca^{2+}]_i$. These findings suggest that nuclear/cytosolic VDRs are not required for the rapid effects of $1{,}25(OH)_2D_3$ on $[Ca^{2+}]_i$. Another report from this laboratory (Sorensen *et al.*, 1993) described an interesting observation that $1{,}25(OH)_2D_3$ increased the concentration of free Ca^{2+} in isolated nuclei of ROS 17/2.8 cells.

An article from Hruska's laboratory (Civitelli *et al.*, 1990) described in detail rapid effects of $1{,}25(OH)_2D_3$ on $[Ca^{2+}]_i$ in ROS 17/2.8 cells. An acute rise in $[Ca^{2+}]_i$ was observed in 40% of the cells following the addition of $1{,}25(OH)_2D_3$ with a threshold concentration 10^{-11} M. In most cases, the $[Ca^{2+}]_i$ rise was transient, with return to baseline within 1 min; less frequently, a more prolonged effect was observed. At high concentrations (10^{-8}–10^{-7} M) of $1{,}25(OH)_2D_3$, the $[Ca^{2+}]_i$ rise was due to both influx of extracellular Ca^{2+} through VDCC and release of Ca^{2+} from intracellular stores, as the effect was only partially inhibited by nifedipine. At low concentrations (10^{-9}–10^{-10} M), the $[Ca^{2+}]_i$ increase was entirely dependent on extracellular Ca^{2+}. $1{,}25(OH)_2D_3$ also increased the production of IP_3 and DAG, at a threshold concentration of 10^{-9} M. A second addition of $1{,}25(OH)_2D_3$ resulted in a $[Ca^{2+}]_i$ transient of higher amplitude than the first; this response amplification was associated with production of IP_3.

More recent studies utilizing the colonic carcinoma cell line Caco-2 (Bissonette *et al.*, 1994; Tien *et al.*, 1993a,b; Wali *et al.*, 1992) have demonstrated that $1,25(OH)_2D_3$, in a concentration-dependent manner, rapidly (seconds to minutes) increased $[Ca^{2+}]_i$, stimulated membrane phosphoinositide turnover, and induced translocation of protein kinase C from the cytosol to the membrane fraction. The pattern of $[Ca^{2+}]_i$ increase was typical for $1,25(OH)_2D_3$ responses in other cell lines; namely, a sharp initial elevation (transient phase) followed by a slower return to a level above baseline (plateau phase). Experiments using Ca^{2+}-free buffer suggested that the rapid rise in $[Ca^{2+}]_i$ was due to Ca^{2+} mobilization from intracellular stores and that the influx of extracellular Ca^{2+} attributed to plateau phase. Structural requirements for the rapid action of $1,25(OH)_2D_3$ analogues on $[Ca^{2+}]_i$ were different from those for the conventional VDR binding and transcriptional regulation, suggesting a role for the membrane VDR in signal transduction. The mechanism of $1,25(OH)_2D_3$-induced increase in $[Ca^{2+}]_i$ included accumulation of IP_3; subsequent activation of PKC-α limited the rise in $[Ca^{2+}]_i$ at a step distal to IP_3 accumulation (Bissonnette *et al.*, 1995, 1994).

$1,25(OH)_2D_3$ induces rapid and sustained increases in $[Ca^{2+}]_i$ in normal human keratinocytes (Jackson and Holick, 1994; Bittiner *et al.*, 1991). A sustained phase of this increase, i.e., elevated basal $[Ca^{2+}]_i$, can be observed over a period of hours (Pillai *et al.*, 1995). An increase in $[Ca^{2+}]_i$ was seen in 80% of cell cultures, with over half the responses occurring within 30 sec of exposure to $1,25(OH)_2D_3$ and the remainder occurring within minutes (Bittiner *et al.*, 1991). Long-term treatment of keratinocytes with $1,25(OH)_2D_3$ resulted in a potentiation of the Ca^{2+} response of these cells to an agonist such as ATP and was associated with an increase in ATP-stimulated IP_3 production (Pillai *et al.*, 1995). A similar effect of $1,25(OH)_2D_3$ on increases in ATP- or IP_3-evoked changes in $[Ca^{2+}]_i$ has been demonstrated in FRTL-5 rat thyroid cells (Ahlstrom *et al.*, 1995; Tornquist *et al.*, 1992) and GH_4C_1 pituitary cells (Tornquist, 1992).

In our recently published studies (Sergeev and Rhoten, 1995a, 1994), fluorescence digital ratiometric video imaging at the single-cell level was utilized for the first time to study the effects of $1,25(OH)_2D_3$ on $[Ca^{2+}]_i$. In pancreatic β-cell cells (rat insulinoma cell line RINr1046-38) equilibrated at a steady-state glucose concentration, $1,25(OH)_2D_3$ (2–20 nM) rapidly, within 5–10 sec, increased $[Ca^{2+}]_i$ and evoked sinusoidal $[Ca^{2+}]_i$ oscillations with concentration-dependent amplitude. The $[Ca^{2+}]_i$ oscillations were acutely dependent on extracellular Ca^{2+}, but not on extracellular glucose. The mechanisms of $1,25(OH)_2D_3$-induced $[Ca^{2+}]_i$ oscillations included activation of Ca^{2+} entry through VICC and VDCC and mobilization of intracellular Ca^{2+} through the ryanodine receptor/Ca^{2+} release channel, but not through activation of the IP_3 receptor. These results are consistent with a role of the plasma and endoplasmic reticulum membrane VDRs coupled to Ca^{2+} channels in mediating rapid effects of $1,25(OH)_2D_3$. Additionally, we postulated that the $1,25(OH)_2D_3$-mediated Ca^{2+} signaling pathway is involved in the regulation of insulin secretion from the pancreatic β-cell. RIN cells cultivated in horizontally rotating vessels on microcarriers (a model for microgravity) responded to $1,25(OH)_2D_3$ with desynchronization of Ca^{2+} oscillations (Sergeev, Rhoten, and Chaudhry, unpublished observations). This may indicate that synchronization of $1,25(OH)_2D_3$-evoked Ca^{2+} oscillations requires intercellular communication and/or mechanotransduction within the cell.

Several other articles report typical $1,25(OH)_2D_3$-induced increases in $[Ca^{2+}]_i$ in hepatocytes (Baran *et al.*, 1990), isolated enterocytes (Lieberherr *et al.*, 1989; de Boland and Norman, 1990b), renal proximal tubule cells (Friedlaender *et al.*, 1995), vascular smooth muscle cells (Shan *et al.*, 1993), and promyelocytic cells (Tasaka *et al.*, 1991).

Our results on the role of $1,25(OH)_2D_3$ in regulation of intracellular Ca^{2+} in breast cancer cells (estrogen receptor-positive MCF-7 and estrogen receptor-negative BT-20 cell lines) showed that the hormone did not evoke rapid rapid changes in $[Ca^{2+}]_i$, but did increase Ca^{2+} entry through VICC (Sergeev *et al.*, 1997). Treatment with $1,25(OH)_2D_3$ for 4–72 hr significantly increased the percentage of cells with an elevated basal $[Ca^{2+}]_i$. The Ca^{2+} responses of cells with an elevated $[Ca^{2+}]_i$ to thapsigargin, a highly specific mobilizer of the ER Ca^{2+} stores, were dramatically attenuated or abolished, indicating depletion of these stores. It is necessary to emphasize that breast cancer cells do not express VDCC and that VICC and the thapsigargin-sensitive ER Ca^{2+} pool are the principal pathways for Ca^{2+} entry and Ca^{2+} mobilization in these cells (Sergeev and Rhoten, Submitted, 1995b).

A summary of the effects of $1,25(OH)_2D_3$ on $[Ca^{2+}]_i$ is presented in Table I. Taken together, these results clearly indicate that $1,25(OH)_2D_3$ induces an increase in $[Ca^{2+}]_i$ due to an influx of extracellular Ca^{2+} through VDCC and VICC and mobilization of Ca^{2+} from intracellular stores. An initial rapid phase of increase in $[Ca^{2+}]_i$ seems to be associated with release of intracellular Ca^{2+} from the ER stores through the IP_3 and/or ryanodine receptor/Ca^{2+} release channels. Ca^{2+} channels located in the plasma membrane provide the pathway for Ca^{2+} entry from the extracellular space, which, in turn, supports a sustained phase of the $[Ca^{2+}]_i$ increase and $[Ca^{2+}]_i$ oscillations. Coordinative regulation of Ca^{2+} entry and Ca^{2+} mobilization by $1,25(OH)_2D_3$ underlies the mechanism of the hormone-specific Ca^{2+} signaling. Importantly, although other steroid hormones may induce rapid increases in $[Ca^{2+}]_i$, such effects seems to be hormone, dose, and cell-type specific (Lieberherr *et al.*, 1994).

Table I
$1,25(OH)_2D_3$ Effects on Intracellular Ca^{2+}

Cell type/line	Onset of $[Ca^{2+}]_i$ increase	Mechanism of $[Ca^{2+}]_i$ increase	References
Osteosarcoma (ROS 17/2.8)	Rapid	VDCC, Ca^{2+} mobilization (IP_3)	Lieberherr *et al.* (1994), Baran (1994), Civitelli *et al.* (1990), Caffrey and Farach-Carson (1989)
Breast carcinoma (MCF-7, BT-20)	Delayed	VICC, Ca^{2+} mobilization (thapsigargin-sensitive Ca^{2+} stores)	Sergeev *et al.* (1997)
Insulinoma (RINr 1046–38)	Rapid (Ca^{2+} oscillations)	VDCC, VICC, Ca^{2+} mobilization (ryanodine receptor)	Sergeev and Rhoten (1995a)
Colonic carcinoma (Caco-2)	Rapid and sustained	VDCC, Ca^{2+} mobilization (IP_3)	Bissonette *et al.* (1994, 1995), Tien *et al.* (1993a,b)
Normal human keratinocytes	Rapid and sustained	Ca^{2+} influx and Ca^{2+} mobilization (IP_3)	Pillai *et al.* (1995), Bittiner *et al.* (1991)
Vascular smooth muscle cells	Rapid	VDCC	Shan *et al.* (1993)
Thyroid (FRTL-5) and pituitary (GH_4C_1) cells	Delayed	Potentiation of agonist-induced Ca^{2+} mobilization	Ahlstrom *et al.* (1995), Tornquist (1992), Tornquist *et al.* (1992)

5. 1,25-DIHYDROXYVITAMIN D_3 AND Ca^{2+} FLUXES

The $1,25(OH)_2D_3$-induced changes in $[Ca^{2+}]_i$ result from the activation of Ca^{2+} fluxes through Ca^{2+} channels located in plasma and ER membranes. However, literature concerning direct effects of $1,25(OH)_2D_3$ on Ca^{2+} channels is limited, and most of the data were obtained using the $^{45}Ca^{2+}$ uptake method.

The $^{45}Ca^{2+}$ uptake induced by $1,25(OH)_2D_3$ has been observed in many cell types and tissues. Prior treatment with $1,25(OH)_2D_3$ increases $^{45}Ca^{2+}$ uptake in intestine, kidney, bone, heart, skeletal and smooth muscle, mammary gland, lung, testis, and yolk sack of the bird (Walters, 1995, 1992; Johnson and Kumar, 1994; Marinissen *et al.*, 1994; Lee and Clark, 1993; Vazquez and de Boland, 1993; Selles and Boland, 1991; Lucas *et al.*, 1989). In most cases, $^{45}Ca^{2+}$ uptake has been characterized as demonstrating $1,25(OH)_2D_3$ specificity and, in some instances, this effect has been abolished by treatment with cycloheximide and/or actinomycin D [thus implying a genomic 1,25(OH)$_2$D$_3$-mediated effect]. The magnitude and time course of the $^{45}Ca^{2+}$ uptake response differ substantially both with respect to time of 1,25(OH)$_2$D$_3$ exposure and sustained versus transient nature of the effect. Generally, the maximum increase in $^{45}Ca^{2+}$ uptake occurs either within minutes or hours after pretreatment with hormone. For example, we have shown (Farach-Carson *et al.*, 1991a,b) that an acute addition of 1,25(OH)$_2$D$_3$ induced a rapid, within 1 min, increase in $^{45}Ca^{2+}$ uptake in rat osteosarcoma ROS 17/2.8 cells which was dose dependent and required structural ligand specificity distinct from that of the nuclear VDR (and thus providing for the first time evidence of the existence of the membrane VDR). Maximum response in primary culture of cardiac muscle cells occurred at 8–24 hr after 1,25(OH)$_2$D$_3$ exposure, and the elevated $^{45}Ca^{2+}$ uptake was sustained throughout a 1-hr $^{45}Ca^{2+}$ incubation period (Walters *et al.*, 1987). Uptake of $^{45}Ca^{2+}$ into mammary gland explants demonstrated an intermediate pattern: detectable increase after treatment with 1,25(OH)$_2$D$_3$ for 30 min, with the response reaching a plateau at 3 hr; remarkably, an inhibitory effect of actinomycin D or cycloheximide was not observed (Mezzetti *et al.*, 1988).

The rapid 1,25(OH)$_2$D$_3$-induced Ca^{2+} transport in the intestine, or transcaltachia, was first described and characterized by Nemere and Norman (1989, 1988). In this model, an initial Ca^{2+} uptake at the brush border of the enterocyte occurs through Ca^{2+} channels via actions of 1,25(OH)$_2$D$_3$ independent of genome activation (de Boland and Norman, 1990b). Subsequent vesicular Ca^{2+} transport associated with the cytoskeletal elements has been suggested (Nemere and Norman, 1990). A review of transcaltachia can be found elsewhere (Norman, 1997; Nemere, 1995; Nemere and Norman, 1991, 1990). Importantly, stimulation of Ca^{2+} influx by 1,25(OH)$_2$D$_3$ in enterocytes required expression of the vitamin D-dependent transport elements. Stimulation of transcaltachia in vitamin D-deficient chicks (Nemere and Norman, 1990) and 1,25(OH)$_2$D$_3$-induced $^{45}Ca^{2+}$ uptake in enterocytes isolated from naturally vitamin D-deficient mole rats (Buffenstein *et al.*, 1994) were absent. Noteworthy, vitamin D deficiency also led to lower $[Ca^{2+}]_i$ in hepatocytes (Gascon-Barre *et al.*, 1994).

Effects of 1,25(OH)$_2$D$_3$ on Ca^{2+} uptake in pituitary cells have been related to a specific biological response (thyrotropin-releasing hormone-stimulated prolactin production) (Tornquist *et al.*, 1989). Pretreatment with 1,25(OH)$_2$D$_3$ for 24 hr resulted in enhanced $^{45}Ca^{2+}$ uptake through dihydropyridine-sensitive L-type VDCC in response to depolarization with K^+ and in potentiation of the increase in $[Ca^{2+}]_i$ by thyrotropin-releasing hormone. This study represents an example (although permissive) of interaction of genomic 1,25$(OH)_2D_3$- and intracellular Ca^{2+}-mediated signaling pathways in regulating specific biological functions.

The Ca^{2+} uptake studies are important in establishing an involvement of 1,25$(OH)_2D_3$ in regulation of Ca^{2+} fluxes, but the utility of such an approach for understanding the mechanisms of the 1,25$(OH)_2D_3$ effects on intracellular Ca^{2+} and, consequently, regulation of Ca^{2+}-dependent biological functions is to some extent limited. Direct measurement of Ca^{2+} fluxes through Ca^{2+} channels, obviously, may provide more specific information. Unfortunately, only a few reports where a patch-clamp technique was used to study the effects of 1,25$(OH)_2D_3$ on Ca^{2+} channels are available.

Using the whole-cell voltage clamp technique, Caffrey and Farach-Carson (1989) demonstrated a rapid activation of L-type VDCC by 1,25$(OH)_2D_3$ in rat osteoblast-like osteosarcoma cell line ROS 17/2.8. At physiologically relevant concentrations (0.1–3 nM), 1,25$(OH)_2D_3$ stimulated an inward Ca^{2+} current and rapid $^{45}Ca^{2+}$ influx by shifting the threshold for activation of the current to more negative and physiological potentials. At higher concentrations (5–10 nM), inhibitory effects became predominant. These opposing effects are functionally similar to those of the dihydropyridine Ca^{2+}-channel agonist Bay K 8644. Further experiments utilizing vitamin D analogues demonstrated that opening of VDCC in ROS cells seems to involve membrane VDRs (Farach-Carson *et al.*, 1991a). Yukihiro *et al.* (1994) confirmed that 1,25$(OH)_2D_3$ (at a concentration of 50 pM) rapidly, within seconds, shifts the threshold for activation of inward Ca^{2+} currents in ROS 17/2.8 cells to more negative potentials and suggested that 1,25$(OH)_2D_3$ does not gate the Ca^{2+} channel itself. They also concluded, by evaluating a series of 1,25$(OH)_2D_3$ analogues, that the opening of VDCC in bone cells is mediated by a signal transduction pathway distinct from the conventional VDR.

Shan *et al.* (1993) reported a rapid activation by 1,25$(OH)_2D_3$ (10–100 nM) of the voltage-dependent Ca^{2+} current in vascular smooth muscle cells. Effects of 1,25$(OH)_2D_3$ appeared by 5–10 min of exposure with a maximum by 15 min and were dose dependent.

Activation of Ca^{2+} channels *via* second messenger pathways, including Ca^{2+} signaling, seems essential for the action of 1,25$(OH)_2D_3$. Activators of protein kinase C (PKC), phorbol esters, and the adenylate cyclase activator forskolin were found to stimulate the rapid intestinal Ca^{2+} transport utilizing Ca^{2+} channels, transcaltachia, to the same extent that physiological levels of 1,25$(OH)_2D_3$ achieved (de Boland and Norman, 1990a; Massheimer *et al.*, 1994). Additionally, 1,25$(OH)_2D_3$-dependent stimulation of Ca^{2+} uptake in the intestine was suppressed by a PKC inhibitor, staurosporine. Similar results were obtained in myocytes (Vazquez *et al.*, 1995; de Boland and Boland, 1994; de Boland *et al.*, 1994; Marinissen *et al.*,

1994). Furthermore, direct stimulation of membrane protein phosphorylation by $1,25(OH)_2D_3$, which may include Ca^{2+} channels, was reported (Selles and Boland, 1991).

It is noteworthy that $1,25(OH)_2D_3$ also rapidly enhanced an outwardly rectifying voltage-dependent Cl^- current (i.e., Cl^- influx) (Zanello and Norman, 1996), an outward K^+ current (Chesnoy-Marchais and Fritsch, 1993), and increased intracellular pH (Jenis *et al.*, 1993) in osteoblastic cells. Taken together, these findings imply that integrated modulatory mechanisms underlying the control of membrane ionic pathways and electrical states of the cell include $1,25(OH)_2D_3$.

6. CALBINDINS AND INTRACELLULAR Ca^{2+}

Calbindins are high-affinity calcium-binding proteins regulated in some organs, e.g., intestine and kidney, by $1,25\text{-}(OH)_2D_3$. Calbindin-D_{28k} (also known as $CaBP_{28k}$), with a molecular mass of about 28 kDa, was first described by Wasserman and Taylor (1966). Calbindin-D_{28k} has a widespread distribution in tissues and is highly conserved phylogenetically (Christakos *et al.*, 1997, 1989). Calbindin-D_{9k}, with a molecular mass of about 9 kDa, has a more narrow tissue distribution and appears to be an evolutionarily more recent protein (Thomasset, 1997; Christakos *et al.*, 1989). Experimental evidence in support of a role for the calbindins in Ca^{2+} homeostasis has accumulated in the areas of Ca^{2+} buffering, Ca^{2+} transport, and Ca^{2+} signaling.

The reference point of genomic regulation of calbindins by $1,25(OH_2)D_3$ may have inhibited the acceptance of these proteins as mediators of rapid Ca^{2+} responses, such as intestinal transport or renal reabsorption, in response to the hormone. Nonetheless, Dupret *et al.* (1987) found a doubling of transcription of the intestinal calbindin-D_{9k} gene in 15 min in response to $1,25(OH_2)D_3$ *in vivo*. In culture, both $1,25(OH_2)D_3$ and extracellular Ca^{2+} increased transcription and translation of calbindin-D_{9k} (Brehier and Thomasset, 1990). Walters (1989) found a threefold increase in ATP-dependent Ca^{2+} uptake into rat duodenal basolateral membrane vesicles with calbindin-D_{9k}. In addition to a stimulatory effect on mammalian intestinal Ca^{2+}-ATPase, calbindin-D_{9k} may facilitate translocation of Ca^{2+} in the enterocyte (Feher *et al.*, 1992; Feher, 1983). Downregulation of calbindins, e.g., as a consequence of the weightlessness of space flight (Sergeev *et al.*, 1996), may contribute to decreased Ca^{2+} absorption in the intestine.

The four or more high-affinity binding sites for Ca^{2+} in calbindin-D_{28k} (Akerfeldt *et al.*, 1996; Leathers *et al.*, 1990) make it particularly attractive for identifying roles of calbindins in Ca^{2+} homeostasis. Although localization of calbindin-D_{28k} in the shell gland of chickens strongly suggests a role for calbindin-D_{28k} in Ca^{2+} secretion, direct experimental evidence is apparently lacking. Perhaps the strongest evidence of calbindin-D_{28k} involvement in Ca^{2+} transport comes from studies with the chicken intestine, where translocation of Ca^{2+} and calbindin-D_{28k} in the enterocyte have been linked with organelles, including vesicles and lysosomes.

Nemere and co-workers in a series of studies have provided evidence linking calbindin-D_{28k} to vesicular transport of Ca^{2+} in chicken intestine (Nemere *et al.*, 1991, 1986; Nemere and Norman, 1989, 1988). In a biochemical study (Nemere *et al.*, 1986), 1,25(OH_2)D_3-stimulated intestinal transport of Ca^{2+} was associated with calbindin-D_{28k} in lysosomes and putative endocytic vesicles. A subsequent time-course study (Nemere and Norman, 1988) followed 1,25$(OH)_2D_3$ stimulation of Ca^{2+} absorption and localization of calbindin-D_{28k} from endocytic vesicles at the brush border initially, to lysosomes, and finally to the basolateral membrane. Morphological studies at the light and electron microscopic level (Nemere *et al.*, 1991) reinforced both the dynamic relationship of calbindin-D_{28k} and intestinal Ca^{2+} translocation, and the involvement of vesicular and lysosomal structures in the translocation process. However, dose–response (Nemere and Norman, 1989) and translocation (Lee *et al.*, 1988) studies support a multiple-pathway view of intestinal Ca^{2+} transport with calbindin-D_{28k} involved in one or more of these pathways.

Calbindin-D_{28k} has been shown to modulate enzymes controlling $[Ca^{2+}]_i$. Morgan *et al.* (1986) reported that calbindin-D_{28k} stimulated erythrocyte plasma membrane Ca^{2+},Mg^{2+}-ATPase. Subsequently, Reisner *et al.* (1992) found that calbindin-D_{28k} enhanced the activity of purified erythrocyte Ca^{2+}-ATPase, and Wasserman *et al.* (1992) reported that calbindin-D_{28k} stimulated ATP-dependent Ca^{2+} transport into basolateral membrane vesicles from chicken intestine. Calbindin-D_{28k} stimulation of ATP-dependent Ca^{2+} uptake into rat duodenal basolateral membrane vesicles and porcine erythrocyte vesicles was shown by Timmermans *et al.* (1995). Bovine brain 3′,5′-cyclic nucleotide phosphodiesterase was also found to be activated, in a study by Reisner *et al.* (1992).

Recent evidence links calbindin-D_{28k} in the kidney to Ca^{2+} reabsorption in the distal tubule. Bouhtiauy *et al.* (1994a,b) reported differential effects of the calbindins on Ca^{2+} handling in membrane vesicles of renal distal and proximal tubules from the rabbit. Calbindin-D_{28k} markedly increased Ca^{2+} uptake into luminal vesicles and slowed Ca^{2+} efflux from them (Bouhtiauy *et al.*, 1994a). Calbindin-D_{9k} increased ATP-dependent Ca^{2+} pumping in basolateral membrane vesicles (Bouhtiauy *et al.*, 1994b).

Roles for calbindin-D_{9k} in maternofetal transfer of Ca^{2+} and in female generational organs remain exciting possibilities (Mathieu *et al.*, 1989; Bruns *et al.*, 1988, 1986).

Much speculation has surrounded the role of calbindin-D_{28k} in the nervous system (Christakos *et al.*, 1989); experimental studies on the nervous system have been especially productive for revealing effects of calbindin-D_{28k} on Ca^{2+} homeostasis. Mattson *et al.* (1991) demonstrated that cultured calbindin-D_{28k}-containing hippocampal neurons were better able to maintain $[Ca^{2+}]_i$ in response to ionophore-induced influx of Ca^{2+}. Resistance to degeneration was enhanced in calbindin-D_{28k}-containing neurons in response to potentially excitotoxic stimulation with glutamate as well as exposure to Ca^{2+} ionophore. Thus, neurons immunoreactive for calbindin-D_{28k} were better able to buffer an elevated $[Ca^{2+}]_i$ and maintain Ca^{2+} homeostasis.

Buffering of Ca^{2+} by calbindin-D_{28k} was shown with direct introduction of calbindin-D_{28k} into cultured dorsal root ganglion cells of the rat (Chard *et al.*, 1993). Calbindin-D_{28k} caused a marked decrease in the rate of rise in $[Ca^{2+}]_i$ and a significant reduction in the $[Ca^{2+}]_i$ maximum in response to depolarization. Calbindin-D_{28k}-containing cells exhibited a monophasic decay in $[Ca^{2+}]_i$ that was slower than unloaded control cells and lacked a fast-decay component. The basal $[Ca^{2+}]_i$ and the rate of inactivation of the Ca^{2+} current were unaffected by the introduction of calbindin-D_{28k}. The latter finding differs from the results of transfection of the calbindin-D_{28k} gene into a pituitary cell line (GH_3) in which a faster inactivation of voltage-dependent L- and T-type Ca^{2+} currents was observed (Lledo *et al.*, 1992). However, the increase in intracellular calbindin-D_{28k} did not affect activation of Ca^{2+} currents. Using a replication-defective, calbindin-D_{28k}-expressing adenovirus construct, Chard *et al.* (1995) found that calbindin-D_{28k} selectively suppressed posttetanic potentiation in cultured hippocampal pyramidal neurons. Evoked, glutamate-mediated, excitatory postsynaptic Ca^{2+} currents were unaffected in calbindin-D_{28k}-expressing neurons. The Ca^{2+}-binding kinetics of calbindin-D_{28k} were believed to account for the differential effect and the authors speculated that calbindin-D_{28k} acted by limiting and disrupting evoked Ca^{2+} signals at the nerve terminal. Such an effect on local Ca^{2+} signaling associated with synaptic transmission is in agreement with the theoretical study of Roberts (1994).

Motoneuron hybrid cells carrying a calbindin-D_{28k}-expressing retrovirus had high levels of calbindin-D_{28k} and increased resistance to amyotrophic lateral sclerosis (ALS) immunoglobulin G (IgG)-mediated toxicity (Ho *et al.*, 1996). Survival of the motoneurons infected with retrovirus carrying calbindin-D_{28k} was significantly decreased when the cells were treated with antisense oligodeoxynucleotides to calbindin-D_{28k}. Levels of calbindin-D_{28k} in the antisense-treated cells were significantly reduced. The authors related the protective effect of calbindin-D_{28k} to their earlier studies showing that ALS IgG increased Ca^{2+} currents, $[Ca^{2+}]_i$, and apoptosis in motoneuron hybrid cells lacking the retroviral calbindin-D_{28k} cDNA.

Airaksinen *et al.* (1997) fabricated mice with a targeted null mutation of the calbindin-D_{28k} gene. Null mutants developed normally, but motor coordination upon challenge was impaired, with ataxia and kinetic tremor in the calbindin-D_{28k} null mutants. Afferent fiber-evoked Ca^{2+} transients in Purkinje cells of null mutants displayed a much greater amplitude of the fast-decay component, and a gene dosage effect was observed. Resting $[Ca^{2+}]_i$ was unchanged. The authors concluded that calbindin-D_{28k} is a fast-acting (100 msec or less), high-affinity Ca^{2+} buffer.

Buffering of Ca^{2+} by calbindin-D_{28k} in magnocellular neurons of the supraoptic nucleus was found to suppress depolarizing afterpotentials and to convert phasic firing to continuous firing (Li *et al.*, 1995). Furthermore, introduction of antibody directed against calbindin-D_{28k} into continuous-firing neurons, which are associated with oxytocin secretion, converted these neurons into phasic-firing ones. Thus, this study (Li *et al.*, 1995) showed not only an effect on neuronal electrical activity of exogenous calbindin-D_{28k}, but also an effect of depletion of endogenous calbindin-D_{28k}.

It appears that intrinsic Ca^{2+}-regulatory mechanisms influence the outcome of calbindin-D_{28k} gene expression in cells. Transfection of the calbindin-D_{28k} gene into murine fibroblasts (NIH3T3 cells) had no effect on the rise or the return to basal levels of $[Ca^{2+}]_i$ in response to serum stimulation (Muir *et al.*, 1993). Nonetheless, the NIH3T3 cells with the transgene for calbindin-D_{28k} buffered ionophore-induced increases in $[Ca^{2+}]_i$.

Our studies with rat insulinoma cells (RINr1046-38, RIN) indicated a large increase in cytoplasmic Ca^{2+} buffering capacity in cells with elevated levels of calbindin-D_{28k} (Rhoten and Sergeev, 1994a,b). RIN cells treated with 2 mM butyrate for 24–72 hr had calbindin-D_{28k} levels about threefold higher than control cells. In treated cells, the $[Ca^{2+}]_i$ response to glucose, K^+-depolarization, a calcium ionophore (ionomycin), and Ca^{2+}-channel agonist Bay K 8644 were dramatically altered. At 24 and 48 hr, $[Ca^{2+}]_i$ peak values were markedly reduced, although basal $[Ca^{2+}]_i$ was unchanged. The increase in $[Ca^{2+}]_i$ in response to mobilization of Ca^{2+} from intracellular stores was also attenuated. Only high concentrations of the ionophore (5 μM) overcame the buffering effect of calbindin-D_{28k} and produced a dramatic rise in $[Ca^{2+}]_i$. Our finding of buffering of Ca^{2+} by calbindin-D_{28k} in RIN cells was recently confirmed in RIN cells overexpressing this protein

(Reddy *et al.*, 1997). Interestingly, these authors found that the calbindin-D_{28k}-transfected and-overexpressing cells had increased insulin mRNA content and insulin release.

In summary, calbindins have been shown to act as mediators of $[Ca^{2+}]_i$. Studies on classical target tissues of vitamin D action, such as intestine and kidney, have provided provocative potential roles for calbindins, e.g., vesicular transport of Ca^{2+} and stimulation of the ATP-dependent Ca^{2+} pump. Calbindin-D_{28k} buffers Ca^{2+} and enhances insulin synthesis and secretion in a pancreatic β-cell line. Investigations involving the nervous system and calbindin-D_{28k} have demonstrated, inter alia, rapid buffering of Ca^{2+}, increased resistance to neurotoxicity, and alteration in firing patterns of neuroendocrine cells. The alterations in Purkinje cell Ca^{2+} transients and behavioral abnormalities in mice with a targeted null mutation of the calbindin-D_{28k} gene presage the breadth and depth of calbindin action that is likely to be revealed in the near future.

7. REGULATION OF CELL FUNCTIONS VIA VITAMIN D–INTRACELLULAR Ca^{2+} SIGNALING

Calcium acts as an intracellular second messenger in a wide variety of cellular responses (Petersen *et al.*, 1994). Short-term responses to Ca^{2+} include excitation–contraction coupling in skeletal and cardiac muscle and stimulus–secretion coupling in endocrine and neuronal cells. Mechanisms underlying these processes involve, in particular, phosphorylation of Ca^{2+}-channel subunits catalyzed by Ca^{2+}-activated kinases. Long-term responses to Ca^{2+}, which require new gene expression, include cell proliferation, differentiation, programmed cell death (apoptosis), and neuronal adaptive responses. Proteins whose expression is regulated by changes in intracellular Ca^{2+} include hormones, neuropeptides, ion channels, and oncogene products (Rosen *et al.*, 1995).

The effects of $1,25(OH)_2D_3$ on Ca^{2+}-dependent pathways may provide the mechanism for rapid effects on phosphorylation and immediate gene expression or activation of alternate gene products. $1,25(OH)_2D_3$ initiates turnover of phospholipids, including phospholipase C, D, and A_2 (Le Mellay *et al.*, 1997; Pillai *et al.*, 1995; de Boland *et al.*, 1994; Lieberherr *et al.*, 1994; Grosse *et al.*, 1993; Tien *et al.*, 1993a,b; Schwartz *et al.*, 1991), stimulates release of DAG (Wali *et al.*, 1992), induces activation/translocation of PKC (Sylvia *et al.*, 1996; Bissonnette *et al.*, 1994; Khare *et al.*, 1994; Marinssen *et al.*, 1994; Simboli-Campbell *et al.*, 1994; Wali *et al.*, 1992), including phospholipase- and DAG-independent mechanisms (van Leeuwen *et al.*, 1992), and enhances specific phosphorylation of protein signaling intermediates at serine/threonine and tyrosine residues (Bhatia *et al.*, 1996; Berry *et al.*, 1996). Stimulation of translocation of PKC-α and PKC-δ isoforms and activation of the tyrosine kinase phosphorylation cascade were attenuated by 1β,25$(OH)_2D_3$, suggesting a nongenomic 1,25$(OH)_2D_3$-mediated signaling pathway (Berry *et al.*, 1996). Activation of the adenylate cyclase–protein kinase A (PKA) pathway (Vazquez *et al.*, 1995; Massheimer *et al.*, 1994; Selles *et al.*, 1994), involvement of guanylate cyclase (Khare *et al.*, 1993), and mitogen-activated protein kinase (Beno *et al.*, 1995) have been also reported. Furthermore, 1,25$(OH)_2D_3$ has been shown to increase the level of IP_3 receptors (Bradford *et al.*, 1993). The sequence of effects of 1,25$(OH)_2D_3$ on intracellular Ca^{2+} and steps which follow in the signal transduction are not always obvious, e.g., whether 1,25$(OH)_2D_3$-induced increase in $[Ca^{2+}]_i$ initiates phospholipid turnover or results from it remains to be elucidated. Nevertheless, the definitive role of 1,25$(OH)_2D_3$ in rapid increase in $[Ca^{2+}]_i$, Ca^{2+} influx, and intracellular Ca^{2+} mobilization strongly implies the subsequent activation of intracellular Ca^{2+} signaling events, leading to modulation in immediate gene expression and rapid functional response. For example, 1,25$(OH)_2D_3$-evoked Ca^{2+} oscillations in pancreatic β cells may induce synchronous pulses of insulin release (Sergeev and Rhoten, 1995a), as demonstrated for other agonists of insulin secretion inducing Ca^{2+}oscillations (Gilon *et al.*, 1993).

The relationship between increased Ca^{2+} influx and Ca^{2+} mobilization after treatment with 1,25$(OH)_2D_3$ and long-term target gene activation is less clear. Our data indicate that 1,25$(OH)_2D_3$-induced increase in $[Ca^{2+}]_i$ in breast cancer cells observed after a 4-hr treatment with the hormone is associated with significant DNA breakage, a hallmark of apoptosis, at 24–72 hr (as measured using terminal deoxynucleotidyl transferase and fluorescein-dUTP) (Sergeev *et al.*, 1997). Vandewalle *et al.* (1995) also suggested that growth inhibition of breast cancer MCF-7 cells is related to increase in $[Ca^{2+}]_i$. Welsh *et al.* (1995) described the role of apoptosis in the genomic (not Ca^{2+}-mediated) growth-inhibitory effects of 1,25$(OH)_2D_3$ in MCF-7 cells. 1,25$(OH)_2D_3$-induced increase in $[Ca^{2+}]_i$ may be an initial signal that triggers formation of the cornified envelope, the terminal event during keratinocyte differentiation (Pillai and Bikle, 1991; Bittiner *et al.*, 1991). However, prodifferentiation effects of 1,25$(OH)_2D_3$ in keratinocytes seem secondary to the effects of extracellular Ca^{2+} (Su *et al.*, 1994).

Baran *et al.* (1992) showed that the 1,25$(OH)_2D_3$-induced rapid increase in $[Ca^{2+}]_i$ is functionally related to activation of the osteocalcin gene in osteoblsts. In ROS 17/2.8 cells, 1β,25$(OH)_2D_3$, which competes with 1α,25$(OH)_2D_3$ for the rapid signaling system (membrane VDRs) and does not appear to affect the binding of the VDR complex to the vitamin D-responsive element of the osteocalcin gene, blocked the 1,25$(OH)_2D_3$-

induced increase in $[Ca^{2+}]_i$ and osteocalcin mRNA transcription at 1 hr and steady-state level in 3 hr.

Khoury *et al.* (1995, 1994) examined the relationship between $1,25(OH)_2D_3$-stimulated Ca^{2+} influx and genomic and nongenomic pathways leading to osteoblast activation. The authors concluded that $1,25(OH)_2D_3$-mediated upregulation of osteopontin and osteocalcin in ROS 17/2.8 cells is independent of Ca^{2+} influx and suggested that stimulation of Ca^{2+} influx by $1,25(OH)_2D_3$ is not required for target gene activation. The L-type VDCC antagonist (nitredipine) and agonist (Bay K 8644) did not affect mRNA accumulation of osteocalcin and osteopontin, whereas the analogue of $1,25(OH)_2D_3$, oxacalcitriol, which had no effect on Ca^{2+} influx in ROS cells, increased mRNA levels of these proteins (Farach-Carson *et al.*, 1993). Because $[Ca^{2+}]_i$ was not measured in these studies, there is a possibility that Ca^{2+} influx through VICC and Ca^{2+} mobilization from intracellular stores could influence the osteocalcin and osteopontin gene expression in ROS cells. Meszaros *et al.* (1996) demonstrated that $1,25(OH)_2D_3$ downregulates, via a genomic pathway, a steady-state level of the L-type α1C-a, but not α1C-d isoform of the Ca^{2+} channel in ROS 17/2.8 cells. This genomic effect of $1,25(OH)_2D_3$ on VDCC may be considered as regulatory, opposing a rapid channel activation by the hormone. It is reasonable to conclude that rapid signals generated by $1,25(OH)_2D_3$ and long-term, direct genomic modulation of cell function by this hormone involve separate receptors and that structural requirements for activation of genomic versus nongenomic pathways by $1,25(OH)_2D_3$ are distinct.

Remarkably, intracellular Ca^{2+} appears to regulate VDRs through feedback effects. Krishnan and Feldman (1991) observed that an elevation in $[Ca^{2+}]_i$ caused by a Ca^{2+} ionophore, resulted in a decrease in VDR levels in mouse fibroblasts, while van Leeuwen *et al.* (1990) found that intracellular Ca^{2+} is involved in the PTH- and cAMP-mediated regulation of VDRs in rat osteosarcoma cells. We have shown that VDRs are carboxylated at the Glu residues in the vitamin K-dependent reaction and that such a modification modulates DNA binding properties of the receptor (Sergeev and Spirichev, 1988; Sergeev and Norman, 1992). Importantly, when γ-carboxylation of VDRs was blocked (in vitamin K–deficient rats) and, therefore, receptors had no Ca^{2+} binding Gla residues, the DNA binding and nuclear localization of VDRs, as well as the biological response to $1,25(OH)_2D_3$ (intestinal Ca^{2+} absorption) were increased (Sergeev and Spirichev, 1989). Recent molecular modeling proposes that Glu-92, which is situated in the T box in DNA binding domain of human VDR, apparently determines selective association of VDR with a vitamin D responsive element in DNA (Rastinejad *et al.*, 1995). Collectively, these findings imply that intracellular Ca^{2+} plays a role in regulation of VDRs, probably directly interacting with the receptor protein.

8. CONCLUSIONS

The reviewed literature clearly allows one to make a definitive conclusion that $1,25(OH)_2D_3$ induces an increase in $[Ca^{2+}]_i$ in a variety of cell types. Rapid onset of such an increase, selectivity of Ca^{2+} responses to specific $1,25(OH)_2D_3$ analogues, and involvement of PKC, PKA, mitogen-activated protein kinase, and tyrosine kinase in signal transduction imply that VDRs located in plasma (and ER?) membranes mediate the effects of $1,25(OH)_2D_3$ on intracellular Ca^{2+}. The nature and precise localization of this receptor have to be determined. Most findings indicate that it is a different form of the VDR (Kim *et al.*, 1996; Norman, 1995; Nemere, 1995; Nemere *et al.*, 1994), although some evidence points to a rapid translocation of the conventional VDR (Barsony *et al.*, 1997, 1994).

Increases in $[Ca^{2+}]_i$ induced by $1,25(OH)_2D_3$ in different cell types are characterized by a rapid transient rise followed by a slower return to a level at or above the initial baseline. Mechanisms of $[Ca^{2+}]_i$ increase include mobilization of Ca^{2+} from the ER stores through IP_3 and ryanodine receptor/Ca^{2+} release channels and Ca^{2+} influx through VDCC and VICC. Coordinative regulation by $1,25(OH)_2D_3$ of the Ca^{2+} entry and Ca^{2+} mobilization pathways is responsible for triggering and maintaining Ca^{2+} oscillations (Sergeer and Rhoten, 1995a). The dependence of the amplitude of these oscillations on the $1,25(OH)_2D_3$ concentration implies that specific signaling information might be conveyed in such a form to control the magnitude of the cellular response.

The precise mechanisms of how $1,25(OH)_2D_3$ induces increase in $[Ca^{2+}]_i$ remain to be elucidated. Cloning of the membrane VDR would provide an opportunity to specify signal transduction events leading to, or dependent on, the $1,25(OH)_2D_3$-triggered $[Ca^{2+}]_i$ response. The type of Ca^{2+}channels stimulated by $1,25(OH)_2D_3$ in different cell types, and the possibility of direct interactions of the hormone with Ca^{2+} channels, as well as Ca^{2+} with VDRs, also remain open to research. Obviously, the nature of intracellular Ca^{2+} signals evoked by $1,25(OH)_2D_3$ is more complex than a simple surge in $[Ca^{2+}]_i$. Ca^{2+} oscillations, Ca^{2+} waves and spirals, propagation of the Ca^{2+} signal along the cytoskeletal elements, and the role of calbindins in such a propagation are possible research areas, which may be examined using confocal Ca^{2+} imaging.

From a practical point of view, the role of $1,25(OH)_2D_3$ in regulation of biological responses via intracellular Ca^{2+} signaling pathways opens exciting opportunities for the search for potent, clinically promising $1,25(OH)_2D_3$ analogues for inducing apoptosis in cancer cells, regulate insulin secretion from pancreatic β-cells or cell functions in bone, and alleviate neurodegenerative diseases, among other applications.

ACKNOWLEDGMENTS. I.N.S. would like to extend special thanks to A. W. Norman for his many scientific insights and encouragement with regard to vitamin D research. This work was supported in part by NIH grant CA 67317 to INS.

9. REFERENCES

Airaksinen, M. S., Eilers, J., Garaschuk, O., Thoenen, H., Konnerth, A., and Meyer, M., 1997, Ataxia and altered dendritic calcium signaling in mice carrying a targeted null mutation of the calbindin-D_{28k} gene, *Proc. Natl. Acad. Sci. USA* **94:**1488–1493.

Ahlstrom, M., Tornquist, K., and Lamberg-Allardt, C., 1995, 1,25-Dihydroxyvitamin D_3 reduces the number of α1-adrenergic receptors in FRTL-5 rat thyroid cells, *Mol. Cell. Endocrinol.* **108:**143–148.

Akerfeldt, K. S., Coyne, A. N., Wilk, R. R., Thulin, E., and Linse, S., 1996, Ca^{2+}-binding stoichiometry of calbindin-D_{28k} as assessed by spectroscopic analyses of synthetic peptide fragments, *Biochemistry* **35:**3662–3669.

Baran, D. T., 1994, Nongenomic actions of the steroid hormone 1α,25-dihydroxyvitamin D_3, *J. Cell. Biochem.* **56:**303–306.

Baran, D. T., and Milne, M., 1986, 1,25-Dihydroxyvitamin D increases hepatocyte cytosolic calcium levels, *J. Clin. Invest.* **77:**1622–1626.

Baran, D. T., Sorensen, A. M., Honeyman, T. W., Ray, R., and Holick, M. F., 1990, 1,25α-Dihydroxyvitamin D_3-induced increments in hepatocyte cytosolic calcium and lysophospatidylinositol: Inhibition by pertussis toxin and 1β,25-dihydroxyvitamin D_3, *J. Bone Miner. Res.* **5:**517–522.

Baran, D. T., Sorensen, A. M., Shalhoub, V., Owen, T., Oberdore, A., Stein, G., and Lian, J., 1991, 1α,25-Dihydroxyvitamin D_3 rapidly increases cytosolic calcium in clonal rat osteosarcoma cells lacking the vitamin D receptor, *J. Bone Miner. Res.* **6:**1269–1275.

Baran, D. T., Sorensen, A. M., Shalhoub, V., Owen, T., Stein, G., and Lian, J., 1992, The rapid nongenomic actions of 1α,25-dihydroxyvitamin D_3 modulate the hormone-induced increments in osteocalcin gene transcription in osteoblast-like cells, *J. Cell. Biochem.* **50:**124–129.

Barsony, J., McKoy, W., Renyi, I., and Liberman, M. E., 1994, A common mechanism for the vitamin D receptor activation and for the nongenomic actions of calcitriol, in *Vitamin D, A Pluripotent Steroid Hormone: Structural Studies, Molecular Endocrinology and Clinical Applications* (A. W. Norman, R. Bouillon, and M. Thomasset, eds.), pp. 345–346, de Gruyter, Berlin.

Barsony, J., Renyi, I., and McKoy, W., 1997, Subcellular distribution of normal and mutant vitamin D receptors in living cells, *J. Biol. Chem.* **272:**5774–5782.

Beno, D. W., Brady, L. M., Bissonnette, M., and Davis, B. H., 1995, Protein kinase C and mitogen-activated protein kinase are required for 1,25-dihydroxyvitamin D_3-stimulated Erg induction, *J. Biol. Chem.* **270:**3642–3647.

Berry, D. M., Antochi, R., Bhatia, M., and Meckling-Gill, K. A., 1996, 1,25-Dihydroxyvitamin D_3 stimulates expression and translocation of protein kinase C-α and C-δ via a nongenomic mechanism and rapidly induces phosphorylation of a 33-kDa protein in acute promyelocytic NB_4 cells, *J. Biol. Chem.* **271:**16090–16096.

Bhatia, M., Kirkland, J. B., and Meckling-Gill, K. A., 1996, 1,25-Dihydroxyvitamin D_3 primes acute promyelocytic cells for TPA-induced monocytic differentiation through both PKC and tyrosine phosphorylation cascades, *Exp. Cell. Res.* **222:**61–69.

Bissonnette, M., Tien, X.-Y., Niedziela, S. M., Hartmann, S. C., Frawley, B. P., Roy, H. K., Sitrin, M. D., Perlman, R. L., and Brasitus, T. A., 1994, $1,25(OH)_2$-Vitamin D_3 activates PKC-α in Caco-2 cells: A mechanism to limit secosteroid-induced rise in $[Ca^{2+}]_i$, *Am. J. Physiol.* **267:**G465–G475.

Bissonnette, M., Wali, R. K., Hartman, S. C., Niedziela, S. M., Roy, H. K., Tien, X. Y., Sitrin, M. D., and Brasitus, T. A., 1995, 1,25-Dihydroxyvitamin D_3 and 12-O-tetradecanol phorbol 13-acetate cause differential activation of Ca^{2+}-dependent and Ca^{2+}-independent isoforms of protein kinase C in rat colonocytes, *J. Clin. Invest.* **95:**2215–2221.

Bittiner, B., Bleehen, S. S., and MacNeil, S., 1991, $1\alpha,25(OH)_2$-Vitamin D_3 increases intracellular calcium in human keratinocytes, *Br. J. Dermatol.* **124:**230–235.

Bouhtiauy, I., Lajeunesse, D., Christakos, S., and Brunette, M. G., 1994a, Two vitamin D_3-dependent calcium binding proteins increase calcium reabsorption by different mechanisms. I. Effect of $CaBP_{28k}$, *Kidney Int.* **45:**461–468.

Bouhtiauy, I., Lajeunesse, D., Christakos, S., and Brunette, M. G., 1994b, Two vitamin D_3-dependent calcium binding proteins increase calcium reabsorption by different mechanisms. II. Effect of $CaBP_{9k}$, *Kidney Int.* **45:**469–474.

Bouillon, R., Okamura, W. H., and Norman, A. W., 1995, Structure–function relationships in the vitamin D endocrine system, *Endocr. Rev.* **16:**200–257.

Bradford, P. G., Jin, Y., and Hui, P., 1993, 1,25-Dihydroxyvitamin D_3 enhances the transcription and expression of the inositol trisphospate receptor gene in HL-60 cells, *Mol. Pharmacol.* **44:**292–297.

Brehier, A., and Thomasset, M., 1990, Stimulation of calbindin-D_{9k} ($CaBP_{9k}$) gene expression by calcium and 1,25-dihydroxycholecalciferol in fetal rat duodenal organ culture, *Endocrinology* **127:**580–587.

Bruns, M. E., Kleeman, E., and Bruns, D. E., 1986, Vitamin D dependent calcium binding protein in mouse yolk sac: Biochemical and immunochemical properties and responses to 1,25-dihydroxycholecalciferol, *J. Biol. Chem.* **261:**7485–7490.

Bruns, M. E., Overpeck, J. G., Smith, G. C., Hirsch, G. N., Mills, S. E., and Bruns, D. E., 1988, Vitamin D dependent calcium binding protein in rat uterus: Differential effects of estrogen, tamoxifen, progesterone and pregnancy on accumulation and cellular localization, *Endocrinology* **122:**2371–2378.

Buffenstein, R., Sergeev, I. N., and Pettifor, J. M., 1994, Absence of calcitriol-mediated nongenomic actions in isolated intestinal cells of the damara mole rat (*Cryptomys damarensis*), *Gen. Comp. Endocrinol.* **95:**25–30.

Caffrey, J. M., and Farach-Carson, M. C., 1989, Vitamin D_3 metabolites modulate dihydropyridine-sensitive calcium currents in clonal rat osteosarcoma cells, *J. Biol. Chem.* **264:**20265–20274.

Civitelli, R., Kim, Y. S., Gunsten, S. L., Fujimori, A., Huskey, M., Avioli, L. V., and Hruska, K. A., 1990, Nongenomic activation of the calcium message system by vitamin D metabolites in osteoblast-like cells, *Endocrinology* **127:**2253–2262.

Chard, P. S., Bleakman, D., Christakos, S., Fullmer, C. S., and Miller, R. J., 1993, Calcium buffering properties of calbindin D_{28k} and parvalbumin in rat sensory neurones, *J. Physiol.* **472:**341–357.

Chard, P. S., Jordan, J., Marcuccilli, C. J., Miller, R. J., Leiden, J. M., Roos, R. P., and Ghadge, G. D., 1995, Regulation of excitatory transmission at hippocampal synapses by calbindin-D_{28k}, *Proc. Natl. Acad. Sci. USA* **92:**5144–5148.
Chesnoy-Marchais, D., and Fritsch, J., 1993, Potassium currents and effects of vitamin D_3 metabolites and cyclic GMP in rat osteoblastic cells, *Biochim. Biophys. Acta* **1148:**239–248.
Christakos, S., Gabrielides, C., and Rhoten, W. B., 1989, Vitamin D-dependent calcium-binding proteins: Chemistry, distribution, functional considerations, and molecular biology, *Endocr. Rev.* **10:**3–26.
Christakos, S., Beck, J. D., and Hyllner, S. J., 1997, Calbindin-D_{28k}, in *Vitamin D* (D. Feldman, F. H. Glorieux, and J. W. Pike, eds.), pp. 209–221, Academic Press: San Diego.
Clapham, D. E., and Sneyd, J., 1995, Intracellular calcium waves, *Adv. Second Messenger Phosphoprotein Res.* **30:**1–24.
de Boland, A. R., and Boland, R. L., 1994, Non-genomic signal transduction pathway of vitamin D in muscle, *Cell. Signal.* **6:**717–724.
de Boland, A. R., and Norman, A. W., 1990a, Evidence for involvement of protein kinase C and cyclic adenosine 3′,5′-monophosphate-dependent protein kinase in the 1,25-dihydroxyvitamin D_3-mediated rapid stimulation of intestinal calcium transport (transcaltachia), *Endocrinology* **127:**39–45.
de Boland, A. R., and Norman, A. W., 1990b, Influx of extracellular calcium mediates 1,25-dihydroxyvitamin D_3-dependent transcaltachia (the rapid stimulation of duodenal Ca^{2+} transport), *Endocrinology* **127:**2475–2480.
de Boland, A. R., Morelli, S., and Boland, R., 1994, $1,25(OH)_2$-Vitamin D_3 signal transduction in chick myoblasts involves phosphatidylcholine hydrolysis, *J. Biol. Chem.* **269:**8675–8679.
Dupret, J. M., Brun, P., Perret, C., Lomri, N., Thomasset, M., and Cuisinier-Gleizes, P., 1987, Transcriptional and post-transcriptional regulation of vitamin D dependent calcium binding protein gene expression in the rat duodenum by 1,25-dihydroxycholecalciferol, *J. Biol. Chem.* **262:**16553–16557.
Farach-Carson, M. C., Sergeev, I., and Norman, A. W., 1991a, Nongenomic actions of 1,25-dihydroxyvitamin D_3 in rat osteosarcoma cells: Structure–function studies using ligand analogs, *Endocrinology* **129:**1876–1884.
Farach-Carson, M. C., Sergeev, I., Zhou, L.-X., and Norman, A. W., 1991b, Nongenomic actions of $1,25(OH)_2D_3$ in rat osteosarcoma cells and chick intestinal epithelial cells: Structure–function studies of ligand agonist analogs, in *Vitamin D: Gene Regulation, Structure–Function Analysis and Clinical Application* (A. W. Norman, R. Bouillon, and M. Thomasset, eds.), pp. 157–158, de Gruyter, Berlin.
Farach-Carson, M. C., Abe, J., Nishii, Y., Khoury, R., Wright, G. C., and Norman, A. W., 1993, 22-Oxacalcitriol: Dissection of $1,25(OH)_2D_3$ receptor-mediated and Ca^{2+} entry-stimulating pathways, *Am. J. Physiol.* **265:**F705–F711.
Feher, J. J., 1983, Facilitated diffusion by intestinal calcium-binding protein, *Am. J. Physiol.* **244:**C303–C307.
Feher, J. J., Fullmer, C. S., and Wasserman, R. H., 1992, Role of facilitated diffusion of calcium by calbindin in intestinal calcium absorption, *Am. J. Physiol.* **262:**C517–C526.
Felder, C. C., Singer-Lahat, D., and Mathes, C., 1994, Voltage-independent calcium channels. Regulation by receptors and intracellular calcium stores, *Biochem. Pharmacol.* **48:**1997–2004.
Friedlaender, M. M., Yagil, Y., Wald, H., and Popovtzer, M. M., 1995, 22-Oxacalcitriol does not interfere with parathyroid hormone-induced phosphaturia or cyclic AMP excretion, *Bone* **17:**301–306.
Galione, A., 1994, Cyclic ADP-ribose, the ADP-ribosyl cyclase pathway and calcium signaling, *Mol. Cell. Endocrinol.* **98:**125–131.

Gascon-Barré, M., Haddad, P., Provencher, S. J., Bilodeau, S., Pecker, F., Lotersztajn, S., and Vallieres, S., 1994, Chronic hypocalcemia of vitamin D deficiency leads to lower intracellular calcium concentration in rat hepatocytes, *J. Clin. Invest.* **93:**2159–2167.

Gilon, P., Shepherd, R. M., and Henquin, J.-C., 1993, Oscillations of secretion driven by oscillations of cytoplasmic Ca^{2+} as evidenced in single pancreatic islets, *J. Biol. Chem.* **268:**22265–22268.

Golovina, V. A., and Blaustein, M. P., 1997, Spatially and functionally distinct Ca^{2+} stores in sarcoplasmic and endoplasmic reticulum, *Science* **275:**1643–1648.

Grosse, B., Bourdeau, A., and Lieberherr, M., 1993, Oscillations in inositol 1,4,5-trisphosphate and diacylglycerol induced by vitamin D_3 metabolites in confluent mouse osteoblasts, *J. Bone Miner. Res.* **8:**1059–1069.

Hannah, S. S., and Norman, A. W., 1994, 1α,25-dihydroxyvitamin D_3-regulated expression of the eukaryotic genome, *Nutr. Rev.* **52:**376–382.

Ho, B.-K., Alexianu, M. E., Colom, L. V., Mohamed, A. H., Serrano, F., and Appel, S. H., 1996, Expression of calbindin-D_{28k} in motoneuron hybrid cells after retroviral infection with calbindin-D_{28k} cDNA prevents amyotrophic lateral sclerosis IgG-medicated cytotoxicity, *Proc. Natl. Acad. Sci. USA* **93:**6796–6801.

Jackson, D. M., and Holick, M. F., 1994, 1α,25-Dihydroxyvitamin D_3 induces a rapid and sustained increase in cytosolic calcium in cultured human keratinocytes, in *Vitamin D, A Pluripotent Steroid Hormone: Structural Studies, Molecular Endocrinology and Clinical Applications* (A. W. Norman, R. Bouillon, and M. Thomasset, eds.), pp. 351–352, de Gruyter, Berlin.

Jenis, L. G., Lian, J. B., Stein, G. S., and Baran, D. T., 1993, 1α,25-Dihydroxyvitamin D_3-induced changes in intracellular pH in osteoblast-like cells modulate gene expression, *J. Cell. Biochem.* **53:**234–239.

Johnson, J. A., and Kumar, R., 1994, Vitamin D and renal calcium transport, *Curr. Opin. Nephrol. Hypertens.* **3:**424–429.

Khare, S., Wilson, D. M., Tien, X.-Y., Dudeja, P. K., Wali, R. K., Sitrin, M. D., and Brasitus, T. A., 1993, 1,25-Dihydroxycholecalciferol rapidly activates rat colonic particulate guanylate cyclase via a protein kinase C-dependent mechanism, *Endocrinology* **133:**2213–2219.

Khare, S., Tien, X., Wilson, D., Wali, R. K., Bissonnette, B. M., Scaglione-Sewel, B., Sitrin, M. D., and Brasitus, T. A., 1994, The role of protein kinase-Cα in the activation of particulate guanylate cyclase by 1α,25-dihydroxyvitamin D_3 in Caco-2 cells, *Endocrinology* **135:**277–283.

Khoury, R., Ridall, A. L., Norman, A. W., and Farach-Carson, M. C., 1994, Target gene activation by 1,25-dyhydroxyvitamin D_3 in osteosarcoma cells is independent of calcium influx, *Endocrinology* **135:**2446–2453.

Khoury, R. S., Weber, J., and Farach-Carson, M. C., 1995, Vitamin D metabolites modulate osteoblast activity by Ca^{2+} influx-independent genomic and Ca^{2+} influx-dependent nongenomic pathways, *J. Nutr.* **125:**1699S–1703S.

Kim, Y. S., MacDonald, P. N., Dedhar, S., and Hruska, K. A., 1996, Association of 1α,25-dihydroxyvitamin D_3-occupied vitamin D receptors with cellular membrane acceptance sites, *Endocrinology* **137:**3649–3658.

Krishnan, A. V., and Feldman, D., 1991, Activation of protein kinase C inhibits vitamin D receptor gene expression, *Mol. Endocrinol.* **5:**605–612.

Leathers, V. L., Linse, S., Forsen, S., and Norman, A. W., 1990, Calbindin-D_{28k}, a 1α,25-dihydroxyvitamin D_3-induced calcium-binding protein, binds five or six Ca^{2+} ions with high affinity, *J. Biol. Chem.* **265:**9838–9841.

Lee, S. K., and Clark, N. B., 1993, Effects of cytochalasin B on calcium transport by $1,25(OH)_2D_3$- or PTH-treated chick embryonic yolk sac *in vitro* and *in vivo, J. Exp. Zool.* **266:**11–18.

Lee, Y. S., Reimers, T. J., Cowan, R. G., Fullmer, C. S., and Wasserman, R. H., 1988, Calcium-dependent translocation of calbindin-D_{28k} from intestine to blood, *Arch. Biochem. Biophys.* **261:**27–34.

Le Mellay, V., Grosse, B., and Lieberherr, M., 1997, Phospholipase Cβ and membrane action of calcitriol and estradiol, *J. Biol. Chem.* **272:**11902–11907.

Li, Z., Decavel, C., and Hatton, G. I., 1995, Calbindin-D_{28k}: Role in determining intrinsically generated firing patterns in rat supraoptic neurones, *J. Physiol.* **488:**601–608.

Lieberherr, M., 1987, Effects of vitamin D_3 metabolites on cytosolic free calcium in confluent mouse osteoblasts, *J. Biol. Chem.* **262:**168–173.

Lieberherr, M., Grosse, B., Duchambon, P., and Drueke, T., 1989, A functional cell surface receptor is required for the early action of 1,25-dihydroxyvitamin D_3 on the phosphoinositide metabolism in rat enterocytes, *J. Biol. Chem.* **262:**403–406.

Lieberherr, M., Grosse, B., Tassin, M.-T., Kachkache, M., and Bourdeau, A., 1994, Transmembrane signal pathways induced by calcitriol, estradiol, testosterone and progesterone in osteoblasts, in *Vitamin D, A Pluripotent Steroid Hormone: Structural Studies, Molecular Endocrinology and Clinical Applications* (A. W. Norman, R. Bouillon, and M. Thomasset, eds.), pp. 315–323, de Gruyter, Berlin.

Lledo, P.-M., Somasundaram, B., Morton, A. J., Emson, P. C., and Mason, W. T., 1992, Stable transfection of calbindin-D_{28k} into the GH_3 cell line alters calcium currents and intracellular calcium homeostasis, *Neuron* **9:**943–954.

Lowe, K. E., Maiyar, A. C., and Norman, A. W., 1992, Vitamin D-mediated gene expression, *Crit. Rev. Eukaryotic Gene Expression* **2:**65–109.

Lucas, P. A., Roulet, C., Duchambon, P., Lacour, B., and Drueke, T., 1989, Rapid stimulation of calcium uptake by isolated rat enterocytes by $1,25(OH)_2D_3$, *Pflügers Arch.* **413:**407–413.

Marinissen, M. J., Selles, J., and Boland, R., 1994, Involvement of protein kinase C in $1,25(OH)_2$-vitamin D_3 regulation of calcium uptake by cultured myocytes, *Cell. Signal.* **6:**531–538.

Massheimer, V., Boland, R., and de Boland, A. R., 1994, Rapid $1,25(OH)_2$-vitamin D_3 stimulation of calcium uptake by rat intestinal cells involves a dihydropyridine-sensitive cAMP-dependent pathway, *Cell. Signal.* **6:**299–304.

Mathieu, C. L., Mills, S. E., Burnett, S. H., Cloney, D. L., Bruns, D. E., and Bruns, M. E., 1989, The presence and estrogen control of immunoreactive calbindin-D_{9k} in the fallopian tube of the rat, *Endocrinology* **125:**2745–2750.

Mattson, M. P., Rychlik, B., Chu, C., and Christakos, S., 1991, Evidence for calcium-reducing and excito-protective roles for the calcium-binding protein calbindin-D_{28k} in cultured hippocampal neurons, *Neuron* **6:**41–51.

Meissner, G., 1994, Ryanodine receptors/Ca^{2+} release channels and their regulation by endogenous effectors, *Annu. Rev. Physiol.* **56:**485–508.

Meszaros, J. G., Karin, N. J., Akanbi, K., and Farach-Carson, M. C., 1996, Down regulation of L-type Ca^{2+} channel transcript by 1,25-dihydroxyvitamin D_3. Osteoblastic cells express L-type α1C Ca^{2+} channel isoforms, *J. Biol. Chem.* **271:**32981–32985.

Mezzetti, G., Monti, M. G., Casolo, L. P., Piccinini, G., and Moruzzi, M. S., 1988, 1,25-Dihydroxycholecalciferol-dependent calcium uptake by mouse mammary gland in culture, *Endocrinology* **122:**389–394.

Morgan, D. W., Welton, A. F., Heich, A. E., and Christakos, S., 1986, Specific *in vitro* activation of Ca,Mg-ATPase by vitamin D-dependent rat renal calcium binding protein (calbindin D_{28k}), *Biochem. Biophys. Res. Commun.* **138:**547–553.

Muir, E. M., Lawson, D. E. M., Sepulveda, F. V., O'Brien, J. A., and Harding, M., 1993, Calbindin D_{28k} expression in transfected mouse NIH3T3 cells, *Cell Calcium* **14:**349–358.

Nemere, I., 1995, Nongenomic effects of 1,25-dihydroxyvitamin D_3: Potential relation of a plasmalemmal receptor to the acute enhancement of intestinal calcium transport in chick, *J. Nutr.* **125:**1695S–1698S.

Nemere, I., and Norman, A. W., 1988, 1,25-Dihydroxyvitamin D_3-mediated vesicular transport of calcium in intestine: Time-course studies, *Endocrinolgy* **122:**2962–2969.

Nemere, I., and Norman, A. W., 1989, 1,25-Dihydroxyvitamin D_3-mediated vesicular calcium transport in intestine: Dose–response studies, *Mol. Cell. Endocrinol.* **67:**47–53.

Nemere, I., and Norman, A. W., 1990, Transcaltachia, vesicular calcium transport, and microtubule-associated calbindin-D_{28k}: Emerging views of 1,25-dihydroxyvitamin D_3-mediated intestinal calcium absorption, *Miner. Electrolyte Metab.* **16:**109–114.

Nemere, I., and Norman, A. W., 1991, Steroid hormone actions at the plasma membrane: Induced calcium uptake and exocytotic events, *Mol. Cell. Endocrinol.* **80:**C165–C169.

Nemere, I., Leathers, V., and Norman, A. W., 1986, 1,25-Dihydroxyvitamin D_3-mediated intestinal calcium transport: Biochemical identification of lysosomes containing calcium and calcium-binding protein (calbindin-D_{28k}), *J. Biol. Chem.* **261:**16106–16114.

Nemere, I., Leathers, V. L., Thompson, B. S., Luben, R. A., and Norman, A. W., 1991, Redistribution of calbindin-D_{28k} in chick intestine in response to calcium transport, *Endocrinology* **129:**2972–2984.

Nemere, I., Zhou, L.-X., and Norman, A. W., 1993, Nontranscriptional effects of steroid hormones, *Receptor* **3:**277–291.

Nemere, I., Dormanen, M. C., Hammond, M. W., Okamura, W. H., and Norman, A. W., 1994, Identification of a specific binding protein for 1α,25-dihydroxyvitamin D_3 in basal-lateral membranes of chick intestinal epithelium and relationship to transcaltachia, *J. Biol. Chem.* **269:**23750–23756.

Norman, A. W., 1979, *Vitamin D: The Calcium Homeostatic Steroid Hormone*, Academic Press, New York.

Norman, A. W., 1995, Pleiotropic actions of 1α,25-dihydroxyvitamin D_3: An overview, *J. Nutr.* **125:**1687S–1689S.

Norman, A. W., 1997, in *Vitamin D* (D. Feldman, F. H. Glorieux and J. W. Pike, eds.), pp. 233–256, Academic Press: San Diego.

Norman, A. W., Koeffler, H. P., Bishop, J. E., Collins, E. D., Sergeev, I., Zhou, L.-X., Nemere, I., Henry, H. L., and Okamura, W. H., 1991, Structure–function relationships in the vitamin Dendocrine system for $1,25(OH)_2D_3$ analogs, in *Vitamin D: Gene Regulation, Structure–Function Analysis and Clinical Application* (A. W. Norman, R. Bouillon, and M. Thomasset, eds.), pp. 146–153, de Gruyter: Berlin.

Norman, A. W., Nemere, I., Zhou, J., Bishop, J. E., Lowe, K. E., Maiyar, A. C., Collins, E. D., Taoka, T., Sergeev, I., and Farach-Carson, M. C., 1992, $1,25(OH)_2$-Vitamin D_3, a steroid hormone that produces biological effects via both genomic and nongenomic pathways, *J. Steroid Biochem. Mol. Biol.* **41:**231–240.

Norman, A. W., Sergeev, I. N., Bishop, J. E., and Okamura, W. H., 1993, Selective biological response by target organs (intestine, kidney and bone) to 1,25-dihydroxyvitamin D_3 and two analogs, *Cancer Res.* **53:**3935–3942.

Petersen, O. H., Petersen, C., and Kasai, H., 1994, Calcium and hormone action, *Annu. Rev. Physiol.* **56:**297–319.

Pillai, S., and Bikle, D. D., 1991, Role of intracellular free calcium in the cornified envelope formation of keratinocytes: Difference in the mode of action of extracellular calcium and 1,25-dihydroxyvitamin D_3, *J. Cell. Physiol.* **146:**94–100.

Pillai, S., Bikle, D. D., Su, M. J., Ratnam, A., and Abe, J., 1995, 1,25-Dihydroxyvitamin D_3 upregulates the phospatidylinositol signaling pathway in human keratinocytes by increasing phospholipase C levels, *J. Clin. Invest.* **96:**602–609.

Rastinejad, F., Perlmann, T., Evans, R. M., and Sigler, P. B., 1995, Structural determinants of nuclear receptor assembly on DNA direct repeats, *Nature* **375:**203–211.

Reddy, D., Pollock, A. S., Clark, S. A., Sooy, K., Vasavada, R. C., Stewart, A. F., Honeyman, T., and Christakos, S., 1997, Transfection and overexpression of the calcium binding protein calbindin-D_{28k} results in a stimulatory effect on insulin synthesis in a rat β cell line (RIN 1046-38), *Proc. Natl. Acad. Sci. USA* **94:**1961–1966.

Reisner, P. D., Christakos, S., and Vanaman, T. C., 1992, *In vitro* enzyme activation with calbindin-D_{28k}, the vitamin D-dependent 28 kDa calcium binding protein, *FEBS Lett.* **297:**127–131.

Rhoten, W. B., and Sergeev, I. N., 1994a, Calbindin-D_{28k} appears to buffer intracellular Ca^{2+} in butyrate-treated rat insulinoma cells, *Endocrine* **2:**989–995.

Rhoten, W. B., and Sergeev, I. N., 1994b, Does calbindin-D_{28k} buffer intracellular free calcium? in *Vitamin D, A Pluripotent Steroid Hormone: Structural Studies, Molecular Endocrinology and Clinical Applications* (A. W. Norman, R. Bouillon, and M. Thomasset, eds.) pp. 412–413, de Gruyter, Berlin.

Roberts, W. M., 1994, Localization of calcium signals by a mobile calcium buffer in frog saccular hair cells, *J. Neurosci.* **14:**3246–3262.

Rosen, L. B., Ginty, D. D., and Greenberg, M. E., 1995, Calcium regulation of gene expression, *Adv. Second Messenger Phosphoprotein Res.* **30:**225–254.

Schwartz, Z., Langston, G. G., Swain, L. D., and Boyan, B. D., 1991, Inhibition of $1,25(OH)_2D_3$-dependent stimulation of alkaline phosphatase activity by A23187 suggests a role for calcium in the mechanism of vitamin D regulation of chondrocyte cultures, *J. Bone Miner. Res.* **6:**159–162.

Selles, J., and Boland, R., 1991, Rapid stimulation of calcium uptake and protein phosphorylation in isolated cardiac muscle by 1,25-Dihydroxyvitamin D_3, *Mol. Cell. Endocrinol.* **77:**67–73.

Selles, J., Bellido, T., and Boland, R., 1994, Modulation of calcium uptake in cultured cardiac muscle cells by 1,25-dihydroxyvitamin D_3, *J. Mol. Cell. Cardiol.* **26:**1593–1599.

Sergeev, I. N., 1989, Metabolism and receptors of vitamin D: An overview, *Vopr. Med. Khimii* **35:**2–11 [in Russian].

Sergeev, I. N., and Norman, A. W., 1992, Vitamin K-dependent γ-carboxylation of the vitamin D receptor, *Biochem. Biophys. Res. Commun.* **189:**1543–1547.

Sergeev, I. N., and Rhoten, W. B., 1994, Video imaging of intracellular calcium in insulinoma cells: Effects of $1,25(OH)_2D_3$, in *Vitamin D, A Pluripotent Steroid Hormone: Structural Studies, Molecular Endocrinology and Clinical Applications* (A. W. Norman, R. Bouillon, and M. Thomasset, eds.), pp. 355–356, de Gruyter, Berlin.

Sergeev, I. N., and Rhoten, W. B., 1995a, 1,25-Dihydroxyvitamin D_3 evokes oscillations of intracellular calcium in a pancreatic β-cell line, *Endocrinology* **136:**2852–2861.

Sergeev, I. N., and Rhoten, W. B., 1995b, Video imaging of calcium and pH in breast cancer cells, *Mol. Biol. Cell* **6:**351a.

Sergeev, I. N., and Rhoten, W. B., 1997, Regulation of intracellular calcium in breast cancer cells, in *Abstracts 10th Workshop on Vitamin D*, Strasbourg, France, p. 187.

Sergeev, I. N., Rhoten, W. B., and Carney, M. D., 1996, Calbindins decreased after space flight, *Endocrine* **5:**335–340.

Sergeev, I. N., and Spirichev, V. B., 1988, Vitamin K role in the interaction of $1,25(OH)_2D_3$ receptors with DNA, in *Vitamin D: Molecular, Cellular and Clinical Endocrinology* (A. W. Norman ed.), pp. 237–239, Walter de Gruyter, Berlin.

Sergeev, I. N., and Spirichev, V. B., 1989, Do 1,25-dihydroxyvitamin D_3 receptors depend on vitamin K? *Nutr. Res.* **9:**725–733.

Sergeev, I. N., Rhoten, W. B., and Norman, A. W., 1997, 1,25-Dihydroxyvitamin D_3, intracellular Ca^{2+} and apoptosis in breast cancer cells, in *Vitamin D: Chemistry, Biology and Clinical Applications of the Steroid Hormone* (A. W. Norman, R. Bouillon and M. Thomasset, eds.), pp. 473–474, Univ. California Printing: Riverside.

Shan, J., Resnick, L. M., Lewanczuk, R. Z., Karpinski, E., Li, B., and Pang, P. K., 1993, 1,25-Dihydroxyvitamin D as a cardiovascular hormone. Effects on calcium current and cytosolic free calcium in vascular smooth muscle cells, *Am. J. Hypertens.* **6:**983–988.

Simboli-Campbell, M., Gagnon, A., Franks, D. J., and Welsh, J., 1994, 1,25-Dihydroxyvitamin D_3 transclocates protein kinase Cβ to nucleus and enhances plasma membrane association of protein kinase Cα in renal cells, *J. Biol. Chem.* **269:**3257–3264.

Sorensen, A. M., Bowman, D., and Baran, D. T., 1993, 1α,25-Dihydroxyvitamin D_3 rapidly increases nuclear calcium levels in rat osteosarcoma cells, *J. Cell. Biochem.* **52:**237–242.

Spirichev, V. B., and Sergeev, I. N., 1988, Vitamin D: Experimental research and its practical application, *World Rev. Nutr. Dietet.* **56:**173–216.

Su, M. J., Bikle, D. D., Mancianti, M. L., and Pillai, S., 1994, 1,25-Dihydroxyvitamin D_3 potentiates the keratinicyte response to calcium, *J. Biol. Chem.* **269:**14723–14729.

Sylvia, V. L., Schwartz, Z., Ellis, E. B., Helm, S. H., Gomez, R., Dean, D. D., and Boyan, B. D., 1995, Nongenomic regulation of protein kinase C isoforms by the vitamin D metabolites $1\alpha,25(OH)_2D_3$ and $24R,25(OH)_2D_3$, *J. Cell. Physiol.* **167:**380–393.

Szego, C. M., 1994, Cytostructural correlates of hormone action: New common ground in receptor-mediated signal propagation for steroid and peptide agonists, *Endocrine* **2:**1079–1093.

Tasaka, T., Tokuda, M., Taoka, T., Itano, T., Matsui, H., Etoh, S., Nishio, H., Miyamoto, O., Irino, S., and Hatase, O., 1991, Mechanism of transient increase in intracellular concentration of free calcium ions in HL-60 cell differentiation induced by vitamin D_3 and phorbol ester, *Biochem. Int.* **23:**137–143.

Thomasset, M., 1997, Calbindin-D_{9k}, 1997, in *Vitamin D* (D. Feldman, F. H. Glorieux and J. W. Pike, eds.), pp. 223–232, Academic Press: San Diego.

Tien, X.-Y., Brasitus, T. A., Qasawa, B. M., Norman, A. W., and Sitrin, M. D., 1993a, Effect of $1,25(OH)_2D_3$ and its analogues on membrane phosphoinositide turnover and $[Ca^{2+}]_i$ in Caco-2 cells, *Am. J. Physiol.* **265:**G143–G148.

Tien, X. Y., Katnik, C., Qasawa, B. M., Sitrin, M. D., Nelson, D. J., and Brasitus, T. A., 1993b, Characterization of the 1,25-dihydroxychoilecalciferol-stimulated calcium influx pathway in Caco-2 cells, *J. Membr. Biol.* **136:**159–168.

Timmermans, J. A. H., Bindels, R. J. M., and Van Os, C. H., 1995, Stimulation of plasma membrane Ca^{2+} pump by calbindin-D_{28k} and calmodulin is additive in EGTA-free solutions, *J. Nutr.* **125:**1981S–1986S.

Tornquist, K., 1992, Pretreatment with 1,25-dihydroxycholecalciferol enhances thyrotropin-releasing hormone- and inositol 1,4,5-trisphosphate-induced release of sequestered Ca^{2+} in permeabilized GH_4C_1 pituitary cells, *Endocrinology* **131:**1677–1688.

Tornquist, K., and Tashjian, A. H., 1989, Dual actions of 1,25-dihydroxycholecalciferol on intracellular Ca^{2+} in GH_4C_1 cells: Evidence for effects on voltage-operated Ca^{2+} channels and Na^+/Ca^{2+} exchange, *Endocrinology* **124:**2765–2776.

Tornquist, K., Stewen, P., and Lamberg-Allard, C., 1992, Regulatory effects of 1,25-dihydroxycholecalciferol on calcium fluxes in thyroid FRTL-5 cells, *Mol. Cell. Endocrinol.* **86:**21–27.

Vandewalle, B., Hornez, L., Wattez, N., Revillion, F., and Lefebvre, J., Vitamin D_3 derivatives and breast tumor cell growth: Effect of intracellular calcium and apoptosis, *Int. J. Cancer* **61:**806–811.

van Leeuwen, J., Birkenhager, J. C., Scilte, J. P., Buurman, C. J., and Pols, H. A., 1990, Role of calcium and cAMP in heterologous up-regulation of the 1,25-dihydroxyvitamin D_3 receptor in an osteoblast cell line, *Cell Calcium* **11:**281–289.

van Leeuwen, J., Birkenhager, J. C., van den Bemd, G., Buurman, C. J., Staal, A., Bos, M. P., and Pols, H., 1992, Evidence for the functional involvement of protein kinase C in the action of 1,25-dihydroxyvitamin D_3 in bone, *J. Biol. Chem.* **267:**12562–12566.

Vazquez, G., and de Boland, A. R., 1993, Stimulation of dihydropyridine-sensitive Ca^{2+} influx in cultured myoblasts by 1,25$(OH)_2$-vitamin D_3, *Biochem. Mol. Biol. Int.* **31:**677–684.

Vazquez, G., Boland, R., and de Boland, A. R., 1995, Modulation by 1,25$(OH)_2$-vitamin D_3 of the adenylyl cyclase/cyclic AMP pathway in rat and chick myoblasts, *Biochim. Biophys. Acta* **1269:**91–97.

Wali, R. K., Baum, C. L., Bolt, M. J., Brasitus, T. A., and Sitrin, M. D., 1992, 1,25-Dihydroxyvitamin D_3 inhibits Na^+–H^+ exchange by stimulating membrane phosphoinositide turnover and increasing cytosolic calcium in Caco-2 cells, *Endocrinology* **131:**1125–1131.

Walters, J. R. F., 1989, Calbindin-D_{9k} stimulates the calcium pump in rat enterocyte basolateral membranes, *Am. J. Physiol.* **256:**G124–G128.

Walters, M. R., 1992, Newly identified actions of the vitamin D endocrine system, *Endocr. Rev.* **13:**719–764.

Walters, M. R., 1995, Newly identified actions of the vitamin D endocrine system: Update 1995, *Endocr. Rev.* **4:**47–56.

Walters, M. R., Ilenchuk, T. T., and Claycomb, W. C., 1987, 1,25-Dihydroxyvitamin D_3 stimulates $^{45}Ca^{2+}$ uptake by cultured adult rat ventricular cardiac muscle cells, *J. Biol. Chem.* **262:**2536–2541.

Wasserman, R. H., and Taylor, A. N., 1966, Vitamin D_3-induced calcium-binding protein in chick intestinal mucosa, *Science* **152:**791–793.

Wasserman, R. H., Chandler, J. S., Meyer, S. A., Smith, C. A., Brindak, M. E., Fullmer, C. S., Penniston, J. T., and Kumar, R., 1992, Intestinal calcium transport and calcium extrusion processes at the basolateral membrane, *J. Nutr.* **122:**662–671.

Welsh, J., Simboli-Campbell, M., Narvaez, C. J., and Tenniswood, M., 1995, Role of apoptosis in the growth inhibitory effects of vitamin D in MCF-7 cells, *Adv. Exp. Med. Biol.* **375:**45–52.

Yukihiro, S., Posner, G. H., and Guggino, S. E., 1994, Vitamin D_3 analogs stimulate calcium currents in rat osteosarcoma cells, *J. Biol. Chem.* **269:**23889–23893.

Zanelo, L. P., and Norman, A. W., 1996, 1α,25$(OH)_2$-vitamin D_3-mediated stimulation of outward anionic currents in osteoblast-like ROS 17/2.8 cells, *Biochem. Biophys. Res. Commun.* **225:**551–556.

Part IV

Vitamin E

Chapter 13

Antioxidant Chemistry of α-Tocopherol in Biological Systems

Roles of Redox Cycles and Metabolism

Daniel C. Liebler

1. INTRODUCTION

Vitamin E is the family name given to a group of tocopherols and tocotrienols that function as the principal lipid-soluble chain-breaking antioxidants in biological membranes and lipoproteins. α-Tocopherol (**1**) and, to a lesser extent, γ-tocopherol (**2**) constitute an essential component of cellular defense against endogenous and exogenous oxidants.

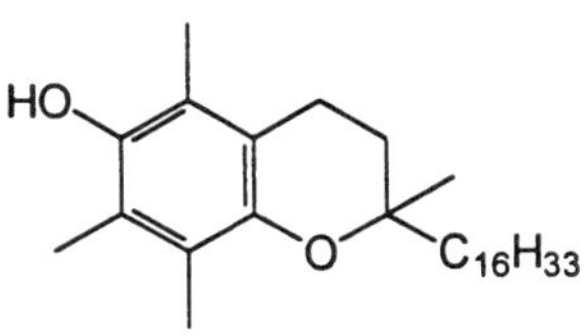

1

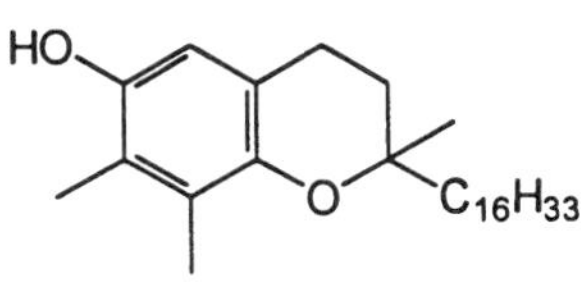

2

Daniel C. Liebler Department of Pharmacology and Toxicology, College of Pharmacy, University of Arizona, Tucson, Arizona 85721-0207.

Subcellular Biochemistry, Volume 30: Fat-Soluble Vitamins, edited by Quinn and Kagan. Plenum Press, New York, 1998.

Many components of cellular antioxidant defense are enzymes, such as superoxide dismutase, or enzyme-dependent systems, such as the glutathione redox cycle. In contrast, the antioxidant reactions of tocopherols do not require the participation of enzymes. The reaction of α-tocopherol with lipid peroxyl radicals or other oxidants in a membrane is a fast, nonenzymatic reaction in which the radical is consumed and α-tocopherol is converted to oxidation products. The principal role of α-tocopherol as an antioxidant is in scavenging lipid peroxyl radicals, which are the chain-carrying species that propagate lipid peroxidation (Liebler, 1993; Burton and Ingold, 1986). However, tocopherols do react with other reactive oxidants, including other oxygen radicals, singlet oxygen, and reactive nitrogen species, and these reactions also contribute to cellular antioxidant defense.

Although tocopherol antioxidant reactions themselves are not catalyzed by enzymes, interactions with other cellular antioxidants are thought to regenerate and maintain tocopherols and thereby to link tocopherol antioxidant function to cellular metabolism. This chapter reviews three aspects of the antioxidant function of tocopherols in cell membranes. First, the antioxidant reactions of tocopherols and analytical approaches to monitoring these antioxidant reactions are discussed. Second, the participation of redox reactions with other antioxidants in maintaining the cellular α-tocopherol status is discussed. Finally, the metabolic reduction of α-tocopherolquinone to α-tocopherolhydroquinone by quinone oxidoreductase is considered, along with the possible contribution of α-tocopherol hydroquinone generated by this route to cellular antioxidant defense.

2. ANTIOXIDANT CHEMISTRY

2.1. Peroxyl Radical Scavenging

The principal antioxidant function of α-tocopherol is as a peroxyl radical scavenger, a role to which it is ideally suited (Burton and Ingold, 1986). Stereoelectronic features of α-tocopherol contribute to the stabilization the tocopheroxyl radical (**3**) and account for the high first-order rate constant for hydrogen transfer from α-tocopherol to peroxyl radicals (Burton and Ingold, 1981; Burton *et al.*, 1985). α-Tocopherol reacts with a peroxyl radical either by concerted hydrogen transfer or by sequential electron, then proton transfer to form a lipid hydroperoxide and the tocopheroxyl radical:

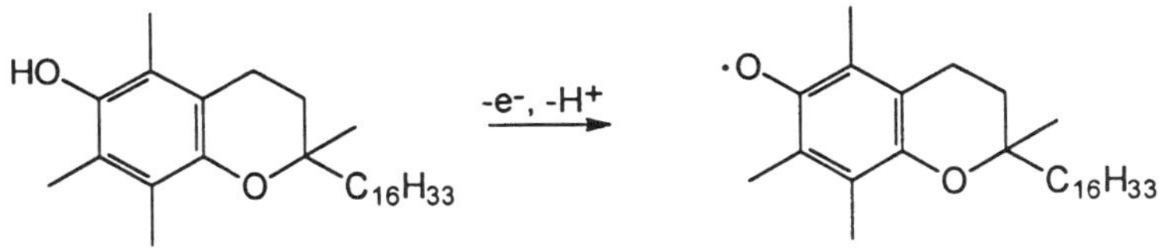

(1)

1 3

The tocopheroxyl radical then may either (1) react with a second peroxyl radical to form nonradical products, (2) undergo reduction back to α-tocopherol by another cellular reductant, or (3) abstract a hydrogen atom from a nonradical molecule (e.g., a polyunsaturated lipid) to reform α-tocopherol and a lipid radical. The first two reactions contribute to antioxidant protection. The latter may facilitate oxidative damage under certain conditions, as is discussed later in this chapter and in considerable detail by Witting *et al.* in Chapter 15 of this volume.

Reaction of the tocopheroxyl radical with a peroxyl radical yields two groups of products (Winterle *et al.*, 1984; Yamauchi *et al.*, 1989a,b; Liebler *et al.*, 1990; Liebler and Burr, 1992) (Fig. 1). The first are 8α-substituted tocopherones, which bear a peroxyl radical-derived adduct or a hydroxyl group at C-8a of the chromanol ring system. The 8α-alkyldioxytocopherones (**4**) hydrolyze to 8α-hydroxytocopherone (**5**), which then rearranges to α-tocopherolquinone (**6**) (Liebler *et al.*, 1989). The second group are the epoxy-8α-hydroperoxytocopherones (**8/9**), which bear an epoxide at either C-4a/C-5 or C-7/C-8 of the chromanol system and a hydroperoxy substituent at C-8a. These hydrolyze and rearrange to epoxyquinones (**10/11**).

Minor products formed during peroxyl radical scavenging may include tocopherol dimers and trimers (Yamauchi *et al.*, 1989a). Spiro dimer (**12**) is formed by dimerization of an orthoquinone methide intermediate (Skinner and Alaupovic, 1963; Suarna *et al.*, 1988), whereas both spiro dimer 12 and dihydroxy dimer (**13**) are formed in UV-induced α-tocopherol photooxidations, where locally high concentrations of tocopherol radicals favor dimerization (Kramer and Liebler, 1997).

Finally, reversible addition of oxygen to C-8a of the tocopheroxyl radical followed by hydrogen abstraction by the resulting 8α-peroxyl intermediate yields 8α-hydroperoxytocopherone (**14**), a minor oxidation product observed during peroxyl radical-mediated oxidations in lipid bilayers and isolated mitochondria (Matsuo *et al.*, 1989; Liebler *et al.*, 1990; Ham and Liebler, 1995).

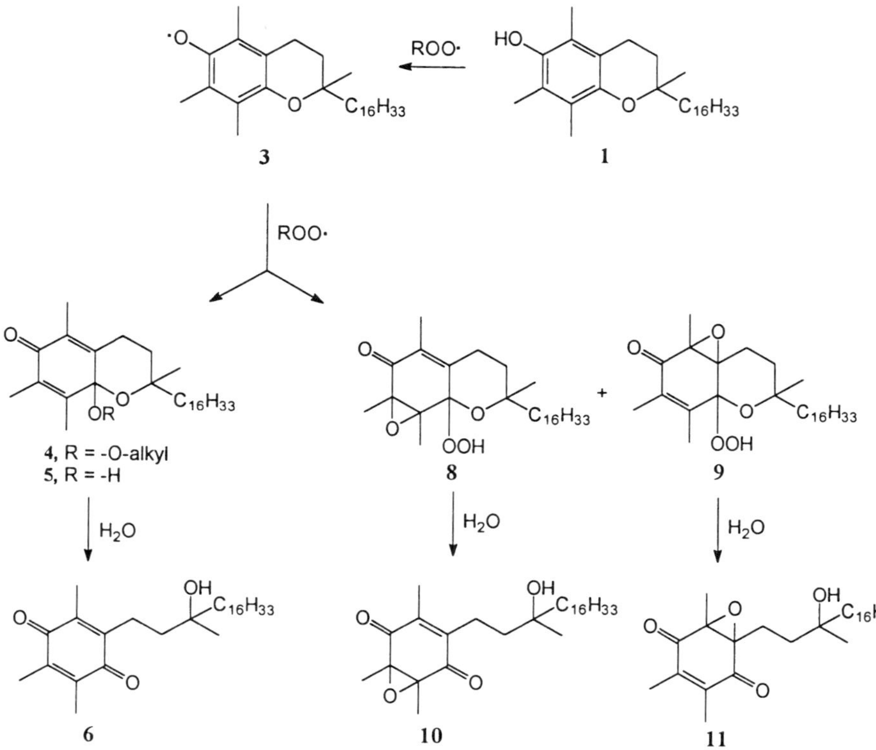

FIGURE 1. Structures of products formed by reaction of α-tocopherol with peroxyl radicals.

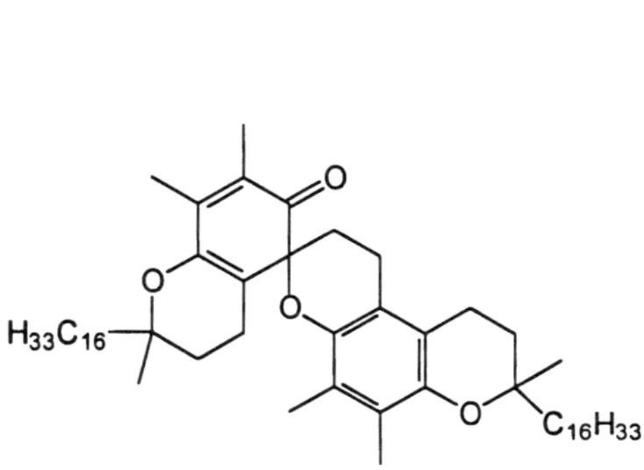

12

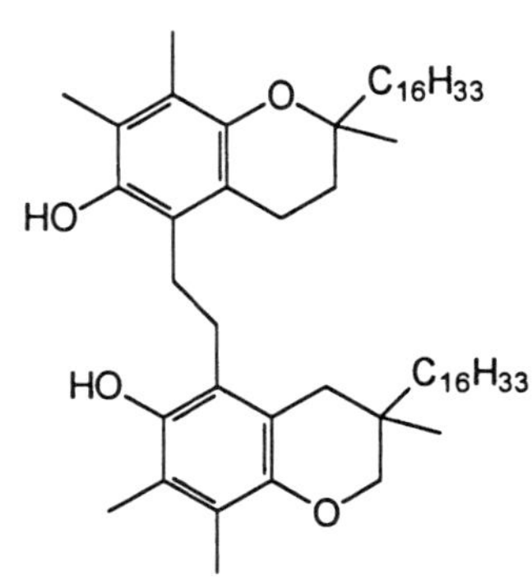

13

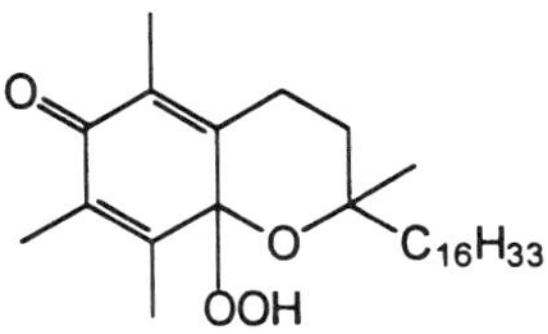

14

2.2. Chemistry of 8a-Substituted Tocopherones

2.2.1. Formation of 8a-Substituted Tocopherones

The 8a-substituted tocopherones appear to be the major end products of peroxyl radical scavenging in biological membranes. Product studies in chemically defined model systems indicate that organic peroxyl radicals derived from azo initiators [e.g., azobis(isobutyronitrile) and azobis(2,4-dimethylvaleronitrile) (AMVN) (Winterle *et al.*, 1984; Yamauchi *et al.*, 1989a,b; Liebler *et al.*, 1990)] form 8a-alkyldioxytocopherones and that analogous products are formed with methyl linoleate peroxyl radicals (Yamauchi *et al.*, 1990) and phospholipid peroxyl radicals (Yamauchi *et al.*, 1994). 8a-Substituted tocopherones also are formed by electron transfer from the tocopheroxyl radical to form a tocopherone action (**15**), which then hydrolyzes to form an 8a-hydroxytocopherone (Durckheimer and Cohen, 1964; Liebler and Burr, 1992).

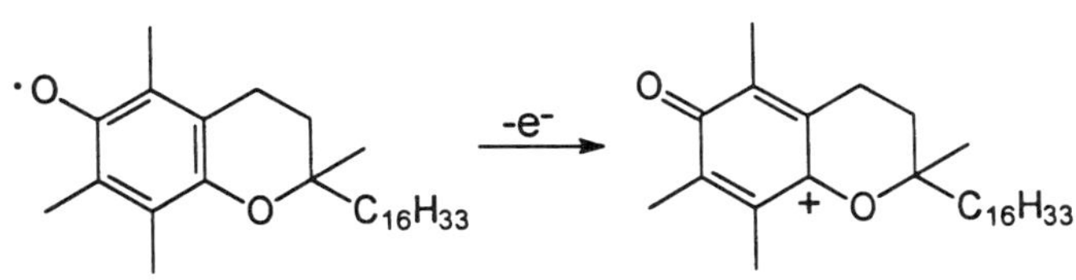

3 **15** (2)

It is not clear how the balance between radical adduction at C-8a and oxidation of the tocopheroxyl radical to tocopherone cation (**15**) is controlled. In biological membranes and lipid bilayer models, both 8a-alkyldioxytocopherones and 8a-hydroxytocopherone have been identified as α-tocopherol oxidation products. In liposomal oxidations initiated by the azo initiator AMVN, 8a-alkyldioxytocopherones derived from both AMVN-derived peroxyl radicals and phospholipid peroxyl radicals were identified (Liebler *et al.*, 1991; Yamauchi *et al.*, 1994). In contrast, lipid-derived 8a-alkyldioxytocopherones were not detected in iron- and superoxide-dependent oxidation of liposomes (Liebler and Burr, 1992) or in mitochondria exposed *in vitro* to azobis(amidinopropane HCl), a water-soluble, peroxyl radical-generating compound (Ham and Liebler, 1995). The yield of 8a-alkyldioxytocopherones derived from phospholipid peroxyl radicals will depend on the propagation of a radical chain in membrane phospholipids, which should be suppressed until α-tocopherol is largely oxidized. These results suggest that peroxyl radicals may react with tocopheroxyl radicals either by adduction at C-8a or by one-electron oxidation. However, the data indicate that the latter process predominates in biological membranes.

2.2.2. Analysis of 8a-Substituted Tocopherones

8a-Alkyldioxytocopherones are stable to extraction and analysis at neutral pH and in the absence of metals and can be analyzed directly in extracts of biological membranes (Liebler, 1994). However, the tocopherones are not fluorescent nor do they undergo facile electrochemical reduction, except in acid, where they decompose. Direct high-pressure liquid chromatography (HPLC) analysis can be done by UV detection with monitoring at 240 nm or by LC-molecular spectrometry (MS). Chemiluminescence has been used to detect 8a-hydroperoxytocopherone (**14**) in reverse-phase HPLC analyses, although this method is not applicable to tocopherones that do not contain a hydroperoxide (Miyazawa *et al.*, 1992). Detection of 8a-substituted tocopherones can be achieved by analysis of α-tocopherolquinone, the end product of tocopherone hydrolysis and rearrangement. Because tocopherones are readily converted to α-tocopherolquinone by acid treatment, measurement of α-tocopherolquinone levels in samples processed with and without acid permits tocopherone content to be calculated by difference (Liebler and Burr, 1992; Liebler, 1994). Recent studies of α-tocopherol oxidation in rat liver microsomes *in vitro* (Liebler *et al.*, 1996) and in the perfused rat liver (Ham and Liebler, 1997) employed this analytical approach. Measured levels of α-tocopherolquinone and its reduction product α-tocopherolhydroquinone were approximately two- to threefold higher in acid-treated samples, which indicates that the majority of oxidized α-tocopherol was present as tocopherone products.

2.2.3. Stability and Fate of 8a-Substituted Tocopherones

In lipid bilayers as neutral pH, 8a-alkyldioxytocopherones undergo gradual hydrolysis to 8α-hydroxytocopherone, and hydrolysis is accelerated by acid (Liebler *et al.*, 1989). The 8a-hydroxytocopherone then undergoes rearrangement to α-tocopherolquinone. Both 8a-alkyldioxytocopherones and 8a-hydroxytocopherone can be reduced to α-tocopherol by ascorbic acid or nordehydroguaretic acid under acidic, but not neutral conditions (Liebler and Burr, 1992). In both cases, the intermediate reduced is the tocopherone cation (**15**), which is formed during 8a-alkyldioxytocopherone hydrolysis and is in hydration/dehydration equilibrium with 8a-hydroxytocopherone (**5**). The oxidation of α-tocopherol to an 8a-substituted tocopherone and subsequent reduction to α-tocopherol completes a two-electron redox cycle. The reduction of tocopherones to α-tocopherol requires a reductant and an acid catalyst and there is no direct evidence that 8a-substituted tocopherones are reduced in biological systems. However, the studies of Chan and colleagues indicated that α-tocopherol was regenerated from reversibly oxidized products in platelet and erythrocyte membranes treated with a metal chelator and either ascorbic acid or reduced glutathione (Chan *et al.*, 1991). Although these investigators hypothesized that the reducible intermediate was the tocopheroxyl radical, the stability of the reducible intermediate(s) suggests that they were 8a-substituted tocopherones (Liebler, 1993).

2.3. Chemistry of Epoxytocopherones

Epoxytocopherones (**8/9**) comprise the other major products of α-tocopherol oxidized by peroxyl radicals in aqueous/organic mixtures and in lipid bilayers (Liebler *et al.*, 1990, 1991; Lieblar and Burr, 1992). These products represent a four-electron oxidation of α-tocopherol, but the mechanism by which they are formed is unknown. Epoxide yield varies considerably with properties of the reaction medium, but the antioxidant stoichiometry of two radicals consumed by each α-tocopherol is unaffected (Liebler and Burr, 1995). This argues against previously proposed mechanisms (Matsuo *et al.*, 1989; Liebler *et al.*, 1990) for epoxytocopherone formation, which do not result in any net peroxyl radical consumption.

Epoxytocopherones are less stable than corresponding 8a-substituted alkyldioxytocopherones. They rapidly hydrolyze and rearrange to epoxyquinones (**10/11**), which are the stable end products of this reaction pathway (Liebler *et al.*, 1990). The hydrolytic instability of epoxytocoherones is the major barrier to direct analysis of these compounds. Reverse-phase HPLC and HPLC-MS analysis of these compounds has been reported (Yamauchi *et al.*, 1989b; Liebler *et al.*, 1990; Kramer and Liebler, 1997), although rearrangement to epoxyquinones (**10/11**) yields mixtures of epoxytocopherones and epoxyquinones during HPLC purification. Epoxytocopherones can be quantified indirectly by measuring levels of epoxyquinones (**10/11**) before and after treatment of samples with dilute acid, which facilitates epoxytocoherone rearrangement to epoxyquinones (Liebler *et al.*, 1996).

The biological significance of epoxytocopherones, if any, is unclear. These oxidation products are not directly reducible to α-tocopherol, as are the 8a-substituted tocopherones, so it seems unlikely that they are intermediates in a tocopherol redox cycle. Aside from any possible biological activities, their principal significance may be as markers for peroxyl radical scavenging reactions of α-tocopherol.

2.4. Other Antioxidant Reactions of Tocopherols

Tocopherols react with a variety of oxidants, including hydroxyl radicals, singlet oxygen (Grams *et al.*, 1971; Clough *et al.*, 1979; d'Ischia *et al.*, 1991), alkoxyl radicals (Suarna *et al.*, 1988; Suarna and Southwell-Keely, 1988; Suarna *et al.*, 1992), peroxynitrite (Hogg *et al.*, 1994; Christen *et al.*, 1997; Hoglen *et al.*, 1997), nitrogen dioxide (Cooney *et al.*, 1995), ozone (Giamvala *et al.*, 1986; Liebler *et al.*, 1993), and superoxide (Nishikimi *et al.*, 1980; Matsuo *et al.*, 1987; Ha and Csallany, 1992; Csallany and Ha, 1992). Several reports of direct α-tocopherol oxidation by nitric oxide (de Groot *et al.*, 1993; Hogg *et al.*, 1993; d'Ischia and Novellino, 1996; Gorbunov *et al.*, 1997) probably reflect oxidations mediated by peroxynitrite or higher oxides of nitrogen formed in the experimental systems employed (Hogg *et al.*, 1996). At physiological levels, tocopherols cannot efficiently scavenge hydroxyl radicals and alkoxyl radicals, which are too highly reactive toward many other biomolecules. However, alkoxyl radicals derived from fatty acids rearrange to epoxyallylic peroxyl radicals (Dix and Marnett, 1985), which then may be scavenged by tocopherols. Tocopherols are effective singlet-oxygen quenchers. Tocopherols quench singlet oxygen both by a physical quenching mechanism and by reaction to form oxidation products (Kaiser *et al.*, 1990). Singlet oxygen oxidation of α-tocopherol yields 8a-hydroperoxytocopherone (**14**) as the principal primary product (Clough *et al.*, 1979).

Tocopherols may be physiologically significant scavengers of reactive nitrogen species. Although nitric oxide itself does not react directly with α-tocopherol, more reactive nitric oxide-derived species including nitrogen dioxide (Cooney *et al.*, 1995) and peroxynitrite (Christen *et al.*, 1997; Hoglen *et al.*, 1997; Hogg *et al.*, 1994) may be scavenged by tocopherols in biological systems. Reaction of α-tocopherol with either species yields α-tocopherolquinone, apparently via an 8a-hydroxytocopherone (Hogg *et al.*, 1994). In contrast, γ-tocopherol reacts with peroxynitrite to form 5-nitro-γ-tocopherol (**16**), a 5,6-chromaquinone ("tocored", **17**), and 8a-hydroxy-γ-tocopherone (**18**) (Christen *et al.*, 1997; Hoglen *et al.*, 1997). The former two products are also produced by reaction of γ-tocopherol with nitrogen dioxide (Cooney *et al.*, 1995).

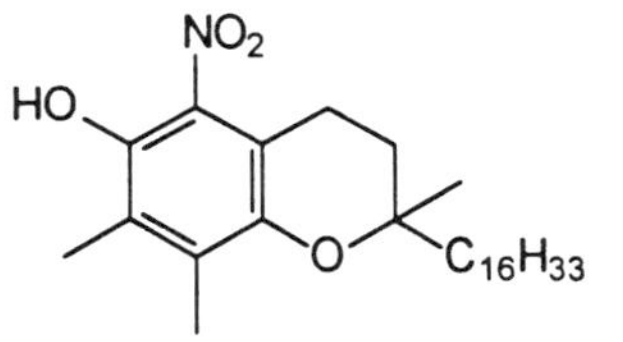

16

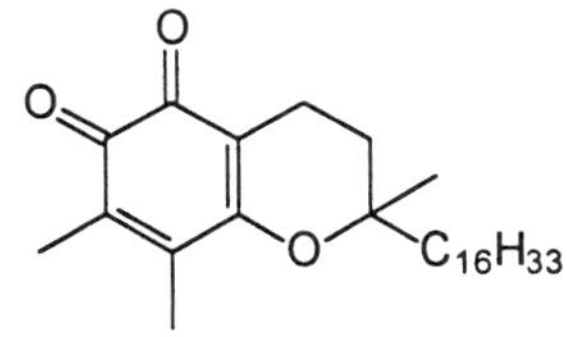

17

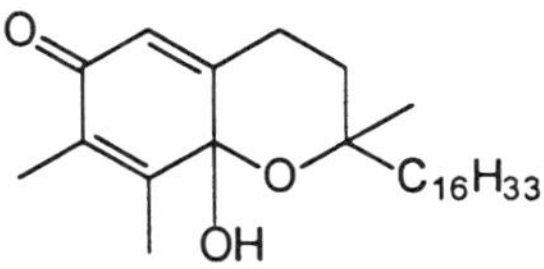

18

γ-Tocopherol may be superior to α-tocopherol as an antioxidant against reactive nitrogen species, as recent reports indicate that γ-tocopherol was more effective than α-tocopherol in inhibiting peroxynitrite-induced lipid peroxidation (Christen *et al.*, 1997) and nitrogen dioxide toxicity to rat insulinoma cells (Cooney *et al.*, 1995).

3. REDOX CYCLES IN α-TOCOPHEROL ANTIOXIDANT FUNCTION

3.1. One-Electron Redox Cycle

The ability of α-tocopheroxyl radicals to undergo reductive recycling by ascorbic acid was first documented by J. E. Packer *et al.* (1979), and numerous reports have since documented the interaction (for reviews, see Liebler, 1993; L. Packer and Kagan, 1993). A one-electron redox cycle (Fig. 2) regenerates α-tocopherol at the expense of another cellular reductant (e.g., ascorbate) and may couple membrane antioxidant defense to cellular metabolism. This metabolic coupling would involve regeneration of ascorbate by cellular glutathione, which in turn in maintained in reduced form enzymatically through the glutathione redox cycle.

Several lines of evidence support the feasibility of a one-electron α-tocopherol redox cycle. First, pulse radiolysis and laser flash photolysis studies have established the temporal relationship between tocopheroxyl radical formation and subsequent tocopherol regeneration accompanied by semidehydroascorbyl radical formation (J. E. Packer *et al.*, 1979; Bisby and Parker, 1995). Second, electron spin resonance studies have established that in tocopherol-containing solutions, micelles, liposomes, and membranes incubated with oxidants the tocopheroxyl radical signal is suppressed by the

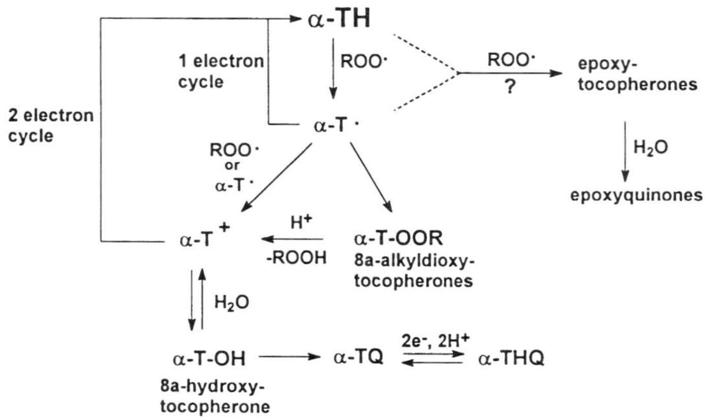

FIGURE 2. Redox cycles of α-tocopherol and its oxidation products.

presence of a coantioxidant, such as ascorbate or ubiquinol (Mehlhorn *et al.*, 1989; Kagan *et al.*, 1990*a*,*b*). Finally, prooxidant effects of the tocopheroxyl radical in lipoproteins are reversed by either ascorbate or ubiquinol acting as reductant to trap the phenoxyl radical and regenerate α-tocopherol (Bowry and Stocker, 1993; Bowry *et al.*, 1995). In this latter system, the participation of a coantioxidant such as ascorbate or ubiquinol appears to be essential to the α-tocopherol-dependent protection of the lipoprotein against oxidative damage. This requirement stems from the fact that lipoprotein particles may contain a single α-tocopheroxyl radical, which, in the absence of another radical, will initiate lipid peroxidation. Tocopheroxyl radicals in lipoprotein particles cannot diffuse to other particles to react with other radicals. Thus, the relatively unfavorable reaction of the tocopheroxyl radical with polyunsaturated lipids becomes the only available reaction pathway. In biological membranes, tocopheroxyl radicals that do not trap a second peroxyl radical may encounter another tocopheroxyl radical. Disproportionation yields α-tocopherol and tocopherone cation (**15**), which hydrolyzes to 8a-hydroxytocopherone (**5**) (Liebler and Burr, 1992). In membranes, regeneration of tocopheroxyl radicals extends the duration of antioxidant protection provided by α-tocopherol and this interaction is thought to account for the synergistic antioxidant effects of ascorbate and α-tocopherol in combination *in vitro*.

What remains unclear is the extent to which a one-electron redox cycle or independent actions of antioxidants contribute to antioxidant protection *in vivo*. Ascorbate may act to regenerate α-tocopherol as described above. However, ascorbate also may react directly with oxidants and thus spare α-tocopherol indirectly (Glascott *et al.*, 1996). The degree to which either mechanism facilitates the antioxidant action of α-tocopherol most likely depends on the nature of the oxidants present and concentrations of both ascorbate and α-tocopoherol. This independent, complementary antioxidant interaction has been proposed to account for the antioxidant synergism between α-tocopherol and glutathione in microsomal membranes *in vitro* (Maiorino *et al.*, 1989). Glutathione was proposed to inhibit lipid peroxidation primarily via phospholipid glutathione peroxidase, which consumes lipid hydroperoxides and indirectly reduces oxidation of membrane α-tocopherol.

Whether α-tocopherol redox cycles function *in vivo* has been a difficult question to address experimentally. One well-designed experiment by Burton *et al.* (1990) employed stable isotope dilution gas chromatography–mass spectrometry to measure the turnover of α-tocopherol in several tissues in guinea pigs fed diets containing either low, normal, or high levels of ascorbate. Because guinea pigs do not synthesize ascorbate, dietary manipulation can modulate tissue ascorbate levels. Analyses of the tocopherol status of tissues of animals on the different diets revealed no difference in kinetics of α-tocopherol turnover between guinea pigs differing in ascorbate status. Had ascorbate been required to regenerate α-tocopherol, the turnover of α-tocopherol should have been faster in ascorbate-deficient animals. Since this was not the case, the authors concluded that ascorbate did not regenerate α-tocopherol *in vivo*. Implicit in this conclusion is the assumption that α-tocopherol turnover results from α-tocopherol oxidation. Although this seems most likely, it is possible that a factor other than ascorbate modulates α-tocopherol regeration. In this experiment, the animals were not subjected to any oxidative stress. It would be interesting to repeat this experiment with animals subjected to oxidative stress, as ascorbate may affect α-tocopherol turnover primarily under conditions where α-tocopherol is rapidly oxidized.

3.2. Two-Electron Redox Cycle

Oxidation of α-tocopherol to 8a-substituted tocopherones may occur via the tocopherone cation (**15**), which may be reduced to α-tocopherol by ascorbate, nordehydroguaretic acid, and possibly other reductants (Liebler *et al.*, 1989; Liebler and Burr, 1992). Moreover, 8a-hydroxytocopherone (**5**), the immediate precursor to α-tocopherolquinone (**6**) is in hydrolytic equilibrium with tocopherone cation (**15**). Reduction of **15** thus offers a potentially useful pathway for α-tocopherol regeneration from nonradical oxidation products. Fukuzawa *et al.* (1993) proposed that the tocopherone cation was the ascorbate-reducible intermediate in α-tocopherol recycling in phosphatidylcholine liposomes, although it was not possible on the basis of the evidence presented to rule out the possibility that the toxopheroxyl radical instead was reduced. Reduction of 8a-substituted tocopherones *in vitro* requires acidic conditions (e.g., pH < 4) to facilitate formation of tocopherone cation (**15**) (Liebler *et al.*, 1989). In biological systems, this would require a catalyst, although none is known to exist. However, previous work by Chan and colleagues indicated that ascorbate, nordehydroguaretic acid, and glutathione regenerated oxidized α-tocopherol in platelet homogenates and that the regeneration by glutathione did not occur in chloroform/methanol-denatured homogenates (Chan *et al.*, 1991). Although the authors suggested that the reducible species was the tocopheroxyl radical, the stability and behavior of the reducible intermediate suggest that tocopherones instead may have been the reducible intermediates. Whether an enzyme mediated the reductions observed remains to be determined. However, the significant yield of 8a-substituted tocopherones in oxidations of α-tocopherol in intact tissues (Ham and Liebler, 1997) suggests that a two-electron redox cycle could be an effective means of maintaining cellular α-tocopherol.

4. REDOX REGULATION OF α-TOCOPHEROLQUINONE AND α-TOCOPHEROLHYDROQUINONE

α-Tocopherolquinone (**6**), a major product of α-tocopherol oxidation in biological systems, may contribute to antioxidant defense via its two-electron reduction product, α-tocopherolhydroquinone (**19**).

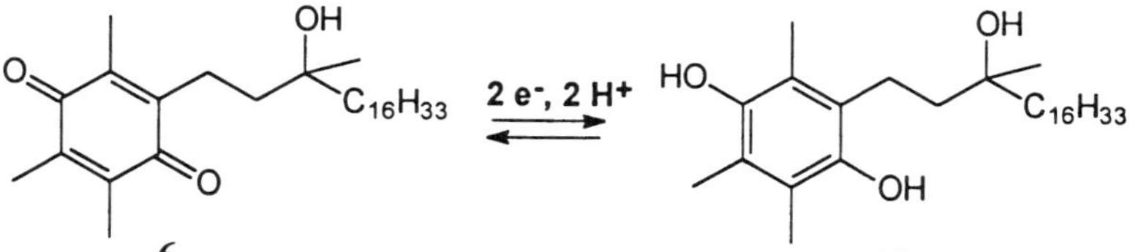

6 **19** (3)

α-Tocopherolquinone displays vitamin E activity *in vivo* in animal models and undergoes reduction to α-tocopherolhydroquinone (**19**) (Mackenzie and Mackenzie, 1958, 1960; Chow *et al.*, 1967). Moreover, a recent report documented the low-yield conversion of α-tocopherolquinone to α-tocopherol in a single human subject (Moore and Ingold, 1997). The reduction of quinone (**6**) to hydroquinone (**19**) has been documented in hepatocytes *in vitro* by Hayashi *et al.* (1992), who reported that both NADH- and NADPH-dependent reduction occurred in rat liver microsomes and mitochondria. Kohar *et al.* (1995) further reported that α-tocopherolquinone was reduced to α-tocopherolhydroquinone by peripheral mononuclear cells *in vitro*. These findings collectively have led to the suggestion that α-tocopherolquinone reduction generates an antioxidant form of vitamin E (α-tocopherolhydroquinone), which acts as a biological antioxidant.

A significant new finding in this regard is the demonstration that α-tocopherolquinone is a substrate for human NAD(P)H:quinone oxidoreductase (EC 1.6.99.2, NQO1). Siegel *et al.* (1997) recently reported that human NQO1 catalyzed the NADH- and NADPH-dependent reduction of α-tocopherolquinone to α-tocopherolhydroquinone in detergent micelles. α-Tocopherolquinone appeared to be a better substrate for NQO1 than was ubiquinone-10 under the conditions employed. α-Tocopherolhydroquinone was the sole product of the reaction as determined by HPLC and gas chromatography–molecular spectrometry (GC-MS). In sonicates of Chinese hamster ovary cells stably transfected with human NQO1, the level of NQO1 activity correlated with the rate of α-tocopherolquinone reduction to α-tocopherolhydroquinone. Cells with higher NQO1 levels also generated and maintained higher levels of α-tocopherolhydroquinone when treated with exogenous α-tocopherolquinone. The ability to reduce α-tocopherolquinone to α-tocopherolhydroquinone conferred greater resistance to cellular lipid peroxidation induced by cumene hydroperoxide. These data suggest that α-tocopherolquinone may be a physiological substrate for NAD(P)H:quinone oxidoreductase and that reductive metabolism of α-tocopherolquinone to α-tocopherolhydroquinone may play an important role in cellular antioxidant protection.

5. CONCLUSION

Much has been learned in the last decade about the chemistry of vitamin E antioxidant function. Application of GC-MS methods now makes it possible to measure α-tocopherol antioxidant reactions via analysis of specific reaction products. Studies with this methodology indicate that 8a-substituted tocopherones and epoxytocopherones are key intermediates in α-tocopherol oxidation in biological membranes. Continuing work on the role of redox cycles in maintaining α-tocopherol status indicates that tocopheroxyl radical recycling is an indispensable component of α-tocopherol antioxidant function in lipoproteins. Reduction of the tocopheroxyl radical in lipoprotein particles prevents the initiation of lipid peroxidation by the tocopheroxyl radical. The feasibility of tocopheroxyl radical recycling in biological membranes is well documented, but the roles of these reactions *in vivo* have been difficult to assess experimentally. A critical role for tocopherol redox cycles in membranes *in vivo* must be considered likely, but unproven. The recent finding that α-tocopherolquinone undergoes metabolic reduction to α-tocopherolhydroquinone has focused renewed interest on this hydroquinone as a bioactive form of vitamin E. Future work should clarify the potential significance of NAD(P)H:quinone oxidoreductase as a physiological modulator of vitamin E antioxidant function.

ACKNOWLEDGMENTS. Work in the author's laboratory described here was supported by USPHS Grants CA 47943, CA 59585, and ES 06694.

6. REFERENCES

Bisby, R. H., and Parker, A. W., 1995, Reaction of ascorbate with the α-tocopheroxyl radical in micellar and bilayer membrane systems, *Arch. Biochem. Biophys.* **317:**170–178.

Bowry, V. W., and Stocker, R., 1993, Tocopherol-mediated peroxidation. The prooxidant effect of vitamin E on the radical-initiated oxidation of human low-density lipoprotein, *J. Am. Chem. Soc.* **115:**6029–6044.

Bowry, V. W., Mohr, D., Cleary, J., and Stocker, R., 1995, Prevention of tocopherol-mediated peroxidation in ubiquinol-10-free human low density lipoprotein, *J. Biol. Chem.* **270:**5756–5763.

Burton, G. W., and Ingold, K. U., 1981, Autoxidation of biological molecules. 1. The antioxidant activity of vitamin E and related chain-breaking phenolic antioxidants *in vitro*, *J. Am. Chem. Soc.* **103:**6472–6477.

Burton, G. W., and Ingold, K. U., 1986, Vitamin E: Application of the principles of physical organic chemistry to the exploration of its structure and function, *Acc. Chem. Res.* **19:**194–201.

Burton, G. W., Doba, T., Gabe, E. J., Hughes, L., Lee, F. L., Prasad, L., and Ingold, K. U., 1985, Autoxidation of biological molecules. 4. Maximizing the antioxidant activity of phenols, *J. Am. Chem. Soc.* **107:**7053–7065.

Burton, G. W., Wronska, U., Stone, L., Foster, D. O., and Ingold, K. U., 1990, Biokinetics of dietary RRR-alpha-tocopherol in the male guinea pig at three dietary levels of vitamin C and two levels of vitamin E. Evidence that vitamin C does not "spare" vitamin E *in vivo*, *Lipids* **25:**199–210.

Chan, A. C., Tran, K., Raynor, T., Ganz, P. R., and Chow, C. K., 1991, Regeneration of vitamin E in human platelets, *J. Biol. Chem.* **266:**17290–17295.

Chow, C. K., Draper, H. H., Csallany, A. S., and Chiu, M., 1967, The metabolsim of C^{14}-alpha-tocopheryl quinone and C^{14}-alpha-tocopheryl hydroquinone, *Lipids* **2:**390–396.

Christen, S., Woodall, A. A., Shigenaga, M. K., Southwell-Keely, P. T., Duncan, M. W., and Ames, B. N., 1997, gamma-Tocopherol traps mutagenic electrophiles such as NO_x and complements α-tocopherol: Physiological implications, *Proc. Natl. Acad. Sci. USA* **94:**3217–3222.

Clough, R. L., Yee, B. G., and Foote, C. S., 1979, Chemistry of singlet oxygen. 30. The unstable primary product of tocopherol photooxidation, *J. Am. Chem. Soc.* **101:**683–686.

Cooney, R. V., Harwood, P. J., Franke, A. A., Narala, K., Sundstrom, A. K., Berggren, P. O., and Mordan, L. J., 1995, Products of gamma-tocopherol reaction with NO_2 and their formation in rat insulinoma (RINm5F) cells, *Free Radic. Biol. Med.* **19:**259–269.

Csallany, A. S., and Ha, Y. L., 1992, α-Tocopherol oxidation mediated by superoxide anion. I. Reactions in aprotic and protic conditions, *Lipids* **27:**195–200.

d'Ischia, M., and Novellino, L., 1996, Nitric oxide-induced oxidation of α-tocopherol, *Bioorg. Med. Chem.* **4:**1747–1753.

d'Ischia, M., Costantini, C., and Prota, G., 1991, Dye-sensitized photooxidation of vitamin E revisited. New 7-oxaspiro[4.5]dec-1-ene-3,6-dione products by oxygenation and ring contraction of α-tocopherol, *J. Am. Chem. Soc.* **113:**8353–8356.

de Groot, H., Hegi, U., and Sies, H., 1993, Loss of α-tocopherol upon exposure to nitric oxide or the syndonimine SIN-1, *FEBS Lett.* **315:**139–142.

Dix, T. A., and Marnett, L. J., 1985, Conversion of linoleic acid hydroperoxide to hydroxy, keto, epoxyhydroxy, and trihydroxy fatty acids by hematin, *J. Biol. Chem.* **260:**5351–5357.

Durckheimer, W., and Cohen, L. A., 1964, The chemistry of 9-hydroxy-alpha-tocopherone, a quinone hemiacetal, *J. Am. Chem. Soc.* **86:**4388–4393.

Fukuzawa, K., Ikebata, W., and Sohmi, K., 1993, Location, antioxidant and recycling dynamics of α-tocopherol in liposome membranes, *J. Nutr. Sci. Vitaminol.* **39:**S9–S22.

Giamvala, D. H., Church, D. F., and Pryor, W. A., 1986, Kinetics of ozonation. 4. Reactions of ozone with alpha-tocopherol and oleate and linoleate esters in carbon tetrachloride and in aqueous micellar solvents, *J. Am. Chem. Soc.* **108:**6646–6651.

Glascott, P. A., Tsyganskaya, M., Gilfor, E., Zern, M., and Farber, J. L., 1996, The antioxidant function of the physiological content of vitamin C, *Mol. Pharmacol.* **50:**994–999.

Gorbunov, N. V., Osipov, A. N., Sweetland, M. A., Day, B. W., Elsayed, N. M., and Kagan, V. E., 1997, NO-redox paradox: Direct oxidation of α-tocopherol and α-tocopherol mediated oxidation of ascorbate, *Biochem. Biophys. Res. Commun.* **219:**835–841.

Grams, G. W., Eskins, K., and Inglett, G. E., 1971, Dye-sensitized photooxidation of alpha-tocopherol, *J. Am. Chem. Soc.* **93:**866–868.

Ha, Y. L., and Csallany, A. S., 1992, α-Tocopherol oxidation mediated by superoxide anion. II. Identification of stable α-tocopherol oxidation products, *Lipids* **27:**201–205.

Ham, A. J. L., and Liebler, D. C., 1995, Vitamin E oxidation in rat liver mitochondria, *Biochemistry* **34:**5754–5761.

Ham, A. J. L., and Liebler, D. C., 1997, Antioxidant reactions of vitamin E in the perfused rat liver: Product distribution and effect of dietary vitamin E supplementation, *Arch. Biochem. Biophys.* **339:**157–164.

Hayashi, T., Kanetoshi, A., Nakamura, M., Tamura, M., and Shirahama, H., 1992, Reduction of α-tocopherolquinone to α-tocopherolhydroquinone in rat hepatocytes, *Biochem. Pharmacol.* **44:**489–493.

Hogg, N., Darley-Usmar, V. M., Wilson, M. T., and Moncada, S., 1993, The oxidation of α-tocopherol in human low density lipoprotein by the simultaneous generation of superoxide and nitric oxide, *FEBS Lett.* **326:**199–203.

Hogg, N., Joseph, J., and Kalyanaraman, B., 1994, The oxidation of α-tocopherol and Trolox by peroxynitrite, *Arch. Biochem. Biophys.* **314:**153–158.

Hogg, N., Singh, R. J., Goss, P., and Kalyanaraman, B., 1996, The reaction between nitric oxide and α-tocopherol: A reappraisal, *Biochem. Biophys. Res. Commun.* **224:**696–702.

Hoglen, N. C., Waller, S. C., Sipes, I. G., and Liebler, D. C., 1997, Reactions of peroxynitrite with gamma-tocopherol, *Chem. Res. Toxicol.* **10:**401–407.

Kagan, V., Serbinova, E., and Packer, L., 1990a, Antioxidant effects of ubiquinones in microsomes and mitochondria are mediated by tocopherol recycling, *Biochem. Biophys. Res. Commun.* **169:**851–857.

Kagan, V. E., Serbinova, E. A., and Packer, L., 1990b, Recycling and antioxidant activity of tocopherol homologs of differing hydrocarbon chain lengths in liver microsomes, *Arch. Biochem. Biophys.* **282:**221–225.

Kaiser, S., Di Mascio, P., Murphy, M. E., and Sies, H., 1990, Physical and chemical scavenging of singlet molecular oxygen by tocopherols, *Arch. Biochem. Biophys.* **277:**101–108.

Kohar, I., Baca, M., Suarna, C., Stocker, R., and Southwell-Keely, P. T., 1995, Is α-tocopherol a reservoir for α-tocopherolhydroquinone? *Free Radic. Biol. Med.* **19:**197–207.

Kramer, K. A., and Liebler, D. C., 1997, UVB induced photooxidation of vitamin E, *Chem. Res. Toxicol.* **10:**219–224.

Liebler, D. C., 1993, The role of metabolism in the antioxidant function of vitamin E, *Crit. Rev. Toxicol.* **23:**147–169.

Liebler, D. C., 1994, Tocopherone and epoxytocopherone products of vitamin E oxidation, *Meth. Enzymol.* **234:**310–316.

Liebler, D. C., and Burr, J. A., 1992, Oxidation of vitamin E during iron-catalyzed lipid peroxidation: Evidence for electron-transfer reactions of the tocopheroxyl radical, *Biochemistry* **31:**8278–8284.

Liebler, D. C., and Burr, J. A., 1995, Antioxidant stoichiometry and the oxidative fate of vitamin E in peroxyl radical scavenging reactions, *Lipids* **30:**789–793.

Liebler, D. C., Kaysen, K. L., and Kennedy, T. A., 1989, Redox cycles of vitamin E: Hydrolysis and ascorbic acid dependent reduction of 8a-(alkyldioxy)tocopherones, *Biochemistry* **28:**9772–9777.

Liebler, D. C., Baker, P. F., and Kaysen, K. L., 1990, Oxidation of vitamin E: Evidence for competing autoxidation and peroxyl radical trapping reactions of the tocopheroxyl radical, *J. Am. Chem. Soc.* **112:**6995–7000.

Liebler, D. C., Kaysen, K. L., and Burr, J. A., 1991, Peroxyl radical trapping and autoxidation reactions of alpha-tocopherol in lipid bilayers, *Chem. Res. Toxicol.* **4:**89–93.

Liebler, D. C., Matsumoto, S., Iltaka, Y., and Matsuo, M., 1993, Reactions of vitamin E and its model compound 2,2,5,7,8-pentamethylchroman-6-ol with ozone, *Chem. Res. Toxicol.* **6:**69–74.

Liebler, D. C., Burr, J. A., Philips, L., and Ham, A. J. L., 1996, Gas chromatography-mass spectrometry analysis of vitamin E and its oxidation products, *Anal. Biochem.* **236:**27–34.

Mackenzie, J. B., and Mackenzie, C. G., 1958, The effect of α-tocopherol, α-tocopherylhydroquinone and their esters on experimental muscular dystrophy in the rat, *J. Nutr.* **67:**223–235.

Mackenzie, J. B., and Mackenzie, C. G., 1960, The antisterility activity of alpha-tocohydroquinone in the female rat, *J. Nutr.* **72:**322–324.

Maiorino, M., Coassin, M., Roveri, A., and Ursini, F., 1989, Microsomal lipid peroxidation: Effect of vitamin E and its functional interaction with phospholipid hydroperoxide glutathione peroxidase,*Lipids* **24:**721–726.

Matsuo, M., Matsumoto, S., and Iltaka, Y., 1987, Oxygenations of vitamin E (alpha-tocopherol) and its model compound 2,2,5,7,8-pentamethylchroman-6-ol in the presence of the superoxide radical solubilized in aprotic solvents: Unique epoxidations and recyclizations, *J. Org. Chem.* **52:**3514–3520.

Matsuo, M., Matsumoto, S., Iltaka, Y., and Niki, E., 1989, Radical scavenging reactions of vitamin E and its model compound, 2,2,5,7,8-pentamethylchroman-6-ol, in a tert-butylperoxyl radical generating system, *J. Am. Chem. Soc.* **111:**7179–7185.

Mehlhorn, R. J., Sumida, S., and Packer, L., 1989, Tocopheroxyl radical persistence and tocopherol consumption in liposomes and in vitamin E-enriched rat liver mitochondria and microsomes, *J. Biol. Chem.* **264:**13448–13452.

Miyazawa, T., Yamashita, T., and Fujimoto, K., 1992, Chemiluminescence detection of 8a-hydroperoxytocopherone in photooxidized α-tocopherol, *Lipids* **27:**289–294.

Moore, A. N. J., and Ingold, K. U., 1997, α-Tocopheryl quinone is converted into vitamin E in man, *Free Radic. Biol. Med.* **22:**931–934.
Nishikimi, M., Yamada, H., and Yagi, K., 1980, Oxidation by superoxide of tocopherols dispersed in aqueous media with deoxycholate, *Biochim. Biophys. Acta* **627:**101–108.
Packer, J. E., Slater, T. F., and Willson, R. L., 1979, Direct observation of a free radical interaction between vitamin E and vitamin C, *Nature* **278:**737–738.
Packer, L., and Kagan, V. E., 1993, Vitamin E: The antioxidant harvesting center of membranes and lipoproteins, in *Vitamin E in Health and Disease* (L. Packer and J. Fuchs, eds.), pp. 179–192, Marcel Dekker, New York.
Siegel, D., Bolton, E. M., Burr, J. A., Liebler, D. C., and Ross, D., 1997, The reduction of α-tocopherolquinone by human NAD(P)H:quinone oxidoreductase: The role of α-tocopherolhydroquinone as a cellular antioxidant, *Mol. Pharmacol.* **52:**300–305.
Skinner, W. A., and Alqaupovic, P., 1963, Oxidation products of vitamin E and its model 6-hydroxy-2,2,5,7,8-pentamethylchroman. V. Studies of the products of alkaline ferricyanide oxidation, *J. Org. Chem.* **28:**2854–2858.
Suarna, C., and Southwell-Keely, P. T., 1988, New oxidation products of alpha-tocopherol, *Lipids* **23:**137–139.
Suarna, C., Craig, D. C., Cross, K. J., and Southwell-Keely, P. T., 1988, Oxidations of vitamin E (alpha-tocopherol) and its model compound 2,2,5,7,8-pentamethyl-6-hydroxychroman. A new dimer, *J. Org. Chem.* **53:**1281–1284.
Suarna, C., Baca, M., and Southwell-Keely, P. T., 1992, Oxidation of the α-tocopherol model compound 2,2,5,7,8-pentamethyl-6-chromanol in the presence of alcohols, *Lipids* **27:**447–453.
Winterle, J., Dulin, D., and Mill, T., 1984, Products and stoichiometry of reaction of vitamin E with alkylperoxy radicals, *J. Org. Chem.* **49:**491–495.
Yamauchi, R., Matsui, T., Kato, K., and Ueno, Y., 1989a, Reaction of alpha-tocopherol with 2,2′-azobis(2,4-dimethylvaleronitrile) in benzene, *Agric. Biol. Chem.* **12:**3257–3262.
Yamauchi, R., Matsui, T., Satake, Y., Kato, K., and Ueno, Y., 1989b, Reaction products of alpha-tocopherol with a free radical initiator, 2,2′-azobis(2,4-dimethylvaleronitrile), *Lipids* **24:**204–209.
Yamauchi, R., Matsui, T., Kato, K., and Ueno, Y., 1990, Reaction products of α-tocopherol with methyl linoleate-peroxyl radicals, *Lipids* **25:**152–158.
Yamauchi, R., Yagi, Y., and Kato, K., 1994, Isolation and characterization of addition products of α-tocopherol with peroxyl radicals of dilinoleoylphosphatidylcholine in liposomes, *Biochim. Biophys. Acta* **1212:**43–49.

Chapter 14

Localization of Vitamin E in Membranes

Peter J. Quinn

1. INTRODUCTION

Vitamin E is a term encompassing a group of eight isomeric molecules which are characterized by a 6-chromanol ring structure and a side chain at the 2 position. The tocopherols possess a 4′,8′,12′-trimethyltridecyl phytol side chain, and the tocotrienols differ by the presence of double bonds at the 3′,7′, and 11′ positions of the side chain. The α, β, γ, and δ isomers of tocopherol and tocotrienols differ in the number and position of the methyl substituents attached to the chromanol ring. A total of eight optical isomers arise from the existence of three asymmetric carbons at the 2 position of the ring and the 4′ and 8′ positions of the phytol chain.

The *RRR*-α isomer of tocopherol predominates in mammalian cells and tissues despite a usually higher dietary intake of the γ isomer. The selection process is not due to differences in absorption of vitamin E isomers from the gut, as all appear to be transported at the same rate (Traber and Kayden, 1989; Meydani *et al.*, 1989). The difference appears to arise during the process of metabolic turnover of lipoproteins (Traber *et al.*, 1990,

---

Peter J. Quinn Division of Life Sciences, King's College London, London W8 7AH, England.

Subcellular Biochemistry, Volume 30: Fat-Soluble Vitamins, edited by Quinn and Kagan. Plenum Press, New York, 1998.

1992). Thus, when tocopherols are delivered to the liver in chylomicrons and circulating lipoproteins, a hepatic tocopherol-binding protein that preferably interacts with the *RRR*-α isomer (Sato *et al.*, 1993; Arita *et al.*, 1995) mainly sequesters this isomer for assembly into very low density lipoprotein (VLDL), which acts to distribute vitamin E to the tissues. The other isomers are secreted in the bile.

Vitamin E is believed to have a multiplicity of effects in animals, but two primary functions in cells are recognized. Its major role is regarded to be as an antioxidant acting to prevent free radical damage to tissues and specifically to unsaturated lipids. The antioxidant action is said to be as a scavenger of active oxygen radicals (Burton and Ingold, 1986; Liebler, 1993; Mukai, 1993) as well as singlet oxygen (Sies *et al.*, 1993; Fukuzawa *et al.*, 1997). While active oxygen species will attack any biomolecule, the primary location of vitamin E in hydrophobic environments means that its principal site of action is in cell membranes and lipoproteins. The second mode of action is subsidiary to its role as an antioxidant in membranes and that is to stabilize the structure of membranes by forming complexes with destabilizing molecules so as to prevent disturbance of the amphipathic balance within the structure.

This chapter is concerned with how vitamin E localizes in subcellular membranes and how it interacts with other membrane components. Much of the experimental evidence of how vitamin E is arranged in membranes is derived from biophysical studies of model membrane systems consisting of phospholipids dispersed in aqueous media. The implications of the results obtained from such studies will be examined in the context of the putative roles of vitamin E as an antioxidant and membrane-stabilizing agent.

2. UPTAKE AND TRANSPORT OF VITAMIN E

Vitamin E is classed as a fat-soluble vitamin because of its solubility in solvents of low polarity. The solubility in aqueous media is therefore relatively low, and vitamin E will tend to either partition into tissue lipids or, in translocation through the body, locate in hydrophobic domains of molecular structures like lipoproteins. Within the cell, vitamin E partitions into the hydrophobic core of the various cell membranes. The relative concentration of vitamin E differs from one membrane to another. This chapter will examine how vitamin E is distributed among the different subcellular membranes and the identity of proteins that appear to specifically bind vitamin E. Brief mention will also be made of the partition of vitamin E into lipoproteins and model membrane systems which are commonly used to characterize how vitamin E interacts with membranes.

2.1. Distribution of Vitamin E in the Body

Dietary uptake of vitamin E has been examined in some detail in the rat (Zhang *et al.*, 1996). When the diet of laboratory rats was supplemented with vitamin E (1 g/kg diet) the concentration of α-tocopherols in the plasma reached a maximum level after 48 hr and the liver was the main organ where the vitamin accumulated. Levels in the kidney, heart, muscle, and brain continued to increase throughout a 6-week period. The level of α-tocopherol decreased only slowly when the supplement was removed from the diet. Studies were undertaken to determine the distribution of α-tocopherol in the various subcellular compartments of the liver. These were found to be similar to those published earlier by Buttriss and Diplock (1988), which are summarized in Table I. Of all the subcellular membrane fractions, the greatest concentrations of α-tocopherol were found in the Golgi membranes and lysosomes. The molar ratio of α-tocopherol:phospholipid in these membranes was on the order of 1:65 phospholipid molecules, which is about an order of magnitude greater than found in the other subcellular membranes.

2.2. Vitamin E-Binding Proteins

Tocopherol-binding proteins are believed to be involved in the transport and metabolism of α-tocopherol in tissues. Several tocopherol-binding proteins have been isolated from the cytosol of mammalian tissues and characterized. A hepatic protein of 30–35 kDa has been identified and sequenced (Sato *et al.*, 1993). The amino acid sequence indicates homology with 11-*cis* retinaldehyde-binding protein of the retina (Wolf, 1994). A similar protein which specifically binds α-tocopherol has been isolated from

Table I
Content of α-Tocopherol in Subcellular Membranes of Rats Fed a Standard Laboratory Diet

Subcellular fraction	α-Tocopherol (mmol/mole lipid P)
Nuclear fraction	0.72 ± 0.16
Mitochondria	1.07 ± 0.27
Inner membrane	1.27 ± 0.32
Outer membrane	1.14 ± 0.35
Lysosomes	14.60 ± 2.33
Microsomes	0.52 ± 0.08

[a]Data from Buttriss and Diplock (1988).

human liver cytosol (Kuhlenkamp *et al.*, 1993) and it appears to exist in two charged isoforms. The tocopherol-binding protein appears to be present only in hepatocytes and has not been detected in an examination of a number of other mammalian tissues. Studies *in vitro* have shown that the tocopherol-binding protein mediates the transfer of tocopherol between microsomes and liposomes loaded with the vitamin (Sato *et al.*, 1991). It was found that the tocopherol binding protein has preferential binding affinities for the different isomers of tocopherol in the series $\alpha = \beta > \gamma > \delta$. Related compounds, including α-tocopherol acetate, tocopherol quinone, and cholesterol, did not appear to compete for the binding of tocopherol isomers.

Another tocopherol-binding protein, which has a molecular weight of about 16kDa, has been identified in rat liver, but is also present in the cytosol of other tissues, including heart (Dutta Roy *et al.*, 1994). The protein has a high specific binding affinity for α-tocopherol and facilitates transfer of α-tocopherol between different subcellular membranes. The protein has been purified from bovine heart cytosol and an analysis of its amino acid composition has been performed (Gordon *et al.*, 1995). The protein did not bind oleic acid and was immunologically distinct from fatty acid-binding protein.

2.3. Lipoproteins

Vitamin E is transported around the body as a component of plasma lipoproteins. Dietary vitamin E is transported via the portal vein to the liver in chylomicrons. In the liver the isomeric forms are sorted and the *RRR*-α-tocopherol isomer is preferentially incorporated into lipoproteins assembled for export from the liver (Traber, 1994). Studies using deuterated isomers of the *RRR*-α-tocopherol and *SRR*-α-tocopherol have shown that this selection process is highly discriminatory. After administration of equal amounts of the two isomers to healthy subjects, the chylomicrons contained approximately the same amount of each isomer after 24hr. The concentration of the *RRR* isomer, however, began to dominate the plasma concentration after this time because it was secreted preferentially in the low-density lipoprotein (LDL) fraction manufactured in the liver for export to the blood (Traber *et al.*, 1994). Further studies of the clearance rates of the two deuterated isomers indicated that the fractional disappearance rates of *RRR*-α-tocopherol were significantly lower than the *SRR*-α-tocopherol, which was said to reflect the higher rate of return of the *RRR* isomer to the circulation. Overall turnover rates showed that the total pool of tocopherol was replaced each day.

The lipoprotein which carries most of the vitamin E is the LDL fraction, although other lipoprotein fractions contain some vitamin E. Normal LDL contains between 6 and 12 molecules of α-tocopherol and lesser amounts of γ-tocopherol (0.5 mole/LDL) (Stocker and Bowry, 1996). The structure of LDL consists of a monolayer of phospholipids and protein, analogous to one leaflet of a biological membrane, on the surface and a core of neutral lipid including cholesterol esters. The tocopherol is believed to be located within the surface monolayer of phospholipid.

3. INTERACTION WITH PHOSPHOLIPID BILAYERS

Dispersions of certain phospholipd classes in aqueous media result in the spontaneous assembly of the molecules into bilayers, which have been extensively used as a model for the lipid matrix of biological membranes. The properties of codispersions of α-tocopherol with phospholipids provide information about how α-tocopherol interacts with membrane phospholipids and where it is located with respect to the phospholipid molecules in the assembly. A number of such studies have been reported using phospholipids with particular acyl chain composition, different phospholipid classes, and mixtures of different phospholipids with tocopherol.

The lipophilic character of α-tocopherol has been demonstrated by electron spin resonance (ESR) probe experiments which showed that the phytol chain anchors the molecule firmly within the phospholipid bilayer (Niki *et al.*, 1985). Intercalation of α-tocopherol between phospholipid molecules in bilayer configuration is consistent with studies of $[5^{13}CH_3]\alpha$-tocopherol dispersed in egg glycerophosphorylcholine, showing splitting and broadening of the ^{13}C resonances in the presence of inorganic salts of Pr^{3+} and Gd^{3+}, respectively (Perly *et al.*, 1985). This is further supported by quenching of the intrinsic fluorescence of α-tocopherol by stearic acid derivatives of nitroxyl or anthroyloxyl groups interpolated into egg glycerophosphorylcholine (Aranda *et al.*, 1989). The orientation and motion of α-tocopherol has been examined in egg glycerophosphorylcholine bilayers over a temperature range 5–50°C using ^{13}C- and ^{2}H-NMR methods (Urano *et al.*, 1987; Salgado *et al.*, 1993; Ekiel *et al.*, 1988). It was concluded that the molecule rotates about an axis perpendicular to the plane of the phospholipid bilayer, and the motion of the rigid chromanol ring is more constrained than the phytol chain, in which the frequency of *gauche*$^{\pm}$ rotamers increases with distance toward the center of the bilayer.

Steady-state (Ohyashiki *et al.*, 1986) and time-resolved (Bisby and Birch, 1989) fluorescence anisotropy measurements of probes interpolated into phospholipid dispersions or membranes indicate that α-tocopherol causes an increase in the order parameter and decrease in the fluidity of the hydrocarbon domain of the structures. ^{13}C-NMR (Srivastava *et al.*, 1983) and ^{2}H-NMR (Wassall *et al.*, 1986) relaxation measurements of α-tocopherol in phospholipids in the fluid phase both indicate a restriction in the motional freedom of both components of the mixture.

The effect of up to 20 mole% incorporation of α-tocopherol on acyl chain order and dynamics in liquid-crystalline phosphatidylcholine bilayer dispersions has been examined as a function of acyl chain unsaturation using ESR of 5-, 7-, 12-, and 16-doxyl spin-labeled stearic acids intercalated into the membrane (Wassall *et al.*, 1991). Order parameters S measured in the upper portion of the hydrocarbon chain (positions 5 and 7) and correlation times τ_C in the lower portion (positions 12 and 16) determined from the ESR spectra indicated that α-tocopherol restricts acyl chain motion within the membrane. The magnitude of the increases in order were dependent upon the extent of unsaturation of the phospholipid, which, in turn, is related to the area occupied by the molecules at the bilayer interface. The effect was greatest (up to 15%) for saturated dimyristoylphosphatidylcholine, which possesses a relatively small area per molecule, as opposed to much smaller increases (less than 3%) for unsaturated phosphatidylcholine bilayers, where the phospholipid occupies a much larger surface area.

Many studies have shown that α-tocopherol influences the phase behavior of phospholipids. Massey *et al.* (1982) showed that the presence of α-tocopherol in aqueous dispersions of dimyristoylphosphatidylcholine decreased the temperature of the gel-to-liquid-crystalline phase transition and reduced the enthalpy of the transition. Similar effects were noted for fully saturated derivatives of other phospholipid classes. Other studies with unsaturated phospholipids (Fukuzawa *et al.*, 1979), which indicated relatively minor effects of α-tocopherol even when present in proportions of 10 mole/100 mole, is in marked contrast with the effect on transitions from the gel phase in saturated molecular species of phospholipid, where, as pointed out above, relatively large perturbations are produced.

An indirect approach has been, for example, to assess the effect of the presence of α-tocopherol on the activity of enzymes that act on phospholipid substrates. Yamamoto *et al.* (1994) noted that the activity of phospholipase D acting on egg-yolk phosphatidylcholine was enhanced twofold by the presence of 18 mole% α-tocopherol in the substrate, suggesting that susceptibility of the phospholipid to hydrolysis was altered by the presence of the vitamin. Comparison with α-tocopherol acetate showed no effect on enzymic activity, implying that the location of the acetate derivative is in a domain that does not perturb the bilayer–water interface. The activation of protein kinase C in smooth muscle cells provoked by low-density lipoproteins is also inhibited by α-tocopherol (Ozer *et al.*, 1993), but it is uncertain whether this is a direct effect on the enzyme or an effect on the membrane site to which the enzyme is bound.

3.1. Effect of Acyl Chain Configuration

Examination of the thermal properties of mixed aqueous dispersions of 1,2-dipalmitoylphosphatidylcholine (DPPC) with α-tocopherol by differential scanning calorimetry (DSC) showed that increasing proportions of α-tocopherol cause a progressive broadening of the gel-to-liquid-crystalline phase transition (decrease in cooperativity of the transition) such that the onset temperature decreased, but the temperature of completion of the transition was unaltered (Massey *et al.*, 1982; DeKruijff *et al.*, 1974). A report by Villalain *et al.* (1986) that the temperature of completion of the transition decreases with increasing proportions of α-tocopherol has not been confirmed in subsequent studies (Quinn, 1995). Proportions of α-tocopherol greater than about 20 mole% resulted in almost complete loss of transition enthalpy. The effect of α-tocopherol on the transition enthalpy of DPPC is shown in Fig. 1. Similar findings have been reported for the dimyristoyl derivative of phosphatidylcholine (McMurchie and McIntosh, 1986). It has been consistently observed in the thermal studies of mixtures of α-tocopherol with saturated phosphatidycholines that the presence of α-tocopherol apparently eliminates the pretransition enthalpy. This does not mean that the pretransition itself is eliminated by the presence of α-

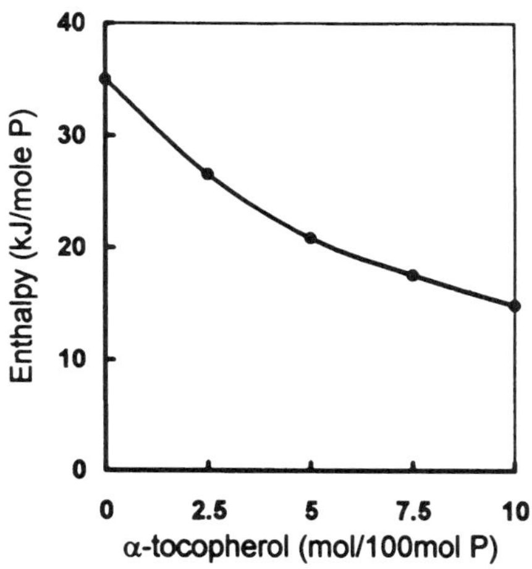

FIGURE 1. Transition enthalpy of the gel-to-liquid-crystalline phase transition of dipalmitoylphosphatidylcholine plotted as a function of the mole% α-tocopherol.

tocopherol, as it is possible that the enthalpies of the two transitions merge in the broad transition produced. Evidence for this comes from synchrotron X-ray diffraction studies of the DPPC–α-tocopherol system, which showed that a diffraction band at about 15 nm persisted up to at least 5 mole% α-tocopherol, and signs of an uncoordinated ripple structure were observed at even higher proportions of α-tocopherol (Quinn, 1995). This suggests that α-tocopherol reduces the cooperativity of the gel-to-liquid-crystalline phase transition, but does not eliminate the pretransition, i.e., the transition from gel to ripple conformation is unaffected, at least in those molecules undergoing the transition. The reduction in overall transition enthalpy indicates that an increasing proportion of the phospholipid molecules is removed from the transition.

The effect of α-tocopherol on the phase behavior of heteroacid phosphatidylcholines (PCs) is complex, with evidence of domain formation. Stillwell *et al.* (1996) reported the creation of at least two transition endotherms in mixtures of α-tocopherol with 18:0/18:1-PC and 18:0/22:6-PC, with transitions at lower temperatures for the more saturated molecular species and transitions at higher temperatures for the polyunsaturated molecular species. Sanchez-Migallon *et al.* (1996b) examined the thermal behavior of α-tocopherol with 1,2-di18-phosphatidylcholines having 18:0 acylated in the *sn*-1 position and 18:0, 18:1, 18:3, or 20:4 acylated at the *sn*-2 position of the glycerol. They constructed partial phase diagrams of the mixed acyl dispersions and concluded that there was fluid-phase immiscibility between the phospholipids and α-tocopherol. They concluded that lateral phase separations of domains containing different amounts of α-tocopherol were formed in the fluid phase. The magnitude of the effect of α-tocopherol on transition enthalpy was found to be dependent on the extent of unsaturation of the hydrocarbon chain located at the *sn*-2 position of the phospholipid. It was suggested that α-tocopherol served to reduce the differences between gel and fluid states of the phospholipids, which, in turn, is related to the molecular shape of the unsaturated phosphatidylcholines.

Fourier-transform infrared studies of α-tocopherol–dipalmitoylphosphatidylcholine mixed aqueous dispersions have shown effects of the presence of the vitamin on acyl chain conformation. The original studies of this mixture (Villalian *et al.*, 1986) showed that the number of *gauche*$^{\pm}$ rotamers in the acyl chains of the phospholipid in the gel phase was decreased by the presence of α-tocopherol. This, however, was not confirmed by subsequent studies (Severcan, 1997), which showed that the number of *gauche*$^{\pm}$ rotamers increased in a manner consistent with ESR (Severcan and Cannistraro, 1990; Wassall *et al.*, 1991; Suzuki *et al.*, 1993) and NMR (Wassall *et al.*, 1986; Suzuki *et al.*, 1993) observations.

3.2. Phospholipid Class

Most of the studies of the effect of α-tocopherol on phospholipid structure and stability have been performed using phosphatidycholines, but more recently studies have been extended to other phospholipid classes. Like phosphatidylcholine, phosphatidylserine is a bilayer-forming phospholipid and the effect of α-tocopherol on the phase transition behavior of this charged lipid in the presence and absence of Ca^{2+} has been reported (Sanchez-Migallon *et al.*, 1996a). It was found that the enthalpy of the gel-to-liquid-crystalline phase transition of dimyristoylphosphatidylserine decreased proportionately with increasing α-tocopherol in the bilayer. No phase transition could be detected when the proportion of α-tocopherol exceeded about 25 mole%.

Unlike the bilayer-forming phospholipids, the effect of α-tocopherol on the phase behavior of non-bilayer-forming lipids appears to be different (Salgado *et al.*, 1995). Ortiz *et al.* (1987) examined the effect of different proportions of α-tocopherol in dipalmitoyl and dilauroyl derivatives of phosphatidylethanolamines, and found that α-tocopherol induces formation of multiple peaks in differential scanning calorimetric curves, which they interpretated as evidence of phase separation within the bilayer.

α-Tocopherol has been found to decrease the temperature of the lamellar-to-hexagonal-II phase transition in dielaidoylphosphatidylethanolamine as well as to induce phase separations of isotropic and hexagonal-II phases in the gel-phase phospholipid as deduced from ^{31}P-NMR spectroscopy (Micol *et al.*, 1990). Similar effects of α-tocopherol on the lamellar-to-nonlamellar phase transition temperature has been found by X-ray diffraction measurements, as shown in Fig. 2. This shows a series of low-angle X-ray diffraction patterns recorded from mixed aqueous dispersions of 1-palmitoyl-2-oleoylphosphatidylethanolamine at three temperatures below the lamellar-to-nonlamellar phase transition temperature (69°C). In the absence of α-tocopherol the phospholipid remains in a lamellar phase at temperatures at least up to 45°C. Increasing proportion of α-tocopherol causes the phase separation of domains of the nonbilayer phase which coexists with the bilayer phase, and at the highest mole ratio of α-tocopherol (10 mole%) the formation of a cubic phase is indicated by the appearance of peaks at scattering angles less than the lamellar repeat spacing.

3.3. Binary Phospholipid Mixtures

Attempts to determine whether α-tocopherol partitions preferentially into fluid or gel phases has been examined using phospholipid mixtures

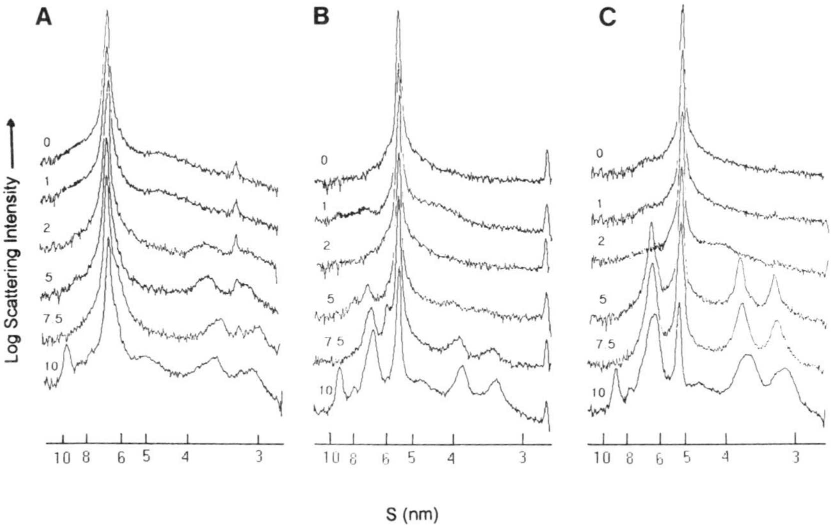

FIGURE 2. Low-angle X-ray scattering patterns recorded from dispersions of 1-palmitoyl-2-oleoylphosphatidylethanolamine containing different amounts of α-tocopherol (as indicated on each scattering curve) at (A) 15°C, (B) 30°C, and (C) 45°C.

differing in acyl chain composition and hence gel-to-liquid-crystalline phase transition temperature. Using mixed dispersions of phosphatidylcholines, zone-refining effects were observed in which α-tocopherol partitions into fluid phases in mixtures of phosphatidylcholines which exhibit gel-phase separation (Villalain *et al.*, 1986; Ortiz *et al.*, 1987). In the case where cocrystallization of the hydrocarbon chains occurs, α-tocopherol was said to distribute randomly throughout the phospholipid bilayer and no preference was observed for either molecular species of phosphatidylcholine or phosphatidylethanolamine.

The effect of α-tocopherol on the phase behavior of mixed acyl phosphatidylcholines has been reported by Stillwell *et al.* (1996). Increasing proportions of α-tocopherol in an equimolar mixture of 18:0/18:1- and 18:0/22:6-phosphatidylcholine produces a two-component transition enthalpy in place of a single endotherm in each of the pure phospholipids. The lower temperature transition is progressively broadened with increasing amount of α-tocopherol, but the higher temperature component is virtually unaffected by the presence of up to 10 mole% α-tocopherol. This effect was said to arise from a phase separation of α-tocopherol within the mixed phospholipid dispersion, with a preference for an association with the more unsaturated molecular species of phospholipid. This was attributed to a steric fit of the chromanol ring structure of α-tocopherol into the delta end of the polyunsaturated fatty acyl chains.

3.4. Orientation and Motion of Vitamin E in Bilayer Membranes

Three possible locations have been considered for the orientation of α-tocopherol in phospholipid bilayer membranes (Fukuzawa *et al.*, 1992). The first, based on paramagnetic resonance and other studies of α-tocopherol incorporated into phospholipid bilayer membranes, was a location in the region of the aqueous interface of the structure (Srivastava *et al.*, 1983; Perly *et al.*, 1985). The second, based on studies reported by Fragata and Bellemare (1980), indicated a location of the chromanol ring of α-tocopherol 1 nm from the aqueous interface. Finally, fluorescence quenching of probes located in different domains of the bilayer place the chromanol nucleus at varying depths within the bilayer structure (Bisby and Ahmed, 1989). A systematic study of the location of α-tocopherol and homologues with side chains of varying lengths using fluorescence quenching and fluorescence energy transfer methods has been reported by Kagan and Quinn (1988). Because the excitation and fluorescence emission wavelengths of α-tocopherol are appropriate for fluorescence energy transfer with anthroyloxystearate (AS) probes, it was possible to obtain fairly precise distance measurements between the interacting species. The efficiency of fluorescence energy transfer between α-tocopherol and different *n*-AS probes differing in the location of the 9-anthroyloxy group in egg phosphatidylcholine bilayers was measured and it was found that the order of efficiency followed the series 7-AS > 2-AS > 9-AS = 12-AS. This led to the conclusion that the chromanol nucleus of α-tocopherol is located in the bilayer in a region between the 9-anthroyloxy group attached to carbon 7 and carbon 2 and preferentially nearest to carbon 7. Water-soluble fluorescence quenching agents were found to have low efficiency in quenching the fluorescence of α-tocopherol, suggesting that the chromanol nucleus does not extend into the lipid–water interface.

The mobility and molecular orientation of α-tocopherol in egg phosphatidylcholine bilayers have been investigated using ^{19}F-NMR methods with homologues of α-tocopherol labeled with fluorine (Urano *et al.*, 1993). The effect of water-soluble lanthanide shift reagents was found to perturb fluorine atoms preferenetially on the chromanol ring, indicating that the chromanol nucleus was oriented between the phospholipid molecules and located in the polar head group region of the bilayer structure. Fluorine labels on the isoprenoid chain were not greatly affected, consistent with their penetration into the hydrophobic interior of the bilayer.

Fluorescence quenching of the α, γ, and δ isomers of tocopherol by homologues of coenzyme Q differing in number of isoprene units in the side chain have been undertaken to investigate the interaction between these two lipids in phospholipid bilayers (F. Piccinetti and P. J. Quinn, unpublished). A summary of the results in the form of a Stern–Volmer plot of the fluorescence quenching of α-tocopherol by coenzyme Q10 and coenzyme Q4 is shown in Fig. 3. This shows that the fluorescence of α-tocopherol is effectively quenched by the coenzyme Q with a side chain of four isoprene units, but not by that with ten isoprene units. Essentially the same result was obtained for each of the three isomers of tocopherol. Studies of the temperature dependence of the quenching indicated that the relationship between the coenzyme Q and tocopherol was unaffected by whether the phospholipid was in the gel or the liquid-crystalline phase. From these data it may be concluded that the short-chain homologue, which is known to be oriented perpendicular to the plane of the bilayer between the phospholipid molecules (Quinn and Katsikas, 1985), is able to interact with the chromanol ring of the tocopherol molecules, but that the benzoquinone ring of the long-chain homologue is not able to access the same domain, or at least in a form in which the benzoquinone substituent can quench the fluorescence of the tocopherol. Recent Fourier-transform infrared spectroscopic data suggest that coenzyme Q10 occupies a separate domain in phospholipid bilayers (Castresana *et al.*, 1992) and its participation in redox cycling with α-tocopherol (Kagan *et al.*, 1994) must involve a relocation of one or other of the molecules within the bilayer.

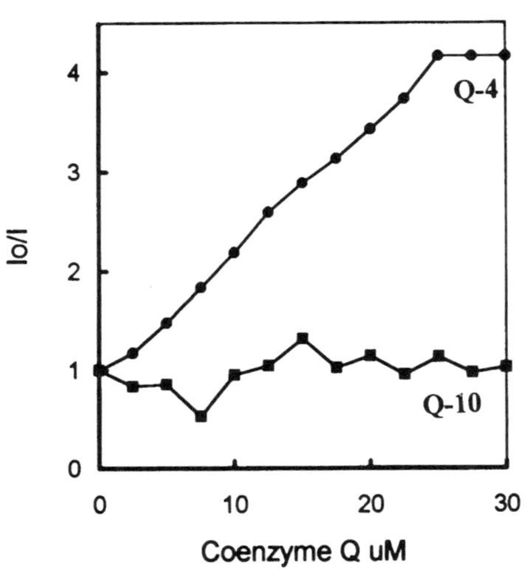

FIGURE 3. Stern–Volmer plot of quenching of fluorescence of α-tocopherol in bilayers of dipalmitoylphosphatidycholine by ubiquinone-10 (■) and ubiquinone-4 (●).

The location of the chromanol nucleus established on the basis of fluorescence energy transfer and quenching measurements is supported by Fourier-transfer infrared spectroscopy measurements (Gomez-Fernandez *et al.*, 1989, 1991; Salgado *et al.*, 1993, 1995). These studies showed evidence of hydrogen bonding of the phenoxyl hydroxy group to either the carbonyl or phosphate oxygens of the phospholipid molecules. The possibility of hydrogen bonding to water cannot be excluded at this stage, but the possibility seems unlikely (see above). Hydrogen bonding of α-tocopherol to the phospholipid is consistent with bilayer permeability experiments reported by Urano *et al.* (1990) which led them to conclude that the phenoxy group of α-tocopherol is hydrogen bonded to the carbonyl group of the ester bond of unsaturated fatty acids in the phospholipids arranged in a bilayer configuration. Dynamic spectroscopic studies of α-tocopherol and α-tocotrienol indicate that the molecules rotate about their long axis perpendicular to the plane of the membrane (Suzuki *et al.*, 1993)

4. TOCOPHEROL–LIPID COMPLEXES

The interaction of tocopherol with other membrane constituents is said to influence the stability of the membrane. Some evidence points to a stabilizing effect through the formation of complexes with membrane components that tend to destabilize bilayer structures, while other evidence suggests that tocopherol destabilizes membranes and promotes fusion.

With respect to membrane fusion, there are conflicting reports on the effects of α-tocopherol. An early observation by Ahkong *et al.* (1972) showed that when α-tocopherol was added to a suspension of avian erythrocytes the cells were induced to fuse, presumably because α-tocopherol destabilized the membrane lipid bilayer. Similar studies with chromaffin granules aggregated with synexin, however, showed that the addition of α-tocopherol did not induce the granules to fuse (Creutz, 1981). Moreover, α-tocopherol was found to inhibit platelet aggregation (Steiner and Anastasi, 1976; Steiner and Mower, 1982).

Studies with pure lipid systems also supported the notion that α-tocopherol stabilizes lipid bilayers and prevents fusion. For example, the aggregation rate of phosphatidylcholine vesicles containing free oleic acid is decreased (Ortiz and Gomez-Fernandez, 1988), and Ca^{2+}-induced fusion of large unilamellar vesicles of phosphatidylserine (Aranda *et al.*, 1996) is inhibited by α-tocopherol. One explanation for the inhibitory effect of α-tocopherol on Ca^{2+}-induced fusion is that α-tocopherol reduces the binding of Ca^{2+} to phosphatidylserine vesicles (Sanchez-Migallon *et al.*, 1996b).

4.1. Fatty Acids and Lysophospholipids

One of the ways in which α-tocopherol is believed to stabilize membranes is to form complexes with membrane lipid components that have a tendency to destabilize the bilayer structure, thereby countering their effects and rendering the membrane more stable. Kagan (1989) used paramagnetic resonance spectroscopy to demonstrate the formation of complexes between free fatty acids and α-tocopherol in phospholipid bilayer membranes. The incorporation of a fatty acid spin probe into bilayers of palmitoyloleoylphosphatidylcholine was found to take place when the phospholipid was in a liquid-crystalline phase. At temperatures below the gel-to-liquid-crystalline phase pransition temperature the spin probe could become incorporated into the bilayer when 15 mole% arachidonic acid was added alone, but not with 5 mole% α-tocopherol. This was interpreted as the formation of a complex between the arachidonic acid and α-tocopherol which did not perturb the phase behavior of the phospholipid to the same extent as the free fatty acid. Similar conclusions were drawn from ^{1}H-NMR studies of dipalmitoylphosphatidylcholine–20 mole% linoleic acid mixed dispersions with and without 5 mole% α-tocopherol. Evidence was also obtained for the formation of complexes between lysophosphatidylcholine and α-tocopherol, which was said to explain the protective effect of α-tocopherol against the action of products of phospholipase A hydrolysis on biological membranes (Mukherjee *et al.*, 1997). The mode of interaction of unsaturated fatty acids with α-tocopherol has been investigated by Urano and Matsui (1989) using fluorescence and NMR methods. They showed that α-tocopherol decreased the fluidity of phospholipid liposomes that were perturbed by the presence of free fatty acids with more than one double bond. Examination of spin–lattice relaxation modes of the methyl carbons led to the conclusion that the three methyl groups attached to the aromatic ring rather than the isoprenoid side chain have the strongest affinity for unsaturated lipids.

4.2. Diacylphospholipids

The formation of molecular complexes of α-tocopherol with membrane constituents implies that the lateral distribution of the constituents is not random. The distribution of stoichiometric complexes, however, may be random within the lateral plane of the bilayer or they could assemble into lateral domains. Preferential interactions between α-tocopherol and phospholipids and indirect evidence for the formation of specific complexes has been obtained from a number of different experimental approaches. The presence of two signals from spin-label studies of mixed dispersions of dipalmitoylphosphatidylcholine–α-tocopherol containing relatively high proportions of α-tocopherol (20 and 30 mole/100 mole) was reported at temperatures over the range 30–50°C (Severcan and Cannistraro, 1988). A relatively immobile component was attributed to an α-tocopherol-rich phospholipid domain coexisting with bilayers of pure phospholipid. Measurements of intrinsic fluorescence of 7 mole/100 mole α-tocopherol in bilayers of dipalmitoylphosphatidylcholine also led to the conclusion that α-tocopherol was closely associated in particular domains of the phospholipid bilayer (Kagan and Quinn, 1988; Aranda *et al.*, 1989). Information about the forces which may operate between phospholipids and α-tocopherol have come from Fourier-transform infrared spectroscopy of dipalmitoylphosphatidylcholine–α-tocopherol mixed dispersions (Villalain *et al.*, 1986). It was found from measurements of bandwidth and frequencies of the CH_2 antisymmetric and symmetric stretch vibrations of the phospholipid acyl chains that the presence of α-tocopherol causes a decrease in the average number of *gauche*$^{\pm}$ chain conformers at temperatures above the gel-to-liquid-crystalline phase transition. Interestingly, the bandwidth and frequency of maximum absorbance do not appear to change in concert with changes in temperature, which was interpreted as evidence for the coexistence of two phases. This conclusion, however, assumes that temperature has identical effects on *trans–gauche* isomerizations, on one hand, and lipid order and mobility on the other.

Clear evidence of phase separations of α-tocopherol within both gel and liquid-crystalline phases of dipalmitoylphosphatidylcholine has been provided from time-resolved low-angle X-ray diffraction and freeze-fracture electron microscopic studies (Quinn, 1995). Multiple low-angle diffraction peaks were observed in mixed aqueous dispersions both above and below the gel-to-liquid-crystalline phase transition temperature of the phospholipid. The identity of one set of low-angle diffraction peaks was assigned on the basis of the *d* spacing of pure dipalmitoylphosphatidylcholine and the temperatures at which transitions take place. Additional peaks observed in mixed dispersions consisted of different lamellar phases superimposed on profiles of the phospholipid which displayed diffraction patterns almost identical to those of $L_{\beta'}$, $P_{\beta'}$, and L_{α} phases of pure dipalmitoylphosphatidylcholine. The reationship between relative X-ray scattering intensity and the proportion of α-tocopherol in the mixture was used to identify reflections from mesophase structures enriched in α-tocopherol. This is illustrated in Fig. 4A, which shows the relationship between integrated X-ray scattering intensity of low-angle diffraction peaks as a function of mole/100 mole α-tocopherol in the dispersions. The relative intensity of scattering of the peaks centered at about 6.4 and 6.7 nm in the gel and liquid-crystalline phases, respectively,

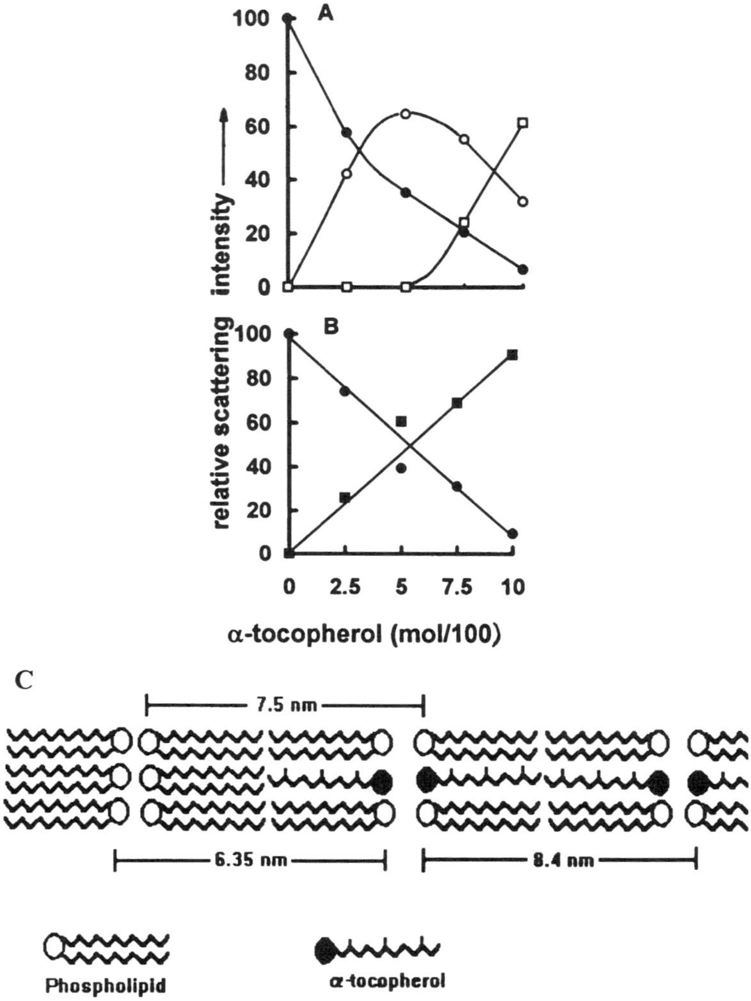

FIGURE 4. Relative X-ray scattering intensities of the first-order lamellar diffraction peaks observed in mixed dispersions of dipalmitoylphosphatidylcholine and α-tocopherol at 25°C (gel phase, A) and 45°C (liquid-crystalline phase, B). (C) Schematic model according to which peaks are assigned. ●, Pure phospholipid; O, 7.5 nm; □, 8.4 nm; ■, 6.2 nm (liquid-crystalline phase).

decreases with increasing proportion of α-tocopherol. There is a corresponding increase in the relative intensity of scattering of peaks centered at about 7.5 and 6.2 nm, respectively, suggesting that these spacings correspond to domains enriched in α-tocopherol. The increase in relative intensity of the peak centered at 8.4 nm at higher proportions of α-tocopherol in dipalmitoylphosphatidylcholine in the gel phase may be due to formation of a unit cell containing a higher proportion of α-tocopherol than that giving rise to the *d* spacing of 7.5 nm. This could be due either to formation of a new phase containing an increased proportion of α-tocopherol or to an arrangement of domains created with lower proportions of α-tocopherol so as produce a longer *d* spacing. Since there are only two low-angle peaks in the fluid phase, there is only one structural coordinate of the α-tocopherol-enriched domain of dipalmitoylphosphatidylcholine.

An expansion of the lamellar repeat spacing of dipalmitoylphosphatidylcholine gel phase due to the formation of stoichiometric complexes with dipalmitin has been shown to be due to an increase in thickness of both the bilayer and the interposing water layer (Quinn *et al.*, 1994). The increase in dimension of the water layer apparently results from an increase in repulsive forces between opposing bilayer surfaces as the phospholipid head groups are separated by interpolated dipalmitin. The insertion of α-tocopherol between molecules of phospholipid may also cause an expansion of the smectic mesophase in this way. A schematic model that explains the existence of two lamellar repeat spacings of α-tocopherol-enriched domains is shown in Fig. 4B. From this scheme it can be seen that the repeat spacing of 7.5 nm results when a domain of α-tocopherol-rich dipalmitoylphosphatidylcholine opposes a bilayer surface of pure dipalmitoylphosphatidylcholine. The spacing increases to 8.4 nm when two α-tocopherol-rich domains are interfaced across the aqueous layer. The increase in scattering intensity from the spacing at 8.4 nm relative to that at 7.5 nm with increasing proportion of α-tocopherol is consistent with the presence of a greater area of domains enriched in α-tocopherol opposed across the aqueous layer. The formation of only one spacing by α-tocopherol-rich domains of dipalmitoylphosphatidylcholine in the fluid phase is said to be due to a change in correlation between opposing monolayers of α-tocopherol-enriched domains, a change in stoichiometry or structure of the domain, or a combination of both. The stoichiometry of α-tocopherol : phospholipid in these enriched domains appears to be relatively constant judging from the consistent and comparatively sharp lamellar repeat spacings they produce.

The calorimetric data suggest that α-tocopherol causes a perturbation of the gel-to-liquid-crystalline phase transition of dipalmitoylphosphatidylcholine and progressive removal of phospholipid molecules from contributing to the transition enthalpy with increasing proportion of α-tocopherol in the mixture (Villalain *et al.*, 1986). While there is a decrease in temperature of the solidus phase boundary with increasing proportion of α-tocopherol, as indicated by the temperature of onset of the transition, there is no significant change in temperature of the liquidus phase boundary as judged by both thermal and wide-angle X-ray diffraction. This is consistent with a phase separation of components in the fluid phase. These data are in agreement with calorimetric studies of mixtures of α-tocopherol and dimyristoylglycerophosphocholine (Massey *et al.*, 1982), but not with those of Villalain *et al.* (1986), who reported a significant shift of the liquidus phase boundary to lower temperatures in mixed dispersions of α-tocopherol and dipalmitoylphosphatidylcholine. No endothermic transitions other than the pre and main transitions of dipalmitoylphosphatidylcholine were detected in the temperature range 25–50°C that would indicate the formation of a compound(s) of fixed stoichiometry between α-tocopherol and dipalmitoylphosphatidylcholine. Both the thermal and X-ray evidence, however, are consistent with the creation of domains within the phospholipid bilayer that are enriched with respect to α-tocopherol.

Estimates of the stoichiometry of the components of the domains enriched with respect to α-tocopherol were obtained from a comparison of the respective changes in transition enthalpy and low-angle X-ray diffraction intensities due to the presence of α-tocopherol in the bilayer. Calorimetric data showed that the presence of α-tocopherol caused a loss in cooperativity in the gel-to-liquid-crystalline phase transition of dipalmitoylphosphatidylcholine and a decrease in transition enthalpy. The values obtained for transition enthalpy in the different mixtures of dipalmitoylphosphatidylcholine and α-tocopherol are presented in Table II.

Table II
Transition Enthalpies and Stoichiometry of α-Tocopherol–Dipalmitoylphosphatidylcholine (DPPC) Clusters Calculated from Thermal and Low-Angle X-Ray Diffraction Data of Quinn (1995)

		DPPC:α-tocopherol stoichiometry		
			By X-ray data	
α-Tocopherol in DPPC (mole/100 mole)	ΔH (kJ/mole DPPC)	By calorimetry data	Gel (25°C)	Fluid (45°C)
0	35.0	(9.6)[a]	—	—
2.5	26.6	8.5	10.5	10.2
5.0	20.9	7.0	9.1	11.5
7.5	17.6	5.6	8.8	8.5
10.0	14.9	4.7	8.3	8.2

[a]Value derived from regression analysis.

Because of the high degree of correlation between the change in excess specific heat capacity and disordering of the hydrocarbon chains, broadening of the transition with increasing α-tocopherol concentration indicates that the normal phase transition of the phospholipid is perturbed. It is not possible, on the basis of the present evidence, to conclude that the loss in enthalpy is due to a decrease in molar enthalpy of the phospholipid or removal of a proportion of the total phospholipid from undergoing a gel-to-liquid-crystalline phase transition. If it is assumed, however, that domains of pure dipalmitoylphosphatidylcholine contribute normally to the transition enthalpy (35 kJ $mole^{-1}$), then the number of molecules of dipalmitoylphosphatidylcholine that do not contribute to enthalpy in mixtures with α-tocopherol can be calculated from the measured transition enthalpies. When the number of dipalmitoylphosphatidylcholine molecules that do not contribute to transition enthalpy, expressed as a ratio of α-tocopherol molecules in the mixture, is plotted as a function of mole/100 mole α-tocopherol in the mixture, a linear regression of the form $y = -0.56x + 9.6$ is obtained, which indicates that increasing proportions of α-tocopherol result in a decrease in efficiency in removing dipalmitoylphosphatidylcholine from an endothermic transition. The intercept with the *y* axis yields a stoichiometry of 9.6 dipalmitoylphosphatidylcholine:α-tocopherol. In order to account for the loss in cooperativity of the gel-to-liquid-crystalline phase transition, a large-scale phase separation of pure phospholipid and α-tocopherol-rich domains of phospholipid would appear to be excluded.

Calculations of stoichiometry of α-tocopherol and dipalmitoylphosphatidylcholine have also been performed using the ratio of integrated intensities of the first-order lamellar repeat X-ray scattering bands recorded in mixtures in the gel phase (25°C) and the fluid phase (50°C). The model presented in Fig. 4C provides a plausible explanation of the appearance of two lamellar *d* spacings for α-tocopherol-rich domains in the gel phase. It can be seen that the gel-phase peak at 7.5 nm is comprised equally of a bilayer of pure dipalmitoylphosphatidylcholine and α-tocopherol-rich domains of phospholipid. The peak at 8.4 nm is observed where α-tocopherol-rich domains are opposed across the aqueous layer. The values presented in Table II for the dipalmitoylphosphatidylcholine: α-tocopherol ratio in the gel phase were obtained from the sum of the integrated X-ray scattering intensity of the band centered at 8.4 nm plus half the integrated intensity of the band centered at 7.5 nm. This complication does not arise in the fluid phase because there is a clear distinction between the scattering attributed to the pure phospholipid phase and that from the α-tocopherol-rich domains. Calculations of dipalmitoylphosphatidylcholine:α-tocopherol ratio in the fluid phase are also presented in Table II.

The stoichiometry of dipalmitoylphosphatidylcholine : α-tocopherol clusters in mixed aqueous dispersions of the two lipids obtained from direct structural parameters and by extrapolation from the thermal analysis is remarkably similar. Furthermore, the values obtained from integrated X-ray scattering intensities in the gel phase (9.2 ± 1.1) and fluid phase (9.6 ± 1.6) suggest that the forces responsible for cohesion of the α-tocopherol-rich domains of phospholipid are not dependent on the phase state of the bilayer in which they are formed. The fact that the two *d* spacings are produced in the gel phase, however, indicates that there may be greater coordination between successive bilayers of α-tocopherol-rich and pure dipalmitoylphosphatidylcholine domains in the fluid phase.

Whether the molecules are arranged in a regular 9.6 : 1 stoichiometric lattice or oligomerization of α-tocopherol produces some higher order structure is unknown. It has been suggested that because there is no persuasive evidence of strong cohesive van der Waals interactions between hydrocarbon chains of the phospholipid and the phytol chain of α-tocopherol, the driving force for formation of α-tocopherol clusters is the reduction in repulsion between adjacent phospholipid molecules when they are separated by interpolating α-tocopherol molecules.

5. IMPLICATIONS AND CONCLUSIONS

Vitamin E, being a lipid-soluble vitamin, partitions into the hydrophobic domains of membranes, lipoproteins, and lipid-storage organelles. Dietary vitamin E is comprised of many sterioisomers, but the α- and γ-tocopherols usually predominate. These are carried to the liver in chylomicrons, where the *RRR*-α-tocopherol isomer is preferentially packaged by tocopherol-specific proteins into serum low-density lipoproteins. In cells, tocopherol distributes between all subcellular membranes, but achieves a tenfold higher concentration in Golgi membranes and lysosomes than in the remaining membrane fractions.

The location and distribution of vitamin E in membranes have been largely modeled by biophysical studies of aqueous dispersions of mixtures of phospholipids and α-tocopherol. The general consensus from these studies is that α-tocopherol is interpolated between phospholipid molecules in a bilayer configuration, with the chromanol ring locating in a region of the polar head groups of the phospholipid molecules. Whether the phenoxyl group forms hydrogen bonds with water is still a moot point, but there is evidence that such interactions take place with phosphate oxygens and/or the ester carbonyl oxygen of the fatty acyl chains. The α-tocopherol molecules are able to diffuse laterally in the plane of the bilayer with a diffusion constant similar to that of the phospholipid molecules. The molecules of α-tocopherol are also seen to rotate about their long axis perpendicular to the bilayer plane.

Vitamin E has been shown to form complexes with free fatty acids and lysolipids in phospholipid bilayer membranes. The formation of these complexes is believed to stabilize the structure which would otherwise be disturbed; this is regarded as an important function of the vitamin in addition to its well-characterized antioxidant role in biological membranes. More recent evidence on the distribution of α-tocopherol in bilayer membranes indicates that it forms stoichiometric complexes with diacylphospholipids in both the gel and the liquid-crystalline configuration in a proportion of about 10:1 phospholipid:α-tocopherol. The significance of these complexes for the functional roles of α-tocopherol is yet to be determined.

ACKNOWLEDGMENTS. Financial support for work described in this chapter was provided by an International Research Collaborative Award from the Wellcome Trust.

6. REFERENCES

Ahkong, Q. F., Fisher, D., Tampion, W., and Lucy, J. A., 1972, The fusion of erythrocytes by fatty acids, esters, retinol and α-tocopherol, *Biochem. J.* **136:**147–155.

Aranda, F. J., Coutinho, A., Berberan-Santos, M. N., Prieto, M. J. E., and Gomez-Fernandez, J. C., 1989, Fluorescence study of the location and dynamics of α-tocopherol in phospholipid vesicles, *Biochim. Biophys. Acta* **985:**26–32.

Aranda, F. J., Sanchez-Migallon, M. P., and Gomez-Fernandez, J. C., 1996, Influence of α-tocopherol incorporation on Ca^{2+}-induced fusion of phosphatidylserine vesicles, *Arch. Biochem. Biophys.* **333:**394–400.

Arita, M., Sato, Y., Miyata, A., Tanabe, T., Takahashi, E., Kayden, H. J., Arai, H., and Inoue, K., 1995, Human α-tocopherol transfer protein—cDNA cloning, expression and chromosomal localization, *Biochem. J.* **306:**437–443.

Bisby, R. H., and Ahmed, S., 1989, Transverse distribution of α-tocopherol in bilayer membranes studied by fluorescence quenching, *Free Radic. Biol. Med.* **6:**231–239.

Bisby, R. H., and Birch, D. J. S., 1989, A time-resolved fluorescence anisotropy study of bilayer membranes containing α-tocopherol, *Biochem. Biophys. Res. Commun.* **158:**386–391.

Burton, G. W., and Ingold, K. U., 1986, Vitamin E: Applications of the principles of physical organic chemistry to the exploration of its structure and function, *Acc. Chem. Res.* **19:**194–201.

Buttriss, J. L., and Diplock, A. T., 1988, The α-tocopherol and phospholipid fatty acid content of rat liver subcellular membranes in vitamin E and selenium deficiency, *Biochim. Biophys. Acta* **963:**61–69.

Castresana, J., Alonso, A., Arrondo, J.-L. R., Goni, F. M., and Casal, H., 1992, The physical state of ubiquinone-10 in pure form and incorporated into phospholipid bilayers. A Fourier-transform infrared spectroscopic study, *Eur. J. Biochem.* **204:**1125–1130.

Creutz, C. C., 1981, *cis*-Unsaturated fatty acids induce the fusion of chromaffin granules aggregated by synexin, *J. Cell Biol.* **91:**247–256.

DeKruijff, B., Van Dijck, P. W. M., Demel, R. A., Schuijff, A., Brants, F., and Van Deenen, L. L. M., 1974, *Biochim. Biophys. Acta* **356:**1–7.

Dutta Roy, A. K., Gordon, M. J., Campbell, F. M., Duthie, G. G., and James, W. P. T., 1994, Vitamin E requirements, transport and metabolism—Role of α-tocopherol-binding proteins, *J. Nutr. Biochem.* **5:**562–570.

Ekiel, I. H., Hughes, L., Burton, G. W., Joval, P. A., Ingold K. U., and Smith, I. C. P., 1988, Structure and dynamics of α-tocopherol in model membranes and in solution. A broadline and high-resolution NMR study, *Biochemistry* **27:**1432–1440.

Fragata, M., and Bellemare, F., 1980, Model for singlet oxygen scavenging by α-tocopherol in biomembranes, *Chem. Phys. Lipids* **27:**93–99.

Fukuzawa, K., Ikeno, H., Tokumura, A., and Tsukatani, H., 1979, Effect of α-tocopherol incorporation on glucose permeability and phase transition of lecithin liposomes, *Chem. Phys. Lipids* **23:**13–22.

Fukuzawa, K., Ikebata, W., Shibata, A., Kumadaki, I., Sakanaka, T., and Urano, S., 1992, Location and dynamics of α-tocopherol in model phospholipid membranes with different charges, *Chem. Phys. Lipids* **63:**69–75.

Fukuzawa, K., Matsuura, K., Tokumura, A., Suzuki, A., and Terao, J., 1997, Kinetics and dynamics of singlet oxygen scavenging by α-tocopherol in phospholipid model membranes, *Free Radic. Biol. Med.* **22:**923–930.

Gomez-Fernandez, J. C., Villalain, J., Aranda, F. J., Ortiz, A., Micol, V., Coutinho, A., Berberan-Santos, M. N., and Prieto, M. J. E., 1989, Localization of α-tocopherol in membranes, *Ann. N. Y. Acad. Sci.* **570:**109–120.

Gomez-Fernandez, J. C., Aranda, F. C., and Villalain, J., 1991, in *Progress in Membrane Biotechnology* (J. C. Gomez-Fernandez, D. Chapman, and L. Packer, eds.), pp. 98–115, Birkhauser, Basel.

Gordon, M. J., Campbell, F. M., Duthie, G. G., and Dutta Roy, A. K., 1995, Characterisation of a novel α-tocopherol-binding protein from bovine heart cytosol, *Arch. Biochem. Biophys.* **318:**140–146.

Kagan, V. E., 1989, Tocopherol stabilizes membrane against phospholipase A, free fatty acids, and lysophospholipids, *Ann. N. Y. Acad. Sci.* **570:**121–135.

Kagan, V. E., and Quinn, P. J., 1988, The interaction of α-tocopherol and homologues with shorter hydrocarbon chains with phospholipid dispersions. A fluorescence probe study, *Eur. J. Biochem.* **171:**661–667.

Kagan, V. E., Stoyanovsky, D. O., and Quinn, P. J., 1994, Integrated functions of coenzyme Q and vitamin E in antioxidant action, in *Free Radicals in the Environment, Medicine and Toxicology* (H. Nohl, H. Esterbauer, and C. Rice-Evans, eds.), pp. 221–248, Richelieu Press, London.

Kuhlenkamp, J., Ronk, M., Yusin, M., Stolz, A., and Kaplowitz, N., 1993, Identification and purification of a human liver cytosolic tocopherol binding protein, *Protein Exp. Purif.* **4:**382–389.

Liebler, D. C., 1993, The role of metabolism in the antioxidant function of vitamin E, *Crit. Rev. Toxicol.* **23:**147–169.

Massey, J. B., She, H. S., and Pownall, H. J., 1982, Interaction of vitamin E with saturated phospholipid bilayers, *Biochem. Biophys. Res. Commun.* **106:**842–847.

McMurchie, E. J., and McIntosh, G. H., 1986, Thermotropic interaction of vitamin E with dimyristoyl and dipalmitoylphosphatidyl choline liposomes, *J. Nutr. Sci. Vitaminol.* **32:**551–558.

Meydani, M., Cohn, J. S., Macauley, J. B., McNamara, J. R., Blumberg, J. B., and Schaefer, E. J., 1989, Postprandial changes in the plasma concentration of α- and γ-tocopherol in

human subjects fed a fat-rich meal supplemented with fat-soluble vitamins, *J. Nutr.* **119:**1252–1258.
Micol, V., Aranda, F. J., Villalain, J., and Gomez Fernandez, J. C., 1990, Influence of vitamin E on phosphatidylethanolamine lipid polymorphism, *Biochim. Biophys. Acta* **1022:**194–202.
Mukai, K., 1993, Synthesis and kinetic study of antioxidant and prooxidant actions of vitamin E derivatives, in *Vitamin E in Health and Disease* (L. Packer and J. Fuchs, eds.), pp. 97–119, Marcel Dekker, New York.
Mukherjee, A. K., Ghosal, S. K., and Maity, C. R., 1997, Lysosomal membrane stabilization by α-tocopherol against the damaging action of *Vipera russelli* venom phospholipase A_2, *Cell. Mol. Life Sci.* **53:**152–155.
Niki, E., Kawakami, A., Saito, M., Yamamoto, Y., Tsuchiya, J., and Kamiya, Y., 1985, Effect of phytol side chain of vitamin E on its antioxidant activity, *J. Biol. Chem.* **260:**2191–2196.
Ohyashiki, T., Ushiro, H., and Mohri, T., 1986, Effects of α-tocopherol on the lipid peroxidation and fluidity of porcine intestinal brush border membranes, *Biochim. Biophys. Acta* **858:**294–300.
Ortiz, A., and Gomez-Fernandez, J. C., 1988, Calcium-induced aggregation of phosphatidylcholine vesicles containing free oleic acid, *Chem. Phys. Lipids* **46:**259–266.
Ortiz, A., Aranda, F. J., and Gomez Fernandez, J. C., 1987, A differential scanning calorimetric study of the interaction of α-tocopherol with mixtures of phospholipids, *Biochim. Biophys. Acta* **898:**214–222.
Ozer, N. K., Palozza, P., Boscoboinik, D., and Azzi, A., 1993, α-Tocopherol inhibits low density lipoprotein induced proliferation and protein kinase C activity in vascular smooth muscle cells, *FEBS Lett.* **322:**307–310.
Perly, B., Smith, I. C. P., Hughes, L., Burton, G. W., and Ingold, K. U., 1985, Estimation of the location of natural α-tocopherol in lipid bilayers by ^{13}C-NMR spectroscopy, *Biochim. Biophys. Acta* **819:**131–135.
Quinn, P. J., 1995, Characterisation of clusters of α-tocopherol in gel and fluid phases of dipalmitoylphosphatidylcholine, *Eur. J. Biochem.* **233:**916–925.
Quinn, P. J., and Katsikas, H., 1985, Thermal characteristics of coenzyme Q and its interaction with model membrane systems, in *Coenzyme Q* (G. Lenaz, ed.), pp. 107–130, Wiley, Chichester.
Quinn, P. J., Takahashi, H., and Hatta, I., 1994, Characterisation of complexes formed in fully hydrated dispersions of dipalmitoyl derivatives of phosphatidylcholine and diacylglycerol, *Biophys. J.* **68:**1374–1382.
Salgado, J., Villalain, J., and Gomez-Fernandez, J. C., 1993, Magic-angle spinning C-13 NMR spin–lattice relaxation study of the location and the effects of α-tocopherol, ubiquinone-10 and ubiquinol-10 in unsonicated model membranes, *Eur. Biophys. J.* **22:**151–155.
Salgado, J., Villalain, J., and Gomez-Fernandez, J. C., 1995, Metastability of dimyristoylphosphatidylethanolamine as studied by FT-IR and the effect of α-tocopherol, *Biochim. Biophys. Acta* **1239:**213–225.
Sanchez-Migallon, M. P., Aranda, F. J., and Gomez-Fernandez, J. C., 1996a, The interaction of α-tocopherol with phosphatidylserine vesicles and calcium, *Biochim. Biophys. Acta* **1281:**23–30.
Sanchez-Migallon, M. P., Aranda, F. J., and Gomez-Fernandez, J. C., 1996b, Interaction between α-tocopherol and heteroacid phosphatidylcholines with different amounts of unsaturation, *Biochim. Biophys. Acta* **1279:**251–255.
Sato, Y., Hagiwara, K., Arai, H., and Inoue, K., 1991, Purification and characterisation of the α-tocopherol transfer protein from rat liver, *FEBS Lett.* **288:**41–45.

Sato, Y., Arai, H., Myata, A., Tokita, S., Yamamoto, K., Tanabe, T., and Inoue, K., 1993, Primary structure of α-tocopherol transfer protein from rat liver. Homology with retinaldehyde-binding protein, *J. Biol. Chem.* **268:**17705–17710.

Severcan, F., 1997, Vitamin E decreases the order of the phospholipid model membranes in the gel phase: An FTIR study, *Biosci. Rep.* **17:**231–235.

Severcan, F., and Cannistraro, S., 1988, Direct electron spin resonance evidence for α-tocopherol-induced phase separation in model membranes, *Chem. Phys. Lipids* **47:**129–133.

Severcan, F., and Cannistraro, S., 1990, A spin-label ESR and saturation transfer ESR study of α-tocopherol containing model membranes, *Chem. Phys. Lipids* **53:**17–26.

Sies, H., Kaiser, S., Di Mascio, P., and Murphy, M. E., 1993, Scavenging of singlet molecular oxygen by tocopherols, in *Vitamin E in Health and Disease* (L. Packer and J. Fuchs, eds.), pp. 141–152, Marcel Dekker, New York.

Srivastava, S., Phadke, R. S., Govil, G., and Rao, C. N. R., 1983, Fluidity, permeability and antioxidant behaviour of model membranes incorporated with α-tocopherol and vitamin E acetate, *Biochim. Biophys. Acta* **734:**353–362.

Steiner, M., and Anastasi, I., 1976, Vitamin E. An inhibitor of the platelet release reaction, *J. Clin. Invest.* **57:**732–737.

Steiner, M., and Mower, R., 1982, Mechanism of action of vitamin E on platelet function, *Ann. N. Y. Acad. Sci.* **393:**289–299.

Stillwell, W., Dallman, T., Dumaual, A. C., Crump, F. T., and Jenski, L. J., 1996, Cholesterol versus α-tocopherol—Effects on properties of bilayers made from heteroacid phosphatidylcholines, *Biochemistry* **35:**13353–13362.

Stocker, R., and Bowry, V. W., 1996, Tocopherol-mediated peroxidation of lipoprotein lipids and its inhibition by co-oxidants, in *Handbook of Antioxidants* (E. Cadenas and L. Packer, eds.), pp. 27–41, Marcel Dekker, New York.

Suzuki, Y., Tsuchiya, M., Wassall, S. R., Govil, G., Kagan, V. E., and Packer, L., 1993, Structural and dynamic membrane properties of α-tocopherol and α-tocotrienol. Implication to the molecular mechanism of their antioxidant potency, *Biochemistry* **32:**10692–10699.

Traber, M. G., 1994, Determinants of plasma vitamin E concentrations, *Free Radic. Biol. Med.* **2:**229–239.

Traber, M. G., and Kayden, H. J., 1989, Preferential incorporation of α-tocopherol vs γ-tocopherol in human lipoproteins, *Am. J. Clin. Nutr.* **49:**517–526.

Traber, M. G., Burton, G. W., Ingold, K. U., and Kayden, H. J., 1990, RRR- and SRR-α-tocopherols are secreted without discrimination in human chylomicrons, but RRR-α-tocopherol is preferentially secreted in very low density lipoproteins, *J. Lipid Res.* **31:**675–685.

Traber, M. G., Burton, G. W., Hughes, L., Ingold, K. U., Hidaka, H., Malloy, M., Kane, J., Hyams, J., and Kayden, H. J., 1992, Discrimination between forms of vitamin E by humans with and without genetic abnormalities of lipoprotein metabolism, *J. Lipid Res.* **33:**1171–1182.

Traber, M. G., Ramakrishnan, R., and Kayden, J. H., 1994, Human plasma vitamin E kinetics demonstrate rapid recycling of plasma RRR-α-tocopherol, *Proc. Natl. Acad. Sci. USA* **91:**10005–10008.

Urano, S., and Matsuo, M., 1989, Membrane stabilizing effect of vitamin E, *Annals NY Acad. Sci.* **570:**524–526.

Urano, S., Iida, M., Otani, I., and Matsuo, M., 1987, Membrane stabilization of vitamin E; interactions of α-tocopherol with phospholipids in bilayer liposomes, *Biochem. Biophys. Res. Commun.* **146:**1413–1418.

Urano, S., Yano, K., and Matsuo, M., 1988, Membrane stabilizing effect of α-tocopherol and its model compounds on fluidity of lecithin liposomes, *Biochem. Biophys. Res. Commun.* **150:**469–475.

Urano, S., Kitahara, M., Kato, Y., Hasegawa, Y., and Matsuo, M., 1990, Membrane stabilizing effect of vitamin E. Existence of a hydrogen bond between α-tocopherol and phospholipids in bilayer liposomes, *J. Nutr. Sci. Vitaminol.* **36:**513–519.

Urano, S., Matsuo, M., Sakanaka, T., Uemura, I., Koyama, M., Kumadaki, I., and Fukuzawa, K., 1993, Mobility and molecular orientation of vitamin E in liposomal membranes as determined by ^{19}F-NMR and fluorescence, *Arch. Biochem. Biophys.* **303:**10–14.

Villalain, J., Aranda, F. J., and Gomez-Fernandez, J. C., 1986, Calorimetric and infrared spectroscopic studies of the interaction of α-tocopherol and α-tocopherol acetate with phospholipid vesicles, *Eur. J. Biochem.* **158:**141–147.

Wassall, S. R., Thewalt, J. L., Wong, L., Gorrisen, H., and Cushley, R. J., 1986, Deuterium NMR-study of the interaction of α-tocopherol with a phospholipid model membrane, *Biochemistry* **25:**319–326.

Wassall, S. R., Wang, L., McCabe, R. C., Ehringer, W. D., and Stillwell, W., 1991, Electron spin resonance study of the interaction of α-tocopherol with phospholipid model membranes, *Chem. Phys. Lipids* **60:**29–37.

Wolf, G., 1994, Structure and possible function of an α-tocopherol-binding protein, *Nutr. Rev.* **52:**97–98.

Yamamoto, I., Mazumi, T., Asai, Y., Handa, T., and Miyajima, K., 1994, Effects of α-tocopherol and its acetate on the hydrolytic activity of phospholipase D in egg-yolk phosphatidylcholine bilayers, *Colloid Polymer Sci.* **272:**598–603.

Zhang, Y., Turunen, M., and Appelkvist, E. L., 1996, Restricted uptake of dietary coenzyme Q is in contrast to the unrestricted uptake of α-tocopherol into rat organs and cells, *J. Nutr.* **126:**2089–2097.

Chapter 15

The Molecular Action of α-Tocopherol in Lipoprotein Lipid Peroxidation

Pro- and Antioxidant Activity of Vitamin E in Complex Heterogeneous Lipid Emulsions

P. K. Witting, J. M. Upston, and R. Stocker

1. INTRODUCTION

Free radicals are inevitable by-products of biological redox reactions involving one-electron transfer processes such as those that occur during the catalytic cycles of some enzymes and electron transfer chains of mitochondria (Halliwell and Gutteridge, 1989). Radicals are also generated in biological systems as a result of exposure to a wide variety of external factors including some drugs, pollutants, metal ions, heat, UV or visible light, and other forms of radiation (Halliwell and Gutteridge, 1989). The generation of these reactive species in an uncontrolled fashion causes significant reversible or irreversible damage to a wide range of biological molecules including DNA, proteins, carbohydrates, and lipids. Many of these processes are chain reactions with a single initiating radical species damaging a

P. K. Witting, J. M. Upston, and R. Stocker Heart Research Institute, Camperdown, Sydney, NSW 2050, Australia.

Subcellular Biochemistry, Volume 30: Fat-Soluble Vitamins, edited by Quinn and Kagan. Plenum Press, New York, 1998.

large number of target molecules (Halliwell and Gutteridge, 1989). There is considerable interest in the reactions of these species, the damage they induce, and their relationship in the physiology and pathology of a variety of diseases and processes including cancer, arthritis, aging, and heart disease (Halliwell and Gutteridge, 1989). The "oxidative theory of atherosclerosis" proposes that oxidation of human low-density lipoprotein (LDL)* is implicated in the development of atherosclerosis (Steinberg *et al.*, 1989; Ylä-Herttuala *et al.*, 1989; Steinbrecher *et al.*, 1990; Witztum and Steinberg, 1991; Ross, 1993), the single largest cause of death in Western societies. The oxidation of different classes of biological macromolecules in isolation is well defined; however, less is known about oxidative processes in complex mixtures of lipids and proteins, such as lipoproteins, and the mechanisms inhibiting these deleterious processes.

α-Tocopherol (α-TOH), biologically and chemically the most active form of vitamin E (Burton and Ingold, 1986), is also the major lipid-soluble antioxidant in extracts prepared from human LDL (Esterbauer *et al.*, 1992) (Table I). Accordingly, most research into inhibition of LDL lipid peroxidation has focused on α-TOH, with comparatively little interest in other LDL antioxidants, such as ubiquinol-10 ($CoQ_{10}H_2$), γ-tocopherol, lycopene, and β-carotene (Table I).

While the oxidative modification of LDL is implicated as an early event in atherogenesis (Steinberg *et al.*, 1989; Ylä-Herttuala *et al.*, 1989; Steinbrecher, 1990; Witztum and Steinberg, 1991; Ross, 1993), the precise mechanism(s) by which this modification occurs remain unclear. Oxidatively modified LDL is taken up by macrophages via the scavenger receptor to form lipid-laden or "foam" cells, indicative of early atherosclerotic le-

*Abbreviations: AAPH, 2,2′-*Azobis*(2-amidinopropane); apo B, apolipoprotein B-100; ArOH, phenol; ArO$^•$, phenoxyl radical; ROO$^•$, aqueous or lipophilic alkyl peroxyl radical; CE, cholesteryl esters; CE-OOH, cholesteryl linoleate hydroperoxides; CE-O(O)H, cholesteryl linoleate hydroperoxides and hydroxides; Cu^{2+}, cupric ions; $CoQ_{10}H_2$, ubiquinol-10; CoQ_{10}, coenzyme Q_{10}; EPR, electron paramagnetic resonance; FFA, free fatty acids; FFA-OO$^•$, free fatty acid peroxyl radical; FFA-OOH, free fatty acid hydroperoxides; F_{TOH}, fractional content of α-TOH; HDL, high-density lipoprotein; HRP, horseradish peroxidase; H_2O_2, hydrogen peroxide; 3HAA, 3-hydroxyanthranilic acid; $^•$OH, hydroxyl radical; $^-$OCl, hypochlorite; LDL, low-density lipoprotein; LO, 15-lipoxygenase; LOO$^•$, lipid peroxyl radical; LH, bisallylic hydrogen-containing lipid; LOOH, lipid hydroperoxides; MPO, myleoperoxidase; NRP, non-radical product; PC, phosphatidylcholine; PC-OOH, phophatidylcholine hydroperoxides; protein$^•$, protein-derived radical; rhLO, recombinant human 15-lipoxygenase; R_i, rate of initiation of lipid peroxidation; R_p, rate of lipid peroxidation; R_{pmax}, maximal rate of lipid peroxidation; SLO, soybean lipoxygenase-1; α-TOH, α-tocopherol; TMP, tocopherol-mediated peroxidation; α-TO$^•$, α-tocopheroxyl radical; α-TQ, α-tocopherol quinone; α-TQH2, α-tocopheryl hydroquinone; L-TyrO$^•$, L-tyrosyl radical; VLDL, very low density lipoprotein.

Table I
Composition of Human LDL[a]

Component	Weight %	Moles/LDL
Protein		
Apolipoprotein B100	22 ± 1.9	1
Lipids		
Fatty acids	—	20–50
Phospholipids		700
Phosphatidylcholine	22.3 ± 3.9	450
Bisallylic hydrogens[b]	—	375
Cholesterol	9.6 ± 0.7	600
Cholesteryl esters	42.2 ± 3.8	1600
Cholesteryllinoleate	—	880
Cholesterlyarachidonate	—	95
Bisallylic hydrogens	—	1165
Triglycerides	5.9 ± 2.7	180
Bisallylic hydrogens	—	50
Total bisallylic hydrogens	—	1590
Antioxidants		
α-Tocopherol	—	6–12
γ-Tocopherol	—	0.5
Ubiquinol-10	—	0.5–0.8[c]
Lycopene	—	0.2–0.7
β-Carotene	—	0.1–0.4

[a]The values in this table are derived from Chance *et al.* (1979), Esterbauer *et al.* (1992), Bowry and Stocker (1993), Reaven *et al.* (1993), Constantinescu *et al.* (1993), and Upston *et al.* (1996).
[b]Bisallylic hydrogens refer to the polyunsaturated fatty acids in LDL (i.e., the most readily oxidizable lipid moieties of the lipioprotein particle. Linoleic and arachidonic acid contain one and three pairs of bisallylic hydrogens, respectively.
[c]This range has been determined in our laboratory and is consistently higher than that reported by others (see, e.g., Esterbauer *et al.* (1992). Reasons for this difference are discussed in Stocker (1994).

sions (Goldstein and Brown, 1977), and there is evidence that at least in certain animals, the scavenger receptor is required for atherosclerosis to develop (Suzuki *et al.*, 1997). Recent interest has focused on mechanisms of LDL lipid oxidation initiated by a variety of both physiological and chemical oxidants as a result of the likely physiological relevance of oxidized LDL.

This review addresses the conventional mode of action of vitamin E in inhibiting lipid peroxidation. It also describes a novel mechanism for lipid oxidation in complex lipoprotein emulsions, where the usually accepted paradigms of vitamin E's "antioxidant" activity do not describe the observed lipid peroxidation. Finally, we summarize recent studies on LDL oxidation induced by peroxidases and lipoxygenases.

2. THE CONVENTIONAL MOLECULAR ACTION OF VITAMIN E (EXCLUSIVE ANTIOXIDANT ACTION)

The mechanism of peroxidation of lipid containing bisallylic hydrogen (LH) in homogeneous solution is well established and takes place via the radical-chain mechanism illustrated by reactions (1)–(4). The process is initiated by formation of a carbon-centered lipid radical ($L^{\cdot}$) which subsequently adds, at near diffusion-limited rates, to oxygen, producing the corresponding lipid peroxyl radical ($LOO^{\cdot}$). Reaction of $LOO^{\cdot}$ (the chain-carrying species) with another LH generates a lipid hydroperoxide (LOOH) and another propagating $LOO^{\cdot}$. In the absence of chain-breaking antioxidants, lipid peroxidation eventually terminates via bimolecular processes [reaction (4)]:

Initiation $\quad \text{Oxidant} + LH \rightarrow \text{Inactive Oxidant} + L^{\cdot} \quad (1)$

Propagation $\quad L^{\cdot} + O_2 \rightarrow LOO^{\cdot} \quad (2)$

Propagation $\quad LOO^{\cdot} + LH \rightarrow LOOH + L^{\cdot} \quad (3)$

Termination $\quad LOO^{\cdot} + LOO^{\cdot}/L^{\cdot} \rightarrow NRP \quad (4)$

If the oxidative theory of atherosclerosis plays a role in atherogenesis, an understanding of the earliest stages of LDL lipid (per)oxidation *in vitro*, and how endogenous and synthetic antioxidants affect this process, will provide useful information in designing strategies to prevent oxidative damage to lipoproteins *in vivo*. Lipid peroxidation is generally considered a necessary event in the sequence of oxidative modification of LDL (to the contrary see Hazell *et al.*, 1994). Elucidating the mechanism(s) underlying LDL lipid (per)oxidation and antioxidation is ongoing and forms a broad area of research. As such, characterizing lipid oxidation products present in human atheroma (Glavind *et al.*, 1952; Brooks *et al.*, 1966, 1971; Gilbert *et al.*, 1969, 1972; Traber *et al.*, 1993; Chisolm *et al.*, 1994; Suarna *et al.*, 1995) and determining whether these products correlate with those formed during LDL peroxidation *in vitro* are of major interest. For example, by using this approach, 15-lipoxygenase (LO) (enzymic)-supported LDL lipid peroxidation is thought to contribute to overall lipid oxidation in human atherosclerotic lesions (Kühn *et al.*, 1992, 1997; Folcik *et al.*, 1995).

Unlike enzymic oxidation of lipid (e.g., by cyclooxygenases and LO), which generates regio- and stereo-specific oxidation products, the free radical-induced, random oxidation of esterified and free LH yields a complex array of products. Antioxidants with hydrogen-donating capacity, such as α-TOH, influence product distribution during LH autoxidation (Porter *et al.*, 1980, 1981; Peers and Coxon, 1983). For example, during linoleate autoxidation, abstraction of a bisallylic hydrogen initially generates 13-*cis*, *trans* (*Z*, *E*) and 9-*trans*, *cis* (*E*, *Z*) conjugated diene $LOO^{\cdot}$. Fragmentation of these $LOO^{\cdot}$ at the β-position affords isomerization to the respective thermodynamically more stable 13- and 9-(*E*, *E*) $LOO^{\cdot}$ isomers. Rapid trapping of the intermediate $LOO^{\cdot}$ by hydrogen transfer from a suitable donor, such as a phenolic compound (ArOH), before β-fragmentation, allows preferrential formation of 13-(*Z*, *E*) and 9-(*E*, *Z*)-LOOH (Porter *et al.*, 1980, 1981). This effect is also observed in aqueous peroxyl radical ($ROO^{\cdot}$)-induced oxidation of liposomes, where the presence of trolox (a water-soluble vitamin E analog) or α-TOH significantly enhances formation of the (*Z*, *E*)- and (*E*, *Z*)-hydroperoxide isomers (Barclay *et al.*, 1995).

α-TOH also influences the distribution of esterified linoleate oxidation products that accumulate in heterogeneous lipid emulsions, such as human high-density lipoprotein (HDL) (Kenar *et al.*, 1996) and LDL (Kenar *et al.*, 1996; Upston *et al.*, 1996) exposed to $ROO^{\cdot}$ or LO (Upston *et al.*, 1996). Thus, so long as (endogenous) α-TOH is present, 13-(*Z*, *E*) and 9-(*E*, *Z*) cholesteryl linoleate oxidation products predominate with relatively little formation of the corresponding (*E*, *E*) isomers (Kenar *et al.*, 1996; Upston *et al.*, 1996). This is explained by α-TOH acting as the hydrogen donor, shifting the equilibrium of product accumulation toward (*Z*, *E*) and (*E*, *Z*) products (see above). Hence the relative ratios of (*E*, *Z*) versus (*E*, *E*) oxidation products of free and esterified linoleate in complex biological lipids such as membranes, lipoproteins, and human arterial intima may be indicative of whether lipid oxidation has occurred in the presence or absence of α-TOH.

In addition to controlling the distribution of LOOH isomers, ArOH can act as a chain-breaking antioxidant removing the chain-propagating $LOO^{\cdot}$ [reaction (5)]; a less reactive phenoxyl radical ($ArO^{\cdot}$) is formed at the expense of one molecule of LOOH. Alternatively, ArOH inhibits by reacting directly with the initiating oxidant [reaction (6)] to arrest $LOO^{\cdot}$ formation. It is generally assumed that $ArO^{\cdot}$ takes no further part in lipid peroxidation, or becomes involved in peroxidation termination reactions [reaction (7)]. Thus, ArOH has a high efficacy toward inhibiting lipid peroxidation by virtue of removing 2 moles of propagating radicals per mole of antioxidant.

Inhibition $\quad ArOH + LOO^{\cdot} \rightarrow ArO^{\cdot} + LOOH \quad (5)$

Inhibition $\quad ArOH + \text{Oxidant} \rightarrow \text{Inactive Oxidant} + ArO^{\cdot} \quad (6)$

Termination $\quad LOO^{\cdot} + ArO^{\cdot} \rightarrow NRP \quad (7)$

In the "conventional" view of vitamin E antioxidant activity (Burton and Ingold, 1986), lipid peroxidation in the presence of the vitamin results in less than 1 mole of LOOH formed per mole of α-TOH consumed, i.e., a lipid peroxidation chain length ν of $\ll$1. Indeed, for lipid peroxidation in homogeneous solutions, a well-defined "inhibited period" is observed while α-TOH is present and during which time $\nu \ll 1$; this period is also termed the "lag time" (Burton and Ingold, 1986). Following consumption of the vitamin, peroxidation proceeds at much higher rates with $\nu > 1$.

A clearly defined, initial period of a low rate of lipid peroxidation (R_p) is also observed for isolated, ascorbate-free LDL oxidized with high concentrations of cupric ion (Cu^{2+}) (i.e., ≥12–16 molecules of Cu^{2+} per LDL particle) (Jürgens *et al.*, 1987; Bowry *et al.*, 1992a). In studies using these oxidizing conditions, Esterbauer and co-workers (Jürgens *et al.*, 1987) observed that relatively little lipid peroxidation is observed during the initial period of incubation in which rapid depletion of α-TOH and other antioxidants occurs, whereas after α-TOH depletion, lipid peroxidation proceeds at higher rates. The similarity of these observations and the action of vitamin E in homogeneous solution prompted Esterbauer and co-workers to suggest that α-TOH acts as a "conventional" antioxidant for LDL *in vitro*. This view of α-TOH's activity is supported by the finding that under conditions of high rates of lipid peroxidation initiation (R_i),* supplementation of LDL with α-TOH or other antioxidants (including low-molecular-weight ArOH) significantly increases the lag time (see, e.g., Jürgens *et al.*, 1987; Cosgrove *et al.*, 1987; Sattler *et al.*, 1991; Reaven *et al.*, 1992; Kleinveld *et al.*, 1994). However, there also exists a clear lack of correlation between endogenous vitamin E levels and lag-time in native LDL (Dieber-Rotheneder *et al.*, 1991; Esterbauer *et al.*, 1992), inconsistent with a "conventional" antioxidant role for α-TOH in LDL. In addition, the suggestion that resistance of LDL lipid peroxidation, R_p, and the overall extent of LDL lipid peroxidation are dependent on the ratio of oleic to linoleic acids (Kleinveld *et al.*, 1993) conflicts with this conventional view (see also Stocker, 1994).

3. α-TOCOPHEROL-MEDIATED PEROXIDATION (TMP) OF LDL LIPID (ANTI- AND PROOXIDANT ACTIONS OF VITAMIN E)

Recent observations show that in the absence of endogenous $CoQ_{10}H_2$ and other low-molecular-weight antioxidants (see below), LDL lipids

* R_i is defined by the rate of α-TOH consumption, i.e., $R_i = -2d(\alpha\text{-TOH})/dt$, according to Boozer *et al.* (1955) and as applied to LDL (Bowry and Stocker, 1993).

peroxidize readily in the presence of α-TOH when the lipoprotein is exposed to low concentrations of initiating oxidant, e.g., aqueous or lipophilic ROO˙, Cu^{2+} ions, hydroxyl radicals (˙OH), horseradish peroxidase (HRP/H_2O_2), or soybean lipoxygenase-1 (SLO). Indeed, under these conditions α-TOH supports peroxidative chain reactions (i.e., $\nu > 1$) (Sato *et al.*, 1990) and acts as a prooxidant (Bowry *et al.*, 1992a, b; Bowry and Stocker, 1993; Witting *et al.*, 1995, 1997; Neuzil *et al.*, 1997a). Thus, R_p is inversely proportional to R_i, is decreased by exchanging α-TOH's phenolic proton with deuterium, is faster in the presence of α-TOH than immediately after its depletion, is enhanced upon enrichment of the lipoprotein with α-TOH, and is retarded when the lipoprotein is depleted of α-TOH (either during the course of the oxidation reaction or prior to exposure oxidant), and the maximal rate of lipid peroxidation (R_{pmax}) occurs in the presence of the vitamin and is independent of R_i. These observations clearly conflict with and cannot be explained by the "conventional" view of α-TOH as a chain-breaking antioxidant (see Bowry and Stocker, 1993, and above). Such observations, however, are explained readily in terms of both the phase transfer activity of α-TOH [reaction (8), Scheme 1] and the chain transfer activity of α-tocopheroxyl radical (α-TO˙) [reaction (9), Scheme 1]. The chain transfer activity is characterized by initiation and propagation of lipid peroxidation via α-TO˙ [reactions (9)–(11)]. The overall process whereby α-TOH facilitates LDL lipid peroxidation is termed tocopherol-mediated peroxidation (TMP) (Bowry and Stocker, 1993).

Unlike the oxidation of LH in homogeneous phase, free radical-mediated lipid peroxidation in lipoprotein emulsions such as LDL occurs within discrete particles separated by an aqueous/lipid interface. Impor-

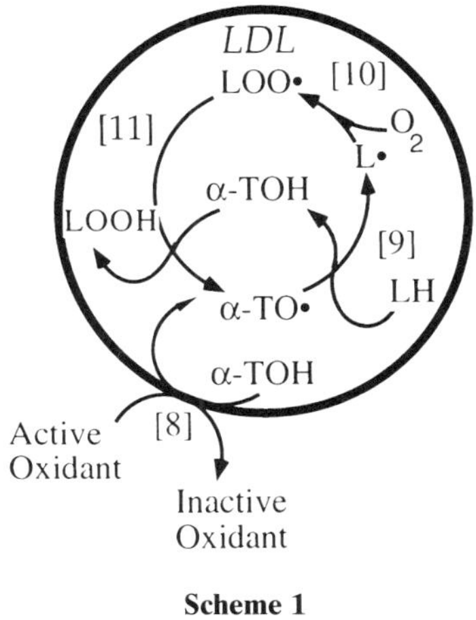

Scheme 1

tantly, the highly hydrophobic nature of the radicals limits their diffusion between particles, and thereby localizes the radicals within the particles. Thus, a single radical such as α-TO˙ in a peroxidizing particle is physically segregated from nonperoxidizing and other (single radical-containing) peroxidizing particles. Therefore, under a low flux of initiating radicals, a peroxidizing lipoprotein particle continues peroxidizing by virtue of the low incidence of bimolecular radical termination reactions (Bowry and Stocker, 1993).

3.1. Chain-Transfer Activity of LDL's Vitamin E

The particle size and α-TOH content significantly influence lipid peroxidation kinetics (Bowry and Stocker, 1993; Ingold *et al.*, 1993). Thus, under conditions of identical R_i, the chain length of lipid peroxidation decreases in the order very low density lipoprotein (VLDL) > LDL > HDL (Mohr and Stocker, 1994; Bowry and Stocker, 1993); i.e., decreasing ratios of particle volume to surface area. This order also reflects the decreasing ability to sustain a single radical per particle by virtue of a relatively higher incidence of termination of α-TO˙ at the surface of the particle under identical R_i. Enrichment of lipoproteins with α-TOH modulates the chain-transfer activity of the vitamin. For example, under comparable R_i, lipids in α-TOH-enriched LDL peroxidize faster than those in the corresponding native lipoprotein particle (Bowry *et al.*, 1992a).

3.2. Phase-Transfer Activity of LDL's Vitamin E

Exchanging vitamin E's phenolic hydrogen with deuterium (Witting *et al.*, 1995) effectively attenuates the ability of α-TOH to transfer radicals from the aqueous compartment into LDL [reaction (5), $k_H/k_D \approx 2$, measured at 37°C]. This exchange results in an overall twofold decrease in both the rates of accumulation of hydroperoxides of cholesteryl ester (CE-OOH) and the consumption of α-TOH in LDL exposed to aqueous ROO˙, despite deuterium-exchange reducing the radical scavenging ability of α-TOH (Ingold and Howard, 1962). Thus, if lipid peroxidation were to proceed in LDL via the "conventional" mechanism of vitamin E antioxidant activity, a reduction in radical scavenging ability is predicted to enhance rather than diminish R_p (Burton and Ingold, 1981). By contrast, the inverse deuterium kinetic isotope effect observed is fully consistent with the reduced phase transfer activity of α-TOD decreasing TMP of LDL lipids.

Vitamin E May Be Required for Effective Initiation of LDL Lipid Peroxidation

The oxidizability of LDL lipids can be *dependent* on the presence and initial levels of α-TOH, *independent* of the one-electron oxidant employed (low-molecular-weight, water- and lipid-soluble ROO˙, Cu^{2+} ions, ˙OH, Ham's F-10 medium, SLO, $MPO/H_2O_2/Cl^-$, and HRP/H_2O_2) to initiate lipid peroxidation (Neuzil *et al.*, 1997a; see below). Most strikingly, the initiation of lipid peroxidation in (hydroperoxide-free) LDL by SLO, Cu^{2+} ions (Fig. 1), or $MPO/H_2O_2/Cl^-$/tyrosine (see below) only occurs in the presence of α-TOH. In the case of Cu^{2+} ions this is true even when Cu^{2+} : LDL ratios of up to 20 : 1 are used. For these and other oxidants employed under conditions of low R_i, CE-OOH and hydroperoxides of phosphatidyl choline (PC-OOH) accumulate at lower rates in *in vitro* α-TOH-depleted LDL

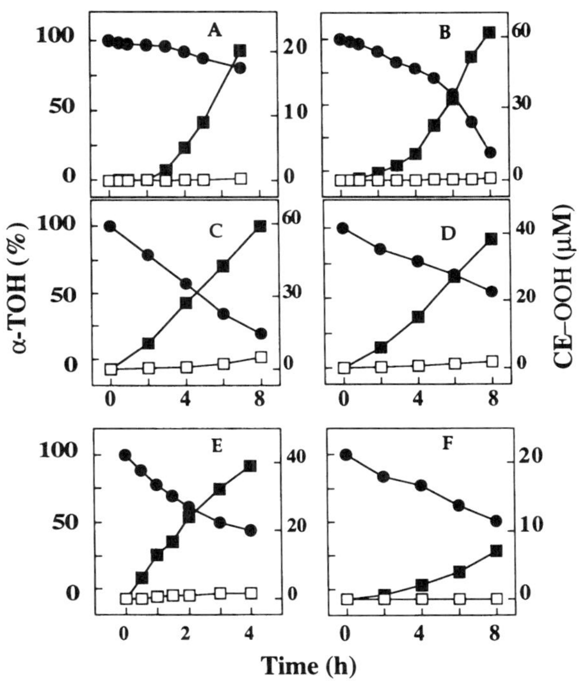

FIGURE 1. Exposure of native (filled symbols), but not α-TOH-depleted (open symbols), LDL to (A) SLO, (B) Cu^{2+}, (C) aqueous ROO˙, (D) lipophilic ROO˙, (E) HRP/H_2O_2, and (F) Ham's F-10 medium results in the time-dependent accumulation of CE-OOH (squares) and consumption of α-TOH (●). Experimental conditions are as detailed in Neuzil *et al.* (1997a). Reproduced from Neuzil *et al.* (1997a) with permission from the publisher.

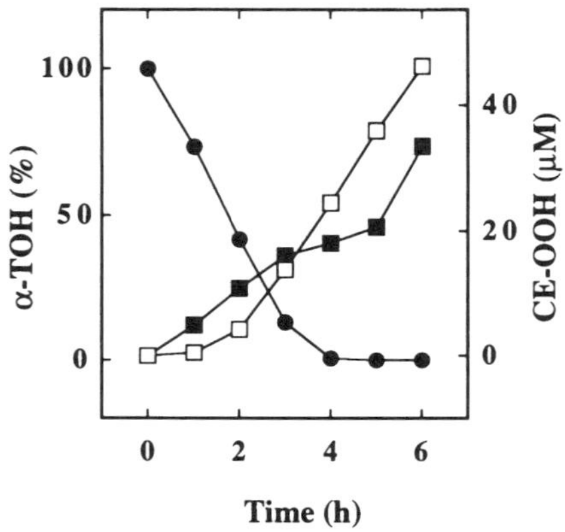

FIGURE 2. *In vitro* α-TOH-depleted LDL (open symbol) is initially more resistant to ˙OH-induced lipid peroxidation compared with native LDL (closed symbol). However, subsequent to α-TOH consumption (●) in the native LDL sample, CE-OOH (squares) accumulate at similar rates in both systems. Experimental conditions are as detailed in Neuzil *et al.* (1997a). Reproduced from Neuzil *et al.* (1997a) with permission from the publisher.

than the corresponding native samples (Neuzil *et al.*, 1997). This applies even to highly reactive ˙OH, where α-TOH-depleted LDL shows higher resistance toward lipid peroxidation in the early stages of oxidation than native LDL, although at later stages, CE-OOH accumulation is greater in the vitamin-depleted than in the native lipoprotein (Fig. 2). The oxidizability of lipids in isolated LDL decreases proportionally with the particle's α-TOH content (Fig. 3). Replenishing the *in vitro* depleted LDL with α-TOH (Esterbauer *et al.*, 1991), but not with α-tocopherol acetate, fully restores both its original content of vitamin E and its lipid oxidizability (Fig. 4). These observations are consistent with α-TOH actively promoting the initiation of LDL lipid peroxidation.

Importantly, *in vivo* α-TOH-depleted plasma and lipoproteins obtained from a patient with familial isolated vitamin E deficiency syndrome (FIVE)* (Traber *et al.*, 1990, 1993) behave similarly to the corresponding *in vitro* α-TOH-depleted samples (Fig. 5). This demonstrates that the observed requirement for α-TOH in the initiation of lipoprotein lipid peroxidation is not the result of an *in vitro* artifact.

*The FIVE syndrome is characterized by a low level of plasma α-TOH unless the patient is supplemented daily with vitamin E. α-TOH-depleted lipoproteins are obtained from plasma after a period of 5 consecutive days without vitamin E supplementation. Plasma and lipoproteins serially repleted with α-TOH (native LDL and HDL) are obtained after the patient is supplemented with 400, 800, and 1200 mg of vitamin E over the ensuing 3 consecutive days.

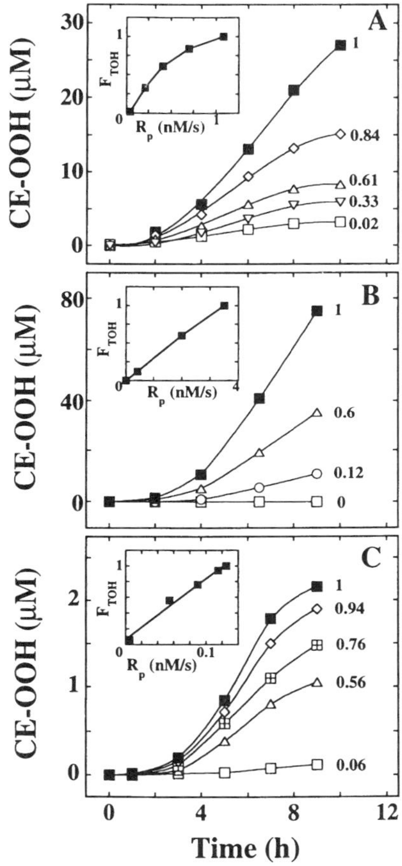

FIGURE 3. LDL lipid oxidizability is dependent on the fractional content F_{TOH} of the lipoprotein's α-TOH. LDL with decreasing F_{TOH} was exposed to (A) aqueous ROO˙, (B) SLO, or (C) Ham's F-10 medium and the time-dependent accumulation of CE-OOH monitored. Insets show the dependence of the maximal rate of lipid peroxidation R_{pmax} on LDL's F_{TOH}. Numbers listed adjacent to lines represent the various F_{TOH} values in LDL for each oxidant, with $F_{TOH} = 1$ corresponding to 100% endogenous α-TOH. Experimental conditions are as detailed in Neuzil *et al.* (1997a). Reproduced from Neuzil *et al.* (1997a) with permission from the publisher.

3.3. Antioxidant Activity of LDL's Vitamin E under Conditions of High Radical Flux

With high fluxes of ROO˙ or ˙OH, the activity of α-TOH in LDL switches from pro- to antioxidant, with the "switching point" for ˙OH observed at a lower flux than that for ROO˙ (Neuzil *et al.*, 1997a). For Cu^{2+} ion-induced initiation of LDL lipid peroxidation, this switching point is achieved at Cu:LDL ratios less than ~3 (Witting *et al.*, 1995; Ziouzenkova *et al.*, 1996). Such switching in α-TOH's action is explained readily by TMP alone (Waldeck and Stocker, 1996), so that complicated explanations involving several different mechanisms for different Cu:LDL ratios (Ziouzenkova *et al.*, 1996) are not needed. As R_i increases, the incidence of termination reactions involving the otherwise chain-carrying α-TO˙ also

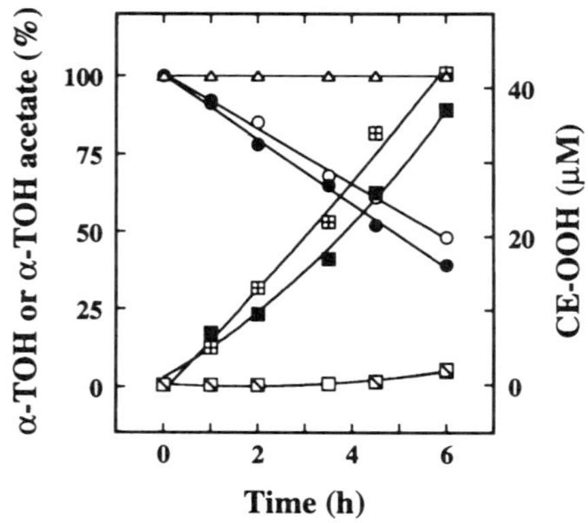

FIGURE 4. Replenishment of *in vitro* vitamin E-depleted LDL with α-TOH, but not α-TOH acetate, restores its lipid peroxidizability to aqueous ROO˙. Symbols listed correspond to α-TOH analogue or CE-OOH for native (●, ■), α-TOH-depleted (none, □), α-TOH-repleted (○, ⊞), or α-TOH acetate-replenished (△, ⧅) LDL, respectively. Experimental conditions are as detailed in Neuzil *et al.* (1997a). Reproduced from Neuzil *et al.* (1997a) with permission from the publisher.

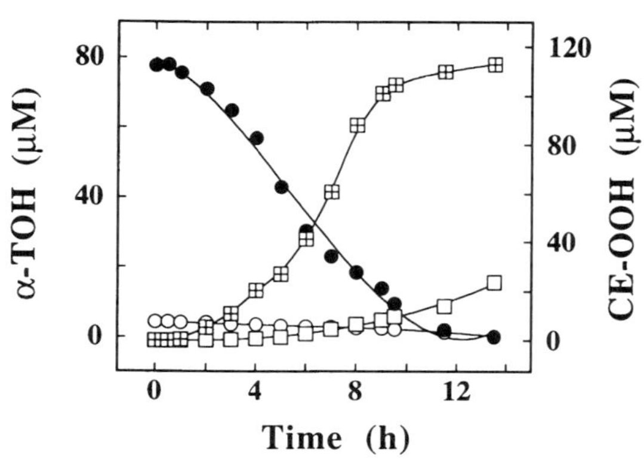

FIGURE 5. Plasma from a vitamin E-deficient patient is remarkably resistant to aqueous ROO˙-induced LDL lipid peroxidation, but its oxidizability is restored upon *in vivo* replenishment with vitamin E. Plasma obtained prior to (○, □) and after (●, ⊞) vitamin E supplementation was oxidized by aqueous ROO˙ under identical conditions and both α-TOH (circles) and CE-OOH (squares) monitored. Experimental conditions are as detailed in Neuzil *et al.* (1997a). Reproduced from Neuzil *et al.* (1997a) with permission from the publisher.

increases, thereby reducing the extent of lipid peroxidation, and allowing vitamin E to appear to act as a "conventional" antioxidant.

The above illustrates that α-TOH, in general, renders lipoproteins more reactive toward one-electron radical oxidants. Whether this results in an overall pro- or antioxidant activity depends, in part, on the specific oxidation conditions employed.

4. ROLE OF VITAMIN E IN ENZYME-MEDIATED LDL LIPID PEROXIDATION

Peroxidase-mediated LDL lipid oxidation is thought to represent an *in vitro* model for transition metal-independent oxidation of LDL (Wieland *et al.*, 1993; Braun and von Bruchhausen, 1994; Panasenko *et al.*, 1994; Savenkova *et al.*, 1994; Santanam and Parthasarathy, 1995), though the precise mechanism for LDL lipid peroxidation initiated by peroxidases is not clear. We first consider two peroxidases, horseradish peroxidase (HRP) as a general model for peroxidases, and myeloperoxidase (MPO).

4.1. Horseradish Peroxidase-Induced LDL Lipid Peroxidation

Upon treatment of LDL with HRP/H_2O_2, conjugated dienes are formed rapidly and concomitantly with the consumption of LDL's α-TOH and the detection of α-TO$^\cdot$ (Kalyanaraman *et al.*, 1995). While these results are consistent with LDL lipid peroxidation proceeding via TMP, additional results suggest a different role for vitamin E in HRP/H_2O_2-induced LDL lipid peroxidation. For example, enrichment with α-TOH has been reported to increase LDL's resistance toward HRP/H_2O_2-induced lipid peroxidation (Kalyanaraman *et al.*, 1995), although several other groups observed that supplementation of LDL with low levels of α-TOH enhances LDL lipid oxidation (Braun and von Bruchhausen, 1994; Santanam and Parthasarathy, 1995; Witting *et al.*, 1997). Also, lipid hydroperoxide-free LDL devoid of α-TOH completely resists oxidation induced by HRP/H_2O_2 (Neuzil *et al.*, 1997), indicating a requirement for α-TOH for the peroxidation of LDL lipids (see above).

One mechanism proposed for the initiation of LDL lipid peroxidation by HRP/H_2O_2 suggests that another free radical (tentatively assigned to an apolipoprotein B-100 radical, apo B$^\cdot$) initiates lipid peroxidation via the following reactions (Kalyanaraman *et al.*, 1995):

$$\text{apo B} + \text{HRP/H}_2\text{O}_2 \rightarrow \text{apo B}^\cdot + \text{Resting State Enzyme} \qquad (12)$$

$$\text{apo B}^\cdot + \text{LH} \rightarrow \text{apo B} + \text{L}^\cdot \qquad (13)$$

Inconsistent with this mechanism, however, lipid peroxidation occurs concomitantly with α-TO$^\cdot$ detection, which itself precedes detection of the apo B$^\cdot$ (Kalyanaraman *et al.*, 1995). Protein radicals (protein$^\cdot$) formed upon exposure of HRP or other peroxidases to H_2O_2 can translocate intra- (Wilks and Ortiz de Montellano, 1992) and intermolecularly (Ortiz de Montellano and Catalano, 1985; Ortiz de Montellano and Grab, 1987; Heinecke *et al.*, 1993a; Moreau *et al.*, 1995). The latter may result in the cooxidation of low-molecular-weight substrates, of which α-TOH is by far the most abundant (Esterbauer *et al.*, 1992) in human LDL. Indeed, direct oxidation of the vitamin by HRP/H_2O_2 occurs readily even when α-TOH is present in nonoxidizable or complex emulsions [reaction (14)] (Kalyanaraman *et al.*, 1995; Witting *et al.*, 1997). Once formed, α-TO˙ may then react with LH to initiate lipid oxidation via TMP [see reaction (9), Scheme 1] (Bowry and Stocker, 1993):

$$\alpha\text{-TOH} + \text{HRP/H}_2\text{O}_2 \rightarrow \alpha\text{-TO}^{\cdot} + \text{Rest State Enzyme} \qquad (14)$$

Several lines of evidence support a role of α-TO˙ rather than apo B˙ in the initiation of LDL lipid peroxidation by HRP/H_2O_2 (Witting *et al.*, 1997). First, the concentration of α-TO˙ in LDL correlates positively with R_p, particularly during the early phase of the oxidation, and R_{pmax} is independent of R_i (Witting *et al.*, 1997). Also, supplementation with α-TOH enhances the HRP/H_2O_2-induced oxidizability of LDL lipids, independent of the presence of preformed CE-O(O)H (Witting *et al.*, 1997). Further, CE-O(O)H continues to accumulate as long as α-TO˙ is detectable, even in the absence of active HRP (Fig. 6). In addition, the ability of urate to eliminate the redox-active forms of HRP (and hence eliminate further initiation of oxidation reactions) combined with its *inability* to both eliminate α-TO˙ and prevent CE-O(O)H accumulation (Figs. 6 and 7) is fully consistent with the "urate paradox" (Bowry and Stocker, 1993). In fact, this is the most direct evidence to support TMP and a role for α-TO˙ as the lipid peroxidation chain-transfer agent in oxidizing LDL.

While these results strongly support a role for TMP in HRP/H_2O_2-induced LDL lipid oxidation, they do not unambiguously prove that this mechanism is operative. Indeed, the rapid initial depletion of α-TOH without CE-O(O)H accumulation at high HRP:LDL ratios could be interpreted as evidence against TMP (Witting *et al.*, 1997). However, as discussed earlier, a rapid initial loss of α-TOH can be explained readily by bimolecular termination reactions under these strongly oxidizing conditions (Bowry and Stocker, 1993; Neuzil *et al.*, 1997). Radical-radical termination reactions occur rapidly, effectively preventing α-TO˙-induced oxidation of LH at the expense of vitamin E consumption (see Section 3.3). Thus, a lack of correlation between α-TOH consumption and LOOH accumulation per se does not argue against TMP (Waldeck and Stocker, 1996).

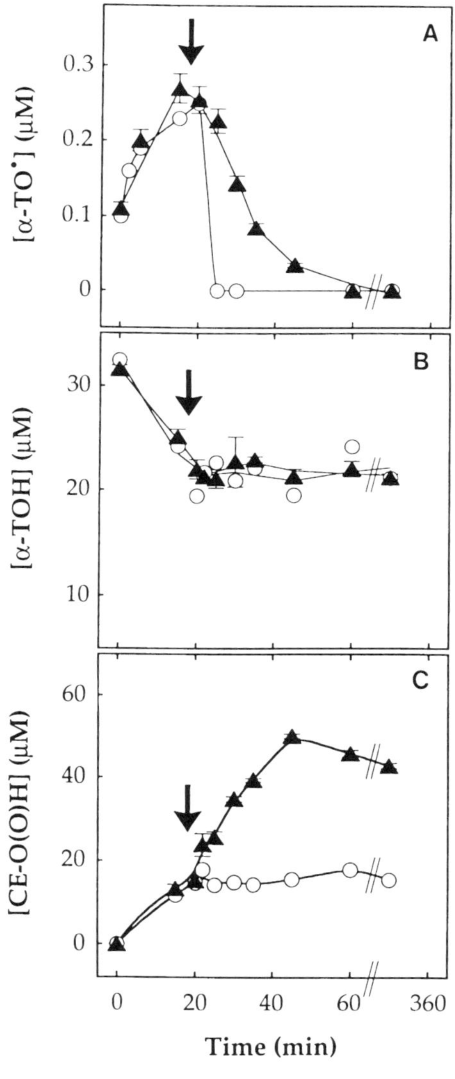

FIGURE 6. Time-dependent changes in (A) α-TO• concentration, (B) consumption of α-TOH, and (C) accumulation of CE-O(O)H in HRP/H_2O_2-oxidizing LDL treated sequentially with catalase followed by either urate (▲) or ascorbate (○). The arrow indicates time of treatment with catalase followed by addition of urate (100 μM) or ascorbate (50 μM). Experimental conditions are as detailed in Witting *et al.* (1997). Reproduced from Witting *et al.* (1997) with permission from the publisher.

4.2. MPO/H_2O_2/Cl^-/Tyrosine-Induced LDL Lipid Peroxidation

MPO is the only human enzyme known to produce hypochlorite ($^-$OCl) under physiological conditions. $^-$OCl, a potent oxidant, bleaches heme proteins, chlorinates proteins and unsaturated lipids, and oxidatively modifies LDL to a form that causes macrophages to accumulate cholesteryl

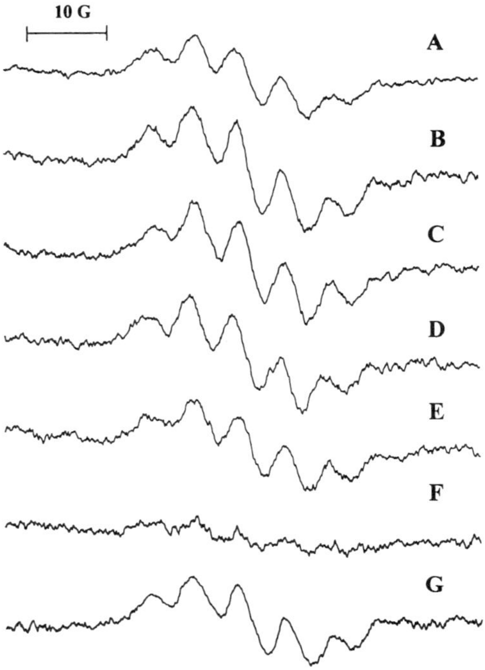

FIGURE 7. Time-dependent changes in α-TO˙ concentration in LDL treated sequentially with HRP/H_2O_2 then either phosphate buffer (control) or catalase followed by urate. LDL was treated with HRP/H_2O_2; after 20 min, catalase was added to the mixture and after a further 2 min incubation, urate (100 μM) was added. At times (A) 5, (B) 15, (C) 20, (D) 25, (E), 30, (F) 45 min, aliquots were removed for EPR spectroscopic analyses. (G) A spectrum taken after 90 min from a control sample treated with phosphate buffer alone. Experimental conditions are as detailed in Witting *et al.* (1997). Reproduced from Witting *et al.* (1997) with permission from the publisher.

esters (CE; see below). However, MPO can also act on L-tyrosine, generating tyrosyl radicals (L-TyrO˙; Heinecke *et al.*, 1993b). L-TyrO˙ can dimerize to dityrosine (Heinecke *et al.*, 1993b), react with protein tyrosyl residues to yield protein-bound dityrosines (Heinecke *et al.*, 1993b), and initiate LDL lipid peroxidation (Savenkova *et al.*, 1994). The latter is proposed to occur via direct reaction of the tyrosyl radical with LH:

$$\text{L-TyrO}^{\cdot} + \text{LH} \rightarrow \text{L-TyrOH} + \text{L}^{\cdot} \tag{15}$$

As pointed out earlier, α-TOH is the most redox-active molecule present on the surface of LDL, and therefore it is expected to compete with LH for reaction with L-TyrO˙. Indeed, LDL lipid peroxidation induced by

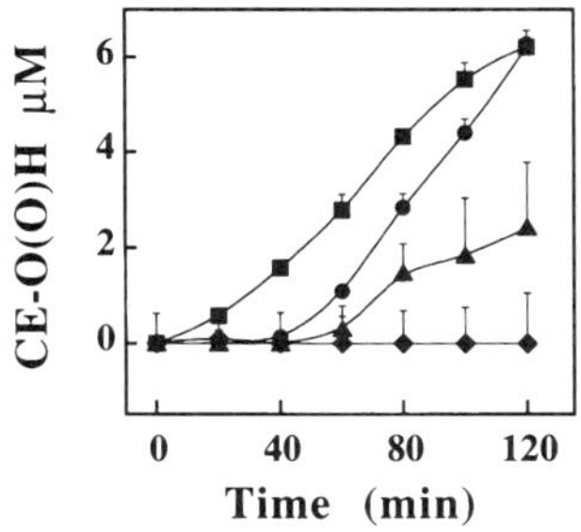

FIGURE 8. Treatment of vitamin E-enriched (■), vitamin E-replenished (▲), or native (●), but not vitamin E-depleted LDL (◆) with $MPO/H_2O_2/Cl^-$/tyrosine results in the initiation of lipid peroxidation. Reaction conditions: MPO (4 nM), glucose oxidase 1×10^{-3} U/ml, glucose (10 μM), and tyrosine (500 μM) were mixed prior to addition of LDL at final concentration (0.2 mg apo B/ml). Lipid peroxidation was measured as the time-dependent accumulation of both hydroperoxides and hydroxides of CE, i.e., CE-O(O)H. The results shown represent the mean ± SD of a triplicate experiment.

$MPO/H_2O_2/Cl^-$/tyrosine shows a clear requirement for α-TOH in that the vitamin-depleted lipoprotein is completely resistant to peroxidation, whereas replenishing α-TOH in previously vitamin E-depleted LDL restores lipid oxidizability (Fig. 8). Thus, α-TOH acts as a phase-transfer agent, aiding the reaction of aqueous L-TyrO˙ with lipid components of LDL. By doing so, α-TOH appears to compete effectively with the bimolecular reaction of L-TyrO˙, as indicated by the significant reduction in dityrosine formation in reaction mixtures containing vitamin E and, in contrast, the enhancement of dityrosine levels in vitamin E-depleted LDL (Fig. 9).

4.3. Oxidation of LDL by 15-Lipoxygenase (15-LO)

Evidence for a putative role of 15-LO in atherogenesis is based on the findings that human atherosclerotic lesions contain 15-LO mRNA, protein (Ylä-Herttuala *et al.*, 1990), and specific lipid oxidation products (Kühn *et al.*, 1992, 1994a, 1997; Folcik *et al.*, 1995). Murine fibroblasts expressing human 15-LO also increase lipid peroxide levels in LDL (Benz *et al.*, 1995; Ezaki *et al.*, 1995), and transfer of the 15-LO gene into rabbit iliac arteries results in the appearance of lipid–protein adducts characteristic of oxidized LDL (Ylä-Herttuala *et al.*, 1995). Furthermore, 15-LO oxidizes LDL's lipids *in vitro* (Parthasarathy *et al.*, 1989; Belkner *et al.*, 1993; Kühn *et al.*, 1993, 1994b; Lass *et al.*, 1996). These findings suggest that 15-LO represents a possible agent of LDL oxidation *in vivo*, raising interest in the underlying mechanism(s) of this lipid peroxidation.

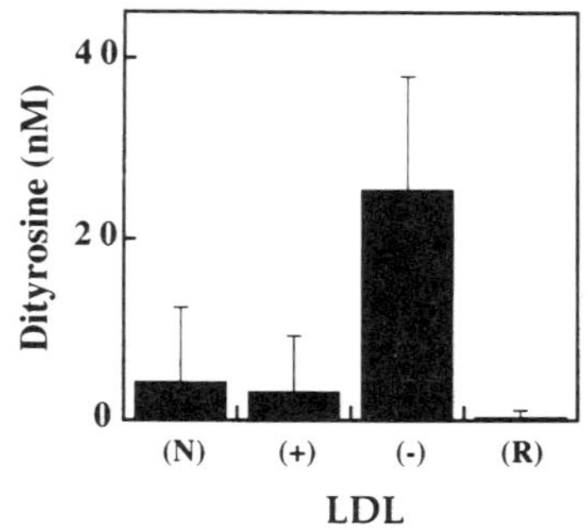

FIGURE 9. Treatment of native (N), vitamin E-enriched (+), or vitamin E-replenished (R), but not vitamin E-depleted (−) LDL with MPO/H_2O_2/Cl^-/tyrosine results in significant accumulation of dityrosine. The results shown are the mean ± SD of four independent experiments. It is noteworthy that for LDL (N) and (+) only one of the four experiments showed the presence of traces of dityrosine, hence the large deviation for these samples. Dityrosine was measured by fluorescence spectroscopy with excitation set at 280 nm and the emission monitored at 410 nm.

Several studies of Kühn and colleagues suggest that the 15-LO-induced peroxidation of LDL's CE and phospholipids (principally those of phosphatidyl choline, PC) to the corresponding CE-OOH and PC-OOH occurs via direct interaction of 15-LO with the surface and core lipids of LDL. Thus, oxidation products accumulate during the earliest stages of LDL lipid peroxidation with regio- and stereospecificity characteristic of enzymic-induced lipid oxidation, although this selectivity is considerably lower than that of 15-LO-mediated peroxidation of free (nonesterified) polyenoic fatty acids (Belkner *et al.*, 1993; Kühn *et al.*, 1993, 1994a; Lass *et al.*, 1996). Also, when used at a fixed enzyme-to-LDL ratio, α-TOH appears to be unable to inhibit lipid peroxidation induced by rabbit reticulocyte 15-LO (Lass *et al.*, 1996).

In apparent contradiction, α-TOH can, however, significantly influence the extent of CE-OOH and PC-OOH accumulation in LDL exposed to recombinant human 15-LO (rhLO) and SLO (Upston *et al.*, 1996; Neuzil *et al.*, 1997a). Thus, in LDL substantially depleted of endogenous α-TOH *in vivo*;* only small amounts of CE-OOH accumulate when the lipoprotein is incubated with rhLO (Fig. 10) or SLO (Neuzil *et al.*, 1997a). By contrast, LDL isolated from the same patient following vitamin E supplementation contains large amounts of α-TOH, and its CE readily peroxidizes under the same conditions (Fig. 10). Similarly, LDL samples depleted of (Neuzil *et al.*, 1997a) or enriched with α-TOH *in vitro* (Esterbauer *et al.*, 1991)

*See note on page 354.

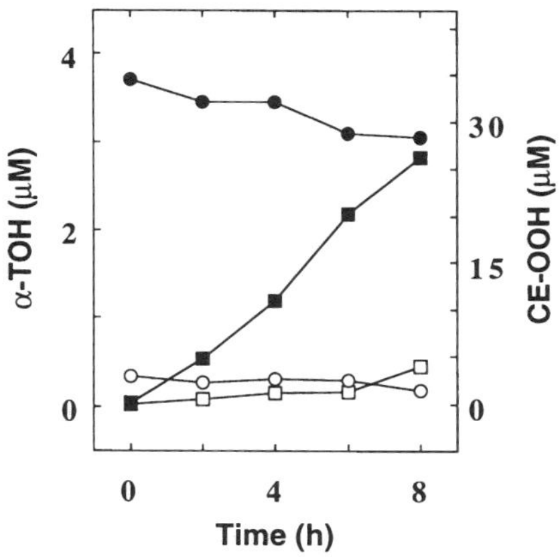

FIGURE 10. rhLO induces CE oxidation (squares) and α-TOH consumption (circles) in α-TOH-containing (filled symbols), but not *in vivo* α-TOH-depleted LDL (open symbols). LDL was obtained from the plasma of a FIVE patient receiving or not receiving oral supplementation with vitamin E. Experimental conditions are as detailed in Upston *et al.* (1996). Reproduced from Upston *et al.* (1996) with permission from the publisher.

show significantly lower or higher lipid oxidizability than native LDL, respectively.

These observations suggest that α-TOH controls 15-LO-induced oxidation of LDL's esterified lipids, and that α-TO˙ acts as the lipid peroxidation chain-propagating radical. Indeed, α-TO˙ is detectable in LO-oxidizing LDL (Kalyanaraman *et al.*, 1990; Thomas *et al.*, 1996b), and ascorbic acid, which completely eliminates the phenoxyl radical (see below), prevents both α-TOH consumption and CE-OOH accumulation in LDL exposed to rhLO when employed at an enzyme:LDL ratio of 3.8:1 (Fig. 11). By contrast, ascorbate is incapable of inhibiting enzymic oxidation of 18:2 by rhLO or SLO (Upston *et al.*, 1996).

The analysis of the geometrical isomers of Ch18:2-OH (which are derived from reduction of Ch18:2-OOH) provides more direct evidence to support a role for TMP in 15-LO-initiated peroxidation of LDL's lipids. Thus, 9-(*E*, *Z*)- and 13-(*Z*, *E*)Ch18:2-OH accumulate at indistinguishable rates during the α-TOH-containing period of aqueous ROO˙ or rhLO-induced LDL oxidation, at least under the enzyme:LDL ratios employed (Figs. 12 and 13). This pattern is identical to that for α-TOH-controlled autoxidation of polyunsaturated lipids (Peers and Coxon, 1983) and work reported by Kenar *et al.* (1996) (see above). Indeed for aqueous ROO˙-oxidizing LDL, significant amounts of 9- and 13-(*E*, *E*)Ch18:2-OH are detected only after complete consumption of α-TOH (Fig. 13). By contrast

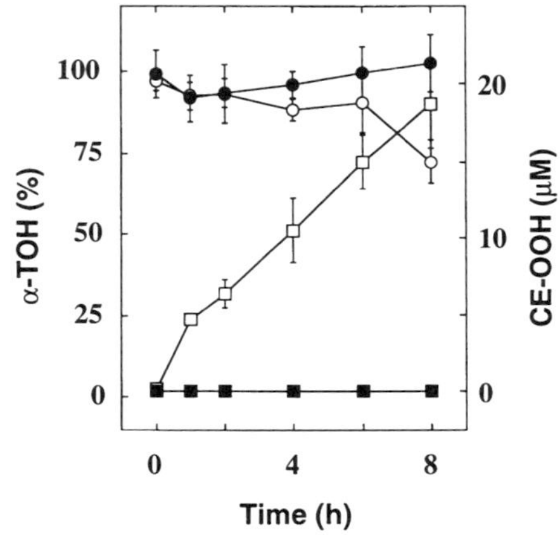

FIGURE 11. Ascorbate inhibits rhLO-initiated peroxidation of LDL's CE. The oxidation of LDL (1.25 ± 0.29 μM in apo B) was initiated by rhLO (0.33 μM) in the presence (closed symbols) or absence (open symbols) of ascorbate. At various times CE-OOH accumulation (squares) and α-TOH consumption (circles) were determined. Experimental conditions are as detailed in Upston *et al.* (1996). Reproduced from Upston *et al.* (1996) with permission from the publisher.

with LDL's CE, and in support of previous work (Sigal *et al.*, 1990), rhLO oxidation of free 18:2 gives 13-(*Z*, *E*)18:2-OOH as the typical enzymic product, with much lower amounts of other regioisomers (Upston *et al.*, 1996).

Unambiguous evidence for nonenzymic oxidation of LDL's CE and PC by 15-LO is obtained by direct analysis of stereospecific oxidation products of Ch18:2-OH. With SLO as the oxidant, the *S*- and *R*-chiral isomers of 13-(*Z*, *E*)Ch18:2-O(O)H accumulate at equimolar amounts from the onset of the oxidation (J. M. Upston *et al.*, 1997). With rhLO, some preferrential formation of the enzymic product, 13(*S*)-(*Z*, *E*)Ch18:2-O(O)H, occurs in the earliest stages of the oxidation, although this preference diminishes as oxidation proceeds (J. M. Upston *et al.*, 1997). These results show some differences in behavior between different types of 15-LO, consistent with earlier studies of Kühn and co-workers (Belkner *et al.*, 1993), although enzymic oxidation of esterified lipids by 15-LO at best represents a minor contribution to the overall lipid peroxidation products detected.

A unified model can be proposed in which 15-LO primarily acts on free fatty acids (FFA) associated with LDL isolated from plasma, generating both stereospecific FFA-OOH and FFA-derived peroxyl radicals (FFA-OO˙). The latter are released as by-products during the enzymic reaction

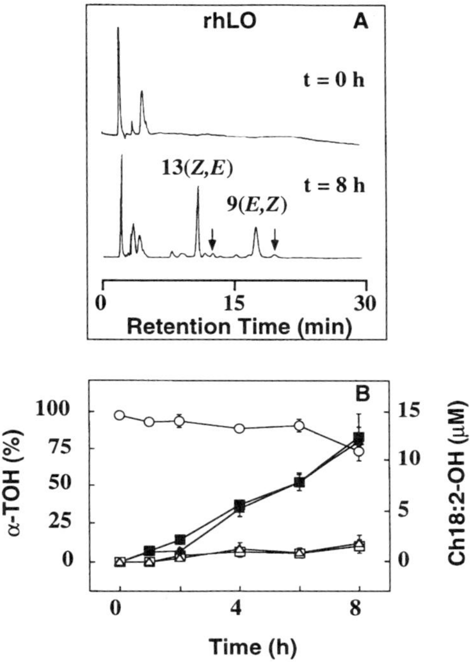

FIGURE 12. Formation of geometric isomers of Ch18:2-OH during LDL oxidation initiated by rhLO. Native LDL was incubated with rhLO, aliquots were removed as indicated, treated with $NaBH_4$ to reduce CE-OOH products to the corresponding Ch18:2-OH, and the lipids extracted and analyzed by HPLC. (A) Chromatographic profile of LDL's Ch18:2-OH isomers present at 0 and 8 hr, with the major oxidation products being the 13(*Z*, *E*)-and 9(*E*, *Z*)-Ch18:2-OH, and the minor products 13- and 9(*E*, *E*)-Ch18:2-OH indicated by the arrows. (B) Accumulation of 13 (*Z*, *E*)- (■), 9(*E*, *Z*)- (▲), 13(*E*, *E*)- (□), and 9(*E*, *E*)-Ch18:2-OH (△) isomers and consumption of α-TOH (○) during rhLO-induced LDL oxidation. Experimental conditions are as detailed in Upston *et al.* (1996). Reproduced from Upston *et al.* (1996) with permission from the publisher.

(Chamulitrat and Mason, 1989). In addition, or alternatively, lipid-derived radicals could result from interaction of 15-LO with FFA-OOH (Davies and Slater, 1987). In either case, these secondary FFA-OO˙ are expected to be scavenged rapidly by α-TOH, giving rise to α-TO˙:

$$\text{FFA-OO}^{\cdot} + \alpha\text{-TOH} \rightarrow \text{FFA-OOH} + \alpha\text{-TO}^{\cdot} \qquad (16)$$

Once formed, α-TO˙ mediates nonenzymic formation of CE-OOH and PC-OOH, by initiating and carrying the chain of lipid peroxidation of LDL's esterified lipids via TMP (Bowry and Stocker, 1993; Waldeck and

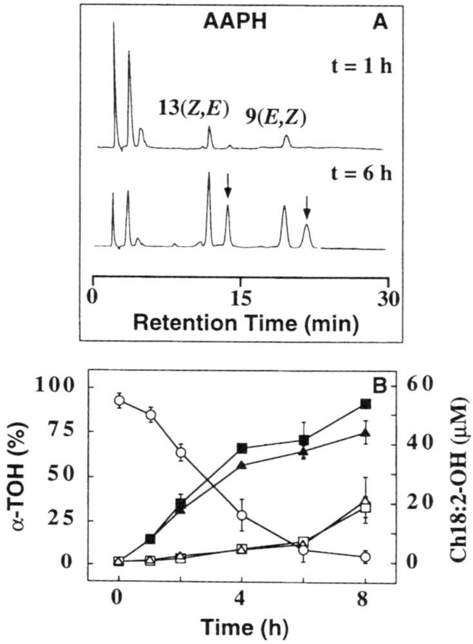

FIGURE 13. Formation of geometric isomers of Ch18:2-OH during LDL oxidation initiated by aqueous ROO˙. Native LDL was oxidized by aqueous ROO˙ and analyzed as in Fig. 12. (A) Representative chromatograms of Ch18:2-OH isomers produced while α-TOH was present (1 hr) and after α-TOH depletion (6 hr). (B) Accumulation of the various Ch18:2-OH isomers during aqueous ROO˙-initiated oxidation of LDL. Symbols are as for Fig. 12. Experimental conditions are as detailed in Upston *et al.* (1996). Reproduced from Upston *et al.* (1996) with permission from the publisher.

Stocker, 1996) (i.e., Scheme 1). Additional evidence for this mechanism includes the fact that LDL contains free fatty acids (Aviram *et al.*, 1988; Upston *et al.*, 1996) and that small amounts of enzymic oxidation products accumulate rapidly upon exposure of LDL to SLO (Fig. 14).

5. OXIDATION OF LDL BY NUCLEOPHILIC OXIDANTS

The above discussion is limited to the use of one-electron oxidants, where the primary target of oxidation in the lipoproteins is α-TOH. Under these conditions, generation of the chain-carrying α-TO˙ occurs and lipoprotein lipid peroxidation proceeds according to TMP. These oxidants, however, are only one group which causes oxidative modification of lipo-

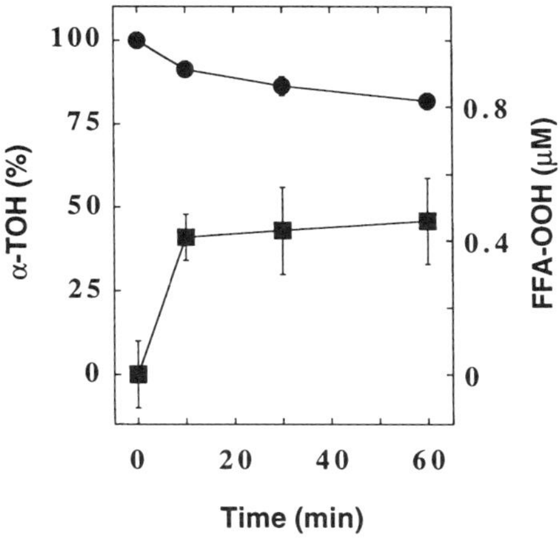

FIGURE 14. Accumulation of FFA-OOH in native LDL during rhLO-induced oxidation. Aliquots of isolated LDL were incubated with rhLO and the time-dependent consumption of α-TOH (●) and accumulation of FFA-OOH (■) monitored. FFA-OOH coelutes with authentic 13(*S*) (*Z*, *E*)-18:2-OOH, as verified by reversed-phase HPLC analysis (Upston *et al.*, 1996). Experimental conditions are as detailed in Upston *et al.* (1996). Reproduced from Upston *et al.* (1996) with permission from the publisher.

proteins. Another group of oxidants includes nucleophilic species giving rise to two-electron oxidation reactions. Of these reagents, hypochlorite ($^{-}$OCl) and peroxynitrite ($ONOO^{-}$) have been used to oxidize LDL *in vitro*. By contrast with one-electron radical oxidants, these two compounds are capable of rapidly oxidizing α-TOH to tocopheryl quinone (α-TQ) in a two-electron process, thereby potentially preventing the formation of the chain-carrying α-TO$^{\cdot}$.

5.1. $^{-}$OCl

At neutral pH, $^{-}$OCl acts primarily on the protein moiety of LDL, largely modifying tryptophan and lysine residues and causing aggregation in the latter case (Hazell *et al.*, 1994) and cross-linking of LDL's apolipoprotein B100 (O'Connell *et al.*, 1994). By contrast, only minor accumulation of CE-OOH occurs in LDL exposed to $^{-}$OCl, and this is largely independent of the $^{-}$OCl concentration employed (Hazell *et al.*, 1994). During this period $CoQ_{10}H_2$ and α-TOH are consumed rapidly. α-TOH supplementation and vitamin depletion of LDL affect neither the extent of $^{-}$OCl-mediated apolipoprotein damage (assessed by the loss of tryptophan and lysine residues and the changes in relative electrophoretic mobility of LDL), nor the damage to the lipid components (assesed by the consumption of the vitamin and the small extent of accumulation of CE-OOH) (Hazell and Stocker, 1997). These latter observations indicate that $^{-}$OCl acts on LDL independently of the level of vitamin E, and further, α-TOH only plays a minor role in modulating either lipid or protein oxidation induced by $^{-}$OCl.

5.2. $ONOO^-$

Authentic $ONOO^-$ also acts on apolipoprotein B-100, with comparatively minor oxidation of LDL's lipids (Graham *et al.*, 1993; Rubbo *et al.*, 1994; S. R. Thomas and R. Stocker, unpublished). However, when used at a molar ratio ($ONOO^-$ to LDL) of <100 μM, the extent of LDL lipid peroxidation increases proportionally with increasing initial α-TOH content, whereas the α-TOH content of LDL does not affect tryptophan or lysine loss, independent of the amounts of $ONOO^-$ added (Thomas *et al.*, 1998). Thus, it appears that for $ONOO^-$, α-TOH can control lipid peroxidation, while protein oxidation proceeds independent of the vitamin. In any case, bicarbonate (which is present at millimolar concentrations in human blood) increases the range of $ONOO^-$ concentration over which α-TOH is prooxidant, although bicarbonate overall decreases the extent of $ONOO^-$-induced LDL lipid and protein oxidation (Thomas *et al.*, 1998). Moncada and co-workers have shown that α-TO$^•$ is generated in LDL exposed to $ONOO^-$, although the majority of α-TOH is converted to tocopheryl quinone (Graham *et al.*, 1993). Together, these results indicate that although reagent $ONOO^-$ primarily undergoes two-electron reactions independent of α-TOH, the oxidant also gives rise to radical reactions, and the latter can result in the induction of LDL lipid peroxidation via TMP. Importantly, such a role of TMP may be physiologically relevant, as it is more pronounced if $ONOO^-$ is generated continuously from superoxide anion and nitric oxide radicals (Thomas *et al.*, 1998).

In summary, although developed initially with ROO$^•$ as oxidant, TMP represents a general model describing LDL lipid peroxidation for other radical oxidants and conditions that promote one-electron oxidations (Table II). While differences exist among the various oxidizing conditions (see, e.g., Lynch and Frei, 1995, for Cu^{2+} vs. Fe^{3+}), the common features are that α-TOH aids the "entry" of aqueous radicals into LDL, and that during the α-TOH-containing period, α-TO$^•$ is likely the major lipid peroxidation chain-carrying radical. Therefore, elimination of α-TO$^•$ may represent a unified *general* strategy to inhibit radical-induced lipid peroxidation in LDL (see below) and other lipoproteins. Where α-TO$^•$ formation is dependent on active enzymes (15-LO, MPO, and, likely, heme-containing proteins), their inhibition represents an alternative and/or supplementary approach.

Table II
Low-Density Lipoprotein Lipid (Per)oxidation: TMP versus Non-TMP

TMP	Non-TMP
$ROO^{\cdot}_{(aq)}$, $ROO^{\cdot}_{(lip)}$	$^{1}O_2$
Cu^{2+}, Fe^{3+}, Ham's F-10 medium, autoxidation	
$^{\cdot}OH$ (plus $O_2^{\cdot -}$)	
SIN-1 (nitric oxide donor), HONOO	(HONOO)
HRP/H_2O_2	
$MPO/H_2O_2/Cl^-$/Tyrosine	$MPO/H_2O_2/Cl^-$ (HOCl)
Recombinant human 15-lipoxygenase, SLO	
Human macrophages	

Notwithstanding this, oxidative modification of LDL may also be achieved by mechanisms independent of TMP and lipid peroxidation, as for ^-OCl (Table II). In this case, effective inhibition of oxidation is unlikely achieved by α-TOH alone and/or the prevention of TMP.

6. COANTIOXIDANTS FOR α-TOH ARE INHIBITORS OF TMP AND PROTECT LDL AGAINST LIPID PEROXIDATION

As mentioned earlier, it is the isolation of a single α-TO$^{\cdot}$ within an oxidizing lipoprotein particle that permits its reaction with LDL's LH, although this reaction is kinetically unfavorable (k_{TMP} [k_9] $\approx$ 0.01–0.1 $M^{-1}sec^{-1}$; Scheme 1; Mukai *et al.*, 1992; Bowry and Stocker, 1993). It follows that prevention of TMP is dependent on the destruction or elimination of α-TO$^{\cdot}$ from the peroxidizing particle. This may occur via a bimolecular termination reaction involving another radical [see reaction (7), Scheme 1] or by direct reduction of α-TO$^{\cdot}$ by a suitable reductant termed "coantioxidant" (XH) (Scheme 2; Doba *et al.*, 1985; Neuzil and Stocker, 1994; Bowry *et al.*, 1995). Importantly, a compound's ability to regenerate α-TOH from α-TO$^{\cdot}$ per se does *not* equate with coantioxidation. Whether or not regeneration of α-TOH results in LDL antioxidation depends on the fate of the resulting coantioxidant-derived radical (X$^{\cdot}$). "Radical export," i.e., the conversion of the *lipophilic* α-TO$^{\cdot}$ into a relatively harmless *aqueous* radical via a "reduction" reaction with the particular coantioxidant, is the process which actually prevents LOOH formation; the concomitant regeneration of the vitamin E may merely extend the period of this activity, depending on both the concentration of the coantioxidant and its redox stability under the conditions employed (Scheme 2).

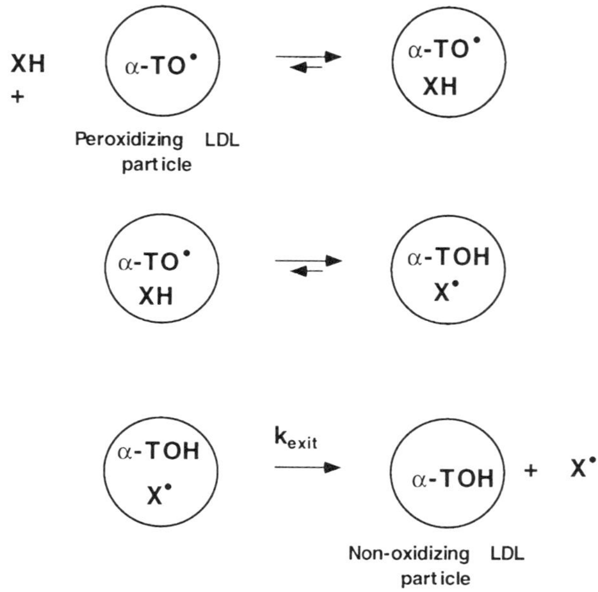

Scheme 2

A variety of natural and synthetic reductants behave as coantioxidants. The natural coantioxidants include $CoQ_{10}H_2$ (Stocker *et al.*, 1991; Mohr *et al.*, 1992; Thomas *et al.*, 1996a), α-tocopheryl hydroquinone (Neuzil *et al.*, 1997), ascorbate (Bowry *et al.*, 1995; Witting *et al.*, 1996), bilirubin (Neuzil and Stocker, 1994), and the tryptophan metabolite 3-hydroxyanthranillic acid (Thomas *et al.*, 1996b); various synthetic coantioxidants are described in Witting *et al.* (1996).

6.1. $CoQ_{10}H_2$

$CoQ_{10}H_2$ is an intrinsic component of mitochondrial electron transport systems, and is also an antioxidant in biological membranes. Freshly isolated LDL (Stocker *et al.*, 1991; Sattler *et al.*, 1994) contains small amounts of $CoQ_{10}H_2$, the reduced form of coenzyme Q_{10}, relative to the lipoprotein's content of α-TOH (Table I). $CoQ_{10}H_2$ is the first antioxidant consumed when ascorbate-free LDL is exposed to aqueous and lipophilic $ROO^•$, transition metal ions (Cu^{2+}, Fe^{3+}), activated human neutrophils, unstimulated monocytes/macrophages, authentic ^-OCl, singlet oxygen, $ONOO^-$, and nitric oxide (Stocker *et al.*, 1991; Jessup *et al.*, 1992; Wagner *et al.*, 1993; Hazell and Stocker, 1993). Although $CoQ_{10}H_2$ itself is an effective lipid-soluble antioxidant (Frei *et al.*, 1990), in LDL it appears to act predominantly as a coantioxidant for α-TOH, as the rate of $CoQ_{10}H_2$ consump-

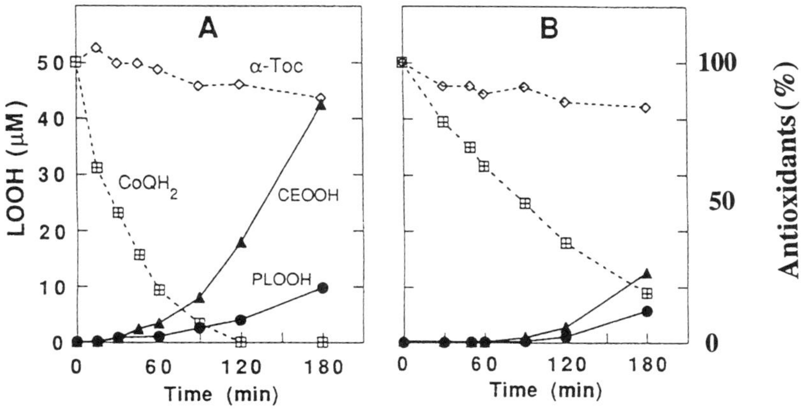

FIGURE 15. Increasing LDL $CoQ_{10}H_2$ levels increases its resistance to lipid peroxidation initiation. Native LDL (A) and $CoQ_{10}H_2$-supplemented LDL (B) were exposed to aqueous $ROO^{\cdot}$ and lipid peroxidation monitored. Levels of $CoQ_{10}H_2$ were 1.9 μM initially (A) or 5.7 μM (B) after 11 days of CoQ_{10} supplementation, while 100% α-TOH levels were ~21 μM (A) and 16 μM (B). Symbols correspond to α-TOH (◇), $CoQ_{10}H_2$ (⊞), CE-OOH (▲), and PCOOH (●), respectively. Experimental conditions are as detailed in Mohr *et al.* (1992). Reproduced from Mohr *et al.* (1992) with permission from the publisher.

tion is independent of R_g (Neuzil *et al.*, 1997a). $CoQ_{10}H_2$ likely eliminates α-$TO^{\cdot}$ indirectly (via the semiquinone radical) resulting in the formation of aqueous superoxide anion radicals which may diffuse from the particle or scavenge another α-$TO^{\cdot}$ (Ingold *et al.*, 1993).

The efficiency of $CoQ_{10}H_2$ as a coantioxidant for LDL is highlighted by supplementation of human subjects with coenzyme Q_{10}. Thus, dietary supplementation for 4–5 days results in an approximately fourfold increase in LDL's $CoQ_{10}H_2$ levels (i.e., from ~0.5–0.8 to ~2.0–2.4 molecules of $CoQ_{10}H_2$ per LDL particle; Mohr *et al.*, 1992; Kontush *et al.*, 1994; Thomas *et al.*, 1996a). This increase in $CoQ_{10}H_2$ content efficiently enhances LDL's resistance to lipid peroxidation, such as that initiated by $ROO^{\cdot}$ (Fig. 15). In addition, $CoQ_{10}H_2$-enriched LDL shows greater resistance toward initiation of lipid oxidation by Ham's F-10 medium than the corresponding native LDL (Thomas *et al.*, 1996a). Furthermore, and perhaps most importantly, coenrichment of LDL with $CoQ_{10}H_2$ and α-TOH (Thomas *et al.*, 1996a) demonstrates that $CoQ_{10}H_2$ also efficiently attenuates the accumulation of CE-OOH compared with both native LDL and LDL enriched with α-TOH alone, thereby negating any prooxidant effect of vitamin E supplementation (Fig. 16).

As the majority of human LDL contains <1 molecule $CoQ_{10}H_2$ per particle, an intriguing question is how such small levels of this or other coantioxidant(s) can provide significant antioxidant protection to LDL lipids. As mentioned earlier, TMP of LDL lipids proceeds via a free radical chain reaction (Sato *et al.*, 1990; Stocker *et al.*, 1991) with 20–40 LOOH being formed per molecule of α-TOH consumed. Each molecule of $CoQ_{10}H_2$ can potentially terminate two radical chains (see above on reactions between ubisemiquinone radicals and molecular oxygen), and therefore may readily cause the rate of the peroxidation to decrease by 40- to 80-fold. In addition, the degree of inhibition decreases with the *square root*

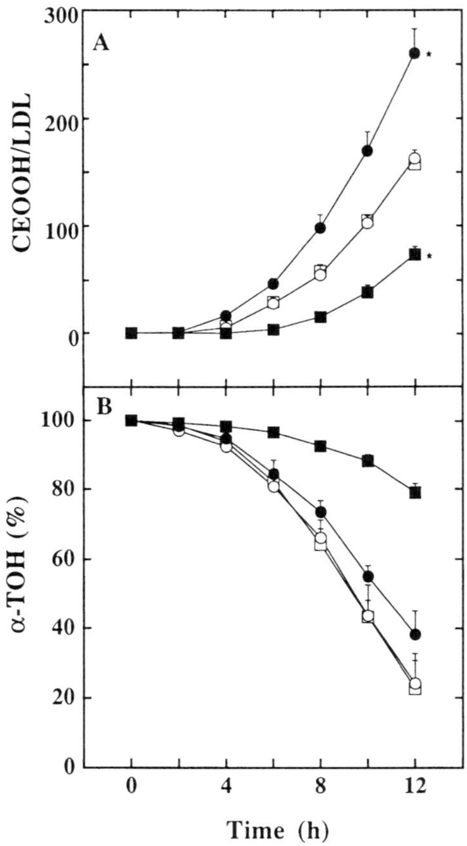

FIGURE 16. *In vivo* cosupplementaion with CoQ_{10} efficiently protects LDL against the prooxidant effects of α-TOH supplementation alone. LDL, either native (○), α-TOH-supplemented (●), stored for 5 days under argon (□), or obtained from plasma of subjects supplemented with CoQ_{10} for 5 days (■), were exposed to Ham's F-10 medium and the accumulation of CE-OOH and consumption of α-TOH monitored. Asterisk indicates significant difference between the lines of CE-OOH values for supplemented versus the corresponding control LDL. Experimental conditions are as detailed in Thomas *et al.* (1996a). Reproduced from Thomas *et al.* (1996a) with permission from the publisher.

of the concentration of the "coantioxidant" [see equation (XXIV) in Bowry and Stocker (1993); also see Bowry *et al.* (1995)]. Together, this offers an explanation of why small amounts of coantioxidants offer substantial protection against LDL peroxidation caused by a relatively low R_i. Also, in the case of $CoQ_{10}H_2$ an increase in the number of molecules per LDL particle from <1 (presupplementation) to >1 (postsupplementation) likely increases the resistance of the lipoprotein toward oxidation, particularly if cellular systems maintain this coantioxidant as the quinol and hence antioxidant form (Stocker and Suarna, 1993).

Interestingly, for human plasma, $CoQ_{10}H_2$ consumption during ROO$^{\cdot}$-induced oxidation shows a dependence on the type of oxidant employed. For example, treatment of freshly isolated plasma with aqueous ROO$^{\cdot}$ results in the consumption of plasma ascorbate prior to consumption of $CoQ_{10}H_2$ (Thomas *et al.*, 1996b). By contrast, oxidation of plasma using a lipophilic ROO$^{\cdot}$ results in $CoQ_{10}H_2$ being consumed prior to ascorbate (Neuzil and Stocker, 1994).

6.2. α-Tocopheryl Hydroquinone

α-Tocopheryl hydroquinone (α-TQH_2) acts as a multifunctional and potent inhibitor of radical-induced LDL lipid peroxidation. Aqueous ROO$^{\cdot}$-induced LDL lipid peroxidation is completely inhibited in the presence of α-TQH_2 (e.g., compare Fig. 17A and Fig. 17B), the hydroquinone being converted stoichiometrically to the corresponding α-tocopherylquinone (α-TQ; Neuzil *et al.*, 1997b). In addition to its potent coantioxidant activity, α-TQH_2 can reduce LDL's CoQ_{10} and thereby main-

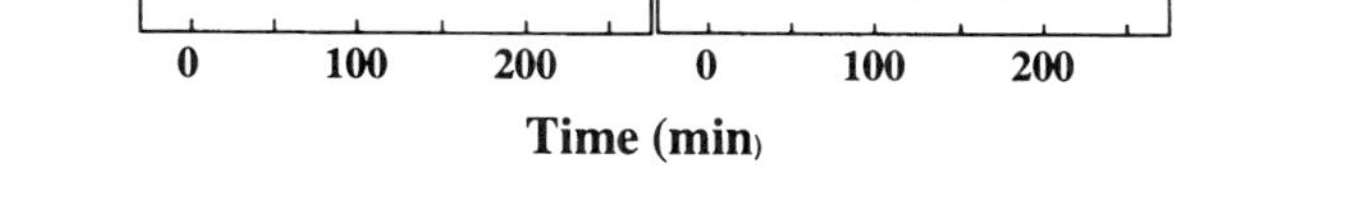

FIGURE 17. α-TQH_2 potently protects LDL cholesteryl esters and lipophilic antioxidants from aqueous ROO$^{\cdot}$-induced oxidation. LDL was treated with either ethanol to serve as control (A) or 10 μM α-TQH_2 (B) on ice, prior to oxidation by aqueous ROO$^{\cdot}$. Symbols represent $CoQ_{10}H_2$ (■), CoQ_{10} (□), α-TQH_2 (▲), α-TQ (△), α-TOH (○), and CE-O(O)H (⊞), respectively. Experimental conditions are as detailed in Neuzil *et al.* (1997b). Reproduced from Neuzil *et al.* (1997b) with permission from the publisher.

tain the coantioxidant activity of $CoQ_{10}H_2$ (Neuzil *et al.*, 1997b). Furthermore, the hydroquinone actively intercepts incipient radicals either in the presence or absence of LDL (Neuzil *et al.*, 1997b). Whether α-TQH_2 is a natural inhibitor of LDL lipid peroxidation is not clear; this antioxidant is not found in LDL isolated from blood of healthy donors. However, the corresponding oxidized form, α-TQ, is a natural oxidation product of α-TOH found in human atherosclerotic plaque (Suarna *et al.*, 1995). Also, supplementation of humans with α-TQ results in detectable levels of α-tocopheryl hydroquinone in plasma (Kohar *et al.*, 1995), indicating that reducing systems exist that can convert α-TQ to the coantioxidant active form.

6.3. Ascorbate

Among the aqueous, biological coantioxidants, the ubiquitous ascorbate (i.e., the reduced, antioxidant form of vitamin C) efficiently inhibits lipid peroxidation (Frei *et al.*, 1989). This activity is likely due to a combination of direct radical interception (where aqueous radicals are involved), and indirect action by acting synergistically with vitamin E as a coantioxidant. The regeneration of α-TOH from α-$TO^{\cdot}$ by acsorbate, with concomitant generation of the ascorbyl radical, is well established (Doba *et al.*, 1985; Packer *et al.*, 1979; Barclay *et al.*, 1985; Kagan *et al.*, 1990, 1992; Sharma and Buettner, 1993). Both ascorbate and ascorbyl palmitate (a lipophilic analogue of ascorbate) directly reduce α-$TO^{\cdot}$ in micellular dispersions, in agreement with a coantioxidant action (Witting *et al.*, 1996), giving rise to ascorbyl radicals, as measured by electron paramagnetic resonance spectroscopy (EPR; Witting *et al.*, 1996). Addition of ascorbate to LDL undergoing oxidation induced by aqueous $ROO^{\cdot}$ results in immediate cessation of α-TOH consumption and lipid oxidation (Fig. 18). Upon removal of ascorbate, via treatment with ascorbate oxidase, consumption of α-TOH and oxidation of cholesteryl esters in LDL resume, attaining the same rates as those prior to the addition of ascorbate. Furthermore, treatment of HRP/H_2O_2-oxidizing LDL with catalase followed by ascorbate results in immediate reduction of the chain-propagating α-$TO^{\cdot}$ and subsequent formation of the relatively nonreactive ascorbyl radical with immediate cessation of lipid peroxidation (Witting *et al.*, 1997). Overall, ascorbate is a phase transfer agent facilitating the export of a radical from within lipid emulsions to the aqueous compartment and thereby preventing lipid peroxidation.

6.4. Bilirubin

Bilirubin, in either its free form or when bound to its physiological carrier protein in extracellular fluids, albumin, efficiently scavenges radicals

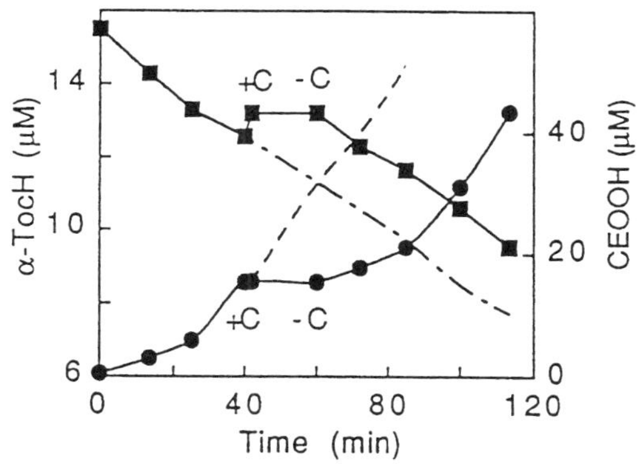

FIGURE 18. Addition and subsequent removal of ascorbic acid modulates LDL lipid peroxidation induced by aqueous ROO˙. LDL lipid peroxidation was initiated by aqueous ROO˙. At 40 min, 300 μM ascorbate was added to the reaction (+C), and the excess ascorbic acid was removed by treatment with ascorbate oxidase, 60 min (−C). CE-OOH accumulation (●) and α-TOH consumption (■) were monitored throughout. Broken lines represent the control sample, where ascorbic acid was not added. Experimental conditions are as detailed in Bowry and Stocker (1993). Reproduced from Bowry and Stocker (1993) with permission from the publisher.

(reviewed in Stocker, 1990; Stocker *et al.*, 1990). Upon addition to α-TOH-containing LDL undergoing oxidation initiated by lipophilic ROO˙, free or albumin-bound bilirubin readily inhibits CE-OOH accumulation with concomitant regeneration of α-TOH (Fig. 19A). In the case of albumin-bound bilirubin, formation of biliverdin indicates the ability of the protein-bound pigment, first, to export the radical reaction from the LDL particle into the aqueous phase and, second, convert α-TO˙ into a NRP, by analogy with ascorbate (Neuzil and Stocker, 1994). Importantly, addition of either free or albumin-bound bilirubin to lipophilic ROO˙ oxidizing LDL, after complete depletion of α-TOH, no longer inhibits CE-OOH accumulation (e.g., Fig. 19B), indicating that the pigment requires interaction with α-TOH to prevent this type of lipid peroxidation. These results exemplify the requirement of an interaction between coantioxidants and α-TOH and the subsequent removal of the radical nature from the lipoprotein particle as important events in LDL coantioxidation. For reasons currently not understood, however, free bilirubin, but not albumin-bound bilirubin, efficiently inhibits aqueous ROO˙-induced plasma lipid peroxidation in the absence of ascorbate and ubiquinol (Thomas *et al.*, 1996b).

6.5. 3-Hydroxyanthranilic Acid

The aminophenol 3-hydroxyanthranilic acid (3HAA) is produced and released by interferon-γ (IFNγ)-activated human monocytes and macroph-

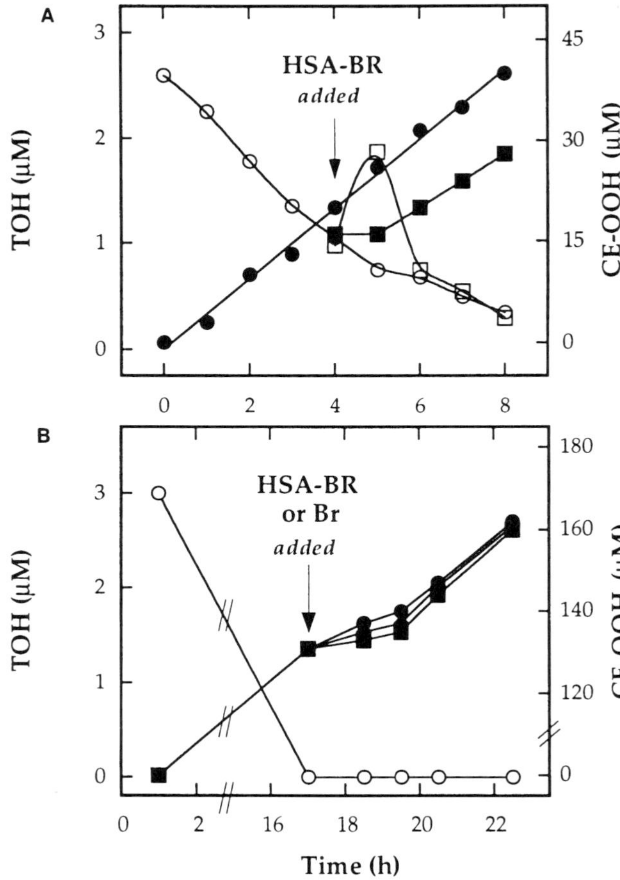

FIGURE 19. (A) Addition of human serum albumin-bound bilirubin (HSA-BR) to lipophilic ROO$^{\cdot}$-induced oxidation of α-TOH-containing LDL results in attenuation of CE-OOH accumulation and concomitant regeneration of α-TOH. (B) By contrast, addition of HSA-BR or free bilirubin to oxidizing LDL after the complete depletion of α-TOH does not inhibit LDL lipid peroxidation, as indicated by parallel rates of CE-OOH accumulation subsequent to addition of HSA-BR (■) or free bilirubin (▲) to the peroxidizing LDL. Experimental conditions are as detailed in Neuzil and Stocker (1994). Reproduced from Neuzil and Stocker (1994) with permission from the publisher.

ages as a metabolite of oxidative tryptophan metabolism along the kynurenine pathway. 3HAA can scavenge ~3 radicals per molecule (Christen *et al.*, 1990, 1992). Release of 3HAA by IFNγ-primed human mononuclear phagocytes is largely responsible for the inhibitory effect of the cytokine on cell-mediated LDL oxidation in transition metal-containing medium (Christen *et al.*, 1994). 3HAA also efficiently inhibits aqueous and lipophilic ROO˙-induced (Bowry *et al.*, 1995; Thomas *et al.*, 1996b) and Cu^{2+}-induced LDL lipid peroxidation; it is also capable of reducing α-TO˙ directly in both positively and negatively charged micellular dispersions (Witting *et al.*, 1996) and LDL (Thomas *et al.*, 1996b). The active moiety of 3HAA is the aminophenol group, as both 2- and 4-aminophenol are capable of reducing α-TO˙, while anthranilic acid lacks this activity (Witting *et al.*, 1996).

3HAA added to freshly isolated plasma acts as a second line of aqueous defense against aqueous ROO˙-induced lipid peroxidation (Thomas *et al.*, 1996b), the order of antioxidant consumption being ascorbate >3HAA > $CoQ_{10}H_2$ > bilirubin > urate (Fig. 20). Although not normally present in plasma, 3HAA may be formed under certain proinflammatory conditions where IFNγ and reactive oxygen species are produced. This offers an intriguing possibility that LDL coantioxidation may be partly under cellular control and oxidative stress may lead to a production of the aminophenol (e.g., Christen *et al.*, 1994).

Overall these findings suggest that biological systems are rich in coantioxidants for LDL's α-TOH. Therefore, lipoprotein antioxidation via inhibition of TMP likely depends on a complex network of interacting low-molecular-weight antioxidants and, perhaps, proteins that eventually result in the reduction of α-TO˙. An important feature of lipid peroxidation inhibition via this pathway is that LOOH is not formed as α-TO˙ is reduced, whereas in the conventional view of α-TOH acting as a chain-breaking antioxidant, LOOH is formed [reaction (5)]. Thus, α-TOH is a more efficient antioxidant in the TMP model as long as coantioxidants are available.

7. POSSIBLE ROLES OF TMP AND ANTI-TMP IN ATHEROGENESIS

Although a large body of evidence implicates oxidatively modified LDL in atherogenesis, the exact mechanism(s) involved in LDL oxidation *in vivo* are not defined. However, as pointed out above, TMP represents a model of LDL lipid peroxidation generally applicable for radical (one-electron) oxidants (Table II) and, as such, the prevention of TMP may be seen as a general approach to prevent *in vivo* lipoprotein lipid peroxidation induced by radical oxidants. Whether, and if so, to what extent, this is relevant to atherogenesis depends on at least three parameters: the nature of the oxidant(s) involved, the rate to which LDL (and other lipoproteins) become exposed to radical oxidant(s), and the extent to which *in vivo* LDL oxidation proceeds.

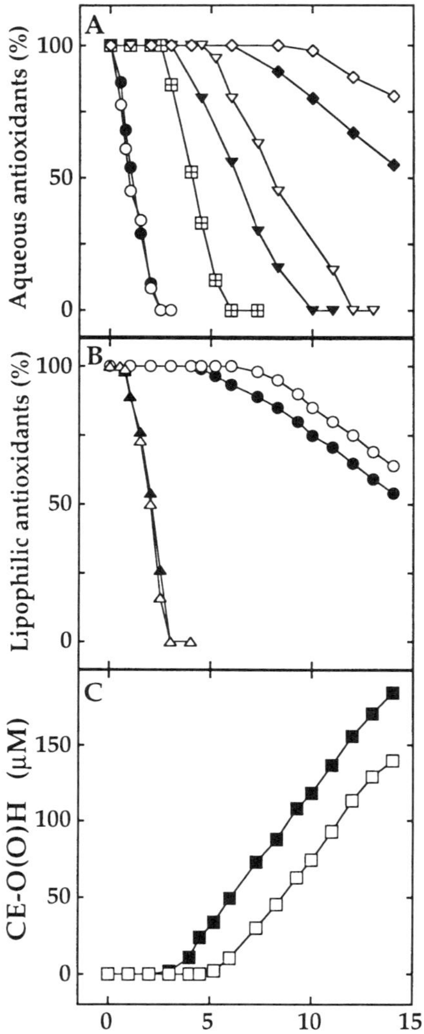

FIGURE 20. 3-Hydroxyanthranilic acid (3HAA) efficiently inhibits aqueous ROO˙-induced lipid peroxidation of human plasma. Consumption of (A) aqueous antioxidants [ascorbate (circles), 3HAA (hatched squares), bilirubin (inverted triangles), and urate (diamonds)] and (B) lipophilic antioxidants [$CoQ_{10}H_2$ (triangles) and α-TOH (circles)] and (C) concomitant accumulation of CE-O(O)H (squares) were monitored in the presence (open symbol) and absence (filled symbol) of 3HAA. Experimental conditions are as detailed in Thomas *et al.* (1996b). Reproduced from Thomas *et al.* (1996b) with permission from the publisher.

7.1. Putative Oxidants Involved in *in Vivo* LDL Oxidation

7.1.1. Transition Metals

Cu^{2+} is the most commonly used oxidant for *in vitro* LDL oxidation (Esterbauer *et al.*, 1989), yet the evidence that (free) transition metals are involved in *in vivo* LDL oxidation remains indirect. The presence of ceroid in human lesion (Parums *et al.*, 1990) provides indirect evidence for a potential involvement of transition metals. Reports on the presence of bleomycin-detectable iron and phenanthroline-detectable copper in gruel samples from advanced lesions (Smith *et al.*, 1992) unfortunately used materials that were first frozen (and thawing of biological materials releases transition metals; Gutteridge *et al.*, 1985), and lacked appropriate control tissue, and the oxidation events did not correlate with the levels of transition metals. Also, the presence of large amounts of keto-derivatives of Ch18:2 (see e.g., Suarna *et al.*, 1995) indirectly supports an involvement of free and/or protein-bound transition metals (Wilcox and Marnett, 1993), although LO can also metabolize LOOH to the corresponding keto-derivatives (Schewe *et al.*, 1986). The presence of LDL modified by other degradation products derived from LOOH, such as malonylaldehyde and 4-hydroxynonenal (e.g., Ylä-Herttuala *et al.*, 1989), relies on the specificity of the antibody used, as well as the specificity of oxidative modification of ε-amino groups for lipid peroxidation. Regarding the latter, reagent ^{-}OCl (Hazell *et al.*, 1994) and aldehydes formed from amino acids exposed to $MPO/H_2O_2/Cl^-$ (Anderson *et al.*, 1997; Hazen *et al.*, 1997) can modify protein amino groups via Schiff base formation, similar to aldehydes derived from lipid peroxidation. Recently, Heinecke and co-workers reported a clear lack of increase in the levels of *o*- and *m*-tyrosine in LDL isolated from early to intermediate lesions obtained from post-mortem aortas (Leeuwenburgh *et al.*, 1997a). The absence of these modified forms of tyrosine from early lesions, previously used as markers for hydroxyl radical-induced oxidative damage catalyzed by free metals (Halliwell and Gutteridge, 1992), indicates that at least for the early stages of atherogenesis, transition metals are not required.

7.1.2. Reactive Nitrogen Intermediates

A potential (free) metal-independent pathway for LDL oxidation involves nitric oxide generated by vessel wall cells. Nitric oxide is a weakly oxidizing radical that fails to substantially oxidize LDL at physiological pH (Jessup *et al.*, 1992). However, nitric oxide reacts rapidly with $O_2^{-\cdot}$ to form $ONOO^-$, which peroxidizes LDL readily (Jessup *et al.*, 1992), and which converts tyrosine to 3-nitrotyrosine (van der Vliet *et al.*, 1995). The recent findings of elevated levels of 3-nitrotyrosine in LDL isolated from vascular lesions supports the notion that nitric oxide and reactive nitrogen intermediates derived from it may participate in *in vivo* LDL oxidation (Leeuwenburgh *et al.*, 1997b). Formation of $ONOO^-$ may not necessarily be required for this, as in the presence of hypochlorous acid, nitrite also gives rise to nitrating species (Eiserich *et al.*, 1996).

7.1.3. Enzymic Modification

Among enzymic oxidants, 15-LO and MPO have received most interest. In addition to the evidence presented earlier (see Section 4.3), support for a role of 15-LO in atherogenesis comes from a recent study showing that administration of a specific inhibitor of 15-LO (which lacks significant antioxidant activity) dramatically lowered lesion formation in New Zealand rabbits fed a high-fat diet containing 0.25% cholesterol (Sendobry *et al.*, 1997). Unfortunately the effect of the enzyme inhibitor on the distribution of chiral 15-LO products could not be carried out in this study. Furthermore, 13-(*Z*, *E*)-hydroxylinoleate was detected as a racemic mixture in the control animals (Sendobry *et al.*, 1997), thereby casting some doubts on the specific involvement of 15-LO in the *in vivo* lipid peroxidation in this animal model of atherosclerosis.

Catalytically active MPO is present in human atherosclerotic lesions, where it colocalizes with lipid-laden macrophages (Daugherty *et al.*, 1994). The enzyme can give rise to both tyrosyl radicals and $^{-}$OCl, both of which oxidize LDL by different pathways (see above). A monoclonal antibody which selectively reacts with $^{-}$OCl-modified proteins recognizes yet unknown epitopes in early human lesions and advanced atheroma and an apolipoprotein B-100-containing fraction derived from $^{-}$OCl-modified LDL, indicating a role for $^{-}$OCl (Hazell *et al.*, 1996). Consistent with this, 3-chlorotyrosine, which is formed when L-tyrosine reacts with $^{-}$OCl or MPO/H_2O_2/Cl^-, is present at elevated levels in LDL isolated from human lesions (Hazen and Heinecke, 1997). The observation that both 3-chlorotyrosine- and $^{-}$OCl-modified proteins are present in early lesions indicates that this type of oxidative modification of LDL may indeed be pathophysiologically relevant; MPO-gene knockout mice may provide more direct support for this notion.

7.2. Frequency of Radical Hits of Intimal LDL

As discussed above, a potentially prooxidant activity of vitamin E *in vivo* is, in part, dependent on the frequency with which intimal LDL may encounter radical oxidants and the availability of effective coantioxidants. Relative to other plasma constituents, LDL accumulates in the intima, and this will decrease the hit frequency for a given radical flux. Unfortunately, information on the rate of radical formation in the subendothelial space is unavailable. On the basis of what is known about the rates of consumption of molecular oxygen through respiration, the maximal fraction of this converted to $O_2^{-\cdot}$ by mitochondria (Chance *et al.*, 1979), and the oxidative damage that radicals cause *in vivo* (Shigenaga *et al.*, 1989; Loft *et al.*, 1994), a mild *in vivo* oxidative stress capable of sustaining TMP in LDL appears feasible.

7.3. Extent of Intimal LDL Oxidation

A putative biological relevance of TMP and anti-TMP crucially depends on the extent of intimal LDL oxidation. If LDL oxidation proceeds substantially beyond α-TOH consumption, TMP may be responsible for a minor proportion of the oxidized lipids, and coantioxidants may merely delay, rather than substantially inhibit or prevent, LDL lipid peroxidation. The ratio of α-TOH to free cholesterol levels varies between arteries, but is consistently low in lesions, especially in lesions rich in macrophage foam cells (Carpenter *et al.*, 1995). However, the levels of free cholesterol greatly increase in lesions, more so than the concentration of oxidizable lipid. When expressed per oxidizable lipids, the levels of α-TOH are *elevated* in homogenates of advanced human lesions (Suarna *et al.*, 1995) and lipoprotein fractions derived from them (V. Zammit, X. Nui, J. May, R. T. Dean, and R. Stocker, unpublished), although large amounts of oxidized lipids are also present in the same materials. While these findings are consistent with intimal lipid peroxidation occurring via TMP, they do not clarify whether the measured lipid oxidation occurs in the presence or the absence of α-TOH, as it remains possible that α-TOH may be temporarily depleted and subsequently replenished as a response to the oxidative stress. As α-TOH largely prevents metal-dependent degradation of LOOH in LDL to secondary lipid oxidation products such as aldehydes (see, e.g., Bowry and Stocker, 1993) the presence of the latter in atherosclerotic lesion may provide indirect information on whether LDL oxidation proceeds beyond the consumption of vitamin E (see, however, Section 7.1.1). Alternatively, such information may be obtained from the distribution of regioisomeric lipid oxidation products (see above). Thus, in the presence of the vitamin (a potent hydrogen donor), the oxidation of Ch18:2 yields primarily *cis*, *trans* (*Z*, *E*) hydroperoxides at positions 13 and 9 on the fatty acid carbon chain. In the absence of α-TOH, the thermodynamically more stable, *trans*, *trans* (*E*, *E*) isomers are formed primarily (Kenar *et al.*, 1996; Upston *et al.*, 1996). Preliminary findings suggest that in homogenates of advanced human plaque the *Z*, *E* regioisomers of Ch18:2-OH are more abundant than the corresponding *E*, *E* isomers (J. M. Upston, A. Terentis, and R. Stocker, unpublished), suggesting that the majority of oxidized lipids may have been formed in the presence of a hydrogen donor, most likely α-TOH. If confirmed, these results would be consistent with a deficiency in coantioxidant activity, rather than α-TOH itself, contributing to intimal LDL lipid oxidation.

8. REFERENCES

Anderson, M. M., Hazen, S. L., Hsu, F. F., and Heinecke, J. W., 1997, Human neutrophils employ the myeloperoxidase–hydrogen peroxide–chloride system to convert hydroxy-amino acids into glyceraldehyde, 2-hydroxypropanal, and acrolein. A mechanism for the generation of highly reactive α-hydroxy and α,β-unsaturaed aldehydes by phagocytes at sites of inflammation, *J. Clin. Invest.* **99:**424–432.

Aviram, M., Lund-Katz, S., Phillips, M. C., and Chait, A., 1988, The influence of the triglyceride content of low density lipoprotein on the interaction of apolipoprotein B-100 with cells, *J. Biol. Chem.* **263:**16842–16848.

Barclay, L. R. C., Locke, S. J., and MacNeil, J. M., 1985, Autoxidation in micelles. Synergism of vitamin C with lipid-soluble vitamin E and water-soluble Trolox, *Can. J. Chem.* **63:**366–374.

Barclay, L. R., Artz, J. D., and Mowat, J. J., 1995, Partitioning and antioxidant action of the water-soluble antioxidant, Trolox, between the aqueous and lipid phases of phosphatidylcholine membranes: ^{14}C tracer and product studies, *Biochim. Biophys. Acta* **1237:**77–85.

Belkner, J., Wiesner, R., Rathman, J., Barnett, J., Sigal, E., and Kühn, H., 1993, Oxygenation of lipoproteins by mammalian lipoxygenases, *Eur. J. Biochem.* **213:**251–261.

Benz, D. J., Mol, M., Ezaki, M., Mori-Ito, N., Zelan, I., Miyanohara, A., Friedmann, T., Parthasarathy, S., Steinberg, D., and Witztum, J. L., 1995, Enhanced levels of lipoperoxides in low density lipoprotein incubated with murine fibroblast expressing high levels of human 15-lipoxygenase, *J. Biol. Chem.* **270:**5191–5197.

Boozer, C. E., Hammond, G. S., Hamilton, C. E., and Jyotirinda, N. S., 1955, Air oxidation of hydrocarbons. The stoichiometry and fate of inhibitors in benzene and chlorobenzene, *J. Am. Chem. Soc.* **77:**3233–3237.

Bowry, V. W., and Stocker, R., 1993, Tocopherol-mediated peroxidation. The prooxidant effect of vitamin E on the radical-initiated oxidation of human low-density lipoprotein, *J. Am. Chem. Soc.* **115:**6029–6044.

Bowry, V. W., Ingold, K. U., and Stocker, R., 1992a, Vitamin E in human low-density lipoprotein. When and how this antioxidant becomes a pro-oxidant, *Biochem. J.* **288:**341–344.

Bowry, V. W., Stanley, K. K., and Stocker, R., 1992b, High density lipoprotein is the major carrier of lipid hydroperoxides in fasted human plasma, *Proc. Natl. Acad. Sci. USA* **89:**10316–10320.

Bowry, V. W., Mohr, D., Cleary, J., and Stocker, R., 1995, Prevention of tocopherol-mediated peroxidation of ubiquinol-10-free human low density lipoprotein, *J. Biol. Chem.* **270:**5756–5763.

Braun, S., and von Bruchhausen, F., 1994, Vitamin E or probucol as donors for oxidation of human low-density lipoprotein by peroxidases/H_2O_2, *Pharmacology* **49:**325–335.

Brooks, C. J. W., Harland, W. A., and Steel, G., 1966, Squalene, 26-hydroxycholesterol and 7-ketocholesterol in human atheromatous plaques, *Biochim. Biophys. Acta* **125:**620–622.

Brooks, C. J. W., Steel, G., Gilbert, J. D., and Harland, W. A., 1971, Lipids in human atheroma. Part 4. Characterisation of a new group of polar sterol esters from human atherosclerotic plaques, *Atherosclerosis* **13:**223–237.

Burton, G. W., and Ingold, K. U., 1981, Autoxidation of biological molecules. 1. The antioxidant activity of vitamin E and related chain-breaking phenolic antioxidants *in vitro*, *J. Am. Chem. Soc.* **103:**6472–6477.
Burton, G. W., and Ingold, K. U., 1986, Vitamin E: Application of the principles of physical organic chemistry to the exploration of its structure and function, *Acc. Chem. Res.* **19:**194–201.
Carpenter, K. L., Cheeseman, K. H., vanderVeen, C., Taylor, S. E., Walker, M. K., and Mitchinson, M. J., 1995, Depletion of alpha-tocopherol in human atherosclerotic lesions, *Free Radic. Res.* **23:**549–558.
Chamulitrat, W., and Mason, R. P., 1989, Lipid peroxyl radical intermediates in the peroxidation of polyunsaturated fatty acids by lipoxygenase. Direct electron spin resonance investigations, *J. Biol. Chem.* **264:**20968–20973.
Chance, B., Sies, H., and Boveris, A., 1979, Hydroperoxide metabolism in mammalian organs, *Physiol. Rev.* **59:**527–605.
Chisolm, G. M., Ma, G., Irwin, K. C., Martin, L. L., Gunderson, K. G., Linberg, L. F., Morel, D. W., and DiCorleto, P. E., 1994, 7β-Hydroperoxycholest-5-en-3β-ol, a component of human atherosclerotic lesions, is the primary cytotoxin of oxidized human low density lipoprotein, *Proc. Natl. Acad. Sci. USA* **91:**11452–11456.
Christen, S., Peterhans, E., and Stocker, R., 1990, Antioxidant activities of some tryptophan metabolites: Possible implication for inflammatory diseases, *Proc. Natl. Acad. Sci. USA* **87:**2506–2510.
Christen, S., Southwell-Keely, P. T., and Stocker, R., 1992, Oxidation of 3-hydroxyanthranilic acid to the phenoxazinone cinnabarinic acid by peroxyl radicals and by compound I of peroxidases or catalase, *Biochemistry* **31:**8090–8097.
Christen, S., Thomas, S. R., Garner, B., and Stocker, R., 1994, Inhibition by interferon-γ of human mononuclear cell-mediated low density lipoprotein oxidation. Participation of tryptophan metabolism along the kynurenine pathway, *J. Clin. Invest.* **93:**2149–2158.
Constantinescu, A., Han, D., and Packer, L., 1993, Vitamin E recycling in human erythrocyte membranes, *J. Biol. Chem.* **268:**10906–10913.
Cosgrove, J. P., Church, D. F., and Pryor, W. A., 1987, The kinetics of the autoxidation of polyunsaturated fatty acids, *Lipids* **22:**299–304.
Daugherty, A., Dunn, J. L., Rateri, D. L., and Heinecke, J. W., 1994, Myeloperoxidase, a catalyst for lipoprotein oxidation, is expressed in human atherosclerotic lesions, *J. Clin. Invest.* **94:**437–444.
Davies, M. J., and Slater, T. F., 1987, Studies on the metal-ion and lipoxygenase-catalysed breakdown of hydroperoxides using electron-spin-resonance spectroscopy, *Biochem. J.* **245:**167–173.
Dieber-Rotheneder, M., Puhl, H., Waeg, G., Striegl, G., and Esterbauer, H., 1991, Effect of oral supplementation with *d*-alpha-tocopherol on the vitamin E content of human low density lipoproteins and resistance to oxidation, *J. Lipid Res.* **32:**1325–1332.
Doba, T., Burton, G. W., and Ingold, K. U., 1985, Antioxidant and co-antioxidant activity of vitamin C. The effect of vitamin C, either alone or in the presence of vitamin E or a water-soluble vitamin E analogue, upon the peroxidation of aqueous multilamellar phospholipid liposomes, *Biochim. Biophys. Acta* **835:**298–303.
Eiserich, J. P., Cross, C. E., Jones, A. D., Halliwell, B., and van der Vliet, A., 1996, Formation of nitrating and chlorinating species by reaction of nitrite with hypochlorous acid. A novel mechanism for nitric oxide-mediated protein modification, *J. Biol. Chem.* **271:**19199–19208.
Esterbauer, H., Striegl, G., Puhl, H., and Rotheneder, M., 1989, Continuous monitoring of *in vitro* oxidation of human low density lipoprotein, *Free Radic. Res. Commun.* **6:**67–75.

Esterbauer, H., Dieber-Rotheneder, M., Striegl, G., and Waeg, G., 1991, Role of vitamin E in preventing the oxidation of low-density lipoprotein, *Am. J. Clin. Nutr.* **53:**314S–321S.

Esterbauer, H., Gebicki, J., Puhl, H., and Jürgens, G., 1992, The role of lipid peroxidation and antioxidants in oxidative modification of LDL, *Free Radic. Biol. Med.* **13:**341–390.

Ezaki, M., Witztum, J. L., and Steinberg, D., 1995, Lipoperoxides in LDL incubated with fibroblasts that overexpress 15-lipoxygenase, *J. Lipid. Res.* **36:**1996–2004.

Folcik, V. A., Nivar-Aristy, R. A., Krajewski, L. P., and Cathcart, M. K., 1995, Lipoxygenase contributes to the oxidation of lipids in human atherosclerotic plaques, *J. Clin. Invest.* **96:**504–510.

Frei, B., England, L., and Ames, B. N., 1989, Ascorbate is an outstanding antioxidant in human blood plasma, *Proc. Natl. Acad. Sci. USA* **86:**6377–6381.

Frei, B., Kim, M., *et al.*, 1990, Ubiquinol-10 is an effective lipid-soluble antioxidant at physiological concentrations, *Proc. Natl. Acad. Sci. USA* **87:**4879–4883.

Gilbert, J. D., Harland, W. A., Steel, G., and Brooks, C. J. W., 1969, The isolation and identification of 5α-cholestan-3β-ol from the human atheromatous aorta, *Biochim. Biophys. Acta* **187:**453–456.

Gilbert, J. D., Brooks, C. J. W., and Harland, W. A., 1972, Lipids of human atheroma. VII. Isolation of diesters of cholest-5-ene-3β,26-diol from extracts of advanced atherosclerotic lesions of human aorta, *Biochim. Biophys. Acta* **270:**149–155.

Glavind, J., Hartmann, S., Clemmesen, J., Jessen, K. E., and Dam, H., 1952, Studies on the role of lipoperoxides in human pathology. II. The presence of peroxidized lipids in the atherosclerotic aorta, *Acta Pathol. Microbiol. Scand.* **30:**1–6.

Goldstein, J. L., and Brown, M. S., 1977, The low-density lipoprotein pathway and its relation to atherosclerosis, *Annu. Rev. Biochem.* **46:**897–930.

Graham, A., Hogg, N., Kalyanaraman, B., O'Leary, V., Darley-Usmar, V., and Moncada, S., 1993, Peroxynitrite modification of low-density lipoprotein leads to recognition by the macrophage scavenger receptor, *FEBS Lett.* **330:**181–185.

Gutteridge, J. M. C., Winyard, P. G., Blake, D. R., Lunec, J., Brailsford, S., and Halliwell, B., 1985, The behaviour of caeruloplasmin in stored human extracellular fluids in relation to ferroxidase II activity, lipid peroxidation and phenanthroline-detectable copper, *Biochem. J.* **230:**517–523.

Halliwell, B., and Gutteridge, J. M. C., 1989, *Free Radicals in Biology and Medicine*, 2nd ed., Clarendon, Oxford.

Halliwell, B., and Gutteridge, J. M. C., 1992, Biologically relevant metal ion-dependent hydroxyl radical generation. An update, *FEBS Lett.* **307:**108–112.

Hazell, L. J., and Stocker, R., 1993, Oxidation of low-density lipoprotein with hypochlorite causes transformation of the lipoprotein into a high-uptake form for macrophages, *Biochem. J.* **290:**165–172.

Hazell, L. J., and Stocker, R., 1997, α-Tocopherol does not inhibit hypochlorite-induced oxidation of apolipoprotein B-100 of low-density lipoprotein, *FEBS Lett.* **414:**541–544.

Hazell, L. J., van den Berg, J. J. M., and Stocker, R., 1994, Oxidation of low-density lipoprotein by hypochlorite causes aggregation that is mediated by modification of lysine residues rather than lipid oxidation, *Biochem. J.* **302:**297–304.

Hazell, L. J., Arnold, L., Flowers, D., Waeg, G., Malle, E., and Stocker, R., 1996, Presence of hypochlorite-modified proteins in human atherosclerotic lesions, *J. Clin. Invest.* **96:**1535–1544.

Hazen, S. L., and Heinecke, J. W., 1997, 3-Chlorotyrosine, a specific marker of myeloperoxidase-catalyzed oxidation, is markedly elevated in low density lipoprotein isolated from human atherosclerotic intima, *J. Clin. Invest.* **99:**2075–2081.

Hazen, S. L., Gaut, S. P., Hsu, F. F., Crowley, J. R., d'Avignon, A., and Heinecke, J. W., 1997, *p*-Hydroxyphenylacetaldehyde, the major product of L-tyrosine oxidation by the myeloperoxidase–H_2O_2–chloride system of phagocytes, covalently modifies ε-amino groups of protein lysine residues, *J. Biol. Chem.* **272:**16990–16998.

Heinecke, J. W., Li, W., Francis, G. A., and Goldstein, J. A., 1993a, Tyrosyl radical generated by myeloperoxidase catalyzes the oxidative cross-linking of proteins, *J. Clin. Invest.* **91:**2866–2872.

Heinecke, J. W., Li, W., Daehnke, H. L., III, and Goldstein, J. A., 1993b, Dityrosine, a specific marker of oxidation, is synthesized by the myeloperoxidase–hydrogen peroxide system of human neutrophils and macrophages, *J. Biol. Chem.* **268:**4069–4077.

Ingold, K. U., and Howard, J. A., 1962, Reaction of phenols with peroxy radicals, *Nature* **195:**280–281.

Ingold, K. U., Bowry, V. W., Stocker, R., and Walling, C., 1993, Autoxidation of lipids and antioxidation by α-tocopherol and ubiquinol in homogeneous solution and in aqueous dispersions of lipids. The unrecognized consequences of lipid particle size as examplified by the oxidation of human low density lipoprotein, *Proc. Natl. Acad. Sci. USA* **90:**45–49.

Jessup, W., Mohr, D., Gieseg, S. P., Dean, R. T., and Stocker, R., 1992, The participation of nitric oxide in cell free- and its restriction of macrophage-mediated oxidation of low-density lipoprotein, *Biochim. Biophys. Acta* **1180:**73–82.

Jürgens, G., Hoff, H. F., Chisolm, G. M., and Esterbauer, H., 1987, Modification of human low density lipoprotein by oxidation—Characterization and pathophysiological implications, *Chem. Phys. Lipids* **45:**315–336.

Kagan, V. E., Serbinova, E. A., and Packer, L., 1990, Recycling and antioxidant activity of tocopherol homologs of differing hydrocarbon chain lengths in liver micrososmes, *Arch. Biochem. Biophys.* **282:**221–225.

Kagan, V. E., Serbinova, E. A., Forte, T., Scita, G., and Packer, L., 1992, Recycling of vitamin E in human low density lipoproteins, *J. Lipid Res.* **33:**385–397.

Kalyanaraman, B., Antholine, W. E., and Parthasarathy, S., 1990, Oxidation of low-density lipoprotein by Cu^{2+} and lipoxygenase: An electron spin resonance study, *Biochim. Biophys. Acta* **1035,3:**286–292.

Kalyanaraman, B., Darley-Usmar, V., Struck, A., Hogg, N., and Parthasarathy, S., 1995, Role of apolipoprotein B-derived radical and α-tocopheroxyl radical in peroxidase-dependent oxidation of low density lipoprotein, *J. Lipid Res.* **36:**1037–1045.

Kenar, J. A., Havrilla, C. M., Porter, N. A., Guyton, J. R., Brown, S. A., Klemp, K. F., and Selinger, E., 1996, Identification and quantification of the regioisomeric cholesteryl linoleate hydroperoxides in oxidized human low density lipoprotein and high density lipoprotein, *Chem. Res. Toxicol.* **9:**737–744.

Kleinveld, H. A., Naber, A. H. J., Stalenhoef, A. F. H., and Demacker, P. N. M., 1993, Oxidation resistance, oxidation rate, and extent of oxidation of human low-density lipoprotein depend on the ratio of oleic acid content to linoleic acid content: Studies in vitamin E deficient subjects, *Free Radic. Biol. Med.* **15:**273–280.

Kleinveld, H. A., Demacker, P. N., and Stalenhoef, A. F., 1994, Comparative study on the effect of low-dose vitamin E and probucol on the susceptibility of LDL to oxidation and the progression of atherosclerosis in Watanabe heritable hyperlipidemic rabbits, *Arterioscler. Thromb.* **14:**1386–1391.

Kohar, I., Baca, M., Suarna, C., Stocker, R., and Southwell-Keely, P., 1995, Is α-tocopherol a reservoir for α-tocopheryl hydroquinone? *Free Radic. Biol. Med.* **19:**197–207.

Kontush, A., Hübner, A., Finckh, C., Kohlschütter, B. A., and Beisiegel, U., 1994, Low density lipoprotein oxidizability by copper correlates to its initial ubiquinol-10 and polyunsaturated fatty acid content, *FEBS Lett.* **341:**69–73.

Kühn, H., Belkner, J., Wiesner, R., Schewe, T., Lankin, V. Z., and Tikhaze, A. K., 1992, Structure elucidation of oxygenated lipids in human atherosclerotic lesions, *Eicosanoids* **5:**17–22.

Kühn, H., Barnett, J., Grunberger, D., Baecker, P., Chow, J., Nguyen, B., Bursztyn-Pettegrew, H., Chan, H., and Sigal, E., 1993, Overexpression, purification and characterization of human recombinant 15-lipoxygenase, *Biochim. Biophys. Acta* **1169:**80–89.

Kühn, H., Belkner, J., Suzuki, H., and Yamamoto, S., 1994a, Oxidative modification of human lipoproteins by lipoxygenases of different positional specificities, *J. Lipid Res.* **35:**1749–1759.

Kuhn, H., Belkner, J., Zaiss, S., Fahrenklemper, T., and Wohlfeil, S., 1994b, Involvement of 15-lipoxygenase in early stages of atherogenesis, *J. Exp. Med.* **179:**1903–1911.

Kühn, H., Heydeck, D., Hogou, I., and Gniwotta, C., 1997, *In vivo* action of 15-lipoxygenase in early stage of human atherogenesis, *J. Clin. Invest.* **99:**888–893.

Lass, A., Belkner, J., Esterbauer, H., and Kühn, H., 1996, Lipoxygenase treatment renders low-density lipoprotein susceptible to Cu^{2+}-catalysed oxidation, *Biochem. J.* **314:**577–585.

Leeuwenburgh, C., Rasmussen, J. E., Fong, F. H., Mueller, D. M., Pennathur, S., and Heinecke, J. W., 1997a, Mass spectrometric quantification of markers for protein oxidation by tyrosyl radical, copper and hydroxyl radical in low density lipoprotein isolated from human atherosclerotic plaques, *J. Biol. Chem.* **272:**3520–3526.

Leeuwenburgh, C., Hardy, M. M., Hazen, S. L., Wagner, P., Oh-ishi, S., Steinbrecher, U. P., and Heinecke, J. W., 1997b, Reactive nitrogen intermediates promote low density lipoprotein oxidation in human atherosclerotic intima, *J. Biol. Chem.* **272:**1433–1436.

Loft, S., Strup, A., Buemann, B., and Poulsen, H. E., 1994, Oxidative DNA damage correlates with oxygen consumption in humans, *FASEB J.* **8:**534–537.

Lynch, S. M., and Frei, B., 1995, Reduction of copper, but not iron, by human low density lipoprotein, LDL, Implications for metal ion-dependent oxidative modification of LDL, *J. Biol. Chem.* **270:**5158–5163.

Mohr, D., and Stocker, R., 1994, Radical-mediated oxidation of isolated human very low density lipoprotein, *Arterioscler. Thromb.* **14:**1186–1192.

Mohr, D., Bowry, V. W., and Stocker, R., 1992, Dietary supplementation with coenzyme Q_{10} results in increased levels of ubiquinol-10 within circulating lipoproteins and increased resistance of human low density lipoprotein to the initiation of lipid peroxidation, *Biochim. Biophys. Acta* **1126:**247–254.

Moreau, S., Davies, M. J., and Puppo, A., 1995, Reaction of ferric leghemoglobin with H_2O_2: Formation of heme–protein cross-links and dimeric species, *Biochim. Biophys. Acta* **1251:**17–22.

Mukai, K., Itoh, S., Itoh, S., and Morimoto, H., 1992, Stopped-flow kinetic study of vitamin E regeneration reaction with biological hydroquinones, reduced forms of ubiquinone, vitamin K, and tocopherolquinone in solution, *J. Biol. Chem.* **267:**22277–22281.

Neuzil, J., and Stocker, R., 1994, Free and albumin-bound bilirubin is an efficient co-antioxidant for α-tocopherol, inhibiting plasma and low density lipoprotein lipid peroxidation, *J. Biol. Chem.* **269:**16712–16719.

Neuzil, J., Thomas, S. R., and Stocker, R., 1997a, Requirement for, promotion, or inhibition by α-tocopherol of radical-induced initiation of plasma lipoprotein lipid peroxidation, *Free Radic. Biol. Med.* **22:**57–71.

Neuzil, J., Witting, P. K., and Stocker, R., 1997b, α-Tocopheryl hydroquinone is an efficient multifunctional inhibitor of radical-initiated oxidation of low-density lipoprotein lipids, *Proc. Natl. Acad. Sci. USA* **96:**7885–7990.

O'Connell, A. M., Gieseg, S. P., and Stanley, K. K., 1994, Hypochlorite oxidation causes cross-linking of Lp(a), Biochim. *Biochim. Acta* **1225:**180–186.

Ortiz de Montellano, P. R., and Catalano, C. E., 1985, Epoxidation of styrene by hemoglobin and myoglobin. Transfer of oxidizing equivalents to the protein surface, *J. Biol. Chem.* **260:**9265–9271.

Ortiz de Montellano, P. R., and Grab, L. A., 1987, Cooxidation of styrene by horseradish peroxidase and phenols: A biochemical model for protein-mediated cooxidation, *Biochemistry* **26:**5310–5314.

Packer, J. E., Slater, T. F., and Willson, R. L., 1979, Direct observation of a free radical interaction between vitamin E and vitamin C, *Nature* **278:**737–738.

Panasenko, O. M., Evgina, S. A., Aidyraliev, R. K., Sergienko, V. I., and Vladimirov, Y. A., 1994, Peroxidation of human blood lipoproteins induced by exogenous hypochlorite or hypochlorite generated in the system of myeloperoxidase + H_2O_2 + Cl^-, *Free Radic. Biol. Med.* **16:**143–148.

Parthasarathy, S., Wieland, E., and Steinberg, D., 1989, A role for endothelial cell lipoxygenase in the oxidative modification of low-density lipoprotein, *Proc. Natl. Acad. Sci. USA* **86:**1046–1050.

Parums, D. V., Brown, D. L., and Mitchinson, M. J., 1990, Serum antibodies to oxidized low-density lipoprotein and ceroid in chronic periaortitis, *Arch. Pathol. Lab. Med.* **114:**383–386.

Peers, K. E., and Coxon, D. T., 1983, Controlled synthesis of monohydroperoxides by α-tocopherol inhibited autoxidation of polyunsaturated lipids, *Chem. Phys. Lipids* **32:**49–56.

Porter, N. A., Weber, B. A., Weenen, H., and Khan, J. A., 1980, Autoxidation of polyunsaturated lipids. Factors controlling the stereochemistry of product hydroperoxides, *J. Am. Chem. Soc.* **102:**5597–5601.

Porter, N. A., Lehman, L. S., Weber, B. A., amd Smith, K. J., 1981, Unified mechanism for polyunsaturated fatty acid autoxidation. Competition of peroxy radical hydrogen atom abstraction, β-scission, and cyclization, *J. Am. Chem. Soc.* **103:**6447–6455.

Reaven, P. D., Parthasarathy, S., Beltz, W. F., and Witztum, J. L., 1992, Effect of probucol dosage on plasma lipid and lipoprotein levels and on protection of low density lipoprotein against *in vitro* oxidation in humans, *Arterioscler. Thromb.* **12:**318–324.

Reaven, P. D., Khouw, A., Beltz, W. F., Parthasarathy, S., and Witztum, J. L., 1993, Effect of dietary antioxidant combinations in humans. Protection of LDL by vitamin E but not beta-carotene, *Arterioscler. Thromb.* **13:**590–600.

Ross, R., 1993, The pathogenesis of atheroslcerosis: A perspective for the 1990's, *Nature* **362:**801–809.

Rubbo, H., Radi, R., Trujillo, M., Telleri, R., Kalyanaraman, B., Barnes, S., Kirk, M., and Freeman, B. A., 1994, Nitric oxide regulation of superoxide and peroxynitrite-dependent lipid peroxidation. Formation of novel nitrogen-containing oxidized lipid derivatives, *J. Biol. Chem.* **269:**26066–26075.

Santanam, N., and Parthasarathy, S., 1995, Paradoxical actions of antioxidants in the oxidation of low density lipoprotein by peroxidases, *J. Clin. Invest.* **95:**2594–2600.

Sato, K., Niki, E., and Shimasaki, H., 1990, Free radical-mediated chain oxidation of low density lipoprotein and its synergistic inhibition by vitamin E and vitamin C, *Arch. Biochem. Biophys.* **279:**402–405.

Sattler, W., Kostner, G. M., Waeg, G., and Esterbauer, H., 1991, Oxidation of lipoprotein Lp(a), A comparison with low-density lipoproteins, *Biochim. Biophys. Acta* **1081:**65–74.

Sattler, W., Mohr, D., and Stocker, R., 1994, Rapid isolation of lipoproteins and assessment of their peroxidation by HPLC postcolumn chemiluminescence, *Meth. Enzymol.* **233:**469–489.

Savenkova, M. L., Mueller, D. M., and Heinecke, J. W., 1994, Tyrosyl radical generated by myeloperoxidase is a physiological catalyst for the initiation of lipid peroxidation in low density lipoprotein, *J. Biol. Chem.* **269:**20394–20400.

Sendobry, S. M., Cornicelli, J. A., Welch, K., Bocan, T., Tait, B., Trivedi, B. K., Colbry, N., Dyer, R. D., Feinmark, S. J., and Daugherty, A., 1997, Attenuation of diet-induced atherosclerosis in rabbits with a highly selective 15-lipoxygenase inhibitor lacking significant antioxidant properties, *Br. J. Pharmacol.* **120:**1199–1206.

Schewe, T., Rapoport, S. M., and Kühn, H., 1986, Enzymology and physiology of reticulocyte lipoxygenase: Comparison with other lipoxygenases, in *Advances in Enzymology and Related Areas of Molecular Biology*, Vol. 58 (A. Meister, ed.), pp. 191–272, Wiley, New York.

Sharma, M. K., and Buettner, G. R., 1993, Interaction of vitamin C and vitamin E during free radical stress in plasma: An ESR study, *Free Rad. Biol. Med.* **14:**649–653.

Shigenaga, M. K., Gimeno, C. J., and Ames, B. N., 1989, Urinary 8-hydroxy-2′-deoxyguanosine as a biological marker of *in vivo* oxidative DNA damage, *Proc. Natl. Acad. Sci. USA* **86:**9697–9701.

Sigal, E., Grunberger, D., Highland, E., Gross, C., Dixon, R. A. F., and Craik, C. S., 1990, Expression of cloned human reticulocyte 15-lipoxygenase and immunological evidence that 15-lipoxygenase of different cell types are related, *J. Biol. Chem.* **265:**5113–5120.

Smith, C., Mitchinson, M. J., Aruoma, O. I., and Halliwell, B., 1992, Stimulation of lipid peroxidation and hydroxyl-radical generation by the contents of human atherosclerotic lesions, *Biochem. J.* **286:**901–905.

Steinberg, D., Parthasarathy, S., Carew, T. E., Khoo, J. C., and Witztum, J. L., 1989, Beyond cholesterol: Modifications of low-density lipoprotein that increase its atherogenicity, *N. Engl. J. Med.* **320:**915–924.

Steinbrecher, U. P., Zhang, H., and Lougheed, M., 1990, Role of oxidatively modified LDL in atherosclerosis, *Free Radic. Biol. Med.* **9:**155–168.

Stocker, R., 1990, Induction of haem oxygenase as a defence against oxidative stress, *Free Radic. Res. Commun.* **9:**101–112.

Stocker, R., 1994, Tracking the daily supplement, *Today Life Sci.* **6:**24–31.

Stocker, R., and Suarna, C., 1993, Extracellular reduction of ubiquinone-1 and -10 by human Hep G2 and blood cells, *Biochim. Biophys. Acta* **1158:**15–22.

Stocker, R., McDonagh, A. F., Glazer, A. N., and Ames, B. N., 1990, Antioxidant activities of bile pigments: Biliverdin and bilirubin, in *Methods in Enzymology*, Vol. 186 (L. Packer and A. N. Glazer, eds.), pp. 301–309, Academic Press, New York.

Stocker, R., Bowry, V. W., and Frei, B., 1991, Ubiquinol-10 protects human low density lipoprotein more efficiently against lipid peroxidation than does α-tocopherol, *Proc. Natl. Acad. Sci. USA* **88:**1646–1650.

Suarna, C., Dean, R. T., May, J., and Stocker, R., 1995, Human atherosclerotic plaque contains both oxidized lipids and relatively large amounts of α-tocopherol and ascorbate, *Arterioscler. Thromb. Vasc. Biol.* **15:**1616–1624.

Suzuki, H., Kurihara, Y., Takeya, M., Kamada, N., Kataoka, M., Jishage, K., Ueda, O., Sakaguchi, H., Higashi, T., Suzuki, T., Takashima, Y., Kawabe, Y., Cynshi, O., Wada, Y., Honda, M., Kurihara, H., Aburatani, H., Doi, T., Matsumoto, A., Azuma, S., Noda, T., Toyoda, Y., Itakura, H., Yazaki, Y., Horiuchi, S., Takahashi, K., Kruijt, J. K., van Berkel, T. J. C., Steinbrecher, U. P., Ishibashi, S., Maeda, N., Gordon, S., and Kodama, T., 1997, A role for macrophage scavenger receptors in atherosclerosis and susceptibility to infection, *Nature* **386:**292–296.

Thomas, S. R., Neuzil, J., and Stocker, R., 1996a, Co-supplementation with coenzyme Q prevents the pro-oxidant effect of α-tocopherol and increases the resistance of low-

density lipoprotein towards transition metal-dependent oxidation initiation, *Arterioscler. Thromb. Vasc. Biol.* **16:**687–696.
Thomas, S. R., Witting, P. K., and Stocker, R., 1996b, 3-Hydroxyanthranilic acid is an efficient, cell-derived co-antioxidant for α-tocopherol, inhibiting human low density lipoprotein and plasma lipid peroxidation, *J. Biol. Chem.* **271:**32714–32721.
Thomas, S. R., Davies, M. J., and Stocker, R., 1998, Oxidation and antioxidation of human low-density lipoprotein and plasma exposed to 3-morpholinosydnonimine and reagent peroxynitrite, *Chem. Res. Toxicol,* **11:**484–494.
Traber, M. G., Sokol, R. J., Burton, G. W., Ingold, K. U., Papas, A. M., Huffaker, J. E., and Kayden, H. J., 1990, Impaired ability of patients with familial isolated vitamin E deficiency to incorporate α-tocopherol, *J. Clin. Invest.* **85:**397–407.
Traber, M. G., Sokol, R. J., Kohlschütter, A., Yokota, T., Muller, D. P., Dufour, R., and Kayden, H. J., 1993, Impaired discrimination between stereoisomers of alpha-tocopherol in patients with familial isolated vitamin E deficiency, *J. Lipid Res.* **34:**201–210.
Upston, J. M., Neuzil, J., and Stocker, R., 1996, Oxidation of LDL by recombinant human 15-lipoxygenase: Evidence for α-tocopherol dependent oxidation of esterified core and surface lipids, *J. Lipid Res.* **37:**2650–2661.
Upston, J. M., Neuzil, J., Witting, P. K., Alleva, R., and Stocker, R., 1997, Oxidation of free fatly acids in low density lipoprotein by 15-lipoxygenase stimulates nonenzymic, α-tocopherol-mediated peroxidation of cholesteryl esters, *J. Biol. Chem.* **272:**30067–30074.
van der Vliet, A., Eiserich, J. P., O'Neill, C. A., Halliwell, B., and Cross, C. E., 1995, Tyrosine modification by reactive nitrogen species: A closer look, *Arch. Biochem. Biophys.* **319:**341–349.
Wagner, J. R., Motchnik, P. A., Stocker, R., Sies, H., and Ames, B. N., 1993, The oxidation of blood plasma and low density lipoprotein components by chemically generated singlet oxygen, *J. Biol. Chem.* **268:**18502–18506.
Waldeck, A. R., and Stocker, R., 1996, Radical-initiated lipid peroxidation in low density lipoproteins: Insights obtained from kinetic modeling, *Chem. Res. Toxicol.* **9:**954–964.
Wieland, E., Parthasarathy, S., and Steinberg, D., 1993, Peroxidase-dependent metal-independent oxidation of low density lipoprotein *in vitro*: A model for *in vivo* oxidation? *Proc. Natl. Acad. Sci. USA* **90:**5929–5933.
Wilcox, A. L., and Marnett, L. J., 1993, Polyunsaturated fatty acid alkoxyl radicals exist as carbon-centered epoxyallylic radicals: A key step in hydroperoxide-amplified lipid peroxidation, *Chem. Res. Toxicol.* **6:**413–416.
Wilks, A., and Ortiz de Montellano, P. R., 1992, Intramolecular translocation of the protein radical formed in the reaction of recombinant sperm whale myoglobin with H_2O_2, *J. Biol. Chem.* **267:**8827–8833.
Witting, P. K., Bowry, V. W., and Stocker, R., 1995, Inverse deuterium kinetic isotope effect for peroxidation in human low-density lipoprotein, LDL: A simple test for tocopherol-mediated peroxidation of LDL lipids, *FEBS Lett.* **375:**45–49.
Witting, P. K., Westerlund, C., and Stocker, R., 1996, A rapid and simple screening test for potential inhibitors of tocopherol-mediated peroxidation of LDL lipids, *J. Lipid Res.* **37:**853–867.
Witting, P. K., Upston, J. M., and Stocker, R., 1997, The role of α-tocopheroxyl radical in the initiation of lipid peroxidation in human low density lipoprotein exposed to horse radish peroxidase, *Biochemistry* **36:**1251–1258.
Witztum, J. L., and Steinberg, D., 1991, Role of oxidized low density lipoprotein in atherogenesis, *J. Clin. Invest.* **88:**1785–1792.
Ylä-Herttuala, S., Palinski, S., Rosenfeld, M. E., Parthasarathy, S., Carew, T. E., Butler, S., Witztum, J. L., and Steinberg, D., 1989, Evidence for the presence of oxidatively modified

low density lipoprotein in atherosclerotic lesions of rabbit and man, *J. Clin. Invest.* **84:**1086–1095.
Ylä-Herttuala, S., Rosenfeld, M. E., Parthasarathy, S., Glass, C. K., Sigal, E., Witztum, J. L., and Steinberg, D., 1990, Colocalization of 15-lipoxygenase mRNA and protein with epitopes of oxidized low density lipoprotein in macrophage-rich areas of atherosclerotic lesions, *Proc. Natl. Acad. Sci. USA* **87:**6959–6963.
Ylä-Herttuala, S., Luoma, J., Viita, H., Hiltunen, T., Sisto, T., and Nikkari, T., 1995, Transfer of 15-lipoxygenase gene into rabbit iliac arteries results in the appearance of oxidation-specific lipid–protein adducts characteristic of oxidized low density lipoprotein, *J. Clin. Invest.* **95:**2692–2698.
Ziouzenkova, O., Gieseg, S. P., Ramos, P., and Esterbauer, H., 1996, Factors affecting resistance of low density lipoproteins to oxidation, *Lipids* **31:**S71–S76.

Chapter 16

Vitamin E
Its Role in Aging

Shiro Urano

1. APPEARANCE AND CAUSE OF AGING

All living organisms undergo irreversible and progressive changes with time in their biological functions. Although some of these changes, namely those associated with growth and development, are not thought of in terms of "aging," other, deteriorative changes are. Thus, one can consider "aging" as the accumulation of changes with the passage of time that increase the risk of death (Upton, 1977; Kohn, 1985; Harman, 1994).

The aging process can be divided into two categories: physiological, that is, normal, aging; and pathological aging, involving the development of one of a number of degenerative diseases, such as cardiovascular disease, diabetes, cancer, dementia, immune system decline, and so on. There is a growing consensus that the pathological aging processes caused by such age-related diseases can be delayed by medical therapy. Many trials have been undertaken for the prevention and treatment of age-related diseases,

Shiro Urano Tokyo Metropolitan Institute of Gerontology, 35-2 Sakaecho, Itabashi-ku, Tokyo 173, Japan; *present address:* Department of Biological Chemistry, Faculty of Industrial Chemistry, Shibaura Institute of Technology, 3-9-14 Shibaura, Minato-ku, Tokyo 108, Japan.
Subcellular Biochemistry, Volume 30: Fat-Soluble Vitamins, edited by Quinn and Kagan. Plenum Press, New York, 1998.

and it is important, in particular, to verify the etiology of such degenerative diseases.

In the last several decades, a number of theories of aging have been proposed, which can be divided into three categories (Table I): (1) genetic theories, (2) dysfunction theories, and (3) oxidative stress theories. The first group considers the process of aging to be genetically controlled and programmed. Since the life span of many organisms is fixed genetically, and since senescence proceeds according to the biological clock, it would seem to be impossible to elongate the intrinsic life span of such organisms.

The second group can be further divided into two additional categories: theories of deterioration of physiological function and theories of degenerative change. The former consider, for example, immunological, endocrinological, and central nervous system (CNS) changes, attributing the cause of aging to deterioration with age of physiological functions. One can also consider the effects of dysfunctions on important reactions *in vivo*, leading to degenerative changes in tissues, taking into account such factors as cross-linking, somatic mutation, wear and tear, and replication and repair errors.

Finally, there are theories based on the hypothesis that chronic oxidative stress acting over a long period of time produces active oxygen species and free radicals in living tissues. This theory, usually attributed to Harman (1956, 1982, 1986), proposes that most changes during aging are caused by free radical reactions and the formation of lipid peroxides in tissues, which lead to age-related damage and eventually to various aging processes and phenomena. In order to better understand this mechanism, it may be insufficient to deal only with an increase in rates of free radical generation under physiological and pathological aging processes. Nevertheless, this theory has been recognized as a reasonable hypothesis, and has been widely supported by many investigations. The starting point of these studies arose from the discovery of superoxide dismutase (SOD), which eliminates super-

Table I
Theories of Aging

	Dysfunction theories		
Genetic theories	Deterioration of functions	Degenerative changes	Oxidation theory
Programmed	Immunological	Cross-linking	Active oxygen species
Biological clock	Endocrinological	Somatic mutation	Free radicals
Metabolic rate	CNS	Error	
	Stress	Wear and tear	
	Hindered control	Waste accumulation	

oxide generated in single-electron transfer metabolic reactions (McCord and Fridovich, 1969), together with the statistical fact that the maximum life expectancy of mammals is inversely correlated with their daily rate of oxygen consumption (Cutler, 1984). The maximum life span of rats is shorter than that of elephants (3.5 years and 70 years, respectively), whereas the relative oxygen consumption of elephants is less than that of rats (0.1 ml/kg body weight/day versus 1.6 ml/kg body weight/day, respectively). This consideration shows that the greater the relative oxygen consumption, the shorter is the life span.

Although there are arguments against the concept that much oxygen consumption actually produces active oxygen species in living tissues, it has been reported that generation of active oxygen species in the lungs of pigs subjected to hyperoxia is proportional to the amount of oxygen absorbed (Turrens *et al.*, 1982), so that it is reasonable to conclude that hyperoxia produces free radicals. Ku *et al.* (1983) also reported that superoxide and hydrogen peroxide in mitochondria of the kidney and heart of seven species including mouse, rabbit, pig, and cow are produced at the site of ubiquinone/cytochrome b in the mitochondrial electron transport system, and that mitochondria of short-lived species produce active oxygen species more significantly than those of long-lived species. Thus, it seems likely that the maximum life span is inversely proportional to the amount of active oxygen species generated in mitochondria (Fig. 1).

2. OXIDATIVE DAMAGE TO BIOLOGICAL SYSTEMS DURING AGING

Oxygen-derived free radicals have been proposed to play a role in cancer (Ames, 1989; Floyd, 1991), cardiovascular disease (Parthasarathy *et al.*, 1990), diabetes (Ghiselli *et al.*, 1992), and aging (Harman, 1992). Work on oxygen-derived free radicals in numerous pathophysiological processes has been summarized in several extensive reviews (Fridovich, 1983; Halliwell and Gutteridge, 1985; Hammond *et al.*, 1985; Pryor, 1982). Oxidative damage caused by free radicals has also been implicated in some aging processes. However, despite many reports on the involvement of free radicals in aging, the relation between oxidative damage caused by free radicals and age-related events in lipid peroxidation processes is unresolved due to conflicting results. Information concerning such an interaction awaits clear documentation of the role of free radicals in age-related events.

It has been recognized that polyunsaturated lipids of biological membranes are a major target of free radicals, and that lipid peroxides produced by free radical reactions disrupt cellular homeostasis, resulting in significant

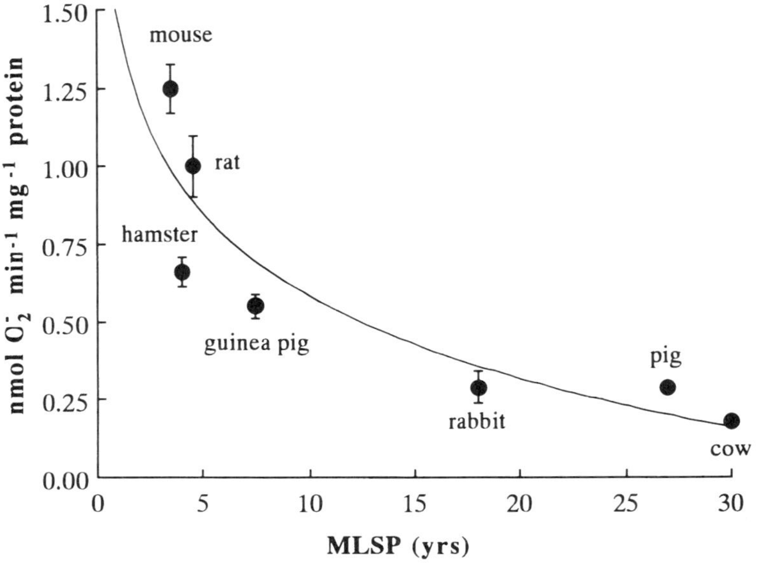

FIGURE 1. Relationship between the rate of mitochondrial superoxide generation in the kidney and the maximum life span potential (MLSP) of seven mammalian species. Reproduced from Ku *et al.* (1993) with permission.

changes in the physiological function of and damage to many organs (van Kujik *et al.*, 1987). Zhang *et al.* (1994) observed that phospholipid hydroperoxide levels in gerbil brain striatum increased significantly during aging. The rate of superoxide generation is enhanced significantly in rat brain mitochondria during aging, accompanied by an increase in lipid peroxides in the brain (Ku *et al.*, 1993; Sawada and Carlson, 1987a,b). Many morphological changes have been seen in rats subjected to hyperoxia, e.g., swollen astrocytes around vessels, deformed nuclei in nerve cells, pigmentation, swollen mitochondria, and quite abnormal accumulation of synaptic vesicles in the swollen nerve terminals (Fig. 2), along with an increase in lipid peroxides and a decrease in vitamin E content of synaptic plasma membranes (Urano *et al.*, 1997). Since these phenomena also have been observed in vitamin E-deficient rats and during aging of normal rats (Urano *et al.*, 1994, 1997), it can be assumed that free radicals caused by oxidative stress may attack synapses, resulting in aging of the nervous system. In human brain, myelin stability significantly decreases beyond age 50; malondialdehyde formation and level of conjugated diene increase with age, indicating an increase in the amount of lipid peroxidation (Chia *et al.*, 1983).

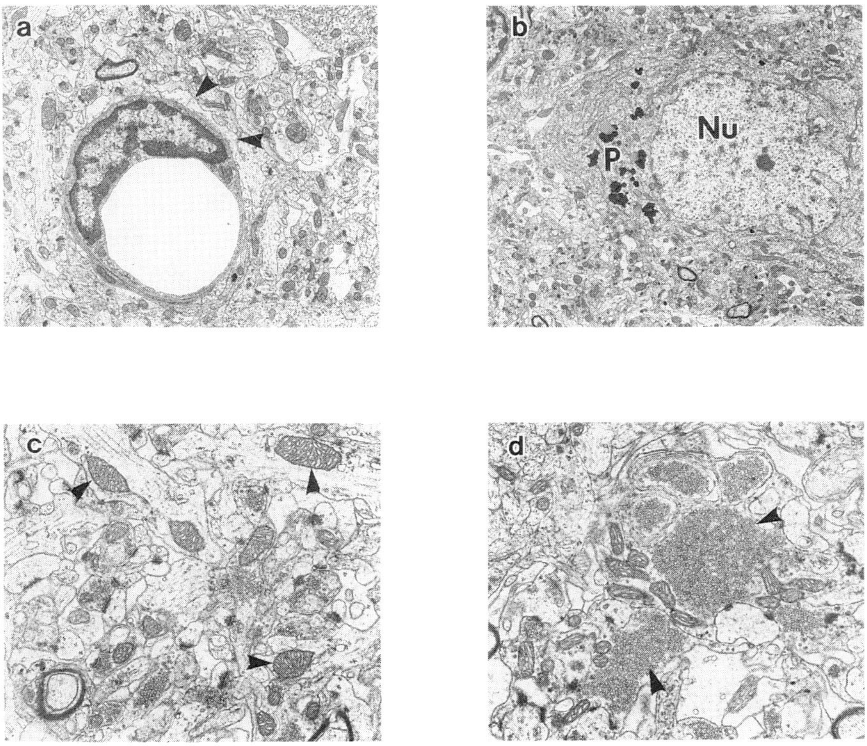

FIGURE 2. Transmission electron micrographs of damaged nerve cell from the brain of a rat exposed to 100% O_2. (a) Swollen astrocytes around a capillary vessel (arrowhead), endotherial cell (Et); (b) deformed nucleus (Nu) in a nerve cell, accumulated pigment (P); (c) swollen mitochondria (arrowhead); (d) abnormally accumulated synaptic vesicles in the swollen nerve terminals (arrowhead). Magnification: (a, c, d) ×14,000; (b) ×7200. From Urano *et al.* (1997).

Lipid peroxides are considered to induce lipofuscin-like age pigment in the brain (Sawada and Carlson, 1987a). Thus, the brain and nervous system are considered to be more susceptible to oxidative damage than other organs due to their high content of polyunsaturated, lipid-rich neural parenchyma, high oxygen consumption (accounting for one-fifth of the total systemic oxygen consumption), low catalase activity, and moderate superoxide dismutase and glutathione peroxidase activities (Halliwell, 1989). There is also considerable evidence that iron accumulates in the rat brain (Francois *et al.*, 1981). Based on these findings, it can be speculated that degenerative oxidative damage to the functioning of the nervous system may be mediated by iron-induced peroxidative processes after dysfunction of antioxidative defense systems.

When antioxidant defense systems are not enough protection against oxidative stress, some DNA is oxidized. It has been estimated that the number of oxidative hits to DNA per cell per day is about 100,000 in rats and about 10,000 in humans (Ames *et al.*, 1993). Damaged DNA is repaired by FAPY-DNA glycosylase, which excises the oxidized lesions from DNA, which are excreted in the urine. It has been reported that oxidized lesions in DNA accumulate with age, and that an old rat has approximately 2 million lesions per cell, which is about twice that in a young rat (Ames *et al.*, 1993). Guanine residues in DNA are considered to be among the most susceptible bases for oxidative stress to produce 8-hydroxy guanine and 8-hydroxy-2′-deoxyguanosine (8-OHdG). These oxidized products increase with age in rat liver, kidney, and intestine. It has been reported that the level of 8-OHdG in the liver of aged (24-month-old) rats is 23.2 ± 1.8 fmol/mg DNA, which represents an increase of 164% over the value observed in young (1-month-old) rats (Fraga *et al.*, 1990). In human heart mitochondria, the content of 8-OHdG in DNA increases exponentially with age after 55 years (Hayakawa *et al.*, 1992).

It is well recognized that during aging, many metabolic enzymes accumulate less active or inactive forms. Although the mechanism of these age-related changes is obscure, it seems likely that the oxidation of amino acid residues, such as histidine, arginine, and proline, in enzymes may be caused by free radicals, leading to enzyme deactivation. Oxidized proteins in many organs have been assessed by the analysis of the content of the carbonyl moiety of oxidized amino acids. The carbonyl content in hepatocytes from old (26-month-old) rats is greater than from young (3-month-old) rats, with the largest incremental increase occurring between 20 and 26 months (Starke-Reed and Oliver, 1989). In cultured human skin fibroblasts from normal donors of different ages the level of oxidatively modified proteins is increased only after the age of 60. The levels in fibroblasts from individuals with progeria or Werner's syndrome are significantly higher than age-matched controls (Fig. 3). In addition, glucose-6-phosphate dehydrogenase is inactivated by oxidative stress (Oliver *et al.*, 1987). Aged gerbils also show a significantly higher level of oxidized protein in the brain, as assessed by carbonyl residues, compared to young gerbils (Carney, *et al.*, 1991). These observations suggest that dysfunction of enzymes during aging may be due partly to modification of protein by oxidative stress.

3. DETERIORATION OF FUNCTION IN VITAMIN E DEFICIENCY

Although vitamin E deficiency has detrimental effects on the steady state of the nervous system and skeletal and cardiac muscles of many

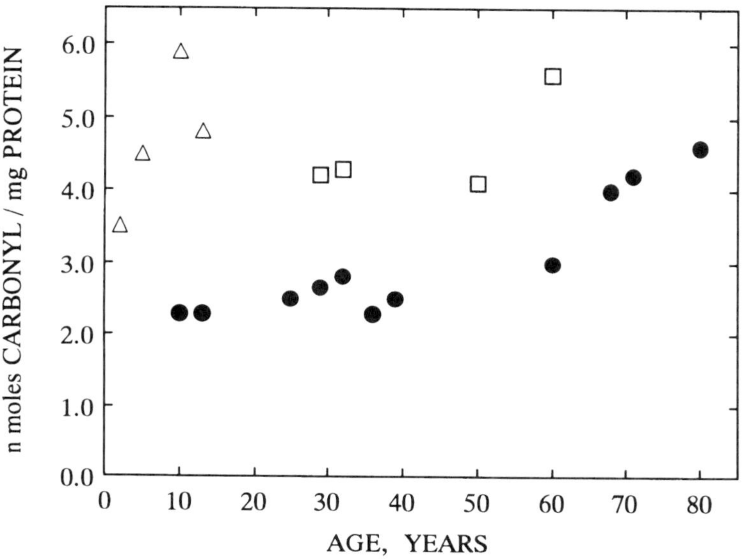

FIGURE 3. Oxidatively modified protein in cultured fibroblasts. ●, Cultures from normal individuals; □, cultures from individuals with Werner's syndrome; △, cultures from individuals with progeria. Reproduced from Oliver *et al.* (1987) with permission.

experimental animals (Nelson, 1980), a similar effect in humans has not been recognized except for rare human diseases with fat malabsorption, such as cystic fibrosis, abetalipoproteinemia, and familial isolated vitamin E deficiency syndrome. Because of the widespread occurrence of vitamin E in food, nutritional deficiency in humans is extremely uncommon. Abetalipoproteinemia caused by fat malabsorption is attributed to the blockage of chylomicron formation, and results in the absence of low-density lipoprotein (LDL) and very low density lipoprotein (VLDL) in blood (Muller *et al.*, 1983). Since lipid-soluble vitamin E is usually absorbed via the chylomicrons, and is transported mostly by LDL and VLDL in liver and blood circulation, in the absence of these lipoproteins vitamin E level is depressed, leading to vitamin E deficiency (Urano, 1993). On the other hand, ataxia patients with isolated vitamin E deficiency have normal gastrointestinal lipid absorption and normal incorporation into chylomicrons, but they have an impaired ability to incorporate vitamin E into VLDL secreted by the liver, a function putatively attributable to the α-tocopherol transfer protein (Ouahchi *et al.*, 1995). Muscle damage and neurodegeneration are observed in both diseases of malabsorption and familial isolated vitamin E deficiency. Although the physiological process is obscure, vitamin E is implicated as a protective agent against neurological damage in these diseases.

Kay *et al.* (1986) proposed that erythrocytes from vitamin E-deficient rats have a senescent cell antigen, regardless of their age, which develops on the surface of aged red cells. The antigen is recognized and binds to the Fab region of an immunoglobulin G (IgG) autoantibody; subsequently the Fc region of IgG is recognized and bound by macrophages, leading to phagocytosis of the old cell. Vitamin E-deficient red cells behave like old erythrocytes from normal rats. These abnormalities are not reversed by the addition of exogenous vitamin E to the assay systems.

Age-related accumulation of substances damaged by oxidative stress in living tissues corresponds to a decline in the efficiency of oxidative defense systems. Vitamin E deficiency in both experimental animals and humans leads to excessive accumulation of an age pigment, lipofuscin, in several organs such as the brain, liver, heart, etc., of rats, monkeys, and humans (Papafrangos and Lyman, 1982; Sokol, 1989) (Fig. 4). Lipofuscin accumulation in many tissues also increased with the age of a normal organism as

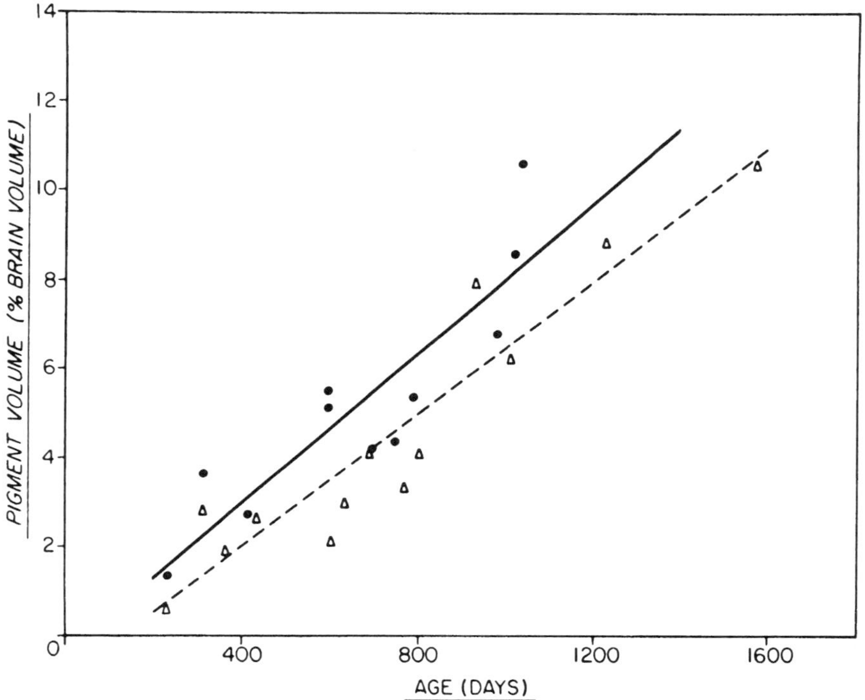

FIGURE 4. Regression of lipofuscin content as a function of age in the brains of nonhibernating hamsters (—, ●) and of hibernating hamsters (---, △). From Papafrangos and Lyman (1982) with permission.

specifically shown in histochemistry. Fluorescent light microscopy showed that a strong correlation was found between age and the amount of lipofuscin in the myocardium of humans and dogs. Interestingly, it has been calculated that lipofuscin accumulates about 5.5 times faster in dogs than humans, whereas the life span of humans is approximately five times longer than that of dogs (Strehler *et al.*, 1959; Munnell and Getty, 1968). A relation between vitamin E deficiency and the aging process has been proposed, leading to the belief that vitamin E supplementation may retard senescence. However, it is ambiguous whether accumulation of lipofuscin in many organs impairs physiological function, or is merely stored in tissue as age-related reaction products of lipid peroxides caused by the vitamin E deficiency.

Although there are numerous reports of significant increases in the lipid peroxide level of several organs in vitamin E deficiency and during aging (Hirai and Yoshikawa, 1970; Shirai *et al.*, 1980; Tappel *et al.*, 1973; Sawada and Carlson, 1987b), most methods for measuring lipid peroxidation in biological samples, such as thiobarbituric acid methods, have been indirect and nonspecific, so that these data probably do not reflect the actual evaluation of lipid peroxides in biological systems. Tokumaru *et al.* (1997) have recently demonstrated by a specific and sensitive method, using 1-naphthyldiphenylphosphine, that the level of lipid hydroperoxides increases significantly in rat liver, lung, and kidney as early as 4 weeks after feeding on a vitamin E-deficient diet, and that during aging in the mouse, lipid hydroperoxide levels also increase significantly in the liver, heart, and kidney (Tokumaru *et al.*, 1996) (Fig. 5). Since lipid hydroperoxides are considered to be detrimental substances which induce age-related diseases such as arteriosclerosis, diabetes, etc., it is likely that changes in lipid hydroperoxides will be an effective indicator for the evaluation of the aging process. Lipid peroxidation in vitamin E deficiency and in the aging process involves a free radical-induced chain reaction which may result in cell membrane, protein, and DNA damage (van Kujik *et al.*, 1987; Davies, 1987; Ames, 1989), finally leading to pathological senescence.

Other biochemical changes in vitamin E deficiency, which are probably independent of an involvement with peroxidative damage caused by free radicals, have been widely demonstrated. Platelet-activating factor (PAF), which is metabolized in polymorphonuclear leukocytes (PMN) by certain stimuli, is a potent phospholipid mediator which causes activation of leukocytes and platelets. Fukuzawa *et al.* (1989) indicated that PAF synthesis is significantly enhanced in vitamin E-deficient rat PMN, and that the activity of acetyl transferase, which transfers an acyl moiety to 2-lyso-PAF to form PAF, is higher in vitamin E-deficient than in vitamin E-supplemented rat

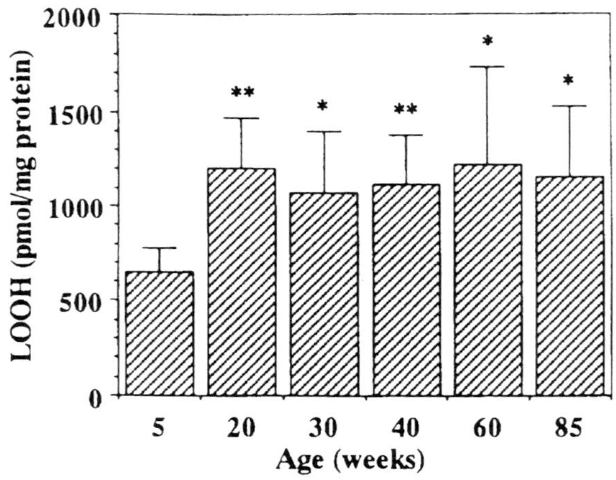

FIGURE 5. Changes in the lipid hydroperoxide (LOOH) level in the kidney. Asterisks indicate significance differences from 5-week-old mice: $*p < .05$; $**p < .01$. From Tokumaru *et al.* (1996) with permission.

PMN. On the basis of these findings, it is concluded that vitamin E may inhibit the activity of phospholipase A_2, which produces 2-lyso-PAF in PMN, rather than inhibiting the activity of acetyl transferase.

An increase in the production of thromboxane A_2 (TXA_2) in platelets leading to platelet aggregation and prostaglandin E_2 (PGE_2) content in serum, and a decrease in synthesis of prostaglandin I_2 (PGI_2) in endothelium of the vessel wall having antiaggregatory and antithrombotic effects, were observed in vitamin E-deficient rats (Machlin *et al.*, 1975; Okuma *et al.*, 1980) (Fig. 6). Since platelet aggregation and its inhibition are regulated under normal conditions by a balance in the biosynthesis of TXA_2 and PGI_2, the ratio of TXA_2/PGI_2 increases in vitamin E deficiency. It is unclear whether the effect of vitamin E deficiency on arachidonic acid metabolism is linked to age-related changes in living tissues.

4. VITAMIN E AND AGING PHENOMENA

Vitamin E levels in human plasma are higher than in plasma of any other mammal. Since there is a positive correlation between maximum life expectancy and plasma vitamin E concentration (Cutler, 1984), vitamin E is considered to be an important factor in influencing aging phenomena in many species.

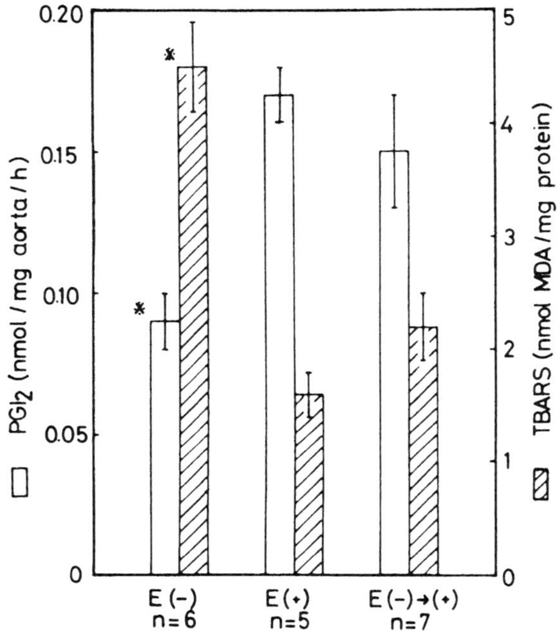

FIGURE 6. Generation of PGI_2-like substance and thiobarbituric acid-reactive substances in the aorta of rats fed vitamin E-deficient diet and/or vitamin E-supplemented diet. $^*p < .001$. From Okuma *et al.* (1980) with permission.

Many investigations have demonstrated that vitamin E deficiency is responsible for the formation of greater amounts of toxic lipid peroxides in experimental animals (Hafeman and Hoekstra, 1977; Dillard *et al.*, 1978). There are many reasons to implicate lipid peroxides and oxidized lipoprotein in the pathogenesis of age-related atherosclerosis (Steinbrecher *et al.*, 1990). Some evidence suggests that oxidized apoprotein B100 caused by lipid peroxidation plays a key role in recognition of LDL, and that oxidatively modified LDL uptake by scavenger receptors in macrophages leads to formation of foam cells and atherosclerotic plaques (Esterbauer *et al.*, 1992; Harberland *et al.*, 1992). It has been shown that vitamin E supplementation decreases the concentration of lipid peroxides in the organs of experimental animals (Takeuchi *et al.*, 1976, 1978; Sawada and Carlson, 1987b). In human trials, serum levels of lipid peroxides in people aged 60–100 years in Poland decreased significantly by a mean of 26% through vitamin E supplementation in doses of 200 mg per day for 1 year (Wartanowicz *et al.*, 1984). A number of *in vitro* investigations have demonstrated that addition of a high concentration of vitamin E into the culture medium prevents the oxidative modification of LDL by cells and conversion into foam cells (Esterbauer *et al.*, 1993). Besides the *in vitro* studies, *ex vivo* studies have shown that vitamin E supplementation (1200 IU/day) has a positive effect on the oxidation resistance of LDL (Esterbauer *et al.*, 1993).

Since endotherial cells are highly susceptible to oxidative stress (Hennig and Chow, 1988), vitamin E is expected to play a role as an antioxidant against the oxidative reaction in endothelium. Age-related injuries in the aortic endothelium of inherited-cataract rats can be reduced by the administration of vitamin E (Masuda *et al.*, 1989). Similar studies of endotherial cell cultures subjected to linoleic acid hydroperoxide as oxidative stress (Hennig *et al.*, 1989) demonstrated that the endotherial barrier function in vessels was damaged by oxidative stress, and that the addition of vitamin E significantly decreased albumin transfer across the cells.

Although it is expected that vitamin E functions as an antiatherogenic agent, in general its action in most pathological conditions is not significant (Farrell, 1980), except for the antioxidant probucol (Kita, 1987), even if there is ample evidence to propose a therapeutic role (Gisinger *et al.*, 1990). Nevertheless, a prospective study on the role of vitamin E in atherogenesis has demonstrated that the risk of major coronary disease in women is decreased by about 40% with vitamin E supplementation in doses of 200 IU per day for 2 years or more (Stampfer *et al.*, 1993) (Table II).

Platelet aggregation and enhancement of PGE_2 production were found to increase in vitamin E-deficient compared to normal subjects (Machlin *et al.*, 1975; Hope *et al.*, 1975), accompanied by an increase in TXA_2 production and a decrease in PGI_2 in vessel walls (Pichard *et al.*, 1982; Okuma *et al.*, 1980). Since these abnormalities are improved by vitamin E supplementation (Creter *et al.*, 1979), vitamin E is thought to inhibit either phospholipase A_2 (PLA_2) activity of arachidonic acid release from platelets or cyclooxygenase activity in the synthesis of TXA_2 and PGI_2.

As shown in Table I, deterioration of immunological function is one hypothesized cause of aging. In addition to its antioxidative effect, several reviews have concluded that vitamin E may have an immunoenhancing effect (Corwin and Shloss, 1980; Meydani *et al.*, 1991). Redox balance is an important determinant of immune cell function. Immune cells are highly sensitive to change in balance because of the high content of polyunsaturated lipids in their plasma membranes. Since immune cells produce active oxygen species in their normal functioning, these cells are also susceptible to the balance. Consequently, in order to maintain their functional integrity, cells of the immune system contain a high concentration of vitamin E (Hatman and Kayden, 1979). By the activation of macrophages, immuno-

Table II
Age-Adjusted Relative Risks of Major Coronary Heart Disease, According to Quintile Group for Total Vitamin E Intake and Intake of Vitamin E from Dietary Sources[a]

Variable	Quintile group for vitamin E intake 1	2	3	4	5	*p* value for trend
Total intake (including supplements)						
Median (IU/day)	2.8	4.2	5.9	17	208	—
Range (IU/day)	1.2–3.5	3.6–4.9	5.0–8.0	8.1–21.5	21.6–1000	—
Age-adjusted relative risk	1.0	0.90	1.00	0.68	0.59	—
95% Confidence interval	—	0.70–1.16	0.78–1.27	0.52–0.89	0.45–0.78	<0.001
Relative risk adjusted for age and smoking	1.0	1.00	1.15	0.74	0.66	—
95% Confidence interval	—	0.78–1.28	0.90–1.48	0.57–0.98	0.50–0.87	<0.001
Dietary intake (without supplements)						
Median (IU/day)	2.6	3.6	4.4	5.4	7.7	—
Range (IU/day)	0.3–3.1	3.2–3.9	4.0–4.8	4.9–6.2	6.3–100	—
Age-adjusted relative risk	1.0	0.97	0.77	0.98	0.79	—
95% Confidence interval	—	0.75–1.26	0.59–1.01	0.77–1.26	0.61–1.03	0.12
Relative risk adjusted for age and smoking	1.0	1.04	0.87	1.14	0.95	—
95% Confidence interval	—	0.80–1.35	0.66–1.14	0.89–1.47	0.72–1.23	0.99

[a]Major heart disease includes nonfatal myocardial infarction and death due to coronary disease. From Stampfer *et al.* (1993) with permission.

suppressive factors such as hydrogen peroxide and PGE_2 are secreted, leading to the depression of lymphocyte proliferation (Metzger *et al.*, 1980). Hence, it is likely that vitamin E deficiency may affect the immune response. PGE_2 production is enhanced during aging in mice. Meydani *et al.* (1986) demonstrated that vitamin E supplementation to old mice (500 μg/g diet) enhances immune response by reducing PGE_2 production (Table III). In this study, an increase in mitogenic response to concanavalin A and lipopolysaccharide, a delayed-type hypersensitivity (DTH) response to 2,4-dinitro-7-fluorobenzene, which is an *in vivo* indicator of cell-mediated immunity, interleukin-2 (IL-2) production, and a decrease in PGE_2 production were found in old mice fed vitamin E. Since the increased activity of cyclooxygenase caused by aging is reduced by vitamin E supplementation, it has been proposed that vitamin E exerts an immunoenhancing effect at least partially by its ability to inhibit cyclooxygenase activity, resulting in decreased PGE_2 production. It is known, however, that vitamin E may also have an inhibitory effect on phospholipase A_2, which can result in decreased PGE_2 production (Douglas *et al.*, 1986). Further investigation is needed to define the mechanism of the immunoenhancing effect of vitamin

Table III
The Effect of Age and Vitamin E Supplementation on *ex Vivo* Synthesis of PGE_2 by Spleen Homogenates (Mean ± SE, n = 9)

Age (months)	Dietary vitamin E (ppm)	PGE_2 (μg/g)
3	30	2.60 ± 0.08[a]
24	30	3.20 ± 0.07
24	500	2.30 ± 0.10[a]

[a]Significantly different from 24-month-old mice fed 30 ppm vitamin E at $p < .05$. From Meydani *et al.* (1986) with permission.

E. Thus, although firm evidence is lacking, it is interesting to consider that vitamin E may play a role in life span elongation partly by its immunoenhancing property.

One of the most obvious signs of the aging process is, as described above, the accumulation of age pigment, lipofuscin, at the cellular level, probably produced by the reaction of lipid peroxides in organs (Papafrangos and Lyman, 1982). Because neuronal lipofuscin accumulation and axonal dystrophy are commonly observed after 30 years of age in normal people, a parallel has been drawn between vitamin E deficiency and the aging process, leading some to believe that vitamin E supplementation in normal people might retard aging (Sokol, 1989). A reduction in the content of lipofuscin pigments has been shown in animals fed a vitamin E-rich diet (Reddy *et al.*, 1973; Kruk and Enesco, 1981; Amemiya, 1981). The mechanism of the reduction of lipofuscin in tissues by vitamin E is unknown, i.e., whether it may be prevented by the antioxidative effect of vitamin E, because chemical characteristics of lipofuscin pigment have not been identified in detail.

Recently, the effect of antioxidants in improving the cognitive performance of aged experimental animals and humans has been demonstrated. Socci *et al.* (1995) showed that aged male Sprague-Dawley rats treated by the spin-trapping compound phenyl-α-*tert*-butylnitrone (PBN) and vitamin E exhibited significantly greater memory and learning retention compared to normal, aged rats in the water maze test. When geriatric patients in Finland were treated with vitamin E and selenium, several mental functions such as depression, anxiety, self-care, mental alertness, emotional lability, interest in the environment, etc., were significantly improved after only 2 months, and the improvement continued until the end of the 1 year study period (Tolonen *et al.*, 1985). These findings support the free radical theory of aging related to brain function, and the prevention of oxidative damage by vitamin E.

5. VITAMIN E AND LIFE SPAN ELONGATION

In the last two decades many attempts have been made to increase the life span of organisms using antioxidants, in accord with the free radical theory of aging. Vitamin E supplementation has been found to extend the life span of many species, including the nematodes *Turbatrix aceti* and *Caenorhabditis elegans* (Kahn and Enesco, 1981; Zuckerman and Geist, 1983), *Drosophila melanogaster* (Miquel *et al.*, 1973), the rotifers *Philodina* and *Asplancha brightwelli* (Enesco and Verdone-Smith, 1980; Sawada and Enesco, 1984), *Paramecium tetraurelia* (Thomas and Nyberg, 1988), and rats (Porta *et al.*, 1980). On the other hand, although Packer and Smith (1974) reported that vitamin E extends the life span of cultured human fibroblasts, these findings have not been confirmed (Packer and Smith, 1977; Sakagami and Yamada, 1977; Balin *et al.*, 1977). Vitamin E has also been found to have no significant effect on life span in mice (Ledvina and Hodanova, 1980; Blackett and Hall, 1981). This discrepancy could be caused either by problems with vitamin uptake and its metabolism by organisms, or by the environmental conditions of the investigation, such as temperature. This conclusion is supported by the finding that the life span of rotifers is primarily a function of metabolic rate, and that it is also extended by lowering the temperature (Verdon-Smith and Enesco, 1982). These findings suggest that vitamin E and other antioxidants which extend life span in *Drosophila* may act by depressing oxygen consumption.

Interestingly, recent reports demonstrate that the chronic spin-trapping compound PBN resulted in greater survival of rats during the treatment period (Sack *et al.*, 1996); this was accompanied by improvement of cognitive performance and decrease in oxidative damage within brain areas important for cognitive function. Although this effect is considered to be, in part, a consequence of the free radical scavenging property of PBN, most likely the major action of PBN is to decrease the initiation rate of adverse free radical reactions. It has been proposed that this is probably accomplished by the joint action of the following: (1) association with mitochondria followed by reaction with superoxide, resulting in nitroxide, then competition with oxygen for electrons from ubisemiquinone of the respiratory chain system; (2) decrease in hydroxyl radical formation by maintenance of cellular iron in the ferric state; (3) nitroxide serving as a

SOD mimic and depressing hydroxyl radical formation inside Cu/Zn SOD (Harman, 1995).

On the other hand, Sohal *et al.* (1984) reported that diethyl dicarbamate, an inhibitor of the antioxidative enzyme superoxide dismutase, increased the life span of house-flies about 40%. Enesco *et al.* (1989) also demonstrated that although vitamin E extended the life span of rotifers, cortisone, which is not considered to be an antioxidant, can also increase their life span. Both compounds increased the length of either the premature developing period or the mature reproductive period, but had no effect on the aging phase. They concluded that the reproductive period is the most labile period and one of the most important variables in rotifer life span extension (Fig. 7). These observations may suggest that vitamin E retards the metabolic rate during the development stage, and stores the energy reserve in the mature stage. On the basis of these results, it is likely that the effect of vitamin E on life span elongation may not always be due to its antioxidant effect.

6. CONCLUSIONS

In conclusion, although the mechanism of aging is still obscure, there is litte doubt that aging may arise from both genetic and environmental factors. Free radicals, which are generated as one environmental factor, may

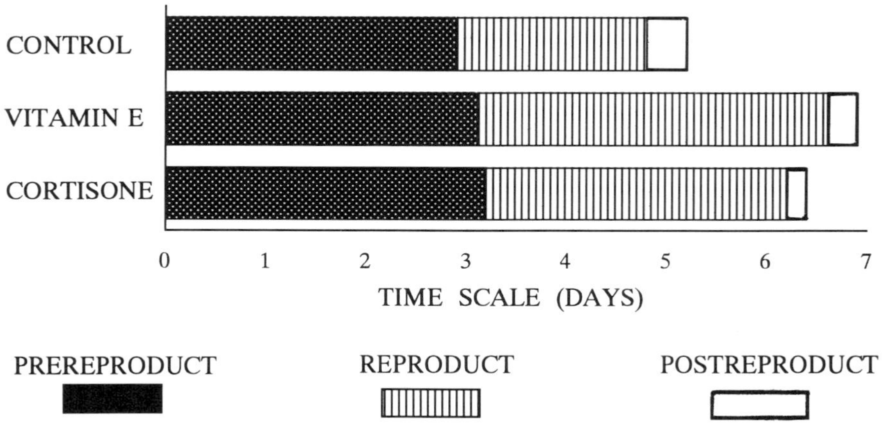

FIGURE 7. The phases of the life span represented graphically to show that agents which extend life span extend both the prereproductive and reproductive periods. Reproduced from Enesco *et al.* (1989) with permission.

attack many organs to induce their dysfunction and result in aging phenomena. It is, however, unknown whether free radicals cause aging or are a result of aging. Since vitamin E is an excellent chain-breaking antoxidant, it is likely that the life span of small species may be extended by the antioxidant effect of vitamin E. As mentioned before, deterioration of immunofunction is considered to induce aging. An increase in maximum life span of several species may be caused by the immunoenhancing effect of vitamin E. The detailed mechanism of life-span elongation by vitamin E is still unclear. Thus, one of the most neglected topics in vitamin E research concerns the role of vitamine E in aging. In order to better define its role, further progress in vitamin E studies on aging is needed.

7. REFERENCES

Amemiya, T., 1981, Effect of vitamin E administration on photoreceptor outer segment and retinal pigment epithelium of vitamin E deficient rats, *Int. J. Vitamin Nutr. Res.* **51:**114–118.

Ames, B. N., 1989, Endogenous DNA damage as related to cancer and aging, *Mutat. Res.* **214:**41–46.

Ames, B. N., Shigenaga, M. K., and Hagen, T. M., 1993, Oxidants, antioxidants, and the degenerative diseases of aging, *Proc. Natl. Acad. Sci. USA* **90:**7915–7922.

Balin A. K., Goodman, D. B. P., Rasmussen, H., and Cristofalo, V. J., 1977, The effect of oxygen and vitamin E on the life span of human diploid cells *in vitro*, *J. Cell. Biol.* **74:**58–67.

Blackett, A. D., and Hall, D. A., 1981, The effect of vitamin E on mouse fitness and survival, *Gerontology* **27:**133–139.

Carney, J. M., Starke-Reed, P. E., Oliver, C. N., Landum, R. W., Cheng, M. S., Wu, J. F., and Floyd, R. A., 1991, Reversal age-related increase in brain protein oxidation, decrease in enzyme activity, and loss in temporal and spatial memory by chronic administration of the spin trapping compound N-tert-butyl-α-phenylnitrone, *Proc. Natl. Acad. Sci. USA* **88:**3633–3636.

Chia, L. S., Thompson, J. E., and Moscarello, M. A., 1983, Changes in lipid phase behavior in human myelin during maturation and aging, *FEBS Lett.* **157:**155–158.

Corwin, L. M., and Shloss, J., 1980, Role of antioxidants in the stimulation of the mitogen response, *J. Nutr.* **110:**2497–2505.

Creter, D., Pavlotzky, F., and Savir, H., 1979, Effect of vitamin E on platelet aggregation in diabetic retinopathy, *Acta Haematol.* **62:**74–77.

Cutler, R. G., 1984, Antioxidants, aging, and longevity, in *Free Radicals in Biology*, Vol. 6 (W. A. Pryor, ed.), pp. 371–428, Academic Press, Orlando, Florida.

Davies, K. J. A., 1987, Protein damage and degeneration by oxygen radical, I. general aspect, *J. Biol. Chem.* **262:**9895–9901.

Dillard, C. J. E., Dumelin, E. E., and Tappel, A. L., 1978, Effects of dietary vitamin E, selenium and polyunsaturated fats on *in vivo* lipid peroxidation in the rat as measured by pentane production, *Lipids* **13:**396–402.

Douglas, C. E., Chan, A. C., and Choy, P. C., 1986, Vitamin E inhibits platelet phospholipase A_2, *Biochim. Biophys. Acta* **876:**639–645.

Enesco, H. E., and Verdone-Smith, C., 1980, Alpha-tocopherol increases life span in the rotifer *Philodina*, *Exp. Gerontol.* **15:**335–338.
Enesco, H. E., Bozovic, V., and Anderson, P. D., 1989, The relationship between life span and reproduction in the rotifer *Asplanchna brightwelli*, *Mech. Ageing Dev.* **48:**281–289.
Esterbauer, H., Gebicki, J., Puhl, H., and Jurgens, G., 1992, The role of lipid peroxidation and antioxidants in oxidative modification of LDL, *Free Radic. Biol. Med.* **13:**341–390.
Esterbauer, H., Puhl, H., Waeg, G., Krebs, A., and Dieber-Rotheneder, M., 1993, The role of vitamin E in lipoprotein oxidation, in *Vitamin E in Health and Disease* (L. Packer and J. Fuchs, eds.), pp. 649–671, Marcel Dekker, New York.
Farrell, P. M., 1980, Deficiency states, pharmacological effect, and nutrient requirements, in *Vitamin E, A Comprehensive Treatise* (L. J. Machlin, ed.), pp. 520–620, Marcel Dekker, New York.
Floyd, R. A., 1991, Role of oxygen free radicals in carcinogenesis and brain ischemia, *FASEB J.* **4:**2587–2597.
Fraga, C. G., Shigenaga, M. K., Park, J.-W., Degan, P., and Ames, B. N., 1990, Oxidative damage to DNA during aging: 8-Hydroxy-2′-deoxyguanosine in rat organ DNA and urine, *Proc. Natl. Acad. Sci. USA* **87:**4533–4537.
Francois, C., Nguyen-Legros, J., and Percheron, G., 1981, Topographical and cytological localization of iron in rat and monkey brains, *Brain Res.* **215:**317–322.
Fridovich, I., 1983, Superoxide radical: An endogenous toxicant, *Annu. Rev. Pharmacol. Toxicol.* **23:**239–257.
Fukuzawa, K, Kurotori, Y., Tokumura, A., and Tsukatani, H., 1989, Vitamin E deficiency increases the synthesis of platelet-activating factor (PAF) in rat polymorphonuclear leucocytes, *Lipids* **24:**236–239.
Ghiselli, A., Laurenti, O., Mattia, G. D., Maiani, G., and Ferro-Luzzi, A., 1992, Salicylate hydroxylation as an early marker of *in vivo* oxidative stress in diabetic patients, *Free Radic. Biol. Med.* **13:**621–626.
Gisinger, C., Watanabe, J., and Colwell, J. A., 1990, Vitamin E and platelet eicosanoids in diabetes millitus, *Prostaglandins Leukotrienes Essential Fatty Acids* **40:**169–176.
Haberland, M. E., Fless, G. M., Scanu, A. M., and Fogelman, A. M., 1992, Malondialdehyde modification of lipoprotein produces avid uptake by human monocyte-macrophages, *J. Biol. Chem.* **267:**4143–4151.
Hafeman, D. G., and Hoekstra, W. G., 1977, Lipid peroxidation *in vivo* during vitamin E and selenium deficiency in the rats as monitored by ethane evolution, *J. Nutr.* **107:**666–672.
Halliwell, B., 1989, Oxidants and the central nervous system: Some fundamental questions, *Acta Neurol. Scand.* **126:**23–33.
Halliwell, B., and Gutteridge, J. M. C., 1985, The importance of free radicals and catalytic metal ions in human diseases, *Mol. Aspects Med.* **8:**189–193.
Hammond, B., Kontos, H. A., and Hess, M. L., 1985, Oxygen radicals in the adult respiratory distress syndrome, in madrigal ischemia and reperfusion injury, and in cerebral vascular damage, *J. Physiol. Pharmacol.* **63:**173–187.
Harman, D., 1956, Aging, A theory based on free radical and radiation chemistry, *J. Gerontol.* **11:**298–300.
Harman, D., 1982, The free-radical theory of aging, in *Free Radicals in Biology*, Vol. 5 (W. A. Pryor, ed.), pp. 255–275, Academic Press, Orlando, Florida.
Harman, D., 1986, Free radical theory of aging: Role of free radicals in the origination and evolution of life, aging and disease process, in *Free Radicals, Aging, and Degenerative Disease* (J. E. Johnson, ed.), pp. 3–49, Alan R. Liss, New York.
Harman, D., 1992, Free radical theory of aging, *Mutat. Res.* **275:**257–266.

Harman, D., 1994, Aging: Prospects for further increases in the functional life span, *Age* **17:**119–146.

Harman, D., 1995, Role of antioxidant nutrients in aging: Overview, *Age* **18:**51–62.

Hatman, L. J., and Kayden, H. J., 1979, A high-performance liquid chromatographic method for the determination of tocopherol in plasma and cellular elements of the blood, *J. Lipid Res.* **20:**639–645.

Hayakawa, M., Hattori, K., Sugiyama, S., and Ozawa, T., 1992, Age associated oxygen damage and mutations in mitochondrial DNA in human hearts, *Biochem. Biophys. Res. Commun.* **189:**979–985.

Hennig, B., and Chow, C. K., 1988, Lipid peroxidation and endotherial cell injury: Implications in atherosclerosis, *Free Radic Biol. Med.* **4:**99–106.

Hennig, B., Boissonneault, G. A., and Wang, Y., 1989, Protective effects of vitamin E in age-related endotherial cell injury, *Int. J. Vitamin Nutr. Res.* **59:**273–279.

Hirai, S., and Yoshikawa, M., 1970, Vitamin E and aging of the nervous system, in *Proceedings International Symposium on Vitamin E* (N. Shimazono and Y. Takagi, eds.), pp. 228–237, Kyoritsu Shuppan, Tokyo.

Hope, W. C., Dalton, C., and Machlin, L. J., 1975, Influence of vitamin E on prostaglandin biosynthesis rat blood, *Prostaglandins* **10:**577–567.

Kahn, M., and Enesco, H. E., 1981, Effect of α-tocopherol on the life span of *Urbatrix aceti*, *Age* **4:**109–115.

Kay, M. M. B., Bosman, G. J. C. G. M., Shapiro, S. S., Bendich, A., and Nassal P. S., 1986, Oxidation as a possible mechanism of cellular aging: Vitamin E deficiency causes premature aging and IgG binding to erythrocytes, *Proc. Natl. Acad. Sci. USA* **83:**2463–2467.

Kita, T., Nagano, Y., Yokode, M., Ishii, K., Kume, N., Ooshima, A., Yoshida, H., and Kawai, C., 1987, Probucol prevents the progression of atherosclerosis in Watanabe heritable hyperlipidemic rabbits, *Proc. Natl. Acad. Sci. USA* **84:**5928–5931.

Kohn, R. R., 1985, Aging and age-related diseases: Normal processes, in *Relation between Normal Aging and Disease* (H. A. Johnson, ed.), pp. 1–44, Raven Press, New York.

Kruk, P., and Enesco, H. E., 1981, Tocopherol reduces fluorescent age pigment levels in heart and brain of young mice, *Experimentia* **37:**1301–1302.

Ku, H. H., Brunk, U. T., and Sohal, R. S., 1993, Relationship between mitochondrial superoxide and hydrogen peroxide production and longevity of mammalian species, *Free Radic. Biol. Med.* **15:**621–627.

Ledvina, M., and Hodanova, M., 1980, The effect of simultaneous administration of tocopherol and sunflower oil on the life span of female mice, *Exp. Gerontol.* **15:**67–71.

Machlin, L. J., Filipski, R., and Kuhn, D. C., 1975, Influence of vitamin E on platelet aggregation and thrombocytopenia in rat, *Proc. Soc. Exp. Biol. Med.* **149:**275–277.

Masuda, M., Ihara, N., Kuriki, H., Komiyama, Y., Nishikado, H., Egawa, H., and Murata, K., 1989, Spontaneous injuries in the aortic endotherium of the inherited cataract rats and their prevention by tocopherol, *Atherosclerosis* **75:**23–30.

McCord, J. M., and Fridovich, I., 1969, Superoxide dismutase, An enzymic function for erythrocuprein, *J. Biol. Chem.* **244:**6049–6055.

Metzger, Z., Hoffeld, J. T., and Oppenheim, J. J., 1980, Macrophage-mediated suppression I. Evidence for participation of both hydrogen peroxide and prostaglandin of suppression of murine lymphocyte proliferation, *J. Immunol.* **124:**983–988.

Meydani, S. N., and Tengerdy, R. P., 1991, Vitamin E and immune response, in *Vitamin E: Biochemical Clinical Applications* (L. Packer and J. Fuchs, eds.), pp. 549–561, Marcel Dekker, New York.

Meydani, S. N., Meydani, M., Verdon, C. P., Blumberg, J. P., and Hayes, K. C. (1986). Vitamin E supplementation suppresses prostaglandin E_2 synthesis and enhances immune response of aged mice, *Mech. Ageing Dev.* **34:**191–201.

Miquel, J., Binnard, R., and Howard, W. H., 1973, Effect of dl-α-tocopherol on the life span of *Drosophila melanogaster*, *Gerontologist* **13:**37.

Muller, D. P. R., Lloyd, J. K., and Wolff, O. H., 1983, Vitamin E and neurological function: Abetalipoproteinaemia and other disorders of fat absorption, in *Ciba foundation Symposium 101, Biology of Vitamin E*, pp. 106–121, Pitman, London.

Munnell, J. F., and Getty, R., 1968, Rate of accumulation of cardiac lipofuscin in the aging canine, *J. Gerontol.* **23:**154–158.

Nelson, J. S., 1980, Phathology of vitamin E deficiency, in *Vitamin E—A Comprehensive Treatise* (L. J. Machlin, ed.), pp. 397–428, Marcel Dekker, New York.

Okuma, M., Takayama, H., and Uchino, H., 1980, Generation of prostacycline-like substance and lipid peroxidation in vitamin E-deficient rats, *Prostaglandin*, **19:**527–536.

Oliver, C. N., Ahn, B.-W., Moerman, E. J., Goldstein, S., and Stadtman, E. R., 1987, Age-related changes in oxidised proteins, *J. Biol. Chem.* **262:**5488–5491.

Ouahchi, K., Arita, M., Kayden, H., Hentati, F., Hanida, M. B., Sokol, R., Arai, H., Inoue, K., Mandel, J.-L., and Koenig, M., 1995, Ataxia with isolated vitamin E deficiency is caused by mutations in the α-tocopherol transfer protein, *Nature Genet.* **9:**141–145.

Packer, L., and Smith, J. R., 1974, Extension of the life span of cultured normal human diploid cells by vitamin E, *Proc. Natl. Acad. Sci. USA* **71:**4763–4767.

Packer, L., and Smith, J. R., 1977, Extension of the life span of cultured normal human diploid cells by vitamin E. A reevaluation, *Proc. Natl. Acad. Sci. USA* **74:**1640–1641.

Papafrangos, E. D., and Lyman, C. P., 1982, Lipofuscin accumulation and hibernation in the Turkish hamster, *Mesocricetes brandti*, *J. Gerontol.* **37:**417–421.

Parthasarathy, S., Khoo, J. C., Miller E., Barnett, J., Witztum, J. L., and Steinberg, D., 1990, Low density lipoprotein rich in oleic acid is protected against oxidative modification: Implications for dietary prevention of atherosclerosis, *Proc. Natl. Acad. Sci. USA* **87:**3894–3898.

Pichard, K. A., Karpen, C. W., and Merola, C. W., 1982, Influence of dietary vitamin C on platelet thromboxane A_2 and vascular prostacycline I_2 in rabbit, *Prostaglandins Leukotrienes Med.* **9:**373–378.

Porta, E. A., Joun, N. S., and Nitta, R. T., 1980, Effects of the type of dietary fat at two levels of vitamin E in Wistar male rats during development and aging, *Mech. Ageing Dev.* **13:**1–39.

Pryor, W. A., 1982, Free radical biology: Xenobiotics, cancer and aging, *Ann. N. Y. Acad. Sci.* **393:**1–22.

Reddy, K., Fletcher, B., Tappel, A., and Tappel, A. L., 1973, Measurement and spectral characteristics of fluorescent pigments in tissues of rats as a function of dietary polyunsaturated fats and vitamin E, *J. Nutr.* **103:**908–915.

Sack, C. A., Socci, D. J., Crandall, B. M., and Arendash, G. W., 1996, Antioxidant treatment with phenyl-a-tert-butylnitron (PBN) improves the cognitive performance and survival of aging rats, *Neurosci. Lett.* **205:**181–184.

Sakagami, H., and Yamada, M., 1977, Failure of vitamin E to extend the life span of a human diploid cell line in culture, *Cell Struct. Funct.* **2:**219–227.

Sawada, M., and Enesco, H. E., 1984, Vitamin E extends life span in the short-lived rotifer *Asplanchna brightwelli*, *Exp. Gerontol.* **19:**179–183.

Sawada, M., and Carlson, J. C., 1987a, Changes in superoxide radical and lipid peroxide formation in the brain, heart and liver during the life span of the rat, *Mech. Aging Dev.* **41:**125–137.

Sawada, M., and Carlson, J. C., 1987b, Association between lipid peroxidation and life modifying factors in rotifers, *J. Gerontol.* **42:**451–456.
Shirai, K., Matsuoka, N., Morisaki, N., Murano, S., Sasaki, N., Shinoyama, N., Saito, Y., Kumagai, A., and Mizobuchi, M., 1980, Effect of tocopherol deficiency on lipid metabolism in the arterial wall of rats on normal and high cholesterol diet, *Artery* **6:**484–506.
Socci, D. J., Crandall, B. M., and Arendash, G. W., 1995, Chronic antioxidant treatment improves the cognitive performance of aged rats, *Brain Res.* **693:**88–94.
Sohal, R. S., Farmer, K. J., Allen, R. G., and Ragland, S. S., 1984, Effects of diethyldithiocarbamate on life span, metabolic rate, superoxide dismutase, catalase, inorganic peroxide and glutathione in the adult male housefly, *Mech. Ageing Dev.* **24:**175–183.
Sokol, R. J., 1989, Vitamin E and neurologic function in man, *Free Radic. Biol. Med.* **6:**189–207.
Stampfer, M. J., Hennekens, C. H., Manson, J. E., Colditz, G. A., Rosner, B., and Willett, W., 1993, Vitamin E consumption and the risk of coronary disease in women, *N. Engl. J. Med.* **328:**1444–1449.
Starke-Reed, P. E., and Oliver, C. N., 1989, Protein oxidation and proteolysis during aging and oxidative stress, *Arch. Biochem. Biophys.* **275:**559–567.
Steinbrecher, U. P., Zhang, H., and Lougheed, M., 1990, Role of oxidatively modified LDL in atherosclerosis, *Free Radic. Biol. Med.* **9:**155–168.
Strehler, B. L., Mark, D. D., Mildvan, A. S., and Gee, M. V., 1959, Rate and magnitude of age pigment accumulation in the human myocardium, *J. Gerontol.* **14:**430–439.
Takeuchi, N., Tanaka, E., Katayama, I., Matsumiya, K., and Yamamura, I., 1976, Effects of tocopherol on thiobarbituric acid reactive substances in serum and hepatic subcellular organelles and lipid metabolism, *Exp. Gerontol.* **11:**179–185.
Takeuchi, N., Iritani, N., Fukuda, E., and Tanaka, F., 1978, Effects of long-term administration and deficiency of α-tocopherol on lipid metabolism of rats, in *Tocopherol, Oxygen and Biomembranes* (C. deDuve and O. Hayaishi, eds.), pp. 257–267, Elsevier, Amsterdam.
Tappel, A. L., Fletcher, B., and Deamer, D., 1973, Effect of antioxidants and nutrients on lipid peroxidation fluorescent products and aging parameters in the mouse, *J. Gerontol.* **28:**415–424.
Thomas, J., and Nyberg, D., 1988, Vitamin E supplementation and intense selection increase clonal life span in *Paramecium tetraurelia*, *Exp. Gerontol.* **23:**501–512.
Tokumaru, S., Iguchi, H., and Kpjo, S., 1996, Changes of the lipid hydroperoxide level in mouse organs on aging, *Mech. Ageing Dev.* **86:**67–74.
Tokumaru, S., Ogino, R., Shiromoto, A., Iguchi, H., and Kpjo, S., 1997, Increase of lipid hydroperoxides in tissues of vitamin E-deficient rats, *Free Radic. Res.* **26:**169–174.
Tolonen, M., Halme, M., and Sarna, S., 1985, Vitamin E and selenium supplementation in geriatric patients. A double-blind preliminary clinical trial, *Biol. Trace Element Res.* **7:**161–168.
Turrens, J. F., Freeman, B. A., and Crapo, J., 1982, Hyperoxia increases H_2O_2 release by lung mitochondria and microsomes, *Arch. Biochem. Biophys.* **217:**411–421.
Upton, A. C., 1977, Pathobiology, in *The Biology of Aging* (C. H. Finch and L. Hayflick, eds.), pp. 513–535, Van Nostrand, New York.
Urano, S., 1993, The role of vitamin E in blood and cellular aging, in *Vitamin E in Health and Disease* (L. Packer and J. Fuchs, eds.), pp. 159–170, Marcel Dekker, New York.
Urano, S., Matsuo, M., Ohtsubo, K., and Izumiyama, N., 1994, Oxidative damage of synaptic membranes in rats during aging, *Biomed. Gerontol.* **18:**378–379.
Urano, S., Asai, Y., Makabe, S., Matso, M., Izumiyama, N., Ohtsubo, K., and Endo, T., 1997, Oxidative injury of synapse and antioxidative defense systems in rats, and its prevention by vitamin E, *Eur. J. Biochem.* **245:**67–70.

van Kujik, F. J. G. M., Sevanian, A., Handelman, F. J., Dratz, E. D., 1987, A new role for phospholipase A_2: Protection of membranes from lipid peroxidation damage, *Trends Biochem. Sci.* **12:**31–34.

Verdon-Smith, C., and Enesco, H. E., 1982, The effect of temperature and dietary restriction on life span and reproduction in the rotifer *Asplanchna brightwelli*, *Exp. Gerontol.* **17:**255–262.

Wartanowicz, M., Panczenko-Kresowska, B., Ziemlanski, S., Kowalska, M., and Okolska, G., 1984, The effect of α-tocopherol and ascorbic acid on the serum lipid peroxide level in elderly people, *Ann. Nutr. Metab.* **28:**186–191.

Zhang, J.-R., Andrus, P. K., and Hall, E. D., 1994, Age-related phospholipid hydroperoxide levels in gerbil brain measured by HPLC-chemiluminescence and their relation to hydroxyl radical stress, *Brain Res.* **639:**275–282.

Zuckerman, B. M., and Geist, M. A., 1983, Effect of vitamin E on the nematode *Caenorhabditis eleganse*, *Age* **6:**1–4.

Part V

Vitamin F

Chapter 17

The Role of Linoleic Acid in Endothelial Cell Gene Expression

Relationship to Atherosclerosis

Michal Toborek and Bernhard Hennig

1. INTRODUCTION: FATTY ACIDS AND ATHEROSCLEROSIS

Although the mortality from coronary heart disease has declined recently, atherosclerosis and related vascular disorders still are the leading causes of death in the United States and other Western countries. The etiology of this disease is multifactorial, with hyperlipidemia, smoking, diabetes mellitus, hypertension, and obesity being well-established risk factors for the development of atherosclerosis. Dietary fat affects plasma lipids and lipoproteins and thus is linked to atherosclerosis (Watkins *et al.*, 1996). Injury to or abnormal mechanisms of the vascular endothelium may be initiating events in the etiology of atherosclerosis.

There are numerous theories for the pathogenesis of atherosclerosis (Haudenshild, 1990). The current trend is to consider atherosclerosis as a response of the vascular wall to a variety of initiating agents and multiple

Michal Toborek Department of Surgery, University of Kentucky Medical Center, Lexington, Kentucky 40536. **Bernhard Hennig** Department of Nutrition and Food Science, University of Kentucky, Lexington, Kentucky 40506-0054.

Subcellular Biochemistry, Volume 30: Fat-Soluble Vitamins, edited by Quinn and Kagan. Plenum Press, New York, 1998.

pathogenic mechanisms (e.g., hyperlipidemia), contributing to the development of atheromatous plaques. It appears that the major participants in the atherosclerotic disease process include an active vascular endothelium, smooth muscle cells, blood-borne cells such as monocytes and macrophages, and circulating lipoproteins (Ross, 1986; Munro and Cotran, 1988; Simionescu and Simionescu, 1993). The result is a multifactorial sequence of events involving endothelial cell injury/dysfunction, uptake of circulating blood monocytes and their differentiation into macrophages, coupled with smooth muscle cell migration and proliferation.

There is ample evidence suggesting that serum cholesterol is a predictor of atherosclerosis and that serum cholesterol concentrations can be modified by varying the composition of dietary fat. Less is known, however, about the role of specific fatty acids in atherosclerosis. The general indication is that saturated fatty acids increase, whereas polyunsaturated fatty acids decrease, serum cholesterol (Hegsted *et al.*, 1993). Also, monounsaturated fatty acids appear to have no specific effect on plasma cholesterol concentrations. However, it is now known that saturated fatty acids are not equally hypercholesterolemic. For example, stearic acid (18:0) and saturated fatty acids with fewer than 12 carbon atoms are thought not to raise serum cholesterol concentrations (Mensink, 1993). This then would suggest that the cholesterol-raising properties of saturated fatty acids should be attributed solely to lauric (12:0), myristic (14:0), and palmitic acid (16:0). However, these three saturated fatty acids appear to have different effects on serum total-cholesterol concentrations as well. Several studies suggest that lauric acid is less, and myristic acid probably more, hypercholesterolemic than palmitic acid (Hegsted *et al.*, 1965; Mensink, 1993). Furthermore, recent human studies suggest that lauric acid raises total serum cholesterol and low-density lipoprotein (LDL) cholesterol concentrations compared with oleic acid (18:1). However, it is not as potent in increasing cholesterol concentrations as is palmitic acid (Denke and Grundy, 1992). On the other hand, in normocholesterolemic men and women, dietary palmitic and oleic acids seemed to exert similar effects on serum cholesterol and lipoprotein profiles (Ng *et al.*, 1992).

Numerous recent studies and biochemical investigations suggest that lipid-oxidation products, ingested with food or produced endogenously, represent a health risk (Esterbauer, 1993). In fact, evidence supports the hypothesis that LDL undergoes an oxidative modification that increases its uptake by macrophages (Steinberg and Witztum, 1990). Dietary antioxidants such as vitamin E might act as an antiatherogenic agent by suppressing oxidative modification of LDL and the recruitment of monocytes into the arterial subendothelium by smooth muscle cells (Janero, 1991). In fact, data from subjects with varying degrees of coronary atherosclerosis support the hypothesis that high serum polyunsaturated fatty acid levels, when insufficiently protected by antioxidants (e.g., vitamin E), may indicate a higher risk of atherosclerosis (Kok *et al.*, 1991). In particular, a positive relationship between linoleic acid (18:2) intake and coronary artery disease was seen in patients undergoing angiography (Hodgson *et al.*, 1993). Recent research with a population from a country with one of the highest dietary polyunsaturated/saturated fat ratios in the world has concluded that diets rich in omega-6 fatty acids may contribute to an increased incidence in atherosclerosis, hyperinsulinemia, and tumorigenesis (Yam *et al.*, 1996).

There is evidence that selected fatty acids, derived from the hydrolysis of triglyceride-rich lipoproteins, may be atherogenic by causing endothelial injury or dysfunction and subsequent endothelial barrier dysfunction (reviewed by Hennig *et al.*, 1996b). In support of this hypothesis, we have shown that exposure of cultured endothelial cells to selected fatty acids, e.g., oleic acid, decreased endothelial barrier function, expressed as an increased transfer of both albumin and LDL across the endothelium (Hennig *et al.*, 1984, 1985). In addition to oleic acid, we also investigated the effects of other fatty acids on endothelial barrier function (Hennig *et al.*, 1990). Albumin-bound palmitic and stearic acid had little effect on endothelial barrier function, but exposure of cell monolayers to linoleic acid produced an even greater increase in albumin transfer than did equal concentrations of oleic acid. The disruption in endothelial barrier function was exacerbated greatly in the presence of small amounts of oxidation derivatives of unsaturated fatty acids (Hennig *et al.*, 1986). Furthermore, when comparing fatty acid extracts derived from different animal fats and plant oils, the fat-induced disruption of endothelial barrier function was related to the amount of linoleic acid present in the fat source (Hennig *et al.*, 1993). The significance of linoleic acid in the development of atherosclerosis is supported by the observations that levels of this fatty acid in the phospholipid fractions of human coronary arteries were increased in cases of sudden cardiac death due to ischemic heart disease (Loustarinen *et al.*, 1993). In addition, concentrations of linoleic acid in adipose tissue were positively correlated with the degree of coronary artery disease (Hodgson *et al.*, 1993). These and other studies suggest that, in general, selected fatty acids (Hennig *et al.*, 1996b), as well as their oxidation derivatives (Louheranta *et al.*, 1996), can cause endothelial cell activation and dysfunction, events critical in the atherosclerotic disease process.

Inflammation and any agents which provoke an inflammatory response may be among the most critical risk factors in the atherosclerotic disease process (Ridker *et al.*, 1997). Little is known about the interaction of nutrition (and in particular fatty acids) and inflammatory mediators, but there is evidence that selected nutrients can provoke an inflammatory response or, on the other hand, provide antiinflammatory or antiatherogenic properties (Grimble, 1996; Hennig *et al.*, 1996b). It appears that only selected omega-3 fatty acids can reduce endothelial expression of adhesion molecules (Kim *et al.*, 1995; De Caterina and Libby, 1996), and that linoleic acid can increase adhesion receptor expression (Pietsch *et al.*, 1995). In fact, our data strongly support the fact that selected unsaturated fatty acids (e.g., linoleic acid) and inflammatory cytokines may cross-amplify vascular endothelial cell activation, an inflammatory response, and atherosclerosis (Toborek *et al.*, 1996).

Even though numerous risk factors, including hyperlipidemia, smoking, and hypertension, seem to contribute to the development of atherosclerosis, it has not been possible to link these risk factors into a common pathogenic mechanism. There is evidence, however, that modulations in the level of activity of a select set of endothelial transcription factors (e.g., endothelial nuclear factor-κB, NF-κB) may provide a mechanism for linking these seemingly diverse processes with the generation of dysfunctional endothelium and the onset of atherosclerotic lesion formation (Berliner *et al.*, 1995; Collins *et al.*, 1995). Stimuli known to activate the NF-κB complex include inflammatory cytokines, activators of protein kinase C, viruses, and oxidants (Baeuerle, 1991; Siebenlist *et al.*, 1994; Barnes and Karin, 1997). One may speculate that selected diet-derived fatty acids may induce endothelial oxidative stress and generate excess reactive oxygen species, which activate NF-κB and modulate endothelial gene expression (Fig. 1). Our data support the hypothesis that linoleic acid may be atherogenic by intensifying cytokine expression and endothelial cell injury in atherosclerosis. We have recently shown that linoleic acid can activate NF-κB and induce NF-κB-dependent transcription in culture endothelial cells (Hennig *et al.*, 1996a; Toborek *et al.*, 1996). All these studies lead one to conclude that excessive consumption of oils and fats rich in linoleic acid may be atherogenic and that the type of fat becomes a less significant component in the pathogenesis of atherosclerosis when one consumes a low-fat diet rich in soluble fibers and natural antioxidants.

2. ROLE OF NF-κB IN FATTY ACID-MEDIATED GENE EXPRESSION

Recent evidence indicates that fatty acid-mediated injury to endothelial cells and atherogenesis may involve induction a variety of genes, including those encoding for adhesion molecules and inflammatory cytokines (Fig. 1) (De Caterina and Libby, 1996). Detailed mechanisms of gene

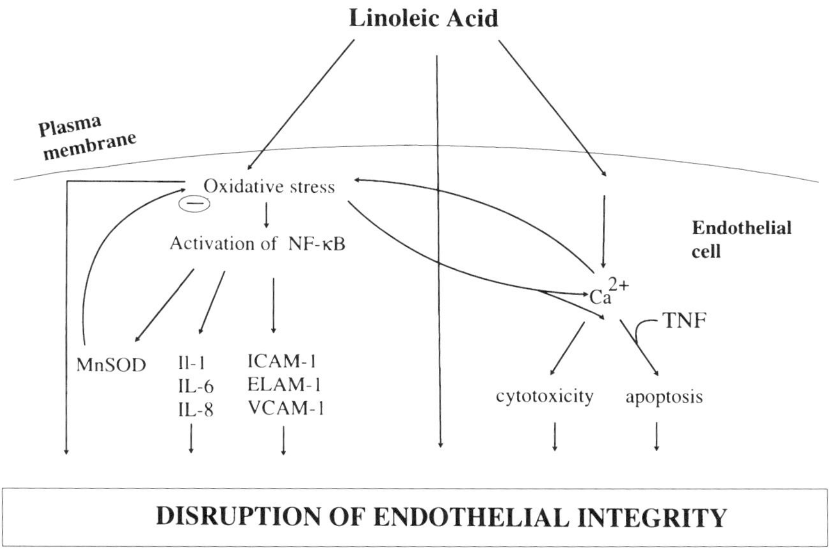

FIGURE 1. Schematic diagram of linoleic acid-mediated alterations in endothelial cell metabolism. NF-κB, Nuclear factor-κB, IL, interleukin; TNF, tumor necrosis factor-α; ICAM-1, intracellular adhesion molecule-1; ELAM-1, endothelial-leukocyte adhesion molecule-1; VCAM-1, vascular cell adhesion molecule-1.

expression in response to exposure to fatty acid are unknown. However, they include, at least in part, activation of NF-κB (Hennig *et al.*, 1996a; Toborek *et al.*, 1996).

The significance of endothelial cell activation of NF-κB on the development of atherosclerosis is related to the stimulation of the inflammatory response through production of inflammatory cytokines or adhesion molecules (Berliner *et al.*, 1995). During atherogenesis, mediators of this process may include fatty acids, oxidized fatty acids, oxidized lipoproteins, and/or cytokines. NF-κB binding sites were identified in the promotor regions of genes encoding for ELAM-1 (endothelial-leukocyte adhesion molecule-1), VCAM-1 (vascular cell adhesion molecule-1), and ICAM-1 (intracellular adhesion molecule-1). Although other transcription factors, such as activating transcription factor-2 (ATF-2) or high-mobility group protein I (Y) are required for expression of ELAM-1 and possibly for other adhesion molecules, NF-κB constitutes an important component of transcriptional regulation of these genes (Collins *et al.*, 1995). Moreover, NF-κB regulates transcription of genes encoding for cytokines (such as interleukin-1 [IL-1], IL-6, IL-8, tumor necrosis factor [TNF]), growth factors, chemokines, and manganese superoxide dismutase (MnSOD). It is interesting that expression of inflammatory cytokines is dependent on activated NF-κB and, in turn, these cytokines stimulate activation of this transcription factor. Thus it appears that inflammatory cytokines use NF-κB to amplify their own signals (Siebenlist *et al.*, 1994). However, no evidence suggests that NF-κB activation may directly participate in fatty acid- or cytokine-induced cytotoxicity (Beyaert and Fiers, 1994). On the other hand, NF-κB stimulates expression of MnSOD, a potent antioxidant enzyme of mitochondrial matrix (Wong, 1995). It is possible that NF-κB-dependent stimulation of MnSOD protects against cytotoxic effects mediated by fatty acids, cytokines, or other inducers of this transcription factor. This notion is supported by the observation that injury to mitochondria and production of superoxides are integral parts of TNF cytotoxicity in endothelial cells (Schulze-Osthoff *et al.*, 1992). Thus, activated NF-κB can lead to both harmful and protective events in endothelial cell metabolism. If induction of inflammatory cytokines and adhesion molecules are most critical events in NF-κB function, activated NF-κB may start an excessive inflammatory response. On the other hand, if stimulation of MnSOD expression plays a main role in NF-κB physiology, NF-κB activation may be critical for cellular antioxidant systems.

Based on this information, one may suggest that in fatty acid-treated endothelial cells, NF-κB emerges as a pleiotropic regulatory activator, capable of induction of two mutually different cellular functions, inflammatory response and antioxidant protection. Regulatory mechanisms which determine the induction of the specific signal(s) are unknown. However, they may depend on the cellular microenvironment, which can be modulated by the presence of fatty acids.

3. SIGNAL TRANSDUCTION MECHANISMS OF FATTY ACID-INDUCED NF-κB

NF-κB consists of several subunits; p50 (NF-κB1), p52 (NF-κB2), p65 (RelA), and inhibitory subunit IκB (Fig. 2). In addition, several forms of IκB have been identified. In quiescent cells, NF-κB remains in the cytoplasm in an inactive state, composed of a p50/p65 complex, bound to IκB-α. Activation of NF-κB is induced by phosphorylation of the IκB-α, followed by dissociation of the complex and proteolytic degradation of IκB. This process induces translocation of the active p50/p65 dimer into the nucleus. Subsequently, this dimer binds to *cis*-acting κB sites in the promoters and enhancers of NF-κB-responsive genes (Collins *et al.*, 1995) (Fig. 2).

Phosphorylation of IκB, the critical process in NF-κB activation, is mediated by protein kinases (Thanos and Maniatis, 1995). Different protein

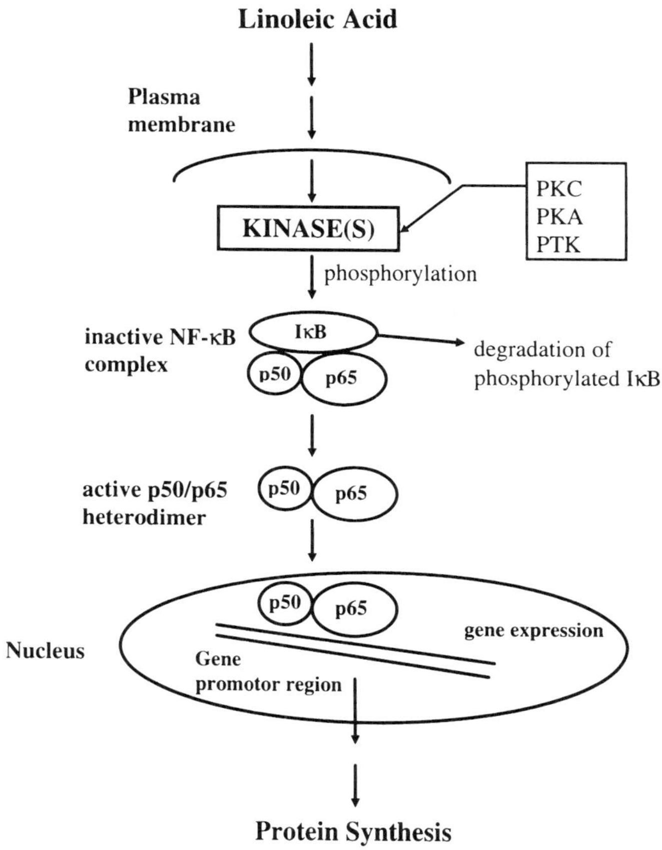

FIGURE 2. Potential mechanisms of linoleic acid-induced activation of NF-κB and gene expression. PKC, Protein kinase C; PKA, cAMP-dependent protein kinase; PTK, protein tyrosine kinase.

kinases appear to participate in this process, depending on stimuli of NF-κB activation and cell types. The specific protein kinase which participates in lipid-induced activation of NF-κB in endothelial cells is unknown. However, the data indicate a potential involvement of either protein kinase C (PKC), cAMP-dependent protein kinase (PKA), protein tyrosine kinase (PTK), or mitogen-activated protein (MAP) kinase in lipid-induced NF-κB activation. This assumption is based on two factors: (a) these kinases can be induced by lipids, and (b) these kinases can regulate NF-κB activation, as indicated in experiments with different nonlipidic modulators. For example, unsaturated lipids, including linoleic acid, but not saturated fats, can activate PKC activity (Graber *et al.*, 1994). PKC-dependent phosphorylation is involved in numerous cellular reactions, including activation of NF-κB (Shirakawa and Mizel, 1989). Endothelial cell cAMP is involved in maintaining endothelial barrier function by phosphorylation processes catalyzed by cAMP-dependent protein kinase. Recent evidence indicates that endothelial cell cAMP and PKA may be involved in activation of NF-κB (Parhami *et al.*, 1993). Protein tyrosine kinases represent a family of several different enzymes. Tyrosine phosphorylation has been linked to several processes, including adhesion molecule expression and NF-κB activation (Thanos and Maniatis, 1995). Both PKC and PTK may induce a signaling cascade which involves phosphorylation of mitogen-activated protein kinases (Koide *et al.*, 1992). Evidence indicates that MAP kinase may also participate in NF-κB activation (Thanos and Maniatis, 1995).

4. FATTY ACID-INDUCED ACTIVATION OF NF-κB AND NF-κB-RELATED TRANSCRIPTION

As mentioned above, several different factors, such as cytokines, including TNF, IL-1, lipopolysaccharides, viruses, double-stranded RNA, hydrogen peroxide (Schreck *et al.*, 1992), or minimally oxidized LDL (Parhami *et al.*, 1993), can activate NF-κB. However, independent of the type of stimulus, activation of NF-κB can be inhibited by cotreatment with antioxidants or metal chelators (Baeuerle, 1991; Schreck *et al.*, 1992; Barnes and Karin, 1997). Therefore, it was suggested that the activation of NF-κB is stimulated by cellular oxidative stress or the cellular redox state (Schreck *et al.*, 1992; Khan *et al.*, 1995).

Selected free fatty acids induce oxidative stress in cultured endothelial cells. Among different methods to measure cellular oxidative stress, 2,7-dichlorofluorescin (DCF) fluorescence was indicated to be extremely sensitive. This method permits measurement of oxidative stress in viable cells and estimates primarily peroxyl radicals and peroxides as indicators of cellular oxidation (Mattson *et al.*, 1995). Using DCF fluorescence assay, we showed that cellular exposures to selected 18-carbon fatty acids differing in degree of unsaturation produce a concentration-dependent increase in cellular oxidation. When endothelial cells were exposed to fatty acids at the concentrations of 60 and 90 μM for 6 hr, only linoleic acid and linolenic acid (18:3) induced oxidative stress. However, all tested fatty acids at the level of 120 μM enhanced cellular oxidation. Independent of concentration, linoleic acid induced endothelial cell oxidative stress most markedly as compared to other 18–carbon fatty acids (Toborek *et al.*, 1996). These results are consistent with the observed fatty acid-mediated increase in lipid hydroperoxides (Hennig *et al.*, 1996a), depletion of intracellular glutathione, as well as alterations in oxidized glutathione and glutathione-dependent enzymes (Toborek and Hennig, 1994). Treatment with fatty acids caused a time-dependent decrease in glutathione concentration with a maximum decrease being observed after a 6-hr exposure. Among tested 18-carbon fatty acids, the most significant depletion of intracellular glutathione was observed in linoleic acid-treated cells (Toborek and Hennig, 1994).

Induction of linoleic acid-mediated cellular oxidative stress was associated with activation of NF-κB. This activation occurred after a 6-hr exposure to linoleic acid, i.e., at the time point when the maximal depletion of glutathione levels due to fatty acid treatment was observed. Both 2- and 4-hr exposures to linoleic acid were not sufficient to activate NF-κB appreciably, and activation of NF-κB was not observed after a 24-hr exposure to this fatty acid. The specificity of the shifted bands was established in a competition analysis performed using nuclear extracts obtained from linoleic acid-treated endothelial cells and molar excess of the unlabeled NF-κB probe. In addition, pretreatment of endothelial cells for 6 hr with *N*-acetyl-L-cysteine or 24 hr with vitamin E inhibited linoleic acid-mediated activation of NF-κB (Hennig *et al.*, 1996a).

Our recent transfection studies with a plasmid (pκB/TK5-CAT) encoding chloramphenicol acetyltransferase (CAT) (Hennig *et al.*, 1996a; Toborek *et al.*, 1996) provided evidence that linoleic acid-activated NF-κB is sufficient to induce transcription in endothelial cells. This plasmid contains a CAT coding sequence driven by a thymidine kinase promoter responsive to NF-κB (Shakhov *et al.*, 1990). Activity of CAT is measured by determination of its products, i.e., acetylated forms of chloramphenicol. Acetylated forms of chloramphenicol indicating CAT activity were elevated in transfected cells exposed to linoleic acid. Preenrichment of cells with vitamin E attenuated NF-κB-related transcription in linoleic acid-treated endothelial cells (Hennig *et al.*, 1996a; Toborek *et al.*, 1996) providing additional evidence that cellular oxidative stress plays a critical role in fatty acid-mediated activation of NF-κB and NF-κB-related transcription. In addition, we now have evidence that exposure of endothelial cells to linoleic acid can result in the induction of adhesion molecules, such as ICAM-1, and inflammatory cytokines, such as IL-8, processes which are controlled by NF-κB (Toborek *et al.*, 1997a).

5. ROLE OF OXIDIZED FATTY ACIDS IN ENDOTHELIAL CELL GENE EXPRESSION

Not only unsaturated free fatty acids, but also their oxidized derivatives may play an important role in endothelial cell injury and the development of atherosclerosis. Polyunsaturated fatty acids can be oxidized both in nonenzymatic reactions of fatty acid peroxidation and in enzyme-mediated reactions, such as in reactions catalyzed by different lipoxygenases (Fig. 3). It appears that both types of fatty acid oxidation are essential in the development of atherosclerosis (Kühn *et al.*, 1994). For example, nonenzymatic oxidation of fatty acids and lipoproteins might be predominant in advanced atherosclerotic lesions. In contrast, enzymatic oxidation of unsaturated fatty acids by lipoxygenases might be more important in the early stages of atherosclerosis (Kühn *et al.*, 1994). In reactions catalyzed by lipoxygenases, linoleic acid is converted into hydroperoxyoctadecadienoic acids (HPODEs) and arachidonic acid into hydroperoxyeicosatetraenoic acids (HPETEs). HOPDE and HPETE are biosynthetic precursors of more stable oxidized metabolites, such as hydroxyoctadecadienoic acids (HODEs) and monohydroxyeicosatetraenoic acids (HETEs), respectively (Fig. 3). A growing body of evidence indicates a role of the 15-lipoxygenase pathway in fatty acid oxidation and in atherogenesis. It was shown that 15-lipoxygenase mRNA and protein were present in atherosclerotic lesions of humans and rabbits, but not in normal vessels. In addition, expression of 15-lipoxygenase was colocalized with epitopes of oxidized LDL (Ylä-Herttuala *et al.*, 1990, 1991). To support the role of 15-lipoxygenase in atherogenesis, HODEs and HETEs were also detected in atherosclerotic lesions (Wang and Powell, 1991; Kühn *et al.*, 1994). It should be noted that 13-HPODE is the major fatty acid oxidation product found in oxidized

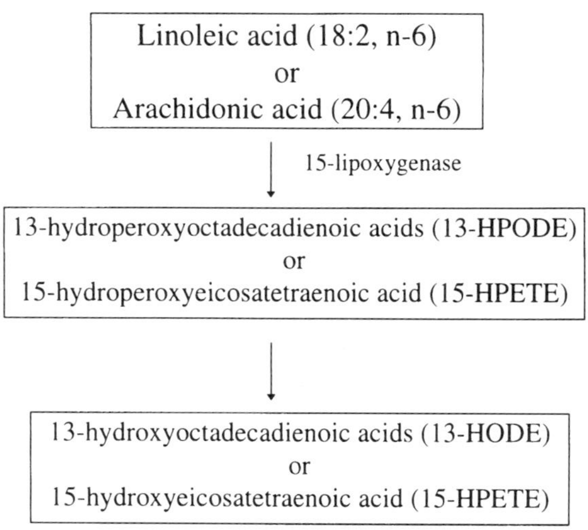

FIGURE 3. Enzymatic conversion of linoleic acid or arachidonic acid to their oxidative derivatives.

LDL (Folcik and Cathcart, 1994). 13-HPODE may also contribute to oxidized LDL-mediated effects on endothelial cell metabolism. It was hypothesized that 13-HOPDE might serve as a specific redox signal in cytokine-mediated expression of adhesion molecules in endothelial cells (Khan *et al.*, 1995).

In addition to their influence on endothelial cell metabolism, linoleic acid and 13-HPODE exert mitogenic effects on vascular smooth muscle cells. Both agents induce DNA synthesis, expression of c-*fos*, c-*jun*, and c-*myc*, as well as activation of mitogen-activated protein kinase. However, cotreatment with nordihydroguaiaretic acid, a potent inhibitor of the lipoxygenase system, but not with indomethacin, an inhibitor of the cyclooxygenase pathway, reduced linoleic acid-mediated mitogenic effects. These results suggest that linoleic acid-induced alterations of cellular growth and prooncogene expression in vascular smooth muscle cells can be mediated by 13-HPODE (Rao *et al.*, 1995). Effects of linoleic acid or 13-HPODE on induction of prooncogenes in endothelial cells remain to be tested.

Despite their several mitogenic effects, oxidized fatty acids might also mediate cellular apoptosis. For example, recent evidence indicates that 15-HPETE and 13-HPODE can induce apoptosis, decrease cellular viability, increase intracellular calcium, and induce DNA fragmentation in different T-cell lines (Sandstrom *et al.*, 1995). Moreover, highly oxidized LDLs, which contain significant amounts of 13-HPODE, can initiate apoptosis in arterial smooth muscle cells, macrophages, and fibroblasts (Bjorkerud and Bjorkerud, 1996). However, their effects on endothelial cell apoptosis remain yet to be determined.

6. INTERACTION BETWEEN FATTY ACIDS AND CYTOKINES AND ENDOTHELIAL CELL DYSFUNCTION

Due to fatty acid-mediated stimulation of production of inflammatory cytokines, endothelial cells may be simultaneously exposed to elevated levels of lipids and cytokines. For example, lipoprotein lipase, an enzyme associated with endothelial cells, not only can release free fatty acids from lipoproteins, but also can induce TNF gene expression and TNF production in macrophages in a protein kinase C signaling pathway (Renier *et al.*, 1994a, b). Monocyte/macrophages and T cells are the main sources of TNF within atherosclerotic vessels (Tracey and Cerami, 1992). Moreover, leukocyte production of TNF is increased in patients with atherosclerosis (Vaddi *et al.*, 1994), and elevated levels of this cytokine are observed in atherosclerotic vessels (Barath *et al.*, 1990; Rus *et al.*, 1991). There is evidence that

TNF can activate vascular endothelial cells (Collins *et al.*, 1995; Read *et al.*, 1995) as well as induce a variety of different cytotoxic effects (Tartaglia *et al.*, 1993).

These facts indicate that fatty acid metabolism may be strongly associated with inflammatory cytokine- or TNF-mediated effects. In fact, we reported that a combined exposure of endothelial cells to linoleic acid and TNF can cross-amplify oxidative injury as measured by DCF fluorescence and depletion of antioxidants, such as glutathione or vitamin E (Toborek *et al.*, 1996). In addition, endothelial cell integrity, determined by transendothelial albumin passage, was more compromised in cells exposed to linoleic acid and TNF as compared to treatment with linoleic acid or TNF alone (Toborek *et al.*, 1996). In contrast, a combined exposure to linoleic acid and TNF did not influence effects mediated by linoleic acid or TNF alone on NF-κB activation. Consistent with these results were alterations in NF-κB-related transcription measured by CAT assay in pκB/TK5-CAT-transfected endothelial cells. Both linoleic acid and TNF increased CAT activity. However, compared to cultures exposed to linoleic acid or TNF alone, no additional effect of the treatment with linoleic acid plus TNF (18:2 + TNF) was observed. In addition, vitamin E, which blocked NF-κB activation, partially inhibited linoleic acid and/or TNF-mediated induction of NF-κB transcription (Toborek *et al.*, 1996).

To further support the hypothesis that the lipid environment and inflammatory cytokines may cross-influence their metabolic effects, it was demonstrated that IL-1 stimulated oxidation of linoleic acid into its oxidative derivatives in cultured endothelial cells (Camacho *et al.*, 1995). In addition, TNF enhanced monocyte or endothelial cell-mediated oxidation of LDL (Maziere *et al.*, 1994), and TNF-mediated cytolysis can be prevented by limiting the availability of arachidonic acid (Hayakawa *et al.*, 1993). Finally, both oxidized LDL and oxidized linoleic acid augmented TNF-induced adhesion molecule expression in human endothelial cells (Khan *et al.*, 1995).

Recently, we also studied effects of linoleic acid on apoptosis of cultured endothelial cells. Although linoleic acid did not induce cell death, it potentiated TNF-mediated apoptosis of endothelial cells (Toborek *et al.*, 1997b). Several mechanisms, including increased oxidative stress, as well as disruption of ion homeostasis, can be responsible for this effect. Cellular oxidation is potentiated in cells exposed 18:2 + TNF as compared to treatment with linoleic acid or TNF alone (Toborek *et al.*, 1996, 1997b) and it is generally accepted that oxidative stress may invoke apoptotic cell death (reviewed by Slater *et al.*, 1995). Different mechanisms, such as direct DNA damage or protein oxidation, may contribute to this process. In addition, oxidative stress may directly regulate gene expression and induce apoptosis by influence on specific transcription factors (Slater *et al.*, 1995). Cellular redox status, regulated by cellular thiol and glutathione, may play a critical role in oxidative stress-induced apoptosis. For example, it was shown that apoptosis is preceded by a decrease in cellular glutathione (Beaver and Waring, 1995) or can be induced by oxidation of thiol groups (Sato *et al.*, 1995). We reported that endothelial cell exposure to linoleic acid or TNF and, in particular, to 18:2 + TNF markedly decreased intracellular glutathione content (Toborek *et al.*, 1996). The disruption of ion homeostasis may be another mechanism by which linoleic acid can increase endothelial cell vulnerability to TNF-mediated apoptosis. We observed a significant increase in intracellular calcium levels $[Ca^{2+}]_i$ in endothelial cells exposed to linoleic acid or TNF and in particular to 18:2 + TNF (Toborek *et al.*, 1996). Elevated $[Ca^{2+}]_i$ might lead to activation of Ca^{2+}-dependent kinases and/or phosphatases with the subsequent induction of gene expression related to apoptosis. In addition, enhanced $[Ca^{2+}]_i$ can activate proteases and/or endonucleases(s) and thus induce DNA damage, a characteristic feature of cellular apoptosis (McConkey and Orrenius, 1996). The role of increased oxidative stress and elevated $[Ca^{2+}]_i$ levels in TNF and/or 18:2 + TNF-induced endothelial cell apoptosis is supported by the observation that BAPTA-AM, a intracellular calcium chelator, and antioxidants, such as vitamin E or *n*-propyl gallate, partially protected against apoptosis in these experimental groups (Toborek *et al.*, 1997b).

Disturbances in endothelial integrity may be an ultimate outcome of endothelial cell apoptosis. In fact, approximately 60% of albumin transfer across normal endothelial cell monolayers occurs via intercellular junctions (Siflinger-Birnboim *et al.*, 1991). In certain experimental settings, aurintricarboxylic acid, which inhibits cellular apoptosis, also can partially protect against the TNF- and 18:2 + TNF-mediated compromised endothelial barrier function (Toborek *et al.*, 1997b). Thus, potentiation of endothelial cell apoptosis due to exposure to 18:2 + TNF may have significant implications in the pathogenesis of atherosclerosis.

7. ROLE OF LINOLEIC ACID IN PROTEOGLYCAN METABOLISM OF THE VASCULAR ENDOTHELIUM

Proteoglycans are a heterogeneous group of extracellular matrix components ubiquitous to tissues of multicellular organisms. These interesting and diverse molecules possess unique physical and chemical properties which largely control their many functions within the interstitium, basement membranes, and cell surface of tissues. Proteoglycans consist of a proteinaceous core to which are covalently bonded one or more large polysaccharide chains, termed glycosaminoglycans. Covalently bonded sulfate molecules are attached to the glycosaminoglycan chains at specific sites. A complete structural classification of the proteoglycans has been described in previous reviews (Hassell *et al.*, 1986; Ruoslahti, 1988; Wight, 1989; Templeton, 1992; Jolles, 1994).

Although proteoglycans are found in all areas of tissues, heparan sulfate proteoglycans appear primarily as components of the cell surface as well as in the basement membrane. On the other hand, the chondroitin and dermatan sulfates are found more often as components of the basement membrane and as interstitial matrix elements. In general, cell surface proteoglycans of the heparin sulfate variety stimulate cell attachment, which is decreased when cells are treated with enzymes that specifically degrade heparan sulfate (Lark, 1984; Farach *et al.*, 1987; Saunders and Bernfield, 1987; Ruoslahti, 1988). In contrast, chondroitin sulfate proteoglycans appear to inhibit cell attachment, probably through interference of the chondroitin sulfate chains with cell surface proteoglycans or masking of the cell attachment sites on matrix molecules, including fibronectin (Ruoslahti, 1988). Interestingly, while decreases in the synthesis of heparan sulfate proteoglycans were observed in cultured pulmonary endothelial cells in response to linoleic acid, there was an increase in a specific, but as-yet-uncharacterized, chondroitin sulfate proteoglycan (Ramasamy *et al.*, 1993). These alterations may inhibit endothelial cell attachment to the underlying matrix, thereby promoting endothelial leakage and invasion of leukocytes. In addition, changes in proteoglycans may reduce the efficacy of reendothelialization *in vivo* by blocking cell migration, which requires cell attachment.

Proteoglycans containing chondroitin, dermatan, and heparan sulfate chains can be produced by vascular endothelial and smooth muscle cells. The vascular smooth muscle produces primarily chondroitin sulfate- and dermatan sulfate-containing proteoglycans while producing only small amounts of heparan sulfate proteoglycans. In contrast, endothelial cells produce primarily heparan sulfate proteoglycan, while chondroitin sulfate proteoglycans represent only a minor fraction of the total proteoglycans synthesized by these cells. Additionally, the specific proteoglycans synthesized in these two cell types can differ widely (Wight *et al.*, 1985). Due to their ubiquitous nature and their important contribution to cellular and tissue function, alterations in proteoglycan synthesis and deposition would be expected to occur in a variety of vascular pathologies in which normal matrix structure and function is altered.

There is evidence that selected lipids can be atherogenic by altering endothelial proteoglycan metabolism (Hennig *et al.*, 1995). Little is known, however, about the role of specific fatty acids in modulating proteoglycan metabolism in endothelial cells. Proteoglycans appear to be important components of endothelial cell stability, and studies with cultured bovine aortic endothelial cells have revealed an increase in survival rate, enhancement of cell density at confluence, and an increase in the nucleus/cytoplasm ratio after *in vitro* administration of heparin or mesoglycan (Tanganelli *et al.*, 1992). There is evidence that specific lipids, such as linoleic acid, can alter normal proteoglycan metabolism, which may be sufficient to impair normal endothelial barrier function (Ramasamy *et al.*, 1993). We have treated vascular endothelial cells with linoleic acid and/or the known proteoglycan synthesis inhibitor β-D-xyloside. Independent exposure to β-D-xyloside or linoleic acid significantly disrupted endothelial barrier function, measured as increased albumin transfer across cultured endothelial monolayers. The linoleic acid-mediated decrease in endothelial barrier function was accompanied by a concurrent inhibition of proteoglycan synthesis, a decrease in anionic properties of heparan sulfate proteoglycans in the cell monolayers, and release of a unique chondroitin sulfate proteoglycan into the culture media (Ramasamy *et al.*, 1993).

The direct mechanism by which linoleic acid alters the synthesis of specific proteoglycans is not clear. It is speculated that linoleic acid might alter the activity of membrane-bound enzymes such as glycosyl transferases and sulfotransferase involved in the proteoglycan synthesis by its effect on membrane fluidity (Chatelain *et al.*, 1985; Tanaka *et al.*, 1988). There is evidence that cell exposure to LDL alters proteoglycan metabolism of cultured endothelial cells (Vijayagopal *et al.*, 1988; Olgemoller *et al.*, 1990). The linoleic acid-induced alterations in proteoglycan molecules may change the architecture or physicochemical properties of the basement membrane (Kleinman *et al.*, 1983; Fujiwara *et al.*, 1984), i.e., reduce the anionic charge density and also affect cell-to-cell associations (Fransson, 1983). These factors could greatly contribute to the disruption of the endothelial barrier function. Because vitamin E can protect against fatty acid-mediated endothelial barrier dysfunction (Hennig *et al.*, 1987), it is conceivable that an increase in oxidative stress can contribute to an abnormal proteoglycan metabolism. In fact, reactive oxygen species have been implicated in the loss of endothelial cell-surface heparan sulfates, which appears to contribute to the increased vascular permeability observed in diabetes mellitus (Wardle, 1994).

Because of our findings that linoleic acid can cross-amplify or potentiate inflammatory cytokine (e.g., TNF)-mediated endothelial barrier dysfunction (Toborek *et al.*, 1996), we also were interested in the effects of TNF on proteoglycan metabolism. We have reported recently that TNF can reduce proteoglycan synthesis in cultured endothelial cells (Radhakrishnamurthy *et al.*, 1990), suggesting that abnormal proteoglycan metabolism may be one of the mechanisms responsible for TNF-mediated endothelial barrier dysfunction. In fact, TNF treatment caused a dose-dependent decrease in mRNA levels for the core proteins of perlecan, a major heparan sulfate proteoglycan, and biglycan in endothelial cultures (Ramasamy *et al.*, 1995). These studies suggest that the decrease in heparan sulfate proteoglycan upon linoleic acid or TNF-mediated injury of the vascular endothelium may play a role in the inflammatory pathogenesis of atherosclerosis.

8. SUMMARY

There is evidence that linoleic acid plays a critical role in gene expression and vascular function as it relates to the pathogenesis of atherosclerosis. The lipid environment, particularly linoleic acid and its derivatives, of the vascular endothelium may profoundly influence the inflammatory response mediated by cytokines. Modulations in the level of activity of a select set of endothelial transcription factors appear to provide a mechanism for linking lipid/cytokine-mediated vessel wall dysfunction, including endothelial cell activation, altered proteoglycan metabolism, and endothelial barrier dysfunction, with the onset of atherosclerotic lesion formation. The activity of endothelial transcription factors is in part regulated by the balance of cellular oxidative stress and antioxidant status. Our data suggest that linoleic acid can activate the vascular endothelium and may thus be an atherogenic fatty acid. Furthermore, nutrients/chemicals with antioxidant properties can protect endothelial cells against lipid-mediated cell injury, suggesting that oxidative stress is a critical component in linoleic acid-mediated gene expression. Our discoveries that linoleic acid can influence significantly the cytokine-mediated inflammatory response may open new fields in dietary intervention of atherosclerosis.

ACKNOWLEDGMENTS. Supported in part by grants from NIH, American Heart Association, Kentucky Affiliate, USDA, University of Kentucky Research Funds, American Cancer Society Institutional Research Grant, and Kentucky Agricultural Experimental Station.

9. REFERENCES

Baeuerle, P. A., 1991, The inducible transcription activator NF-κB: Regulation by distinct protein subunits, *Biochim. Biophys. Acta* **1072:**63–80.

Barath, P., Fishbein, M. C., Cao, J., Berenson, J., Helfant, R. H., and Forrester, J. S., 1990, Detection and localization of tumor necrosis factor in human atheroma, *Am. J. Cardiol.* **65:**297–302.

Barnes, P. J., and Karin, M., 1997, Nuclear factor-κB. A pivotal transcription factor in chronic inflammatory diseases, *N. Engl. J. Med.* **336:**1066–1071.

Beaver, J. P., and Waring, P., 1995, A decrease in intracellular glutathione concentration precedes the onset of apoptosis in murine thymocytes, *Eur. J. Cell Biol.* **68:**47–54.

Berliner, J. A., Navab, M., Fogelman, A. M., Frank, J. S., Demer, L. L., Edwards, P. A., Watson, A. D., and Lusis, A. J., 1995, Atherosclerosis: Basic mechanisms. Oxidation, inflammation, and genetics, *Circulation* **91:**2488–2496.

Beyaert, R., and Fiers, W., 1994, Molecular mechanisms of tumor necrosis factor-induced cytotoxicity. What we do understand and what we do not, *FEBS Lett.* **340:**9–16.

Bjorkerud, B., and Bjorkerud, S., 1996, Contrary effects of lightly and strongly oxidized LDL with potent promotion of growth versus apoptosis on arterial smooth muscle cells, macrophages, and fibroblasts, *Arterioscler. Thromb. Vasc. Biol.* **16:**416–424.

Camacho, M., Godessart, N., Anton, R., Garcia, M., and Vila, L., 1995, Interleukin-1 enhances the ability of cultured human umbilical vein endothelial cells to oxidize linoleic acid, *J. Biol. Chem.* **270:**17279–17286.

Chatelain, P., Laurel, R., and Gillard, M., 1985, Effect of amiodarone on membrane fluidity and Na+/K+ ATPase activity in rat-brain synaptic membranes, *Biochem. Biophys. Res. Commun.* **129:**148–154.

Collins, T., Read, M. A., Neish, A. S., Whitley, M. Z., Thanos, D., and Maniatis, T., 1995, Transcriptional regulation of endothelial cell adhesion molecules: NF-κB and cytokine-inducible enhancers, *FASEB J.* **9:**899–909.

De Caterina, R., and Libby, P., 1996, Control of endothelial leukocyte adhesion molecules by fatty acids, *Lipids* **31:**S57–S63.

Denke, M. A., and Grundy, S. M., 1992, Comparison of effects of lauric and palmitic acid on plasma lipids and lipoproteins, *Am. J. Clin. Nutr.* **56:**895–898.

Esterbauer, H., 1993, Cytotoxicity and genotoxicity of lipid-oxidation products, *Am. J. Clin. Nutr.* **57:**779S–786S.

Farach, M. C., Tang, J. P., Decke, G. L., and Carson, D. D., 1987, Heparin/heparan sulfate is involved in attachment and spreading of mouse embryos *in vitro*, *Dev. Biol.* **123:**401–410.

Folcik, V. A., and Cathcart, M. K., 1994, Predominance of esterified hydroperoxy-linoleic acid in human monocyte-oxidized LDL, *J. Lipid Res.* **35:**1570–1582.

Fransson, L. A., Carlstedt, I., Coster, L., and Malmstrom, A., 1983, Proteoheparan sulfate from human skin fibroblasts. Evidence for self-interaction via the heparan sulfate side chains, *J. Biol. Chem.* **258:**14342–14345.

Fujiwara, S., Wiedemann, H., Timpl, R., Lustig, A., and Engel, J., 1984, Structure and interactions of heparan sulfate proteoglycans from a mouse tumor basement membrane, *Eur. J. Biochem.* **143:**145–157.

Graber, R., Sumida, C., and Nunez, E.A., 1994, Fatty acids and cell signal transduction, *J. Lipid Med.* **9:**91–116.

Grimble, R. F., 1996, Interaction between nutrients, pro-inflammatory cytokines and inflammation, *Clin. Sci.* **91:**121–130.

Hassell, J. R., Kimura, J. H., and Hascall, V. C., 1986, Proteoglycan core protein families, *Annu. Rev. Biochem.* **55:**539–567.

Haudenshild, C. C., 1990, Pathogenesis of atherosclerosis: state of the art, *Cardiovasc. Drugs Ther.* **4:**993–1004.

Hayakawa, M., Ishida, N., Takeuchi, K., Shibamoto, S., Hori, T., Oku, N., Ito, F., and Tsujimoto, M. J., 1993, Arachidonic acid-selective cytosolic phospholipase A2 is crucial in the cytotoxic action of tumor necrosis factor, *J. Biol. Chem.* **268:**11290–11295.

Hegsted, D. M., McGandy, R. B., Myers, M. L., and Stare, F. J., 1965, Quantitative effects of dietary fat on serum cholesterol in man, *Am. J. Clin. Nutr.* **17:**281–295.

Hegsted, D. M., Ausman, L. M., Johnson, J. A., and Dallal, G. E., 1993, Dietary fat and serum lipids: An evaluation of the experimental data, *Am. J. Clin. Nutr.* **57:**875–883.

Hennig, B., Shasby, D. M., Fulton, A. B., and Spector, A. A., 1984, Exposure to free fatty acid increases the transfer of albumin across cultured endothelial monolayers, *Arteriosclerosis* **4:**489–497.

Hennig, B., Shasby, D. M., and Spector, A. A., 1985, Exposure to fatty acid increases human low density lipoprotein transfer across cultured endothelial monolayers, *Circ. Res.* **57:**776–780.

Hennig, B., Enoch, C., and Chow, C. K., 1986, Linoleic acid hydroperoxide increases the transfer of albumin across cultured endothelial monolayers, *Arch. Biochem. Biophys.* 248: 353–357.

Hennig, B., Enoch, C., and Chow, C. K., 1987, Protection by vitamin E against endothelial cell injury by linoleic acid hydroperoxides, *Nutr. Res.* **7:**1253–1260.

Hennig, B., Alvarado, A., Ramasamy, S., Boissonneault, G. A., Decker, E., and Means, W. J, 1990, Fatty acid induced disruption of endothelial barrier function in culture, *Biochem. Arch.* **6:**409–417.

Hennig, B., Ramasamy, S., Alvarado, A., Shantha, N. C., Boissonneault, G. A., Decker, E. A., and Watkins, B. A, 1993, Selective disruption of endothelial barrier function in culture by pure fatty acids and fatty acids derived from animal and plant fats, *J. Nutr.* **123:**1208–1216.

Hennig, B., Lipke, D. W., Boissonneault, G. A., and Ramasamy, S., 1995, Role of fatty acids and eicosanoids in modulating proteoglycan metabolism in endothelial cells, *Prostaglandins Leukotrienes Essential Fatty Acids* **53:**315–324.

Hennig, B., Toborek, M., Joshi-Barve, S., Barger, S. W., Barve, S., Mattson, M. P., and McClain, C. J., 1996a, Linoleic acid activates NF-κB and induces NF-κB-dependent transcription in cultured endothelial cells, *Am. J. Clin. Nutr.* **63:**322–328.

Hennig B., Toborek, M., McClain, C. J., and Diana, J. N., 1996b, Nutritional implications in vascular endothelial cell metabolism, *J. Am. Coll. Nutr.* **15:**345–358.

Hodgson, J. M., Wahlqvist, M. L., Boxall, J. A., and Balazs, N. D., 1993, Can linoleic acid contribute to coronary artery disease? *Am. J. Clin. Nutr.* **58:**228–234.

Janero, D. R., 1991, Therapeutic potential of vitamin E in the pathogenesis of spontaneous atherosclerosis, *Free Radic. Biol. Med.* **11:**129–144.

Jolles, P., ed., 1994, *Proteoglycans*, Birkhauser, Basel.

Khan, B. V., Parthasarathy, S. S., Alexander, R. W., and Medford, R. M., 1995, Modified low density lipoprotein and its constituents augment cytokine-activated vascular cell adhesion molecule-1 gene expression in human vascular endothelial cells, *J. Clin. Invest.* **95:**1262–1270.

Kim, D. N., Eastman, A., Baker, J. E., Mastrangelo, A., Sethi, S., Ross, J. S., Schmee, J., and Thomas, W. A., 1995, Fish oil, atherosclerosis, and thrombogenesis, *Ann. N. Y. Acad. Sci. USA* **748:**474–480.

Kleinman, H. K., McGarvey, M. L., Hassell, J. R., and Martin, G. R., 1983, Formation of a supramolecular complex is involved in the reconstitution of basement membrane components, *Biochemistry* **22:**4969–4974.

Koide, M., Kawahara, Y., Tsuda, T., Ishida, Y., Shii, K., and Yokoyama, M., 1992, Endothelin-1 stimulates tyrosine phosphorylation and the activities of two mitogen-activated protein kinases in cultured vascular muscle cells, *J. Hypertens.* **10:**1173–1182.

Kok, F. J., van Poppel, G., Melse, J., Verheul, E., Schouten, E. G., Kruyssen, D. H. C. M., and Hofman, A., 1991, Do antioxidants and polyunsaturated fatty acids have a combined association with coronary atherosclerosis? *Atherosclerosis* **31:**85–90.

Kühn, H., Belkner, J., Zaiss, S., Fahrenklemper, T., and Wohlfeil, S., 1994, Involvement of 15-lipoxygenase in early stages of atherogenesis, *J. Exp. Med.* **179:**1903–1911.

Lark, M. W., and Culp, L. A., 1984, Turnover of heparan sulfate proteoglycans from substratum adhesion sites of murine fibroblasts, *J. Biol. Chem.* **259:**212–217.

Louheranta, A. M., Porkkala-Sarataho, E. K., Nyyssonen, M. K., Salonen, R. M., and Salonen, J. T., 1996, Linoleic acid intake and susceptibility of very low density and low density lipoproteins to oxidation in men, *Am. J. Clin. Nutr.* **63:**698–703.

Loustarinen, R., Boberg, M., and Saldeen, T., 1993, Fatty acid composition in total phospholipids of human coronary arteries in sudden cardiac death, *Atherosclerosis* **99:**187–193.

Mattson, M. P., Barger, S. W., Begley, J. G., and Mark, R. J., 1995, Calcium, free radicals, and excitotoxic neuronal death in primary cell culture, *Meth. Cell. Biol.* **46:**187–215.

Maziere, C., Auclair, M., and Maziere, J. C., 1994, Tumor necrosis factor enhances low density lipoprotein oxidative modification by monocytes and endothelial cells, *FEBS Lett.* **338:**43–46.

McConkey, D. J., and Orrenius, S., 1996, The role of calcium in the regulation of apoptosis, *J. Leukocyte. Biol.* **59:**775–83.

Mensink, R. P., 1993, Effects of the individual saturated fatty acids on serum lipids and lipoprotein concentrations, *Am. J. Clin. Nutr.* **57:**711S–714S.

Munro, J. M., and Cotran, R. S., 1988, The pathogenesis of atherosclerosis: Atherogenesis and inflammation, *Lab. Invest.* **58:**249–261.

Ng, T. K. W., Hayes, K. C., DeWitt, G. F., Jegathesan, M., Satgunasingam, N., Ong, A. S. H., and Tan, D., 1992, Dietary palmitic and oleic acids exert similar effects on serum cholesterol and lipoprotein profiles in normocholesterolemic men and women, *J. Am. Coll. Nutr.* **11:**383–390.

Olgemoller, B., Schleicher, E. D., Schwaabe, S., Guretzki, H. J., and Gerbitz, K. D., 1990, High concentrations of low density lipoprotein decrease basement membrane-associated heparan sulfate proteoglycan in cultured endothelial cells, *FEBS Lett*, **264:**37–39.

Parhami, F., Fang, Z. T., Fogelman, A. M., Andalibi, A., Territo, M. C., and Berliner, J. A., 1993, Minimally modified low density lipoprotein-induced inflammatory responses in endothelial cells are mediated by cyclic adenosine monophosphate, *J. Clin. Invest.* **92:**471–478.

Pietsch, A., Weber, C., Goretzki, M., Weber, P. C., and Lorenz, R. L., 1995, N-3 but not N-6 fatty acids reduce the expression of the combined adhesion and scavenger receptor CD 36 in human monocyte cells, *Cell Biochem. Funct.* **13:**211–216.

Radhakrishnamurthy, B., Srinivasan, S. R., Vijayagopal, P., and Berenson, G. S., 1990, Arterial wall proteoglycans—Biological properties related to pathogenesis of atherosclerosis, *Eur. Heart J.* **11**(Suppl. E)**:**148–157.

Ramasamy, S., Boissonneault, G. A., Lipke, D. W., and Hennig, B., 1993, Proteoglycans and endothelial barrier function: Effect of linoleic acid exposure to porcine pulmonary artery endothelial cells, *Atherosclerosis* **103:**279–290.

Ramasamy, S., Lipke, D. W., McClain, C. J., and Hennig, B., 1995, Tumor necrosis factor reduces proteoglycan synthesis in cultured endothelial cells, *J. Cell Physiol.* **162:**119–126.

Rao, G. N., Alexander, R. W., and Runge, M. S., 1995, Linoleic acid and its metabolites, hydroperoxyoctadecadienoic acids, stimulate c-*Fos*, c-*Jun*, and c-*Myc* mRNA expression, mitogen-activated protein kinase activation, and growth in rat aortic smooth muscle cells, *J. Clin. Invest.* **96:**842–847.

Read, M. A., Neish, A. S., Luscinskas, F. W., Palombella, V. J., Maniatis, T., and Collins, T., 1995, The proteasome pathway is required for cytokine-induced endothelial-leukocyte adhesion molecule expression, *Immunity* **2:**493–506.
Renier, G., Olivier, M., Skamene, E., and Radzioch, D., 1994a, Induction of tumor necrosis factor alpha gene expression by lipoprotein lipase requires protein kinase C activation, *J. Lipid Res.* **35:**1413–1421.
Renier, G., Skamene, E., DeSanctis, J. B., and Radzioch, D., 1994b, Induction of tumor necrosis factor α gene expression by lipoprotein lipase, *J. Lipid Res.* **35:**271–278.
Ridker, P. M., Cushman, M., Stampfer, M. J., Tracy, R. P., and Hennekens, C. H., 1997, Inflammation, aspirin, and the risk of cardiovascular disease in apparently healthy men, *N. Engl. J. Med.* **336:**973–979.
Ross, R., 1986, The pathogenesis of atherosclerosis—An update, *N. Engl. J. Med.* **314:**488–500.
Ruoslahti, E., 1988, Structure and biology of proteoglycans, *Annu. Rev. Cell Biol.* **4:**229–255.
Rus, H. G., Niculescu, F., and Vlaicu, R., 1991, Tumor necrosis factor-alpha in human arterial wall with atherosclerosis, *Atherosclerosis* **89:**247–254.
Sandstrom, P. A., Pardi, D., Tebbey, P. W., Dudek, R. W., Terrian, D. M., Folks, T. M., and Buttke, T. M., 1995, Lipid hydroperoxide-induced apoptosis: Lack of inhibition by Bcl-2 overexpression, *FEBS Lett.* **365:**66–70.
Sato, N., Iwata, S., Nakamura, K., Hori, T., Mori, K., and Yodoi, J., 1995, Thiol-mediated redox regulation of apoptosis. Possible roles of cellular thiols other than glutathione in T cell apoptosis, *J. Immunol.* **154:**3194–3203.
Saunders, S., and Bernfield, M., 1987, Cell surface proteoglycan binds mouse mammary epithelial cells to fibronectin and behaves as an interstitial matrix receptor, *J. Cell Biol.* **106:**423–430.
Shakhov, A. N., Collart, M. A., Vassalli, P., Nedospasow, S. A., and Jongeneel, C. V., 1990, κB-Type enhancers are involved in lipopolysaccharide-mediated transcriptional activation of the tumor necrosis factor α gene in primary macrophages, *J. Exp. Med.* **171:**35–47.
Schreck, R., Albermann, K., and Baeuerle, P. A., 1992, Nuclear factor κB: An oxidative stress-responsive transcription factor of eukaryotic cells (A review), *Free Radic. Res. Commun.* **17:**221–237.
Schulze-Osthoff, K., Bakker, A. C., Vanhaesebroeck, B., Beyaert, R., Jacob, W. A., and Fries, W., 1992, Cytotoxic activity of tumor necrosis factor is mediated by early damage of mitochondrial functions. Evidence for the involvement of mitochondrial radical generation, *J. Biol. Chem.* **267:**5317–5323.
Shirakawa, F., and Mizel, S. B., 1989, *In vitro* activation and nuclear translocation of NF-kappa B catalyzed by cyclic AMP-dependent protein kinase and protein kinase C, *Mol. Cell. Biol.* **9:**2424–2430.
Siebenlist, U., Franzoso, G., and Brown, K., 1994, Structure, regulation and function of NF-κB, *Annu. Rev. Cell Biol.* **10:**405–455.
Siflinger-Birnboim, A., Schnitzer, J., Lum, H., Blumenstock, F. A., Shen, C.-P. J., Del Vecchio, P. J., and Malik, A. E., 1991, Lectin binding to gp60 decreases specific albumin binding and transport in pulmonary artery endothelial monolayers, *J. Cell Physiol.* **149:**575–584.
Simionescu, M., and Simionescu, N., 1993, Proatherosclerotic events: Pathobiochemical changes occurring in the arterial wall before monocyte migration, *FASEB J.* **7:**1359–1366.
Slater, A. F., Nobel, C. S., and Orrenius, S., 1995, The role of intracellular oxidants in apoptosis, *Biochem. Biophys. Acta* **1271:**59–62.
Steinberg, D., and Witztum, J. L., 1990, Lipoproteins and atherosclerosis, *JAMA* **264:**3047–3052.

Tanaka, T., Hidaka, T., Ogura, R., and Sugiyama, M., 1988, Changes of membrane fluidity and Na+,K+-ATPase activity during cellular differentiation in the guinea pig epidermis, *Arch. Dermatol. Res.* **280:**29–32.

Tanganelli, P., Bianciardi, G., Carducci, A., Palummo, N., Simoes, C., Tarabochia, B., Weber, G., Verzuri, M. S., and Auteri, A., 1992, Updating on *in-vivo* and *in-vitro* effects of heparin and other glycosaminoglycans (mesoglycan) on arterial endothelium: A morphometrical study, *Int. J. Tiss. React.* **14:**149–153.

Tartaglia, L. A., Rothe, M., Hu, Y.-F., and Goedde, D. V., 1993, Tumor necrosis factor's cytotoxic activity is signaled by the p55 TNF receptor, *Cell* **73:**213–216.

Templeton, D. M., 1992, Proteoglycans in cell regulation, *Crit. Rev. Clin. Lab. Sci.* **29:**141–184.

Thanos, D., and Maniatis, T., 1995, NF-ϰB: A lesson in family values, *Cell* **80:**529–532.

Toborek, M., and Hennig, B., 1994, Fatty acid-mediated effects on the glutathione redox cycle in cultured endothelial cells, *Am. J. Clin. Nutr.* **59:**60–65.

Toborek, M., Barger, S. W., Mattson, M. P., Barve, S., McClain, C. J., and Hennig, B., 1996, Linoleic acid and TNF-α cross-amplify oxidative injury and dysfunction of endothelial cells, *J. Lipid Res.* **37:**123–135.

Toborek, M., Kaiser, S., Mattson, M. P., and Hennig, B., 1997a, Fatty acid-mediated injury and endothelial cell autocrine TNF production, *Atherosclerosis* **134:**284.

Toborek, M., Blanc, E. M., Kaiser, S., Mattson, M. P., and Hennig, B., 1997b Linoleic acid potentiates TNF-mediated oxidative stress, disruption of calcium homeostasis and apoptosis of cultured vascular endothelial cells, *J. Lipid Res.* **38:**2155–2147.

Tracey, K. J., and Cerami, A., 1992, Tumor necrosis factor and regulation of metabolism in infection: Role of systemic versus tissue levels, *Proc. Soc. Exp. Biol. Med.* **200:**233–239.

Vaddi, K., Nicolini, F. A., Mehta, P., and Mehta, J. L., 1994, Increased secretion of tumor necrosis factor-α and interferon-γ by mononuclear leukocytes in patients with ischemic heart disease. Relevance in superoxide anion generation, *Circulation* **90:**694–699.

Vijayagopal, P., Srinivasan, S. R., Dalferes, E. R., Radhakrishnamurthy, B., and Berenson, G. S., 1988, Effect of low density lipoproteins on the synthesis and secretion of proteoglycans by human endothelial cells in culture, *Biochem. J.* **255:**639–646.

Yam, D., Eliraz, A., and Berry, E. M., 1996, Diet and disease, the Israeli paradox: Possible dangers of a high omega-6 polyunsaturated fatty acid diet, *Isr. J. Med. Sci.* **32:**1134–1143.

Ylä-Herttuala, S., Rosenfed, M. E., Parthasarathy, S., Glass, C. K., Sigal, E., Witztum, J. T., and Steinberg, D., 1990, Colocalization of 15-Lipoxygenase mRNA and protein with epitopes of oxidized low density lipoprotein in macrophage-rich areas of atherosclerotic lesions, *Proc. Natl. Acad. Sci. USA* **87:**6959–6963.

Ylä-Herttuala, S., Rosenfed, M. E., Parthasarathy, S., Sigal, E., Särkioia, T., Witztum, J. T., and Steinberg, D., 1991, Gene expression in macrophage-rich human atherosclerotic lesions. 15-Lipoxygenase and acetyl low density lipoprotein receptor messenger RNA colocalize with oxidation specific lipid–protein adducts, *J. Clin. Invest.* **87:**1146–1152.

Wang, T., and Powell, W. S., 1991, Increased levels of monohydroxy metabolites of arachidonic acid and linoleic acid in LDL and aorta from atherosclerotic rabbits, *Biochem. Biophys. Res. Commun.* **1084:**129–138.

Wardle, E. N., 1994, Vascular permeability in diabetics and implications for therapy, *Diabetes Res. Clin, Pract.* **23:**135–139.

Watkins, B. A., Hennig, B., and Toborek, M., 1996, Diet and health, in *Bailey's Industrial Oil and Fat Products*, Vol. 1, *Edible Oil and Fat Products: General Applications* 5th ed. (Y. H. Hui, ed.), pp. 159–214, Wiley, New York.

Wight, T. N, 1989, Cell biology of arterial proteoglycans, *Arteriosclerosis* **9:**1–20.

Wight, N., Kinsella, M. G., and Potter-Perigo, S., 1985, Proteoglycans synthesized and secreted by cultured vascular cells, in *Extracellular Matrix: Structure and Function* (A. Riddi ed.), pp. 321–332, Alan R. Liss, New York.

Wong, G. H., 1995, Protective roles of cytokines against radiation: Induction of mitochondrial MnSOD, *Biochim. Biophys. Acta* **1271:**205–209.

Chapter 18

Lipidic Signaling and Peroxisomal Function

Colin Masters

1. INTRODUCTION

While the important role of lipids in relation to cellular signaling is firmly established and in general well reviewed (Exton, 1994; Barrett, 1992), several novel aspects of the subject have only recently come to light, which involve peroxisomal function, and which appear deserving of further commentary.

Part of this reassessment of the significance of peroxisomal input to cellular signaling relates to the parallel reassessment of the metabolic role of the peroxisome in recent years. Far from the early perception that this organelles's activities were limited to minor substrates (De Duve and Baudhuin, 1966), it is now recognized that the peroxisome is a vital organelle implicated in a broad spectrum of metabolic activities and especially prominent in the area of lipid metabolism (Masters, 1996a; Masters and Crane, 1996). Again, it is now recognized that the peroxisome possesses a number of unique signaling characteristics which exert a significant influence on regulatory signaling, within and without the area of direct peroxi-

Colin Masters The School of Biomolecular and Biomedical Science, Griffith University, Brisbane QLD 4111, Australia.

Subcellular Biochemistry, Volume 30: Fat-Soluble Vitamins, edited by Quinn and Kagan. Plenum Press, New York, 1998.

somal metabolism and at subcellular, intracellular, and intraorganellar levels (Masters, 1996b).

At a dietary level, too, it may be recalled that while historically, linoleic acid (ω-6) was long considered the main essential fatty acid (EFA), more recently, the ω-3 polyunsaturated fatty acids derived from α-linolenic acid have become the focus for much biomedical research because of their wide-ranging clinical significance (Hansen, 1994; Budowski, 1988), and the distinctive role of the peroxisome in the regulation of metabolism in this area has become recognized (Masters, 1996a).

This review is concerned with recent advances in lipidic signaling in all of the above areas.

2. LIPID METABOLISM AND THE PEROXISOME

The first of the recent developments that needs to be covered in this chapter is the extent of peroxisomal involvement in cellular lipid metabolism.

It has been known for some time, for example, that the β-oxidation of the common long-chain fatty acids may take place in the peroxisomes as well as in mitochondria (De Duve, 1983). There are several points of difference between the mechanisms of β-oxidation in peroxisomes and mitochondria, and for the details of these differences, the reader is referred to more specialized texts (Masters and Crane, 1995a). For the purposes of this chapter, it is necessary to highlight the fact that in the peroxisome the fatty acid is dehydrogenated to an enoyl CoA by an FAD-requiring fatty acyl–CoA oxidase with the concomitant production of hydrogen peroxide:

$$\text{Fatty Acyl–CoA} + O_2 \rightarrow \Delta^2\text{-Enoyl-Acyl–CoA} + H_2O_2$$

Other differences in the mechanisms which are of significance in the present context relate to the fact that peroxisomal β-oxidation is capable of acting on a much wider range of substrates than mitochondrial β-oxidation, and generally does not proceed to completion. Rather, peroxisomal β-oxidation is mainly concerned with the chain shortening of long-chain fatty acids (Masters and Crane, 1995a, 1984). These differences hold substantial implications in relation to signaling and the regulation of lipid metabolism.

As regards the aspect of substrate specificity, it may be noted that the peroxisome generally shows a marked preference for long-chain fatty acids, with short-chain fatty acids being oxidized little if at all. There is also a marked preference of the peroxisomal system toward the oxidation of unsaturated fatty acids. Indeed, the peroxisomes appear to be the primary site of oxidation of most polyunsaturated fatty acids (Osmundsen *et al.*, 1991). Hence, much of the metabolism of the fatty acid components of complex lipids may be attributable to the peroxisomes rather than the commonly assumed mitochondria.

In addition to this role of chain shortening and catabolism of common fatty acids, the peroxisome is also involved in a broad range of oxidative processes with other lipid substances. The peroxisome is now recognized as the major site of oxidation of very long chain fatty acids, for example, and is also involved in the oxidation of dicarboxylic acids, bile acids, and a number of lipid xenobiotics (Masters and Crane, 1996). It has also been established that the peroxisomes play a major role in the metabolism of leukotrienes and prostaglandins, as well as the biosynthesis of ether lipids, cholesterol, and dolichol (Masters and Crane, 1995a; Table I). Furthermore, studies of the mechanism of peroxisome proliferation have demonstrated an intimate relationship between this process, lipid metabolism, and the cellular signaling pathways (Masters, 1996b; Bieri, 1993).

In summary, then, and as background to the topic of the involvement of the peroxisome with lipidic messengers, it is important to appreciate that it is now established that this organelle has a major, wide-ranging involvement in lipid metabolism, and that this role flows into an associated involvement with the signaling functions of lipids in the cell.

Table I
Peroxisomal Involvement in Lipid Metabolism

Oxidation of long-chain fatty acids (saturated and unsaturated)
Formation of β-hydroxy fatty acyl–CoA
Dehydrogenation of 3-hydroxy fatty acyl–CoA
Thiolysis of 3-oxo-fatty acyl–CoA
Transfer of acyl groups to and from carnitine
Dehydrogenation of glycerol-3-phosphate
Oxidation of very long chain fatty acids, long-chain dicarboxylic acids, and branched–chain fatty acids
Oxidation of prostaglandins, leukotrienes, eicosanoids, and xenobiotics
Synthesis of *n*-3 and *n*-6 polyunsaturated fatty acids
Synthesis of long-chain fatty alcohols and ether glycerolipids
Synthesis of cholesterol, dolichol, isoprenoids, and bile acids
Remodeling of membrane lipids
Contributions to lipid homeostasis

3. UNIQUE SIGNALING FEATURES OF THE PEROXISOME

With its broad involvement in lipid metabolism, the peroxisome possesses a multifaceted potential for regulation, and one of the most striking regulatory features of this organelle is the capacity to vary its numbers remarkably in response to a wide variety of chemicals, including many lipids (Masters and Crane, 1995a; Reddy *et al.*, 1988). This unique property clearly affords an extraordinary ability to correlate peroxisomal flux with its many roles in cellular metabolism, and there is logic in beginning this discussion of peroxisomal involvement with cellular signaling by consideration of this particular characteristic.

To take a typical example of peroxisomal proliferation, it has been demonstrated that the substance clofibrate (ethyl *p*-chlorophenoxyisobutyrate) when fed to mice or rats in appropriate doses may cause the peroxisomal volume in hepatocytes to increase by a factor of ten or more (Reddy *et al.*, 1988). More generally, a wide range of chemicals have been shown to act in a similar manner as peroxisome proliferators (Masters and Crane, 1995a).

As to the mechanism of this singular phenomenon, attention has been drawn in recent years to the close relationship with that of steroid hormone/thyroid hormone/vitamin D receptor superfamily system (Green, 1992). In each case, a signaling substance enters the cell, and binds to a receptor in the cytoplasm, with the resultant complex moving across the nuclear membrane, binding to the nuclear DNA, and giving rise to transcriptional enhancement (Masters and Crane, 1995a; Fig. 1). With both systems, then, signals from outside the cell (agonists) are able to influence the content of individual proteins in the cell by causing an increase in their synthesis. In the case of peroxisome proliferation, for example, the content of acyl CoA oxidase in the hepatocytes may be increased 10–20 times, with consequent significant effects on fatty acid oxidation and signaling.

In the broader perspective, it is noteworthy that there are strong structural similarities between the peroxisome proliferator receptor and the receptors for steroid hormones, glucocorticoids, vitamin D, and retinoids. All of these receptors recognize the same response sequence in the nuclear DNA (AGGTCA) (Maksymowych *et al.*, 1992). As regards function, the peroxisome proliferator-activated receptors (PPARs) are now recognized as transcription factors that regulate gene expression of several enzymes associated with lipid homeostasis (Motojima, 1993). Three types of PPAR have been identified (α, β/δ, γ), and the functions of PPARs have been shown mainly to degrade long-chain fatty acids and detoxify xenobiotics (Von Schacky *et al.*, 1993; Lemberger *et al.*, 1996)

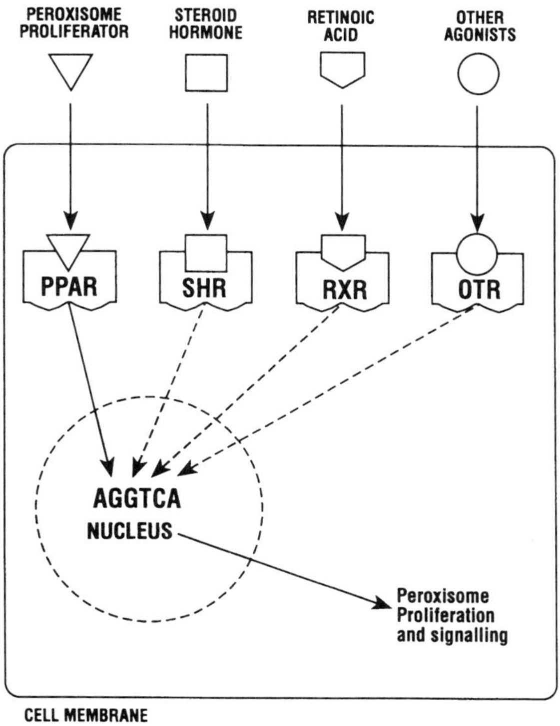

FIGURE 1. A representation of the relationships between the peroxisome proliferator-activated receptor (PPAR) and receptors of the steroid hormone superfamily. All of these receptors recognize the same response sequence (AGGTCA) in the nuclear DNA. SHR, Steroid hormone receptor; RXR, retinoic acid receptor; OTR, other receptors for superfamily members.

Overall, then, the facts that the peroxisome proliferator mechanism shows such a close similarity to that of the wide-ranging steroid hormone/thyroid hormone/vitamin D receptor superfamily, and has been maintained as a regulatory package throughout mammalian evolution, should in themselves stress the considerable significance of the peroxisomal contribution to cellular signaling.

4. FURTHER REGULATORY SIGNALS

Peroxisome proliferators have been shown to interact with many of the major elements in the cellular signaling matrix (e.g., calcium levels, sodium–hydrogen ion exchange, protein kinase C, and protooncogenes), and the involvement of such multilevel control is apparently necessary as an integratory framework for the common pleiotrophic effects involved in peroxisome proliferation (Bieri, 1993). In addition to these general signaling relationships, though, there are a number of special relationships between peroxisomal functioning and cellular signaling, and prominent among these may be listed the lipidic messengers associated with the hydrolysis of phosphatidylcholine, some of the essential fatty acids, and messengers associated with the metabolism of hydrogen peroxide.

All of these messenger relationships are described in the subsequent pages of this review, but first the different possibilities of phosphatidylcholine hydrolysis should be clarified. It has now become evident that there is a special relationship between two aspects of the hydrolysis of this phospholipid and peroxisomal function.

The first of these to be considered is that catalyzed by phospholipase A_2 and leading typically to the formation of lysophosphatidylcholine and an unsaturated fatty acid, such as arachidonic acid. The extent of this hydrolysis has been shown to be increased during peroxisome proliferation, hence establishing a link between peroxisomal function and the signaling possibilities of these hydrolysis products (Crane and Masters, 1996b).

Another special relationship between peroxisomal function and the hydrolysis of phosphatidylcholine occurs at the level of the hydrolysis catalyzed by phospholipase D, and leading to the operation of the phosphatidylcholine cycle (Fig. 2).

These two separate hydrolytic possibilities for phosphatidylcholine and their relationships to peroxisomal function and cellular signaling are considered next.

4.1. Phospholipase A_2 Hydrolysis

As indicated previously, the hydrolysis of phosphatidyl choline by phospholipase A_2 produces lysophosphatidic acid and an unsaturated fatty acid such as arachidonic acid. Considerable signaling significance is possessed by each of these molecules.

In the context of the peroxisomal membrane, for example, it has been demonstrated that a central effect of peroxisome proliferators is to cause a significant increase in the percentage content of lysophosphatidylcholine (Crane *et al.*, 1990; Crane and Masters, 1986). Because of the molecular structure and shape characteristics of this compound, this has the effect of causing marked changes in the permeability characteristics of the peroxisomal membrane. Such effects on target membranes can, of course, be exceedingly important in cell physiology (Barritt, 1992). At another level, too, lysophosphatidylcholine is known to be particularly effective in enhancing

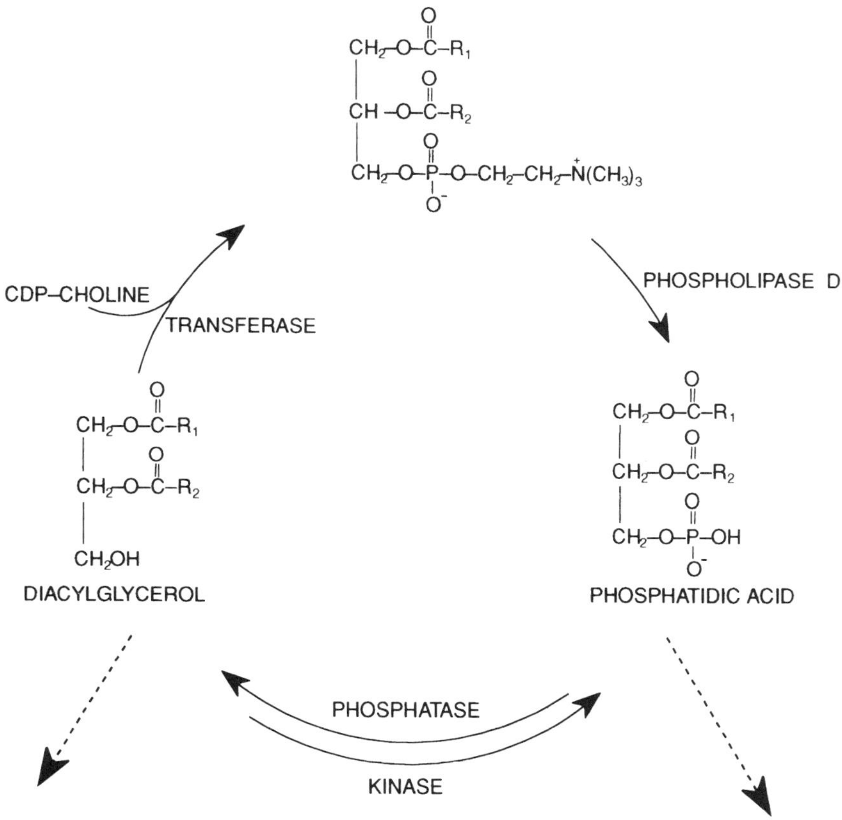

FIGURE 2. The phosphatidylcholine cycle, and the enzymic reactions therein; dashed arrows represent associated signaling pathways.

and sustaining the action of protein kinase C in such long-term physiological responses as cell-to-cell communication and proliferation (Asaoka *et al.*, 1992).

Additionally, lysophosphatidylcholine may be further hydrolyzed to lysophosphatidic acid, which has been demonstrated also to be involved in signaling at the level of adenylate cyclase action, calcium ion elevation, phosphatidylinositol hydrolysis, and the activation of phospholipase D (Bieri, 1993). Lysophosphatidylcholine also plays an essential role in the inhibition of platelet aggregation by secretory phospholipase A_2 (Yuan *et al.*, 1995)

Another important signaling correlation of phospholipase A_2 is associated with the common release of arachidonic acid. This fatty acid, of course, is of primary importance in the production of prostaglandins, leukotrienes, and other eicosanoids. In addition, arachidonic acid is known to activate adenyl cyclase, protein kinase C, and phospholipase C, as well as causing an increased release of calcium ions (Barritt, 1992).

It may be concluded, then, that the hydrolysis of phosphatidylcholine by phospholipase A_2 is a reaction which is of enormous versatility in relation to cellular signaling and one which is closely associated with peroxisomal function.

4.2. The Phosphatidylcholine Cycle

The phosphatidylcholine cycle is well established as a very important element in cellular signaling (Exton, 1994). It forms a major physiological alternative to the phosphatidylinositol cycle, and possesses many advantageous features (e.g., sustainability and enzymic simplicity) over that system. In the case of the phospholipase D hydrolysis of phosphatidylcholine, the major regulators of activity are the phosphokinase C systems, G-proteins, sphingosine, and several receptor systems (Barritt, 1992).

Bringing this system within the realm of especial peroxisomal influence are the recent demonstrations that lysophosphatidic acid and hydrogen peroxide stimulate phospholipase D activity (Kiss and Anderson, 1994). The connection between peroxisomal function and lysophosphatidic acid has been outlined in a previous section of this review; and in relation to hydrogen peroxide, the peroxisome is of course the major organelle involved in hydrogen peroxide metabolism within the cell (Masters and Crane, 1995a).

Phosphatidic acid, the primary product of phospholipase D action, acts as a signaling molecule at several levels. It is known to be a potent mitogen, for example, and has been reported to interact at the levels of adenyl cyclase activity, protein kinases, and actin polymerization (Exton, 1994; Tronchere *et al.*, 1994).

Another level of signaling involvement associated with the phospholipase D system occurs when phosphatidic acid is further hydrolyzed to diacylglyceride, which is an important messenger in its own right (Barritt, 1992). Diacylglycerides induce a prolonged activation of protein kinase C, and in this way are involved in several long-term actions of growth factors and other important agonists. Generally, an increase in the concentration of diacylglycerol in the plasma membrane enhances the association between protein kinase C and the membrane, activates the enzyme, and causes the phosphorylation of target proteins such as membrane receptors, cytoskeletal proteins, regulatory enzymes, and DNA-binding proteins (Barritt, 1992).

As may be seen from the above relationships, then, there is a strong connection between peroxisomal function and the widespread cellular effects of the phosphatidylcholine signaling system.

5. ESSENTIAL FATTY ACIDS

It is well known that the essential fatty acids make a significant contribution to cellular signaling in mammals, but the peroxisomal involvement in these regulatory processes is not so widely recognized.

Much of the physiological importance of the essential fatty acids, of course, relates to their roles as precursors of the eicosanoids. Linoleic acid, for example, may be converted to prostaglandins, thromboxanes, and prostacyclins of the 1 and 2 series, and leukotrienes of the 4 series, while fatty acids of the ω-3 (e.g., α-linolenic acid) series may be transformed into prostaglandins of the 3 series and leukotrienes of the 5 series (Hansen, 1994; Budowski, 1988; Fig. 3).

In relation to the production of these important physiological mediators, too, it is important to realize that there is often competition between ω-6 and ω-3 fatty acids for some of the same enzymes involved in the eicosanoid transformations (Masters, 1996a).

While the eicosanoids, then, are fatty acid derivatives which act as potent short-range messengers on many body functions, it needs to be realized that the families of eicosanoids which derive from linoleic and α-linolenic acids induce markedly different physiological effects and these differences are further intensified by the involvement of peroxisomal metabolism (Masters, 1996a).

As one point, and as previously noted, recent studies have shown that the cellular oxidation of polyunsaturated fatty acids is generally initiated in the peroxisome, and that the peroxisomal contribution to their oxidation may be considerably greater than that of the mitochondria (Osmundsen *et al.*, 1991).

For another point, it has also been shown recently that the peroxisome exhibits a specific involvement in the synthesis of certain polyunsaturated fatty acids which are critical to the balance of the signaling processes derived from the essential fatty acids. Docosahexanoic acid (c22:6, *n*-3), for example, and its *n*-6 analogue are now recognised as having their formation dependent on the proper function of the peroxisome (Masters, 1966a).

Another interesting signal connection that has been established recently in this area is that between leukotriene-B4 (LTB4)—a potent chemotactic agent that initiates and coordinates the inflammatory response—and the α-form of the peroxisome proliferator activated receptor (PPAR-α)

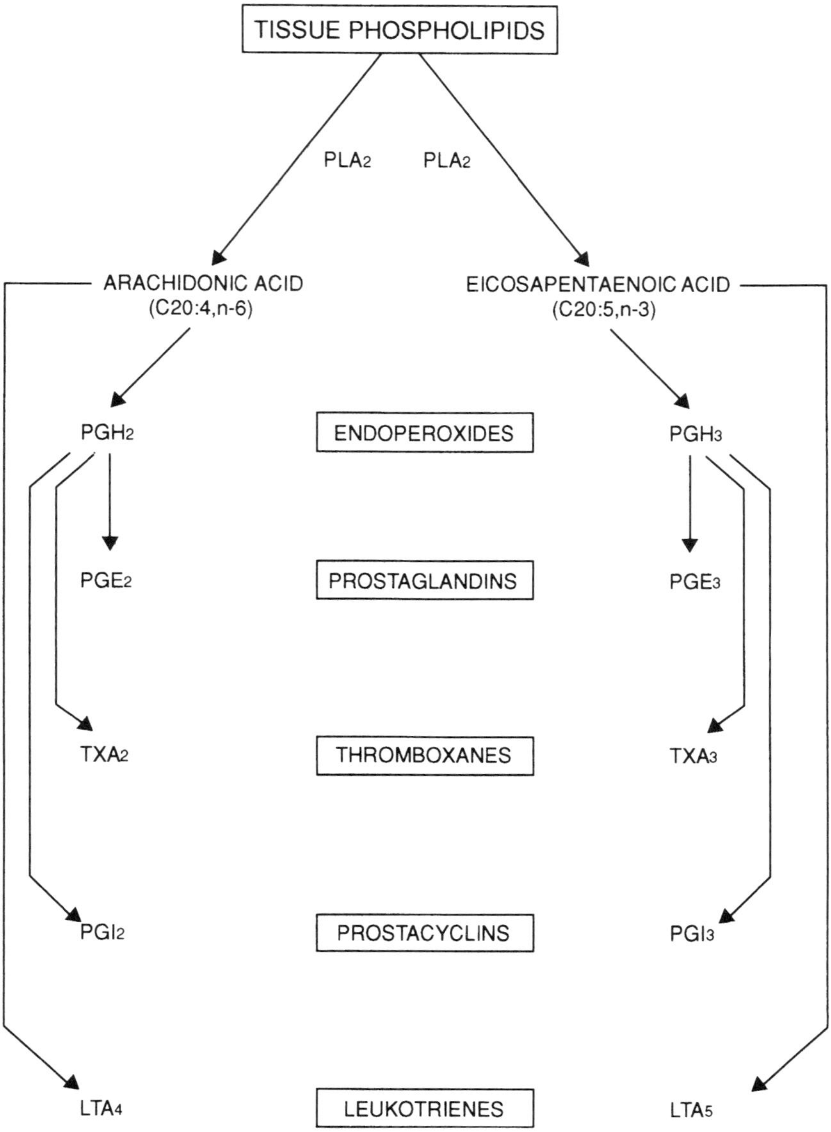

FIGURE 3. An outline of the major differences between the synthesis of eicosanoid families from arachidonic and eicosapentanoic acids. PLA_2, Phospholipase A_2.

(Devchand, 1966). In many inflammatory disease states, LTB4 induces a complex cascade of cellular events that ultimately recruit cells from the immune system to the site of injury and produce an inflammation. As to PPAR-α, as has already been mentioned, this is one of a group of transcription factors that regulate the gene expression of a wide range of enzymes associated with lipid homeostasis, so the establishment of LTB4 as an activating ligand for PPAR-α opens up many new opportunities for signaling between bioactive lipids and gene expression.

Clearly, then, the peroxisomes are intimately involved in the metabolism of the essential fatty acids and in the balance of production of individual polyunsaturated fatty acids and their derivative eicosanoids; in this way they significantly influence the nature of the widespread and varied cellular signaling of these compounds.

6. INVOLVEMENT OF OXYGEN FREE RADICALS AND THE PEROXISOME IN CELLULAR SIGNALING

Another major aspect of the peroxisome in cellular signaling, and one which is intimately integrated with those aspects of lipidic signaling which have been discussed in the previous pages, is the position of this organelle as the major pathway for the removal of oxygen free radicals in the cell (Masters and Crane, 1995b). Oxygen free radicals are well recognized as playing an important role in many aspects of long-term physiological regulation such as aging and degenerative disease (Ames *et al.*, 1993; Chance *et al.*, 1979); they have also been designated as second messengers in the short term, with the oxygen radical-mediated modulation of biomolecules underlying a general mechanism for regulating signal transduction pathways (Masters and Crane, 1995b; Stadtman, 1990).

Oxygen radical modulation of critical biomolecules may occur at many levels. In case of lipids, free radical attack is generally directed at the polyunsaturated fatty acid components of membranes, exerting a marked influence on membrane stability and function, and leading to serious regulatory dysfunction within cells and tissues (Masters and Crane, 1995b; Barritt, 1992). Oxygen free radicals may also cause the oxidation of amino acid residues in proteins, and in so doing lead to a substantial alteration of their normal physiological functions (Stadtman, 1990). As one regulatory protein which is sensitive to the action of free radicals, for example, protein kinase C may be cited, along with its involvement in such diverse cellular processes as receptor function, ion transport, movement and shape changes, metabolism, and DNA transcription (Masters and Crane, 1995b).

Overall, then, it should be clear that the millions of free radicals produced every second by most eukaryotic cells are capable of exerting a considerable influence on the cellular signaling systems, and the peroxisome is a major factor in modifying and ameliorating such influences.

Another point in regard to hydrogen peroxide metabolism is that the peroxisome is the main site of H_2O_2 production in the cell, and that hydrogen peroxide is known to stimulate phospholipase D. As already mentioned, phospholipase D has an important signaling function in relation to many significant cellular activities. Thus the activation of this enzyme may be considered to extend the involvement of the peroxisome and hydrogen peroxide metabolism even further into the cellular regulatory processes (Fig. 4).

7. CONCLUDING COMMENTS

In the overall context of signaling, it should be mentioned that there is much evidence in the literature to indicate that individual signaling pathways do not operate in isolation, but rather are intimately connected with other signaling systems to form a complex regulatory network (Barritt, 1992). Appreciation of this fact, along with the broad involvement of peroxisome signaling that has been outlined in this review, reinforce the conclusion that the signaling roles of this organelle should not be relegated to a peripheral designation, but instead deserve recognition as a major element in cellular signaling.

It is of interest, too, in the present context to note some of the subtle complementarities of some of the peroxisomal involvements in cellular signaling. Considering the increased hydrolysis of phosphatidylcholine in the peroxisomal membrane during proliferation, for example, there is evidence of a beneficial association of the diverse effects of this activity. As well as responding to the potential vulnerability of the nascent membranes to oxidative attack by facilitating the removal of oxidized fatty acid residues via phospholipase A_2 hydrolysis (Van Kuijk *et al.*, 1987), the same mechanism alters the permeability of the peroxisomal membrane and facilitates the release of peroxisomal catalase into the cytoplasm (Crane *et al.*, 1990; Crane and Masters, 1986). This latter aspect then reinforces and complements the first aspect, by supplying an additional defense against oxidative attack to the membranes from oxygen free radicals in the cytoplasm.

Again, while most of the previous connections between lipidic signaling and peroxisomal metabolism have related to catabolic aspects of this organellar function, it should also be emphasized that the unique anabolic capabilities of this organelle may often bear significantly on cellular signal-

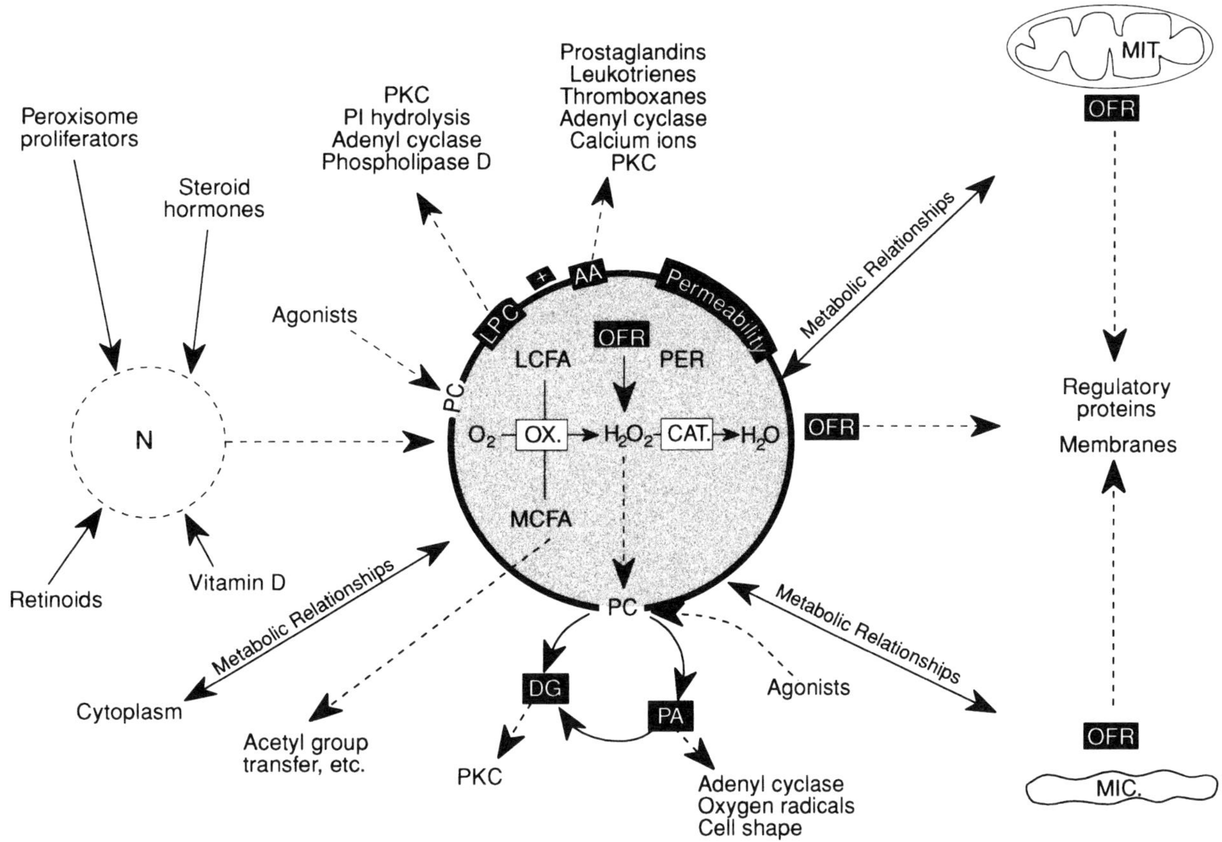

FIGURE 4. A schematic representation of some of the major lipidic signaling roles of peroxisome. Dashed arrows indicate signaling relationships. Solid arrows indicate metabolic relationships. PER, Peroxisomes; N, nucleus; MIT, mitochondria; MIC, microsomes; PKC, protein kinase C; PI, phosphatidylinositol; LCFA, long-chain fatty acids; MCFA, medium-chain fatty acids; OX, peroxisomal oxidases; CAT, catalase; OFR, oxygen free radicals; PA, phosphatidic acid.

ing. The most potent lipidic mediator, for example, namely platelet activating factor, is by structure a plasmalogen and a basal product of peroxisomal synthesis (Masters and Crane, 1995a). Platelet activating factor is, of course, involved as a regulator in many important pathophysiological situations such as inflammation, the allergic response, and reproduction (Lehninger *et al.*, 1993).

It would seem, too, that a comment on the putative relationships between peroxisomal metabolism and the fat-soluble vitamins would be in order in the context of this book. First, it may be noted that the efficient absorption of the fat-soluble vitamins (A, E, K, D) is dependent on an adequate bile secretion, and that peroxisomal oxidation plays an important role in bile acid metabolism (Masters and Crane, 1996). The initial series of reactions in bile acid formation, where cholesterol is converted into di- and trihydroxycoprostanoic acids, for example, has been shown to be localized in the peroxisomes rather than in mitochondria, as previously thought (Pederson, 1993). In addition, it has already been mentioned that the peroxisome is the major organelle involved in the removal of oxygen free radicals—a function which overlaps with the antioxidant function of vitamin E—and it should also be brought to attention that the peroxisome displays a significant role in the metabolism of cholesterol (hence influencing vitamin D formation) and isoprenoid compounds (with putative influence on the biosynthesis of vitamins A, E, and K) (Masters and Crane, 1995a).

In the general context of signaling, another major aspect of peroxisomal metabolism should be mentioned—that of the detoxification of xenobiotics (Masters and Crane, 1998). While this topic is too broad to be satisfactorily covered in this chapter, some of the more relevant aspects may be indicated. For one thing, many of the xenobiotics which are detoxified by the peroxisomal mechanisms are often hydrophobic in character and catabolized by pathways which have already been mentioned here. For another thing, signaling is generally involved, both at the level of the initiation of metabolism and the level of the type of metabolism. To take one example, the industrial plasticizer di(2-ethylhexyl) phthalate (DEHP) is a common xenobiotic detoxified by the peroxisomal pathway. Ingestion of DEHP signals a need for peroxisomal proliferation in the first instance, with a subsequent signal from the peroxisome to the microsome indicating an associated requirement for ω-hydroxylase activity via a subfamily of the cytochrome P450 system. This subfamily is specifically induced by peroxisome proliferators and has a substrate specificity which is mainly directed toward the ω-hydroxylation of fatty acids, prostaglandins, and xenobiotics (Gibson, 1989; Black and Coon, 1987). By this combination and integration of pathways, then, a sequential action of ω- and β-oxidation is allowed, which provides a very efficient overall pathway for the detoxification of aliphatic constituents. The process of β-oxidation has been described in previous pages as directed toward the chain shortening of long-chain, monocarboxylic fatty acids; with ω-oxidation, long-chain fatty acids (or structures containing these elements) can be converted to short-chain dicarboxylic acids in the peroxisome. Both of these oxidation processes with their associated chain shortening decrease the hydrophobicity of the original xenobiotic element and facilitate its excretion.

Overall, then, the role of the peroxisome in lipidic signaling should be viewed as being varied in character and extensive in nature. As well as providing a unique imprint on the signaling network of the cell, this organelle should be regarded as a major contributor to several aspects of these communication relationships, and as such deserving of much greater emphasis in future investigations into the regulatory processes in cells and tissues.

8. REFERENCES

Ames, B. N., Shigenaga, M. K., and Hagen, T. M., 1993, Oxidants, antioxidants and the degenerative diseases of ageing, *Proc. Natl. Acad. Sci. USA* **90:**7915–7922.

Asaoka, Y., Nakamura, S., Yoshida, K., and Nishizuka, Y., 1992, Protein kinase C, calcium and phospholipid degradation, *Trends Biochem. Sci.* **17:**414–448.

Barritt, G. J., 1992, *Communication within Animal Cells*, Oxford University Press, Oxford.

Bieri, F., 1993, Peroxisome proliferators and cellular signalling pathways. A review, *Biol. Cell* **77:**43–46.

Black, S. D., and Coon, M. J., 1987, P450 cytochromes: Structure and function, *Adv. Enzymol.* **60:**35–87.

Budowski, P., 1988, Omega-3 fatty acids in health and disease, *World Rev. Nutr. Dietet.* **57:**214–274.

Chance, B., Sies, H., and Boveris, A., 1979, Hydroperoxide metabolism in mammalian organs, *Physiol. Rev.* **59:**527–605.

Crane, D., and Masters, C., 1986, The effect of clofibrate on the phospholipid composition of peroxisomal membranes in mouse liver, *Biochim. Biophys. Acta* **876:**256–263.

Crane D., Zamattia, J., and Masters, C. J., 1990, Alterations in the integrity of peroxisomal membranes in livers of mice treated with peroxisome proliferators, *Mol. Cell. Biochem*, **96:**153–161.

De Duve, C., 1983, Microbodies in the living cell, *Sci. Am.* **248:**74–84.

De Duve, C., and Baudhuin, P., 1966, Peroxisomes (microbodies and related particles), *Physiol. Rev.* **46:**323–357.

Devchand, P. R., Keller, H., Peters, J. M., Vazquez, M., Gonzalez, F. J., and Wahli, W., 1996, The PPARα–leukotriene B4 pathway to inflammation control, *Nature* **384:**39–43.

Exton, J., 1994, Phosphatidyl choline breakdown and signal transduction, *Biochim. Biophys. Acta* **1212:**26–42.

Gibson, G. G., 1989, Comparative aspects of the mammalian cytochrome P450 IV gene family, *Xenobiotica* **19:**1123–1148.

Green, S., 1992, Receptor-mediated mechanisms of peroxisome proliferators, *Biochem. Pharmacol.* **43:**393–401.

Hansen, H. S., 1994, New biological and clinical roles for the ω-6 and ω-3 fatty acids, *Nutr. Rev.* **52:**162–167.

Kiss, Z., and Anderson, W. H., 1994, Hydrogen peroxide regulates phospholipase D mediated hydrolysis of phosphatidylethanolamine and phosphatidylcholine by different mechanisms in NIH 3T3 fibroblasts, *Arch. Biochem. Biophys.* **311:**430–436.

Lehninger, A. L., Nelson, D. L., and Cox, M. M., 1993, *Principles of Biochemistry*, Worth, New York.

Maksymowych, A. B., Hsu, T. C., and Litwack, G., 1992, A novel highly conserved structural motif is present in all members of the steroid hormone receptor superfamily, *Receptor* **2:**225–240.

Masters, C., 1996a, Omega-3 fatty acids and the peroxisome, *Mol. Cell. Biochem.* **145:**83–93.

Masters, C., 1996b, Cellular signalling: The role of the peroxisome, *Cell Signal.* **8:**197–208.

Masters, C., and Crane, D., 1984, The role of the peroxisomes in lipid metabolism, *Trends Biochem. Sci.* **9:**314–317.

Masters, C., and Crane, D., 1995a, *The Peroxisome—A Vital Organelle*, Cambridge University Press, Cambridge.

Masters, C., and Crane, D., 1995b, On the role of the peroxisome in ontogeny, aging and degenerative disease, *Mech. Ageing* **80:**69–83.

Masters, C., and Crane, D., 1996, Recent developments in peroxisome biology, *Endeavour* **20:**68–73.

Masters, C., and Crane, D., 1998, On the role of the peroxisome in the metabdism of drugs and xenobiotics, *Biochem. Pharmacol.* (in press).

Motojima, K., 1993, Peroxisome proliferator activated receptor (PPAR)—Structure, mechanisms of action and diverse function, *Cell Struct. Funct.* **18:**267–277.

Osmundsen, H., Bremer, J., and Pedersen, J. I., 1991, Metabolic aspects of β-Oxidation, *Biochim. Biophys. Acta* **1085:**141–158.

Pederson, J. I., 1993, Peroxisomal oxidation of the steroid side chain in bile acid formation, *Biochimie* **75:**159–165.

Reddy, J. K., Usuda, N., and Rao, M. S., 1988, Hepatic peroxisome proliferation: An overview, *Arch. Toxicol. Suppl.* **12:**207–216.

Stadtman, E. R., 1990, Metal ion catalysed oxidation of proteins: Biochemical mechanism and biological consequences, *Free Radic Biol. Med.* **9:**315–325.

Tronchere, H., Record, M., Terce, F., and Chap, H., 1994, Phosphatidyl choline cycle and regulation of phosphatidyl choline biosynthesis by enzyme translocation, *Biochim. Biophys. Acta* **1212:**137–151.

Van Kuijk, F. J. G. M., Sevanian, A., Handelman, G. J., and Dratz, E. A., 1987, A new role for phospholipase A_2: Protection of membranes from lipid peroxidation damage, *Trends Biochem. Sci.* **12:**31–33.

Von Schacky, C., Kiefl, R., Marcus, A. J., Breckman, M. J., and Kaminsky, W. E., 1993, Dietary n-3 fatty acids accelerate catabolism of leukotriene B4 in human granulocytes, *Biochim. Biophys. Acta* **1166:**20–24.

Yuan, Y., Jackson, S. P., Newnham, H. H., Mitchell, C. A., and Salem H. H., An essential role for lysophosphatidylcholine in the inhibition of platelet aggregation by secretory phospholipase, *Blood* **86:**4166–4171.

Part VI

Vitamin K

Chapter 19

Vitamin K Metabolism

Paul Newman and Martin J. Shearer

1. DISCOVERY AND EARLY HISTORY OF VITAMIN K

1.1. The Discovery of a Novel Essential Nutrient

The discovery of vitamin K can be largely attributed to the Danish scientist Henrik Dam, whose work on sterol metabolism in chicks required the feeding of carefully controlled diets, some of which were depleted in lipids (Dam, 1929). This often caused internal hemorrhages and other symptoms similar to scurvy, but which Dam showed to be incurable by large doses of vitamin C. The finding of the lack of influence of cod liver oil concentrates (as a source of vitamins A and D), but the protective effect of cereals and seeds, prompted Dam to suggest that the bleeding syndrome was caused by a lack of an essential dietary component that was different from vitamins A, D, and C (Dam and Schoenheyder, 1934; Dam, 1934). Further work by Dam with experimental diets showed the factor was fat soluble, heat stable and occurred in various animal and plant tissues, hog liver fat being one of the best sources. In 1935, after ruling out vitamin E (hog liver fat was many times as active as wheat-germ oil), Dam proposed that the antihemorrhagic factor was a new fat-soluble vitamin that he called "Koagulations vitamin"

Paul Newman and Martin J. Shearer Vitamin K Laboratory, The Haemophilia Centre, St. Thomas's Hospital, London SE1 7EH, England.

Subcellular Biochemistry, Volume 30: Fat-Soluble Vitamins, edited by Quinn and Kagan. Plenum Press, New York, 1998.

or vitamin K (Dam, 1935a,b). Other rich sources of vitamin K were found to be green leaves such as alfalfa. Another surprisingly rich souce was bran or fishmeal which had become putrified by the action of bacteria (Almquist and Stokstad, 1935; Almquist *et al.*, 1938), suggesting that microbial action as well as plant biosynthesis is capable of producing the factor. Improvements in the assay method based on the restoration of normal clotting time in hemorrhagic chicks led to the proposal that a lack of vitamin K caused a deficiency in prothrombin activity and that vitamin K might be some kind of coenzyme or prosthetic group required for active prothrombin (Schoenheyder, 1936). However, subsequent work showed that prothrombin precipitates were active after all lipids had been removed and that vitamin K concentrates could not induce blood to clot (Dam *et al.*, 1936).

1.2. Structural Determination and Organic Synthesis

Improvements in the methods used to extract vitamin K (Dam and Schoenheyder, 1936; Dam and Lewis, 1937) and to determine its activity (Almquist and Stokstad, 1937; Dam and Glavind, 1938) led to the isolation of an active compound from alfalfa which was named vitamin K_1 (Binkley *et al.*, 1939) and another from putrified fish meal which was named vitamin K_2 (McKee *et al.*, 1939). Both were thought to be quinoids (McKee *et al.*, 1939) and their ultraviolet spectra suggested they were 2,3-disubstituted 1,4-naphthoquinones (Ewing *et al.*, 1939). In the same year, phylloquinone was synthesized in two independent laboratories in Switzerland and the United States and was found to have identical biological properties to natural vitamin K_1 (Karrer *et al.*, 1939; Fieser, 1939). The identity of vitamin K_1 was confirmed as 2-methyl-3-phytyl-1,4-naphthoquinone (MacCorquodale *et al.*, 1939). The following year, the structure of K_2 was found to differ only in its 3-substituted side chain, which was believed at the time to be a difarnesyl moiety (Binkley *et al.*, 1940), although subsequent work showed that the substance extracted by Binkley and co-workers was probably a compound with a geranylfarnesyl side chain, i.e., of seven isoprene units (Isler *et al.*, 1958). Since the discovery and characterization of the vitamin K compounds, the IUPAC-IUB have recommended that vitamin K_1 (Fig. 1) should be named phylloquinone and the K_2 compounds (Fig. 1) should be called menaquinones (abbreviated to MK) followed by the number of isoprene units in the side chain, e.g., the structure proposed by Binkley with a difarnesyl chain would be called menaquinone-6, or MK-6. Other bacterial sources later yielded menaquinones-8 to -13, some containing the occasional saturated unit in an otherwise polyprene chain. The specific menaquinone content of bacteria

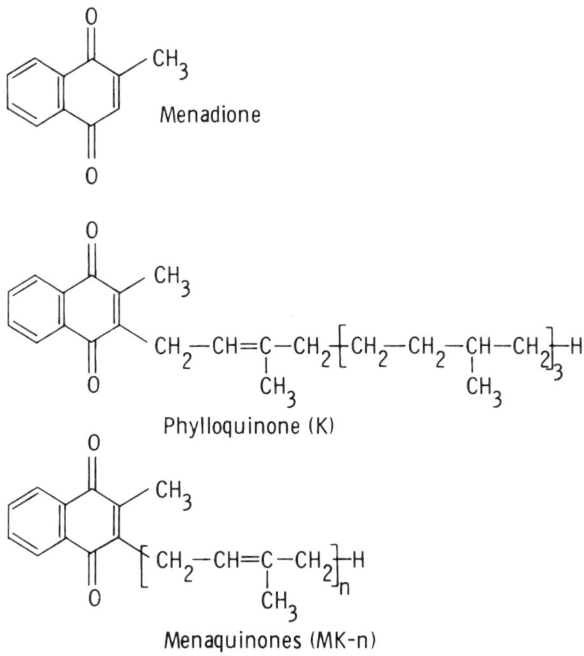

FIGURE 1. Structural formulas of vitamins K.

has been found useful by microbiologists as a biochemical marker in establishing the taxonomy of certain families which would be difficult to distinguish by any other criteria (H. N. Shah and Collins, 1983).

1.3. Dietary Sources of Vitamin K and Nutritional Requirements

The early bioassays used to determine vitamin K were time-consuming, only semiquantitative, and not suitable for reliable assays of dietary sources covering a wide range of concentrations. However, the use of high-performance liquid chromatography (Shearer *et al.*, 1980) has enabled a large number of foodstuffs to be assayed, and a fairly comprehensive body of assay data has been accumulated (Booth *et al.*, 1993; Shearer *et al.*, 1996). Phylloquinone is found in all green vegetables with dark-green leaves, with kale, turnip greens, and spinach being particularly rich sources. Intermediate contents are found in less-green or nonleafy vegetables and certain vegetable oils. The phylloquinone contents of fish and animal skeletal meats are low, but measurable. Menaquinones are less widely distributed than phylloquinone, but are found in significant amounts in animal livers and in foods whose preparation involves a bacterial fermentation, such as cheese (Shearer *et al.*, 1996). Menaquinones are also produced by bacteria resident in the human gut, but the most common have long prenyl side chains (MK-7 to -13) and are extremely lipophilic, which may restrict their absorption by the host (Conly and Stein, 1992). Nevertheless, the vitamin K stores in human liver consist of around 90% of menaquinones (Shearer, 1992), with a similar distribution between the different forms as in the gut bacteria (Shearer *et al.*, 1996). One menaquinone which is not common in bacteria but is found in mammalian and avian tissues is MK-4. Birds and mammals can both synthesize MK-4 from menadione (Fig. 1), although this compound is not thought to occur in the diet or be made by animal tissues and it may be that MK-4 is synthesized from phylloquinone (Thijssen and Drittij-Reijnders, 1994).

The bioavailability of vitamin K from various sources is likely to vary considerably depending on the lipid content of the source, the lipophilicity of the forms of vitamin K it contains, and the presence of fat in the food preperation (Vermeer *et al.*, 1995). These variables make it difficult to establish the requirements for vitamin K, although attempts to do so based on sensitive blood clotting assays agree that around 1 μg per kg of body weight per day is sufficient to maintain a normal prothrombin time (National Research Council, 1989). However, it has recently been suggested that the Gla proteins synthesized by bone require greater dietary intakes of vitamin K to ensure full carboxylation than the Gla coagulation proteins synthesized by the liver (Vermeer *et al.*, 1995) and that this may be of particular importance to certain individuals at risk of diseases such as osteoporosis. This will be discussed in further detail in Section 4.

2. ABSORPTION, TRANSPORT, AND METABOLISM

2.1. Intestinal Absorption

Dietary vitamin K is absorbed chemically unchanged from the proximal intestine after solubilization into mixed micelles composed of bile salts and the products of pancreatic lipolysis (Shearer *et al.*, 1974). In healthy adults the efficiency of absorption of phylloquinone in its free form is about 80% (Shearer *et al.*, 1974), but this efficiency drops to less than 10% from green leafy vegetables such as spinach (Gijsbers *et al.*, 1996). This poor bioavailability of phylloquinone from leafy vegetables is due to its tight association with the thylakoid membranes of chloroplasts.

2.2. Plasma Transport

Within the intestinal mucosa the vitamin is incorporated into chylomicrons, secreted into the lymph, and enters the blood via the lacteals. Once in the circulation, vitamin K is rapidly cleared at a rate consistent with its continuing association with chylomicrons and the chylomicron remnants that are produced by lipoprotein lipase hydrolysis at the surface of capillary endothelial cells. After an overnight fast, more than half of the circulating phylloquinone is still associated with triglyceride-rich lipoproteins, with the remainder being equally distributed between low-density and high-density lipoproteins (Kohlmeier *et al.*, 1996). Dietary vitamin K is delivered to the liver and possibly other tissues, including bone marrow, in the form of chylomicron remnants (Kohlmeier *et al.*, 1996).

2.3. Storage

The liver, the site of synthesis of the vitamin K-dependent coagulation proteins, has a limited capacity for storing vitamin K. Thus the hepatic reserves of vitamin K are several thousandfold lower than those of vitamin A despite the dietary intakes being only about tenfold lower (Shearer *et al.*, 1996). With respect to the different molecular forms of the vitamin K family, the liver stores of the plant form, phylloquinone, represent only about 10% of total stores; the remainder are comprised of bacterial menaquinones, mainly MK-7 to -13. The presence of these very long chain MKs in human liver contrasts with the molecular spectrum found in bone lipid, where phylloquinone appears to be the major form with lower proportions of MK-4 to -7; the concentration of each menaquinone in bone lipid decreases with the side chain length and no higher homologues than MK-8 are detectable (M. J. Shearer, unpublished). An unanswered question is the relative physiological importance and availability of different molecular forms of vitamin K for the microsomal γ-carboxylation reaction in different tissues.

2.4. Catabolism and Turnover

The liver plays an exclusive role in the metabolic transformations leading to the elimination of vitamin K from the body. Doses of phylloquinone in the range of 45 μg to 1 mg are extensively metabolized with about 20% being excreted in the urine within 3 days and about 40–50% excreted in the feces via the bile (Shearer *et al.*, 1974). Since the lower dose of 45 μg approximates to an amount that could be easily obtained from a single meal, this extensive catabolism by the liver suggests that some 60–70% of dietary intakes of phylloquinone are ultimately lost to the body by excretion. This is also illustrated by the rapid depletion of hepatic reserves of phylloquinone seen in surgical patients after they had been placed on a low-phylloquinone diet (Usui *et al.*, 1990). These results suggest that the body stores of vitamin K are being constantly replenished.

The route of hepatic catabolism leading to the urinary excretion of vitamin K proceeds by oxidative degradation of the phytyl side chain, probably involving the same enzymes used for the ω-methyl and β-oxidation of fatty acids, steroids, and prostaglandins. Two major metabolites or aglycones have been identified; they are carboxylic acids with five- and seven-carbon-atom side chains (Fig. 2) that are excreted in the urine as glucuronide conjugates (Shearer *et al.*, 1974). The biliary metabolites have not been clearly identified, but are initially excreted as water-soluble conju-

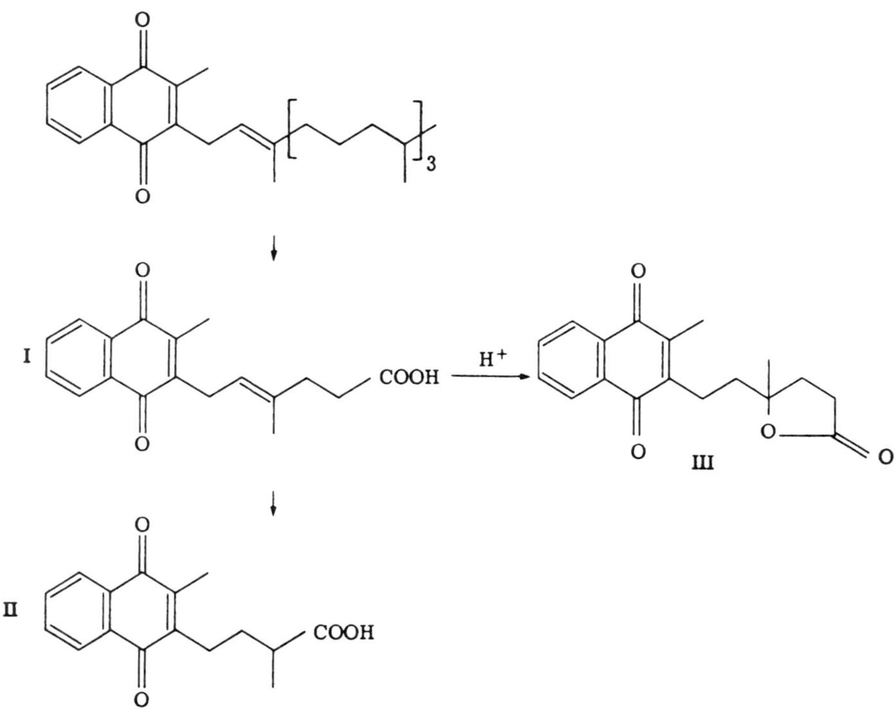

FIGURE 2. Chemical structures of the two major urinary aglycone metabolites of phylloquinone (I and II). Metabolite I is relatively unstable and may become cyclized to phylloquinone γ-lactone (III). (I) 2-methyl-3-(5′-carboxy-3′-methyl-2′-pentenyl)-1,4-naphthoquinone, (II) 2-methyl-3-(3′-carboxy-3′-methylpropyl)-1,4-naphthoquinone, (III) 2-methyl-3-(5′-carboxy-3′-hydroxy-3′-methylpentyl)-1,4-naphthoquinone lactone.

gates and become lipid soluble during their passage through the gut, probably through deconjugation by the gut flora. There is no evidence for body stores of vitamin K being conserved by an enterohepatic circulation. Vitamin K itself is too lipophilic to be excreted in the bile and the side-chain-shortened carboxylic acid metabolites are not biologically active.

3. THE MOLECULAR BASIS OF VITAMIN K ACTION

3.1. Vitamin K-Dependent Protein Modification

After the discovery and isolation of vitamin K in the 1930s it was to take around four more decades to elucidate its biochemical function. Evidence that vitamin K participated in a posttranslational modification began to accumulate when it was found that liver homogenates and cultured cells could still generate active prothrombin when treated with the vitamin in the presence of protein synthesis inhibitors (Goswami and Munro, 1962; Babior, 1966). A molecule similar to prothrombin, but without clotting activity was found to be produced in patients in the absence of vitamin K or in the presence of a vitamin K antagonist which was thought to be an inactive precursor of prothrombin (Hemker *et al.*, 1963), for which the term PIVKA (protein induced by vitamin K absence or antagonists) was later introduced (Hemker *et al.*, 1968). The abnormal protein in humans and other species were found to have impaired calcium-binding properties and could not be precipitated by barium sulfate in the normal way (Ganrot and Nilehn, 1968; Nelsestuen and Suttie, 1972). Further studies showed that it had approximately the same molecular weight as normal prothrombin and similar or identical amino acid composition, and was immunochemically similar, though not identical (Nelsestuen and Suttie, 1972; Malhotra, 1972).

A molecular explanation for the way in which vitamin K participates in active prothrombin synthesis was not provided until 1974 with the independent discovery by three different laboratories that peptides generated by enzymatic digestion of prothrombin contained an unusual amino acid residue, γ-carboxyglutamic acid (Fig. 3), now often abbreviated Gla (Stenflo *et al.*, 1974; Nelsestuen *et al.*, 1974; Magnusson *et al.*, 1974). Stenflo used a peptide from close to the N-terminus containing glutamates 7 and 8, while Nelsestuen used amino acids 33–34, both having previously found that the amino-terminal regions contained the site or sites of vitamin K-dependent activity. Magnusson and co-workers then showed that all ten of the Glu residues in the N-terminal region were modified to Gla (Magnusson *et al.*, 1974).

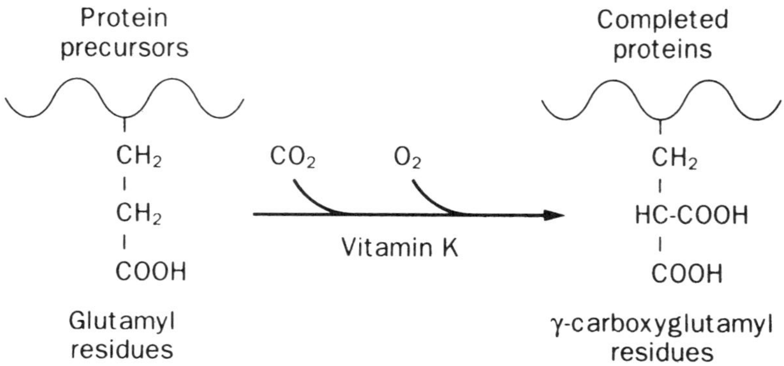

FIGURE 3. The vitamin K-dependent carboxylation reaction.

3.2. Vitamin K-Dependent γ-Glutamyl Carboxylase

3.2.1. Characterization of γ-Glutamyl Carboxylase

Studies of the molecular mechanism behind the γ-carboxy glutamyl modification showed that liver microsomes prepared from rats which had been deprived of vitamin K could produce active prothrombin when vitamin K was added (D. V. Shah and Suttie, 1974). Such microsomes could incorporate $H^{14}CO_3^-$ into the Glu residues of prothrombin already present in the preparations (Esmon *et al.*, 1975a), proving that they must contain a γ-glutamyl carboxylase enzyme or enzymes capable of the modification. Microsomes could also incorporate radioactive label into the small peptide substrate Phe-Leu-Glu-Glu-Val (FLEEV) (Suttie *et al.*, 1976), which enabled the carboxylase to be readily assayed, thereby facilitating its purification. Other studies using peptides helped to define which regions of the substrate are involved with recognition by the carboxylase.

3.2.2. Substrate Recognition

Like all other secreted proteins, the Gla-proteins are synthesized with a signal sequence that is removed by a signal peptidase enzyme immediately following translocation into the lumen of the rough endoplasmic reticulum. In addition, they also have a propeptide that functions as a recognition element for the carboxylase and is later removed. Comparison of the amino acid sequences of such propeptides reveals several areas of homology. All of them contain an Arg at the −1 position and one or two other basic residues at −2 to −5, hydrophobic ones at −6, −7, and −17, and Phe at −16. The importance of these propeptide sequences for substrate recognition by the carboxylase, carboxylation of the protein, and removal of the propeptide has been realized by the study of naturally occurring mutations in the propeptide region, by expression of cDNAs in cultured cells, and by carboxylation reactions using peptide substrates *in vitro*. Three hemophilia type B phenotypes have been found which are caused by mutations in the factor IX propeptide. Factor IX Cambridge (Arg −1 to Ser) and factor IX San Dimas and Oxford 3 (both Arg −4 to Gln) are all incompletely carboxylated and lack coagulation activity. They have been found to retain their propeptides, showing that the Arg-rich sequence is important for propeptide cleavage. However, experiments with a peptide representing the region −18 to +10 of normal profactor IX and others containing the Arg −1 and Arg −4 mutations proved that all were equally good substrates for γ-carboxylase *in vitro* (Hubbard *et al.*, 1989) suggesting that these residues do not constitute part of the γ-carboxylase recognition sequence. The reason for lack of carboxylation in the hemophilia B mutations remains unclear. The propeptide of matrix Gla-protein has an Arg at the −1 position but none in the −2 to −5 region. It undergoes γ-carboxylation, but not propeptide cleavage (Vermeer, 1990), supporting the notion that the −2 to −5 Arg-rich region is required for processing, but not for recognition by the carboxylase. Experiments in which human factor IX (Kaufman *et al.*, 1986) and protein C (Foster *et al.*, 1987) were expressed in cells showed that deletion of the propeptide eliminated carboxylation and that the N-terminal part (−18 to −11) was of particular importance. Studies with peptides showed that carboxylation of peptide substrates such as FLEEV was enhanced substantially by the presence of a second peptide representing a propeptide sequence from a blood coagulation factor and that this enhancement is manyfold stronger if the peptides are covalently linked (Ulrich *et al.*, 1988). In addition to homology within their propeptide sequences, Gla-proteins also contain a well-conserved Gla-Xaa-Xaa-Xaa-Gla-Xaa-Cys motif in the region +17 to +23 (Price *et al.*, 1987). A peptide fragment of descarboxyprothrombin containing this motif, but lacking a propeptide sequence is a substrate for carboxylase (Soute *et al.*, 1981), as is decarboxylated osteocalcin which has had its propeptide removed (Vermeer *et al.*, 1984), strongly suggesting that the conserved residues contribute another substrate recognition site in addition to the N-terminus of the propeptide, possibly due to the cysteine forming a temporary covalent bond with an essential thiol group of the enzyme (Vermeer, 1990).

The propeptide is finally removed by furin, a calcium-dependent endoproteinase which targets the basic region and cleaves at the −1 position (Stanton and Wallin, 1992). This event takes place only after the carboxylation reaction has occurred in the endoplasmic reticulum (Bristol *et al.*, 1996).

3.2.3. The Purification of γ-Glutamyl Carboxylase

The purification of the carboxylase enzyme has proved difficult, but has been achieved using two sources of material. Warfarin-treated bovine liver contains high levels of descarboxy clotting factors, particularly factor X, which remain attached to the carboxylase, enabling purification with anti-factor X antibodies immobilized on Sepharose (De Metz *et al.*, 1981). The problem with this methodology was that the carboxylase could not easily be eluted from the substrate other than by adding vitamin K and allowing the carboxylation reaction to proceed, yielding only an unstable product. However, the complexed carboxylase was useful for certain structural studies and for work on the inhibition of the enzyme by cyanide (De Metz *et al.*, 1982). Another approach used was to prepare detergent-solubilized extracts of microsomes derived from normal livers and then to purify the carboxylase by gel-filtration chromatography (Soute *et al.*, 1987). This gave a more stable product and one which was not attached to its substrate, but it was still not homogeneous. The best preparation has been achieved by use of an affinity column of an immobilized 59-amino acid peptide representing residues −18 to 41 of factor IX with both of the above-mentioned mutations at −4 and −1, which alter the Arg residues to Gln and Ser, respectively (Wu *et al.*, 1991). This peptide has a high affinity for the carboxylase and has enabled purification to near homogeneity.

3.3. The Carboxylation Reaction Mechanism

Vitamin K occurs in tissues in three major metabolic forms differing in their degree of oxidation (Fig. 4). γ-Glutamyl carboxylation is always accompanied by the conversion of vitamin K hydroquinone to vitamin K 2,3 epoxide, which is considered the source of energy for the carboxylation. However, epoxide formation is possible without carboxylation *in vitro* (Sadowski *et al.*, 1977), implying that the two reactions do not have to be coupled. The amount of vitamin K required in the diet is extremely small compared to the molar amount of Gla produced and must therefore be recycled back to the hydroquinone form for reuse as a cofactor by γ-glutamyl carboxylase. Three enzymatic reactions have been identified which constitute a vitamin K cycle (see Fig. 5), plus a fourth activity which can provide an alternative catalytic mechanism for one of the steps. The only compound known to function as a cofactor for γ-glutamyl carboxylase is vitamin K hydroquinone, but dietary vitamin K is found in the oxidized quinone form and needs to be reduced before acting as enzyme cofactor. Similarly, the epoxide generated in the carboxylation reaction must be reconverted into vitamin K hydroquinone for reuse. This recycling pro-

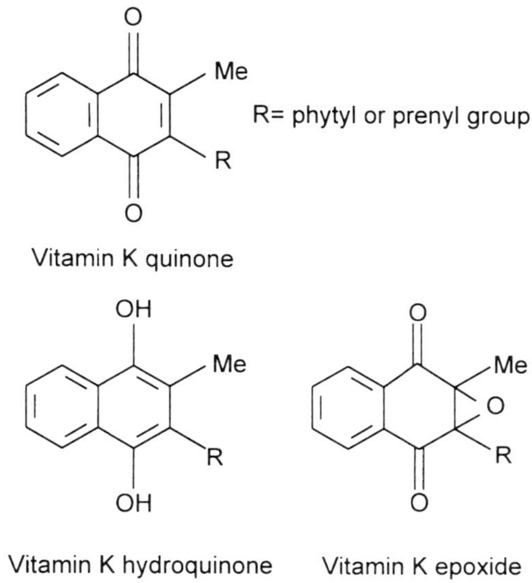

FIGURE 4. Three forms of vitamin K are normally found in cells in which vitamin K-dependent carboxylase is active. Only vitamin K hydroquinone is an active cofactor for the carboxylase.

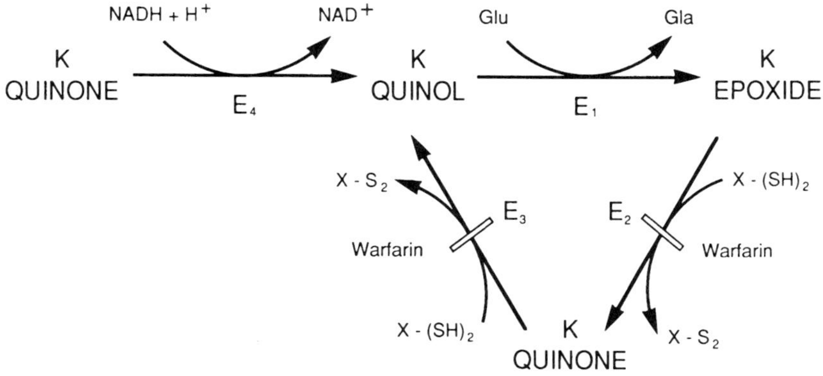

FIGURE 5. The vitamin K–epoxide cycle. Scheme shows the cyclic metabolism of vitamin K in relation to the conversion of glutamate residues (Glu) to γ-caboxyglutamate (Gla) residues in vitamin K-dependent proteins. Known enzyme activities are denoted by E_1, E_2, E_3, and E_4. The active form of vitamin K needed for carboxylation is the reduced form, vitamin K quinol. The carboxylation reaction is driven by a vitamin K-dependent carboxylase activity (E_1) which simultaneously converts vitamin K quinol to vitamin K 2,3-epoxide. Vitamin K 2,3-epoxide is reduced back to the quinone by vitamin K epoxide reductase (E_2). The cycle is completed by the reduction of recycled vitamin K quinone by a vitamin K reductase activity (E_3). Both E_2 and E_3 reductase activities are dithiol dependent [X-(SH_2) and X-S_2 denote reduced and oxidized dithiols, respectively], and are inhibited by coumarin anticoagulants such as warfarin. Dietary vitamin K may enter the cycle via an NAD(P)H-dependent vitamin K reductase activity (E_4) that is not inhibited by warfarin.

ceeds in two reaction steps in which the epoxide is first reduced to vitamin K quinone and then further reduced to the hydroquinone. Both reactions can be catalyzed by reductase activities which depend upon a dithiol cofactor to serve as an electron donor, but several pieces of evidence suggest that the two reactions are performed by the same reductase (Vermeer, 1990). No reductase activity has been found associated with the carboxylase (Morris *et al.*, 1993). The dithiol is usually provided by DTT in *in vitro* reactions, but is likely to be a protein such as thioredoxin *in vivo*, and thioredoxin has indeed been found to serve this role when it has been added to *in vitro* reactions. Both reductase activities are strongly inhibited by coumarin-type vitamin K antagonists, which block the cycle, resulting in an accumulation of vitamin K epoxide. One such reductase activity has been partially purified from rat liver microsomes and seems to consist of multiple components of which a protein of mol. wt. 30,000 binds warfarin (Wallin *et al.*, 1995). A further reductase enzyme is an NADH-dependent activity which can reduce vitamin K quinone to vitamin K hydroquinone, but cannot reduce vitamin K epoxide to the quinone. It is not sensitive to coumarins and therefore makes possible the treatment of coumarin poisoning by the administration of large doses of vitamin K, each molecule being used as cofactor only once, as the epoxide cannot be reduced by the dithiol-dependent reductase if its activity is completely blocked by coumarin antagonists.

The chemical mechanism for the carboxylation event has attracted hypotheses based on either a free radical pathway or the formation of a carbanion intermediate. The latter is currently in favor and an elaborate scheme has been drawn up to explain how a sufficiently strong base can be generated for the carbanion to be formed at the γ-carbon of the Glu (Dowd *et al.*, 1995). The hypothesis is that vitamin K hydroquinone is somewhat acidic and can easily form the monoanion or dianion, both of which can react with molecular oxygen to form intermediates which are believed to be sufficiently strong bases to form the carbanion on the Glu (Fig. 6a). The hypothesis has been tested in a series of model experiments which aim to create reactions involving similar molecules, but in the absence of an enzyme. Such experiments showed that potassium 2,4-dimethyl-α-naphthoxide (Fig. 6b) is not able to cause the condensation of a substrate, diethyl adipate, without the presence of O_2. When O_2 is added the reaction yielded a keto epoxy alcohol and a condensed cyclic product, seen as the result of an intramolecular carboxylation. This was shown to proceed via a dioxetane intermediate because of the *cis* conformation of the epoxide and —OH groups in the product. This structure was verified by X-ray studies and by using radiolabeled O_2 in the reaction, demonstrating that both oxygen atoms acquired by the product are from the same molecule of oxygen in accordance with the hypothesis. It has been suggested that the chemical principles of the model system can be extrapolated to the enzyme system because experimental reactions using rat liver microsomes, phylloquinone, and $^{18}O_2$ have been found to yield an epoxide product that contains two labeled oxygen atoms, suggesting that a similar intermediate is involved in both systems. The term "base strength amplification" has been used to describe the process that generates the strong base required to form the carbanion on the target Glu by using the energy released by the oxidation of vitamin K hydroquinone (Dowd *et al.*, 1995).

4. VITAMIN K ANTAGONISTS

The study of vitamin K has progressed relatively slowly compared to that of other fat-soluble vitamins, but progress would have been virtually blocked were it not for the discovery and development of vitamin K antagonists. The earliest known were the coumarin derivatives, one of which (dicumarol) was initially discovered as a natural microbial product before synthetic derivatives were developed; all are thought to work by inhibiting vitamin K epoxide reductase. Another group of antagonists, which appear to work in the same way, are the indandiones, while certain other compounds are thought to be competitive inhibitors of vitamin K, including some halogen-substituted forms of the molecule. Other compounds have been found with vitamin K antagonist activity, but often with an ill-defined mechanism.

4.1. The Vitamin K Reductase Inhibitors

The first vitamin K antagonist to be discovered was found as a result of investigations into a disease which afflicted cattle in North America from the 1920s onward, characterized by a reduction in the clotting properties of blood resulting in fatal hemorrhage (reviewed by Link, 1959). The cause was traced to the feeding to cattle of hay made from sweet clover, a newly introduced crop intended to overcome problems associated with the growing of more traditional feedstuffs in certain areas. Under some conditions the hay could become spoiled and when fed to cattle it caused the bleeding disorder. Eventually the compound responsible was extracted, identified as 3,3′-methylenebis (4-hydroxycoumarin) (Fig. 7a), and synthesized (Campbell and Link, 1941). The compound was also named dicumarol. Clinical studies showed that dicumarol was useful as an anticoagulant (Butt *et al.*, 1941) and it was used as such for a while, but the synthesis of other substituted 4-hydroxycoumarins led to the development of

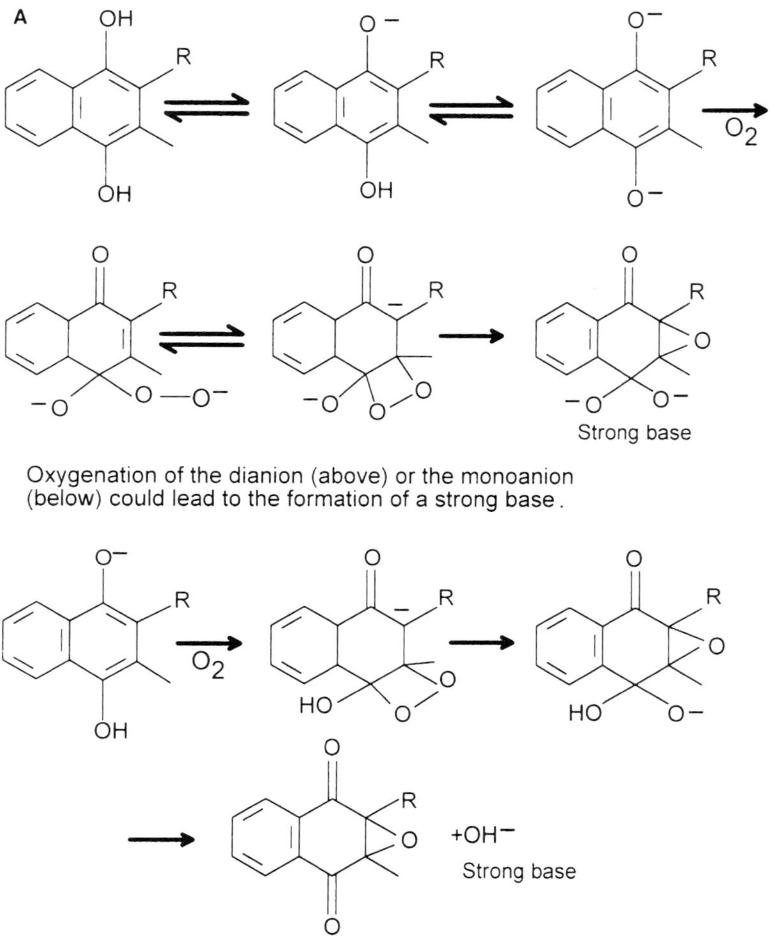

FIGURE 6. (A) Proposed schemes for the formation of a sufficiently strong base to remove a proton from a Glu residue. Vitamin K hydroquinone is able to ionize to both the mono- or dianion. Either is thought to be capable of reaction with molecular oxygen to form very strong bases (Dowd *et al.*, 1995). (B) Model for "intramolecular carboxylation" in which diethyl adipate undergoes C—C bond formation linked to the oxidation of 2,4-dimethyl-α-naphthoxide, a model for the carboxylation of Glu, which is linked to the oxidation of vitamin K hydroquinone.

compounds with better pharmacological properties of which probably the most widely used has been 3-(α-acetonyl-benzyl)-4-hydroxycoumarin, or warfarin (Fig. 7b), which remains in widespread use today.

The mechanism by which warfarin exerted its effects was not understood for some time, but began to be apparent as the metabolism of vitamin

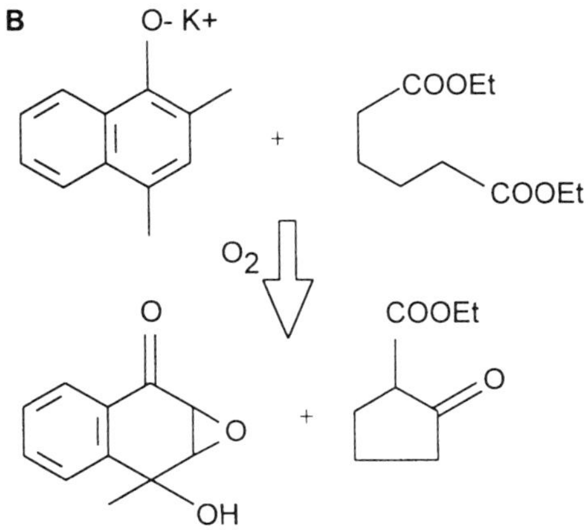

FIGURE 6. (*Continued*).

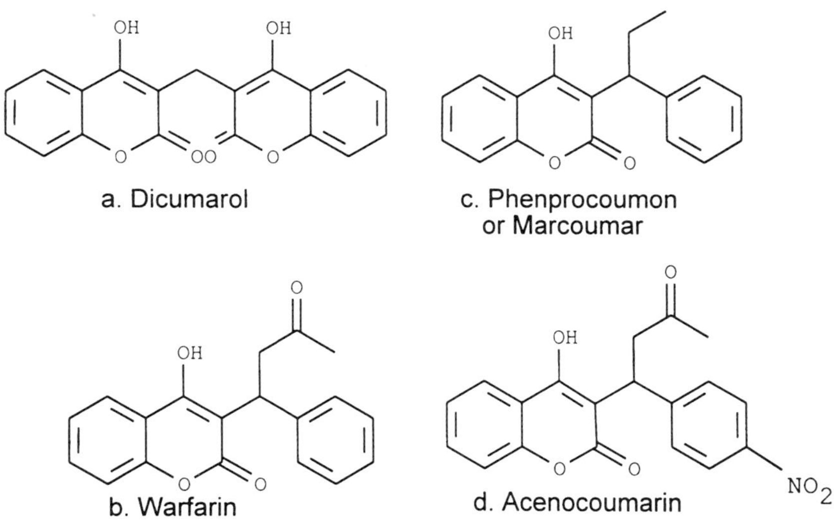

FIGURE 7. Vitamin K antagonists.

K became better understood. Warfarin treatment was found to increase the amount of vitamin K epoxide relative to vitamin K (Bell and Matschiner, 1972), which was initially interpreted to mean that the epoxide was inhibiting the as-yet ill-defined action of vitamin K. After the discovery of vitamin K epoxide reductase (see Section 3) it was found that this enzyme is inhibited in warfarin-sensitive but not in warfarin-resistant rats and that the enzyme was still sensitive to a related compound, difenacoum, that is effective against warfarin-resistant rats (Whitlon *et al.*, 1978). This and other findings have provided strong evidence that the vitamin K epoxide reductase is the binding site for warfarin, thereby interrupting the cellular recycling of vitamin K (Thijssen, 1995).

Several related compounds have been used clinically over the years and although they vary in their ease of absorption and half-life in the bloodstream, they are all thought to work by inhibiting vitamin K epoxide reductase. These include 3-(α-ethylbenzyl)-4-hydroxycoumarin (or Marcoumar) (Fig. 7c) and 3-(α-acetonyl-*p*-nitrobenzyl)-4-hydroxycoumarin (or acenocoumarol) (Fig. 7d), which are widely used in continental Europe and elsewhere, although warfarin remains the most widely prescribed drug of its kind in the United States and the United Kingdom.

Warfarin has also been used very effectively as a domestic rodent poison because it is tasteless and requires multiple doses to kill an animal, thereby reducing the chance of accidental poisoning of humans or wildlife. Since the 1960s warfarin-resistant strains of animals have appeared and the rodenticide industry has responded with the development of new coumarin derivatives which are effective against these animals. This has been a success, but the new compounds are more hazardous to humans, increasing the likelihood of accidental poisoning.

Indandiones are a group of compounds which were found to have vitamin K antagonist activity soon after the discovery of the 4-hydroxycoumarins, and many were synthesized, although their toxicity precluded widespread medical use. They are also thought to inhibit vitamin K epoxide reductase (Ren *et al.*, 1974).

4.2. Vitamin K Derivatives

The 2-chloro derivative of vitamin K_1 (chloro-K) is also an effective vitamin K antagonist, as are certain 2-substituted 3-hydroxy-1,4-naphthoquinones (Smith *et al.*, 1946). Chloro-K does not behave like a reductase inhibitor and may be a true competitive inhibitor of vitamin K (Suttie, 1985).

5. VITAMIN K-DEPENDENT PROTEINS

Very few proteins have been found to contain Gla residues. For the purpose of discussion these can be divided into five groups; (1) plasma proteins of the blood coagulation system, (2) the Gla-proteins of extracellular matrix, (3) the protein Gas6 associated with cellular growth arrest, (4) Gla-containing venom proteins synthesized by certain snakes and the Conidae family of marine molluscs otherwise known as cone shells, and (5) other Gla-proteins, most of which have been only poorly characterized.

5.1. Blood Coagulation Proteins

The first protein to be shown to be vitamin K dependent was prothrombin, as demonstrated by Dam and co-workers (Dam *et al.*, 1936) and by Quick (1937), but the characterization of the chemical nature of the modification was to come almost 40 years later (see Section 3). Once the true nature of the modification was known and could be assayed, attention turned to other proteins known to have protein sequence homologies with prothrombin, and coagulation factors VII, IX, and X were all found to be Gla-proteins. The presence of the Gla residues enables the proteins to bind calcium ions which are required for conformational changes that confer membrane-binding properties vital to the processing of these proteins. In addition to the four vitamin K-dependent clotting factors, three other plasma Gla proteins have been discovered, which have a degree of sequence homology to prothrombin. Two of these have a role in anticoagulation, while the function of the third remains unknown.

5.1.1. Hemostasis

The hemostatic system has been reviewed extensively (Colman *et al.*, 1994) and a detailed account will not be given here; however, a brief description is appropriate due to the dependence of hemostasis upon vitamin K. In a healthy individual the system enables blood to remain fluid under normal conditions, but in the event of an injury to a vessel, it will clot, minimizing bleeding and generating suitable conditions for wound repair. The hemostatic system consists of platelets, the clot-forming protein fibrinogen, and a cascade of protease precursors and associated precofactors which become activated upon partial proteolysis catalyzed by the previous enzyme in the cascade (summarized in Fig. 8). The cascade begins when a signal is generated by vascular trauma, which releases thrombogenic proteins stored beneath and within the endothelium, principally tissue factor, which trigger the clotting processes. The pathway leads to the generation of thrombin, which can cleave soluble fibrinogen to release insoluble fibrin monomers which spontaneously polymerize into a fibrous material together with platelets to form the clot. The polymerized fibrin clot is strengthened by the formation of cross-links between polymers catalyzed by factor XIIIa, a transglutaminase which is also activated by thrombin. The

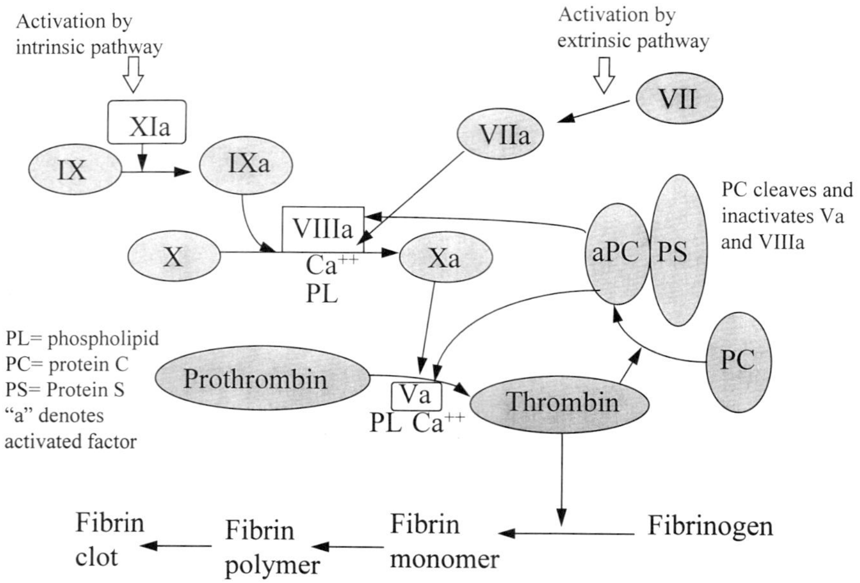

FIGURE 8. An abbreviated diagram of the blood coagulation system. Vitamin K-dependent proteins (shaded) undergo activation upon cleavage by a serine protease in a calcium-dependent association with phospholipid. Protein C, associated with its cofactor protein S, can inactivate factors Va and VIIIa, controlling the coagulation system.

multiple steps of the system, each involving a catalytic cleavage, enables a small thrombogenic signal at a site of injury to be amplified sufficiently to form rapidly a clot large enough to seal the wound. In the absence of injury, the endothelium which lines the luminal surfaces of blood vessels prevents the initiation of the clotting cascade.

5.1.2. Gla-Dependent Activation of Plasma Proteins

Prothrombin has been the prototype plasma protein for study of the vitamin K-dependent activation processes. Prothrombin is made in the liver as a 622-amino acid prepropolypeptide which undergoes processing, as described in Section 2, including carboxylation of ten Glu residues (6, 7, 14, 16, 19, 20, 25, 26, 29, and 32) in the N-terminal region, followed by removal of the 43-residue prepropeptide. The Gla residues in prothrombin enable binding to Ca^{2+} ions (Nelsestuen and Suttie, 1968), which in turn causes a conformational change in the molecule allowing it to bind to membranes which contain acidic phospholipids (Nelsestuen, 1976). When bound to such membranes the molecule becomes a suitable substrate for activation through cleavage by the prothrombinase complex, so for this reason the Gla residues are necessary for generation of thrombin (reviewed by Mann, 1994). Undercarboxylated prothrombin lacking as few as two of the essential Gla residues is not subject to the necessary conformational changes and consequent membrane binding required for activation (Malhotra *et al.*, 1985). Very similar conformational changes are thought to take place during the activation of factors VII and X and protein C (Church *et al.*, 1989). The thrombin released upon activation by the prothrombinase complex is a serine protease known as α-thrombin, held in a noncovalent complex with another fragment of the prothrombin molecule to which it has a high affinity. α-Thrombin is structurally related to other serine proteases and has around 30% homology with α-chymotrypsin (Elion *et al.*, 1977).

α-Thrombin has a number of substrates within the hemostatic system, including fibrinogen, cleavage of which results in clot formation and the activation of factor XIII, which stabilizes the clot by cross-linking fibrin polymers. In addition to its well-known role in hemostasis, it can also activate a specific receptor. The thrombin receptor is activated by means of a proteolytic cleavage leading to a number of cellular effects including modulation of gene expression and proliferation, and was the first receptor found to be activated by such a mechanism (Vu *et al.*, 1991). The receptor contains a tethered ligand which only becomes available to interact with the ligand-binding region after cleavage by thrombin. Studies using transgenic mice with a disrupted thrombin receptor gene have suggested that the receptor plays a role during development. Platelets prepared from the blood of such mice were found to respond strongly to thrombin, implying that a different thrombin receptor is expressed in platelets (Connolly *et al.*, 1996).

Factor Xa has also been found to exhibit cellular effects, being mitogenic for smooth muscle cells (Gasic *et al.*, 1992) and lymphocytes (Altieri and Stamnes, 1994). In addition to its recognition of membrane-bound factor V/Va and related molecules, factor Xa also binds to a receptor on leukocytes known as effector cell protease receptor 1 (EPR-1; Altieri and Edgington, 1990; Altieri, 1995). The physiological function remains unknown, but EPR-1 may play an important role in growth regulation of vascular cells.

5.1.3. Proteins C, S, and Z

In addition to the coagulation factors which contain Gla residues, two Gla proteins having anticoagulant properties have been found. Protein C was discovered in bovine plasma (Stenflo, 1976) and was so named because it was the third protein to elute from a chromatogram, but the same protein had been reported years earlier and referred to as autoprothrombin IIa (Mammen *et al.*, 1960). The blood coagulation system is regulated by means of a set of proteins which constitute a control mechanism to prevent excessive coagulation. Two of these, protein C and protein S, are vitamin K dependent. Protein C is a circulating zymogen which upon cleavage by thrombin or factor Xa is converted to an active serine protease with the potential to inactivate factors Va and VIIIa, again by proteolytic cleavage (Stenflo, 1976; Kisiel *et al.*, 1977; Marler *et al.*, 1982). The activation of protein C requires a cofactor called thrombomodulin, a transmembrane protein found on the surface of endothelial cells where the processing takes place (Esmon and Owen, 1981; Owen and Esmon, 1981). The activated protein C released can cleave and inactivate factors Va and VIIIa. The cleavage is carried out in a reaction complex containing protein S, which acts as a cofactor (Walker, 1981). Human protein C, mol. wt. 62,000, consists of two polypeptide chains joined by a disulfide bond . The light chain contains nine Gla residues which are thought to participate in interactions with calcium ions, thereby inducing a particular structure essential for anticoagulant activity (Zhang and Castellino, 1992).

Protein S is a single-chain molecule of mol. wt. 69,000 (DiScipio and Davie, 1979) made in the liver and endothelial cells (Fair *et al.*, 1986) and possibly in megakaryocytes (Schwartz *et al.*, 1985). The precise nature of the anticoagulation cofactor activity is not known, but protein S is certainly involved in the interaction of protein C with factor Va (Solymoss *et al.*, 1988). More than half of the protein S in the circulation is noncovalently bound to C4b-binding protein and is not active as a cofactor for protein C (Dahlback, 1983). A purpose for this interaction has not been found, but it has been proposed that protein S may have a role in controlling the complement system (Dahlback, 1984). Protein S is a mitogen for smooth muscle cells (Gasic *et al.*, 1992; Benzakour *et al.*, 1995) and binds to a cell surface receptor (Benzakour *et al.*, 1995).

A further vitamin K-dependent plasma glycoprotein is protein Z. The N-terminal half of the protein has considerable homology with other vitamin K-dependent plasma proteins, but the C-terminal half has much less. There are 13 Gla residues in bovine protein Z, which are probably involved in calcium-induced phospholipid binding, as the Gla domain shares structural similarity with the other plasma Gla-proteins, but protein Z does not have the characteristics of a protease. Its function is unknown, but bovine protein Z has been found to interact with thrombin (Hogg and Stenflo, 1991a), although the interaction is less strong between the human proteins (Hogg and Stenflo, 1991b).

5.2. Gla-Proteins in Extracellular Matrix

Evidence for the existence of Gla-proteins outside the coagulation system came with the discovery of osteocalcin in bone (Hauschka *et al.*, 1975) followed by matrix Gla-protein (MGP) in the same tissue several years later (Price *et al.*, 1983). Expression of osteocalcin appears to be restricted to calcified tissues, while MGP or its mRNA has since been found to be very much more widespread (Fraser and Price, 1988) and vitamin K-dependent γ-glutamylcarboxylase has been found in almost every type of tissue examined (Vermeer, 1990). Bone contains at least three Gla-proteins—osteocalcin, MGP, and protein S—believed to be identical or very similar to the anticoagulation cofactor mentioned above (Maillard *et al.*, 1992). The ability of Gla-proteins to bind mineralized calcium suggests a possible structural role for these proteins in bone, but evidence continues to accumulate that Gla-proteins are important to osteoclast recruitment and bone development, and both osteocalcin and MGP have been implicated in these processes.

5.2.1. Osteocalcin

Osteocalcin, also known as bone Gla-protein, was first observed in chick bone extracts (Hauschka *et al.*, 1975) as a small protein capable of binding $BaSO_4$. The bovine homolog was later isolated, found to have a molecular weight of 5700, and when sequenced was found to have 49 amino acid residues, of which 3 are Gla (Price *et al.*, 1976). The vitamin K-dependent hemostatic proteins require Gla residues for calcium ion binding (Esmon *et al.*, 1975b) and osteocalcin requires its Gla residues for the ability to bind to mineralized matrix with high affinity (Poser and Price, 1979). Osteocalcin is the most abundant noncollagen protein of bone, accounting for about 1% of all bone protein and 15–20% of noncollagen protein. When in contact with a surface of mineralized calcium the molecule is thought to undergo a conformational change during which it adopts a structure that includes around 40% α-helix arranged in two antiparallel domains in which the Gla residues are aligned along one side of one domain in a structure termed the Gla-helix (Hauschka and Carr, 1982). In this way the distance between the Gla residues is about the same as that of calcium ions in the hydroxyapatite lattice.

About 20% of the osteocalcin synthesized in bone is released into the bloodstream, enabling it to be assayed quite easily; the level of osteocalcin in the serum is regarded as a useful measure of the rate of bone formation (Price *et al.*, 1980). Osteocalcin has three potential Gla residues, but can exist in bone in undercarboxylated forms (ucOC) with fewer than three Gla residues (Cairns and Price, 1994), though ucOC is deficient in mineral-binding function.

5.2.1a. Osteocalcin and Osteoporosis. Several studies have shown a correlation between elevated levels of serum ucOC or impaired vitamin K status and bone health (reviewed by Vermeer *et al.*, 1995). Elevated levels of ucOC have been found to correlate with both increased fracture risk (Szulc *et al.*, 1993) and decreased bone mineral density (BMD) (Szulc *et al.*, 1994), suggesting that undercarboxylation of osteocalcin leads to more-fragile bones, possibly due to a reduction of mineral-binding capacity. Levels of circulating vitamin K have been found to be lower in elderly women who have suffered hip fracture compared to women of the same age who have not (Hodges *et al.*, 1993). The percentage of serum osteocalcin which binds hydroxyapatite (i.e., carboxylated osteocalcin) is lower in postmenopausal than premenopausal women (Knapen *et al.*, 1989; Plantalech *et al.*, 1991), but this can be restored to normal by treatment with high doses of vitamin K_1 (Knapen *et al.*, 1989, Plantalech *et al.*, 1990), suggesting that carboxylation of osteocalcin is sensitive to circulating vitamin K levels. The dietary intake of vitamin K necessary to maintain adequate hemostatis is considered to be about 1 μg per kg of body weight per day (National Research Council, 1989), but this may not be sufficient to ensure that the Gla-proteins of bone are in a carboxylated state, and it has been suggested that the requirements of bone are greater than those of the coagulation system to maintain normal hemostasis (Vermeer *et al.*, 1995).

5.2.1b. Osteocalcin and Osteoclast Recruitment. Osteocalcin appears to be involved in the induction of differentiation of osteoclast progenitors. *In vivo* studies in which bone particles derived from donor animals were implanted into recipients demonstrated that osteocalcin-depleted particles obtained from warfarin-treated animals were less capable of recruiting osteoclasts than those from normal animals (Glowacki and Lian, 1987). Later *in vitro* experiments which excluded the effects of stromal cells or osteoblasts suggested that osteocalcin worked by complementing the effects upon proliferation and differentiation of osteoclast progenitors by macrophage colony stimulating factor and granulocyte-macrophage colony stimulating factor (Liggett *et al.*, 1994) and osteocalcin has been reported to act as a chemoattractant for osteoclast-like cells (Mundy and Poser, 1983; Chenu *et al.*, 1994). Transgenic mice deficient in osteocalcin have been reported to have elevated bone mass with no change in osteoblast number or degree of mineralization and an elevated number of osteoclasts (Ducy *et al.*, 1996). Clearly, much remains to be learned about the role of osteocalcin in bone.

5.2.2. Matrix Gla-Protein

The purification technique used to extract osteocalcin from bone also yielded a second Gla-protein named matrix Gla-protein (MGP; Price *et al.*, 1983), which has a molecular weight of 9600 but appears to have a higher molecular weight of 15,000 when examined by SDS polyacrylamide gel electrophoresis. Like osteocalcin, MGP is found in bone matrix, though in smaller amounts. A clue as to the function of MGP arose from the observation that deformed fetal bones resulting from maternal use of oral anticoagulants have aberrant calcium deposition (Pettifor and Benson, 1975) and warfarin treatment of pregnant animals causes abnormally short bones of the limbs due to premature ossification of the epiphysial plate (Price *et al.*, 1982), suggesting that a Gla-protein is responsible for normal development. Osteocalcin is not thought to be expressed in cartilage, suggesting that MGP could be responsible for controlling ossification. Recent work with MGP-deficient transgenic mice has demonstrated that a lack of this protein causes inappropriate calcification of soft tissue extracellular matrix (Luo *et al.*, 1997), suggesting that MGP is somehow involved in control of calcification.

5.2.3. Protein S

Protein S is best known for its function as a cofactor for protein C in the anticoagulation system. However, it was noted that two cases of congenital protein S deficiency were accompannied by severe osteopenia and low bone mineral density (Pan *et al.*, 1990). A subsequent study demonstrated that cultured human osteoblasts synthesized and secreted protein S (Maillard *et al.*, 1992), and the authors speculate that it may have a role in maintaining normal bone mass and turnover.

5.3. Gas6

A further and potentially vital Gla-protein has been found through efforts to discover genes which control growth regulation. Several genes were found to be highly expressed during serum starvation of cultured cells (Schneider *et al.*, 1988), one of which, known as growth arrest-specific gene 6 (*gas6*), was found to have 44% sequence identity to the DNA sequence of human protein S (Manfioletti *et al.*, 1993). Most homology is in the N-terminal Gla domain, in three of the four epidermal growth factor (EGF)-like domains, and in the C-terminal sex hormone-binding domain, but the thrombin-sensitive region of protein S is not present in the Gas6 protein, making it unlikely that it can interact with protein C as protein S can.

Expression of *gas6* is greatly elevated after growth arrest and decreases quickly after cells restart the cell cycle. Gas6 was found to bind to a receptor variously known as Rse, Tyro 3, or a few other names (Godowski *et al.*, 1995). Rse/Tyro3 is a receptor with a tyrosine kinase activity which itself becomes phosphorylated upon binding Gas6. Human protein S does not seem to be a ligand for human Rse/Tyro3, but Gas6 does activate other receptors related to Rse/Tyro3 (Godowski *et al.*, 1995). Gas6 has been reported to be a growth-potentiating factor which can enhance the mitogenic effects of thrombin and EGF in smooth muscle cells (Nakano *et al.*, 1995) and was recently found to be a mitogen for serum-deprived cells, which it can protect from undergoing apoptosis (Goruppi *et al.*, 1996). However, the physiological role of Gas6 remains largely unknown and a great deal more remains to be learned about the functions of this interesting ligand and its receptors.

5.4. Conotoxins and Snake Venoms

Conotoxins are a family of small peptides synthesized by species of the marine gastropod family Conidae, often known as cone shells, and are components of their venom. The venom ducts of the *Conus* molluscs are the richest known natural source of Gla, although relatively little is yet known of the functions of the Gla-containing toxins. The first such toxin to be discovered and characterized was conotoxin GV, a 17-amino acid peptide which contains five Gla residues and was the first Gla-protein to be found in an invertebrate (McIntosh *et al.*, 1984). The Gla residues are arranged in a Gla-helix structure similar to the motifs found in osteocalcin and other mammalian Gla-proteins. Conotoxin GV is one of a class of neurotoxins known as ω-toxins, a family of diverse toxins found both in many species of cone shells and in the funnel-web spider, which block certain types of Ca^{2+} channels. However, since other ω-toxins do not contain Gla residues, for instance, the spider toxins, it seems unlikely that the Gla residues play any part in blocking Ca^{2+} channels. For reasons which remain unclear, conotoxin GV induces sleep when injected into mice, but does not affect fish, which are the natural prey of the species from which it was isolated.

At least two Gla-proteins have been found in venoms of different species of snake (Tans *et al.*, 1985; Speijer *et al.*, 1986). Both are activators of prothrombin and exhibit Ca^{2+}-mediated phospholipid binding in which the Gla residues may be involved, though this has not been proven.

5.5. Other Gla-Proteins

A number of other proteins have been found which contain Gla, but most remain incompletely characterized. A protein of approximately

28,000–30,000 mol. wt. which is rich in Gla residues has been found in human spermatozoa (Soute *et al.*, 1985), but its function is not known. Human urine has been found to contain Gla-proteins (Lian *et al.*, 1977), and a protein has been purified from urine with around two Gla residues per molecule. It inhibits the growth of calcium oxalate crystals (Nagakawa *et al.*, 1983). A number of Gla-proteins made by lungs have been found which are thought to be pulmonary surfactants (Rannels *et al.*, 1987). Kidney microsomes synthesize Gla-proteins *in vitro* (Hauschka *et al.*, 1979), which demonstrates that Gla found in kidney does not necessarily arise from plasma proteins which arrive via the bloodstream. Other vitamin K-dependent proteins have been found in atherosclerotic tissue, including one which has been named atherocalcin, molecular weight around 80,000, which is thought to contain about 1.9 % Gla (Levy *et al.*, 1979). This may be of significance to the development of such vascular lesions, as their Gla content has been found to increase as the lesion develops from a fatty streak to a fibrous plaque and rises dramatically when the tissue becomes calcified (Levy *et al.*, 1980).

5.6. Vitamin K and the Sphingolipid Pathway

Vitamin K has been demonstrated to be necessary for the synthesis of sphingolipids. This was first observed in the bacterium *Bacteroides melaninogenicus*, a vitamin K-dependent species in which the synthesis of phosphosphingolipids requires vitamin K (Lev and Milford, 1972). Later studies showed that the sphingolipid metabolism of mammals is modulated by vitamin K or warfarin, the activity of galactocerebroside sulfotransferase being inhibited by warfarin, but restored by the addition of vitamin K (Sundaram and Lev, 1990). The means by which vitamin K influences sphingolipid synthesis are not known and it cannot be assumed that Gla-proteins are involved, although recent experiments suggest that a phosphorylation step is involved (Sundaram *et al.*, 1996).

6. NONCARBOXYLATION FUNCTIONS OF VITAMIN K

In addition to the now well-defined function of vitamin K in the γ-carboxylation of glutamate residues, the K_2 vitamins have other functions which appear to be due to a different, as-yet-undefined mechanism. Menaquinone-4 in particular causes specific effects which are not observed upon treatment with vitamin K_1 and these have now been observed in several cell culture systems, predominantly derived from bone. This is a relatively new and developing field of research which has implications for the treatment and prevention of bone disease, osteoporosis in particular, and possibly for chemotherapeutic treatment of certain leukemias. Most of the work in this area has been done in Japanese laboratories, and MK-4, usually known as menatetrenone in Japan, has already been used for clinical trials there.

6.1. Vitamin K_2 and Bone

In addition to containing at least three Gla-proteins, and therefore reliant upon vitamin K-dependent carboxylation, bone is also a target organ for menaquinone-4 (MK-4), which has been found to affect bone homeostasis. Rats which have undergone ovariectomy experience a loss of bone mass, but this is reduced when they are treated with MK-4 as judged by bone strength, dry weight, hydroxyproline content, or calcium content (Akiyama *et al.*, 1993). In another model designed to investigate steroid-induced bone loss with the drug prednisolone, rats which were treated with MK-4 experienced a smaller loss of calcium, bone density, and Gla content (Hara *et al.*, 1993a). This was thought to be due in part to the inhibition of osteoblastic production of prostglandin E_2, a known inducer of bone resorption (Hara *et al.*, 1993b; Koshihara *et al.*, 1993). Later studies suggested that the effects of MK-4 could be at least partly explained by the prevention of osteoclast formation (Akiyama *et al.*, 1994) and, remarkably, that its effects in a cultured cell system could be mimicked by the alcohol derivative of its side chain, geranylgeraniol (Hara *et al.*, 1995). MK-4 has since been found to cause apoptosis of osteoclasts (Kameda *et al.*, 1996) although this has yet to be corroberated by other labs and geranylgeraniol is known to induce apoptosis in various tumor-derived cell lines (Ohizumi *et al.*, 1995). The mechanisms by which these effects are caused and whether or not the actions of MK-4 and geranylgeraniol are similar remain unknown. One possibility is that these compounds can interact with receptors in target cells which mediate their effects, although to our knowledge no evidence for the existence of a menaquinone receptor has been published.

MK-4 has been evaluated as a drug for the treatment of patients with osteoporosis (Orimo *et al.*, 1992) and was found to increase bone mineral density over a 48-week period, but was used at a very high dose of 45 mg/day. Wider studies will be needed to evaluate its worth as a drug to counter loss of bone mass.

6.2. Vitamin K and Leukemia

Preliminary work on cultured human leukemic cells has demonstrated that MK-3 to -5 and geranylgeraniol all cause apoptosis in cells derived from patients with certain types of leukemia and that the compounds have a synergistic effect with all-*trans* retinoic acid (Yaguchi *et al.*, 1996). Phylloquinone has no such effect. This offers the possibility that menaquinones or related compounds may have a use in the treatment of certain leukemias, and although investigations are at an early stage, this may be a promising area of research in the future.

7. REFERENCES

Akiyama, Y., Hara, K., Ohkawa, I., and Tajima, T., 1993, Effects of menatetrenone on bone loss induced by ovariectomy in rats, *Japan. J. Pharmacol.* **62:**145–153.

Akiyama, Y., Hara, K., Tajima, T., Murota, S., and Morita, I., 1994, Effect of vitamin K_2 (menatatrenone) on osteoclast-like cell formation in mouse bone marrow cultures, *Eur. J. Pharmacol.* **263:**181–185.

Almquist, H. J., and Stokstad, E. L. R., 1935, Hemorragic chick disease of dietary origin, *J. Biol. Chem.* **111:**105–113.

Almquist, H. J., and Stokstad, E. L. R., 1937, Assay procedure for vitamin K (anti-hemorrhagic vitamin), *J. Nutr.* **14:**235–240.

Almquist, H. J., Pentler, C. F., and Mecchi, E., 1938, Synthesis of the antihemorrhagic vitamin by bacteria, *Proc. Soc. Exp. Biol. Med.* **38:**336–338.

Altieri, D. C., 1995, Xa receptor EPR-1, *FASEB J.* **9:**860–865.

Altieri, D. C., and Edgington, T. S., 1990, Identification of effector cell protease receptor-1: A leukocyte-distributed receptor for the serine protease factor Xa, *J. Immunol.* **145:**246–253.

Altieri, D. C., and Stamnes, S. J., 1994, Protease-dependent T-cell activation: Ligation of effector cell protease receptor-1 (EPR-1) stimulates lymphocyte proliferation, *Cell. Immunol.* **155:**372–383.

Babior, B. M., 1966, The role of vitamin K in clotting factor synthesis. I. Evidence for the participation of vitamin K in the conversion of a polypeptide precursor to factor VII, *Biochem. Biophys. Acta* **123:**606–610.

Bell, R. G., and Matschiner, J. T., 1972, Warfarin and the inhibition of vitamin K activity by an oxide metabolite, *Nature* **237:**32–33.

Benzakour, O., Formstone, C., Rahman, S., Kanthou, C., Dennehy, U., Scully, M., Kakkar, V., and Cooper, D. N., 1995, Evidence for a protein S receptor(s) on human vascular smooth muscle cells, *Biochem. J.* **308:**481–485.

Binkley, S. B., MacCorquodale, D. W., Thayer, S. A., and Doisy, E. A., 1939, The isolation of vitamin K_1, *J. Biol. Chem.* **130:**219–234.

Binkley, S. B., McKee, R. W., Thayer, S. A., and Doisy, E. A., 1940, The constitution of vitamin K_2, *J. Biol. Chem.* **133:**721–729.

Booth, S. L., Sadowski, J. A., Weihrauch, J. L., and Ferland, G., 1993, Vitamin K1 (phylloquinone) content of foods: A provisional table, *J. Food Comp. Anal.* **6:**109–120.

Bristol, J. A., Ratcliffe, J. V., Roth, D. A., Jacobs, M. A., Furie, B. C., and Furie, B., 1996, Biosynthesis of prothrombin: Intracellular localization of the vitamin K-dependent carboxylase and the sites of carboxylation, *Blood* **88:**2585–2593.

Butt, H. R., Allen, E. V., and Bollman J. L., 1941, A preparation from spoiled sweet clover [3,3′-methylene-bis-(4-hydroxycoumarin)] which prolongs coagulation and prothrombin time of the blood; preliminary report of experimental and clinical studies, *Proc. Staff Meetings Mayo Clinic* **16:**388–395.

Cairns, J. R., and Price, P. A., 1994, Direct demonstration that the vitamin K-dependent bone Gla protein is incompletely γ-carboxylated in humans, *J. Bone Miner. Res.* **9:**1989–1997.

Campbell, H. A., and Link, K. P., 1941, Studies on the hemorrhagic sweet clover disease; the isolation and crystallization of the hemorrhagic agent, *J. Biol. Chem.* **138:**21–33.

Chenu, C., Colucci, S., Grano, M., Zigrino, P., Barattolo, R., Zambonin, G., Baldini, N., Vergnaud, P., Delmas, P. D., and Zallone, A. Z., 1994, Osteocalcin induces chemotaxis, secretion of matrix proteins, and calcium-mediated intracellular signalling in human osteoclast-like cells, *J. Cell Biol.* **127:**1149–1158.

Church, W. R., Boulanger, L. L., Messier, T. L., and Mann, K. G., 1989, Evidence for a common metal ion-dependent transition in the 4-carboxyglutamic acid domains of several vitamin K-dependent proteins, *J. Biol. Chem.* **264:**17882–17887.

Colman, R. W., Marder, V. J., Salzman, E. W., and Hirsh, J., 1994, Overview of hemostasis, in *Hemostasis and Thrombosis*, 3rd ed. (R. W. Colman, J. Hirsh, V. J. Marder, and E. W. Salzman, eds.), pp. 3–18, Lippincott, Philadelphia.

Conly, J. M., and Stein, K., 1992, Quantitative and qualitative measurements of K vitamins in human intestinal contents, *Am. J. Gasteroenterol.* **87:**311–316.

Connolly, A. J., Ishihara, H., Kahn, M. L., Farese, R. V., and Coughlin, S. R., 1996, Role of the thrombin receptor in development and evidence for a second receptor, *Nature* **381:**516–519.

Dahlback, B., 1983, Purification of human C4b-binding protein and formation of its complex with vitamin K-dependent protein *S, Biochem. J.* **209:**847–856.

Dahlback, B., 1984, Interaction between vitamin K-dependent protein S and the complement protein, C4b-binding protein. A link between coagulation and the complement system, *Semin. Thromb. Hemostas.* **10:**139–148.

Dam, H., 1929, Cholesterinstoffwechsel in Hühnereiern und Hühnchen, *Biochem. Z.* **215:**475–492.

Dam, H., 1934, Haemorrhages in chicks reared on artificial diets: A new deficiency disease, *Nature* **133:**909–910.

Dam, H., 1935a, The antihaemorrhagic vitamin of the chick, *Nature* **135:**652–653.

Dam, H., 1935b, The antihaemorrhagic vitamin of the chick, *Biochem. J.* **29:**1273–1285.

Dam, H., and Glavind, J., 1938, Determination of vitamin K by the curative blood-clotting method, *Biochem. J.* **32:**1018–1023.

Dam, H., and Lewis, L., 1937, The chemical concentration of vitamin K, *Biochem. J.* **31:**17–21.

Dam, H., and Schoenheyder, F., 1934, A deficiency disease in chicks resembling scurvy, *Biochem. J.* **28:**1355–1359.

Dam, H., and Schoenheyder, F., 1936, The occurrence and chemical nature of vitamin K, *Biochem. J.* **30:**897–901.

Dam, H., Schoenheyder, F., and Tage-Hansen, E., 1936, Studies on the mode of action of vitamin K, *Biochem. J.* **30:**1075–1079.

De Metz, M., Vermeer, C., Soute, B. A. M., Van Scharrenberg, G. J. M., Slotboom, A. J., and Hemker, H. C., 1981, Partial purification of bovine liver vitamin K-dependent carboxylase by immunospecific adsorption onto antifactor X, *FEBS Lett.* **123:**215–218.

De Metz, M., Soute, B. A. M., Hemker, H. C., and Vermeer, C., 1982, The inhibition of vitamin K-dependent carboxylase by cyanide, *FEBS Lett.* **137:**253–256.

DiScipio, R. G., and Davie, E. W., 1979, Characterization of protein S, a γ-carboxyglutamic acid containing protein from bovine and human plasma, *Biochemistry* **18:**899–904.

Dowd, P., Hershine, R., Ham, S. W., and Naganathan, S., 1995, Vitamin K and energy transduction: A base strength amplification mechanism, *Science* **269:**1684–1691.

Ducy, P., Desbois, C., Boyce, B., Pinero, G., Story, B., Dunstan, C., Smith, E., Bonadio, J., Goldstein, S., Gundberg, C., Bradley, A., and Karsenty, G., 1996, Increased bone formation in osteocalcin-deficient mice, *Nature* **382:**448–452.

Elion, J., Downing, M. J., Butkowski, R. J., and Mann, K. G., 1977, Structure of human thrombin: Comparison with other serine proteases, in *Chemistry and Biology of Thrombin* (R. L. Lundblad, J. W. Fenton, and K. G. Mann, eds.), pp. 97–111, Ann Arbor Science Publishers, Ann Arbor, Michigan.

Esmon, C. T., and Owen, W. G., 1981, Identification of an endothelial cell cofactor for thrombin-catalyzed activation of protein C, *Proc. Natl. Acad. Sci. USA* **78:**2249–2252.

Esmon, C. T., Sadowski, J. A., and Suttie, J. W., 1975a, A new carboxylation reaction: The vitamin K-dependent incorporation of $H^{14}CO^{-}_{3}$ into prothrombin, *J. Biol. Chem.* **250:**4744–4748.

Esmon, C. T., Suttie, J. W., and Jackson, C. M., 1975b, The functional significance of vitamin K action. Difference in phospholipid binding between normal and abnormal prothrombin, *J. Biol. Chem.* **250:**4095–4099.

Ewing, D. T., Vanderbilt, J. M., and Kamm, O., 1939, The ultraviolet absorption of vitamins K_1, K_2 and some related compounds, *J. Biol. Chem.* **131:**345–356.

Fair, D. S., Marlar, R. A., and Levin, E. G., 1986, Human endothelial cells synthesize protein S, *Blood* **67:**1168–1171.

Fieser, L. F., 1939, Synthesis of vitamin K_1, *J. Am. Chem. Soc.* **61:**3467–3475.

Foster, D. C., Rudinski, M. S., Schach, B. G., Berkner, K. L., Kumar, A. A., Hagen, F. S., Sprecher, C. A., Insley, M. Y., and Davie, E. W., 1987, Propeptide of human protein C is necessary for γ-carboxylation, *Biochemistry* **26:**7003–7011.

Fraser, J. D., and Price, P. A., 1988, Lung, heart and kidney express high levels of mRNA for the vitamin K-dependent matrix Gla protein, *J. Biol.Chem.* **263:**11033–11036.

Ganrot, P. O., and Nilehn, J. E., 1968, Plasma prothrombin during treatment with Dicumarol. II. Demonstration of an abnormal prothrombin fraction, Scand. *J. Clin. Lab. Invest.* **22:**23–28.

Gasic, G. P., Arenas, C. P., Gasic, T. B., and Gasic, G. J., 1992, Coagulation factors X, Xa, and protein S as potent mitogens of cultured aortic smooth muscle cells, *Proc. Natl. Acad. Sci. USA* **89:**2317–2320.

Gijsbers, B. L. M. G., Jie, K.-S. G., and Vermeer, C., 1996, Effect of food composition on vitamin K absorption in human volunteers, *Br. J. Nutr.* **76:**223–229.

Glowacki, J., and Lian, J. B., 1987, Impaired recruitment and differentiation of osteoclast progenitors by osteocalcin-depleted bone implants, *Cell Differ.* **21:**247–254.

Godowski, P. J., Mark, M. R., Chen, J., Sadick, M. D., Raab, H., and Hammonds, R. G., 1995, Reevaluation of the roles of protein S and Gas6 as ligands for the receptor tyrosine kinase Rse/Tyro 3, *Cell* **82:**355–358.

Goruppi, S., Ruaro, E., and Schneider, C., 1996, Gas6, the ligand of the Axl tryosine kinase receptor, has mitogenic and survival activities for serum starved NIH3T3 fibroblasts, *Oncogene* **12:**471–480.

Goswami, P., and Munro, H. N., 1962, The role of ribonucleic acid in the formation of prothrombin activity by rat-liver microsomes, *Biochim. Biophys. Acta* **55:**410–412.

Hara, K., Akiyama, Y., Ohkawa, I., and Tajima, T., 1993a, Effects of menatetrenone on prednisolone-induced bone loss in rats, *Bone* **14:**813–818.

Hara, K., Akiyama, Y., Tajima, T., and Shiraki, M., 1993b, Menatetrenone inhibits bone resorption partly through inhibition of PGE2 synthesis *in vitro*, *J. Bone Miner. Res.* **8:**535–542.

Hara, K., Akiyama, Y., Nakamura, T., Murota, S., and Morita., 1995, The inhibitory effect of vitamin K_2 (menetetrenone) on bone may be related to its side chain, *Bone* **16:**179–184.

Hauschka, P. V., and Carr, S. A., 1982, Calcium-dependent α-helical structure in osteocalcin, *Biochemistry* **21:**2538–2547.

Hauschka, P. V., Lian, J. B., and Gallop, P. M., 1975, Direct identification of the calcium-binding amino acid γ-carboxyglutamate in mineralized tissue, *Proc. Natl. Acad. Sci. USA* **72:**3925–3929.

Hauschka, P. V., Friedman, P. A., Traverso, H. P., and Gallop, P. M., 1979, Vitamin K-dependent γ-carboxyglutamic acid formation by kidney microsomes *in vitro*, *Biochem. Biophys. Res. Commun.* **71:**1207–1213.

Hemker, H. C., Veltkamp, J. J., Hensen, A., and Loeliger, E. A., 1963, Nature of prothrombin biosynthesis: Preprothrombinaemia in vitamin K-deficiency, *Nature* **200:**589–590.

Hemker, H. C., Veltkamp, J. J., and Loeliger, E. A., 1968, Kinetic aspects of the interaction of blood clotting enzymes. III. Demonstration of an inhibitor of prothrombin conversion in vitamin K deficiency, *Thromb. Diath. Haemorrh.* **19:**346–363.

Hodges, S. J., Akesson, K., Vergnaud, P., Obrant, K., and Delmas, P. D., 1993, Circulating levels of vitamins K1 and K2 decreased in elderly women with hip fracture, *J. Bone Miner. Res.* **8:**1241–1245.

Hogg, P. J., and Stenflo, J., 1991a, Interaction of vitamin K-dependent protein Z with thrombin, *J. Biol. Chem.* **266:**10953–10958.

Hogg, P. J., and Stenflo, J., 1991b, Interaction of human protein Z with thrombin: Evaluation of the species difference in the interaction between bovine and human protein Z and thrombin, *Biochem. Biophys. Res. Commun.* **178:**801–807.

Hubbard, B. R., Jacobs, M., Ulrich, M. M. W., Walsh, C., Furie, B., and Furie, B. C., 1989, Vitamin K-dependent carboxylation, *J. Biol. Chem.* **264:**14145–14150.

Isler, O., Rüegg, R., Chopard-dit-Jean, L. H., Winterstein, A., and Wiss, O., 1958, Synthese und isolierung von vitamin K_2 und isoprenalogen verbindungen, *Helv. Chim. Acta* **41:**786–807.

Kameda, A., Miyazawa, K., Mori, Y., Yuasa, M., Nakamura, Y., Mano, H., Hakeda, Y., Kameda, A., and Kumegawa, M., 1996, Vitamin K_2 inhibits osteoclastic bone resorption by inducing osteoclast apoptosis, *Biochem. Biophys. Res. Commun.* **220:**515–519.

Karrer, P., Geiger, A., Ruegger, A., and Salomon. H., 1939, Uber Noe-a-phyllochinon (Nor-vitamin-K1) und ahnliche verbindungen, *Helv. Chim. Acta* **22:**1513–1516.

Kaufman, R., Wasley, L., Furie, B. C., Furie, B., and Shoemaker, C., 1986, Expression, purification and characterization of recombinant γ-carboxylated human factor IX synthesized in chinese hamster ovary cells, *J. Biol. Chem.* **261:**9622–9628.

Kisiel, W., Canfield, W. M., Ericsson, L. H., and Davie, E. W., 1977, Anticoagulant properties of bovine plasma protein C following activation by thrombin, *Biochemistry* **16:**5824–5831.

Knapen, M. H. J., Hamulyak, K., and Vermeer, C., 1989, The effect of vitamin K supplementation on circulating osteocalcin (bone Gla protein) and urinary calcium excretion, *Ann. Intern. Med.* **111:**1001–1005.

Kohlmeier, M., Salomon, A., Saupe, J., and Shearer, M. J., 1996, Transport of vitamin K to bone in humans, *J. Nutr.* **126:**1192S–1196S.

Koshihara, Y., Hoshi, K., and Shiraki, M., 1993, Vitamin K_2 (menatetrenone) inhibits prostaglandin synthesis in cultured human osteoblast-like periosteal cells by inhibiting prostaglandin H synthase activity, *Biochem. Pharmacol.* **46:**1355–1362.

Lev, M., and Milford, A. F., 1972, Effects of vitamin K depletion and restoration on sphingolipid metabolism in *Bacteriodes melaninogenicus*, *J. Lipid Res.* **13:**364–370.

Levy, R. J., Lian, J. B., and Gallop, P., 1979, Atherocalcin, a γ-carboxyglutamic acid containing protein from atherosclerotic plaque, *Biochem. Biophys. Res. Commun.* **91:**41–49.

Levy, R. J., Zenker, J. A., and Lian, J. B., 1980, Vitamin K-dependent calcium binding proteins in aortic valve calcification, *J. Clin. Invest.* **65:**563–566.

Lian, J. B., Glimcher, M. J., and Gallop, P. M., 1977, Identification of γ-carboxyglutamic acid in urinary proteins, in *Calcium-Binding Proteins and Calcium Function* (R. H. Wasserman, ed.), pp. 379–381, Elsevier/North-Holland, Amsterdam.

Liggett, W. H., Lian, J. B., Greenberger, J. S., and Glowacki, J., 1994, Osteocalcin promotes differentiation of osteoclast progenitors from murine long-term bone marrow cultures, *J. Cell. Biochem.* **55:**190–199.
Link, K. P., 1959, The story of Dicumarol and its sequels, *Circulation* **19:**97–107.
Luo, G., Ducy, P., McKee, M. D., Pinero, G. J., Loyer, E., Behringer, R. R., and Karsenty, G., 1997, Spontaneous calcification of arteries and cartilage in mice lacking matrix Gla protein, *Nature* **386:**78–81.
MacCorquodale, D. W., Binkley, S. B., Thayer, S. A., and Doisy, E. A., 1939, On the constitution of vitamin K_1, *J. Am. Chem. Soc.* **61:**1928–1929.
Magnusson, S., Sottrup-Jensen, L., Petersen, T. E., Morris, H. R., and Dell, A., 1974, Primary structure of the vitamin K-dependent part of prothrombin, *FEBS Lett.* **44:**189–193.
Maillard, C., Berruyer, M., Serre, C. M., Dechavanne, M., and Delmas, P. D., 1992, Protein S, a vitamin K-dependent protein, is a bone matrix component synthesized and secreted by osteoblasts, *Endocrinology* **130:**1599–1604.
Malhotra, O. P., 1972, Atypical prothrombin in purified preparations from Dicumarol-treated steers, *Life Sci.* **11:**901–907.
Malhotra, O. P., Nesheim, M. E., and Mann, K. G., 1985, The kinetics of activation of normal and γ-carboxyglutamic acid-deficient prothrombins, *J. Biol. Chem.* **260:**279–287.
Mammen, E. F., Thomas, W. R., and Seegers, W. H., 1960, Activation of purified prothrombin to autoprothrombin I or autoprothrombin II (platelet cofactor II) or autoprothrombin II-A, *Thromb. Diath. Haemorrh.* **5:**218–250.
Manfioletti, G., Brancolini, C., Avanzi, G., and Schneider, C., 1993, The protein encoded by a growth arrest-specific gene (*gas6*) is a new member of the vitamin K-dependent proteins related to protein S, a negative coregulator in the blood coagulation cascade, *Mol. Cell. Biol.* **13:**4976–4985.
Mann, K. G., 1994, Prothrombin and thrombin, in *Hemostasis and Thrombosis: Basic Principles and Cinical Practice*, 3rd ed. (R. W. Colman, J. Hirsh, V. J. Marder, and E. W. J. B. Salzman, eds.), pp. 184–199, Lippincott, Philadelphia.
Marlar, R. A., Kleiss, A. J., and Griffin, J., 1982, Mechanism of action of human activated protein C, a thrombin-dependent anticoagulant enzyme, *Blood* **59:**1067–1072.
McIntosh, J. M., Olivera, B. M., Cruz, L. J., and Gray, W. R., 1984, γ-Carboxyglutamate in a neuroactive toxin, *J. Biol. Chem.* **259:**14343–14346.
McKee, R. W., Binkley, S. B., Thayer, S. A., MacCorquodale, D. W., and Doisy, E. A., 1939, The isolation of vitamin K_2, *J. Biol. Chem.* **131:**327–344.
Morris, D. P., Soute, B. A. M., Verneer, C., and Stafford, D. W., 1993, Characterization of the purified vitamin K-dependent γ-glutamyl carboxylase, *J. Biol.Chem.* **268:**8735–8742.
Mundy, G. R., and Poser, J. W., 1983, Chemotactic activity of gamma-carboxyglutamic acid containing protein in bone, *Calcif. Tiss. Int.* **35:**164–168.
Nagakawa, Y., Abram, V., Kezdy, F. J., Kaiser, E. T., and Coe, F. L., 1983, Purification and characterization of the principal inhibitor of calcium oxalate monohydrate crystal growth in human urine, *J. Biol. Chem.* **258:**12594–12600.
Nakano, T., Higashino, K., Kikuchi, N., Kishino, J., Nomura, K., Fujita, H., Ohara, O., and Arita, H., 1995, Vascular smooth muscle cell-derived, Gla-containing growth-potentiating factor for Ca^{2+}-mobilizing growth factors, *J. Biol. Chem.* **270:**5702–5705.
National Research Council, 1989, *Recommended Dietary Allowances*, 10th ed. National Academy Press, Washington D.C.
Nelsestuen, G. L., 1976, Role of γ-carboxyglutamic acid: An unusual protein transition required for the calcium-dependent binding of prothrombin to phospholipid, *J. Biol. Chem.* **251:**5648–5656.

Nelsestuen, G. L., and Suttie, J. W., 1968, Mode of action of vitamin K. Calcium binding properties of bovine prothrombin, *Biochemistry* **11:**4961–4964.

Nelsestuen, G. L., and Suttie, J. W., 1972, The purification and properties of an abnormal prothrombin protein produced by Dicumarol-treated cows. A comparison to normal prothrombin, *J. Biol. Chem.* **247:**8176–8182.

Nelsestuen, G. L., Zytkovicz, T. H., and Howard, J. B. 1974, The mode of action of vitamin K. Identification of γ-carboxyglutamic acid as a component of prothrombin, *J. Biol. Chem.* **249:** 6347–6350.

Ohizumi, H., Masuda, Y., Nakajo, S., Ohsawa, S., and Nakaya, K., 1995, Geranylgeraniol is a potent inducer of apoptosis in tumor cells, *J. Biochem.* **117:**11–13.

Orimo, H., Fujita, T., Onomura, T., Inoue, T., Kushida, K., and Shiraki, M., 1992, Clinical evaluation of Ea-0167 (Menatetrenone) in the treatment of osteoporosis: Phase III multicenter double blind comparative study with alfacalcidol, *Clin. Eval.* **20:**45–100.

Owen, W. G., and Esmon, C. T., 1981, Functional properties of an endothelial cell cofactor for thrombin catalysed activation of protein C, *J. Biol. Chem.* **256:**5532–5535.

Pan, E. Y., Gomperts, E. D., Millen, R., and Gilsanz, V., 1990, Bone mineral density and its association with inherited protein S deficiency, *Thromb. Res.* **58:**221–231.

Pettifor, J. M., and Benson, R., 1975, Congenital malformations associated with the administration of oral anticoagulants during pregnancy, *J. Pediat.* **86:**459–462.

Plantalech, L. C., Chapuy, M. C., Guillaumont, M., Chapuy, P., Leclerq, M., and Delmas, P. D., 1990, Impaired carboxylation of serum osteocalcin in elderly women: Effect of vitamin K_1 treatment, in *Osteoporosis* (C. Christiansen and K. Overgaard, eds.), pp. 345–347, Osteopress, Copenhagen, Denmark.

Plantalech, L. C., Guillaumont, M., Vergnaud, P., Leclercq, M., and Delmas, P. D., 1991, Impairment of gamma carboxylation of circulating osteocalcin (bone Gla protein) in elderly women, *J. Bone Miner. Res.* **6:**1211–1216.

Poser, J. W., and Price, P. A., 1979, A method for decarboxylation of γ-carboxyglutamic acid in proteins, *J. Biol. Chem.* **254:**431–436.

Price, P. A., Poser, J. W., and Raman, N., 1976, Primary structure of the γ-carboxyglutamic acid-containing protein from bovine bone, *Proc. Natl. Acad. Sci. USA* **73:**3374–3375.

Price, P. A., Parthemore, J. G., and Deftos, L. J., 1980, A new biochemical marker for bone metabolism, *J. Clin. Invest.* **66:**878–883.

Price, P. A., Williamson, M. K., Haba, T., Dell, R. B., and Jee, W. S., 1982, Excessive mineralization with growth plate closure in rats on chronic warfarin treatment, *Proc. Natl. Acad. Sci. USA* **79:**7734–7738.

Price, P. A., Urist, M. R., and Otawara, Y., 1983, Matrix Gla protein, a new γ-carboxyglutamic-containing protein which is associated with the organic matrix of bone, *Biochem. Biophys. Res. Commun.* **117:**765–771.

Price, P. A., Fraser, J. D., and Metz-Virca, G., 1987, Molecular cloning of matrix Gla protein: Implications for substrate recognition by the vitamin K-dependent γ-carboxylase, *Proc. Natl. Acad. Sci. USA* **84:**8335–8339.

Quick, A. J., 1937, The coagulation defect in sweet clover disease and in the hemorrhagic chick disease of dietary origin, *Am. J. Physiol.* **118:**260–271.

Rannels, S. R., Gallaher, K. J., Wallin, R., and Rannels, D. E., 1987, Vitamin K-dependent carboxylation of pulmonary surfactant-associated proteins, *Proc. Natl. Acad. Sci. USA* **84:**5952–5956.

Ren, P., Laliberte, R. E., and Bell, R. G., 1974, Effects of warfarin, phenylindanedione, tetrachloropyridinol, and chloro-vitamin K_1 on prothrombin synthesis and vitamin K metabolism in normal and warfarin-resistant rats, *Mol. Pharmacol.* **10:**373–380.

Sadowski, J. A., Schnoes, H. K., and Suttie, J. W., 1977, Vitamin K epoxidase: Properties and relationship to prothrombin synthesis, *Biochemistry* **16:**3856–3863.

Schneider, C., King, R. M., and Philipson, L., 1988, Genes specifically expressed at growth arrest of mammalian cells, *Cell* **67:**469–481.

Schoenheyder, F., 1936, The quantitative determination of vitamin K, *Biochem. J.* **30:**890–896.

Schwartz, H. P., Heeb, M. J., Wencel-Drake, J., Burstein S. A., and Griffin, J. H., 1985, Localization of protein S, an antithrombotic protein, in human platelets and megakaryocytes, *Blood* **66:**353a.

Shah, D. V., and Suttie, J. W., 1974, The vitamin K-dependent, *in vitro* production of prothrombin, *Biochem. Biophys. Res. Commun.* **60:**1397–1402.

Shah, H. N., and Collins, M. D., 1983, Genus *Bacteroides*, A chemotaxonomical perspective, *J. Appl. Bacteriol.* **55:**403–416.

Shearer, M. J., 1992, Vitamin K metabolism and nutriture, *Blood Rev.* **6:**92–104.

Shearer, M. J., McBurney, A., and Barkham, P., 1974, Studies on the absorption and metabolism of phylloquinone (vitamin K_1) in man, *Vitamins Hormones* **32:**513–542.

Shearer, M. J., Allan, V., Haroon, Y., and Barkham, P., 1980, Nutritional aspects of vitamin K in the human, in *Vitamin K Metabolism and Vitamin K-Dependent Proteins* (J. W. Suttie, ed.), pp. 317–327, University Park Press, Baltimore.

Shearer, M. J., Bach, A., and Kohlmeier, M., 1996, Chemistry, nutritional sources, tissue distribution and metabolism of vitamin K with special reference to bone health, *J. Nutr.* **126:**1181S–1186S.

Smith, C. C., Fradkin, R., and Lackey, M. D., 1946, A hemorrhagic syndrome induced by derivatives of 3-hydroxy-1,4-naphthoquinone, *Proc. Soc. Exp. Biol. Med.* **61:**398–403.

Solymoss, S., Tucker, M. M., and Tracy, P. B., 1988, Kinetics of inactivation of membrane-bound factor Va by activated protein C. Protein S modulates factor Xa protection, *J. Biol. Chem.* **263:**14884–14890.

Soute, B. A. M., Vermeer, C., de Metz, M., Hemker, H. C., and Lijnen, H. R., 1981, *In vitro* prothrombin synthesis from a purified precursor protein. III. Preparation of an acid-soluble substrate for vitamin K-dependent carboxylase by limited proteolysis of bovine descarboxyprothrombin, *Biochim. Biophys. Acta* **676:**101–107.

Soute, B. A. M., Mueller-Esterl, W., de Boer-van den Berg, M. A. G., Ulrich, M., and Vermeer, C., 1985, Discovery of a γ-carboxylic acid-containing protein in human spermatozoa, *FEBS Lett.* **190:**137–141.

Soute, B. A. M., Ulrich, M. M. W., and Vermeer, C., 1987, Vitamin K-dependent carboxylase: Increased efficiency of the carboxylation reaction, *Thromb. Haemostas.* **57:**77–81.

Speijer, H., Govers-Riemslag, J. W. P., Zwaal, R. F. A., and Rosing, J., 1986, Prothrombin activation by an activator from the venom of *Oxyuranus scutellatus* (Taipan snake), *J. Biol. Chem.* **261:**13258–13267.

Stanton, C., and Wallin, R, 1992, Processing and trafficking of clotting factor X in the secretory pathway. Effects of warfarin, *Biochem. J.* **284:**25–31.

Stenflo, J., 1976, A new vitamin K-dependent protein, purification from bovine plasma and preliminary characterization, *J. Biol. Chem.* **251:**355–363.

Stenflo, J., Fernlund, P., Egan W., and Roepsdorff, P., 1974, Vitamin K-dependent modifications of glutamic acid residues in prothrombin, *Proc. Natl. Acad. Sci. USA* **71:**2730–2733.

Sundaram, K. S., and Lev, M., 1990, Regulation of sulfatide metabolism by vitamin K in mouse brain, *Arch. Biochem. Biophys.* **277**:109–113.

Sundaram, K. S., Fan, J.-H., Engelke, J. A., Foley, A. L., Suttie, J. W., and Lev, M., 1996, Vitamin K status influences brain sulfatide metabolism in young mice and rats, *J. Nutr.* **126:**2746–2751.

Suttie, J. W., 1985, Vitamin K, in *Fat-Soluble Vitamins: Their Biochemistry and Applications* (A. T. Diplock, ed.), pp. 225–311, Heinemann, London.

Suttie, J. W., Hageman, J. M., Lehrman, S. R., and Rich, D. H., 1976, Vitamin K-dependent carboxylase. Development of a peptide substrate, *J. Biol. Chem.* **251:**5827–5830.

Szulc, P., Chapuy, M.-C., Meunier, P. J., and Delmas, P. D., 1993, Serum undercarboxylated osteocalcin is a marker of the risk of hip fracture in elderly women, *J. Clin. Invest.* **91:**1769–1774.

Szulc, P., Arlot, M., Chapuy, M.-C., Duboef, F., Meunier, P. J., and Delmas, P. D., 1994, Serum undercarboxylated osteocalcin correlates with hip bone mineral density in elderly women, *J. Bone Miner. Res.* **9:**1591–1595.

Tans, G., Govers-Riemslag, J. W. P., van Rijn, J. L. M. L., and Rosing, J., 1985, Purification and properties of a prothrombin activator from the venom of *Notechis scutatus scutatus*, *J. Biol. Chem.* **260:**9366–9322.

Thijssen, H. H. W., 1995, Warfarin-based rodenticides: Mode of action and mechanism of resistance, *Pestic. Sci.* **43:**73–78.

Thijssen, H. H. W., and Drittij-Reijnders, M. J., 1994, Vitamin K distribution in rat tissues: Dietary phylloquinone is a source of tissue menaquinone-4, *Br. J. Nutr.* **72:**415–425.

Ulrich, M. M. W., Furie, B., Jacobs, M., Vermeer, C., and Furie, B. C., 1988, Vitamin K-dependent carboxylation: A synthetic peptide based upon the γ-carboxylation recognition site sequence of the prothrombin propeptide is an active substrate for the carboxylase *in vitro*, *J. Biol. Chem.* **263:**9697–9702.

Usui, Y., Tanimura, H., Nishimura, N., Kobayashi, N., Okanoue, T., and Ozawa, K., 1990, Vitamin K concentrations in the plasma and liver of surgical patients, *Am. J. Clin. Nutr.* **51:**846–852.

Vermeer, C., 1990, γ-Carboxyglutamate-containing proteins and the vitamin K-dependent carboxylase, *Biochem J.* **266:**625–636.

Vermeer, C., Soute, B. A. M., Hendrix, H., and de Boer-van den Berg, M. A. G., 1984, Decarboxylated bone Gla-protein as a substrate for hepatic vitamin K-dependent carboxylase, *FEBS Lett.* **165:**16–20.

Vermeer, C., Jie, K.-S. G., and Knapen, M. H. J., 1995, Role of vitamin K in bone metabolism, *Annu. Rev. Nutr.* **15:**1–22.

Vu, T.-K. H., Hung, D. T., Wheaton, V. I., and Coughlin, S. R., 1991, Molecular cloning of a functional thrombin receptor reveals a novel proteolytic mechanism of receptor activation, *Cell* **64:**1057–1068.

Walker, F. J., 1981, Regulation of activated protein C by protein S, *J. Biol. Chem.* **256:**11128–11131.

Wallin, R., Guenthner, T. M., and Higson, T. S., 1995, Warfarin sensitive vitamin K 2,3-epoxide reduction in rat liver involves epoxide hydrolase and a second 30 kDa protein which appears to carry the warfarin binding site, *FASEB J.* **9:**A726, abstract 4206.

Whitlon, D. S., Sadowski, J. A., and Suttie, J. W., 1978, Mechanism of coumarin action: Significance of vitamin K epoxide reductase inhibition, *Biochemistry* **17:**1371–1377.

Wu, S.-M., Morris, D. P., and Stafford, D. W., 1991, Identification and purification to near homogeneity of the vitamin K-dependent carboxylase, *Proc. Natl. Acad. Sci. USA* **88:**2236–2240.

Yaguchi, M., Miyazawa, K., Katagiri, T., Nishimaki, J., Iwase, O., Kizaki, M., Tohyama, K., and Toyama, K., 1996, Vitamin K_2 and its derivatives induce apoptosis in laekemia cells and have a synergistic effect with all *trans* retinoic acid, *Blood* **88**(Suppl. 1)**:**270-I.

Zhang, L., and Castellino, F. J., 1992, Influence of specific gamma-carboxyglutamic acid residues on the integrity of the calcium-dependent conformation of human protein C, *J. Biol. Chem.* **267:**26078–26084.

Part VII

Coenzyme Q

Chapter 20

Role of Coenzyme Q and Superoxide in Vitamin E Cycling

Valerian E. Kagan, Yulia Y. Tyurina, and Eric Witt

1. INTRODUCTION AND SUMMARY

Coenzyme Q, an indispensable component of membrane electron transport chains, can also function, in its reduced form, as an antioxidant in membranes and lipoproteins. It decreases lipid peroxidation in liposomes, mitochondria, cells, and low-density lipoprottein (LDL), which explains its ubiquitous presence in biological systems, even where no energy transduction occurs (e.g., LDL). Electron transport reduces coenzyme Q to antioxidant forms, but it is unknown if it is reduced in the absence of electron transport. In an aprotic environment (dimethylsolfoxide, DMSO), in aqueous environments, and in liposomes, we found that superoxide reduces coenzyme Q, and that this reduced coenzyme Q can, in turn, regenerate vitamin E from its radical form, thus "recycling" vitamin E. In HL-60 cells,

Valerian E. Kagan Department of Environmental and Occupational Health, and Department of Pharmacology, University of Pittsburgh, Pittsburgh, Pennsylvania 15238.
Yulia Y. Tyurina Department of Environmental and Occupational Health, University of Pittsburgh, Pittsburgh, Pennsylvania 15238; on leave from the Institute of Evolutionary Physiology and Biochemistry, Russian Academy of Science, St. Petersburg, 194223 Russia.
Eric Witt Department of Environmental and Occupational Health, University of Pittsburgh, Pittsburgh, Pennsylvania 15238.
Subcellular Biochemistry, Volume 30: Fat-Soluble Vitamins, edited by Quinn and Kagan. Plenum Press, New York, 1998.

coenzyme Q protected membrane phospholipids against peroxidation induced by a lipid-soluble azo-initiator of peroxyl radicals, 2,2′-azobis (2,4-dimethylvaleronitrile) (AMVN), only when superoxide was generated in the incubation system. Thus, coenzyme Q is reduced by superoxide, and the reduced forms of coenzyme Q exert antioxidant effects, either directly or via recycling of the vitamin E radical. Hence, under conditions of high superoxide production (i.e., high oxidative stress), coenzyme Q's antioxidant power is multiplied manifold by its continuous reduction by the very superoxide which causes oxidative stress.

The discovery of coenzyme Q by Crane *et al.* (1957) four decades ago greatly advanced our understanding of membrane redox processes. Membrane electron transport systems contain coenzyme Q, which participates in redox reactions (Nohl *et al.*, 1986, Kalen *et al.*, 1987). In mitochondria, coenzyme Q is a mobile link between NADH dehydrogenase, succinate dehydrogenase, and cytochrome b-c1 in electron transport (Gutman, 1980). Coenzyme Q also participates in the protonmotive Q cycle (Mitchell, 1975a,b).

For many years, studies of coenzyme Q's function in mitochondrial respiration overshadowed persistent reports of coenzyme Q in other membrane fractions in which its function was not readily apparent—it is found in plasma membranes, in all intracellular membranes, and in lipoproteins (Crane and Morre, 1977; Crawford *et al.*, 1983; Elmberger *et al.*, 1989). Its concentration is very high in Golgi membranes and in lysosomal membranes (higher, in fact, than in mitochondria), where its function cannot be rationalized in terms of energy-transducing activity (Crane and Morre, 1977). It is also found in lipoproteins, despite the fact that it is manufactured intracellularly and need not be transported between cells.

2. COENZYME Q AS AN ANTIOXIDANT

There is now a substantial amount of experimental data showing that coenzyme Q, in addition to its role in electron transport, functions as an antioxidant in its reduced forms in various biological membranes and in low-density lipoproteins (LDL) (for reviews see Ernster and Dallner, 1995; Beyer, 1990).

2.1. Intracellular Antioxidant Activity (Membranes)

In 1962, Lea and Kwietney reported that coenzyme Q functioned as an antioxidant (Lea and Kwietney, 1962). While some experiments suggest that both ubiquinones and ubiquinols might function as antioxidants (Landi *et al.*, 1985), other studies have shown that very high concentrations of ubiquinone are required to exhibit significant antioxidant activity (Landi *et al.*, 1990a; Mellors and Tappel 1966a), and it is the reduced form of the compound that is considered to be the major antioxidant. In pioneering studies Mellors and Tappel showed that ubiquinol 6 was efficient in inhibiting lipid peroxidation and that electron transport-driven reduction of ubiquinone to ubiquinol resulted in pronounced inhibition of lipid peroxidation in mitochondria (Mellors and Tappel, 1966a,b). Numerous subsequent studies in liposomes (Booth *et al.*, 1982; Yamamoto *et al.*, 1990; Frei *et al.*, 1990; Landi *et al.*, 1985, 1990a,b, 1992; Cipollone *et al.*, 1994), mitochondria, microsomes, and submitochondrial particles (Ernster *et al.*, 1992; Beyer, 1988; Norling *et al.*, 1974; Landi *et al.*, 1984, Takeshige *et al.*, 1980; Kagan *et al.*, 1990a,b; Forsmark *et al.*, 1991; Jakobsson-Borin *et al.*, 1994; Forsmark-Andree *et al.*, 1995) and cells (Takahashi *et al.*, 1988) have established beyond a doubt that ubiquinol is capable of inhibiting lipid peroxidation in biological membranes, and that electron transport, by a number of membrane (Takeshige *et al.*, 1980; Maguire *et al.*, 1992) and cytosolic (Frei *et al.*, 1990) systems, can reduce ubiquinone back to ubiquinol. This has led to the model of coenzyme Q antioxidant activity in membranes shown in Fig. 1.

2.2. Extracellular Antioxidant Activity (Low-Density Lipoproteins)

Oxidation of LDL is a step in the genesis of atherosclerosis in the coronary arteries (Esterbauer *et al.*, 1991; Esterbauer and Ramos, 1995). LDLs contain antioxidants, including vitamin E, carotenoids, and ubiquinol, which collectively protect it from oxidation (Niki *et al.*, 1993). Many reports indicate that ubiquinol is the first LDL antioxidant consumed during induction of lipid peroxidation, thus acting as the first "line of defense" (Stocker *et al.*, 1991; Mohr *et al.*, 1992; Tribble *et al.*, 1994). Bowry and Stocker (1993) suggested that vitamin E acts as a prooxidant in LDL when peroxyl radical flux is low, and that the additional presence of

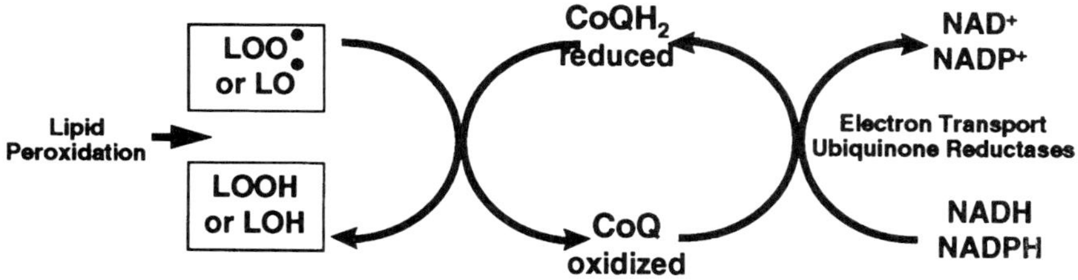

FIGURE 1. Model 1 for antioxidant action of coenzyme Q.

ubiquinol in LDL is important in preventing this prooxidant effect. Even if this is not so, there is a consensus that ubiquinol is an effective antioxidant in LDL.

However, in LDL, there are no electron transport systems to regenerate ubiquinol from its oxidized form, ubiquinone. Estimates of the ubiquinol content of LDL range from 0.1 molecule/LDL particle (Tribble *et al.*, 1994) to 1 molecule per particle (Stocker *et al.*, 1991). Even with the highest of these estimates, the antioxidant capacity of ubiquinol in LDL, in the absence of a means of reduction, is trivial compared to that of vitamin E, which is present in a concentration of about 6 molecules per LDL particle (Esterbauer *et al.*, 1992). If vitamin E does, indeed, exert a prooxidant effect in LDL in the absence of ubiquinol, the protective effect of ubiquinol will be quite limited unless there exists some other mechanism than electron transport to reduce ubiquinone to its antioxidant form. Hence, it is puzzling as to how ubiquinol can be more than a minor antioxidant component in LDL, and how it could be protective against possible prooxidant effects of vitamin E in LDL.

3. DOES COENZYME Q RECYCLE VITAMIN E?

In addition to coenzyme Q, another ubiquitous antioxidant in membranes and LDL is vitamin E, which acts as a lipid peroxidation chain-breaking antioxidant by reducing peroxyl radicals and forming the vitamin E radical (a relatively unreactive radical). In virtually every location that coenzyme Q is found, vitamin E is also found (Frei *et al.*, 1990). There is no doubt that these two compounds both exert antioxidant effects. However, evidence indicates that when reduced coenzyme Q and vitamin E coexist, coenzyme Q will act as an antioxidant indirectly, by regenerating vitamin E from the vitamin E radical, thus recycling vitamin E for another round of lipid peroxidation chain-breaking. The question of whether coenzyme Q's antioxidant function in biological systems is the result of direct or indirect (vitamin E recycling) effects has become the topic of vigorous discussion.

3.1. Direct Radical Scavenging Effects of Coenzyme Q

Ubiquinol acts as a direct chain-breaking antioxidant by donating a hydrogen atom to reduce peroxy and/or alkoxy radicals:

$$CoQH_2 + ROO^{\cdot} \rightarrow CoQH^{\cdot} + RO_2H \qquad (1)$$

where $CoQH_2$ = ubiquinol and $CoQH^{\cdot}$ = ubisemiquinone radical.

In liposomes which do not contain vitamin E (Frei *et al.*, 1990; Landi *et al.*, 1990a, 1992; Yamamoto *et al.*, 1990; Cipollone *et al.*, 1994), as well as in biological membrane systems in which vitamin E has been removed by mild nondenaturing extraction with pentane (Forsmark *et al.*, 1991), ubiquinol exerts antioxidant effects against lipid peroxidation. These results demonstrate that direct antioxidant action of ubiquinol in biological systems is certainly possible.

However, the chemical reactivity of ubiquinols with peroxyl radicals in organic solvents is about one tenth ($k = 3.4 \times 10^5 M^{-1} sec^{-1}$) the reactivity of tocopherols ($k = 33.0 \times 10^5 M^{-1} sec^{-1}$) (Naumov and Khrapova, 1987; Mukai *et al.*, 1990, 1993). Since concentrations of ubiquinols in membranes roughly equal to those of α-tocopherol (Aberg *et al.*, 1992; Lang *et al.*, 1986) and their concentration in LDL is roughly 10–20% of that of α-tocopherol (Esterbauer and Ramos, 1995), it is likely that *in vivo* vitamin E is more efficient as a chain-breaking antioxidant, i.e., as a direct peroxyl radical scavenger. Indeed, the antioxidant activity of ubiquinols in membranes *in vitro* is significantly lower than that of vitamin E (Kagan *et al.*, 1990a).

3.2. Reduction of Vitamin E Radicals by Ubiquinols

Mellors and Tappel (1966a) suggested a second possibility for the antioxidant function of coenzyme Q, namely that it could recycle vitamin E. The one-electron redox potential for ubiquinol [$E_{7.0}$ ($QH_2/Q^{\cdot}$) = −0.24 V] is more negative than that for tocopherol [$E_{7.0}$ (T-OH/T-O$^{\cdot}$) = 0.48 V] (Neta and Steenken, 1982), suggesting that ubiquinol may reduce the vitamin E phenoxyl radical (formed when vitamin E quenches a peroxyl radical), thus regenerating vitamin E:

$$\mathrm{CoQH_2 + T\text{-}O^{\cdot} \rightarrow CoQH^{\cdot} + T\text{-}OH} \qquad (2)$$

where T-O$^{\cdot}$ = vitamin E phenoxyl radical and T-OH = vitamin E.

In organic solvent, the interaction of ubiquinols with tocopheroxyl radicals is very efficient: the rate constant for this reaction is about $10^6 M^{-1} sec^{-1}$ (Mukai *et al.*, 1990), i.e., higher than the rate constant for the reaction of ubiquinol with peroxyl radicals (see above). Thus, ubiquinols should preferentially reduce tocopheroxyl radicals rather than peroxyl radicals.

We have previously demonstrated that electron carriers in rat liver microsomes, mitochondria, and submitochondrial particles, as well as in mitochondrial complexes integrated into liposomes, regenerate α-tocopherol from its phenoxyl radical by *ubiquinone-dependent reduction* (Kagan *et al.*, 1990b; Kagan and Packer, 1993; Maguire *et al.*, 1992). This suggests that redox interactions of coenzyme Q with α-tocopherol are important in the antioxidant protection of electron transport membranes.

Hence, there is little doubt that coenzyme Q exerts direct antioxidant effects and will also, in the presence of vitamin E, recycle the vitamin E radical back to vitamin E, leading to a more complete mechanism of the antioxidant action of coenzyme Q in biological systems (Model 2; see Fig. 2).

However, the exact contribution of each of these pathways to antioxidant effects of coenzyme Q in biological systems is unknown. Knowledge of the exact mechanism of interaction of these two antioxidants is important in determining their relative roles in health and possible preventive and therapeutic roles in disease. A major goal of our research is to determine the balance of these mechanisms in a variety of physiological milieus.

4. THE ROLE OF SUPEROXIDE IN THE ANTIOXIDANT ACTION OF UBIQUINOL

Electron transport in membranes is always accompanied by generation and release of superoxide, whose level may become very high under some conditions (e.g., during the oxidative burst of phagocytic cells, or in the presence of redox-cycling drugs) (Crawford and Schneider, 1983; Nohl and Jordan, 1986). Overproduction of superoxide creates prooxidant conditions, resulting in depletion of antioxidants and development of oxidative stress.

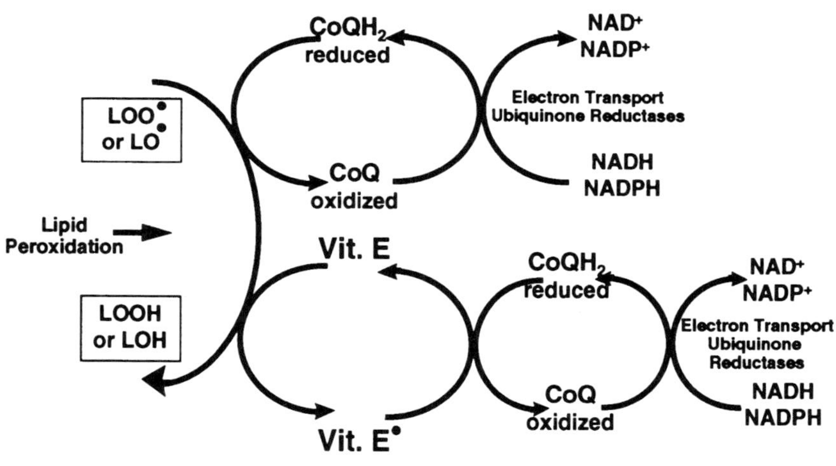

FIGURE 2. Model 2 for antioxidant action of coenzyme Q.

Superoxide readily reduces coenzyme Q to form ubisemiquinone and ubiquinol (directly or via disproportionation of ubisemiquinone radicals) (Nohl and Jordan, 1986, Sugioka *et al.*, 1988):

$$2CoQ + 2O_2^{-\cdot} \rightarrow 2CoQH^{\cdot} + 2O_2 \tag{3}$$

$$2CoQH^{\cdot} + 2H^{+} \rightarrow CoQH_2 + CoQ \tag{4}$$

$$CoQ + 2O_2^{-\cdot} + 2H^{+} \leftrightarrow CoQH^{\cdot} + 2O_2 \tag{5}$$

where CoQ = ubiquinone, $CoQH^{\cdot}$ = ubisemiquinone, and $CoQH_2$ = ubiquinol.

Reaction (3) is reversible: ubisemiquinone can undergo autooxidation yielding superoxide radicals (Boveris *et al.*, 1976; Turrens *et al.*, 1985). This suggests that the ubiquinone/ubisemiquinone redox couple may exert either antioxidant or prooxidant effects depending on the steady-state concentrations of oxidized and reduced coenzyme Q, oxygen, and superoxide.

During many periods of high oxidative stress, conditions would exist to shift reaction (5) to the right. This is especially true in oxidative stress conditions involving high levels of superoxide. Superoxide causes lipid peroxidation (through dismutation to hydrogen peroxide and reaction with transition metals to form hydroxyl radicals, which initiate peroxidation), which causes formation of the vitamin E radical during the quenching of lipid peroxidation by vitamin E; superoxide also interacts directly with vitamin E to form the vitamin E radical (Fukuzawa and Gebicki, 1983; Ozawa and Hanaki, 1985; Nani *et al.*, 1985; Beyer *et al.*, 1987; Nohl, 1990; Noack *et al.*, 1994). Ubiquinol levels are decreased, either by its direct antioxidant lipid peroxidation chain breaking, or by recycling of vitamin E radicals, or both. This, in turn increases ubiquinone levels. Thus, reactants in reaction (5) increase, and products decrease, pulling it to the right.

Hence, under conditions of high superoxide production, superoxide itself may serve as a means to reduce ubiquinone, multiplying coenzyme Q's antioxidant potency by providing another pathway besides electron transport for converting ubiquinone to ubiquinol. We proposed this as a novel pathway for the reduction of ubiquinone in biological systems (Stoyanovsky *et al.*, 1995). This has led us to postulate a new model for the biological action of coenzyme Q (see Fig. 3).

Thus, a major factor in the antioxidant function of coenzyme Q is superoxide-dependent reduction of ubiquinone to ubisemiquinone and ubiquinol, which can act as antioxidants either directly or by reducing the

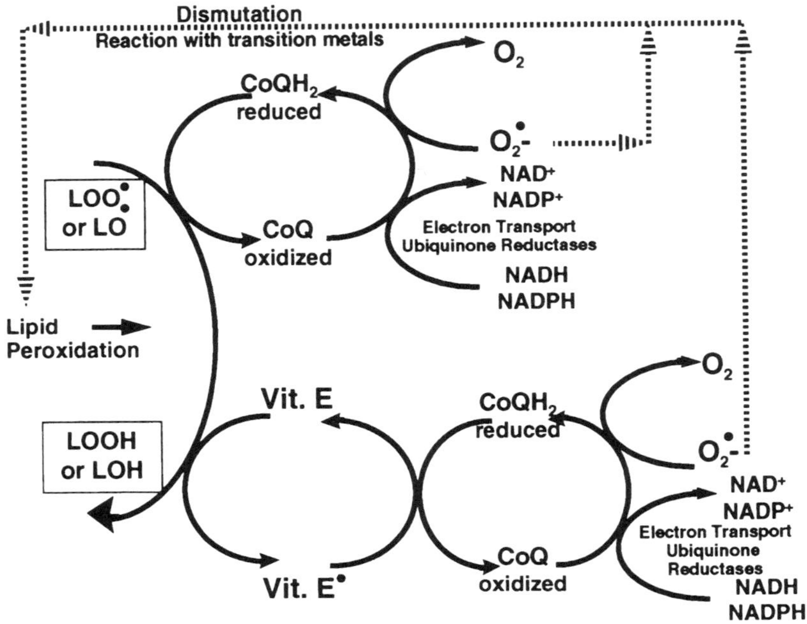

FIGURE 3. Model 3 for antioxidant action of coenzyme Q.

vitamin E radical, resulting in regeneration of vitamin E. Hence, under conditions of high superoxide production (i.e., high oxidative stress), coenzyme Q's antioxidant power is multiplied manifold by its continuous reduction by the very superoxide which causes oxidative stress.

4.1. Ubisemiquinone Protects Vitamin E against Superoxide-Driven Oxidation

Our experiments in an aprotic medium (DMSO) demonstrated that superoxide-driven reduction of ubiquinone occurs, and that it serves as a source of reducing power to regenerate vitamin E from its radical (Stoyanovsky *et al.*, 1995). When KO_2 (a source of superoxide) and ubiquinone-10 were mixed, ubisemiquinone-10 radicals were formed which could be detected by their characteristic EPR spectrum. Addition of α-tocopherol caused a concentration-dependent decrease of the ubisemiquinone-10 radical steady-state concentration. This is consistent with generation of both tocopheroxyl and ubisemiquinone radicals by superoxide and subsequent reaction between them, leading to a reduction in ubisemiquinone radical levels and protection of tocopherol. Our high-performance liquid chromatography (HPLC) measurements showed that ubiquinone-10 protected α-tocopherol against KO_2 oxidation in a concentration-dependent manner. We concluded that one-electron reduction of ubiquinone-10 by superoxide ion resulted in the formation of ubisemiquinone-10 radicals, which donated an electron to the α-tocopherol phenoxyl radical (formed from interaction with superoxide). Thus redox cycling of α-tocopherol from its radical form maintains α-tocopherol concentrations when both superoxide and ubiquinone are present.

We further investigated redox reactions between superoxide radicals, a water-soluble vitamin E homologue, Trolox, and a water-soluble coenzyme Q homologue, Qo, using EPR spectroscopy. The amplitude of the electron paramagnetic resonance (EPR) signal of Qo semiquinone radical recorded on addition of increasing concentrations of Qo to the superoxide-generating system showed a linear relationship over the concentration range 0–0.8 mM Qo. The signal amplitude of the Qo radical was decreased in a concentration-dependent manner by addition of Trolox. Thus, superoxide radicals reduce Qo to semiquinone radicals and the concentration of semiquinone radicals is decreased when Trolox is added to the superoxide-generating system. Using HPLC, we demonstrated that Trolox oxidation induced by superoxide (xanthine/xanthine oxidase) was almost completely prevented by Qo at a concentration about half that of the concentration of Trolox.

4.2. Superoxide-Dependent Reduction of Coenzyme Q Protects Cell Phospholipids against Oxidative Stress

Finally, we used a cell model to demonstrate that superoxide-driven reduction of coenzyme Q can protect lipids against oxidative stress. In this model, human leukemia HL-60 cells in which membrane phospholipids were metabolically prelabeled with oxidation-sensitive fluorescent fatty acid, *cis*-parinaric acid (PnA), were exposed to oxidative stress. The cells were supplemented with α-tocopherol (20 nmol/10^6 cells) and exposed to a lipid-soluble azo-initiator of peroxyl radicals, AMVN, in the presence and in the absence of superoxide (generated by xanthine/xanthine oxidase). As shown in Fig. 4, AMVN alone and superoxide alone caused a significant oxidation of major phospholipids in membranes of HL-60 cells—phosphatidylcholine, phosphatidylethanolamine, phosphatidylinositol, and phosphatidylserine. Combination of AMVN and superoxide produced additively stronger oxidation of phospholipids. Importantly, the cells supplemented with coenzyme Q10 (20 nmol/10^6 cells) exerted protection against

(AMVN + superoxide)-induced peroxidation (Fig. 4) which was not observed in AMVN-treated cells.

Combined, our results indicate that a major factor in the antioxidant function of coenzyme Q is superoxide-dependent reduction of ubiquinone to ubisemiquinone and ubiquinol, which can act as antioxidants either directly or by reducing the vitamin E radical, resulting in regeneration of vitamin E. Hence, under conditions of high superoxide production (i.e., high oxidative stress), coenzyme Q's antioxidant power is multiplied manifold by its continuous reduction by the very superoxide which causes oxidative stress.

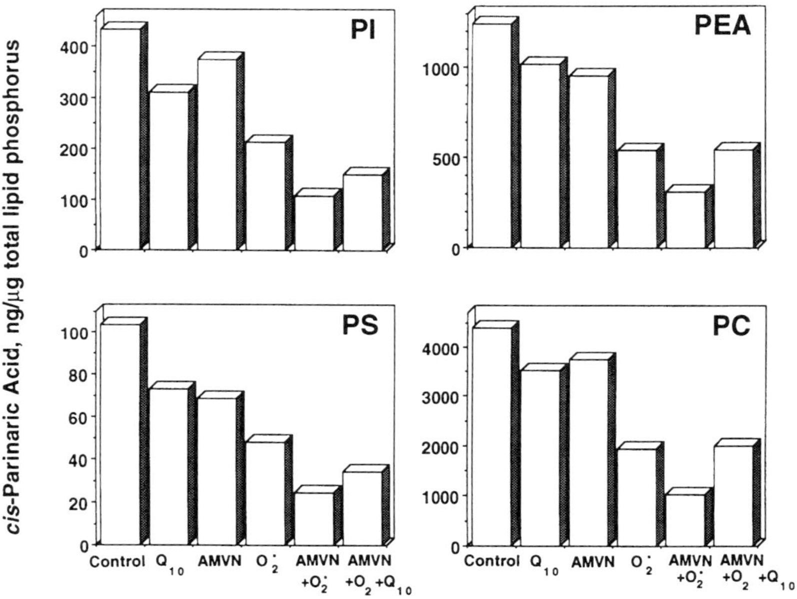

FIGURE 4. Effect of superoxide-driven reduction of coenzyme Q on lipid peroxidation in HL-60 cells supplemented with α-tocopherol. HL-60 cells were grown in RPMI 1640 medium supplemented with 10% fetal bovine serum and at 37°C in 5% CO_2 atmosphere. Cells diluted to a density of 0.5×10^6 cells/ml in RPMI medium containing 10% fetal bovine serum were treated with α-tocopherol (20 nmol/10^6 cells) predisssolved in tetrahydrofuran (THF). Cells were incubated in CO_2 atmosphere for 24 hr at 37°C. After incubation, cells were washed with L1210 buffer. *cis*-Parinaric acid was metabolically incorporated into HL-60 cells supplemented with α-tocopherol as described earlier (Tyurina *et al.*, 1997). PI, Phosphatidylinositol; PS, phosphatidylserine; PEA, phosphatidylethanolamine; PC, phosphatidylcholine. The data presented are means from three measurements; standard error was <10%.

5. ANTIOXIDANTS, OXIDATIVE STRESS, AND HUMAN HEALTH

The importance of antioxidants in human health and aging has now been overwhelmingly established. Laboratory, epidemiological, and clinical data make it clear that antioxidants manufactured by the body (e.g., glutathione and the antioxidant enzymes superoxide dismutase, catalase, glutathione peroxidase, and glutathione reductase) as well those received in the diet (e.g., vitamin E and vitamin C) are crucial to the well-being and longevity of humans (Manson *et al.*, 1993; Ames *et al.*, 1993; Cutler, 1991; Halliwell, 1996). However, mechanisms of interactions among the antioxidants, which have the potential to increase or decrease their effects manifold, are still unclear.

5.1. General Health Significance of Coenzyme Q

The emergence of coenzyme Q as an antioxidant has special significance in this regard. It is the only lipid-soluble antioxidant so far known that is manufactured in the body (Ernster and Dallner, 1995). It is easily converted from its oxidized form back to its reduced, antioxidant form, by well-established electron transport pathways. In its reduced form, it is capable of recycling vitamin E, which is a more potent lipid peroxidation chain-breaking antioxidant. The existence of a mechanism of reduction of coenzyme Q via interaction with superoxide opens the possibility of a greatly expanded antioxidant role for coenzyme Q, because its antioxidant function would no longer be restricted to areas where electron transport exists. This may be important not only in membranes, but in nonmembrane systems such as LDL particles.

Because superoxide represents a previously unrecognized source of reducing power for coenzyme Q that is present in virtually every area of the body, and which increases during periods of oxidative stress, and because vitamin E and coenzyme Q also coexist in virtually every lipid environment, the proposed mechanism provides a new link between metabolism and oxidative stress on one hand, and clearly established antioxidant mechanisms on the other. Elucidation of this link will provide a clearer understanding of oxidant/antioxidant interactions both in normal physiology and during oxidative stress, which will, in turn, lead to a greater ability to manipulate the balance in favor of protection.

5.2. Special Considerations with Low-Density Lipoprotein

This mechanism represents an especially important new dimension in LDL. Although, as outlined above, there is much evidence that coenzyme

Q is one of the most important antioxidants in LDL (Stocker *et al.*, 1991; Mohr *et al.*, 1992; Tribble *et al.*, 1994) and, in fact, vitamin E may act as a prooxidant in LDL in the absence of coenzyme Q (Bowry and Stocker, 1993), it has long been a mystery as to how the low levels of coenzyme Q in LDL could be protective *in vivo*. Although some studies have indicated that the oxidizability of LDL is proportional to it coenzyme Q content, or to its total [coenzyme Q + vitamin E] content, and that supplementation of LDL with coenzyme Q decreases its susceptibility to oxidation (Stocker *et al.*, 1991; Tribble *et al.*, 1994; Thomas *et al.*, 1996; Kontush *et al.*, 1994), other studies have reported no effect of added coenzyme Q (Kaikkonen *et al.*, 1997). However, as the oxidation system used in such studies does not involve the production of superoxide, in view of the new proposed model of coenzyme Q/superoxide/vitamin E interactions it is not surprising that little protective effect is sometimes found.

In the physiological milieu, however, superoxide production may occur during many of the processes related to the development of atherosclerosis, e.g., inflammation (Alexander, 1994; Ridker *et al.*, 1997). Such a source of superoxide may represent a factor that converts coenzyme Q in LDL from a relatively minor component into one that may represent a major source, perhaps *the* major source, of *in vivo* protection. Because coenzyme Q levels in LDL are easily bolstered through dietary supplementation, this represents a potentially useful therapy for atherosclerosis, especially in those in whom inflammatory and other superoxide-producing processes are more prominent. In this regard, it might augment or replace the effects of aspirin, which has recently been shown to be especially effective in inhibiting atherosclerosis in individuals in whom inflammation is prominent (Ridker *et al.*, 1997); in contrast to aspirin, supplementation with coenzyme Q has few or no toxic effects.

Certain widely prescribed medications affect coenzyme Q levels, which may alter antioxidant conditions within the LDL particles and elsewhere. The β-hydroxy-β-methylglutamyl-CoA (HMG-CoA) reductase inhibitors are a major class of drugs used to treat hypercholesterolemia. HMG-CoA reductase inhibitors block an early step in the synthesis of cholesterol, in a pathway which is shared by cholesterol synthesis, synthesis of dolichol, and synthesis of coenzyme Q (Goldstein and Brown, 1990). Decreases in tissue and plasma levels of coenzyme Q have been documented during treatment with HMG-CoA reductase inhibitors (Bargossi *et al.*, 1994; Willis *et al.*, 1990; Elmberger *et al.*, 1991; Ghirlanda *et al.*, 1993; Marinari *et al.*, 1995). Hence, these drugs may have two simultaneous and opposing effects on atherogenesis: decreasing plasma LDL cholesterol levels while at the same time increasing the oxidizability of LDL. Clinical data clearly show that the balance is tipped toward the former effect, since these drugs dramatically lower the incidence of myocardial infarction. However, their effectiveness may be increased if the missing coenzyme Q is replaced, or if another drug which blocks cholesterol synthesis at a step beyond the branch point for coenzyme Q (e.g., squalestatin 1; see Baxter *et al.*, 1992) is used. Supplementation with coenzyme Q may suggest a means of further increasing the effectiveness of HMG-CoA reductase inhibitors in reducing atherogenesis, of potential benefit to hundreds of thousands of people currently using these drugs.

6. REFERENCES

Aberg, F., Appelkvist, E. L., Dallner, G., and Ernster, L., 1992, Distribution and redox state of ubiquinones in rat and human tissues, *Arch. Biochem. Biophys.* **295:**230–234.

Alexander, R. W., 1994, Inflammation and coronary artery disease, *N. Engl. J. Med.* **331:**468–469.

Ames, B. N., Shigenaga, M. K., and Hagen, T. M., 1993, Oxidants, antioxidants, and the degenerative diseases of aging, *Proc. Natl. Acad. Sci. USA* **90:**7915–7922.

Bargossi, A. M., Grossi, G., Fiorella, P. L., Gaddi, A., Di Giullo, R., and Battino, M., 1994, Exogenous CoQ10 supplementation prevents plasma ubiquinone reduction induced by HMG-CoA reductase inhibitors, *Mol. Aspects Med.* **15**(Suppl.)**:**s187–s193.

Baxter, A., Fitzgerald, B. J., Hutson, J. L., McCarthy, A. D., Motteram, J. M., Ross, B. C., Sapra, M., Snowden, M. A., Watson, N. S., and Williams, R. J., 1992, Squalestatin 1, a potent inhibitor of squalene synthase, which lowers serum cholesterol *in vivo*, *J. Biol. Chem.* **267:**11705–11708.

Beyer, R. E., 1988, Inhibition by coenzyme Q of ethanol- and carbon tetrachloride-stimulated lipid peroxidation *in vivo* and catalyzed by microsomal and mitochondrial systems, *Free Radic. Biol. Med.* **5:**297–303.

Beyer, R. E., 1990, The participation of coenzyme Q in free radical production and antioxidation, *Free Radic. Biol. Med.* **8:**545–565.

Beyer, R. F. G., Nordenbrand, K., and Ernster, L., 1987, The function of coenzyme Q in free radical production and as an antioxidant: A review, *Chem. Scripta* **27:**145–153.

Booth, R. F., Galanopoulou, D. G., and Quinn, P. J., 1982, Protection by ubiquinone and ubiquinol against lipid peroxidation in egg yolk phosphatidylcholine liposomes, *Biochem. Int.* **5:**151–156.

Boveris, A., Cadenas, E., and Stoppani, A. O. M., 1976, Role of ubiquinone in the mitochondrial generation of hydrogen peroxide, *Biochem. J.* **156:**435–444.

Bowry, V. W., and Stocker, R., 1993, Tocopherol-mediated peroxidation. The prooxidant effect of vitamin E on the radical-initiated oxidation of human low density lipoprotein, *J. Am. Chem. Soc.* **115:**6029–6044.

Cipollone, M., Fiorentini, D., Galli, M. C., Sechi, A. M., and Landi, L., 1994, Autooxidation and antioxidant activity of ubiquinol homologues in large unilamellar vesicles, *Chem. Phys. Lipids* **69:**87–94

Crane, F. L., and Morre, D. J., 1977, Evidence for coenzyme Q function in Golgi membranes, in *Biomedical and Clinical Aspects of Coenzyme Q* (K. Folkers and Y. Yamamura, eds.), pp. 3–14, Elsevier, Amsterdam.

Crane, F. L., Hatefi, Y., Lester, R. L., and Widmer, C., 1957, Isolation of a quinone from beef heart mitochondria, *Biochim. Biophys. Acta* **25:**220–221.

Crawford, D. R., and Schneider, D. L., 1983, Ubiquinone content and respiratory burst activity of latex-filled phagolysosomes isolated from human neutrophils and evidence for the probable involvement of a third granule, *J. Biol. Chem.* **258:**5363–5367.

Cutler, R. G., 1991, Antioxidants and aging, *Am. J. Clin. Nutr.* **53:**373S–379S.

Elmberger, P. G., Kalen, A., Brunk, U. T., and Dallner, G., 1989, Discharge of newly-synthesized dolichol and ubiquinone with lipoproteins to rat liver perfusate and to the bile, *Lipids* **24:**919–930.

Elmberger, P. G., Kalen, A., Lund, E., Reihner, E., Eriksson, M., Berglund, L., Angelin, B., and Dallner, G., 1991, Effects of pravastatin and cholestyramine on products of the mevalonate pathway in familial hypercholesterolemia, *J. Lipid Res.* **32:**935–940.

Ernster, L., and Dallner, G., 1995, Biochemical, physiological and medical aspects of ubiquinone function, *Biochim. Biophys. Acta* **1271**(1)**:**195–204.

Ernster, L., Forsmark, P., and Nordenbrand, K., 1992, The mode of action of lipid-soluble antioxidants in biological membranes. Relationship between the effects of ubiquinol and vitamin E as inhibitors of lipid peroxidation in submitochondrial particles, *J. Nutr. Sci. Vitaminol.* **548:**41–46.

Esterbauer, H., and Ramos, P., 1995, Chemistry and pathophysiology of oxidation of LDL, *Rev. Physiol. Biochem. Pharmacol.* **127:**31–64

Esterbauer, H., Dieber-Rothenderm, M., Striegel, G., and Waeg, G., 1991, Role of vitamin E in preventing the oxidation of low-density lipoproteins, *Am. J. Clin. Nutr.* **53:**S314–S321.

Esterbauer, H., Gebicki, J., Puhl, H., and Juergens, G., 1992, The role of lipid peroxidation and antioxidants in oxidative modification of LDL, *Free Radic. Biol. Med.* **13:**341–390.

Forsmark, P., Aberg, F., Norling, B., Nordenbrand, K., Dallner, G., and Ernster, L., 1991, Inhibition of lipid peroxidation by ubiquinol in submitochondrial particles in the absence of vitamin E, *FEBS Lett.* **285:**39–43.

Forsmark-Andree, P., Dallner, G., and Ernster, L., 1995, Endogenous ubiquinol prevents protein modification accompanying lipid peroxidation in beef heart submitochondrial particles, *Free Radic. Biol. Med.* **19:**749–757.

Frei, B., Kim, M. C., and Ames, B. N., 1990, Ubiquinol-10 is an effective lipid-soluble antioxidant at physiological concentrations, *Proc. Natl. Acad. Sci. USA* **87:**4879–4883.

Fukuzawa, K., and Gebicki, J. M., 1983, Oxidation of alpha-tocopherol in micelles and liposomes by the hydroxyl, perhydroxyl, and superoxide free radicals, *Arch. Biochem. Biophys.* **226:**242–251.

Ghirlanda, G., Oradei, A., Manto, A., Lippa, S. Uccioli, L., Caputo, S., Greco, A. V., and Littaru, G. P., 1993, Evidence of plasma CoQ10-lowering effect by HMG-CoA reductase inhibitors: A double-blind, placebo-controlled study, *J. Clin. Pharmacol.* **33:**226–229.

Goldstein, J. L., and Brown, M. S., 1990, Regulation of the mevalonate pathway, *Nature* **343:**425–430.

Gutman, M., 1980, Electron flux through the mitochondrial ubiquinone, *Biochim. Biophys. Acta* **594:**53–84.

Halliwell, B., 1996, Antioxidants in human health and disease, *Annu. Rev. Nutr.* **16:**33–50.

Jakobsson-Borin, A., Aberg, F., and Dallner, G., 1994, Lipid peroxidation of microsomal and mitochondrial membranes extracted with *n*-pentane and reconstituted with ubiquinol, dolichol and cholesterol, *Biochim. Biophys. Acta* **1213:**159–166.

Kagan, V. E., and Packer, L., 1993, Antioxidative function of vitamin E and ubiquinols, in *Methods in Toxicology*, Vol. 2 (L. Lash and D. Jones, eds.), pp. 277–285, Academic Press, New York.

Kagan, V. E., Serbinova, E. A., and Packer, L., 1990a, Recycling and antioxidant activity of tocopherol homologues of differing hydrocarbon chain length in liver microsomes, *Arch. Biochem. Biophys.* **282**(2)**:**221–225.

Kagan, V. E., Serbinova, E. A., and Packer, L., 1990b, Antioxidant effects of ubiquinones in microsomes and mitochondria are mediated by tocopherol recycling, *Biochem. Biophys. Res. Commun.* **169:**851–857.

Kaikkonen, J., Nyyssonen, K., Porkkala-Sarataho, P., Poulsen, H. E., Metsa-Ketela, T., Hayn, M., Salonen, R., and Salonen, J. T., 1997, Effect of oral coenzyme Q10 supplementation on the oxidation resistance of human VLDL + LDL fraction: Absorption and antioxidative properties of oil and granule-based preparations, *Free Radic. Biol. Med.* **22:**1195–1202.

Kalen, A., Norling, B., Appelkvist, E. L., and Dallner, G., 1987, Ubiquinone synthesis in the microsomal fraction of rat liver, *Biochim. Biophys. Acta* **926:**79–80.

Kontush, A., Hubner, C., Finckh, B., Kohlschutter, A., and Beisiegel, U., 1994, Low density lipoprotein oxidizability by copper correlates to its initial ubiquinol-10 and polyunsaturated fatty acid content, *FEBS Lett.* **341:**69–73

Landi, L., Cabrini, L., Sechi, A. M., and Pasquali, P., 1984, Antioxidant effect of ubiquinones on mitochondrial membranes, *Biochem. J.* **222:**463–466.

Landi, L., Cabrini, L., Tadolini, B., Sechi, A. M., and Pasquali, P., 1985, Incorporation of ubiquinones into lipid vesicles and inhibition of lipid peroxidation, *Ital. J. Biochem.* **34:**356–363.

Landi, L., Fiorentini, D., Stefanelli, C., Pasquali, P., and Pedulli, G. F., 1990a, Inhibition of autooxidation of egg yolk phosphatidylcholine in homogeneous solutions and in liposomes by oxidized ubiquinone, *Biochim. Biophys. Acta* **1028:**223–228.

Landi, L., Cabrini, L., Fiorentini, D., Sechi, A. M., Sartor, G., Pasquali, P., and Masotti, L., 1990b, Coenzyme Q-3 as an antioxidant. Its effect on the composition and structural properties of phospholipid vesicles, *Cell Biophys.* **16:**1–12.

Landi, L., Cabrini, L., Fiorentini, D., Stefanelli, C., and Pedulli, G. F., 1992, The antioxidant activity of ubiquinol-3 in homogeneous solution and in liposomes, *Chem. Phys. Lipids* **61:**121–130.

Lang, J., Gohil., K., and Packer, L., 1986, Simultaneous determination of tocopherols, ubiquinols and ubiquinones in blood, plasma, tissue homogenates and subcellular fractions, *Anal. Biochem.* **1567:**106–116.

Lea, C. H., and Kwietney, A., 1962, The antioxidant action of ubiquinones and related compounds, *Chem. Ind.* (London) **24:**1245–1246.

Maguire, J. J., Kagan, V. E., Serbinova, E. A., Ackrell, B. A., and Packer, L., 1992, Succinate-ubiquinone reductase-linked recycling of alpha-tocopherol in reconstituted systems and mitochondria: Requirement for reduced ubiquinone, *Arch. Biochem. Biophys.* **292:**47–53.

Manson, J. E., Gaziano, J. M., Jonas, M. A., and Hennekens, C. H., 1993, Antioxidants and cardiovascular disease: A review, *J. Am. Coll. Nutr.* **12:**426–431.

Marinari, U. M., Pronzato, M. A., Dapino, D., Gazzo, P., Traverso, N., Cottalasso, D., and Odetti, P., 1995, Effects of simvastatin on liver and plasma levels of cholesterol, dolichol and ubiquinol in hypercholesterolemic rats, *Ital. J. Biochem.* **44:**1–90.

Mellors, A., and Tappel, A. L., 1966a, The inhibition of mitochondrial peroxidation by ubiquinone and ubiquinol, *J. Biol. Chem.* **241:**4353–4356.

Mellors, A., and Tappel, A. L., 1966b, Quinones and quinols as inhibitors of lipid peroxidation, *Lipids* **1:**282–284.

Mitchell, P., 1975a, Protonmotive redox mechanisms of cytochrome b-c1 complex in the respiratory chain: Protonmotive ubiquinone cycle, *FEBS Lett.* **45:**1–6.

Mitchell, P., 1975b, The protonmotive Q cycle: A general formulation, *FEBS Lett.* **59:**137–139.

Mohr, D., Bowry, V. W., and Stocker, R., 1992, Dietary supplementation with coenzyme Q10 results in increased levels of ubiquinol-10 within circulating lipoproteins and increased resistance of human low-density lipoprotein to the initiation of lipid peroxidation, *Biochim. Biophys. Acta* **1126:**247–254.

Mukai, K., Kikuchi, S., and Urano, S., 1990, Stopped-flow kinetic study of the regeneration reaction of tocopheroxyl radical by reduced ubiquinone-10 in solution, *Biochim. Biophys. Acta* **1035:**77–83.

Mukai, K., Morimoto, H., Kikuchi, S., and Nagaoka, S., 1993, Kinetic study of free radical scavenging action of biological hydroquinones (reduced forms of ubiquinone, vitamin K and tocopherol quinone) in solution, *Biochim. Biophys. Acta* **1157:**313–317.

Nani, A. E., Jr., Stallings, M., and Sawyer, D., 1985, Does superoxide ion oxidize catechol, α-tocopherol, and ascorbic acid by direct electron transfer? *J. Am. Chem. Soc.* **102:**4481–4485.

Naumov, V. V., and Khrapova, N. G., 1987, Chemiluminescence study of ubiquinone and ubiquinol interaction with peroxide radicals, *Biophysics USSR* **28:**730–735.

Neta, P., and Steenken, S., 1982, One electron redox potentials of phenols, hydroxy and aminophenols and related compounds of biological interest, *J. Phys. Chem.* **93:**7654–7659.

Niki, E., Noguchi, N., and Gotoh, N., 1993, Inhibition of oxidative modification of low density lipoprotein by antioxidants, *J. Nutr. Sci. Vitaminol.* **39:**S1–S8.

Noack, H., Kube, U., and Augustin, W., 1994, Relations between tocopherol depletion and coenzyme Q during lipid peroxidation in rat liver mitochondria, *Free Radical Res.* **20:**375–386.

Nohl, H., 1990, Is redox-cycling ubiquinone involved in mitochondrial oxygen activation? *Free Radic. Res. Commun.* **8:**307–315.

Nohl, H., and Jordan, W., 1986,The mitochondrial site of superoxide formation, *Biochem. Biophys. Res. Commun.* **138:**533–539.

Nohl, H., Jordan, W., and Youngman, R. J., 1986, Quinones in biology: Functions in electron transfer and oxygen activation, *Adv. Free Radic. Biol. Med.* **2:**211–279.

Norling, B., Glazek, E., Nelson, B. D., and Ernster, L., 1974, Studies with ubiquinone-depleted submitochondrial particles. Quantitative incorporation of small amounts of ubiquinone and its effects on the NADH and succinate oxidase activities, *Eur. J. Biochem.* **47:**475–482.

Ozawa, T., and Hanaki, A., 1985, Spectroscopic studies on the reaction of superoxide ion with tocopherol model compound, 6-hydroxy-2,2,5,7,8-pentamethylchroman, *Biochem. Biophys. Res. Commun.* **126:**873–878.

Ridker, P. M., Cushman, M., Stampfer, M. J., Tracy, R. P., and Hennekens, C. H., 1997, Inflammation, aspirin, and the risk of cardiovascular disease in apparently healthy men, *N. Engl. J. Med.* **336:**973–979.

Stocker, R., Bowry, V. W., and Frei, B., 1991, Ubiquinol-10 protects human low density lipoprotein more efficiently against lipid peroxidation than does α-tocopherol, *Proc. Natl. Acad. Sci. USA* **88:**1646–1650.

Stoyanovsky, D. A., Osipov, A. N., Quinn, P. J., and Kagan, V. E., 1995, Ubiquinone-dependent recycling of vitamin E radicals by superoxide, *Arch. Biochem. Biophys.* **323:**343–351.

Sugioka, K., Nakano, M., Totsune-Nakano, H., Minakami, H., Tero-Kubota, S., and Ikegami, Y., 1988, Mechanism of O^-_2 generation in reduction and oxidation cycle of ubiquinones in a model of mitochondrial electron transport systems, *Biochim. Biophys. Acta* **936:**377–385.

Takahashi, K., Mayumi, T., and Kishi, T., 1988, Influence of coenzyme Q10 on doxorubicin uptake and metabolism by mouse myocardial cells in culture, *Chem. Pharmaceut. Bull.* **36:**1514–1518.

Takeshige, K., Takayanagi, K., and Minakami, S., 1980, Reduced coenzyme Q10 as an antioxidant of lipid peroxidation in bovine heart mitochondria, in *Biomedical and Clinical Aspects of Coenzyme Q*, Vol. 2 (Y. Yamamura, K. Folkers, and Y. Ito, eds.), pp. 15–26, Elsevier/North-Holland, Amsterdam.

Thomas, S. R., Neuzil, J., and Stocker, R., 1996, Cosupplementation with coenzyme Q prevents the prooxidant effect of alpha-tocopherol and increases the resistance of LDL to transition metal-dependent oxidation initiation, *Arteriosler. Thromb. Vasc. Biol.* **16:**687–696.

Tribble, D. L., van den Berg, J. J. M., Motchnik, P. A., Ames, B. N., Lewis, D. M., Chait, A., and Krauss, R. M., 1994, Oxidative susceptibility of low density lipoprotein subfractions is related to their ubiquinol-10 content, *Proc. Natl. Acad. Sci. USA* **91:**1183–1187.

Turrens, J. F., Alexandre, A., and Lehninger, A. L., 1985, Ubisemiquinone is the electron donor for superoxide formation by complex III of heart mitochondria, *Arch. Biochem. Biophys.* **237:**408–414.

Tyurina, T. Y., Tyurin, V. A., Carta, G., Quinn, P. J., Scor, N. F., and Kagan, V. E., 1997, Direct evidence for antioxidant effect of Bcl-2 in PC12 rat pheochromocytoma cells, *Arch. Biochem. Biophys.* **344:**413–423.

Willis, R. A., Folkers, K., Tucker, J. L., Ye, C.-Q., Xia, L.-J., and Tamagawa, H., 1990, Lovastatin decreases coenzyme Q levels in rats, *Proc. Natl. Acad. Sci. USA* **87:**8928–8930.

Yamamoto, Y., Komuro, E., and Niki, E., 1990, Antioxidant activity of ubiquinol in solution and phosphatidylcholine liposome, *J. Nutr. Sci. Vitaminol.* **36**(5)**:**505–511.

Chapter 21

Prooxidant Functions of Coenzyme Q

Hans Nohl, Lars Gille, and Andrey V. Kozlov

1. BIOLOGICAL FUNCTIONS AND SIDE FUNCTIONS OF COENZYME Q

Ubiquinone (oxidized coenzyme Q, Q, Q_{10}) shares its biological role in membrane-associated redox reactions with a variety of other redox carriers such as dehydrogenases, non-heme-iron proteins, and cytochromes. The cooperation of this biological quinone with respiratory enzymes is often emphasized through the synonymous name "coenzyme Q." Peculiarities arise from the lack of transition metals, which in contrast to the electron carriers mentioned above, do not participate in redox-shuttle activities of coenzyme Q. Another peculiarity is the lipophilicity of coenzyme Q, which allows free movement between reductants and oxidants of a membrane. The chemistry of ubiquinone reduction and ubiquinol (reduced coenzyme Q, QH_2, $Q_{10}H_2$) oxidation requires the stepwise acceptance and transfer of two single electrons associated with the addition or release of two single protons. These special qualities are widely used in biological membranes for linear electron transfer and transmembrane proton translocation. In mitochondria, under certain conditions linear electron transfer from the semireduced form of coenzyme Q (ubisemiquinone radical(s), SQ) to

Hans Nohl, Lars Gille, and Andrey V. Kozlov Institute of Pharmacology and Toxicology, Veterinary University of Vienna, A-1210 Vienna, Austria.

Subcellular Biochemistry, Volume 30: Fat-Soluble Vitamins, edited by Quinn and Kagan. Plenum Press, New York, 1998.

native oxidants of the respiratory chain (Rieske iron sulfur protein and cytochromes) may run out of control, thereby establishing a permanent source of oxygen radical release (Fig. 1A).

Such an event has not been reported in other biological membranes with even higher coenzyme Q contents than found in mitochondria. This may be due to redox rates which are too low to establish detectable amounts of superoxide radicals. It should be mentioned that in mitochon-

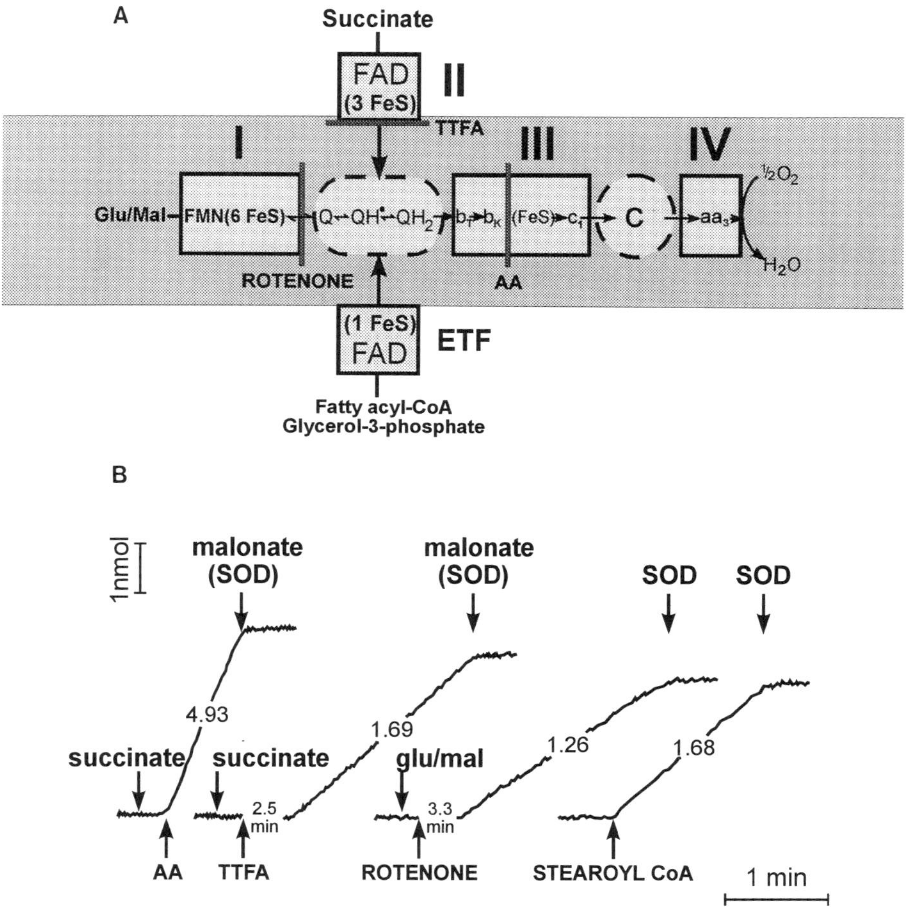

FIGURE 1. Demonstration of $O_2^{\cdot-}$ radical formation through various electron transfer complexes of the respiratory chain. (A) Schematic representation of the respiratory chain demonstrating inhibitors used to reduce selectively the various complexes under study. (B) $O_2^{\cdot-}$ release (nmol/min/mg protein) followed at 37°C. AA, Antimycin A; TTFA; trifluorotheonylacetone; glu/mal, glutamate/malate. For details see Nohl (1982).

dria electron transfer to dioxygen out of sequence requires a particular treatment with inhibitors and uncouplers of the respiratory chain. As shown in Fig. 1, various sections of the respiratory chain, where Q is not involved, can be made directly to divert single electrons to dioxygen when electron flow to the native oxidants is prevented by the use of adequate inhibitors.

These findings reveal that the respiratory chain of mitochondria contains electron carriers operating in complexes I, II, and III which all can be made to establish a direct redox couple with dioxygen, allowing one-electron transfer to the latter.

Nevertheless, it is generally assumed that coenzyme Q is mainly involved in mitochondrial $O_2^{\cdot -}$ generation and that mitochondria represent the major source of $O_2^{\cdot -}$ radicals under physiological and various pathophysiological conditions (Nohl and Hegner, 1978). The ever-increasing application of coenzyme Q as an antioxidant for the prophylaxis and treatment of a great variety of functional disorders, including senescence, has considerably stimulated the interest in the potential prooxidative potency of this natural electron carrier. In this chapter we discuss whether and under which conditions coenzyme Q will contribute to toxic activation of molecular oxygen. We also show that antioxidant-derived reaction products of ubiquinol may stimulate oxygen activation to reactive prooxidants.

2. IS REDUCED COENZYME Q AUTOXIDIZABLE?

Although redox potentials of coenzyme Q redox couples reported in the literature are not homogeneous and are even conflicting, conformity seems to exist that one-electron transfer from ubisemiquinone to dioxygen is thermodynamically more favored than autoxidation of ubiquinol to ubisemiquinone and $O_2^{\cdot -}$.

Experimental evidence comes from recent studies by Kozlov *et al.* (1998). In these studies it was observed that ubiquinol oxidation only proceeds in alkaline (ethanol) solutions, while ubiquinol is stable at physiological pH values. Assuming that in alkaline solutions ubiquinol exists mainly in the deprotonated dianion form (Q^{2-}), one has to consider the following reaction pathways leading to oxidation of Q^{2-}:

$$Q^{2-} + O_2 \rightarrow SQ^{\cdot -} + O_2^{\cdot -} \quad (1)$$

$$SQ^{\cdot -} + O_2 \rightarrow Q + O_2^{\cdot -} \quad (2)$$

$$SQ^{\cdot -} + SQ^{\cdot -} \rightleftharpoons Q^{2-} + Q \quad (3)$$

Reactions (2) and (3) yield Q, which is recovered when oxygen is in the reaction system (Fig. 2A). To assess the pathway by which Q^{2-} oxidation proceeds, oxygen consumption was followed under conditions where the reverse reaction of reaction (3) (comproportionation) was stimulated upon the addition of Q (Fig. 2B). The results support a major role of SQ in oxygen consumption, but do not exclude a contribution of autoxidizing Q^{2-} as well. When the same type of experiment was performed with ubiquinol at physiological pH values the neutral form of ubiquinol was predominant. Under these conditions reoxidation of QH_2 did not occur and consequently oxygen consumption was not observed (Figs. 2C, 2D).

These observations contradict a direct interaction of O_2 with QH_2, thus excluding SQ formation from autoxidizing QH_2 at physiological pH values.

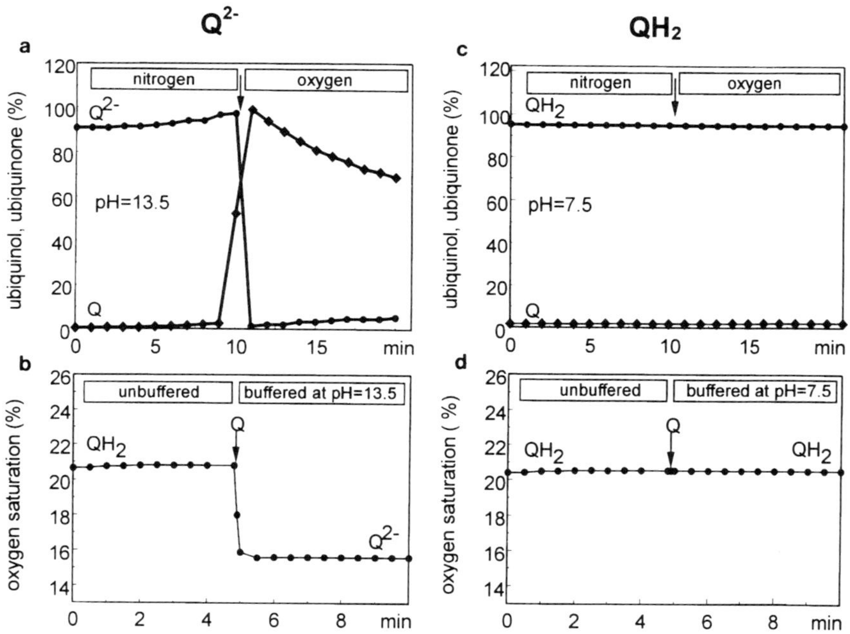

FIGURE 2. Effect of pH on the oxidation of ubiquinol in the presence of molecular oxygen. Oxidation of the ubiquinol dianion and oxygen consumption following the addition of oxygen (a) and the addition of oxidized ubiquinone (b) at pH 13.5. (c, d) Analogous experiments performed at pH 7.5. Experimental conditions: (a, c) QH_2 (final concentration 300 mM) was mixed with ethanol, containing 1% H_2O and 2.5 mM Tris (pH = 7.5 or 13.5) in anaerobic conditions. To start ubiquinol oxidation the solution was equilibrated with air oxygen (indicated by an arrow). (b, d) Unbuffered ethanol solution of QH_2 (final concentration 300 mM) was mixed (indicated by an arrow) with Q (final concentration 300 mM) buffered at pH 7.5 or 13.5 in the presence of air oxygen. For details see Kozlov *et al.* (1998).

Considering that standard redox potentials disregard the effect of steady-state concentrations of redox couples as well as kinetic peculiarities and pH values, which all have effects on the direction of biochemical reactions, experimental evidence is required for the assessment whether or not univalently reduced coenzyme Q may function as a prooxidant in biological systems.

In the early 1970's Chance and co-workers observed the release of the $O_2^{\cdot -}$ dismutation product H_2O_2 from uncoupled and antimycin A-poisoned heart mitochondria (Loschen *et al.*, 1971, 1973; Boveris *et al.*, 1972; Boveris and Cadenas, 1975). Extraction of Q from the respiratory chain totally inhibited H_2O_2 release, while reincorporation of Q resulted in the reestablishment of the mitochondrial H_2O_2 source. Based on these observations the authors concluded that redox-cycling Q was responsible for electron leakage to oxygen out of sequence (Cadenas *et al.*, 1977; Boveris *et al.*, 1976). Disregarding possible serious uncertainties, which question the credibility of this concept, the idea that redox-cycling Q represents a permanent source of cellular $O_2^{\cdot -}$ formation is still adopted as a paradigm in many reports.

The following considerations elucidate why conclusions drawn from Q extraction/reincorporation experiments may be misleading. Removal of Q from the respiratory chain prevents electron flux to electron carriers at the oxidant site of Q; e.g., low potential cytochrome b, which is thermodynamically an even more proper one-electron donor for dioxygen (Chance *et al.*, 1970; Erecinska and Wilson, 1976) than semireduced Q, remains oxidized in Q-extracted mitochondria (Nohl and Hegner, 1978).

It was also demonstrated that extraction of Q from the inner mitochondrial membrane affects the stability and activity of respiratory complexes which normally interact with Q (Lester and Fleischer, 1961; Gutman *et al.*, 1971). On the other hand, reincorporation of Q may form charge-transfer complexes at the aqueous/lipid phase border which channel electrons from the respiratory chain to oxygen. This may explain the phenomenon observed by Chance and co-workers that H_2O_2 formation could be linearly titrated with the amount of Q added for reincorporation (Cadenas *et al.*, 1977; Boveris *et al.*, 1976). Starting from thermodynamic considerations, midpoint potentials of mitochondrial SQ/Q couples also seem to exclude the establishment of a redox couple with dioxygen in the aprotic environment of the phospholipid membrane. However, a prediction as to whether or not $O_2^{\cdot -}$ formation can be expected from autoxidizing SQ must also take into account that midpoint potentials are biologically not relevant because they may be changed electrostatically through membrane potentials and kinetically through equilibrium shifts (pushed on mass action effects). A good example in this respect is the transient formation of a highly unstable semiquinone intermediate (SQ_o), which has an estimated Em_7 value between −300 and −400 mV, under conditions of oxidant-induced reduction (in the presence of an excess of oxidants and antimycin A) (De Vries *et al.*, 1981; Ohnishi and Trumpower, 1980; Ding *et al.*, 1995). The latter semiquinone species was suggested as the site responsible for mitochondrial $O_2^{\cdot -}$ release.

However, SQ_o was not detected under steady-state conditions and thus it remains unclear whether this semiquinone species is formed at all during normal (unpoisoned) mitochondrial respiration.

3. CONDITIONS ALLOWING AUTOXIDATION OF MITOCHONDRIAL SEMIQUINONES

An experimental approach for a better evaluation of physicochemical conditions required for autoxidation of SQ came from *in vitro* studies of SQ and oxygen in aprotic solvents. This concept takes into account that redox-cycling SQ in mitochondria operate in the aprotic environment of the phospholipid membrane. When Q was dissolved in acetonitrile, which was purified from water by treating the solvent with a molecular sieve, $O_2^{\cdot -}$ released from KO_2 in crown ether caused the formation of the stable one-electron reduction product (Fig. 3). The SQ species was identified by electron spin resonance (ESR) characteristics. The presence of oxygen was not critical for the ESR signal intensity unless H_2O was inserted into the reaction system. With increasing amounts of water added, the SQ-related ESR signal decreased and $O_2^{\cdot -}$ radicals were detected by means of spin trapping

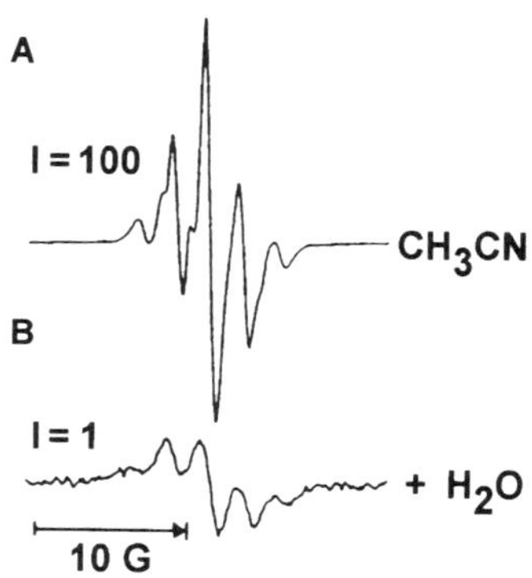

FIGURE 3. ESR signal intensity of ubisemiquinone radicals ($SQ_0^{\cdot -}$) in the absence (A) and presence (B) of H_2O (2.5%) added to purified acetonitrile. For details see Nohl and Stolze (1992).

with 5,5-dimethyl-1-pyrroline *N*-oxide (DMPO) (not shown) (Nohl, 1990; Nohl and Stolze, 1992).

The requirement of water for autoxidation of SQ is supported by standard redox potentials reported for SQ in the literature. These data exclude a single electron shuttle from mitochondrial SQ anions to oxygen on thermodynamic grounds (Ilan *et al.*, 1976; Maride and Hodgeson, 1965; Wood, 1974). Since the mitochondrial membrane is not freely permeable for water, the autoxidation of redox-cycling SQ of the respiratory chain is not to be expected.

In contrast, membrane alterations, which disturb the molecular order of phospholipids such that water has access, destabilize SQ in the presence of oxygen. Figure 4 demonstrates a linear correlation between the molecular disorder of phospholipids of the mitochondrial membrane and superoxide radical release by mitochondria (Nohl *et al.*, 1996).

Since semiquinones are one-electron donors and O_2 is a one-electron acceptor, $O_2^{\cdot -}$ radicals from autoxidizing SQ are generated. Accordingly, the oxygen sensitivity of SQ and the $O_2^{\cdot -}$ formation resulting from this interaction are correlated (Fig. 5). Combining the results of Figs. 4 and 5, it is becoming clear that $O_2^{\cdot -}$ release from the respiratory chain is stimulated when the inner mitochondrial membrane becomes permeable to H_2O. Indirect evidence on structural membrane alterations was also assessed from the coupling decrease of respiratory-linked ATP synthesis during $O_2^{\cdot -}$ radi-

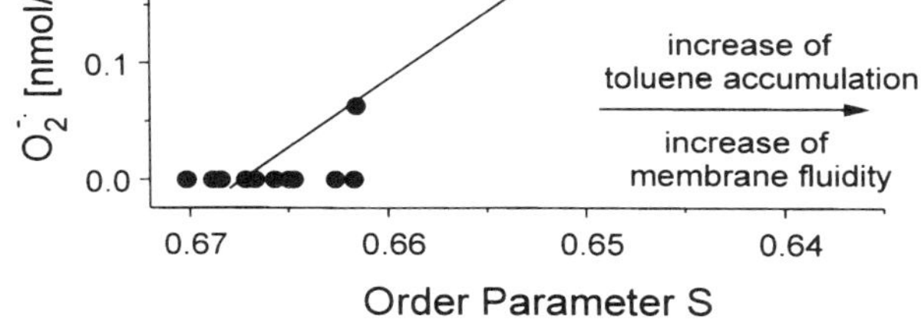

FIGURE 4. Correlation between toluene-induced fluidity changes (order parameter *S*) of the inner mitochondrial membrane and $O_2^{\cdot -}$ release as a side product of respiration. Final toluene concentrations were 1.77 mM (84.4 nmol/mg protein), 3.08 mM (200 nmol/mg protein), and 3.66 mM (223.4 nmol/mg protein), respectively. Preincubation time was 25 min; the controls were treated in the same way, without, however, the addition of toluene. For details see Nohl *et al.* (1996).

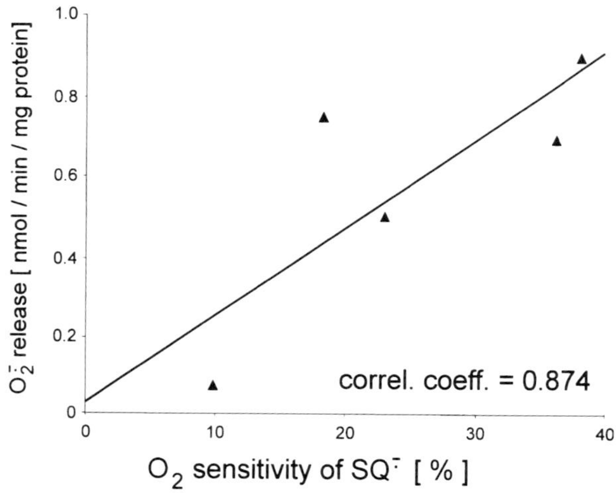

FIGURE 5. Correlation between oxygen sensitivity of redox-cycling ubisemiquinones and superoxide radical release. The O_2 sensitivity of mitochondrial SQ was elevated by increasing proton/water conductivity through incorporation of increasing amounts of toluene. For details see Nohl *et al.* (1997b).

cal formation (Stolze and Nohl, 1994). Various xenobiotics were found to establish this bioenergetic change of mitochondrial function when accumulated in the inner mitochondrial membrane (Fig. 6). Among these lipophilic compounds were environmental pollutants such as toluene or the insecticides atrazine and lindane. In all cases the efficiency of energy production linked to respiration was reduced (reflected by respiratory control (RC) values), indicating an increased proton/H_2O conductivity of the inner mitochondrial membrane. Redox-cycling SQ in this membrane were partially transformed to $O_2^{\cdot-}$ radical-generating sites, which became visible from the O_2 sensitivity of a fraction of the mitochondrial SQ pools (Nohl *et al.*, 1993). Xenobiotics establish this direct redox couple between redox-cycling SQ and dioxygen by abolishing the diffusion barrier for protons/H_2O from the cytosol to SQ implicated in regular electron transfer of the respiratory chain.

Autoxidation of SQ was also observed in unchanged mitochondrial membranes when redox-cycling activities occurred close to the aqueous phase of the membrane (Nohl *et al.*, 1996). This unusual electron transfer pathway was found to be established when reducing equivalents from extramitochondrial (cytosolic) NADH were introduced into the respiratory chain via the exogenous NADH dehydrogenases of heart mitochondria.

It is assumed that this electron transfer pathway is only active when the cytosolic NADH level increases over normal, e.g., as a result of anaerobic

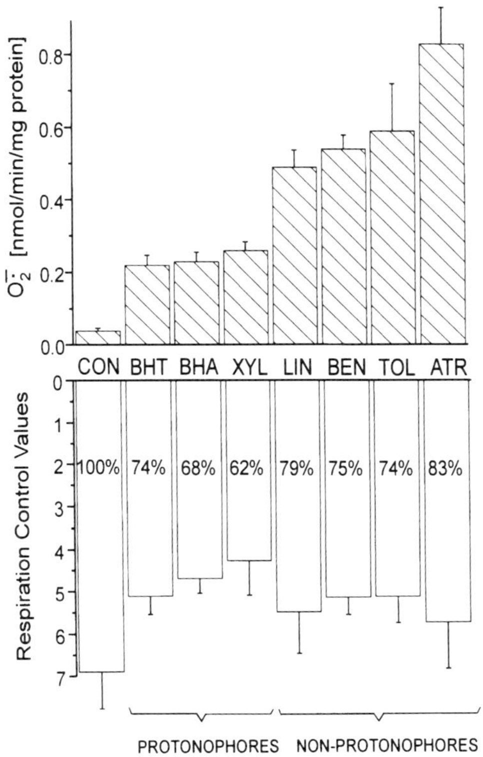

FIGURE 6. Incorporation of xenobiotics into the inner mitochondrial membrane affects both $O_2^{\cdot-}$ release and the impairment of respiratory control values of succinate-respiring mitochondria. Superoxide generation rates were inferred from superoxide dismutase-sensitive cooxidation of epinephrine. BHT, Butylated hydroxytoluene; BHA, butylated hydroxyanisole; XYL, xylenol; LIN, lindane; BEN, benzene; TOL, toluene; ATR, atrazine; CON, controls. For details see Stolze and Nohl (1994).

glycolysis due to a transient decrease of oxygen supply during ischemia. The SQ species implicated in respiration of cytosolic NADH exhibited peculiarities which favor autoxidation. One peculiarity is the impairment of physical interaction with the respective redox partners. This became visible from the decrease of thermal relaxation in the magnetic field of the ESR spectrometer (Fig. 7).

In contrast to SQ implicated in regular electron transfer activities of succinate- or glutamate-respiring mitochondria, the microwave power required for maximal ESR signal intensity was rather low ($\approx$2 mW), suggesting a higher mobility of this SQ species in the lipid phase of the membrane (Fig. 7).

Another peculiarity was the localization of this SQ species close to the aqueous phase of the inner mitochondrial membrane. This was assessed from the physical interaction of the strong hydrophilic chromium complex dissolved in the surrounding aqueous phase with the SQ radical implicated in the NADH-related electron transfer pathway from the cytosol. Due to the high spin density of this compound, spin–spin interaction with paramag-

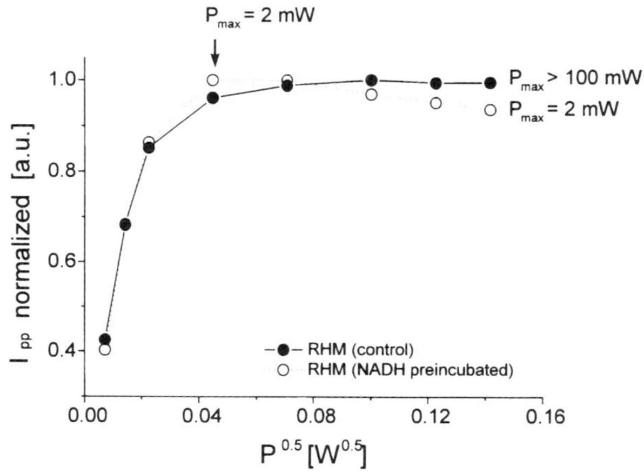

FIGURE 7. Power saturation behavior of the ubisemiquinone ESR signal in succinate/fumarate-respiring rat heart mitochondria (RHM) after preincubation with exogenous NADH (NADH preincubated). Control mitochondria were preincubated analogously without NADH (control). Parameters: Peak-to-peak intensities I_{pp} (normalized) versus the square root of microwave power P; the intensities of the individual curves were normalized based on the maximum intensity of each individual curve for comparison of the saturation behavior at the same level of intensity. Final concentrations: RHM 16.85 mg protein/ml, NADH 10 mM (preincubation; 25 min at 4°C), succinate 20 mM, fumarate 4 mM. For details see Nohl *et al.* (1996).

netic species of the mitochondrial membrane close to the aqueous phase affects the relaxation properties such that the respective ESR signal intensity increases under conditions of power saturation (Fig. 8).

The proximity of this particular SQ to the aqueous border of the inner mitochondrial membrane suggests the accessibility of water from the cytosol, which was shown to be required for autoxidation of SQ. It is assumed, therefore, that the formation of $O_2^{\bullet -}$ radicals reported by many research groups to occur as a side effect of respiration of heart mitochondria subjected to ischemia reperfusion results from autoxidizing SQ involved in the oxidation of cytosolic NADH (Fig. 9).

4. PROOXIDANT ACTIVITY OF COENZYME Q USED AS ANTIOXIDANT

SQ operating as a regular electron carrier in the respiratory chain can be stabilized by accepting a second electron from physiological electron donors or by transferring the odd electron to physiological oxidants of

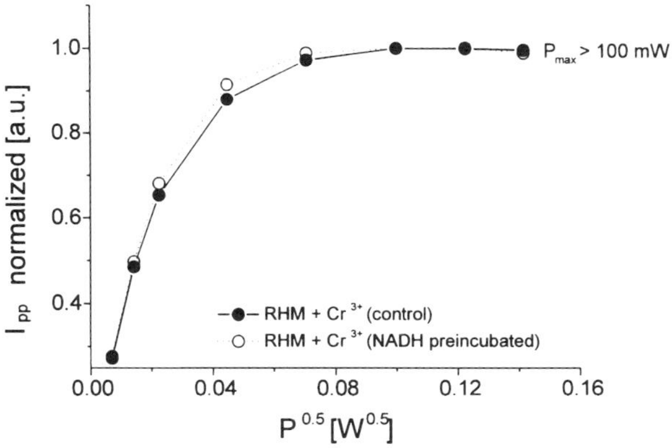

FIGURE 8. Power saturation behavior of the ubisemiquinone ESR signal in rat heart mitochondria (RHM) in the presence of 50 mM Cr^{3+} under otherwise identical conditions as in Fig. 7. Parameters: Peak-to-peak intensities I_{pp} versus the square root of microwave power P; the intensities for the individual curves were normalized as in Fig. 7. For details see Nohl *et al.* (1996).

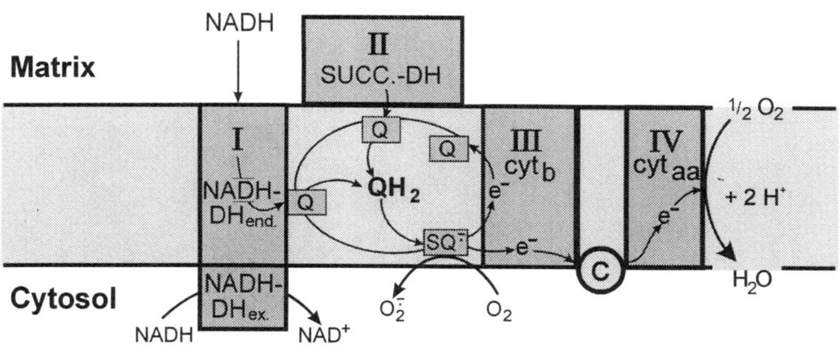

FIGURE 9. Putative scheme of electron diversion to dioxygen from redox-cycling ubisemiquinone of the mitochondrial Q-cycle.

complex III (Fig. 9). On the other hand, SQ formed out of sequence from an uncontrolled interaction of QH_2 with radicals emerging from lipid peroxidation (LPO) are unlikely to range in the regular electron transfer pathway of respiration.

The particular suitability of coenzyme Q in preventing oxygen radical-related biological degradation was suggested to result from (i) its accumulation in membranes, which often are the primary locus of radical formation and radical attack, (ii) its stoichiometric excess in mitochondria, and (iii) the *in vitro* demonstration of α-tocopherol recycling, which increases the

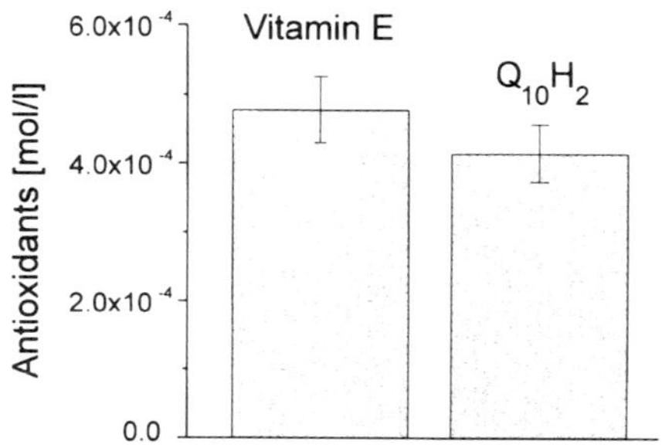

FIGURE 10. Correlation between peroxyl radical scavenging activities of reduced coenzyme Q ($Q_{10}H_2$) and vitamin E in a homogeneous system. The concentration of half-maximal quenching was calculated from the decay of the respective ESR signal intensity of the peroxyl radical spin adduct (DMPO-OOR) versus antioxidant concentration. Concentrations: DMPO, 0.6 mM; vitamin E, 0–3 mM; ubiquinol, 0–3 mM. For details see Nohl *et al.* (1997a).

antioxidant activity of coenzyme Q. Recent studies provide further evidence on the reactivity of QH_2 as antioxidant (Thomas *et al.*, 1996). As shown in Fig. 10, ubiquinol was found to be as effective as vitamin E in scavenging peroxyl radicals in a homogeneous system. It was calculated from these curves that 1 mole of QH_2 provides the same number of antioxidant equivalents as 1 mole of vitamin E. This positive role of coenzyme Q disregards the formation and the fate of the antioxidant-derived reaction products. According to the reaction

$$AOH + XOO^{\cdot} \rightarrow AO^{\cdot} + XOOH \tag{4}$$

which describes the general chemistry of an antioxidant with a radical, semiquinones are expected to be formed when reduced coenzyme Q is the antioxidant. Mitochondrial SQ were suggested to be mainly responsible for cellular development of oxidative stress (Cadenas *et al.*, 1977).

In contrast, the likely existence and reactivity of prooxidant products evolving from the antioxidative activity of coenzyme Q have not been demonstrated. When QH_2 was associated with the membrane of lipid vesicles, Q was recovered as a result of lipid peroxidation (Fig. 11). While oxidation of QH_2 occurred, the transient existence of SQ was observed by means of ESR spectroscopy (Fig. 11, inset). SQ species were not generated prior to starting lipid peroxidation. The presence or absence of oxygen was not critical for the lack of SQ in the controls, excluding both a comproportionation reaction and autoxidation of QH_2 as the source of SQ in peroxidizing lipid membranes. This observation strongly confirms the

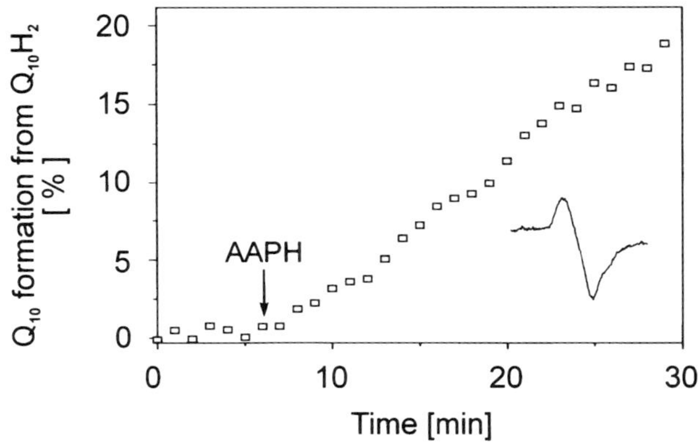

FIGURE 11. Oxidation of reduced coenzyme Q ($Q_{10}H_2$) in liposomes exposed to azo-bis-amidinopropane hydrochloride (AAPH) at 37°C for lipid peroxidation. Inset: SQ-related ESR signal observed during LPO. For methodological reasons LPO from AAPH was initiated by UV irradiation at 77°K to stabilize SQ in the frozen matrix possibly formed. Under these conditions SQ were not obtained when $Q_{10}H_2$ was irradiated with UV in the absence of LPO initiator AAPH and lipids. For details see Nohl *et al.* (1997a).

existence of reaction (4) as the pathway responsible for SQ formation. Thus, QH_2, when present in peroxidizing lipid membranes, forms SQ as the first antioxidant-derived reaction product.

5. AUTOXIDATION OF UBISEMIQUINONES DERIVED FROM RADICAL-SCAVENGING UBIQUINOL

The instability of the SQ species was suggested to result from autoxidation of the latter. This pathway was earlier shown to depend on the accessibility of water (Nohl and Stolze, 1992; Nohl *et al.*, 1996). Since lipid membranes are impermeable to water, the autoxidation can be expected when SQ exists in the outer membrane phase which borders the aqueous phase of the suspension.

The access of QH_2-derived SQ radicals to the aqueous phase was examined by spin–spin interactions with the strongly hydrophilic Gd^{3+} salt. This paramagnetic probe may affect the SQ-related ESR signal through spin exchange if SQ were close to the aqueous phase. In that case increased relaxation occurs, which can be recognized by the requirement of higher microwave power irradiation to achieve ESR signal saturation.

Figure 12 reveals that the presence of the Gd^{3+} ions in the aqueous buffer system significantly stimulated relaxation of membrane-associated

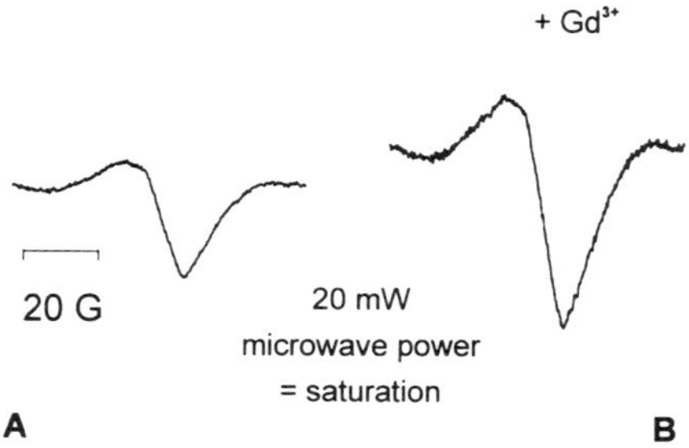

FIGURE 12. Saturation behavior of the SQ-related ESR signal generated during LPO in liposomes from an interaction of lipid radicals with incorporated reduced coenzyme Q ($Q_{10}H_2$). In the presence of Gd^{3+} (B) relaxation of the SQ radical was enhanced resulting in an increased ESR signal which was not seen with the control samples (A) under saturation conditions. For details see Nohl *et al.* (1997a).

semiquinones. This indicates the proximity of a fraction of SQ to the aqueous phase, also suggesting the accessibility of water shown to be essential for superoxide raαical formation through autoxidation (Nohl and Stolze, 1992; Nohl *et al.*, 1996). This assumption was proved by following $O_2^{\cdot-}$ formation of QH_2-free and QH_2–preloaded lipid vesicles during lipid peroxidation. For methodological reasons H_2O_2, the dismutation product of $O_2^{\cdot-}$ was measured. The release of H_2O_2 from peroxidizing lipid membranes was dependent on the existence of QH_2 as shown in Fig. 13. This was also the case in nonperoxidizing liposomes, which, however, always contain small amounts of lipid peroxides as a result of the preparation method using ultrasonication. The $O_2^{\cdot-}$ radicals of these membranes, therefore, stem from SQ in contact with the aqueous phase.

6. UBISEMIQUINONES DONATE ELECTRONS TO PEROXIDES

H_2O_2 existing in the proximity of SQ gives rise to hydroxyl radical ($HO^{\cdot}$) formation, which was demonstrated by characteristic ESR spectra ($a_N = 14.9\,G$, $a_H = 14.9\,G$, in accord with Buettner, 1987) in the presence of the spin trap DMPO.

Single-electron transfer from semiquinones to H_2O_2 was also reported by Winterbourn *et al.* (1981). The presence or absence of diethylentriaminpentaacetic acid to prove the possible role of iron was not critical for the reductive homolytic cleavage of H_2O_2. For methodological reasons the existence of this electron transfer pathway following H_2O_2 accumulation from autoxidizing SQ was inferred from a model system where SQ

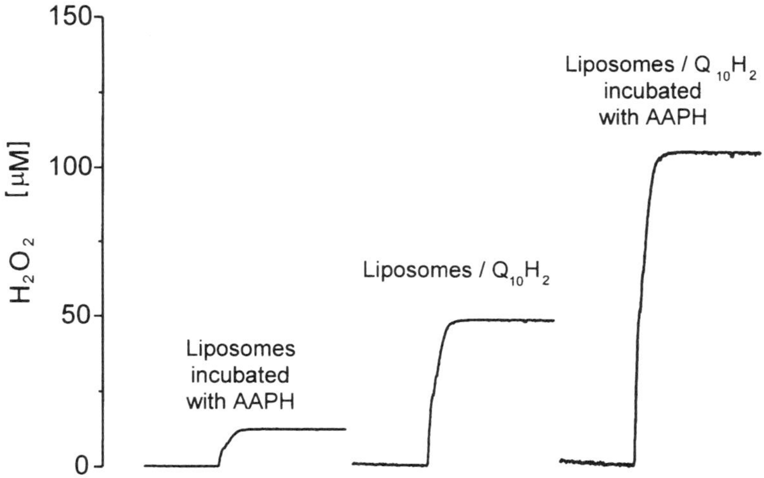

FIGURE 13. H_2O_2 formation as by-product of the antioxidant activities of reduced coenzyme Q ($Q_{10}H_2$) in liposomes subjected to conditions of LPO. Control samples were incubated without the LPO initiator AAPH. The amount of H_2O_2 was assessed from the catalase-sensitive fluorescence decay of the scopoletine horseradish peroxidase compound (Loschen *et al.*, 1971)

were united with H_2O_2. In analogy to its reaction with H_2O_2, also lipid hydroperoxides formed during membrane peroxidation also react with SQ. The expected alkoxyl radicals were not detected; instead DMPO adducts (RC-DMPO$^{\cdot}$: $a_N = 23.7\,G$, $a_H = 14.8\,G$; RC(O)-DMPO$^{\cdot}$: $a_N = 16.9\,G$, $a_H = 13.7\,G$) were found, which indicate the presence of characteristic degradation products of the highly unstable alkoxyl radical (Davies and Slater, 1987).

7. THE AMBIGUOUS ROLE OF COENZYME Q WHEN USED AS ANTIOXIDANT

Direct experimental study of the involvement of QH_2-derived prooxidants in the propagation of oxidative membrane deterioration has to consider many methodological uncertainties.

However, indirect evidence comes from comparative studies with QH_2 and vitamin E in peroxidizing lipid vesicles. Since both compounds were found to be equally effective in scavenging lipid radicals in a homogeneous system, this antioxidant potency should also be visible in peroxidizing lipid membranes.

Scavenging of lipid radicals by antioxidants associated with the lipid membrane delays propagation of lipid peroxidation. A quasilinear correlation was found when the amount of vitamin E was increased. In contrast, QH_2 widely deviated from this correlation and its overall protection capacity was clearly below the effect observed with vitamin E.

From these findings it appears that the antioxidant-derived reaction products stimulate lipid peroxidation. The latter effect gains more significance if more QH_2 is present for chain-breaking interactions with lipid radicals. Physical membrane effects such as dilution or micelle formation can be excluded as being responsible for the inclination of the antioxidant capacity of QH_2 since deviation from linearity becomes visible far below the critical QH_2 content (Lenaz and Esposti, 1985; Kagan *et al.*, 1996).

The scheme in Fig. 14 summarizes conditions under which Q_{10} develops prooxidant functions in biological membranes. SQ involved in redox cycling of mitochondrial respiration are stabilized upon binding to the respective redox partners. In addition, stabilization occurs through the regular one-electron transfer from or to SQ pools. Oxygen may serve as a direct electron acceptor when redox-cycling SQ approach the aqueous phase of the membrane. This can be expected both by an alteration of the molecular order of the phospholipid bilayer and activation of an unusual electron transfer pathway from cytosolic NADH. SQ pools supplemented from chain-breaking activities of QH_2 may destabilize upon one-electron transfer to dioxygen and peroxides. Cycling of odd electrons from lipid radicals to lipid peroxides is mediated by the first antioxidant-derived reaction product of Q_{10}.

Based on this schematic presentation, it is possible to predict conditions favoring the antioxidant activity of coenzyme Q over its prooxidant potency and vice versa; e.g., the probability by which QH_2 will stimulate prooxidant formation is small if SQ emerging from the antioxidant activity of QH_2 is stabilized upon protein binding or if SQ can be recycled back to

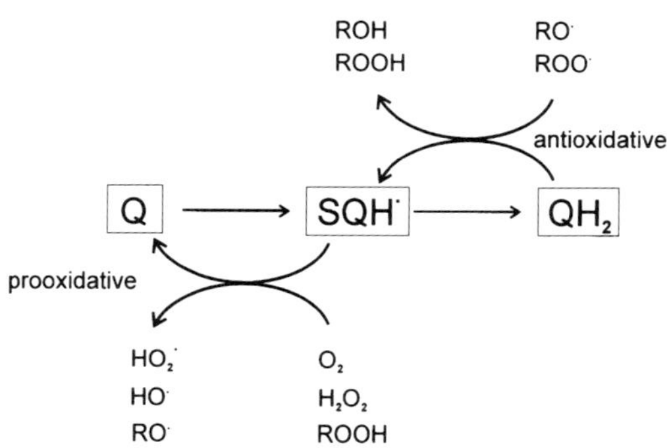

FIGURE 14. Putative scheme of the compulsory formation of prooxidants from reduced coenzyme Q ($Q_{10}H_2$) when operating as antioxidant in lipid membranes.

QH_2. In contrast, in organelles lacking these activities the prooxidant function of Q_{10} is expected to increase with the propagation of oxidative degradation of membrane lipids.

8. REFERENCES

Boveris, A., and Cadenas, E., 1975, Mitochondrial production of superoxide anions and its relationship to the antimycin insensitive respiration, *FEBS Lett.* **54:**311–314.

Boveris, A., Oshino, N., and Chance, B., 1972, The cellular production of hydrogen peroxide, *Biochem. J.* **128:**617–630.

Boveris, A., Cadenas, E., and Stoppani, A. O. M., 1976, Role of ubiquinone in the mitochondrial generation of hydrogen peroxide, *Biochem. J.* **156:**435–444.

Buettner, G. R., 1987, Spin trapping: ESR parameters of spin adducts. *Free Radic. Biol. Med.* **3:**259–303.

Cadenas, E., Boveris, A., Ragan, C. I., and Stoppani, A. O. M., 1977, Production of superoxide radicals and hydrogen peroxide by NADH ubiquinone reductase and ubiquinol cytochrome c reductase from beef heart mitochondria, *Arch. Biochem. Biophys.* **180:**248–257.

Chance, B., Wilson, D. F., Dutton, P. L., and Erecinska, M., 1970, Energy-coupling mechanisms in mitochondria: Kinetic, spectroscopic, and thermodynamic properties of an energy-transducing form of cytochrome b, *Proc. Natl. Acad. Sci. USA* **66:**1175–1182.

Davies, M. J., and Slater, T. F., 1987, Studies on the metal-ion and lipoxygenase-catalysed breakdown of hydroperoxides using electron-spin-resonance spectroscopy, *Biochem. J.* **245:**167–173.

De Vries, S., Albracht, S. P. J., Berden, J. A., and Slater, E. C., 1981, A new species of bound ubisemiquinone anion in QH_2: cytochrome c oxidoreductase, *J. Biol. Chem.* **256:**11996–11998.

Ding, H., Moser, C. C., Robertson, D. E., Tokito, M. K., Daldal, F., and Dutton, P. L., 1995, Ubiquinone pair in the Qo site central to the primary energy conversion reactions of cytochrome bc1 complex, *Biochemistry* **34:**15979–15996.

Erecinska, M., and Wilson, D. F., 1976, The effect of antimycin A on cytochromes b561, b566, and their relationship to ubiquinone and the iron–sulfur centers S-1 (+N-2) and S-3, *Arch. Biochem. Biophys.* **174:**143–157.

Gutman, M., Coles, C. J., Singer, T. P., and Casida, J. E., 1971, On the functional organization of the respiratory chain at the dehydrogenase–coenzyme Q junction, *Biochemistry* **10:**2036–2043.

Ilan, Y. A., Czapski, G., and Meisel, D., 1976, The one electron transfer redox potentials of free radicals. 1. The oxygen/superoxide system, *Biochim. Biophys. Acta* **430:**209–224.

Kagan, V. E., Nohl, H., and Quinn, P. J., 1996, Coenzyme Q: Its role in scavenging and generation of radicals in membranes, in *Handbook of Antioxidants* (E. Cadenas and L. Packer, eds.), pp. 157–201, Marcel Dekker, New York.

Kozlov, A. V., Nohl, H., and Gille, L., 1998, Identification of the ubiquinone species involved in superoxide radical formation, *Biochem. Biophys. Res. Commun.*, Submitted.

Lenaz, G., and Esposti, M. D., 1985, Physical properties of ubiquinones in model systems and membranes, in *Coenzyme Q* (G. Lenaz, ed.), pp. 83–105, Wiley, Chichester.

Lester, R. L., and Fleischer, S., 1961, Studies on the electron-transport system. XXVII. The respiratory activity of acetone extracted beef-heart mitochondria, role of coenzyme Q and other lipids. *Biochim. Biophys. Acta* **71:**355–363.

Loschen, G., Flohe, L., and Chance, B., 1971, Respiratory chain linked H_2O_2 production in pigeon heart mitochondria, *FEBS Lett.* **18:**261–264 .
Loschen, G., Azzi, A., and Flohe, L., 1973, Mitochondrial H_2O_2 formation at site II, *Hoppe Seyler Z. Physiol. Chem.* **354:**791–794.
Maride, D. L., and Hodgson, W. G., 1965, Reduction of oxygen to superoxide anion in aprotic solvents, *Anal. Chem.* **37:**1562–1565.
Nohl, H., 1982, The biochemical mechanism of the formation of reactive oxygen species in heart mitochondria, in *Advances in Studies on Heart Metabolism* (C. M. Caldera and P. Harris, eds.), pp. 413–421, CLUEB, Bologna.
Nohl, H., 1990, Is redox-cycling ubiquinone involved in mitochondrial oxygen activation? *Free Radic. Res. Commun.* **8:**307–315.
Nohl, H., and Hegner, D., 1978, Do mitochondria produce oxygen radicals *in vivo*? *Eur. J. Biochem.* **82:**563–567.
Nohl, H., and Stolze, K., 1992, Ubisemiquinones of the mitochondrial respiratory chain do not interact with molecular oxygen, *Free Radic. Res. Commun.* **16:**409–419.
Nohl, H., Koltover, V., and Stolze, K., 1993, Ischemia/reperfusion impairs mitochondrial energy conservation and triggers $O_2^{\cdot -}$ release as a byproduct of respiration, *Free Radic. Res. Commun.* **18:**127–137.
Nohl, H., Gille, L., Schoenheit, K., and Liu, Y., 1996, Conditions allowing redox-cycling ubisemiquinone in mitochondria to establish a direct redox couple with molecular oxygen, *Free Radic. Biol. Med.* **20:**207–213.
Nohl, H., Gille, L., and Staniek, K., 1997a, Endogenous and exogenous regulation of redox-properties of coenzyme Q, *Mol. Aspects Med.* **18:**s33–s40.
Nohl, H., Staniek, K., and Gille, L., 1997b, Imbalance of oxygen activation and energy metabolism as a consequence or mediator of aging, *Exp. Gerontol.* **32:**485–500.
Ohnishi, T., and Trumpower, B. L., 1980, Differential effects of antimycin on ubisemiquinone bound in different environments in isolated succinate cytochrome c reductase complex, *J. Biol. Chem.* **255:**3278–3284.
Stolze, K., and Nohl, H., 1994, Effect of xenobiotics on the respiratory activity of rat heart mitochondria and the concomitant formation of superoxide radicals, *Environ. Toxicol. Chem.* **13:**499–502.
Thomas, S. R., Neuzil, J., and Stocker, R., 1996, Cosupplementation with coenzyme Q prevents the prooxidant effect of alpha-tocopherol and increases the resistance of LDL to transition metal-dependent oxidation initiation, *Arterioscler. Thromb. Vasc. Biol.* **16:**687–696.
Winterbourn, C. C., 1981, Evidence for the production of hydroxyl radicals from the adriamycin semiquinone and H_2O_2, *FEBS Lett.* **136:**89–94.
Wood, P. M., 1974, The redox potential of the system oxygen–superoxide, *FEBS Lett.* **44:**22–24.

Index

Activator protein-1 (AP-1), 10
  transcription factor, 40
Adenovirus transcription factor, 12
Adriamycin, 130
AIDS-related retinol depletion, 149
Amphibian embryos, 92
Anatomical abnormalities, 97
Androgen receptors, 11
Animals fed β-carotene-rich diets, 210
Anterior transformation of vertebrae, 96
Anteroposterior embryo axis, 92
Anticarcinogenic effects of carotenoids, 195
Antiestrogen-resistant MCF-7 cells, 256
Antiestrogens, 245
Antioxidant
  action of coenzyme Q, 492, 493
  activity in membranes, 136
  activity of vitamin A, 113
  chemistry of α-tocopherol, 301
  effectiveness of vitamin A, 125
  properties of β-carotene, 210
  protection, 193
  reactions of tocopherols, 308
  role of carotenoids, 183
Antioxidants in LDL, 502
Antiproliferative effecs of 1,25$(OH)_2D_3$, 257
Anurans, 82
β-apo-carotenals, 164
β-apo-carotenic acids, 174
Apo
  and holo retinol and retinoic-acid binding proteins, 59
  structures of CRPB II and CRABP I, 62, 64
Apocarotenals, 37
Apoptosis, 245
  in cancer therapy, 246
Apoptotic mediators, 258
Appearance and cause of aging, 391
Ascorbate, 374
  and tocopherol, 310
Astaxanthin, 223
Atherogenesis, 132, 377
Atherosclerosis, 348
Atherosclerotic lesions, 380
Autoxidation
  of mitochondrial semiquinones, 514
  of polyunsaturated lipids, 363
  of ubiquinol, 511
  of ubisemiquinones, 521

β-barrel protein, 33
Bilirubin, 374, 376
Bimolecular termination reactions, 369
Binary lipid mixtures with α-tocopherol, 327
Bioavailability of vitamin K, 458
Bioconversion of β-carotene into retinoids, 159
Biogenesis of retinoic acids, 36
Biomarkers for heart disease, 188
Biophysical properties of carotenoids, 219
Biosynthesis of vitamin A, 161
Birth defects and retinoids, 151
Blastemal cells, 86
Blood coagulation proteins, 471

C3H 10T1/2 transformation assay, 200

Calbindin-D_{28k}
-expressing retrovirus, 283
gene, 284
Calbindins, 281
in calcium homeostasis, 282
Calcium
-activated kinases, 285
channels, 280
reabsorption in the kidney, 282
regulation in breast cancer cells, 277
transport in the intestine, 279
uptake in pituitary cells, 279
Calreticulum, 10
c-AMP-mediated regulation of VDRs, 287
Canthaxanthin, 197
Carnitine palmitoyl transferase I, 170
β-carotene
in HIV patients, 147
integrated into HL-60 cells, 212
and oxidative DNA base damage, 192
and oxidative stress, 209
Carotenes and DNA stability, 181
Carotenoid
monolayers, 221
structure and antioxidant activity, 181
structures, 182
supplementation, 192
Carotenoids
and cultured cells, 197
in human plasma, 190
in membranes, 237
Catabolism and turnover of vitamin K, 459
CD4 cells, 150
Cell lines resistant to antiestrogens and vitamin D_3, 254
Cellular calcium homeostasis and signaling, 273
Cellular proliferation and differentiation, 172
Cellular retinoic acid-binding proteins (CRABP) I and II, 4, 61
Cellular retinol-binding proteins (CRBP) I and II, 4, 61
Cellular uptake of retinol, 33, 34
Central double-bond cleavage of β-carotene, 162
Chain-breaking antioxidants, 116, 349
Chick limb bud, 88
Chimeric receptors, 87
Chlorophyll photooxidation, 237
Choroid plexus lipocalin 1, 56
9-*cis*-β-carotene, 175
Citral effect on retinoic acid production, 171
Clathrin-coated pits and vesicles, 35
Clot formation in blood, 472
CNS
defects, 100
phenotypes and retinoic acid, 98
Coantioxidants, 370
Coenzyme Q
as antioxidant, 523
health significance of, 501
and vitamin E cycling, 491
Collapse pressures of monolayers, 231
Colonic carcinoma cell line, 275
Complexes of α-tocopherol and free fatty acids, 332
Composition of human LDL, 347
Condensing effect of xanthophylls, 229
Conotoxins, 478
Content of vitamin E in membranes, 321
Control of RAR
activity, 18
transcription, 16
Conversion of β-carotene into retinoids, 161, 172
Copper-induced oxidation, 132
CoQ_{10} and prooxidant effects of α-tocopherol, 372
$CoQ_{10}H_2$, 370
Corepressors, 8
Coronary heart disease, 403
CRABP mutant, 66
Cumulative antioxidant index, 188
β-cryptoxanthin, 223
Cyclic-AMP, 16
response element binding protein, 10
Cyclic-GMP, 53
Cyclin E, 44
Cyclooxygenases, 129, 348
Cytokine-fatty acid interactions, 425
Cytoplasmic dioxygenase, 163

Delayed-type hypersensitivity, 403
Detoxification of xenobiotics, 450
Development
in absence of retinoic acid, 104
of chick limb bud, 89
Dietary sources
of carotenoids, 190
of vitamin K, 457
Differential splicing and promoter usage, 14
Disruption of retinoic acid signaling, 102

Distribution of β-carotene, 166
Distribution of vitamin E
in the body, 321
in membranes, 338
DNA
repair in human lymphocytes, 194
stability and carotenoids, 181
Donation of electrons to peroxisomes by ubisemiquinones, 522
Downstream effector genes of retinoic acid, 101
Doxorubicin, 124

Effect of β-carotene and α-tocopherol on phospholipid oxidation, 215
Effect of carotenoids on human and animal cells, 198, 199
Effect of retinoids on regenerating limbs, 83
Effect of α-tocopherol
on lipid bilayer dynamics, 324
on phospholipid structure, 327
Effector genes of retinoic acid, 87, 91
Effects of 1,25-dihydroxyvitamin D_3 on intracellular calcium, 274
Effects of linoleic acid on apoptosis, 426
Eicosanoids, 446
Eicosapentanoic acids, 446
Elevated serum cholesterol, 186
Endogenous retinoic acid in the early embryo, 102
Endogenous retinoids, 85
Endogenous α-tocopherol, 133
Endothelial-leucocyte adhesion molecule-1, 419
Endothelial nuclear factor-χB, 418
Enrichment of LDL with coenzyme Q, 371
Epidermal growth factor receptor (EGFR), 43
Epididymal retinoic acid-binding protein (E-RABP), 55
Epoxytocopherones, 307
Essential fatty acids, 445
Estradiol, 17
Estrogen-independent breast cancer, 263
Estrogen receptor, 12
Excentric double-bond cleavage of β-carotene, 164
Expanded-liquid monolayers, 227
Extent of intimal LDL oxidation, 381
Extracellular matrix Gla-proteins, 475

Fatty acid-binding protein aP2, 69
Fatty acids and atherosclerosis, 415
Flow cytometric analysis of MCF-7 cells, 251
Foods rich in carotenoids, 184
Forebrain/midbrain structures, 99
Formation of 8α-substituted tocopherones, 305
Formation of prooxidants from reduced coenzyme Q, 524
Formation of retinoids from β-carotene, 163
Fourier-transform infrared spectroscopy, 333
Free radical-mediated peroxidation, 186
Frequency of radical hits of intimal LDL, 380
Fungal lipase, 69
Furin, 463

Gas6, 477
Gastrulating mesodermal cells, 96
Genomic regulation of calbindins, 281
Gla-dependent activation of plasma proteins, 472
Gla proteins, 473
Glucocorticoid, 11
Glucuronidation of retinoic acid, 169
γ-glutamyl carboxylase, 462
γ-glutamyl carboxylation, 464
Glutathione, 312
Glycosylation, 40
Growth
inhibition by vitamin D_3, 259
of MCF-7 cells, 252

Head-to-tail axis, 92
Hemostasis, 471
Hensen's node, 96
Hepatic reserves of vitamin K, 459
Heptad repeats, 73
Homeobox genes, 87
Horseradish peroxidase (HRP), 357
Hox family, 92
homeobox genes, 39
transcription factors, 44
Human keratinocytes, 276
Human pancreatic lipase, 69
Hydroperoxyeicosatetraenoic acids, 424
Hydroperoxyoctadecadienoic acids, 424
3-hydroxyanthranilic acid, 375
8α-hydroxytocopherone, 306
Hypochlorite, 367
in the central nervous system, 97

Induction of limb duplication, 90
Inhibition
of endogenous retinoids, 90
of retinoic acid formation, 171

Initiation of LDL lipid peroxidation by vitamin E, 353
Inner mitochondrial membrane, 517
Insect fatty-acid binding protein, 67
Interaction of α-tocopherol with phospholipids, 323, 332
Interactions
between vitamin D_3 and estrogens, 249
between vitamins A and E, 133
Interleukin-1, 419
Intestinal absorption of vitamin K, 458
Intracellular retinoid-binding proteins, 60
Intramolecular carboxylation, 468
IP_3 receptor, 276
Ischemic heart disease, 417
Isoforms of retinoic acid receptors, 71
Isoprenoids, 38

β-lactoglobulin, 57
β-lactone ring, 58
Langmuir–Blodgett films, 222
Lecithin:retinol acyltransferase (LRAT), 30
Leucine zipper interface, 74
Leukemia and vitamin K, 480
Ligand-activated transcription factors, 257
Ligand-binding
domain of holo RAR, 72
sites, 65
Linoleate oxidation, 349
Linoleic acid
-induced activation of NF-χB, 421
and proteoglycan metabolism, 427
Lipid
hydroperoxides, 127
metabolism and the peroxisome, 438
oxidation in low-density lipoproteins, 117
peroxidation, 121, 192
and carotenoids, 181
reduction of coenzyme Q, 500
Lipidic signaling
and peroxisomal function, 437
roles of peroxisomes, 449
Lipocalin family, 33
Lipocalin β-lactoglobulin, 57
Lipocalins, 55
Lipofuscin, 398
-like age pigment, 395
Lipoperoxyl radical scavenging, 123
by retinoids in membranes, 128
by vitamin A, 120
Lipoperoxyl radicals, 125
Lipoxygenase, 129, 167
15-lipoxygenase, 348
Liver fatty acid-binding protein, 68
Localization of vitamin E in membranes, 319
Low-density lipoproteins (LDL), 131
lipid oxidizability, 355

Major carotenoid food sources, 189
Malabsorption, 145
Malignant transformation, 201
Malnutrition, 145
Mammary cells and vitamin D_3, 248
Manganese superoxide dismutase, 419
Matrix Gla protein, 477
MCF-7 tumor xenografts, 262
Mechanisms of LDL lipid oxidation, 347
Melanocyte-stimulatng hormone, 16
Membrane lipid peroxidation, 130, 194
Membranes, 127
Menaquinones, 456
3-Mercaptopropionic acid, 170
Metabolic pathways of β-carotene, 173
Metabolism
of β-apo-carotenoic acid, 170
of retinoic acid, 61
of retinol, 32
Metabolites
of β-apo-carotenoic acids, 168
of phylloquinone, 460
β_2-microglobulin, 152
Miscibility and the phase rule, 230
Mitochondrial superoxide generation, 394
Mixed acyl phosphatidylcholines, 328
Mixed monolayers
of carotenoids and lipids, 227, 234
of zeaxanthin and phospholipid, 232
Modulation
of apoptosis, 265
of lipid chain order by α-tocopherol, 325
Molecular basis of vitamin K action, 461
Molecular epidemiology, 187
Molecular model of β-cryptoxanthin, 226
Monolayers of β-cryptoxanthin, 224
Myelin protein P2, 65
Myeloperoxidase, 359

NAD(P)H:quinone oxidoreductase, 313
Natural carotenoids, 160
Neoplastic transformation, 200
Neural development, 103

NF-χB related transcription, 422
Noncarboxylation functions of vitamin K, 479
Nordehydroguaretic acid, 312
Nuclear hormone receptors, 30
Nuclear receptor corepressor, 9
Nuclear receptors for retinoic acid, 71
Nuclear retinoid receptors, 5
Nutritional antioxidant intakes, 187
Nutritional formulations, 236
Nutritional requirements of vitamin K, 457

$O_2^{\cdot -}$ radical formation, 510
$O_2^{\cdot -}$ radicals from autoxidising semiquinones, 515
Odorant-binding protein (OBP), 60
  of mesoderm, 96
Oncogenes, 42
One-electron radical oxidants, 356
$ONOO^-$, 368
Orientation and motion of vitamin E in membranes, 329
Orientation of carotenoid molecules, 220
Ornithine decarboxylase (ODC), 43
Osteoblast activation, 287
Osteocalcin, 475
  gene in osteoblasts, 286
Osteoclast recruitment, 476
Osteoporosis, 476
β-oxidation in peroxisomes, 438
Oxidation
  of β-carotene, 213
  of LDL, 493
    by 15-lipoxygenase, 361
    by nucleophilic oxidants, 366
    transition metals, 378
  of reduced coenzyme Q, 521
  of ubiquinol, 512
  of unsaturated lipids, 116
Oxidative damage, 115
  to biological systems during aging, 393
Oxidative modification of LDL, 346
Oxidative stress, 402, 500
  and defense, 115
  in membrane phospholipids, 211
Oxidatively modified protein of fibroblasts, 397
Oxygen free radicals, 447

Patterning of the CNS, 94, 95
Patterns of expression of *Hox* genes, 101
PC 12 rat pheochromocytoma cells, 130
Peroxidation of lipid: *see* Lipid peroxidation
Peroxisomal involvement in lipid metabolism, 439
Peroxisomal proliferators, 441
Peroxisome proliferator-activated receptor, 441
Peroxyl radical
  scavenging, 302
  spin adduct, 520
Phase behavior of phospholipids, 326
Phase diagrams of β-cryptoxanthin and phospholipid, 231
Phases of life span, 406
Phosphatidylcholine
  cycle, 443, 444
  liposomes, 123
Phospholipase D, 444
Phospholipid
  peroxyl radicals, 305
  turnover, 285
Phosphorylation of RARβ1, β3, 12
Phototoxicity, 201
Phylloquinone, 456
Physiological effects of retinoids, 41
Plasma
  lipoproteins, 322
  retinol-binding protein, 54
  vitamin E in HIV patients, 148
Platelet-activating factor, 399
Polyunsaturated lipids, 127
Polyunsaturated/saturated fat ratios, 417
Pro- and antioxidant activity of vitamin E, 345
Profactor K, 463
Promyelocytic leukemia, 159
Prooxidant
  activity of coenzyme Q used as antioxidant, 518
  functions of coenzyme Q, 509
Prostaglandin E_2, 400
Prostaglandin I_2, 400
Protection by vitamin A of doxorubicin damage, 124
Protein
  carboxylation mediated by vitamin K, 462
  kinase A, 13
  kinase C, 13
  S, 477
  synthesis, 18
Proteoglycans, 428
Prothrombin synthesis, 461
Provitamin A carotenoids, 160
Purification of γ-glutamyl carboxylase, 464

Radical-radical termination reactions, 358
Radical scavenging
effects of coenzyme Q, 494
ubiquinol, 521
RAR
activity, control of, 18
family, 5
gene, 6
and RXR ligand-binding domains, 73
transcription, control of, 16
Reaction of tocopheroxyl radical with peroxyl radical, 303
Reaction pathways of retinoids during lipid peroxidation, 122
Reactive nitrogen intermediates, 379
Reactivity of retinoids
with lipoperoxyl radicals, 126
with radicals, 119
Redox cycles of α-tocopherol, 310
Redox cycling ubisemiquinones, 516
Redox regulation of α-tocopherolquinones, 312
Reduction of vitamin E radicals by ubiquinols, 495
Regenerating amphibian tail, 85
Regenerating newt limb, 86
Regeneration of amphibian limb, 82, 84
Regulation
between central and excentric pathway of β-carotene, 169
of gene expression, 39
of viral genes, 41
Retinal endocrinology, 29
Retinoic acid (RA)
-binding protein, 4, 36
-metabolising enzymes, 70
receptors (RARs), 85
α and β, 56
response element, 86
secretion in bile, 169
synthesis, 167
Retinoid
binding proteins, 53
-derived radicals, 121
nuclear receptor, 3
signaling pathways, 38
transport and metabolism, 30
Retinoids
and embryo development, 81
in embryogenesis, 41
and oncogenesis, 41
Retinol
-binding protein (RBP), 30, 55
Retinol (*cont.*)
dehydrogenase, 37
dynamics, 131
endocytosis, 35
Retinoylation, 40
Retinyl
acetate, 85
esters, 37
palmitate, 83
Role
of linoleic acid in gene expression, 415
of oxidized fatty acids in gene expression, 423
of superoxide in antioxidant action of ubiquinol, 496
of vitamin E
in aging, 391
in LDL lipid peroxidation, 357
RRR-α-tocopherol, 319
RXR family, 7

Saturated fatty acids, 416
Serum cholesterol, 416
Serum levels of β-carotene, 184
Signal transduction by fatty acid-induced NF-χB, 420
Signaling features of peroxisomes, 440
Single-component monolayers, 229
Singlet oxygen, 118, 370
Snake venoms, 478
Sonic hedgehog, 91
Soybean lipoxygenase, 364
Sphingolipid pathway and vitamin K, 479
Stability of lipid membranes, 227
Steroid hormone receptors, 8
Structural formulas of vitamins K, 457
Structure of carotenoid monolayers, 224
Structures
of apo-hetic RXR and holo-RAR, 74
of β-carotene and retinoids, 114
of syntretinoids, 120
8α-substituted tocopherones, 306
SUM159 tumors, 264
Superoxide
anions, 118
-dependent reduction of coenzyme Q, 499
and vitamin E cycling, 491
Supplementation with β-carotene, 152
Surface-active properties of carotenoids, 223
Synergism between vitamins A and E, 134
Synthesis of PGE_2 by spleen homogenates, 404
Synthetic vitamin D_3 analogs, 258

Terminal transferase-mediated fluorescein dUTP nick end labeling, 263
Terpenoid biosynthesis, 31
Theories of aging, 392
Therapeutic preparations, 236
Thermodynamics of mixed monolayers, 228
Thiobarbituric acid-reactive substances, 401
THR uncoupling protein, 9
Thromboxane A_2, 400
Thyroid
hormone, 33
receptor (THR), 87
-stimulating hormone (TSH), 14
Thyroxine receptor, 71
α-tocopherol
clusters in phosphatidylcholines, 336
enriched HL-60 cells, 214
-mediated LDL lipid peroxidation, 350
radicals, 351
-rich domains in phospholipid bilayers, 337
α-tocopherolquinone, 306
α-tocopheryl hydroquinone, 373
γ-tocopherol, 309
Tocopherol
antioxidant function, 302
-lipid complexes, 331
regeneration, 309
Tocopheroxyl radical, 309
Transcription
factors, 10
rate of RAR genes, 17
regulators, 17
Transformed cells, 197
Transforming growth factor α (TGFα), 43
Transthyretin (TTR), 30, 54
Trials of vitamin D_3 analogs in breast cancer, 260
Trigeminal and facial ganglia, 100
Tumor
growth inhibition by vitamin D_3 analogs, 261
necrosis factor, 419
promoter, 17
Tumorigenesis, 195
Two-electron redox cycle, 311
type I and II, 61
Tyrosyl radicals, 360

Ubisemiquinone
and protection of vitamin E, 498
radicals, 514
Upregulation of p21, 257
Uptake and transport of vitamin E, 320
Urodele amphibians, 82

Vitamin A
deficiency in HIV patients, 150
dietary intakes, 147
and E depletion, 149
excess toxicity, 131
levels in HIV patients, 146
and pregnancy, 151
and retinoids as antioxidants, 117
Vitamin D
and intracellular calcium, 271
-mediated signal transduction pathways, 272
Vitamin D_3
and apoptosis, 245
and calcium fluxes, 278
and cell proliferation/differentiation, 247
endocrine system, 247
receptor (VDR), 247
signaling pathway, 253
Vitamin E
and aging, 400
-binding proteins, 321
and chain-transfer in LDL, 352
deficiency, 396
and elongation of life span, 405
Vitamin K
antagonists, 467, 469
cycle, 464
-dependent γ-glutamyl carboxylase, 462
-dependent plasma glycoprotein, 474
-dependent protein modification, 461
-dependent proteins, 470
derivatives, 470
-epoxide cycle, 465
hydroquinone, 466
metabolism, 455
quinone, 466
reductase inhibitors, 467
Vitamin K2 and bone, 480
Vitamin/mineral supplements, 185
Vitamins A and E in HIV patients, 145

Warfarin, 468
Whole-cell voltage clamp, 280

Xenobiotics, 517

Zeaxanthin, 223
Zone of polarizing activity (ZPA), 88